3rd International Conference on Automation, Control and Robotics Engineering (CACRE 2018)

IOP Conference Series: Materials Science and Engineering
Volume 428

Chengdu, China
19 – 22 July 2018

ISBN: 978-1-5108-7337-7
ISSN: 1757-8981

Printed from e-media with permission by:

Curran Associates, Inc.
57 Morehouse Lane
Red Hook, NY 12571

CURRAN ASSOCIATES INC. proceedings.com

Some format issues inherent in the e-media version may also appear in this print version.

Printed by Curran Associates, Inc. (2018)

Institute of Physics
Dirac House, Temple Back
Bristol BS1 6BE UK

Phone: 44 1 17 929 7481
Fax: 44 1 17 920 0979

techtracking@iop.org

Additional copies of this publication are available from:

Curran Associates, Inc.
57 Morehouse Lane
Red Hook, NY 12571 USA
Phone: 845-758-0400
Fax: 845-758-2633
Email: curran@proceedings.com
Web: www.proceedings.com

Table of contents

Volume 428

3rd International Conference on Automation, Control and Robotics Engineering

(CACRE 2018) 19–22 July 2018, Chengdu, China

Accepted papers received: 6 September 2018
Published online: 1 October 2018

Preface

Papers

Automation Technology and Sensing Methods

Intelligent Robotics and Mechatronic Systems

IOP Conf. Series: Materials Science and Engineering **428** (2018) 011001 doi:10.1088/1757-899X/428/1/011001

PREFACE

CACRE 2018 was held in Chengdu, China during Jul.19-22, 2018. CACRE 2018 was organized by Sichuan University and HKSME. The conference provides a useful and wide platform both for displaying the latest research and for exchanging of research results and thoughts in broad Automation, Control, Robotics Engineering, Artificial Intelligence and other topics. The participants of the conference were from almost every part of the world, with background of either academia or industry, even well-known enterprise. The success and prosperity of the conference is reflected by high level of the papers received.

The proceeding is a compilation of the accepted papers which represents an interesting outcome of the conference. This book covers 3 chapters: Automation Technology and Sensing Methods; Advanced Control and Intelligent Algorithms; Intelligent Robotics and Mechatronic Systems.

We would like to acknowledge all of those who supported CACRE 2018. Each individual and institutional help was very important for the success of this conference. Especially we would like to thank the organizing committee for their valuable advices in the organization and helpful peer review of the papers.

We sincerely hope that CACRE 2018 will be a forum for excellent discussions that will put forward new ideas and promote collaborative researches. We are sure that the proceedings will serve as an important research source of references and the knowledge, which will lead to not only scientific and engineering progress but also other new products and processes.

Songyi Dian

Prof. Songyi Dian, Sichuan University, China

IOP Conf. Series: Materials Science and Engineering **428** (2018) 011002 doi:10.1088/1757-899X/428/1/011002

COMMITTEES

General Chair

Prof. Songyi Dian, Sichuan University, China

Conference Committee Co-Chairs

Prof. Xiaoping Liu, Carleton University, Canada

Prof. Du Ruxu, The Chinese University of Hong Kong, Hong Kong

Program Committee Co-chairs

Prof. Tao Zhao, Sichuan University, China

Prof. Xuechao Duan, Xidian University, China

Advisory Committee

Prof. Aleksander Zgrzywa, Wrocław University of Technology, Czech

Prof. Tshilidzi Marwala, University of Pretoria, South Africa

International Technical Committee

Prof. J. W. Yan, Jinan University, China

Prof. Cho-Pei Jiang, National Formosa University, Taiwan

Prof. Shinn-Liang Chang, National Formosa University, Taiwan

Prof. Meng-Hui Hsu, Kun Shan University, Taiwan

Prof. Robert Babuska, Delft University of Technology, Netherlands

Prof. Nan-Chyuan Tsai, National Cheng Kung University, Taiwan

Prof. Genci Capi, Hosei University, Japan

Prof. Spilios Fassois, the University of Patras, Greece

Dr. Ali GhaffarianHoseini, AUT University, New Zealand

Dr. Darwin Lau, the Chinese University of Hong Kong

Dr. Giuseppe Carbone, University of Cassino and South Latium, Italy

Dr. Xiang Ping, City University of Hong Kong

Dr. Felix Pasila, Petra Christian University, Indonesia

Dr. Zhauyt Algazy, Satbayev University, Kazakhstan

IOP Conf. Series: Materials Science and Engineering **428** (2018) 011003 doi:10.1088/1757-899X/428/1/011003

Peer review statement

All papers published in this volume of *IOP Conference Series: Materials Science and Engineering* have been peer reviewed through processes administered by the proceedings Editors. Reviews were conducted by expert referees to the professional and scientific standards expected of a proceedings journal published by IOP Publishing.

IOP Conf. Series: Materials Science and Engineering **428** (2018) 012001 doi:10.1088/1757-899X/428/1/012001

Interface Byte Stream Parsing Algorithm for Black Box Test Cases Automatic Generation

Cao Jianping*, **Sun Wenzhu, Zhou Xiuzhi, Zhu Guotao and Hu Wenting**

Naval Aviation University Qingdao Branch, 266041, QingDao, China

*jp_cao2016@126.com

Abstract. Traditional black box test cases are mostly generated by manual methods and are difficult to automate. This is caused by the diversification of software module's input interface. In this paper, an intelligent black box test case automatic generation algorithm referred to as Interface Byte Stream Parsing (IBSP) is proposed. In this algorithm, the input interface of the software module to be tested is firstly abstracted as the number of bytes occupied by the input parameters of the module, and a section of dynamic memory is applied, then a binary sequence is generated in the dynamic memory by using Dynamic Memory Increment (DMI) or Random Sampling (RS) method. Follow that, the type information contained in the interface is inputted through the software module to be tested. The binary sequence is parsed into the input data of the software module to be tested, and the input data is filtered by using the filter conditions. Finally, the test expectations given by the user and the filtered input data constitute the test cases. For improving the performance of the algorithm, a testing experiment is conducted. The experimental result shows that the algorithm facilitates the automatic generation of test cases in software testing, improves the efficiency of test case generation, and has been initially applied.

1. Introduction

Software testing is an important means for ensuring software quality. An automated testing process can increase test efficiency and reduce cost. The black box test is a commonly used software testing method. It regards the program to be measured as a black box that cannot be opened. Without considering the internal structure and characteristics of the program, the test is performed from the correspondence between the input data and the output data[1]. During the black box automated testing process, automatic acquisition of test cases is a key and difficult point. So, it is of great significance to research automatic black box test cases generation.

It is generally considered that the exhaustive testing of input data is a task that cannot be completed. For this reason, traditional black box testing methods including equivalent division method, boundary value analysis method, causality graph analysis method and error guessing method usually select some representative input samples as test cases with the assistance of computer or artificially. These methods cannot be defined as test case automatic generation methods.

The automatic generation of test cases is the research hotspot currently. Most of the researches focus on the white-box testing method and generate test paths for overlay programs, for example genetic algorithms [2], particle swarm optimization algorithm[3] and other intelligent optimization algorithms. But the studies about black-box test cases generation are less.

In terms of applications, Nubian Corporation has published a tree-based black box test case generation method [4]. It first obtains n interface parameters of the software module to be measured and constructs them into an inverted tree structure. Finally, a path from the root node to child node forms a test case. Based on this method, valid use cases with higher coverage can be searched and system defects can be located quickly. A patent from Harbin Institute of Technology[5] proposes a method for generating embedded black box test cases based on dynamic models, which focuses more on the management of black box testers and data. But the methods given above are strongly dependent on the interface parameters of the software module to be tested, if the interface parameter types and numbers change, the test cases need to be recreated. From this perspective, these methods have not achieved the goal of universal and automatic test case generation.

This paper proposes a novel black box test case automatic generation algorithm referred to as Interface Byte Stream Parsing (IBSP). In this algorithm, the input interface of the software module to be tested is firstly abstracted as the number of bytes occupied by the module input parameters, and a corresponding byte of dynamic memory is applied. Then, a binary sequence is generated in the dynamic memory. Finally, the binary sequence is parsed into the input data of the software module to be tested through the type information contained in the input interface of the software module, and the input data may be filtered by using the filter condition. The test expectations given by the user and the filtered input data constitute the test cases. The binary sequences mentioned above can be generated by using Dynamic Memory Increment (DMI) or Random Sampling (RS) method which is explained in chapter 3.2 and 3.3. A testing experiment is conducted to assess the performance of IBSP algorithm. The experimental result shows that the algorithm facilitates the automatic generation of test cases in programming and improves the efficiency of test case generation.

2. Theoretical basis

For computer systems, the information can be expressed by "bit + context." All information in a computer system including disk files, programs in memory, user data stored in memory, and data transmitted over the network are represented by a string of bits [6].

The data in the computer memory is expressed by bit sequences composed of 0 and 1. The same data can be interpreted in different meanings for different application backgrounds. For example, for the 8byte binary sequence 0x43AE044534EE0443 (in hexadecimal notation), if it resolves to {int p1, int p2}, its value is {888013891, 1135477829}; if it resolves to {double p3}, its value is 1.0814649073392767e18. Therefore, according to the corresponding context of the input interface parameter set, the generated binary sequence can be parsed into the input sample.

According to the current requirements, the input interfaces of the module under test are diversified, but the same parameter can be abstracted from the input interface parameter set of the module under test M, that is, the number of bytes occupied by the input interface parameters. For example, if the input interface parameter set of the module M1 to be tested is {int p1, int p2}, the number of bytes is 8; the input interface parameter set of the module M2 to be tested is {double p3}, and the number of bytes is also 8.

According to the byte number of the module M, a corresponding size of the dynamic memory is applied in the computer, and a certain method is used to generate a binary sequence in the dynamic memory independently. The binary sequence is then parsed using the context contained in the input interface parameter set to obtain test samples. In this process, the difficulty of test cases automatic generation is how to generate a binary sequence on a given dynamic memory. By researching, two generation methods are given in this paper, one is Dynamic Memory Increment (DMI) method and the other is Random Sampling (RS) method. The DMI method is to give a specific size of memory block and initialized to a special value, then increase the value by a fixed value to generate test cases. The RS method is to randomly generate a binary data generation test on a given memory.

3. Algorithm implementation

3.1. Basic process of IBSP

The steps of IBSP algorithm are as follows:

Step1: The input Interface Parameter Set represented as IPS of the software module M to be tested is given by the user.

Step2: According to the IPS, the size of its byte number represented as C is obtained, the Dynamic Memory represented as DM with the size C is requested, and the first address of DM is obtained, the first address of dynamic memory is represented as DA.

Step3: According to DA, we can address to the DM, and then a Binary Sequence represented as BS is generated.

Step4: Use the IPS to parse the BS, and the Instantiated Test Sample Data (ITD) is obtained.

Step5: A filter condition represented as FC is provided by the user to determine whether the ITD is in conformity with the FC, if so, go to Step6, otherwise go to Step3;

Step6: A set of Test Case (TS) is composed of the ITD and the Test Expectation (TE) which is given by user in advance.

The flow chart of IBSP algorithm is shown in Figure 1.

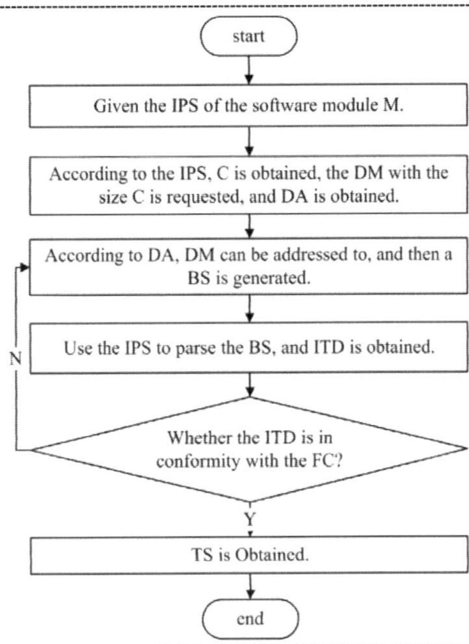

Figure 1. Flow chart of IBSP algorithm

In step 3, the BS can be generated by DMI method or RS method, which will be discussed separately below.

3.2. DMI method

The DMI (dynamic memory increment) method first apply for a dynamic memory block, next specify the memory block as a data type, and then continue to add a specific value to the memory block to generate a binary sequence. When the initial value is 0 and the step size is 1, the value of dynamic memory can be traversed. This method is equivalent to increasing a large integer until overflow. In this method, recursive algorithm is applied. The DMI method steps are as follows:

Step1: The Dynamic Memory (DM) is divided into a continuous sub dynamic memory space DM_i which represents the value stored in the memory of the address in DA+i, the size is 1 byte, and the i represents the offset of the sub memory space in bytes relative to the first address, as an unsigned integer, from 0 to C-1;

Step2: Incremental Step represented as IV is given by the user to obtain the DA of DM, set the Current Incremental Step (CIV) to IV, set the current offset i to 0, and IV is an unsigned integer, and its value is no less than 1 and no more than 255;

IOP Conf. Series: Materials Science and Engineering **428** (2018) 012001 doi:10.1088/1757-899X/428/1/012001

Step3: The Current Dynamic Memory Address represented as CA is obtained from the equation of CA=DA+i, in which i is mentioned in Step1, and through CA addressing to obtain the Current Value (CV) stored in the byte, and the value CV is assigned to the Register Value (RV);

Step4: Add the CIV value to the CV value and assign it to CV, judge whether the following conditions are satisfied: the CV is less than the CIV and the memory address offset i is less than C-1, if the content is satisfied, it enters the Step5, and if not, it is ended and exited;

Step5: Sets the increment step size CIV = (CIV + RV) /256, the memory address offset i=i+1, and turns to step2.

Obviously, step 2 to step 5 is a recursive process. The flow chart of DMI method is shown in Figure 2.

3.3. RS method

The random sampling (RS) method first divide the dynamic memory into N parts, then fill in the sub Dynamic Memory (DM) with pseudo random number, and finally generate Binary Sequence (BS). According to the theory of multidimensional random variable distribution, the distribution of the generation result based on this method is uniform distribution. The method steps are as follows:

Step1: The dynamic memory DM is divided into a continuous sub dynamic memory space DM(i) which represents the value stored in the memory of the address in DA+i, the size of 1 byte, and the i represents the offset of the sub memory space in bytes relative to the first address, as an unsigned integer, from 0 to C-1;

Step2: Generating a random number seed;

Step3: Define the counter represented as CT and initialize it to 0;

Step4: Judge whether the counter CT is less than C-1, if it satisfies, turn to Step6, otherwise turn to Step5.

Step5: Use program to produce random number and copy the random number to the sub dynamic memory DM (CT), and add the counter then turn to Step4.

Step6: Output the BS in DM, and the program ends.

The flow chart of RS methods is shown in Figure 3.

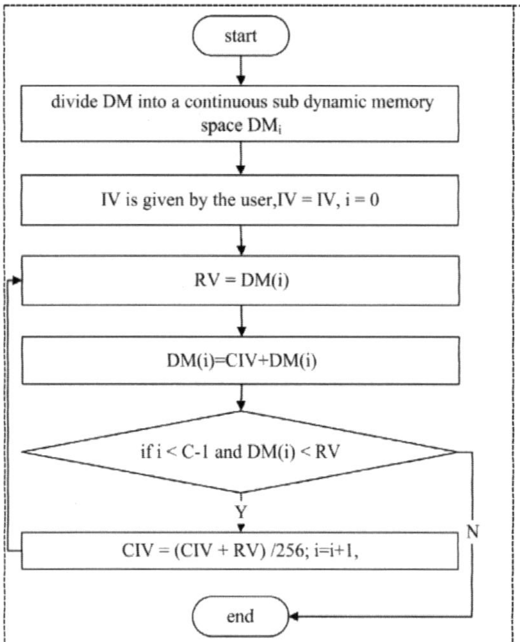

Figure 2. Flow chart of DMI method

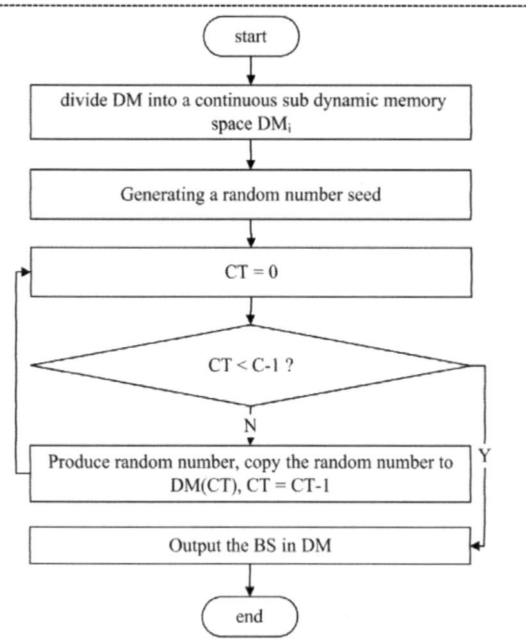

Figure 3. Flow chart of RS method

The test case samples generated by DMI method are concentrated in a certain area, which is suitable for testing interfaces with sensitive boundaries. Under the premise of not knowing, it is preferable to use RS method to generate test cases.

4. Application examples

4.1. Coding of the IBSP algorithm

IBSP algorithm is realized by C++ language programming. First, a DataBlock class is designed to represent the memory block of the test case. There are two member variables, the character pointer which points to the memory data block and the integer variable representing the size of the data block. Secondly, the + + symbol for the incremental operation is overloaded; finally, the two friendly function "Traverse()" and "RandomSample()" are defined to complete the traversal and random sampling of the data blocks. It is worth noting that there is a callback function type in Traverse() and RandomSample() parameters, which can return the result of each test by using a user-defined callback function.

4.2. Example

The following simple functional module is tested by IBSP algorithm.

int Min(int a, int b)

The function of the module is to find a smaller one from two integers. First, the callback function is defined in Table 1.

The callback function saves every test result to the local file. The memory data block is then defined in Table 2.

Next, use the memory increment method, call the function Traverse (db, testAdd) and generated 1000 tests results shown in Figure 4:

Table 1. The definition of callback function

```
struct minS
    {
        unsigned int a;
        unsigned int b;
    };
void testMin(unsigned char *pD, long long times)
    {
        addS *pV = (addS*)pD;
        *ffile<<(int)pV->a<<"\t"<<(int)pV->b<<endl;
    }
```

Table 2. The definition of memory data block

```
DataBlock db;
    addS* pvv = new addS();
    db.BlockPtr = (unsigned char *)pvv;
    db.Len = sizeof(addS);
```

RS method can also be used to call RandomSample (db, 1000, testAdd) to generate 1000 test cases, and the generated test results are shown in Figure 5:

IOP Conf. Series: Materials Science and Engineering **428** (2018) 012001 doi:10.1088/1757-899X/428/1/012001

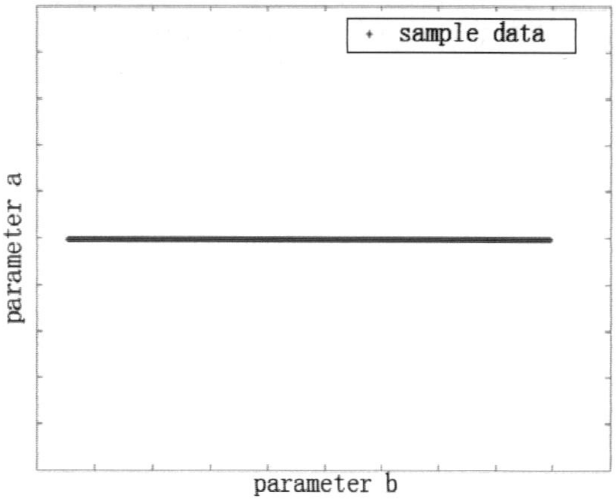

Figure 4. Test cases generated based on DMI method

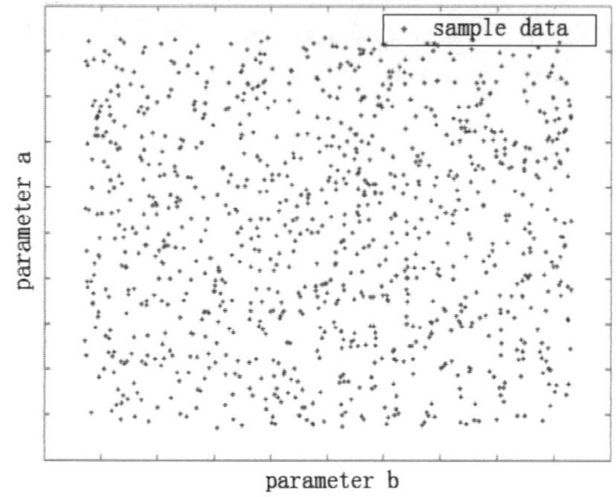

Figure 5. Test cases generated based on RS method

From the comparison between the two, it can be seen that the test cases generated based on the RS method have a wider coverage, so the application is more extensive; and the test cases generated based on the DMI method are more sensitive to the boundary, so this method is more suitable for generating test cases near a certain value.

5. Conclusion

In this paper, a novel algorithm for automatic generation of black box test cases referred to as interface byte stream parsing is proposed, and two methods for generating binary sequence named dynamic memory increasing method and random sampling method are given. This method abstracts the input interface parameters of the diversified input module into the number of bytes used, so only the number of bytes can be given to generate the related test cases, which is beneficial to the automatic generation of test cases by programming and can improve the generation efficiency of test cases.

References

[1] Ron Patton 2006, *Software Testing (2nd Edition)*(Paul Boger): p3
[2] Shurong Liu 2015, *Research on Automatic Generation of Test Data Based on Hierarchic Genetic Algorithm* (Beijing Institute of Technology)
[3] Mao Chengying, Yu Xinxin and Xue Yunzhi 2014, Algorithm design and empirical analysis for particle swarm optimization-based test data generation, *Journal of Computer Research and Development*. 51(4): p824-837
[4] Nubia Technology Co., Ltd. *A black box test case generation method and system based on tree type* (CN201610244017.0)
[5] Harbin Institute of Technology *Single Parameter Event Sequence Test Data Generation Method of Embedded Software*(CN201610663751.0)
[6] Randal E.Bryant, David R.O'Hallaron 2010, *Computer System A Programmer's Perspective (Second Edition)* Beijing (China Machine Press): p1-3

IOP Conf. Series: Materials Science and Engineering **428** (2018) 012002 doi:10.1088/1757-899X/428/1/012002

A Graphical Authentication System Controlled by Eye Movement

Zhanheng Li, Chenwei Xi, and Qiang Zhang and Xiujuan Zheng*

(College of Electrical Engineer and Information Techonolgy, Sichuan University, Chengdu 610065, China)

*Email: xiujuanzheng@scu.edu.cn

Abstract. Graphical password is a potential alternative to the traditional alphanumerical password due to its advantages of recognition and memory in the security system. Eye movement, a natural interaction modality, is a promising tool to replace the keyboard and the mouse in authentication. This paper proposes a graphical authentication system controlled by eye movement. In this keyboard-less authentication system, the user's eye movements were recorded by eye-tracker as the control signal to generate recognition-based graphical passwords. The user test study has been conducted to verify the availability and effectiveness of this keyboard-less graphical authentication system. It's also shown that the efficiency could be improved as the user getting familiar with this system.

1. Introduction

Human is often considered as the weakest link in a computer security system. Many security problems often derived from bad interactions between humans and systems. For human, the pictures are generally easier to recognize and remember than text [1] regardless the race, cultural, and educational backgrounds. In recent years, graphical password has been proposed as a potential alternative to the traditional alpha-numerical password for promoting more memorable passwords against guessing [2-3]. If the possible pictures used in graphical password system is sufficiently large, it will has the advantage for preventing knowledge-based guess and exhaustive search. Furthermore, if the password space can be customized by the users, a great contribution can be provided for the enhancing the password strength. Thus, it seems that the graphical passwords could has a better resistance to brute-force and dictionary attacks than text-based passwords. In light of these advantages, graphical password has been attracted more and more attentions in various authentication applications including workstation, web log-in systems, mobile devices as well as ATM machines [2]. However, the more pictures or details used for generating the graphical password, the larger image interfaces are required. By using the click-based interaction, the graphical password system was more vulnerable to shoulder surfing problems [3]. Some researcher have been proposed several shoulder-surfing resistant algorithms to improve the click-based graphical password schemes [2, 4].

Eye movement is a natural human-computer interaction modality [5]. By using eye tracking technology, eye movement can be collected and analyzed to obtain the eye's behavioral and attentional features. The eye's behavioral features, such as fixation and saccades, were usually utilized in real time human-computer interaction. The eye movement can be treated as a directly control tool in the interaction. This technique has been used as a non-touchable mouse pointer for disables [6]. Eye movement is collected by an eye tracker and the eye's features are hard to directly extract by human observers. So that, eye movement is exactly suitable for using in authentication systems because of it high resistance to the shoulder surfing attacks.

Eye tracking technology has been explored to collect and measure eye movements precisely and conveniently. It could provide a new interaction mechanism to strengthen the existing graphical password schemes. Some researchers made attempts to develop eye-movement based graphical password systems by combining the advantages of graphical and eye movement together. Maeder et al. [7] and Hoanca et al. [8] has both proposed graphical password systems based on user's gaze. Moreover, Dunphy et al. [9] implemented Passfaces (http://www.realuser.com) equipped by eye tracking to a real ATM password system.

This paper aimed to propose a graphical authentication system controlled by eye movement. In this authentication system, the merits of eye movement were fully utilized and integrated with a recognition-based graphical password scheme. Furthermore, the availability and effectiveness of this system was validated by a user test study.

2. Method

In this paper, the keyboard-less graphical authentication system was designed with eye tracker. Thus, the eye tracking technology and the design of graphical authentication system would be respectively introduced in this part.

2.1. Eye Tracking Technology

Eye tracking technology has been developed to study of the mechanisms and dynamics of eye rotation and used in researches on psychology, psycholinguistics, human factors engineering, et al. However, the goal of eye tracking is most often to estimate gaze direction. This could promote the application of eye tracking technology in the field of human-computer interaction. The video-based eye tracker, such as the solutions form Tobii Technology AB, is the most widely used current design that is commonly used cameras to focus eyes and record their movements. When using video-based eye tracking technique, the center of the pupil and infrared/near-infrared non-collimated light can be implemented to create corneal reflections. The vector between the pupil center and the corneal reflections can be used to compute the point of regard on surface or the gaze direction. Before using the eye tracker, a calibration of the individual user is required [10].

For the features of eye movement, fixation is the eye movement to stabilize the retina over a stationary object of area of interest. Saccades are the discrete movements that quickly change the orientation of the eyes. Fixation shows the static characteristics of the human vision system, while saccades reflect the dynamic characteristics. The fixation-based (sometimes named as "gaze-based") interaction and saccades-based interaction are two available modalities for human-computer interactions using eye tracking technology [5]. The fixation-based interaction was adopted for the design of graphical authentication system in this paper.

2.2. Graphical Authentication System

The graphical authentication system was designed based on graphical password mechanism equipped with eye tracking technology. A set of single-object pictures was used to generate the personal identification images (PIIs) of the graphical password. The recognition-based technique was used for authentication. The user was required to recognize and identify their preselected PIIs in the registration stages in order to be authenticated. In the porotype, there were 24 single-object images shown in the interface by 4×6 or 5×5 tiles (as shown in Fig. 1), in order to avoid large image interface and reduce the risk of shoulder surfing attacks. These pictures would be randomly sorted for each 5 seconds. The time interval was defined as a refresh frame. While using eye tracking, the user is asked to fixate on the prescribed pictures among the decoy ones for authentication. In this process, the user's fixation were extracted and utilized as a pointing device, like a mouse, by using Tobii EyeX controller (Tobii Technology AB). The fixation used in this study was defined as the total duration and the average location of a series of fixations within the one picture. By fixating eyes for at least 4 seconds at a certain picture, the user can activate the command for selecting this picture as a PII of graphical password. This process can be repeated for several refresh frames (usually 2~10 frames) until the system determined the picture selected by the user. An example is given in Fig. 2.

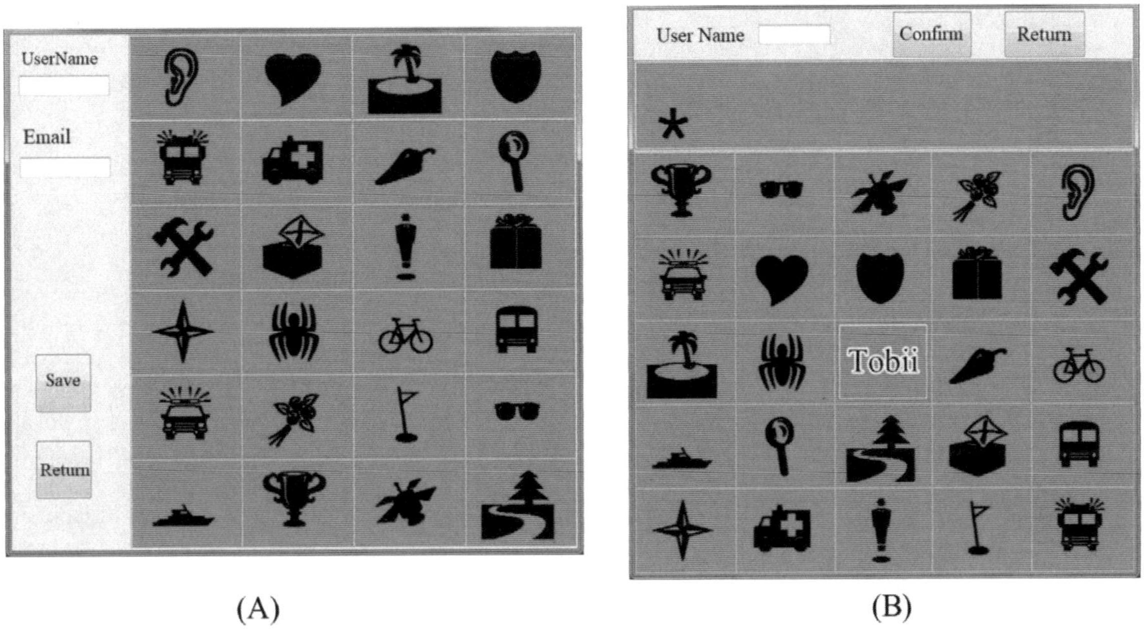

Figure 1. The prototype of the graphical authentication system. (A) the interface for inputting passwords in registration stage; (B) the interface for inputting passwords in login stage.

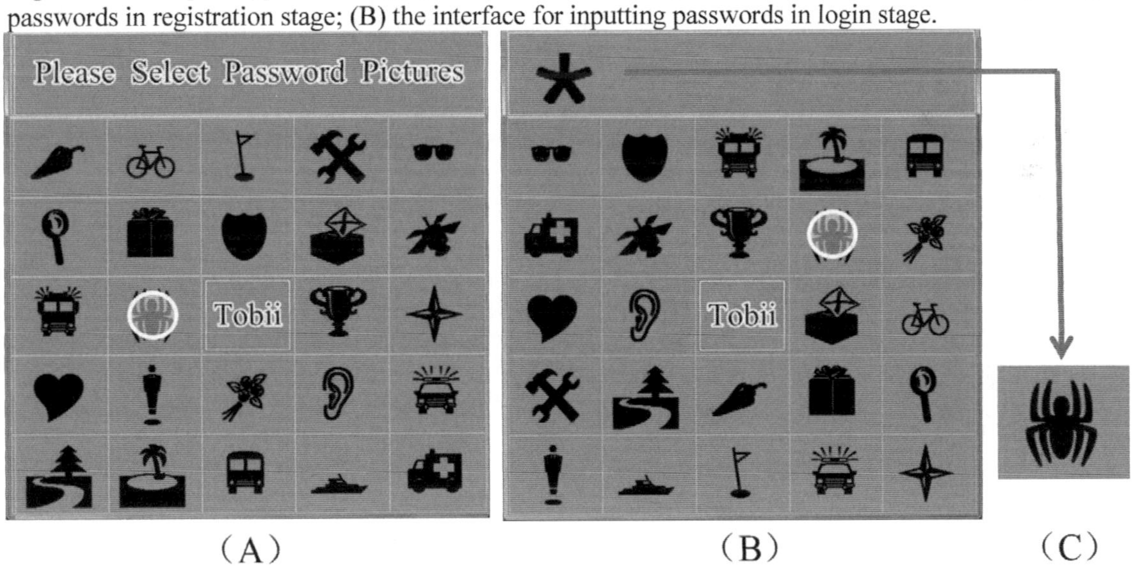

Figure 2. An example plot of extracted fixations for inputting one password in two refresh frames. (A) the first 5 second fresh frame; (B) the following 10 second fresh frame; (C)The selected personal identification image (spider).

To ensure the integrity of the authentication system, there are three main stages in the proposed graphical authentication system: registration, login and password reset. The user can created their own accounts through the stage of registration. The username and password can be both generated by inputting PIIs via the interface shown in Fig.1 (A). These information were then transformed into encrypted information and stored in a database. The graphical password can be rest in the stage of password reset. The framework of this system was developed using Microsoft Visual Studio 2012 and Access 2013.

2.3. User Test Study

The user test study has been conducted to verify the availability and effectiveness of the proposed graphical authentication system. Total six participants, 3 male and 3 female, were recruited for the test study. The ages of the participants ranged from 20 to 24 with an average of 22.5 years. Three of the participants had normal vision and the rest participants had corrected normal vision by wearing the glasses. All of the participants were experienced computer users. None of them had any previous experience with eye-tracking based graphical authentication.

The participant tested all three stages of authentication individually and repeated for five times. For the 5 trails, the participants were asked for performing every trail with an interval of ten minutes and completing whole test within 2 hours. In each trail, the participant sequentially tested the stages of registration, login and password reset. For equal comparisons, the lengths of username were both set as 2. That is, there were only two PIIs in one username or one graphical password. The time spent in every stage was recorded for each trail and each participant. Meanwhile, the error input of PII at login stage were also recorded for each participant.

Finally, the participants were ask for answering a questionnaire about the usability of graphical authentication system.

3. Results

The setup of the proposed graphical authentication system is demonstrated in Fig. 3. The Tobii EyeX controller was mounted at the bottom of a 21-inch monitor and connected with a computer (i7-4790 3.6GHz, 4G RAM) through USB3.0. The participant sat in front of the screen with his/her comfortable position. If needed, the participant can wear the glasses to have normal vision.

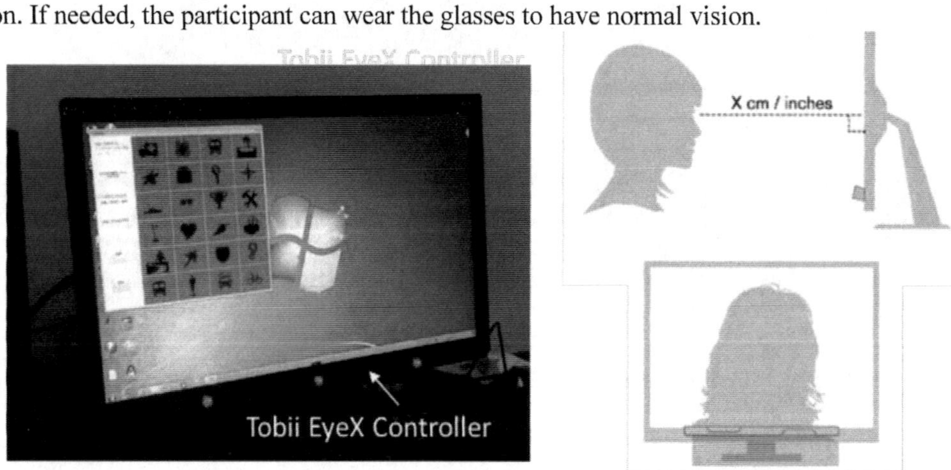

(A) (B)

Figure 3. The demonstration for the setup of the proposed graphical authentication system. (A) the installation of Tobii EyeX Controller in this study; (B) the position of the participant (reproduced from http://www.tobii.com/).

After given a brief introduction of the graphical authentication system, the participant began with a calibration for using Tobii EyeX controller, and then completed the three stages of registration, login and password reset. At the stage of registration, the participant generated a graphical username corresponding with a graphical password. The information of account was encrypted and saved in a database. The process of registration is illuminated in Fig. 4. The process of password reset is similar with registration.

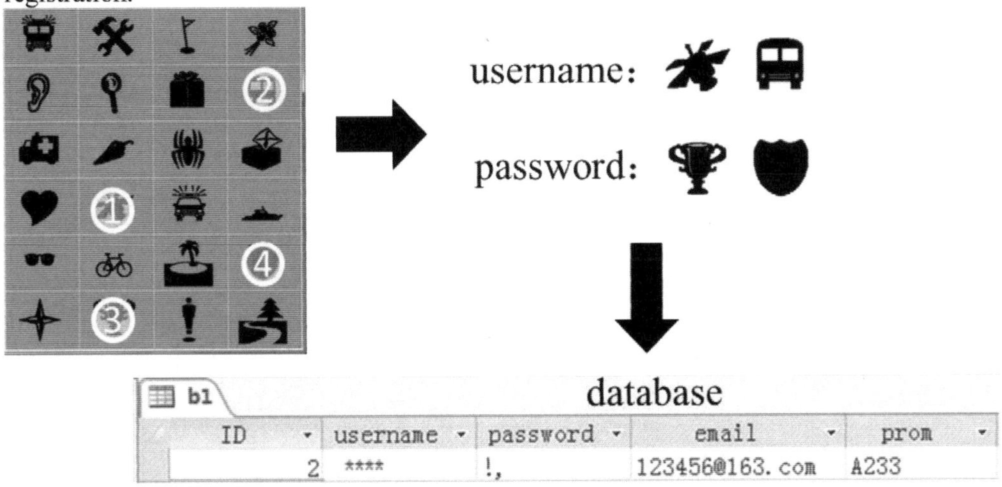

Figure 4. An illumination of the stage of registration.

The average time spent on every stage is calculated and illuminated in Table 1. The results shows that the total time to complete one whole trail ranged from 145 to 180 seconds, and about 50% time was spent on the password reset stage. The login stage, which is most frequently used in authentication system, consumed the least time compared with the other two stages.

Table 1. The results of time (mean±std) spent on every stage in graphical authentication system*.

participants	registration	login	password reset	total
P1	50.6±5.3	28.0±3.2	86.4±6.3	165.0±10.9
P2	51.8±4.1	29.6±3.1	91.6±6.6	173.0±11.9
P3	47.0±4.4	24.2±2.0	84.2±3.4	155.4±9.4
P4	51.8±4.0	28.8±4.0	87.4±6.2	168.0±13.6
P5	50.4±4.0	25.0±2.8	86.8±5.3	162.2±11.7
P6	48.0±3.7	24.6±2.9	86.4±5.5	159.0±11.6

*The unit is second.

The time spent on every stages were recorded and compared for the five trails for all participants. The results are illuminated in Fig. 5. From Fig.5 (A)~(D), the apparent declines of the time spent on authentication could be observed with fluctuant standard deviations. The tendencies reflected by the results shows that all the participants spent more time for the first use no matter in which stage. The time would be decreased when the participants getting familiar with the authentication system. It could indicate that the length of time spent for each stage was highly correlated with the user's experience of the graphical authentication system. Among five trails, one participant had two errors for PII inputting while two participants had one error. The rest three participants had no error during login. It is generally required additional time to remove and fix the error PIIs, so that the time spent on the same stage would have a large standard deviation due to the errors. This is a reasonable explanation for the fluctuant standard deviations of the login time in Fig. 5 (B).

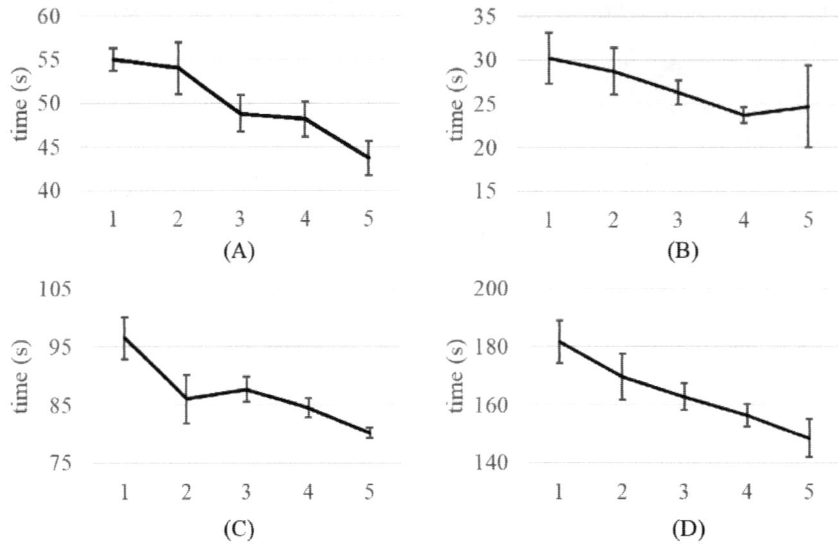

Figure 5. The changes of the average time spent on every stage of authentication. The standard deviations are given by error bars. (A) time spent on registration stage; (B) time spent on login stage; (C) time spent on password reset stage; (D) the total time for authentication.

All of the participants succeeded in the authentication using eye tracking. From the questionnaire, all the participants pointed out that the eye movement based approaches required more operating time than click based approaches. Most of them showed the interests in this system and gave a positive affirmation of the usability. All participants expressed a willingness to use the graphical authentication system based on eye tracking in the future. However, the user test were still limited to give a comprehensive evaluation for the proposed authentication system. In the future work, the strategies of PII determination based on the fixations at the refresh frames could be further explored to improve the efficiency and usability of the proposed authentication system.

4. Conclusions

A prototype of the graphical authentication system controlled by eye movement has been developed and tested by 6 users. The results of user test study have validated the availability and effectiveness of the proposed graphical authentication system. The operating time would be decreased, when the users were getting familiar with the system especially the use of eye tracker. In a conclusion, this paper has proposed a potential solution for the shoulder-surfing resistant authentication system with keyboard-less real-time interactions.

5. Acknowledgment

The author would like to acknowledge an insightful discussion regarding eye tracking technology with Dr. Yun Zhang at the Eye Tracking Lab, Electronic Information and Engineering School, Xi'an Jiaotong University, China.

References

[1] R. N. Shepard, 'Recognition memory for words, sentences, and picutres', Journal of Verbal Learning and Verbal Behavior, 6: 156-163, 1967.

[2] X. Suo, Y. Zhu and G.S. Owen, "Graphical Passwords: A Survey", 21st Annual Computer Security Applications Conference (ACSAC'05), Tucson, AZ, 2005, pp.463-472.

[3] K.V. Renaud, "Guidelines for designing graphical authentication mechanism interfaces". Int. J. Inf. Comput. Secur., 3(1):60-85,2009.

[4] D. Hong, S. Man, B. Hawes, and M. Mathews, "A password scheme strongly resistant to spyware," Processing of International conference on security and management. Las Vagas, NV, 2004.
[5] Y. Zhang, X. Mou, "Survey on Eye Movement based Authentication Systems", Proceeding of Chinese Conference on Computer Vision 2015, Xi'an China, Spet 2015, pp. 144-159.
[6] A. T. Duchowski, Eye Tracking Methodology: Theory and Practice. Secaucus, NJ, USA: Springer-Verlag New York, Inc, 2003, pp. 319.
[7] A. J. Maeder, C. B. Fookes, and S. Sridharan,"Gaze based user authentication for personal computer applications", Proceedings of International Symposium on Intelligent Multimedia, Video and Speech Processing, 2004. October 20-22, 2004, Hong Kong.
[8] B. Hoanca and K. Mock. Secure graphical password system for high traffic public areas. In ETRA '06: Proceedings of the 2006 symposium on Eye tracking research \& applications, San Diego, California, Anonymous ACM, New York, NY, USA, 35-35.
[9] P. Dunphy, a. Fitch and P. Oliver, P. 2008. Gaze-Contingent Passwords at the ATM. In Prague, Czech Republic, September 2-3, pp, 50-62.
[10] D. W. Hansen, Q. Ji ,"In the Eye of the Beholder: A Survey of Models for Eyes and Gaze". IEEE Trans. Pattern Anal. Mach. Intell. 32 (3): 478–500, 2010.

IOP Conf. Series: Materials Science and Engineering **428** (2018) 012003 doi:10.1088/1757-899X/428/1/012003

Research of UHF External Sensor Performance Calibration Test

YAN Jiangbao[1, 2, *], TAN Xiangyu[2], ZHANG Changsheng[1], ZHOU Lei[2,3] and ZHANG Shaoquan[2]

[1]College of Information Engineering and Automation, Kunming University of Science and Technology, Kunming, 650500, China
[2]Yunnan Electric Power Research Institute, Kunming, 650217, China
[3]Department of Automation, North China Electric Power University, 071003, China

[*]Email:120958764@qq.com

Abstract: UHF sensor is one of the most important parts of GIS equipment partial discharge detection system, and its performance directly affects the accuracy of fault diagnosis results. UHF sensor lacks uniform standard, performs equivalent height test and sensitivity test to calibrate its performance, and builds a UHF sensor performance calibration test platform based on GTEM cell for testing. This paper mainly tests the equivalent height of different types of UHF external sensors at different installation angles. With the UHF detection system, the sensitivity test and performance comparison of each sensor are used to select the better sensor.

1.Introduction

Gas Insulated Switchgear (GIS) has the advantages of miniaturization, high reliability, good safety, prevention of adverse impact to the outside, short installation period, convenient maintenance and long maintenance period. [1].As GIS is more and more widely used in power system, the accident rate in GIS increases. Therefore, it is necessary to carry out partial discharge test to detect the defects in GIS. At present, the main methods for detecting partial discharge of GIS include Ultra High Frequency (UHF) detection, pulse current method, TEV (Transient Earth Voltage) detection and ultrasonic testing[2].The UHF detection method can well shield the electromagnetic interference which is not easy to be identified, overcomes the shortcoming of low frequency and narrow frequency band of the traditional pulse current method, and is safe and reliable. It has been well used in Partial discharge detection of GIS, cable motor, combined electrical apparatus and transformer[3].

UHF sensor in the entire testing system is crucial, the performance of a direct impact on the entire testing system fault diagnosis accuracy, it is necessary UHF sensor performance calibration test.Based on the UHF detection platform of partial discharge in GTEM cell, the UHF external sensor is tested for its effective altitude and sensitivity, compared with the performance of different models to choose a better sensor.

2 UHF Sensor

UHF sensor by the antenna, UHF amplifier, detector, high-pass filter, the coupler and the shield shell composition, in addition to the antenna surface is epoxy board for receiving the discharge signal, everything else is the use of metal shielding material to prevent external signal interference [4].UHF sensor internal and physical structure shown in Figure 1:

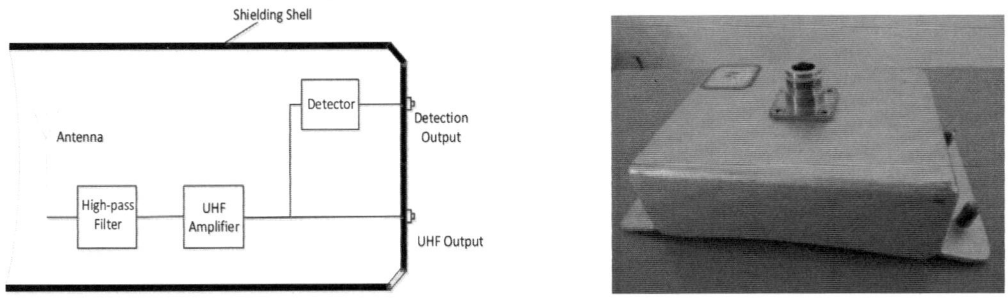

Figure 1. UHF sensor internal structure and physical map

In partial discharge UHF detection method, UHF sensor is divided into two types of built-in and external [5].At present, most of GIS devices with high voltage level at home and abroad are equipped with built-in sensors pre-installed in production, which have high sensitivity and can well collect partial discharge signals generated in GIS devices. However, the introduction of built-in sensors is to a certain extent will change the internal structure of GIS equipment, may change the internal electric field distribution, causing breakdown.A large number of medium-voltage GIS equipment and GIS equipment which has been put into operation for a long time without installing built-in sensors, and external sensors are required to be installed on the inside of the GIS equipment for online monitoring [5].

The external sensor belongs to the antenna detection method and is installed on the pot insulator. The partial discharge condition is judged by detecting the electromagnetic wave signal transmitted from the insulating material. The external sensor is easy to use and disassembled and is widely used in the power system at present [6].

3 GTEM cell testing platform

Due to the low available frequency range of traditional transverse electromagnetic (TEM) cells, which is only a few hundred megahertz, it can not effectively cover the working frequency band of UHF sensors, so the application has been limited[7].Therefore, a Gigahertz transverse electromagnetic wave (GTEM) cell is proposed to transform the TEM cell into a semi-conical coaxial structure to prevent the reflection of indoor electromagnetic waves [8].The size of the GTEM cell is 4×2×2.5(L×W×H). The voltage standing wave ratio is <1.5 in the 80M-2000MHz range and the input impedance is 50Ω.A test window was opened in the middle of the top of the GTEM cell and covered with a teflon cover. The sensor under test was placed on the cover to receive the signal.Window opening in the GTEM chamber near the terminal 1/3 of the area, where there is open space and the field strength distribution is more uniform [9].The calibration signal is injected into the GTEM chamber through a standard calibration pulse source, and a pulsed electromagnetic field is established in the GTEM chamber. The UHF detection platform based on the GTEM cell is composed of a standard pulse voltage source, a measurement and control computer, a high-speed oscilloscope, a GTEM cell, a unipolar standard probe (reference sensor), and measurement and control analysis software, as shown in figure 2.

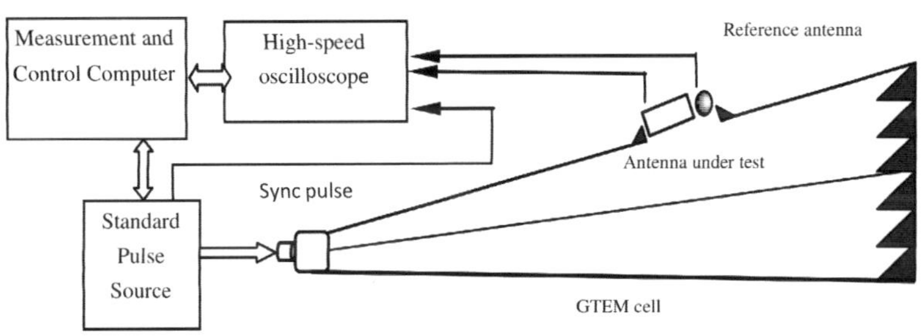

Figure 2.UHF detection system based on GTEM chamber test calibration platform

The test platform software (control/analysis) system mainly includes six parts: (1) Measurement project data management; (2) Through the serial port control signal source to generate periodic pulse calibration signal of a certain amplitude and repetition frequency;(3) Through the USB or Ethernet port control high-speed oscilloscope, the calibration signal, the reference sensor and the test sensor output signal synchronization acquisition and preservation;(4) According to the pulse time domain measurement data and reference method principle, the frequency response characteristics of the sensor under test are analyzed and calculated;(5) The control signal source scans the calibration signal amplitude, according to the signal to noise ratio (SNR) is greater than or equal to 3 principle to determine the sensitivity response of the detection system.;(6) The test data is reported output.

Specific software workflow shown in Figure 3.

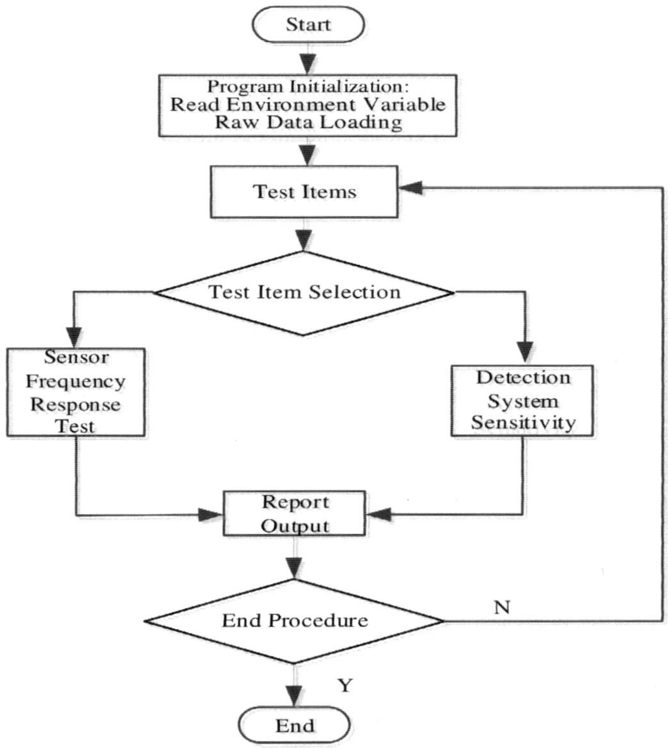

Figure 3. System software work flow chart

The GTEM cell test platform built uses an external sensor and is installed on the top of the GTEM cell window.This approach minimizes the effect of the antenna under test on the field inside the GTEM cell,flexible and does not affect its operation,not only close to the actual installation of high-voltage equipment, but also measurement results more accurate.

4 UHF external sensor equivalent height test

4.1 Equivalent height

Judd M D et al proposed using the equivalent height of the frequency domain to define the coupling performance of the sensor [10].A pulse voltage U1 is fed to the GTEM cell via a standard pulsed voltage source and a field strength E1 is generated inside the GTEM.The output voltage of the reference sensor and the sensor under test are respectively Ur, Us; GTEM cell, reference sensor, the transfer function of the sensor under test are Hcell, Href, Hsens respectively, and the transfer characteristic of the measurement system is Hsys, The sensor's measurement output is respectively

IOP Conf. Series: Materials Science and Engineering **428** (2018) 012003 doi:10.1088/1757-899X/428/1/012003

$$\begin{cases} U_r = U_1 \cdot H_{cell} \cdot H_{ref} \cdot H_{sys} \\ U_s = U_1 \cdot H_{cell} \cdot H_{sens} \cdot H_{sys} \end{cases} \quad (1)$$

By subtracting Href and Hsens from the upper and lower expressions of equation (1), the expression of the transfer function of the sensor to be measured is expressed by the transfer function of the reference sensor.

$$H_{sens} = \frac{U_s}{U_r} H_{ref} \quad (2)$$

By formula (2), it can be known that when the reference transfer function Href, and the voltage response of the reference sensor and the tested sensor to the injected pulse signal are known, the transfer function Hsens of the tested sensor can be obtained.

The test sensor is placed on the test window of the GTEM cell, that is, the antenna under test. The electric field at this position is $E_0(f)$, and the voltage signal output by the antenna is U0(t).It is converted into frequency domain signals $E_0(f)$, $U_0(f)$ after FFT transformation . The unit of electric field is V/mm, and the unit of voltage is V. According to the relationship between the incident electric field and the output voltage, the transfer function of the sensor to be measured is obtained:

$$H_{sens}(f) = \frac{U_0(f)}{E_0(f)} \quad (3)$$

From Equation (3), the unit of Hsens(f) is mm, which is known as the effective height in the frequency domain.When $E_0(f)$ is constant, the higher the output voltage $U_0(f)$ of the sensor is, the stronger its coupling ability is, ie, the greater the equivalent height. It can be seen that the size of the equivalent height reflects the receiving capability of the sensor and is used to characterize the relevant performance of the sensor. In the test band of 0.3 ~ 1.5 GHz, the cumulative height of the equivalent height of each frequency point of the sensor is called the average equivalent height [11].

4.2 Equivalent height test of different types of UHF external sensors

In this paper, six UHF external sensors from three vendors were tested from different models and different installation angles, including two models of PDS-620W, GWA, and SPM-2/GPD. The test result screenshot is shown in Figure 4, and the statistical data results are shown in Table 1.

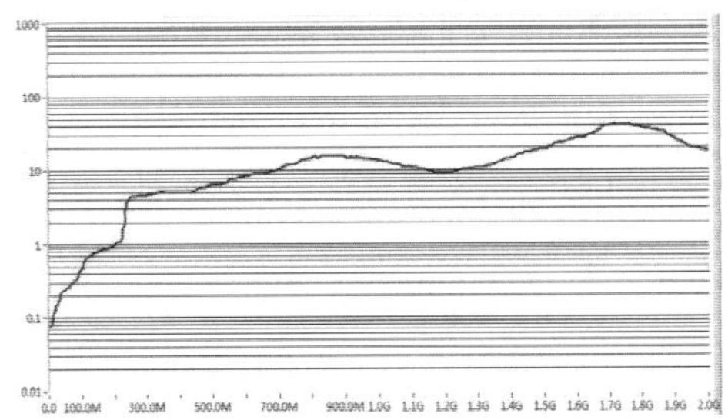

(a)PDS-620W#5 type

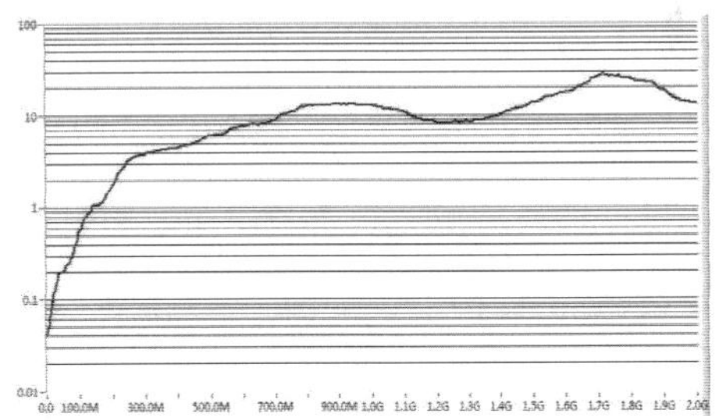

(b)PDS-620W#6 type

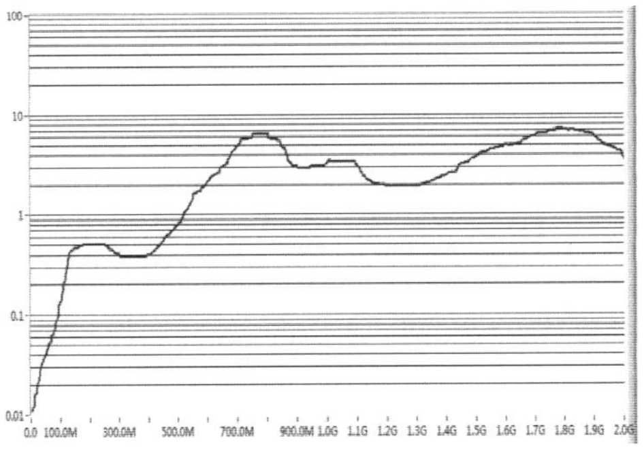

(c)GWA-#1 type

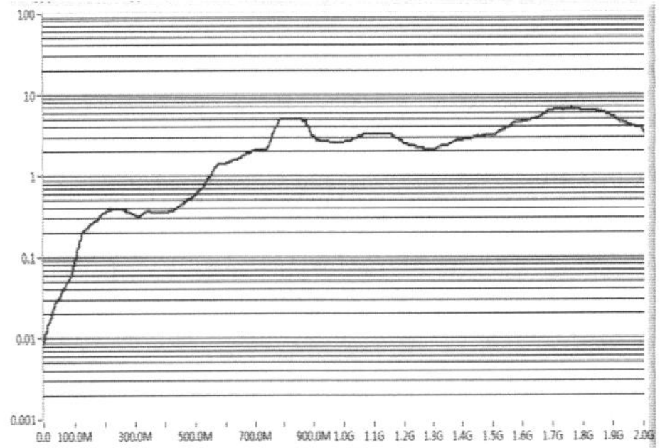

(d)GWA-#2 type

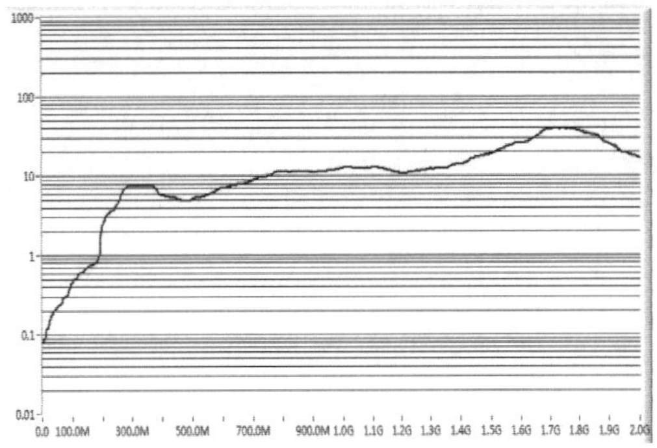

(e)SPM-2/GPD #8 type

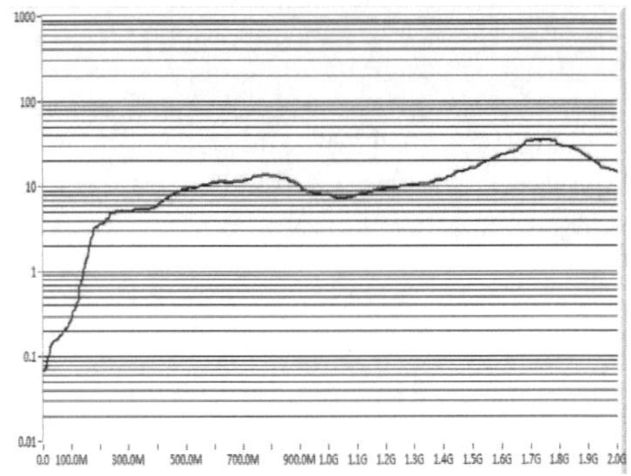

(f)SPM-2/GPD#9 type

Figure 4. different product types UHF external sensor frequency response curve

Table 1. Equivalent height of different models of sensors

Model	Average equivalent height/mm
PDS-620W-#5	11.85
PDS-620W-#6	10.22
GWA-#1	3.14
GWA-#2	2.81
SPM-2/GPD #8	11.44
SPM-2/GPD#9	11.89

When the incident field intensity is the same, the higher the output signal of the GTEM cell antenna, the stronger the coupling performance of the sensor, ie, the greater the equivalent height.From the above results, it can be seen that there is a significant difference in the equivalent height of different types of sensors, and the average equivalent height of sensors in the same factory exhibits a certain degree of consistency.The average equivalent height of the GWA model sensor is obviously lower than that of the other two types of sensors by the above-mentioned frequency response curves of the sensor and the average equivalent height data. The average equivalent height difference between them is 3 to 5 times, which indicates that the GWA model sensor is relatively Sensor coupling performance is poor with the PDS-620W model and the SPM-2/GPD model.

4.3 UHF external sensors with different mounting angles

Since the sensor is not ideally mounted to various fixtures during actual installation, UHF external sensors with different mounting angles are tested. The test result data is shown in Table 2.

Table 2. different mounting angle UHF external sensor equivalent height

Model	Average equivalent height/mm		
	0°	90°	180°
PDS-620W-#5	11.85	1.66	11.57
PDS-620W-#6	10.22	3.9	10.82
GWA-#1	3.14	2.43	4.56
GWA-#2	2.81	3.02	2.61
SPM-2/GPD #8	11.44	1.54	10.22
SPM-2/GPD#9	11.89	3.7	12.61

The data in Table 2 shows that the equivalent heights of the sensors at different mounting angles are different, indicating that different mounting angles have an impact on the detection performance of UHF external sensors. The effective height of the sensor is basically the same at 0° and 180°, and the effective height at 90° is obviously lower than 0° and 180° about 3 to 6 times. Among them, the performance of the GWA sensor is relatively poor compared to other models. The conclusion shows that the different mounting angles of UHF external sensors have a great influence on the acceptance performance (coupling performance) of the sensors.

5. Seneitivity testing

The UHF detection system measures the electric field strength of electromagnetic waves radiated by partial discharges. In order to unify the unit problem of the UHF detection system, the transient peak pulse electric field intensity is used to express the partial discharge UHF detection result. The sensitivity is defined as the peak value of the minimum pulsed electric field strength that the UHF detection system can recognize, that is, the field strength corresponding to the minimum output voltage that the detection system can detect. The smaller the field intensity peak that the UHF detection system can recognize, the better the sensitivity.

The field strengths corresponding to different output voltages are calibrated by standard sensors with known frequency response curves:

$$E_i(t) = IFFT[E_I(f)] = IFFT\left[\frac{U_O(f)}{H_{sens}(f)}\right] \tag{4}$$

The sensor under test is placed on the GTEM cell close to the top window of the upper 1/3 of the terminal, connected to the relevant measurement equipment, and a standard pulse signal is injected into the GTEM cell to adjust the output voltage of the signal source, until under test system detects no voltage signal.The $E_i(t)$ value at this time is the minimum pulse field strength value, that is, the sensitivity of the system under test.The output voltage of the signal source is continuously increased until a certain value makes the signal received by the detection system reach the upper limit. At this time, $E_i(t)$ is the maximum pulse field strength peak value, that is, the maximum measurable value of the detection system.

Through the above method, three different types of sensor sensitivity of the UHF detection system were tested. The test data is shown in Table 3:

Table 3. Sensitivity Test Data Sheet

Model	Minimum Injection Voltage U(V)	Minimum Injection Voltage Corresponding to the Peak Field Strength μV/m	Voltage Injection Upper Limit U(V)	Injection Upper Limit Voltage Corresponding to the Peak Field Strength μV/m
PDS-620W	5	0.0538	40	0.43
GWA	6	0.1038	10	0.208
SPM-2/GPD	6	0.0645	8	0.086

The peak value of the field strength corresponding to the minimum output voltage of the signal source of the PDS-620W type sensor detection system in the table is 0.0538 μV/m,that is the sensitivity is 0.0538 μV/m;When the received signal of this detection system reaches the upper limit, the maximum output voltage of the signal source is 40V, and the corresponding peak field strength is 0.43 μV/m maximum.The dynamic range of the detection system can be derived from the above minimum field strength peak and maximum field intensity peak.The smaller the sensitivity of the UHF sensor, the better the performance. Compared with the above three sets of data, the PDS-620W model sensor has the best sensitivity and can detect the maximum dynamic range of the system, so its performance is best.Based on the above test data, a better sensor can be selected and used in the actual system.

6. Conclusions

In this paper, UHF external sensors based on GTEM cell partial discharge UHF detection system are tested to test the performance. The test results show that the test data show that the coupling performance of UHF sensors of different types is significantly different. The greater the equivalent height, the better the coupling performance.Different installation angles also affect the acceptance performance (coupling performance) of the sensor to a certain extent; the smaller the UHF sensor sensitivity, the better the performance. Through the above performance tests, compare the test data of different sensors,better sensor can be selected according to the actual demand.

The evaluation platform can evaluate the internal and external sensor characteristics, system sensitivity, and partial discharge ultra-high frequency dynamic range characteristics. It can evaluate the overall performance of different manufacturers of different types of UHF detection instruments to solve quantitative calibration issues and comprehensive evaluation issues of the GIS partial discharge detection system, provide a good reference for establishing a complete and standardized test calibration system. UHF sensor performance test can be used to evaluate the performance of the entire UHF partial discharge detection system, improve the accuracy of GIS fault diagnosis results, and ensure the normal operation of the system.

Acknowledgments

The author would like to thank Mr. Tan Xiangyu for his useful guidance and advice on the calibration test of UHF sensor performance during the experiment.

References

[1] Shu Shengwen and Chen Jinxiang et al 2017 Alignment test of effective height of GIS partial discharge UHF sensors *J. Electric Power Automation Equipment* 37(07): 166-170.

[2] Wang Weidong and Zhao Xianping et al 2012 *Analysis and study of partial discharge detection method for GIS J. High Voltage Apparatus* 48(08): 13-17.

[3] Zhang Lei and Lin Qun et al 2011 Simulation Analysis of Partial Discharge Propagation Characteristics of UHF Signals in GIS J . High Voltage Engineering 37(03): 726-731.

[4] Shao Xianjun and He Wenlin et al 2015 Experimental Study on Partial Discharge Detection Characteristics of UHF Sensor GIS J. High Voltage Apparatus 51(01):46-55.

[5] Li Zhili 2005 Research on internal and external sensor detection methods in ultrasonic testing of transformer partial discharge D.North China Electric Power University (Beijing) North China Electric Power University pp 15-20.

[6] Guo Hongfu and Fu Mi, et al 2014 External Ultra High Frequency Partial Discharge Detection Sensor J .World of Sensors 20(08):45-48.

[7] Chen Jun and Wan Fayu, et al 2016 Design of a new type of broadband transverse electromagnetic chamber J .Journal of Hefei University of Technology(Natural Science) 39(07):938-942.

[8] Yang Chao and Meng Cui, et al 2016 Standard device for electromagnetic field of mirror single-cone TEM cell J .High Voltage Engineering 42(05):1476-1482.

[9] Zhao Yang and Yan Wei,et al 2010 Design of multi-functional Ghz transverse electromagnetic chamber J . Journal of Nanjing Normal University(Engineering & Technology Edition) 10(01):1-4.

[10] Judd M D and Yang L. Hunter Lan B 2005 B.partial discharge monitoring for power transformers using UHF sensors part1: sensors and signal interpretation J . IEEE Electrical Insulation Magazine 21(2): 5-14.

[11] Liu Guangbiao and Wang Ke, et al. 2014 Equivalent Height Parameters Test of External Ultra High Frequency Sensors J . Journ*al of Yunnan electric power* 42(6): 81-85.

IOP Conf. Series: Materials Science and Engineering **428** (2018) 012004 doi:10.1088/1757-899X/428/1/012004

Bagging of Xgboost Classifiers with Random Under-sampling and Tomek Link for Noisy Label-imbalanced Data

Luo Ruisen[1], Dian Songyi[1], Wang Chen[1, 2 *], Cheng Peng[1], Tang Zuodong[1], Yu YanMei[1] and Wang Shixiong[3]

[1] College of Electrical Engineering and Information Technology, Sichuan University, 24 South Section 1, One Ring Road, Chengdu, China, 610065

[2] Department of Computer Science, University College London, Gower Street, London, United Kingdom, WC1E 6BT

[3] School of Electronics and Information, Northwestern Polytechnical University, 1 Dongxiang Road, Chang'an District, Xi'an, China, 710129

[*]chen.wang@ucl.ac.uk

Abstract. Fitting label-imbalanced data with high level of noise is one of the major challenges in learning-based intelligent system design. In this paper, for the two-class problem, we propose a bagging-based algorithm with Xgboost classifier (Gradient Boosting Machine) and under-sampling approaches to overcome the challenge. To avoid model misspecification caused by imbalanced data, random sampling with replacement is employed to obtain several balanced training sets; and to mitigate the problem of misleading information produced by noise, Tomek Link method is introduced to eliminate the cross-class overlapped instances, which are the primal sources of noise. And to obtain robust individual learners, we utilize Xgboost, a novel Gradient Boosting Machine-based classifier with convenient parameter tuning interface, to fit each component of the bagging ensemble. The performance of the proposed method is tested with Mandarin radio records (MFCC features) with the task of keywords recognition, and experimental results show that the new method could outperform single Xgboost classifier, verified the rationality and effectiveness of the bagging scheme. The method proposed in the paper could offer a novel solution to the challenge of noisy imbalanced data classification, and the implementation of Xgboost in this area could also serve as an innovative work.

1. Introduction

Data-driven learning algorithms are grasping increasing popularities in the designing of automation and intelligent systems with their record-breaking capacities for various tasks [1]. While this branch of algorithms has shown state-of-the-art performances, learning with highly-noisy class-imbalanced data imposes a significant challenge for the further development of them. Specifically, for classification tasks in intelligent and automation system designing, most of the learning algorithms are based on the assumption of a roughly class-balanced dataset with minor noise; in practice, however, data collected from industrial applications could be highly skewed in label distribution and associated with significant noise. Thus, when encountering label-skewed noisy data in automation and intelligent system, the performances of the advanced learning techniques often drop drastically [2].

There exist previous research endeavours to address the problem, and they could be roughly categorized into four types: surrogate loss functions, re-sampling methods, ensemble learning and special class feature representation (so-called one-class learning) [3]. Among the above categories, the re-sampling approaches, which intend to produce a balanced dataset by oversampling the minority class and/or sub-sampling the majority class, and ensemble learning methods, which aims to stack multiple learners working together to get a more satisfying prediction, could be regarded as approaches from a data-level perspective. The major advantage of these two types of methods is that it does not demand the modification of the structure of the classifiers, thus they could be flexibly plugged into different types of powerful individual algorithms. Previous research mostly utilizes re-sampling and ensemble method separately, and it naturally comes to an intuition that a hybrid of the two methods might have a better performance, which constitute the inspiration of this paper.

In this paper, following the above inspirations of sub-sampling and ensemble learning, we design a novel algorithm to process class-imbalanced noisy data for two-class classification. Specifically, for a dataset considered as label unbalanced and significantly noisy-affected, the algorithm will firstly produce several balanced sub-datasets by randomly sampling the majority with replacement and combining them with the minority. The number of sub-datasets are specified manually, and this process this much alike conventional bagging method. After obtaining the datasets, we perform sub-sampling method, more specifically Tomek Link elimination [4], on them to mitigate the noise by removing the overlapped data between the two classes. Subsequently, we could fit one classifier per subset to get a bagging algorithm result. In our method, the individual classifier is selected as Xgboost Classifier, which is an advanced implementation of the popular-employed Gradient Boosting Machine [5]. And finally, when predicting a new example, the algorithm could weight each individual voter equally and output a probability, which will be its confidence in predicting the instance as majority/minority.

To this end, the rest of the paper is arranged as follows: section 2 will be reviewing related work in the area and discussing their relationships with our work; section 3 will be adopted as detailed introduction of our proposed algorithm; the experimental results will be illustrated in section 4 together with the setting of the experiment and the discussions regarding the result; and finally, a general conclusion will be demonstrated in section 5.

2. Related Work

Re-sampling has been served as one of the conventional approaches to process imbalance data classification [6], and major publications have discussed both over-sampling and under-sampling algorithms. However, both over- and under- sampling methods could incorporate bias to the data and thus adversely affect the results [7]. And despite the abundant number of publications, it still remains unclear whether over- or under- sampling will be more preferable regarding the overall performance [8]. In this paper, under-sampling is considered for the purpose of ensemble. Popular under-sampling algorithms include NearMiss [9], Condensed Nearest Neighbour [10], as well as the Tomek Link method employed in our algorithm. [11] summarized various under-sampling methods and provided related APIs, which are adopted in the program of the proposed algorithm. Notice that although all the algorithms in this branch are all labelled as 'under-sampling', the intuitions behind them could vary to a large extent. For instance, [12] method aims to produce several distinct representations for the majority class, while Tomek Link, the method utilized in our algorithm, primarily serves as an algorithm to reduce noise.

In addition to re-sampling, ensemble learning has also been adopted as an effective method in tackling label-skewed data classification [13]. Both bagging and boosting ensemble algorithms are included in previous literatures [14], and it is noteworthy that when training a bagging ensemble model on label-skewed data, the requirement of random sampling with replacement could be regarded as a form of under-sampling. And if we only under-sample the majority class, it will be possible to regard the under-sampling techniques as part of ensemble and merge the two branches of approaches. [15] elaborated on the above idea and argued that simply combining random under-sampling and bagging could achieve a satisfying performance. However, the authors did not consider the situation when the noise level is high. [16] discussed a combined method of an over-sampling strategy with bagging of SVMs, which is similar to our approach but focusing on another aspect of re-sampling.

Regarding the Xgboost classifier utilized in our algorithm, there also exists plentiful literatures discussing the technique. The classifier is based on Gradient Boosting Machine and it was firstly introduced by the Chen et al. in 2016 [5]. It illustrates impressive performances on various Machine Learning challenges [17] and has been implemented to different scientific and engineering sectors, including predictive finance [18], symptom analysis [19] and biostatistics [20]. However, to the best of the authors' knowledge, there has not been any research outcomes focusing on the bagging of Xgboost models to get better predictions. As a boosting ensemble method, Xgboost tends to more likely to be overfitting [21], thus it is intuitive to think that applying bagging beyond Xgboost models could improve the overall performance. Meanwhile, by removing the overlapped instances with Tomek Link, we could maximize the advantage of reducing bias in boosting while avoid potential overfitting of noise.

3. Methodology

The holistic algorithm is composed of four major parts: random sampling with replacement of the majority data, Tomek Link elimination, individual Xgboost classifier fitting, and bagging of different classifiers for prediction. In this section, the details of each part of the algorithm will be discussed and we will show the intuitions and rationalities of the adopted techniques.

3.1. Random Sampling with Replacement

Random sampling with replacement is the standard technique utilized in bagging algorithm to produce sub-datasets, and the difference in our algorithm is that we only perform this operation on the majority data, which constitute our primal under-sampling algorithm. In our method, to avoid sampling the same instance for multiple times in one subset, we directly sample p distinct instance by getting a random permutation of the index of the majority dataset, and select the first p instances of the permutated data.

Random sampling with replacement could produce datasets sufficiently representing the variance of the data. On the contrary, more complicated under-sampling methods, such as clustering-based technique, tend to better represent the bias which we demand the classifiers to learn. In our algorithm, since we would like to focus on reducing variance during the bagging procedure, it is preferable to directly utilize random sampling with replacement. The comparison between simple random sampling with replacement and the complex clustering-based under-sampling could be shown in figure 1.

Since we intend to obtain a set of instance-balanced data, ideally, we would set the sampling size $p = m_k$, where m_k is the size of the data lying in minority class. However, as we will see in the following paragraphs, the randomly sampled data will be further processed with Tomek Link Elimination, which means, the instances of majority which overlapped with the minorities will be removed. Thus, we would like to add some additional instances to the majority class. In our algorithm, we set $p = \lfloor 1.5 * m_k \rfloor$.

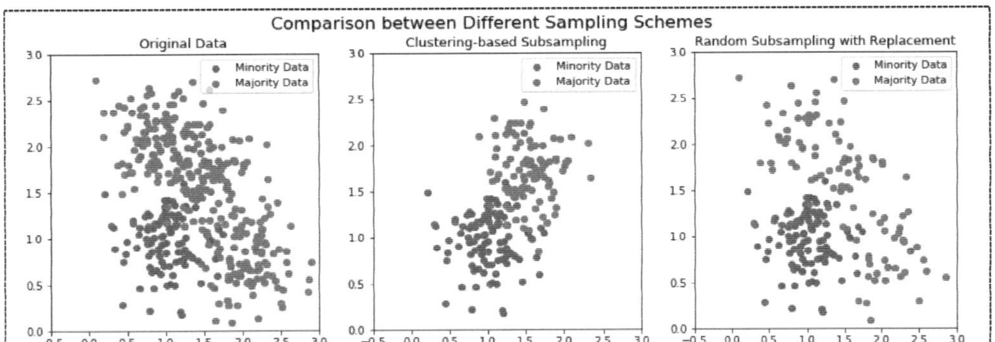

Figure 1. Comparison between clustering-based under-sampling and random sampling with replacement.

3.2. Tomek Link Elimination

Tomek Link Elimination is a widely-applied conventional under-sampling technique. Unlike other under-sampling counterparts in the field, which will remove a large portion of the majority data to produce a nearly-balanced subset, Tomek Link will only remove the instances that overlapped with other items with different labels, which it determines as 'noise'. The idea for Tomek Link to find overlapped instances is to compute 'neighbour pairs', which means for two data x_i and x_j, if they are both the other's nearest neighbour in terms of some distance measures (say, Euclidean distance), then the pair of instances will be called 'Tomek link'. If the class labels of Tomek link are different, then we could remove the majority or the minority instance, or both. Formally, the algorithm could be described as:

For two instances x_i and x_j, if for any $x_k \in X \setminus \{x_i, x_j\}$, we could have $dist(x_i, x_j) < dist(x_i, x_k)$ and $dist(x_i, x_j) < dist(x_j, x_k)$, then x_i and x_j are called Tomek link. If the instances of a pair of Tomek link belongs to different classes, then we could remove one or both of them.

In our proposed algorithm, since the objective is to tackle imbalance data classification, we only remove the instance from the majority class with Tomek Link. The setting reflects our priority on the spotting the minority data, as this is usually the more important part in class-imbalanced data. And since we have multiple Xgboost classifiers, a sufficiently complicated decision boundary for the majority class could still be obtained. And regarding the distance measurement of Tomek Link, Euclidean distance could usually lead to satisfying performance, thus in our method this straightforward metric is implemented. Tomek Link Elimination is arranged after random under-sampling in our overall algorithm procedure, and each subset will be processed with this technique respectively. Notice that this procedure will result in a smaller size of the post-sampled majority class data, thus it justifies our setting of the p value in section 3.1. The effects of Tomek Link could be illustrated as Figure 2.

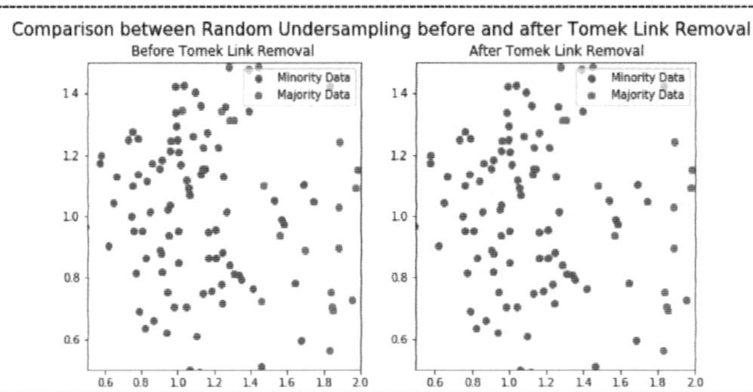

Figure 2. Demonstration of the effect of Tomek Link Elimination.

3.3. Xgboost

Xgboost is based on the algorithm of Gradient Boosting Machine (GBM), which is a popular Machine Learning technique firstly proposed in [22]. As a boosting ensemble method, it follows the idea to learn from the mistakes of the previous steps. Specifically, in GBM we use the gradient of the loss function with respect to the existed model to represent a 'pseudo-residual' between the existed predictions and the true labels/classes. Formally, at step K, suppose we already have a model of:

$$F_K(x) = \sum_{k=1}^{K} \alpha_k f_k(x; \theta_k) \tag{1}$$

Where $f_k(x; \theta_k)$ is the sub-model of the k-th step and α_k is the weight of the corresponding model. Then the 'pseudo-residual' of the $K+1$ step will be:

$$r_{K+1} = -\frac{\partial L(y, F_K(x))}{\partial F_K(x)} \tag{2}$$

Where $L(.,.)$ is the loss function that should be differentiable. Then the model of the $K+1$ step will be designed to fit this gradient as residual:

$$\theta_{K+1} = \arg\min_{\theta} L(r_{K+1}, f_{K+1}(x;\theta)) \tag{3}$$

And after obtaining the new additive model at the current step, we would like to add it back to the overall model $F_K(x)$ with a weighting parameter α obtained by line search:

$$\alpha_{K+1} = \arg\min_{\alpha} L(y, F_K(x) + \alpha f_{K+1}(x;\theta_{K+1})) \tag{4}$$

And finally, the holistic model at this step could be denoted by:

$$F_{K+1}(x) = F_K(x) + \alpha_{K+1} f_{K+1}(x;\theta_{K+1}) \tag{5}$$

And we could iteratively grow the model for given steps, or continue the additive fitting procedure until we get performances satisfying some specific metrics. However, when using the second paradigm in training, one should be aware of potential risks of overfitting. And in practice, like other Machine Learning algorithms, a regularisation term, typically l_2 norm of the parameter, will be added to the model to prevent overfitting.

3.4. Bagging of Xgboost and the Holistic Algorithm

As a bagging approach, in our algorithm each Xgboost model will be fitted with the post-Tomek Link under-sampled data respectively, and this will result in T separate classifiers, where T is the number of under-sampling subsets. Conventionally, for bagging of classifiers with two classes, the combining approach is to set the output of each sub-model $h(x) \in \{-1, 1\}$ and sum the output labels, where the final output will follow the judgement of the majority. This kind of hard-threshold combination, however, will increase the opportunity of misclassification. A better approach would be to obtain a probability for instances to be classified to majority/minority, as we might want the instance to be easier to be determined as a majority/minority and more difficult for the opposite.

Our approach to get this probability is quite straightforward: since our topic is limited to two-class classification, we could get a weighted output with weights sum up to 1, and the final output will be the weights of the classifiers that predict the instance as 1 (minority class, in our settings). And this output could be naturally regarded as the probability of the instance to be in the minority class. Formally, this could be denoted as:

$$p(y_i = 1 | x_i) = \sum_{t=1}^{T} \alpha_t H_t(x_i) \tag{6}$$

In practice, we found that setting α values uniformly with $\alpha_t = 1/T$ could achieve a satisfying performance, and this could also be interpreted as the result of random sampling with replacement. With a probability output, we could use a sign function as follows to determine the class of an instance:

$$f(x_i) = \begin{cases} 1, & p(y_i = 1 | x_i) >= \delta \\ 0, & otherwise \end{cases} \tag{7}$$

And this setting enables us to tune the parameter δ with cross-validation and our emphasis on majority/ minority. A lower δ will lead to a higher recall of minority data, which could be sometimes crucial (say, the diagnosis of cancer, which a false negative will be life-threatening); and a higher δ value will lead to a higher recall of the majority data. The holistic procedure of the algorithm could be described as Figure 3.

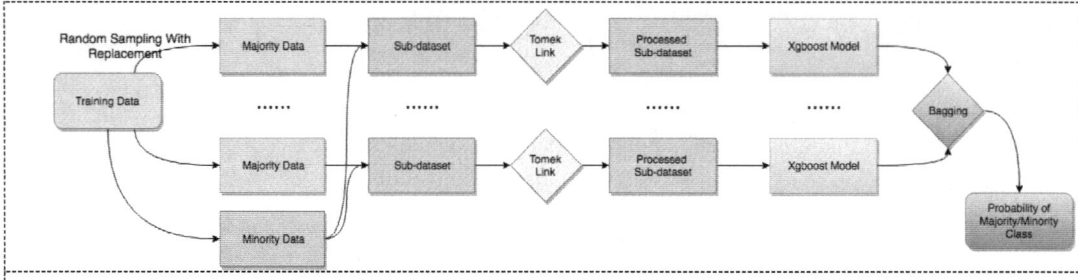

Figure 3. Process of the Holistic Algorithm

4. Experimental Results

In this section, we will be illustrating and discussing the experimental results of the proposed algorithm on a set of Mandarin radio broadcast data. Our task is to recognize if a keyword is presented in a record of radio. In this section, we will show that the proposed algorithm could considerably improve the performance of the recognition task under this dataset.

4.1. Data Composition and Preprocessing

The original data is composed of 133 records of radio broadcast, with few of them containing the keyword 'Beijing Time' (in Mandarin). After pre-processing them into 5-seconds short pieces and labelling the pieces containing the full record of keyword as 'keyword record', we obtained 6906 records in total, with 197 of them containing the keyword. Clearly, this is a label-imbalanced dataset with significant noise, thus it is an excellent example to test our algorithm. The original wave-signal type of data is rarely used in learning-based algorithms. Instead, following the conventional procedure, we transferred the records into MFCC features. And our goal is to build an algorithm to classify instances by their MFCC features.

4.2. Evaluation Metrics and Experiment

Since the classification of label-skewed data could achieve a high accuracy by simply predicting all the instance as majority, ordinary accuracy metric could not sufficiently represent the quality of algorithm. Alternatively, in label-imbalanced data classification, metrics of precision and recall will usually be employed. Using TP, FP, TN, FN to denote the classification results determined as True Positive, False Positive, True Negative and False Negative, then precision and recall of the positive class could be calculated by:

$$\begin{cases} precision = \dfrac{TP}{TP+FP} \\ recall = \dfrac{TP}{TP+FN} \end{cases} \tag{8}$$

And vice versa for the negative class. For specifically the positive or the negative class, we could compute the F1 score of the score:

$$F_1 = 2*(\frac{precision*recall}{precision+recall}) \tag{9}$$

In our experiment, we focus on four evaluation metrics: overall classification accuracy, recall of the majority class, recall of the minority class, and the F1 score of the minority class. We adopt the F1 score of the minority class as the overall evaluation metric, as in our task is to spot the minority (keyword). And as discussed in section 3, the output of our model is the probability to classify the instance as minority (keyword, with label 1), and we could tune the value of δ to get the optimal prediction. In our experiments, the value of δ is tested from 0 to 1 (open interval) with a precision length of 0.05. The change of the classification accuracy, majority (non-keyword) recall and minority(keyword) recall are collected as illustrated as figures. Moreover, the optimal behaviour of different types of algorithms are listed in a table.

There are 4 types of models tested in the experiments, with all of them based on Xgboost classifier and the parameters are best-tuned via validation. The benchmark model is a single Xgboost classifier with the full label-imbalanced data as the training set, and as we could see in figure 4, the problem of overfitting is of great significance. In the figure, the x-axis denotes the value of δ, and the y-axis represents the value of accuracy/recall as shown in the plots. Notice that for the single-classifier situation the output comes back to a hard-threshold decision problem, thus there are large flat-curve areas in the plots.

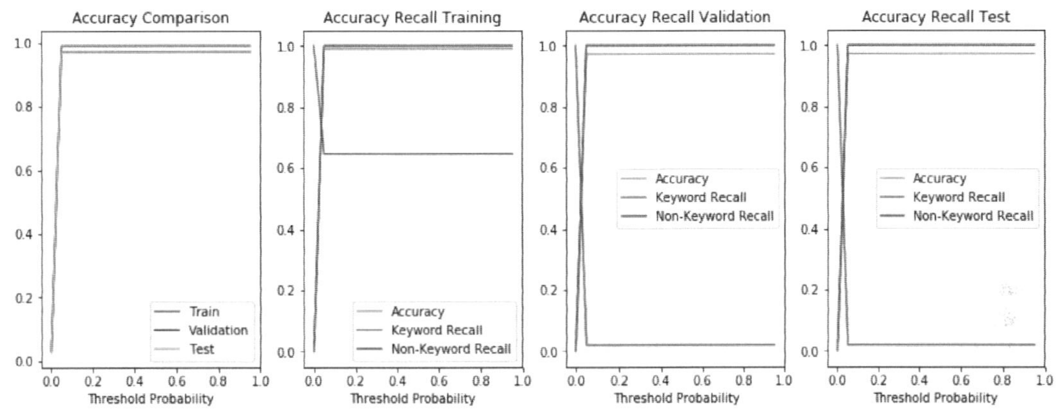

Figure 4. Performances of the benchmark single Xgboost with full dataset.

The second model we tested is the single Xgboost classifier with our under-sampling scheme. This method could be interpreted as 'single-model bagging', which could utilize the advantages of under-sampling and Tomek Link but does not enjoy the merits of bagging of classifiers. The performance of this kind of algorithm could be shown in figure 5. From the figure, it could found that the overfitting problem has been mitigated, and the classifier stop to predict most of the instances as majority. And although there is a decrease in the recall of the majority data, the overall performance has been improved.

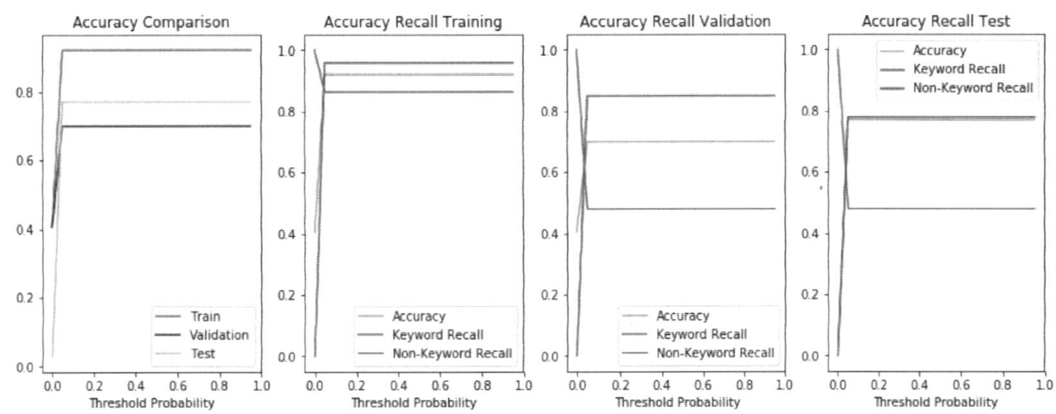

Figure 5. Performances of single Xgboost with the proposed under-sampling scheme.

And finally, the bagging of 5 and 10 Xgboost classifiers following the proposed paradigm are tested with the same dataset. The results of these two methods are illustrated in figure 6 and 7. Two major improvements could be found in the two figures: Firstly, the overall performance considering the hybrid of majority and minority recall has been significantly improved. We could find that the values of the average of the recalls of majority and minority in Figure 6 and 7 could reach significantly higher levels. Secondly, as more than 1 classifiers are added to the holistic model, the differences between the varying values of δ begins to matter. And as we could see from the figure, the choice of δ could be accomplished via validation, and we could get the optimal output together with the prediction confidence of each instance.

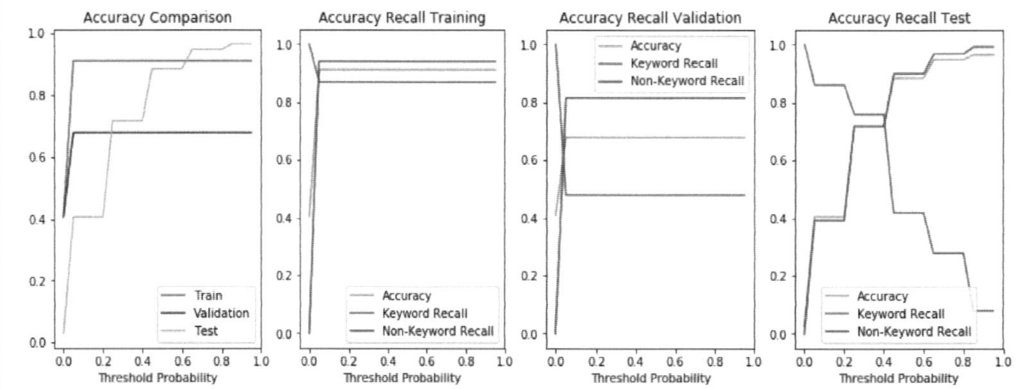

Figure 6. Performances of 5 Xgboost bagging with the proposed algorithm.

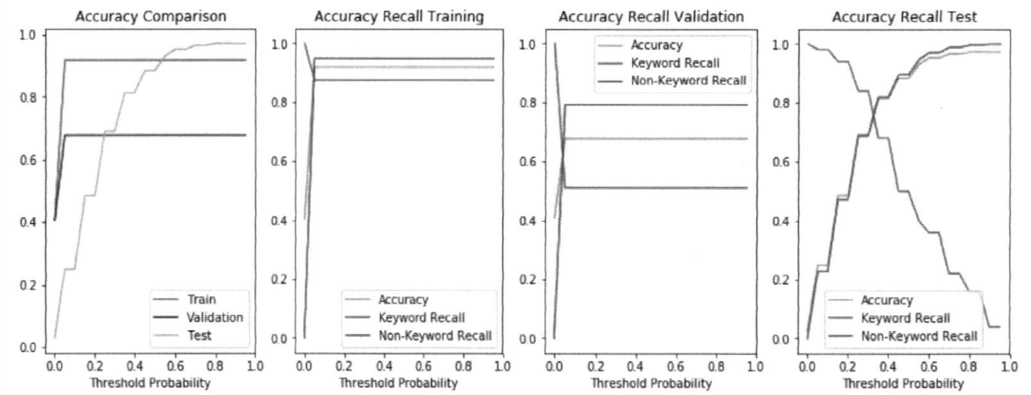

Figure 7. Performances of 10 Xgboost bagging with the proposed algorithm.

Table 1 shows the best F1 score of the test minority (keyword) data and the accuracy/ recall information at the level. There is an additional metric of 'balanced F1 score', which means to assume the minority and majority are in the same size to compute the F1 score. This metric could put additional emphasis on the minority data and increase the retrieved recall of minority (keyword).

4.3. Discussion

From the above experiments, we could found that the proposed bagging of Xgboost model could considerably improve the performance of label-imbalanced data classification. And in addition to accuracy improvement, there also exist other advantages from different aspects in our algorithm, which worth further discussions.

Table 1. Performance Comparison between Different Models

	Best F1 Score	Balanced F1 Score	Minority Recall	Majority Recall	Accuracy	δ
Benchmark	0.0392	0.0392	0.02	**1.0**	**0.9716**	-
1 Xgboost	0.1081	0.5645	0.48	0.7795	0.7708	-

5 bagging Xgboost	0.1300	-	0.32	0.8927	0.8762	0.45
	-	0.6871	0.68	0.7008	0.7002	0.25
10 bagging Xgboost	**0.3077**	-	0.36	**0.9708**	**0.9531**	0.6
	-	**0.7803**	**0.84**	0.6871	0.6915	0.25

The first remark is the preferable time complexity property of our algorithm. With the under-sampling procedure, the sample size of each training set for individual classifiers decreases, which brings the bonus advantage of less demanded training time. Notice that as a boosting method, the training time of Xgboost will considerably raise when the size of the training set is large. In our experiments, the benchmark classifier, which was trained on the full dataset and achieved the worst performance, actually takes the longest time to train. And for the bagging model, the time complexity should be $O(K \bullet M)$, where K is the number of classifiers and M is the time complexity of individual classifier. This is a favourable property, as we could efficiently fit a model with linear time complexity.

The second point to argue is the probabilistic output, which could lead to benefits more than the simple 'soft output' and 'threshold flexibility'. Probabilistic output could reflect the level of confidence, which could be of great importance under some circumstance. The probabilistic output could also bring benefits in evaluation, as we could compute the cross-entropy loss between the prediction and the true labels. And in the two-class scenario, the log-likelihood is equivalent to the negative cross-entropy, thus we could possibly perform model selection with metrics like AIC and/or BIC.

5. Conclusion

In this paper, a novel algorithm based on bagging of Xgboost classifiers is proposed for label-imbalanced noisy data classification. The bagging procedure is designed with random under-sampling with replacement and Tomek Link elimination to generate data and prevent noise, and each classifier is trained independently with the specific subset of data. After obtaining multiple models, we could combine them with uniformly-distributed weights and transform the output into the probability for the instances to be the majority/minority. The performance of the method has been tested on Mandarin radio broadcast data for keyword recognition task, and experimental results have shown that the proposed algorithm could outstrip simple Xgboost, and adding amounts of bagging groups could improve the performance.

To this end, the paper made the following major contributions. Firstly, it designed a novel algorithm to tackle the classification task for label-imbalanced data with high level of noise. The designed model could outperform existed models and the efficiency is in a high grade. Secondly, the paper applied Xgboost, a recently-proposed GBM-based toolkit, to the problem of imbalance data classification and explored its capability in dealing with such problems. The innovative work could provide copious information for further research with the same technique. And finally, the paper elaborated on the idea to use under-sampling and bagging for imbalanced data classification and made major contribution to the development of this branch of algorithms.

Acknowledgments

The authors would like to thank the assistances and encouragements from colleagues, and the support from National Science Foundation of China (Grant: NSFC51475391), University-Enterprise Cooperation Project (17H1199).

Reference

[1] Russell, S J and Norvig P 2016 *Artificial intelligence: a modern approach* (Malaysia: Pearson Education Limited)

[2] Krawczyk B 2016 Learning from imbalanced data: open challenges and future directions *Progress in Artificial Intelligence* pp 221-232

[3] Ali A, Siti M S and Anca L R 2015 Classification with class imbalance problem: a review *Int. J. Advance Soft Compu. Appl*
[4] Qiong G, Cai Z, Zhu L and Huang B 2008 Data mining on imbalanced data sets *In Advanced Computer Theory and Engineering. ICACTE'08. International Conference on Soft* (IEEE:2008) pp. 1020-1024
[5] Chen T and Carlos G 2016 Xgboost: A scalable tree boosting system *Proceedings of the 22nd acm sigkdd international conference on knowledge discovery and data mining* (ACM:2016)
[6] Branco P, Luís T and Rita P R 2016 A survey of predictive modeling on imbalanced domains *ACM Computing Surveys (CSUR)* p 31
[7] Burnaev E, Erofeev P and Papanov 2015 A Influence of resampling on accuracy of imbalanced classification *In: Eighth International Conference on Machine Vision (ICMV 2015), volume 9875* (International Society for Optics and Photonics) p. 987521.
[8] Weiss G M, McCarthy K and Zabar B 2007 Cost-sensitive learning vs. sampling: Which is best for handling unbalanced classes with unequal error costs? *DMIN* pp.35-41
[9] Mani I and Zhang I 2003 KNN approach to unbalanced data distributions: a case study involving information extraction. *In Proceedings of workshop on learning from imbalanced datasets*
[10] Hart P 2006 The condensed nearest neighbor rule (corresp.) *IEEE Trans. Inf. Theor. (vol 14).* pp 515–516
[11] Lemaître G, Fernando N and Christos K A 2017 Imbalanced-learn: A python toolbox to tackle the curse of imbalanced datasets in machine learning *Journal of Machine Learning Research 18.17* pp. 1-5
[12] Yen S and Lee Y 2009 Cluster-based under-sampling approaches for imbalanced data distributions *Expert Systems with Applications* pp 5718-5727
[13] Sobhani P, Viktor, H and Matwin S 2014 Learning from imbalanced data using ensemble methods and cluster-based undersampling *In International Workshop on New Frontiers in Mining Complex Patterns* (Springer, Cham.) pp. 69-83
[14] Woźniak M, Graña M and Corchado E 2014 A survey of multiple classifier systems as hybrid systems *Information Fusion* pp. 3-17
[15] Galar M, Fernandez A, Barrenechea E, Bustince H and Herrera F 2012 A review on ensembles for the class imbalance problem: bagging-, boosting-, and hybrid-based approaches *IEEE Transactions on Systems, Man, and Cybernetics, Part C (Applications and Reviews)* pp. 463-484
[16] Wang Q, Luo Z, Huang J, Feng Y and Liu Z 2017 A novel ensemble method for imbalanced data learning: bagging of extrapolation-SMOTE SVM. *Computational intelligence and neuroscience*
[17] Omar KBA 2018 XGBoost and LGBM for Porto Seguro's Kaggle challenge: A comparison *Preprint Semester Project* pub.tik.ee.ethz.ch
[18] Song Y 2018 Stock Trend Prediction: Based on Machine Learning Methods *Master's thesis, UCLA*
[19] Torlay L, Perrone-Bertolotti M, Thomas E and Baciu M 2017 Machine learning–XGBoost analysis of language networks to classify patients with epilepsy *Brain informatics* pp. 159-169
[20] Wang C, Wang S, Shi F and Wang Z 2018 Robust Propensity Score Computation Method based on Machine Learning with Label-corrupted Data *preprint* arXiv:1801.03132
[21] Bühlmann P 2012 Bagging, Boosting and Ensemble Methods *Handbook of Computational Statistics* (Springer: Berlin, Heidelberg). pp. 985-1022.
[22] Friedman J H 2001 Greedy function approximation: a gradient boosting machine *Annals of statistics* pp. 1189-1232

The Design of Intelligent Grasping Control System for a Special Operation Manipulator

Songbo Deng*, **He Cai, Ke Li, Tongtong Li, Tao Deng and Yanbo Wang**

Beijing Research Institute of Precise Mechatronics and Controls, No.1 South Da Hongmen Road, Dong Gaodi, Feng Tai District, Beijing, China

*d531477298@163.com

Abstract. In this paper, the control system of intelligent grasping for a special operation manipulator is designed to complete disaster relief, fire fighting, explosive disposal and the like for replacing human beings. The control system of the special operation manipulator is needed to have the characteristics of multiple terminal functions, accurate end accuracy and high system reliability and redundancy within outdoor environment. It puts forward higher requirements for the manipulator performance and operator. After analyzing the work requirements of the manipulator, the D-H parameters and motion space is calculated for further research. The forward and inverse kinematics models are built to make the control system into reality. The target grasping is achieved by using the force and position hybrid control and single joint/end control methods. The intelligent recognition system based on machine learning is constructed. The simulation training based on machine learning is carried out by using the back image of the end camera of the manipulator to help the operator locate the target quickly and accurately. Finally the special operation manipulator is used to carry out the target intelligent grasping experiment based on visual guidance. The test results show the effectiveness of the control system design. Thereby it is suitable to be promoted to other robot or manipulator control system design.

1. Introduction

The terrorist attack presents a new situation with complex and changeable environment, diversified forms and high scientific and technological content. Traditional means could not fully meet the needs of anti-terrorism security. Special operation manipulator could be applied to reduce casualties and improve mission success rate dealing with different types of tasks and environments. Special operations manipulator could shed a vital role in the field of disaster detection, rescue, biochemical detection, road clearance and car bomb detection [1-3].

The research of special operation manipulator is extensive, including executing agency, measuring and positioning technology, intelligent control technology, multi-sensor information fusion technology, visual guidance technology and so on. Some special operation manipulators suitable for certain situations have been developed by some famous companies and research institutes, such as Hunter developed by British researcher, the TEODOR produced by Telerob in Germany, RMI-9WT robot from Pedsco company in Canada, Lizard from Shenyang Institute of Automation Chinese Academy of Sciences in China, Super-DII-type developed by Shanghai jiaotong university in China and the like [4-7].

In this paper, a kind of special operation manipulator control system is studied. After analyzing the work requirements of the manipulator, the D-H parameters and motion space is calculated for further research in section 2. The forward and inverse kinematics models are built to make the control system into reality in section 3. In section 4, the target grasping is achieved by using the force and position hybrid control and single joint/end control methods. At the same time, the intelligent recognition system based on machine learning is constructed, as well as the simulation training based on machine learning is carried out by using the back image of the end camera of the manipulator to help the operator locate the target quickly and accurately. Finally the special operation manipulator is used to carry out the target intelligent grasping experiment based on visual guidance.

2. Development of work

For satisfying the application requirement of special operation manipulator, the selected mechanism contains 4 degrees of freedom, two of which is used to change the attitude of the manipulator while the other two is the direction and open angle freedom of the ender. The mechanism brief diagram is shown in figure 1which is without the open angle freedom of the ender [8]. In figure 1, XY is the world coordinate system and X_0Z_0 manipulator coordinate system.

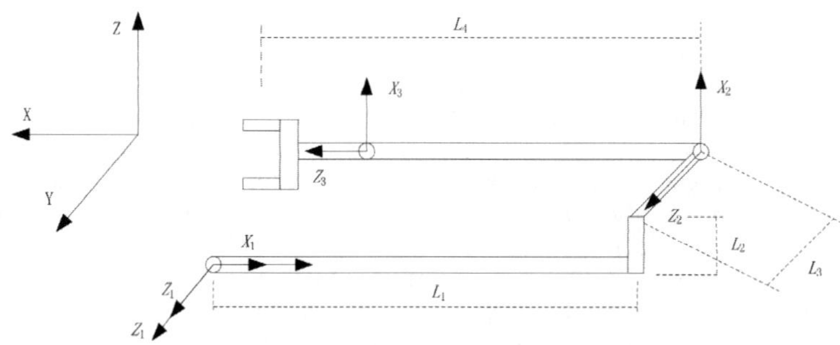

Figure 1. The mechanism brief diagram

Where:

$L_1 = 400mm$, $L_2 = 4mm$, $L_3 = 65.6mm$, $L_4 = 410mm$

Afterward the D-H coordinate system is set up using the standard D-H method shown in Table.1.

Table 1. D-H parameters

i	α_{i-1}	a_{i-1}	d_i	θ_i
1	0	0	0	$\theta_1 + 0.573$
2	0	L_1	$-L_3$	$\theta_2 + 90 - 0.573$
3	-90	0	L_2	θ_3

Relying on the above analysis, the working space of the manipulator is calculated with Matlab simulation shown in figure 2. That indicates the design could fully meet the extension length requirement of operation manipulator application [9].

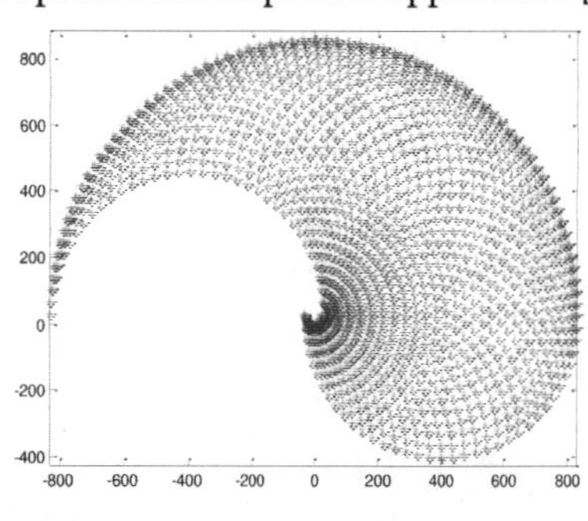

Figure 2. The working space of manipulator

3. Forward and inverse kinematics modeling

3.1. Forward kinematics modeling

The zero position of manipulator is the starting position defined as follows in figure 3. When the manipulator is fully expanded, the joint 1 is 120degree, at the same time the joint 2 is -60 degree and joint 3 remains zero degree.

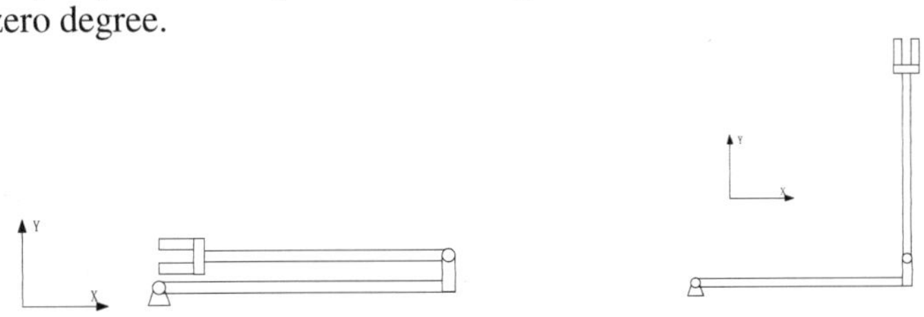

(a) The electrical zero position (b) Kinematics zero position

Figure 3. Zero position definition

While the operator needs to get the end position information of the manipulator, it is necessary to calculate the forward kinematics according to the real-time position of the first two joints obtained by reading the encoders of the joints. The real time angles of joints is named as θ_1 and θ_2. The coordinates of the end of the manipulator is as follows.

$$x = L_1 \times \cos(\theta_1 + \theta) - L_2 \times \sin\alpha_2 \times \cos(\theta_1 + \theta) \times \sin(\frac{\pi}{2} + \theta_2 - \theta) + \sin(\theta_1 + \theta) \times \cos(\frac{\pi}{2} + \theta_2 - \theta) \quad (1)$$

$$y = L_1 \times \sin(\theta_1 + \theta) - L_2 \times \sin\alpha_2 \times \sin(\theta_1 + \theta) \times \sin(\frac{\pi}{2} + \theta_2 - \theta) - \cos(\theta_1 + \theta) \times \cos(\frac{\pi}{2} + \theta_2 - \theta) \quad (2)$$

Where: $\theta = \frac{0.573\pi}{180}$, $L_1 = \sqrt{l_1^2 + l_1^2}$, $L_2 = l_4$.

Thus the end position x and y of the manipulator could be obtained by inputting the angle value.

3.2. Inverse kinematics modeling

It is crucial to pay attention to the initial posture and rotation direction of each joint during the inverse kinematic calculation. It is decouple of the joint 3 from the inverse kinematics of joint 1and2. The inverse kinematics of the manipulator is solved by geometrical method. Inverse kinematics is solved in two cases in coordinate system$\{X_0Y_0Z_0\}$: the end point in second quadrant and in third quadrant. The mechanism brief diagrams of the above two situations are shown in figure 4.

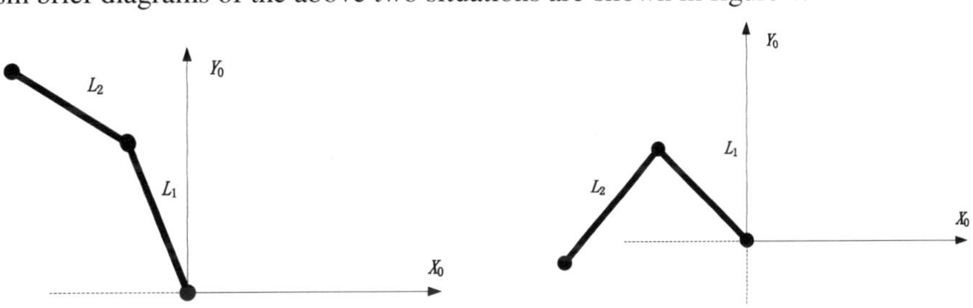

(a) The end point in second quadrant (b) The end point in third quadrant

Figure 4. Inverse kinematics modelling situations

IOP Conf. Series: Materials Science and Engineering **428** (2018) 012005 doi:10.1088/1757-899X/428/1/012005

The actual position of joints driven by the corresponding motors is as follows based on the kinematics modelling and the electrical zero definition when $y > 0$.

$$\theta_1' = a\tan 2(y, x) - \arccos\frac{L_1^2 - L_2^2 + x^2 + y^2}{2L_1\sqrt{x^2 + y^2}} - \theta \tag{3}$$

$$\theta_2' = -\arccos\frac{L_1^2 + L_2^2 - x^2 - y^2}{2L_1L_2} + \theta \tag{4}$$

Where: $\theta = \dfrac{0.573\pi}{180}$.

While $y < 0$, the corresponding position of joints is as below.

$$\theta_1' = 2\pi + a\tan 2(y, x) - \arccos\frac{L_1^2 - L_2^2 + x^2 + y^2}{2L_1\sqrt{x^2 + y^2}} - \theta \tag{5}$$

$$\theta_2' = -\arccos\frac{L_1^2 + L_2^2 - x^2 - y^2}{2L_1L_2} + \theta \tag{6}$$

3.3. Inverse kinematic feasibility analysis

When the end position is operated in Cartesian coordinate system, the end position is calculated by the current joint position and then adjusted according to the instructions issued by the operator. Through inverse kinematics, the angles of each joint are calculated and executed. The feasibility analysis mainly includes two aspects. The one is whether the given end position has exceeded the scope of work space. The other one is whether the joint position obtained is beyond the range of joint motion. For the first case, it could be detected by calculating the distance between the end position and the origin of base coordinate system. The inverse kinematics would be handled when the end position (x, y) meets the limits as behind. Otherwise the feedback information would be given that the manipulator is out of work range [10].

$$\sqrt{x^2 + y^2} < 0.75m\,(x < 0, y > -0.1m) \tag{7}$$

For the second case, the angle of each joint is checked before execution. If it is not beyond the scope of work, the output will be executed. Or else the message will be given that joint is out of the scope of work.

4. Intelligent grasping control system design

4.1. Single joint and end control system design

The single joint and end control technology is mainly switched according to the different task requirements of the special operation manipulator. In this paper, it is adopted with the force and position hybrid grasping control technology and single joint/end control technology. The end of manipulator could be moved forward and backward in X direction as well as up and down in Z direction for the two degrees of freedom of manipulator. The movement in the Z direction is limited while the target is on the uneven surface. The hardware of manipulator or the target will be caused damaged if all of the joints are forced to approach the target. Therefore, it could be carried out position control in the Y direction under the manipulator coordinate system while the force control in Z direction for grasping the target with two sets of parallel complementary feedback [11]. The force and position hybrid control principle is showed in figure 5.

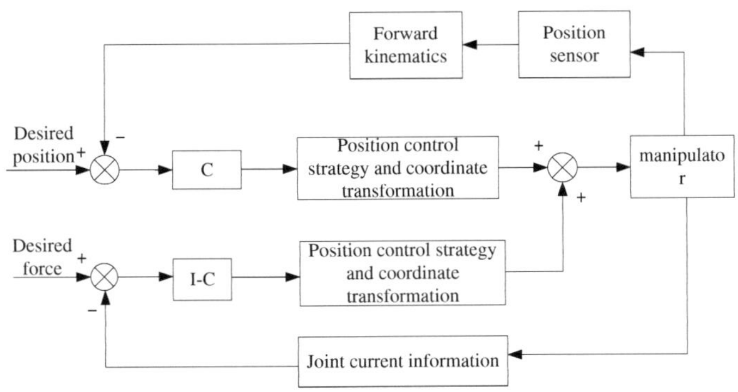

Figure 5. The force and position hybrid control principle diagram

The force and position control is controlled by different control strategies for force and position respectively. The task space is divided into force control space and position control space complementarily. By selecting the matrix to determine the current contact force and position control direction, force and position feedback information are got back to the corresponding loop respectively for force and position control simultaneously in the restricted motion. Each operation of the manipulator could be broken down into several sub-jobs. For each sub-job, a constraint coordinate system{ c} could be used to define a generalized surface. It is defined as the position constraint along the normal direction of the surface accordingly the force constraint the tangent direction. Generally { c} have the following characteristics:

- { c} is a rectangular coordinate system to describe of the operation conveniently.
- Depending on the assignment, { c} may be moved with the end of the manipulator or fixed with base.
- { c} has two degrees of freedom to be decomposed into force and position control along each of the degree at any time.

4.2. Intelligent identification system design

The intelligent identification system based on machine learning is proposed in this paper. The machine learning model is trained to analyze whether there is a target in the field of view, thus helping the operator to locate the target quickly and accurately. The end of the special operation manipulator is equipped with a camera to obtain training data. Firstly, Gaussian blur algorithm is used to remove noise and make the image smooth and easy to deal with. Then, through Canny edge detection algorithm, the edge of the target is identified, which is retained as candidate target. By using the HoughLine algorithm to calculate the target angle, the feature of the target is retained by machine learning algorithm [12]. Finally, the target is grasped with the special operation manipulator. The identification process is shown in figure 6.

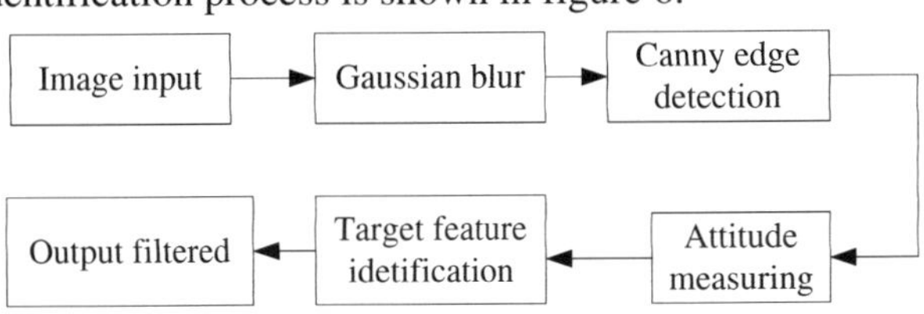

Figure 6. Identification process

The edge gradient is calculated by the blow algorithm.

$$\begin{cases} G_x(x,y) \approx [S(x,y+1) - S(x,y) \\ \qquad + S(x+1,y+1) - S(x+1,y)]/2 \\ G_y(x,y) \approx [S(x,y) - S(x+1,y) \\ \qquad + S(x,y+1) - S(x+1,y+1)]/2 \end{cases} \tag{8}$$

According to the gradient of x and y direction, the gradient amplitude and Angle of the image can be calculated. The edge selection is implemented based on the double threshold method, thus Canny edge detection is completed.

$$\begin{cases} G(x,y)=\sqrt{G_x^2(x,y)+G_y^2(x,y)} \\ \theta(x,y)=\tan^{-1}(G_y(x,y)/G_x(x,y)) \end{cases} \tag{9}$$

Through the Hough transform line detection, a series of straight lines are obtained to determine the general direction of the target. Because of the complex texture region and noise influence, there are many different linear directions, so it is necessary to find the main direction of the target. In this paper, the pixel space of 360 degrees is divided into *m* parts, and then the number of straight lines in each interval is counted. The maximum angle range is set as the main direction. Due to the complexity of environmental background, the candidate target obtained by edge detection is necessary to be identified by machine learning in figure 7.

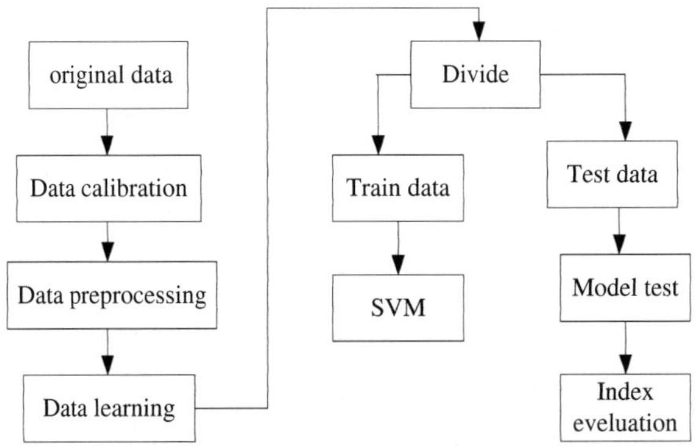

Figure 7. Machine learning process

The special operation manipulator with camera at the end is carried on the mobile platform. At the beginning of the task, the mobile platform firstly moves to the target to make the target located within the grasp of the manipulator. Using a single camera, the collected images are transmitted back to the mobile vehicle platform. After the machine learning and calculation, the target position and grasping path are obtained. The framework of visual processing system is shown in figure 8.

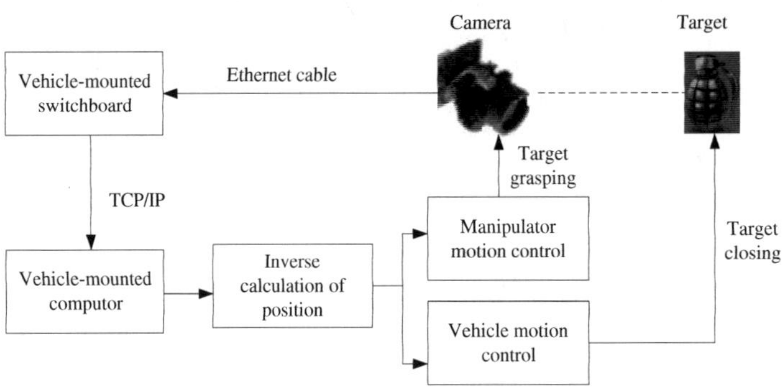

Figure 8. The framework of visual processing system

5. Results and validation

Using the special operation manipulator to grasp the target, the position, speed and current curve of each joint were monitored, as well as the current changes before and after grasping the target. The position, velocity and current curves are shown in figure 9.

IOP Conf. Series: Materials Science and Engineering **428** (2018) 012005 doi:10.1088/1757-899X/428/1/012005

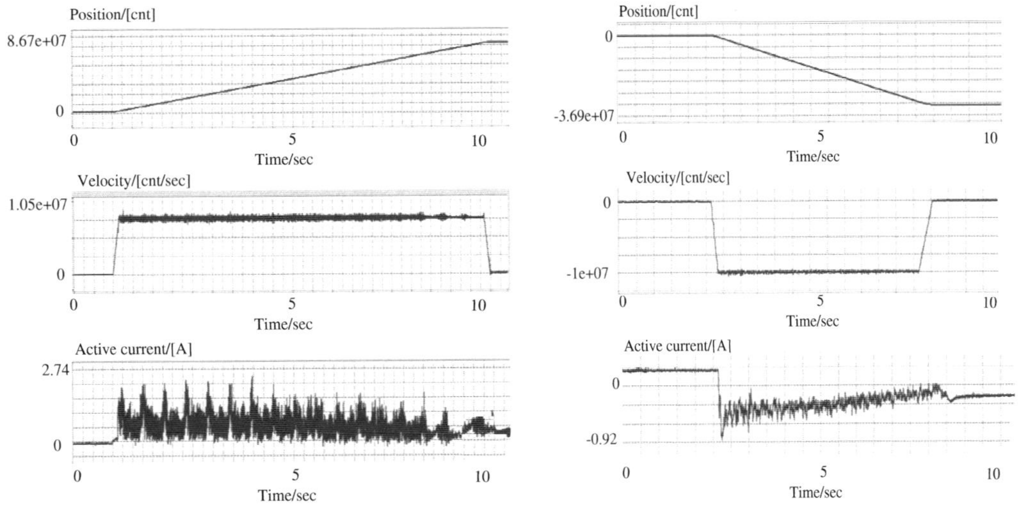

(a) Actual measured position, velocity and current curves of joint 1

(b) Actual measured position, velocity and current curves of joint 2

Figure 9. Actual measured position, velocity and current curves

Current control mode is adopted to grasp the target. In excess of the set current constant value, the claw will maintain constant force grasping to ensure the safety of the target. The change curves of the position, speed and current are shown in figure 10.

Compared with the actual performance curve and design parameters of the special operation manipulator, the design of the control system of the special operation manipulator is proved effectively. It is suitable to be promoted to other robot or manipulator control system design.

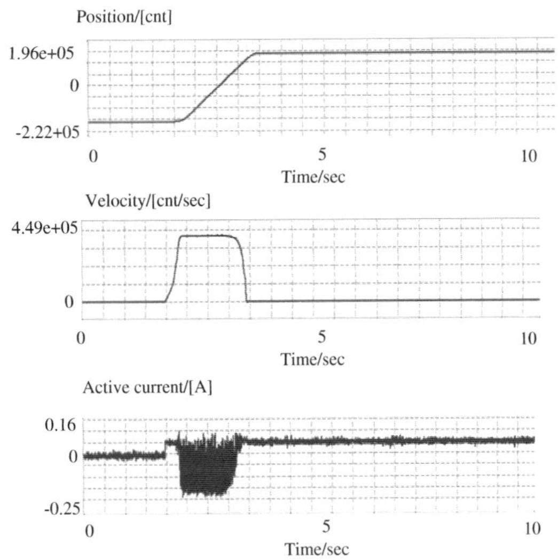

Figure 10. The current control mode grasping

6. Conclusion

In this paper, the control system of intelligent grasping for a special operation manipulator is designed to replacing human beings. After analyzing the work requirements of the manipulator, the D-H parameters and motion space is calculated. Then the forward and inverse kinematics models are built to make the control system into reality. After that, the target grasping is achieved by using the force and position hybrid control and single joint/end control methods. The intelligent recognition system based on machine learning is constructed. The simulation training based on machine learning is carried out to help the operator locate the target. Finally the special operation manipulator is used to carry out the target intelligent grasping experiment based on visual guidance. The test results show the effectiveness of the control system design. Thereby it is suitable to be promoted to other robot or manipulator control system design.

The further research is needed to make the algorithms more efficient for the complex environment, as well as the real time performance of the algorithms and control system is also needed to be improved in future research.

Acknowledgments
Thanks for the team members for their outstanding contributions to the project.

References

[1] Shin Y J, Lee H J, Kim K S, et al. A Robot Finger Design Using a Dual-Mode Twisting Mechanism to Achieve High-Speed Motion and Large Grasping Force[J]. IEEE Transactions on Robotics, 2012, 28(6):1398-1405.

[2] Palli G, Natale C, May C, et al. Modeling and Control of the Twisted String Actuation System[J]. IEEE/ASME Transactions on Mechatronics, 2013, 18(2):664-673.

[3] Wilson J M, Joy J M, Lowery R P, et al. Effects of oral adenosine-5′-triphosphate supplementation on athletic performance, skeletal muscle hypertrophy and recovery in resistance-trained men[J]. Nutrition & Metabolism, 2013, 10(1):57.

[4] Song G. Self-Localization System of the Explosive Ordnance Disposal Robot based on Danger Model Immune Wavelet Neural Network[J]. International Journal of Digital Content Technology & Its Applications, 2013, 7(1):643-651.

[5] Li Zenggang, ADAMS entry Explanation and examples [M], National Defense Industry Press,2006.

[6] R.L. Norton,Design of Machinery, McGraw Hill, 2011.

[7] W. Zhang, J. Yuan, J. Li, Z. Tang, "The Optimization Scheme for EOD Robot Based On Supervising Control Architecture," Proc. of the 2008 IEEE International Conference on Robotics and Biomimetics, Bangkok, Thailand, February, 2009.

[8] K. Massey, J. Sapp, and E. Tsui, "Improved Situational Awareness and Mission Performance for Explosive Ordnance Disposal Robots," Unmanned Systems Technology XI, edited by G.R. Gerhart, D.W. Gage, C.M. Shoemaker, Proc. of SPIE Vol. 7332, 2009.

[9] R. krogh, J. Peng Jin. dynamics of structures[M].Bei Jing, Higher Education Press, 2010.

[10] Wang Y, Li K, Zhou H, et al. Dynamic analysis and co-simulation ADAMS-SIMULINK for a space manipulator joint[C]// International Conference on Fluid Power and Mechatronics. IEEE, 2015:984-989.

[11] N. Pouliot, P. Latulippe and S. Montambault, and S. Tremblay," Reliable and Intuitive Teleoperation of LineScout: a Mobile Robot for Live Trans-mission Line Maintenance," The 2009 IEEE/RSJ International Conference on Intelligent Robots and Systems, St. Louis, Missouri, October 2009.

[12] Yanbo Wang, Ke Li, Haiping Zhou, Songbo Deng, Jian Xu and Jiayu Liu , Dynamic Analysis and Co-simulation ADAMS-SIMULINK for a Spatial robotic manipulator Joint [J], IEEE International Conference on Fluid Power and Mechatronics, 2015.

A Design of Medical Information Sharing Model Based on Blockchain Technology

Hao Jiang, Hao Peng and Songyi Dian*

College of Electrical Engineering and Information Technology, Sichuan University, Chengdu, China

*scudiansy@mail.scu.edu.cn

Abstract. Through the studying of the status quo of medical industry, we find some problems with the sharing of medical data. The medical data in traditional medical institution is stored on severs and its backup database, resulting in lack of communication among medical institutions and ineffective use of medical resources. In order to solve this problem, we propose a medical information sharing model based on blockchain technology. In this model, we build a p2p network to connect different medical institutions and other nodes. Meanwhile, we establish two associated blockchain to store medical data digests and requests. Each node in the p2p network can search the medical information in the distributed medical institutions. In this way, we have integrated medical data between medical institutions and improved the transparency and sharing of medical data. Also, we improve the traditional POW (power of work) algorithm, by using the double-chain structure, we reduce the confirmation time of a block.

1. Introduction

The blockchain was proposed by Nakamoto in 2009, it was first used in bitcoin. It is a decentralized distributed public ledge, which mainly applied to the financial field [1] [2] [3]. In the bitcoin system, transactions can be found in all nodes, and the records are completed by all nodes. With the increase of nodes and blocks, all the transactions in blockchain are hard to be changed. In a broad sense, we can treat the blockchain as a natural distributed system, the data is stored on distributed nodes. Compared with traditional database, the data in blockchain is safer and is widely backed up on distributed nodes. However, there are inherent deficiencies in data search speed and data formatted storage.

We generally think the data in blockchain is very safe as blockchain is composed of interconnected blocks. Figure 1 shows the general structure of a block [3][4].

In Figure 1, A block mainly contains two parts. In block head, it contains some information of the current block and the related information of the previous block. In the block body, it contains many transactions. Each block body records every transaction and encrypts it with a hash algorithm, finally, this information was recorded in block as a merkle root. Every ten minutes, a block may be created, then it will be added to the chain. Due to each block is related to the previous block, all the blocks will be connected in series to form a chain. Each block is created by all nodes, and all nodes reach to a consensus status by POW algorithm. In the process of consensus, each single node searches a proof by a hash calculation, and the node finds the proof first which can generate a block and get profits [5]. Since the blockchain was appeared in 2009, many blockchain applications has emerged. Practical tests found that blockchain has strong security.

IOP Conf. Series: Materials Science and Engineering **428** (2018) 012006 doi:10.1088/1757-899X/428/1/012006

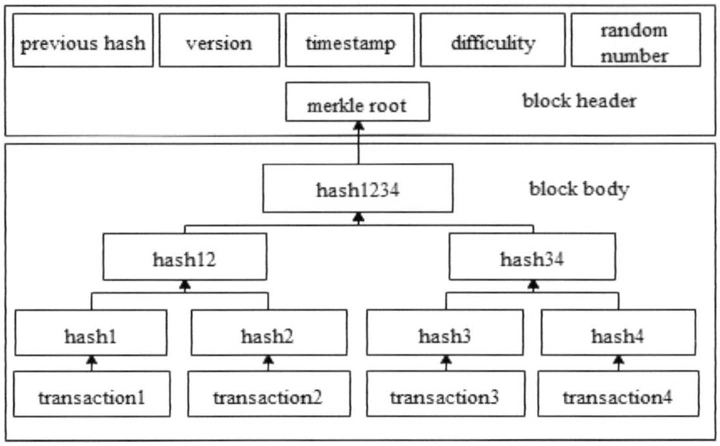

Figure 1. The general structure of a block

The traditional database is a warehouse which was built on a computer storage device, and usually organized by data structure. Due to the complex structure of the data, we can easily store complex data and search for data in the database. The security of database depends on many factors, including the reliability of the system, the security of the network, the stability of the database itself, etc. Therefore, an actual database faces many uncertainties. At present, the security of the database is achieved by many aspects, including identity verification, data verification, access control, etc.

In traditional medical institution, the medical data was stored in serves and its backup database. The data may be tampered maliciously, and the sharing between different institution is difficult [5]. With the development of medical technology, the sharing of medical data is also facing new requirements. We use the knowledge of blockchain, database, cryptography, and using their advantages to establish a medical information sharing system. This system is similar to a widely distributed resource search engine. We put different medical institution onto a p2p network, then we use a public ledger to store the medical data digests base on blockchain technology. Any node in the p2p network can search medical information on public ledge [6] [7]. The method increases the searching speed of medical information. Meanwhile, in order to increase the transparency of medical data query, we build a secondary chain to store requests.

In this paper, we use the POW as the consensus algorithm. As we know, the POW needs many time to confirmation a transaction. We propose a double-chain structure to store medical data digests and requests. By using the double-chain structure, we can greatly reduce the confirmation time.

2. The model of the medical information sharing system

In this medical information sharing model, in order to realize the distributed storage of medical data digests, we build a p2p network between each node. Figure 2 shows the structure of the p2p network.

The p2p network has 4 characters, the characters include doctor, patient, medical institution and searching database. These four types of characters have different functions in the network.

For medical institution nodes in p2p network, it usually contains some databases [6] [7]. Whenever we want to store a medical data, the medical institution node will extract the key information of the medical data at the same time, and then the key information will be transformed as a transaction. The institution node will submit the transaction to the p2p network and stores it in the corresponding database. By using our algorithm, the transaction will be recorded on the main blockchain, so any node can view the medical data digest. In this way, medical data becomes more opening and shareability.

In this model, some non-medical institution nodes also have databases which are called searching database. These nodes convert the medical data digest in the main chain into a corresponding database operation, and then operate on database. These nodes have an API (application programming interface), and we can access the searching database and search the medical data in the p2p network by using the API.

For doctor nodes, this type of node doesn't contain a database, which is mainly used to submit the requests by doctors. When the doctor wants to query some data in database, he will send a request to the p2p network. When the requests reach a consensus by each node, medical institution nodes will check its own database according to the doctor's request. If there is corresponding medical data, it will be send to the doctor node. In addition, we build a secondary chain to store request information to distinguish the information in the main chain.

The last type of node is patient nodes. This type of nodes also doesn't contain a specific database, it is mainly used to submit some authorized information. For some private data, the patient will encrypt and store it into the medical institution's database. In order to access this data for doctor, the patient can submit the doctor's proxy re-encryption key to the p2p network. The corresponding medical institution will use the proxy re-encryption key to handle the encrypted information, then send the re-encrypted data to the corresponding doctor [8]. Similarly, we store these proxy re-encryption keys in the secondary chain.

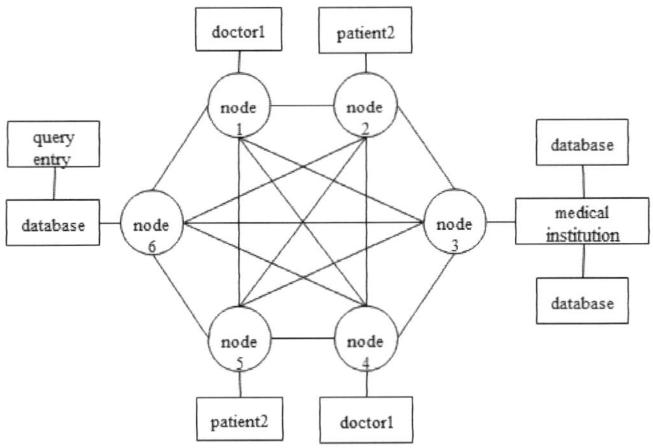

Figure 2. The structure of the p2p network

3. Medical data storage Model

We divide medical data into two types: medical institution's data and patient private data. Figure 3 shows the storage method of medical data.

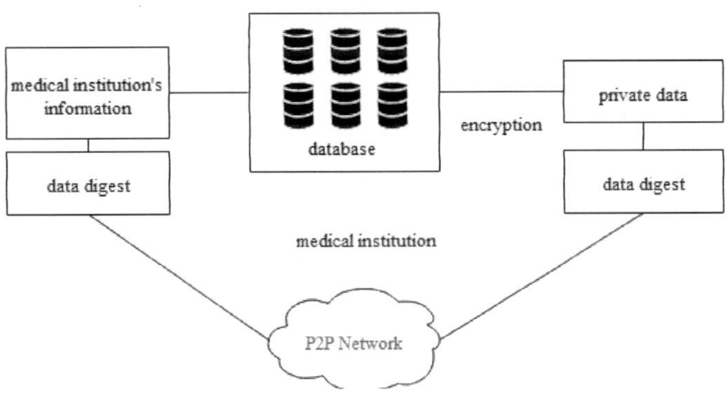

Figure 3. The storage method of medical data

The medical institution's data belongs to medical institution. we store this type of data in the database and submit the data digests to the p2p network.

IOP Conf. Series: Materials Science and Engineering **428** (2018) 012006 doi:10.1088/1757-899X/428/1/012006

For private data, its ownership belongs to the patient. we use the owner's public key to encrypt it, then store ciphertext in the database. Also, we submit the private data digests to the p2p network.

4. An improved pow algorithm

Bitcoin system uses the POW (power of work) as the consensus algorithm [1] [2] [8] [9]. The algorithm generates a block about 10 minutes, people usually think a transaction can be confirmed after 6 blocks, so a transaction needs 1 hour to be confirmed, which greatly reduces efficiency. In the medical information sharing model, we have adopted a double-chain structure to reduce the confirmed time of a transaction. Figure 4 shows the double-chain structure.

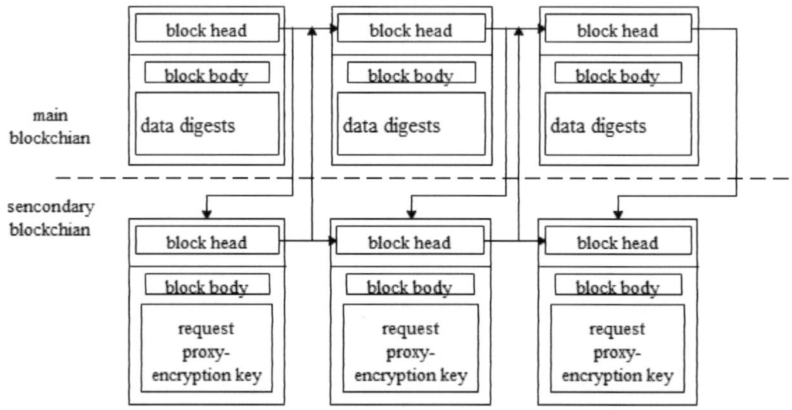

Figure 4. The double-chain structure

We design a double-chain structure for the medical information sharing model. For the data in the p2p network, there are mainly three types: data digest, request, proxy re-encryption key. We store the data digests in a main chain, which uses the traditional POW algorithm, and a block is generated about 10 minutes. For the remaining request data and the proxy re-encryption key, we build a secondary chain to store them. The secondary chain is generated basing on the main chain. When the main chain is generated, the difficulty level will reduce, and all nodes will continue to calculate proof on the basis of the proof in the main chain. When the next proof is found, the secondary block will be generated, and the information in the main block will be added in the secondary block. In this way, the two blockchains will form a strong coupling relationship and the data in the two blockchains is hard to be changed.

Due to the difficulty is different in the two blocks, so that the generate time will be diverse, and the main chain will use more times than the secondary chain. Considering the requests is less than transactions, we control generated time to 2 minutes of the secondary block. It is worth noting that when the main block is generated, the secondary block will start to be generated immediately. Therefore, in some time, the main block and secondary block are generated synchronously.

Through the design of the improved POW algorithm, there are several advantages:

- Reduce the confirmation time of a transaction. Through the connection between the main chain and the secondary chain, we reduce the confirmation time from 1hour to 36 minutes.
- We store different data in different chain, which will be more convenient for data processing.
- The double-chain structure makes the data more secure.

5. Medical data encryption and sharing

For medical data, we divide it into two types: medical institution's data and private data. For medical institution's data, we mainly store it on the database and submit its digest to the main blockchain, and people can freely access these data through the request to the p2p network.

For private data, it is encrypted by the owner's public key and then stored in the medical institution's database, we use the proxy re-encryption to achieve the data sharing [8].

When the patient authorizes the doctor to query his information, the proxy re-encryption key will be sent to the p2p network [10]. After it is verified by the consensus algorithm, and the medical institution will complete the re-encryption operation according to the ciphertext and the proxy re-encryption key, and the ciphertext will be sent to the doctor, by using his private key, so that the doctor can see the specific data. The model uses the LWE (learning with errors) [9] to finish the proxy re-encryption.

1) Generate a random matrix $A \in Z_q^{n*n}$, select safe integer q, n

2) Generate the public key and the private key: public key $pk = P, P = R - A, S \in Z_q^{n*1}$, where R, S are gaussian parameters, then the private key is $sk = S$

3) Encryption algorithm:

$$c = (c_1, c_2) = (e_1 A + e_2, e_1 P + e_3 + m \cdot \left\lfloor \frac{q}{2} \right\rfloor) \tag{1}$$

Where e_1, e_2, e_3 are error parameter.

4) Decryption algorithm:

$$m = c_1 S + c_2 \tag{2}$$

Judging from the distance from 0 to select 0 or 1

5) Re-encryption key generation algorithm:

$$rk_{a \to b} = (P_B, Q)$$
$$Q = \begin{bmatrix} X & -XS_B + E + S_A \\ 0 & I \end{bmatrix} \tag{3}$$

Where X is random matrix and the I is noise.

6) Re-encryption algorithm:

$$(c_1', c_2') = h_1(A, P_B) + (h_2, h_3) + (c_1, c_2) \cdot Q \tag{4}$$

Where h_1, h_2, h_3 are from the error distribution.

6. Conclusion

In the medical information sharing system, we combine the knowledge of blockchain, database and cryptography to build a medical information sharing system. By using the system, we can enhance the utilization of medical data and promote the sharing of medical data. It will be greatly beneficial for the development of medicine.

The model also has some disadvantages. Due to the use of the POW, we will waste some computing ability, when the system become bigger, it will waste a lot of resources. In addition, the encode of the medical date digest need the help of the medical professionals.

Acknowledgments

This work is supported by Fundamental Research Special Funds for Central Universities (20826041A4133).

Reference

[1] Nakamoto S. Bitcoin: A peer-to-peer electronic cash system [OL]. (2008-10-31) [2016-11-02]. https://bitcoin.org/bitcoin.pdf

[2] D. Vujičić, D. Jagodić and S. Ranđić, "Blockchain technology, bitcoin, and Ethereum: A brief overview," 2018 17th International Symposium INFOTEH-JAHORINA (INFOTEH), East

Sarajevo, Bosnia and Herzegovina, 2018, pp. 1-6
[3] E. Yavuz, A. K. Koç, U. C. Çabuk and G. Dalkılıç, "Towards secure e-voting using ethereum blockchain," 2018 6th International Symposium on Digital Forensic and Security (ISDFS), Antalya, Turkey, 2018, pp. 1-7.
[4] L. Mertz, "(Block) Chain Reaction: A Blockchain Revolution Sweeps into Health Care, Offering the Possibility for a Much-Needed Data Solution," in IEEE Pulse, vol. 9, no. 3, pp. 4-7, May-June 2018.
[5] Yuan Yong, Wang Fei-Yue. Blockchain: the state of the art and future trends. Acta Automatica Sinica, 2016, 42(4): 481-494
[6] Merkle R C. A digital signature based on a conventional encryption function. In: proceedings of the 1987 conference on the Theory and Applications of Cryptographic Techniques. Berlin Heidelberg, Germany: Springer. 1987. 369-378
[7] L. Mertz, "Hospital CIO Explains Blockchain Potential: An Interview with Beth Israel Deaconess Medical Center's John Halamka," in IEEE Pulse, vol. 9, no. 3, pp. 8-9, May-June 2018.
[8] Xue Teng-Fei, Fu Qun-Chao, Wang Cong, Wang Xin-Yan. Medical data sharing model via blockchain. Acta Automatica Sinica, 2017, (9):1555-1562
[9] Aono Y, Boyen X, Phong L T, Wang L. Key-private proxy re -encryption under LWE. In: proceedings of the 2013 International Conference on Cryptology in India. Charm, Germany: Springer International Publishing, 2013. 1-18
[10] R. A. Sarıtekin, E. Karabacak, Z. Durgay and E. Karaarslan, "Blockchain based secure communication application proposal: Cryptouch," 2018 6th International Symposium on Digital Forensic and Security (ISDFS), Antalya, Turkey, 2018, pp. 1-4.

IOP Conf. Series: Materials Science and Engineering **428** (2018) 012007 doi:10.1088/1757-899X/428/1/012007

Overview of Maximum Power Point Tracking Control Method for Wind Power Generation System

Yu Li and Li Chengxin

School of Electrical Engineering and Information, Sichuan University, Chengdu, 610065, China

meyuli@outlook.com

Abstract. The maximum power point tracking (MPPT) is one of the most important control issues in the wind power, which aims to maintain the maximum power output as the wind speed is lower than the rated one and thus improve the utilization efficiency of the wind energy. The principle of the MPPT is firstly introduced based on the characteristics of the wind turbine. Next, several MPPT control methods are reviewed. Then, the advantages and disadvantages as well as their respective improvements for each method are analysed. Finally, the development trend and research prospect of MPPT have been pointed out.

1. Introduction

With the rapid development of the modern industry, the energy problem has attracted more and more attention from the world. The wind power has been developed rapidly in all kinds of new energy generations, and the proportion of wind power generation in the total global power generation is becoming higher and higher[1]. The wind energy is renewable, clean and widely distributed. China is one of the world's richest countries with the wind energy. Actively developing wind energy meets the direction of energy reform in China. It can reduce the consumption of the fossil energy while satisfying the energy needs of the people, and then create an environmentally friendly society.

The basic principle of wind power generation is to capture and convert the wind energy into the mechanical energy by using the wind turbine blades first, and then to convert the mechanical energy into the electrical energy by using the wind turbine generator (WTG). The output power of the WTG is directly connected to the power grid or directly supplied the electricity to the consumers.

The WTG is mainly divided into two types: constant speed constant frequency (CSCF) and variable speed constant frequency (VSCF). The speed and frequency of the CSCF are kept constant, and the structure is simple, but the efficiency of the wind energy utilization is particularly low. The speed of the VSCF can be changed with the wind speed by the control strategy to guarantee the maximum coefficient of the wind energy utilization, so the conversion rate of the wind energy is greatly improved. In recent years, with the improvement of the power generation efficiency and the fully-controlled power electronic devices, the VSCF has gradually replaced the CSCF and become the main modern WTG[2]. At present, there are two main types of large WTG including the doubly fed induction generator (DFIG) and the direct drive permanent magnet synchronous generator (PMSG).

The control problems in the WTG primarily include the maximum power point tracking (MPPT), the voltage ride through and the grid-connected control etc[3]. The MPPT control method is the focus of this article. The MPPT is one of the key problems of the VSCF system control. Its function is, under the rated wind speed, to design control strategy to change the rotating speed of the WTG as the wind speed changes so as to keep the tip speed ratio (TSR) always at the best value, then the output power of the WTG is guaranteed to be maximum. The MPPT methods mainly include the optimal TSR method, the power signal feedback (PSF) method and the hill climbing search (HCS) method, etc. The rest of this paper is as follows: A review of the basic principle of the MPPT is given in Section 2. In Section 3 several MPPT control methods are discussed. Then, the advantages and the disadvantages as well as their respective improvements for each method are analysed. The development trend and the research prospects of the MPPT are presented in Section 4.

2. The characteristics of the WTG and the MPPT

2.1 The operation characteristics of the WTG

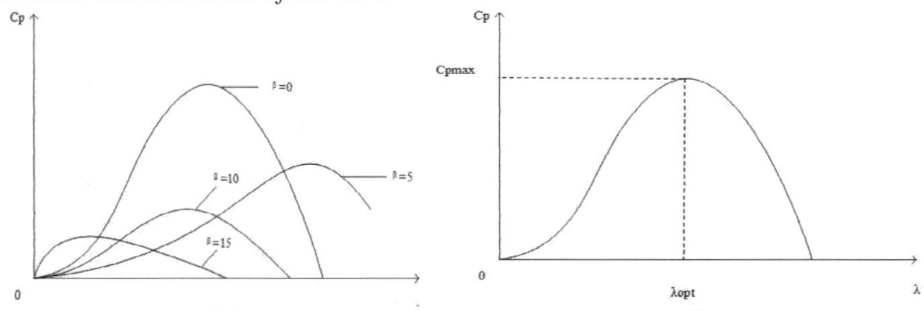

(a) (b)

Figure 1. Relationship between C_p, β and λ

The mechanical power captured by the blades of the WTG can be expressed as:

$$P_m = \frac{1}{2}\rho\pi R^2 C_p(\lambda,\beta)v^3 \tag{1}$$

$$\lambda = \frac{R\omega}{v} \tag{2}$$

Where ρ is the air density, R is the blade radius, C_p is the wind energy utilization coefficient relating to a nonlinear function affected by the blade pitch angle β and the tip-speed ratio λ, v is the wind speed, ω is the wind turbine rotor rotating speed.

According to the Betz' law, no turbine can capture more than 16/27 (59.3%) of the kinetic energy in wind. The object of the MPPT is to find this ideal working point to maintain the highest energy conversion efficiency when the wind speed changes.

The output power regulation scheme of the WTG can be described as: when the actual wind speed is larger than the rated one, the WTG reduces the output power by increasing β and thus maintains its output power at the rated power; otherwise, the WTG keeps β fixed and finds the optimal TSR by adjusting the rotating speed, as shown in figure 1 and thus obtains the maximum power output at the current wind speed.

2.2 The MPPT principle

As can be seen from section 2.1, when the actual wind speed is lower than the rated wind one, the essence of the MPPT is to find the maximum C_p at the current wind speed to maintain the maximum output power. Because the C_p is non-linear with the pitch angle β and the tip-speed ratio λ, for the sake of simplicity, β is fixed and the task is simplified to find the optimal λ to obtain the maximum C_p in practice. It can be seen from equation (2) that λ is linear with the rotating speed ω because of the known wind speed v. The maximum power-rotating speed curve (hereinafter referred to as the optimal power curve) can be obtained by connecting the maximum power points at different wind speeds, as shown in figure. 2. Therefore, the essence of the MPPT is to regulate the rotating speed of the WTG according to the optimal power curve to maintain the maximum output power[4].

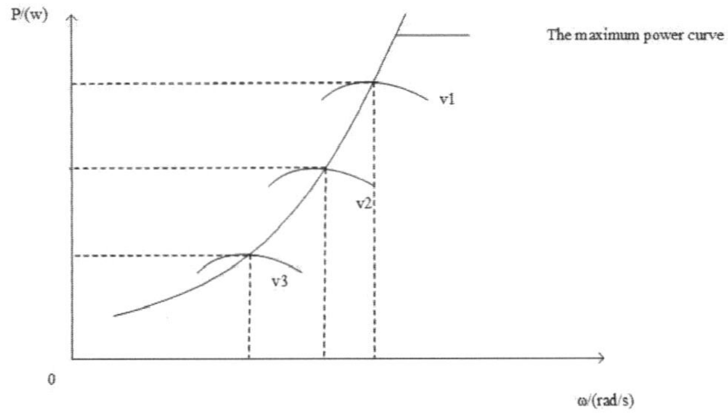

Figure 2. output power of wind turbine under different wind speed and rotating speed

3. Analysis of the MPPT control methods

The MPPT control methods include the PSF method, the optimal torque control (OTC) method, the HCS method, the optimal TSR method, and the extremum search control (ESC) method, as well as some nonlinear control methods. In this section, the above control methods are introduced, and their advantages and disadvantages as well as some relevant improvement strategies are extended.

3.1 The PSF method and its improvements

3.1.1 The method principle. The core of the PSF method is the optimal power curve. The expectation of the control system is the optimal power corresponding to the current wind speed on the optimal power curve. Then the error is obtained by detecting the actual output power of the WTG. Therefore the MPPT is realized by regulating the rotating speed of the WT based on the power signal feedback[5]. This method is stable in operation and suitable for the large wind turbines, but it is necessary to obtain the optimal power curve in advance. Furthermore, the environmental factors will change the actual track of the curve, and thus the precision of the curve will be reduced. Therefore, the efficient wind energy capture ability cannot be guaranteed.

3.1.2 Improvements. In the wind power system, the bandwidth refers to the frequency of the wind speed that the WTG can respond to. In order to increase the tracking bandwidth, an error feed-forward signal of the aerodynamic power and the output power of the WTG is added to the control system[6]. The result is that the tracking speed of the WT becomes obviously quicker.

Since the constant bandwidth control method will increase the machine load, which affects the operating life of the machine[7]. To solve this problem, the response surface analysis method is firstly applied to fit the statistical relationship between the optimal bandwidth and the turbulence wind speed characteristic index (average wind speed, turbulence frequency, turbulence standard deviation), and then the optimal bandwidth is set on line according to the wind speed characteristics[8]. This method realizes the comprehensive optimization of the wind energy capture and the load. However, the load loss coefficient in practical application needs further confirmation.

There exists a beginning speed determination in the control strategy of the WTG. When the rotating speed of the WTG is bigger than the beginning speed, the system regulates the output power according to the optimal power curve. In this method, the optimal compensation coefficient is obtained by the neural network to update the beginning speed, and then the output power of the generator is adjusted according to the optimal power curve, so as to achieve MPPT[9]. This strategy accelerates the tracking process of the WTG to the gradual gust, and provides with great applicability for the different wind regimes and the turbines.

3.2 The OTC method and its improvements

3.2.1 The method principle. Due to the fact that the torque of the WTG is the quadratic function of its rotating speed, the OTC method firstly detects the rotating speed and calculates the corresponding torque as the reference value of control (the optimal torque). Then, it forms the feedback control system with the electromagnetic torque as feedback. That is to say, the WTG regulates the electromagnetic torque to change the rotating speed as the wind speed changes, so that the speed of the WT and the wind speed are satisfied with the optimal TSR of the blade, thus the WT can be operated at the maximum power point. This method does not need the real-time wind speed information. In addition, the power output fluctuation of the WTG is small. However, it is only suitable for the small and the medium turbines[10]. For the large turbines, the rotating speed changes too slow, which results in a longer response time and lower efficiency.

3.2.2 Improvements. Liu designs the optimal torque compensator by the gradient estimation to compensate the optimal given torque value[11]. It reduces the influence of the rotating inertia on the speed regulation. This method can increase the acceleration of the rotating speed of the WT and make system move faster to a new steady working point. The compensator only acts in transient process, which enables the system to go through the transient process faster and effectively speeds up the response speed.

Chen employs the sliding mode variable structure control idea to design a variable structure controller[12]. When the state of the system reaches the sliding surface, the sliding mode motion will be carried out and the equilibrium point is asymptotically stable. It takes into account the dynamics of the WT and makes system reach sliding mode stability point within a limited time, which is equivalent to increasing the unbalanced torque. The average wind energy capture efficiency is raised compared with the traditional method.

3.3 The HCS method and its improvements

3.3.1 The method principle. The power- rotating speed curve of the WTG is the convex function. The HCS method is also called the perturbation observation (P&O) method, which means to observe the change of the output power by continuously applying the rotating speed disturbance to the WTG. If the output power increases, the direction of the disturbance is kept constant, otherwise the direction will be reversed so as to achieve MPPT[13]. The advantage of the HCS is that the accurate parameters of the WT and the wind speed are not required. But the determination of the disturbance step length is a difficult point in the algorithm. If the fixed speed disturbance is selected, the maximum output power will be always in the fluctuating state.

3.3.2 Improvements. Jia presents a variable step length empirical algorithm[14], which does not depend on the turbine parameters and the wind speed. The disturbance step length is adjusted with the change of two continuous sampling power, and the optimal rotating speed point can be automatically searched. As the power increases, the step length decreases to zero. The algorithm effectively reduces the power fluctuation at the maximum power point.

The accurate wind speed measurement is an important but difficult task in the wind power system[15]. Liu first utilizes the support vector machine to estimate the wind speed, then uses the firefly algorithm to optimize the speed, and adjusts the rotating speed to the optimal value according to the TSR method. Finally, the HCS method is adopted to track the maximum output power by small step length[16]. This method not only saves the wind speed measurement devices, but also raises the accuracy of the wind speed estimation. But at the same time, it also brings the complicated control to the system.

The method in [17] combines the TSR method with the HCS method, which is similar to the above method. When the wind speed changes more than 1 m/s, the TSR method is used to control the rotating speed of the WT to quickly track the reference speed, then to determine whether the error between the real-time speed and the reference speed is less than the predetermined threshold value. If so, the HCS method is selected and used as long as the wind speed changes less than 1m/s.

Tian proposes a variable step length control method based on the mechanical power current variation curve[18]. The perturbation direction of this algorithm is determined by the positive and negative slope of the curve. Moreover, the disturbance step length is exponentially decaying at the maximum power point to improve the tracking efficiency. In addition, during each cycle, the system will initialize the step length when the change of the wind speed exceeds the predefined threshold. This method is flexible in changing the step length, which enhances the accuracy of tracking and the applicability of dealing with the abrupt wind speed.

3.4 The Optimal TSR method and its improvements

3.4.1 The method principle. The core idea of the optimal TSR method is to control the TSR of the wind turbine to its optimal value according to the detected real-time wind speed[19]. So the energy utilization coefficient can reach the maximum value and thus the purpose of MPPT is achieved. The advantage of the optimal TSR method is that the control structure is simple and the system response speed is fast, but it depends on the accurate wind speed measurement. It is difficult to obtain the accurate wind speed due to the fact that there is a certain difference in the wind speed around the blades. So the stability and the accuracy of the system are not guaranteed[20].

3.4.2 Improvements. Han puts forward a double closed loop control structure, which is composed of the power outer loop and the current inner loop[21]. The method utilizes the upper computer to simulate the direct drive wind turbine. According to the current wind speed and the corresponding optimal TSR, the optimal rotating speed is calculated as the speed control instruction to realize MPPT. The double closed loop control strategy improves the robustness and the stability of the traditional TSR method.

In view of the uncertainties, the multiple disturbances, the low efficiency and the strong nonlinearity of the wind power system, Li proposes a method which combines the auto disturbances rejection control (ADRC)[22]. First, the overall disturbance of the system is estimated by the observer, and then the feedback controller is employed to compensate the disturbance. Finally, MPPT is realized with the self-disturbance rejection control and the PID controller. The robustness and the capability to fast response of the system are obviously improved.

3.5 The ESC method and its improvements

3.5.1 The method principle. The ESC method mainly makes use of the principle of calculus. As mentioned above, the power- rotating speed curve of the WTG is the convex function. There must be a unique rotating speed that maximizes the output power. Therefore, the maximum power point can be obtained by taking the derivative of the function with respect to the rotating speed, so that maximal wind energy can be captured[23]. But the ESC algorithm needs to detect the differential of the output power of the system, which will cause the system to be very sensitive to the high frequency signal and the dynamic performance of the system will be affected.

3.5.2 Improvements. To eliminate the influence of the differential, the sliding mode (SM) control and ESC can be combined[24] (SMESC). The key of the SMESC method is two switching functions. One is used to generate the output reference values, and the other is applied to generate the output variables. When the function is cut into the sliding surface under the function of the input variables, the system tracks the given power signal. If the disturbance occurs, the sliding surface is cut out, and then the system runs to the new stable point by changing the rotating speed. The control block diagram is shown in figure 3. The SMESC method not only eliminates the measurement steps of the wind speed and the turbine parameters, but also avoids the signal loss phenomenon caused by the differential. The reliability and the tracking accuracy of the system are both increased effectively.

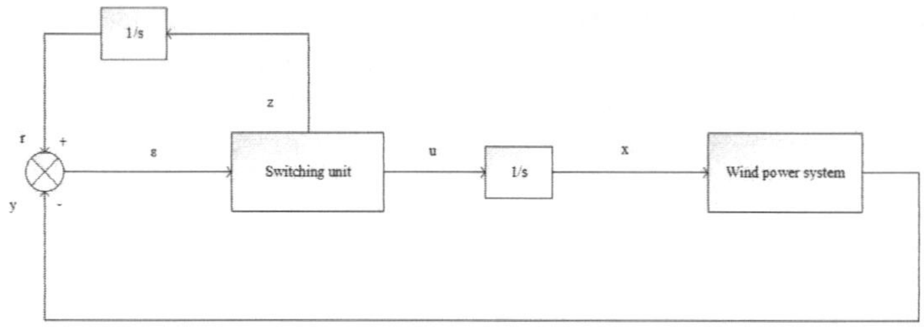

Figure 3. Control block diagram of SM ESC

In [25], first, Fourier transform is used to obtain the phase difference between the TSR and the wind energy utilization coefficient. Next, the softening function is adopted to determine the best reference value of the TSR. Lastly, the PI controller is employed to track the optimal speed. This method does not require the specific system models, and also reduces the operation load of the machine.

3.6 Nonlinear control methods

Since the WTG system is a typical nonlinear complex system, more and more nonlinear modern control algorithms have been proposed by the scholars in recent years[26], such as the speed model predictive control based on the least squares support vector machines[27], the design of nonlinear controllers based on the feedback linearization principle[28], the internal model control design using the exact linearization method and the inverse system theory[29], and the second-order SM control strategy[30], etc. Compared with the traditional linear control methods, the nonlinear controllers improve the response speed and the anti-interference ability of the WTG system, and thus they are more practical in the applications.

In addition, there is a passive MPPT method that changes the system circuit and the WTG structure[31]. The maximum output power of the WTG is reached by matching the generator input power curve and the optimal power curve of the WT.

4. Summary and prospect

The MPPT technology ensures the maximum output power of the WTG under the rated wind speed. It plays an important role in improving the efficiency of the wind energy utilization. In this paper, the existing MPPT control methods are reviewed and summarized, and then the objective evaluations of the various improved strategies as well as their advantages and disadvantages are made. Based on the research of the WTG system and various methods, it can be seen that combining with the artificial intelligence is an overall development trend of the MPPT control strategy. The following future research direction of the MPPT can be expected from two aspects: the basic conditions of the control and the control strategy optimization.

a) Obtaining the accurate wind speed is a prerequisite for several MPPT methods. However, the wind speed is now mostly measured by the wind speed sensor which is not only expensive but also its measurement accuracy cannot be guaranteed. Therefore, the combined prediction model can be used to predict the wind speed. For example, it is viable to combine the artificial intelligence with the physical model as well as the big data analysis to improve the accuracy of prediction.

b) The optimal power curve is also a base for several MPPT methods. However it will change with the change of running time and the external parameters. So the online correction the optimal power curve is one of the important improvement measures to achieve the MPPT.

c) The advantage of the HCS approach is that it does not depend on the system model. The variable step length perturbation method is the mainstream of the current research in the HCS[32]. A searching algorithm with faster tracking capability needs further research.

d) At present, there have been a number of the effective sliding mode variable structure MPPT control methods. But how to avoid the system buffeting caused by sliding mode structure will be still a research hotspot.

e) The ADRC technology has developed rapidly in recent decades[33]. It will be a feasible research scheme to apply the modified ADRC to the MPPT.

References

[1] Cheng Ming, Zhang Yunqian and Zhang Jianzhong 2009 *Journal of Electric Power Science and Technology* **24** 2-9
[2] Xiu Chunbo, Liu Xinting, Zhang Xin and Yu Tingting 2013 *Power System Protection and Control* **41** 14
[3] Li Hui and He Bei 2008 *Acta Energiae Solaris Sinica* **29** 797-803
[4] Johnson K E, Fingersh L J and Balas M J 2004 *Journal of Solar Energy Engineering* **126** 1092-100
[5] Tan K and Island S 2004 *IEEE Transactions Energy Conversion* **19** 392-9
[6] Chen Jiawei, Chen Jie and Gong Chunying 2012 *Proceeding of the CSEE* **32** 32-8
[7] Camblong H, Vechiu I and Guillaud X 2014 *Renewable Energy* **63** 37-45
[8] Zhou Lianjun, Yin Minghui, Zhou Qian, Chen Zaiyu, Wang Chenggen and Zou Yun 2017 *Power System Technology* **41** 64-71
[9] Yin Minghui, Zhang Xiaolian, Zou Yun and Zhou Lianjun 2014 *Power System Technology* **38** 2180-85
[10] Raza Kazmi S M and Goto H 2011 *IEEE Transactions on Industrial Electronics* **58** 29-36
[11] Liu Jizhen, Meng Hongmin and Hu Yang 2014 *Proceeding of the CSEE* **35** 2367-74
[12] Chen Zaiyu, Yin Minghui, Cai Chenxiao, Zhang Baoyang and Zou Yun 2015 *Acta Automatica Sinica* **41** 2047-57
[13] Raju A B, Chatterje K and Fernandes B G 2003 *Proceeding of IEEE PESC* **2** 748-53
[14] Jia Yaoqin, Cao Binggang and Yang Zhongqin 2004 *Acta Energiae Solaris Sinica* **25** 171-6
[15] Wang Shengtie, Zhang Runhe and Tian Lixin 2006 *Acta Energiae Solaris Sinica* **27** 828-34
[16] Liu Weiliang, Liu Changliang, Lin Yongjun, Chen Wenying and Ma Liangyu 2017 *Acta Energiae Solaris Sinica* **38** 631-9
[17] Li Xianshan, Xu Hao and Du Yulong 2015 *Power System Protection and Control* **43** 66-71
[18] Tian Bing, Zhao Ke, Sun Dongyang, Duan Jiandong and Sun Li 2016 *Transactions of China Electrotechnical Society* **31** 226-33
[19] Liu Qihui, He Yikang and Zhao Rende 2003 *Automation of Electric Power System* **27** 62-7
[20] Ma Hongfei, Zhang Wei and Li Weiwei 2007 *Acta Energiae Solaris Sinica* **28** 1278-83
[21] Han Kun, Li Jun, Li Yuling and Chen Guozhu 2010 *Acta Energiae Solaris Sinica* **31** 1497-1502
[22] Li Juan, Zhang Kezhao, Li Shengquan and Liu Chao 2015 *Electric Machines and Control* **19** 94-100
[23] Navid Ghaffarzadeh and Sepher Bijani 2016 *IET Renewable Power Generation* **10** 611-22
[24] Pan Tinglong and Shen Yanxia 2012 *Acta Energiae Solaris Sinica* **33** 2193-97
[25] Zhao Liang, Lv Jianghong and Xiang Wenguo 2011 *Power System Technology* **35** 171-6
[26] Boukhezzar B and Siguerdidjane H 2011 *IEEE Transactions on Energy Conversion* **26** 149-62
[27] Liu Jihong, Xu Daping and Lv Yuegang 2011 *Power System Technology* **35** 159-63
[28] Wang Libing, Mao Chengxiong, Lu Jiming and Wang Dan 2011 *Transactions of China Electrotechnical Society* **26** 1-6
[29] Guo Jiahu, Cai Xu and Gong Youmin 2009 *Control Theory & Applications* **26** 958-64
[30] Liu Xiangjie, Wang Chengcheng and Han Yaozhen 2017 *Acta Energiae Solaris Sinica* **43** 1434-42
[31] Bai Yinru, Kou Baoquan and Chen Qingquan 2017 *Jouanal of Central South University* **48** 370-80
[32] Zhong Qinhong, Ruan Yi and Zhao Meihua 2013 *Power System Protection and Control* **29** 32-8
[33] Yang Jinming, Wu Jie and Ynag Junhua 2004 *Acta Energiae Solaris Sinica* **25** 525-9

IOP Conf. Series: Materials Science and Engineering **428** (2018) 012008 doi:10.1088/1757-899X/428/1/012008

Effect of Quartz Crystal Form on the Measurement Performance of Multi-dimensional Force Sensor

Yingjun Li[1,*], Guicong Wang[1], Shu Huang[1], Xuan Sun[1] and Zhiquan Feng[2]

[1]School of Mechanical Engineering, University of Jinan, Jinan 250022, PR China
[2]School of Information Science and Engineering, University of Jinan, Jinan 250022, PR China

[*]me_liyj@ujn.edu.cn

Abstract. Quartz wafer is an important part of the force sensor of piezoelectric multi-dimensional force sensor. Its structural form affects the overall measurement performance of force-sensitive components and sensors. In this paper, the shape of the quartz crystal group was first analysed, and then two different types of quartz crystal group models were established using finite element software, and their simulation analysis was performed. The corresponding relationship between the deformation of the two different shapes of quartz crystal set under the same external load and the limitation of external space size was studied. The same axial load and tangential load were applied to two different types of quartz crystals respectively, and the corresponding maximum equivalent stress and stress distribution regions were obtained. This paper provides an important reference for the fabrication of quartz crystal arrays and the selection of quartz wafer shape for piezoelectric multi-dimensional force sensors.

1. Introduction

The multi-dimensional force sensor can sense all the information of the six-dimensional external forces and moments in space. Through the monitoring of the direction and the size of the changing force/torque in the work, complete the more complex force feedback pattern recognition for the operation object, and then realize the force feedback control [1, 2]. It has been widely used in important fields such as intelligent robots, aerospace and military, and has become increasingly prominent. The multi-dimensional force sensor measurement principle mainly includes strain type, piezoelectric type, capacitive type and grating type [3, 4].

Among them, piezoelectric sensor is a kind of self-generating electricity type sensor, with high natural frequency (200 KHz), especially suitable for dynamic measurement. Compared with other measurement methods, piezoelectric measurement is the only measurement technology that meets the requirement of high natural frequency. It is an important development direction of multi-dimensional force sensors [5]. Piezoelectric multi-dimensional force sensors mostly use a four-point three-way force quartz crystal group for combined measurement. Ren Zongjin [6] studied the force vector measurement model based on piezoelectric sensors and designed a four-branch three-way force piezoelectric sensor parallel balance. Zhang Jun [7] developed a new four-point piezoelectric quartz parallel shaft-type six-axis force sensor capable of accurately measuring the six-dimensional force in the axial space. Jia Zhenyuan [8] studied the four-point support type large-range six-dimensional force sensor based on the KISTLER Company's research thinking, which is large and difficult to realize the miniaturization of the sensor. Li Dong [9] developed a small-scale integrated piezoelectric six-axis force sensor. Liu Jun [10] proposed a flat plate piezoelectric six-dimensional force sensor program, which achieved multi-dimensional dynamic force measurement in a narrow space.

The current research work is all about the overall performance of six-dimensional force sensors. The piezoelectric six-dimensional force sensor uses a piezoelectric quartz crystal set as the sensor force sensor, the quartz crystal group is the core of the multi-dimensional force sensor, but there are few related researches. The quartz crystal group is composed of quartz crystals. Although the performance of piezoelectric quartz is better than that of other piezoelectric materials, the quartz wafer is limited by the cut type, the shape of the wafer is also limited. The shape will affect the entire sensor assembly process and measurement performance. For this reason, this paper mainly studied the effect of the shape of the quartz wafer on the measurement performance.

2. Quartz crystal shape analysis

According to the piezoelectric effect of the quartz crystal set, the sensitivity of the crystal is related to the unique piezoelectric coefficient and external load of the crystal. Quartz crystal shape analysis is mainly to improve the rigidity and life of the quartz crystal. When the crystal groups of different shapes are subjected to the same load, the crystal group is subjected to a smaller stress and the service life is longer. At present, the shape of the quartz crystal group mainly includes two types of structures: Round with hole and a square with a hole. In order to increase the service life of the quartz crystal set and obtain a suitable crystal shape, the ansys15.0 software was used to model and simulate the analysis. Square wafer size is 18*18mm, thickness 1mm, middle hole diameter Φ12mm; round wafer outer diameter Φ18mm, thickness 1mm, inner diameter Φ12mm. With the same external space installation size constraints, the circular wafer has a smaller bearing area.

2.1. Deformation analysis

Constraints are imposed on the lower surfaces of the two different shapes of the crystal group, and the upper surface is applied with the same magnitude and direction of axial force, and the load sizes are: 400N, 800N, 1200N, 1600N, 2000N, Then, the stress and strain of the crystal group and the transverse shear stress generated when subjected to the axial load are compared. The maximum deformation of the two crystal groups when subjected to the same axial force is shown in Table 1. Figure 1 shows the displacement cloud of two different types of quartz crystals when a 2000N load is applied.

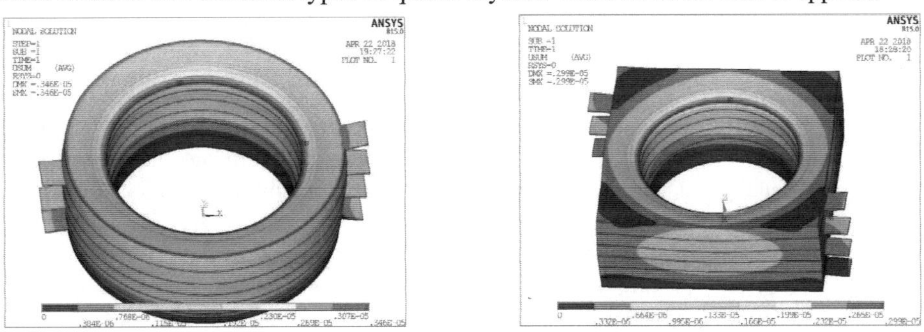

(a) The deformation cloud of Circular crystal; (b) The deformation cloud of Square crystal

Figure 1. Simulation of deformation of quartz crystal at axial load of 2000N

Table 1. Simulation data of crystal deformation(10^{-4}mm)

Normal force /N	0	400	800	1200	1600	2000
Square crystal group	0	5.97	11.9	17.9	23.9	29.9
Circular crystal group	0	6.91	13.8	20.7	27.6	34.6

From Table 1, it can be seen that as the applied stress increases, the deformation of the two crystal groups gradually increases, and the force and the deformation amount become strictly linear, and the

IOP Conf. Series: Materials Science and Engineering **428** (2018) 012008 doi:10.1088/1757-899X/428/1/012008

deformation of the circular crystal group is greater than that of the square crystal group. When the applied load is 2000N, the maximum deformation of the square crystal group is 29.9E-4mm, and that of the round crystal group is 34.6E-4mm. The amount of deformation of the crystal group indicates that the crystal is relatively rigid, and it is verified that the piezoelectric force sensor belongs to a micro-displacement type sensor.

2.2. Stress analysis under axial load

The axial force of the same size and direction is applied to the upper surface of two different shapes of the crystal group. The load sizes are: 400N, 800N, 1200N, 1600N, 2000N. The maximum equivalent stress simulation data for two different shapes of crystal groups can be obtained, as shown in Table 2. Figure 2 shows the equivalent stress cloud and the maximum equivalent stress distribution for the two different shapes of the crystal group when the crystal group is subjected to an axial load of 2000N.

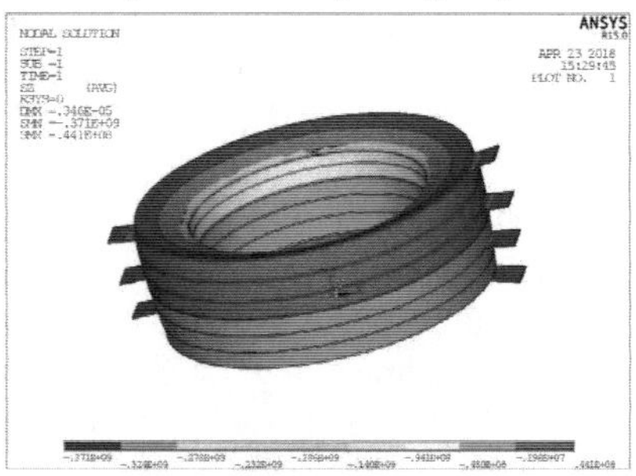

(a) Maximum stress distribution in circular crystal groups

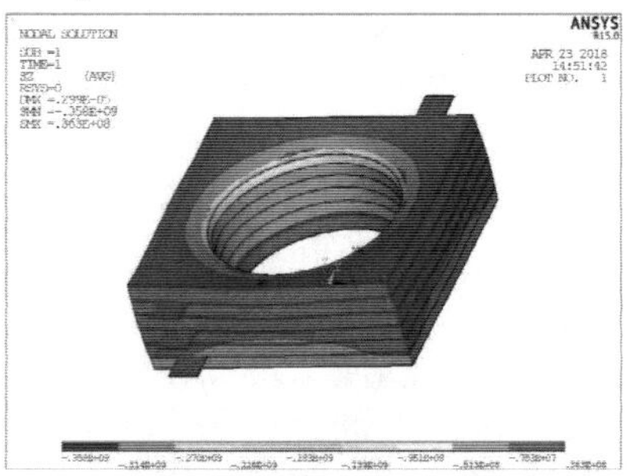

(b) Maximum stress distribution in square crystal groups

Figure 2. Simulation of equivalent stress of crystal group under axial load 2000N

Table 2 . Simulation data of equivalent tensile stress in crystal group

Axial force /N	400	800	1200	1600	2000
Square crystal group(Pa)	7.16E+07	1.43E+08	2.15E+08	2.86E+08	3.58E+08
Circular crystal group(Pa)	7.41E+07	1.48E+08	2.22E+08	2.96E+08	3.71E+08
Stress difference(Pa)	2.50E+06	5.00E+06	7.00E+06	1.00E+07	1.30E+07

From the data in Table 2, it can be seen that under the same load, the stress of the circular crystal group is greater than that of the square crystal group, and the sensitivity of the circular crystal group is greater than that of the square crystal group. From the viewpoint of improving sensitivity, the circular crystal group is suitable. It can be seen from the stress cloud diagram of the crystal in Figure 2 that the maximum stress region of the circular crystal group is distributed widely, while the stress region of the square crystal group is distributed more widely. Compared to the circular crystal group, the square crystal group can withstand greater force.

From the piezoelectric properties of the crystal group, it can be seen that when the normal force is applied, the unit crystal group that measures the tangential force has a corresponding charge output due to the shear force .The presence of shear stress will produce a certain amount of lateral interference to the sensor measurement, affecting the tangential measurement accuracy. In the case of axial forces applied to the crystal stack, the resulting coupled shear stress simulation data is shown in Table 3. Figure 3 shows the coupled shear stress cloud in the F_y direction for an axial load of 2000N. When the crystal group is subjected to axial load, there is a corresponding stress output in the shear direction, and the coupled shear stress generated by the square crystal group is smaller than that of the circular crystal group. According to the characteristics of the unit crystal group, the smaller coupled shear stress generated by the crystal group is, the better the transverse interference ability between the crystal groups can be reduced, and the measurement accuracy of the crystal group can be improved.

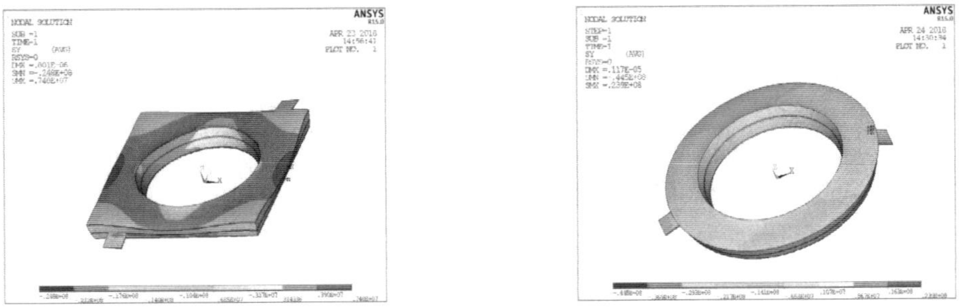

(a) The shear stress cloud of Circular crystal group (b) The shear stress cloud of Square crystal group

Figure 3. Simulation of coupled shear stress of quartz crystal in F_y direction

Table 3. Maximum Coupling Shear Stress Simulation Data for Crystal Groups

Axial force /N	400	800	1200	1600	2000
Square crystal group(Pa)	0.495E+07	0.990E+07	0.149E+08	0.198E+08	0.248E+08
Circular crystal group(Pa)	0.891E+07	0.178E+08	0.267E+08	0.356E+08	0.445E+08

2.3. Stress analysis under tangential loading

In the crystallographic model and invariable parameters, the end face of the crystal group is fixed, and the end face only bears the equivalent stress generated under the shear force condition. When 200N, 400N, 600N, 800N, and 1000N tangential forces are applied, the maximum equivalent stress distribution regions and equivalent stress cloud diagrams for the two crystal groups can be obtained. Figure 4 shows the shear stress cloud of the two crystal groups and the maximum stress distribution of the measured crystal groups at a shear load of 1000 N.

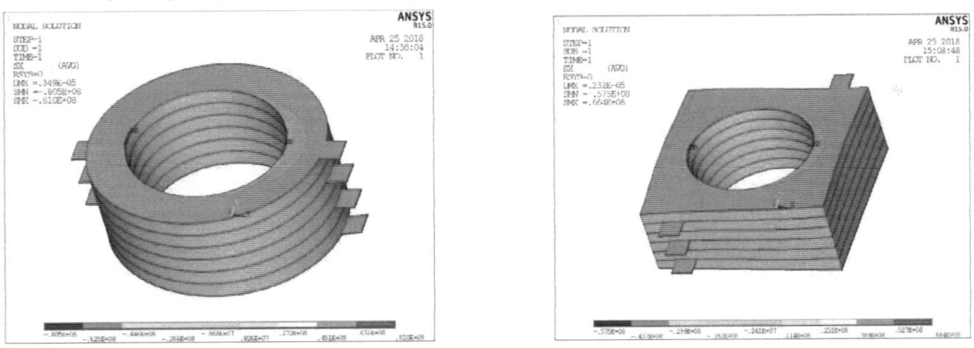

(a) The shear stress cloud of Circular crystal group (b) The shear stress cloud of Square crystal group

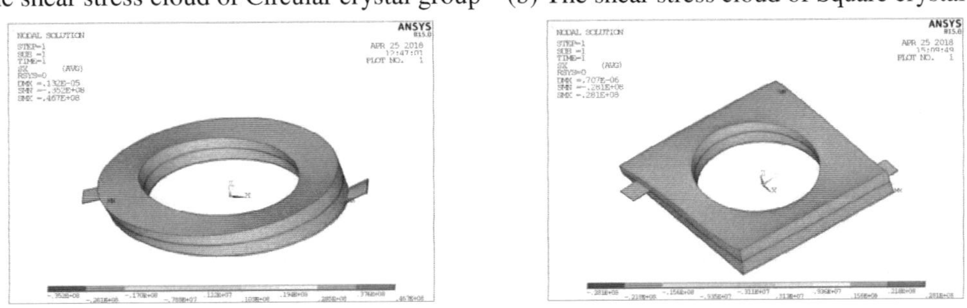

(c) Measurement of the maximum stress distribution of crystal groups in the F_x direction

(d) Measurement of the maximum stress distribution of crystal groups in the Fx directio

Figure 4. Simulation analysis of crystal shear force

IOP Conf. Series: Materials Science and Engineering **428** (2018) 012008 doi:10.1088/1757-899X/428/1/012008

Table 4. Simulation data of shear equivalent stress(Pa)

Shear force /N	0	200	400	600	800	1000
Square equivalent crystal group(Pa)	0	0.133E+08	0.266E+08	0.399E+08	0.531E+08	0.664E+08
Circular equivalent crystal group(Pa)	0	0.162E+08	0.324E+08	0.486E+08	0.648E+08	0.810E+08
Square measuring crystal group(Pa)	0	0.561E+07	0.112E+08	0.168E+08	0.225E+08	0.281E+08
Circular measuring crystal group(Pa)	0	0.934E+07	0.187E+08	0.280E+08	0.374E+08	0.467E+08

The maximum stress values are shown in Table 4. Table 4 shows that when the two crystal groups are subjected to a shear force, the resulting equivalent stress values and the applied load are strictly linear, and the equivalent stress of the circular crystal group is much larger than that of the square crystal group. When the shear strength limit of the crystal is reached, the round crystal group crystal first appears crushing phenomenon, and the square crystal group has better bearing capacity than the round crystal group.

When subjected to the shear load, the unit crystal group for measuring the axial force will have corresponding tensile stress. The existence of the tensile and compressive stress will generate the corresponding charge output, which will cause some coupling interference to the sensor measurement and affect the tangential measurement accuracy. In the case of axial forces applied to the crystal stack, the resulting coupled shear stress simulation data is shown in Table 5. Figure 5 shows the F_z-direction coupled shear stress cloud when the shear load is 1000N. When the crystal group is subjected to a tangential load, there is a corresponding tensile and compressive stress output in the axial direction. Coupling tensile stress generated by the square crystal group is smaller than that of the circular crystal group. According to the characteristics of the unit crystal group, the smaller the coupled tensile and compressive stress generated by the crystal group, the better. It can reduce the horizontal interference between the crystal groups and improve the measurement accuracy of the crystal group.

Table 5. Maximum Coupling Tensile and Compressive Stress Simulation Data for Crystal Groups

Shear force /N	200	400	600	800	1000
Square crystal group(Pa)	0.164E+08	0.328E+08	0.492E+08	0.656E+08	0.820E+08
Circular crystal group(Pa)	0.205E+08	0.408E+08	0.612E+08	0.816E+08	0.102E+08

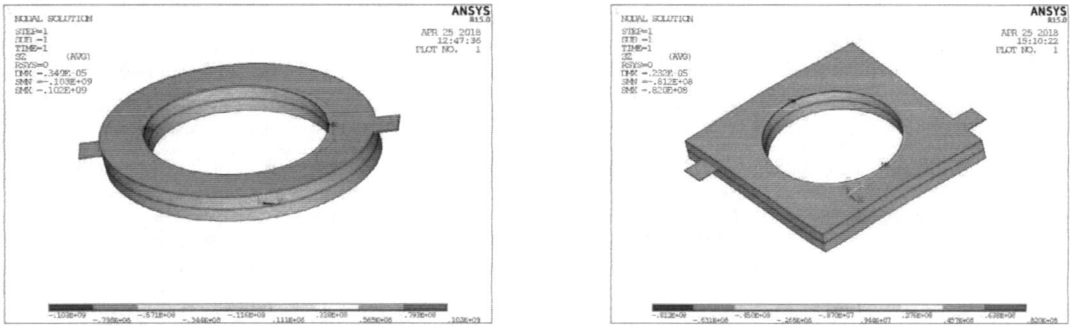

(a) Measurement of the F_z direction of the circular crystal group

(b) Measurement of the F_z direction of the square crystal group

Figure 5. Coupled tensile and compressive stress simulation of quartz crystal in F_z direction

3. Conclusion

In summary, through the simulation analysis of two different crystal groups, it can be concluded that under the same load, the square crystal group deformation is smaller.

When the axial load is applied, the maximum stress region of the circular crystal group is distributed widely, and the circumference of the fixed end of the crystal group is the region where the stress should be maximum, and the stress distribution is more concentrated. However, the maximum equivalent stress of the square crystal group is smaller, and the distribution of the stress region of the square crystal group is more dispersed. When the ultimate compressive strength is reached, the circular crystal group is prone to cracking or crushing, and the crystal group is damaged. When the axial stress is applied, the coupled shear stress generated by the square crystal group is also small, and the transverse interference between the unit crystal groups is reduced during the measurement.

When the tangential load is applied, the equivalent shear stress of the circular crystal group is much larger than that of the square crystal group. Although the larger the stress will increase the sensitivity of the crystal group, the crystal group can not be used normally when the larger force value is measured.

Therefore, choosing a square shape for the crystal group not only reduces the damage of the crystal group, but also effectively improves the service life of the crystal group. It also reduces the lateral interference and improves the measurement accuracy of the crystal group to some extent.

Acknowledgments

This work is supported by a grant from National Natural Science Foundation of China (GrantNo.51875250), Shandong Provincial Natural Science Foundation, China (Grant No. ZR2017QEE012), Key research and development plan of Shandong Province (Grant No. 2018GGX103028), and the National Natural Science Foundation of China (GrantNo.51705200).

References

[1] Liu. X. J, Yu. J. J, Wang. G. B, Lai. Y. N, He. B. Y. Research trend and scientific challenge of robotics. Bulletin of National Nature Science Foundation of China, 2016(5):425-431

[2] Gao. B, Wang. C, Qu. W. X, Man L. The Accuracy and Stiffness Comprehensive Optimal Design of the Heavy Load Stewart Six-axis Force Sensor Based on Partial Decoupling Method. Robot, 2017, 39(06):838-843.

[3] Y. J. Li, C. Yang, G. C. Wang, H. Zhang. Research on the parallel load sharing principle of a novel self-decoupled piezoelectric six-dimensional force sensor, ISA Transactions, 2017, 70:447-457.

[4] Li Y J, Wang G C, Yang X, *et al.* Research on static decoupling algorithm for piezoelectric six axis force/torque sensor based on LSSVR fusion algorithm. Mechanical Systems & Signal Processing, 2018, 110: 509-520.

[5] Yao J, Cui P, Zhu J, et al. Fault-tolerant measurement mechanism and calibration experimental study of pre-stressed parallel six-axis force sensor. Journal of Mechanical Engineering, 2016,52(08):58-66.

[6] Li. P, Yang. R, Ren. Z. J, Qian. W. Design and error analysis of a piezoelectric balance for the measurement of aerodynamic loads on an all-movable rudder. Journal of Harbin Institute of Technology,2018,50(01):181-185

[7] Zhang J, Cheng-Shuai X U, Ren Z J, et al. Research on online calibration of plume force/torque measuring system. Transducer & Microsystem Technologies, 2017,36(01):13-15+19.

[8] Jia Z, Li. Y. J, Zhang. J, Qian. M, Liu. W. Axial Piezoelectric 6-Component Force/torque Sensor Based on Parallel Structure. Journal of Mechanical Engineering, 2010, 46(11):62-68.

[9] Li. D. Research on the Measuring Principle and Performance of the Integrated Piezoelectric 6-axis Force Sensor Dalian University of Technology, 2015.

[10] Li M, Liu J, Qin L, Liu J C. Research of Parallel Piezoelectric Six-axis Force/Torque Sensor's Static Characteristics. Journal of Mechanical Engineering, 2014,50(06):1-7.

IOP Conf. Series: Materials Science and Engineering **428** (2018) 012009 doi:10.1088/1757-899X/428/1/012009

Application of BeiDou Navigation Satellite System in Anti-Drowning System

Yong Lei[1], Minghao Chen[1,3], Tianran Sun[1], Wenrui Li[2], Wensha Gou[1] and Yong Qin[1]

[1]College of Electrical Engineering and Information Technology, Sichuan University
[2]Wu YuzhangHonors College of Sichuan University

[3]chenminghaoscu@163.com

Abstract. Due to the high frequent drowning accident, a new type of drowning-preventing device established on the base ofBeiDou Navigation Satellite System, using the compatibility of GPS tech, GSM, GPRS and other Internet of Things technology as well as Vital Signs Detection technology is designed. The situation whether a person is in drowning is judged by the device monitoring the heart rate and oxygen saturation. And the device utilizes Global Positioning System module to find exact location and uses GPRS to send the information of life signal and position. Based on the experiment, this system can decrease the rate of the drowning accident tosome extent.

1. Introduction

Recently, drowning accident which causes death to have occurred frequently. By the statistics, seventy two thousand people lose their life annually because of drowning, occupying 7% of all deaths related injury[1].Given the events of drowning, nowadays only the propaganda is made for preventing drowning in school, government unit and so on in China. And the effect is very limited due to the fact that it does not decrease the rate of drowning accidents effectively.

The core of control is microcontroller STM32 in this device. It also utilizes Beidou system and GPS, GSM communication, GPRS data transformation, humidity sensor and vital signs detection and other relevant technologies. The situation whether the person is near drowning or drowning is estimated by the device automatically. SOS will be sentautomatically when it happens. The real-time heart beat rate, oxygen content of blood and positioning information are also contained in the SOS to make sure it could decrease the rate of drowning to the minimum.

2. System scheme design

This system is divided into two portions: satellite positioning communication system and vital signs detection system. When the humidity sensor in the device monitor that the device is below the water surface and the humidity is above a certain value, the BeiDou positioning system will start to locate the device user, process the coordinates of the location and estimate whether the user is in the lake and stream or not. When the user is located in water, the vital signs detection system will start to work and to make real-time monitoring of users' vital signs and it also sends the first warning to the outside world. The user's family is able to see the user's real-time heart rate, blood oxygen content and position through this platform. When the vital signals of the user start to show abnormal fluctuation or have a tendency to go in the wrong way, the second warning will give a warning to the user through a buzzer, alerting him to get back to the shore immediately. The user's status will send to their families at once so they could ask for help from the closest fire station and the hospital if necessary. During this process, the vital signs detection system is responsible for the detection of user's vital signal test and judging if the user is in drowning or not while the satellite positioning communication system is in charge of locating the position of human body and makes sure the message will be conveyed effectively. The combination of these two will help drowning people can be quickly known to the outside world and can be effectively given first aid.

2.1. Satellite positioning communication system

*2.1.1. BeiDou satellite positioning technology.*Satellite positioning technology of BeiDou is a Chinese independent strategic satellite navigation system which is aiming for the safety of people and the progress of China's economy. The service , such like positioning, timing and communication all day long and so on, can be provided. The Beidou system can provide multiple fixed frequency navigation signals and use multi-frequency signal combination to improve the service accuracy. BeiDou system also has the characteristic of rapid positioning, short message service and precise timing. Position system of Beidou has been developed rapidly in recent years and the positioning accuracy, power loss and other aspects have been improved greatly. The map of BeiDou is expected to enter our life in May this year and the unique compatibility of BeiDou system can improve the accuracy of positioning through compatibility with other satellite systems.[2] Emergency message can be sentby the short message service of BeiDou system, when the communication stations where the site is used are no longer available or the location is not covered by a communication base. This means that the system can improve the efficiency of information transmission through two-way satellite signals information transmission. We can predict that the short message service will show excellent performance in remote ponds at rural areas andincrease the applicability and dependability of this device.

2.1.2. GSM TECH. Belonging to the second generation cellular mobile communication technology. GSM can run on multiple radio frequencies, which is a cellular network. The advantages of GSM system are the large capacity of the network, strong anti-interference, information sensitivity and high interoperability, etc.[3] But it often suffers from weak signals or even has no signal and therefore the information cannot be transmitted. In this case, GSM and the short message service of BeiDou system should be combined for information transmission.

2.1.3. GPRS TECH. GPRS denotes the generic packet radio service technology. GPRS is able to furnish packet switching. It is achieved packet switching through adding additional hardware entities and modifying the existing base to some extent. The carrier of its data transmission is packet and the cost of using the technology is relatively cheap compared to GSM. The transmission rate of GPRS can reach 56-114Kbps.[4]

2.2. Vital signs detection system

Using the vital signs module to measure the heart rate and the oxygen content is the basic principle of vital signs detection system. The human body suffocates when it is drowning and the lack of oxygen leads to a decrease in blood oxygen content and in the end you will have hypoxemia. Meanwhile, the heart rate increases sharply and the heart appearto have the reflective asystole until heart failure due to asphyxia. Hence, heart rate and blood oxygen content are used as evidences to determine whether the body is drowning.

Vital signs module is a noninvasive test method using Photo PlethyamoGraphy by calculating the residual light intensity after a certain intensity of light absorption through living tissue, determining the blood physiological information. The volume of blood in the skin is periodic pulsating with the systolic and diastolic cycles of the heart. This variation will tell us cardiac function, blood flow and other physiological information that can be used for the measurement of heart rate and oxygen blood content.[5]

3. System hardware design

3.1. Hardware design of Satellite Positioning and Communication system

The hardware design of the satellite positioning communication system is based on the MC20 chip, STM32 single-chip microcomputer as the core of the BDS+GPS dual-mode antenna positioning and communication system.

MC20 module is a multi-functional wireless module which integrates high performance GNSS engine and four-band GSM/GPRS engine. You can choose the All-in-one scheme or the Stand-alone scheme.

This design mainly adopts the Stand-alone scheme, as shown in Figure 1, which can achieve information transmission of GSM and GPRS, etc. The Stand-alone scheme is that GSM part communicates with MCU via main serial port, such as sending AT command and transmitting GSM data. The GNSS part communicates with MCU through GNSS serial port, such as sending PMTK command and output of NMEA statement[6].

MC20 integrates location, GSM, GPRS and other modules, which greatly miniatures the device and leaves the device more portable.

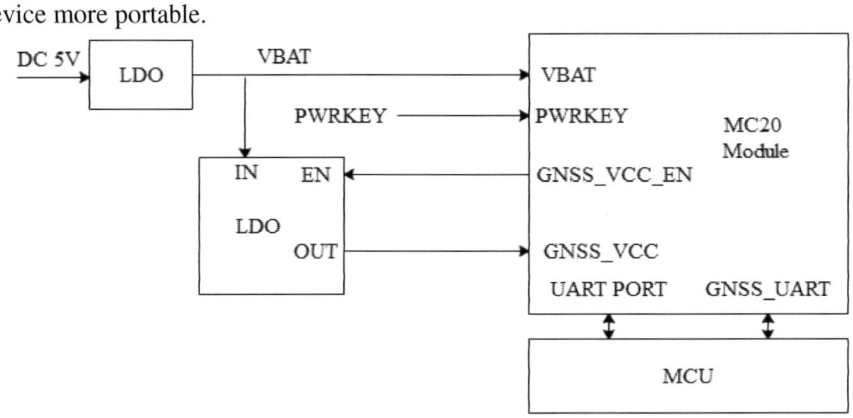

Figure 1. Stand-alone scheme function block diagram.

3.2. Hardware design of Vital Signs Measurement system

In the hardware design of the vital sign measurement system, the core of the detection is Max30102 blood oxygen detection sensor and STM32 single-chip microcomputer. These two important hardwares and the small display screen and the power module constitute the whole detection system.

Max30102 is an integrated sensor for detecting heart rate and blood oxygen content. It can give out and receive red or infrared light. Their peak LED wavelengths are 660nm and 880nm respectively. After the Max30102 receives the light intensity signal, the signal is amplified and digitally filtered, then the output is stored in the internal first-in-first-out(FIFO) memory to complete the acquisition of the original pulse wave[7]. Finally, the Arduino processor can read the data through the standard IIC compatible communication interface on the module and calculate both heart rate and oxygen saturation.

4. Design of system software

4.1. Design of satellite positioning communication systemr

What the main principle of operation of the positioning system isthat it powers up the GNSS (Global Navigation Satellite System)in the MC20 and is controlled by different microcontrollers pin on different circuit boards, and then acquires all the data through the command. Finally, the system stores information about latitude and longitude afterpositioning. The system is verified by software such as u-Center or GNSS_Viewer, whether the positioning software's data is accurate.

The system mainly uses TCP transparent mode to transmit data on GNSS. When the system starts sending messages by the MC20, the module's transparent mode needs to be started up first. When the system exits the transparent mode, it needs to be set for a certain time delay.

The system can configure the module to transmit after launching the projection. Initially, we should close the TCP connection and scene recognition. Then we need to use module of IP SYTATUS to determine if the module is initializing. Finally, we can allocate QIMODE=1.

We can set the data, measured by the vital signs measurement system and positioned by positioning system, to wait for transmission, after starting the data transmission function of GPRS.

4.2. Design of Vital Signs Measurement System

4.2.1. Software Design of Heart Rate Measurement. When the system measures the heart rate, the heart rate value is obtained fromthe PPG signal using the time domain method. Firstly we use the Max30102's FIFO memory to get the intensity signal of infrared light. Then we calculate the average value of the light intensity and use it as the DC signal for the entire sampled waveform. We eliminated the DC signal of the intensity signal of infrared light at sampling points, making them become AC signals that can pass through zero point. We average the AC signal from every four adjacent sampling points. Then we add all the averages and take the average again to get the threshold of AC signal at the minimum peak. The threshold is used as a demarcation point and then we take the AC signals that are greater than or equal to the threshold. We sort the AC signal at each sampling point from the largest to the smallest. We discard the AC signals of some sampling points according to the adjacent range of these sampling points, so that the influence of sharp clutter around the peaks in the same cycle will be reduced. Finally, we arrange the sampling points of the AC signals that are used as peaks based on the time of acquisition, in order to calculate the average of the time between adjacent peaks, which are used as the cycle for each beat of the heartbeat. We obtain the heart rate according to the calculation formula of heart rate[8]. The program flow is shown in Figure 2.

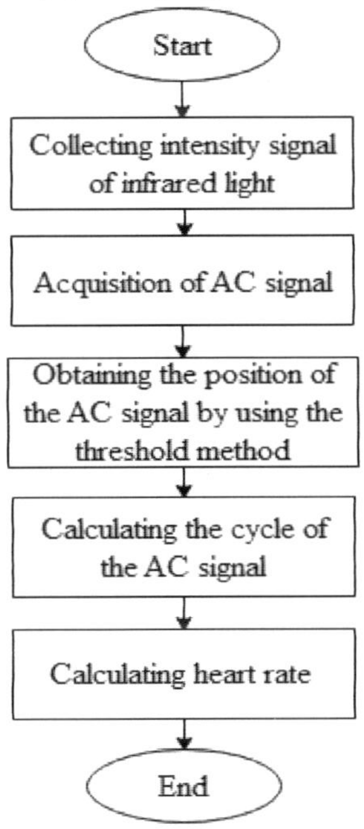

Figure 2. Flowchart formeasuring heart rate.

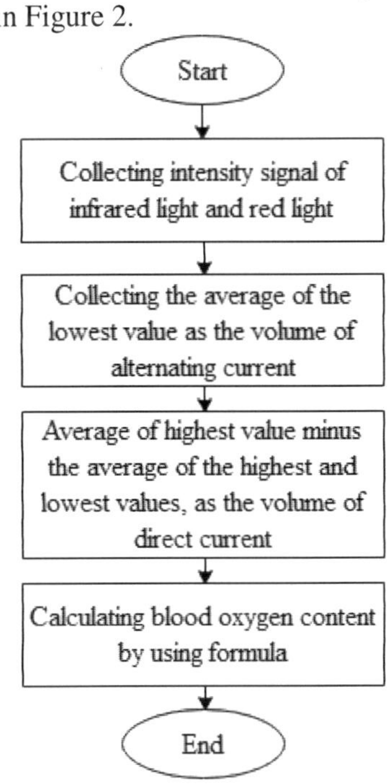

Figure 3. Flowchart for measuring blood oxygen content.

4.2.2. Software Design of Measuring Blood Oxygen Content. Our formula measuring blood oxygen content is$SaO_2 = A_1 \cdot R^2 + B_1 \cdot R + C_1$.The values of A_1,$B_1$andC_1 were determined experimentally. There are currently four ways to get the calibration curve: 1.Experimental data is obtained directly from the human body and used to draw calibration curves. 2.We acquire data from a simulated circulation system with a simulated tissue model in the laboratory. 3.Get data from animal experiments. 4.Pulseoximeter emulator[9]. We chose the empirical value from the official in the experiment.$A_1 = -45.060, B_1 = 30.354, C_1 = 94.845$ *and* $R = \frac{I_{AC}^{\lambda 1} \cdot I_{DC}^{-\lambda 1}}{I_{AC}^{\lambda 2} \cdot I_{DC}^{-\lambda 2}}$. λ_1 signifies red light and λ_2 signifies infrared light.We can calculate the value of R through the AC signal and DC signal of the intensity signal of red light and infrared light.

To determine the alternating amount, we employ the peak method. That is, we select the peaks and valleys of each periodic waveform in the 100 data after sampling for 4s. To get the volume of alternating current, they are subtracted from each other. We calculate the average of the highest and lowest values for all cycles, which is used as the alternating amount of the entire waveform. We use the average of the highest value of the entire waveform to subtract the volume of AC , to get the volume of DC. We can calculate the value of R according to the formula, and then calculate the content of hemorrhagic oxygen[9]. The program flow is shown in Figure 3.

5. Experimental result

5.1. The measuring result of BeiDou module

To use the BeiDou module to locate experimental location and corresponding latitude and longitude information was obtained. Through the MC20 module, we could know the BeiDou positioning data, GPS positioning data and joint positioning final data respectively. Because the static PPP positioning accuracy and convergence of dual mode positioning were better than the single mode positioning[10], this experiment empolyed the information of the former. Figure 4 reveals the measured latitude and longitude information.

When the longitude and latitude information was automatically imported into the upper computer, the corresponding location could be automatically located. The result was shown in figure 5.

It could be seen from the figure above that the measured latitude and longitude information was correct and it could effectively track users' position. It was credible.

```
10:57:48  $GNGLL,3038.0583,N,10405.0035,E,105748.000,A,D*4D
10:57:50  $GNRMC,105750.000,A,3038.0583,N,10405.0035,E,0.00,307.23,080318,,,D*74
10:57:50  $GNVTG,307.23,T,,M,0.00,N,0.00,K,D*23
10:57:50  $GNGGA,105750.000,3038.0583,N,10405.0035,E,2,9,0.96,492.8,M,-31.9,M,,*60
10:57:50  $GPGSA,A,3,01,09,07,193,27,17,30,,,,,,1.21,0.96,0.74*38
10:57:50  $BDGSA,A,3,13,09,,,,,,,,,,,1.21,0.96,0.74*16
10:57:50  $GPGSV,4,1,14,07,81,273,21,11,73,149,13,30,55,312,17,01,48,147,34*75
10:57:50  $GPGSV,4,2,14,08,47,034,18,42,33,120,31,28,30,304,,09,18,212,34*70
10:57:50  $GPGSV,4,3,14,17,12,236,30,27,09,048,25,193,09,137,27,13,02,323,*4D
10:57:50  $GPGSV,4,4,14,23,01,180,,22,01,136,*70
10:57:50  $BDGSV,3,1,11,10,73,014,,08,72,158,,07,64,081,,03,55,167,26*60
10:57:50  $BDGSV,3,2,11,13,47,194,32,02,47,215,31,01,38,124,28,17,37,072,*60
10:57:50  $BDGSV,3,3,11,05,29,242,32,04,21,108,,09,07,200,17*57
10:57:50  $GNGLL,3038.0583,N,10405.0035,E,105750.000,A,D*44
10:57:52  $GNRMC,105752.000,A,3038.0583,N,10405.0035,E,0.00,307.23,080318,,,D*76
10:57:52  $GNVTG,307.23,T,,M,0.00,N,0.00,K,D*23
10:57:52  $GNGGA,105752.000,3038.0583,N,10405.0035,E,2,10,0.96,492.8,M,-31.9,M,,*5A
10:57:52  $GPGSA,A,3,01,09,11,07,193,27,17,30,,,,,1.18,0.96,0.70*36
10:57:52  $BDGSA,A,3,13,09,,,,,,,,,,,1.18,0.96,0.70*18
10:57:52  $GPGSV,4,1,14,07,81,273,20,11,73,149,14,30,56,311,15,01,48,147,34*71
10:57:52  $GPGSV,4,2,14,08,47,034,18,42,33,120,31,28,30,304,,09,18,212,34*70
10:57:52  $GPGSV,4,3,14,17,12,236,30,27,09,048,24,193,09,137,27,13,02,323,*4C
10:57:52  $GPGSV,4,4,14,23,01,180,,22,01,136,*70
10:57:52  $BDGSV,3,1,11,10,73,014,,08,72,158,,07,64,081,,03,55,167,26*60
10:57:52  $BDGSV,3,2,11,13,47,194,32,02,47,215,31,01,38,124,28,17,37,072,*60
10:57:52  $BDGSV,3,3,11,05,29,242,32,04,21,108,,09,07,200,17*57
10:57:52  $GNGLL,3038.0583,N,10405.0035,E,105752.000,A,D*46
```

Figure 4. Latitude and longitude information map of BeiDou test.

IOP Conf. Series: Materials Science and Engineering **428** (2018) 012009 doi:10.1088/1757-899X/428/1/012009

Figure 5. The image from upper computer.

5.2. The experimental results of basic vital signs measurement

By using the Max30102 sensor, we test heart rate and blood oxygen levels of testers. The heart rate was measured by Max30102 and the Mi Band respectively, for the same part of the wrist, and each time the measurement results were obtained by taking the average value for a certain period of time. The blood oxygen content was measured by Max30102 and YK-83C finger sandwich oximeter. In order to compare the difference between wrist and finger blood oxygen content, we decided to test oxygen content of blood at the wrist and fingers through Max30102. Meanwhile, it was also measured by finger clip oximeter and the measurement results were obtained by taking the average value for a certain period of time. The results were shown in table 1 and table 2.

Tab

It could be seen from table 1 that the relative error of the latter was usually around 1.2%, compared the measured heart rate of both Mi band and Max30102. This indicated that not only the heart rate measurement was relatively exactbut also the error of heart rate was around one-beat. As can be seen from table 2, there is about 1% difference in the blood oxygen content measured by Max30102 at wrist and fingertip. Comparing results of Max30102 with results of YK-83C finger clip-type oximeter, we can see that the absolute error was usually 2%, which indicated that the blood oxygen content was faithful. To sum up, the results obtained by the signs measurement system were reliable and stable.

Table 1. Comparison of heart rate test

	1st test	2nd test	3rd test	4th test	5th test	6th test	7th test	8th test
Mi Band (bpm)	76	79	77	79	83	84	82	87
Max30102(bpm)	76	80	77	80	87	84	83	86
Relative error	0%	1.26%	0%	1.26%	4.81%	0%	1.22%	1.15%

IOP Conf. Series: Materials Science and Engineering **428** (2018) 012009 doi:10.1088/1757-899X/428/1/012009

Table 2. Comparison of blood oxygen content

	1st test	2nd test	3rd test	4th test	5th test	6th test
The blood oxygen content measured by the clip-type oximeter (YK-83C)	98%	98%	97%	96%	98%	95%
The blood oxygen content measured at the wrist (Max30102)	100%	100%	97%	99%	100%	96%
The blood oxygen content measured at the fingertips (Max30102)	99%	99%	98%	97%	100%	95%

5.3. Experimental results of information communication

Whether the specific drowning signal detection device of a single chip computer could effectively transmit information, the experimental results were shown in figure 6.

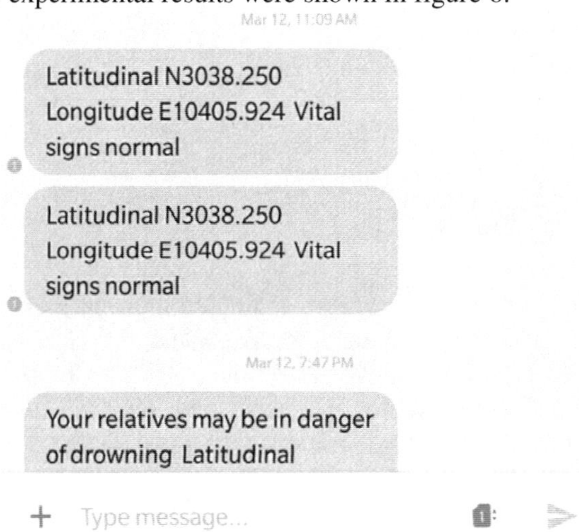

Figure 6. Communication display information.

From the above figure we could know that the information sent included the user's location information (the latitude and longitude information) and the wearer's vital signs. Hence, you only need to automatically import the latitude and longitude information into location program in practice, which will transform these into map position. So the user's actual location information can be triumphantly displayed.

6. Conclusions

In this article, a device built on the BeiDou Navigation Satellite System is designed. The device effectively combines information such as humidity,geographic position, heart rate and blood oxygen content, and sets multiple warnings to enhance the reliability of the device. And contrasted with the traditional anti-drowning methods, this device hasfeatures of convenience, accuracy, real-timeand so on, which makes the risk probability of drowning accidentsreduceeffectively.

References

[1] World Health Organization 2014WHO|WHO highlights devastating global impact of drowningZ. *World Health Organization*
[2] FengqiYe andXiaoleiGao 2018 Beidou Satellite Navigation SystemJ. *SpaceInternational*114-7

[3] TongWu 2012 The Embedded Smart Home System based on QT and GSM*D. Hebei University*
[4] HongSun andYongjunPan 2015 Application of Zigbee in Drowning Prevention System*J. Electronic Sci. &Tech.***28(6)** 72-4
[5] QiaoshouLiu andSenWang 2018Research and Design of Open-ended Heart Rate Device based BLE *J. ElectronicsWorld***3**100-4
[6] MinghuCheng 2016 Hardware Design Manual *S. Quectel Wireless Solutions Co.,Ltd.*
[7] PeifengBai, QiangLiu, FeiboDuan, N FRooij, Nan Li and GuofuZhou 2017 Wearable blood oxygen saturation detection system based on Max30102*J. Laser &Infrared***47(10)**1276-80
[8] MinLi 2016 Research on Motion-resistant Heart Rate and Oxygen Saturation Extracting Algorithm Based on Photoplethysmography*D. Beijing Institute of Technology*
[9] BinZhang 2009 The Study of Blood Oxygen Saturation Measurement Based on Pulse Wave*D. Beijing Jiaotong University*
[10] ChangshengLiu, ChenglinCai, LiangliangWang, YiQin and GangLi 2017 Analysis of positioning accuracy andconvergence of BDS/GPS dual-mode static PPP*J.Journal of HefeiUniversity of Technology***40(4)**481-5

IOP Conf. Series: Materials Science and Engineering **428** (2018) 012010 doi:10.1088/1757-899X/428/1/012010

Test on Stress Distribution of Bionic C-leg Wheel-soil Interaction with Its Data Processing

Gang-dun Hu [1,2,3], Cai-ling He [1], Hui-Ying Li [1] and Lei Sun [1]

[1] School of physics and Electronic-Electrical Engineering, Ningxia University, Yinchuan, China
[2] Ningxia Key Laboratory of Intelligent Sensing for Desert Information, Yinchuan, China

[3] E-mail address: hhggdd@sina.com

Abstract. To study the relationship between the ground support force and drawbar pull of C-leg and the normal and shear stress on the wheel flange, while the C-leg of the six-legged bionic sand-based robot travels on the sand base, the integral model of the above relationship are given on the basis of the designed C-leg loop experimental platform. To solve the limitation of sensors, a "position weighted average numerical interpolation method" is proposed. The normal stress and shear stress sensor on the wheel flange obtained corresponding data by experiment. A number of values are inserted between the measured values of the sensors by the "position weighted average numerical interpolation method". The ground support force and drawbar pull are calculated by the given integral model. At the same time, the three-dimensional force sensor installed on C-leg axle measured ground support force and drawbar pull through the test. The relative error of above force calculated from the integral model is 4.69% and 1.16% respectively compared with the mean value measured by the three-dimensional force sensor. The relative error is less than 5%, which verifies the correctness of the integral model and the feasibility of "position weighted average interpolation method".

1. Introduction

In recent years, in order to meet the needs of adventure, rescue, military reconnaissance, lunar surface and Mars surface detection, domestic and foreign scholars have been engaged in extensive research on robots and planets running on soft ground. Among them, the establishment of mathematical model of wheel-soil interaction is one of the hot topics. Shibly et al. [1] gave a simplified equation of the normal stress and shear stress of planetary exploration rover's wheel-soil interaction and an integral model that can calculate its drawbar pull, the ground support force and the wheel-axle resistance torque verified with good accuracy through the single test bench in the dry sand. Considering the limitation of the Bekker model, Meirion and Spenko[2] propounded an improved pressure–sinkage model for a small rigid wheel. Senatore and Iagnemma[3] installed the force sensor on the wheel flange to measure the movement rule between the normal stress force and shear stress force while wheel travelling on dry sand. The above methods used to analyze the wheel-soil interaction belong to the means of terramechanics. In March 2013, a paper on the Journal of Science [4] attracted the attention of academics to a bionic C-shaped six-legged robot.The authors propose the concept and method of terradynamics, which uses the 6-DOF manipulator to manipulate the movement of a small aluminum plate (3.81 × 2.54cm^2) in the granular media by means of a three-dimensional force sensor and a torque sensor to test drag force (horizontal direction) and lifting force (vertical direction) of the unit area per unit area at different attack angles and different intrusion angles, and thus calculated the ground support force and the drawbar pull of the C-leg by the wheel-soil interaction model. The authors also show that the C-legs have the best performance on the soft granular media in various heterogeneous robot leg types by calculation and experiment.

This paper will briefly introduce the C-leg loop experiment platform based on the theory of terramechanics, and according to the tested normal stress and shear stress of wheel-soil interaction to study its data processing method. The results of this paper can be used to provide theoretical and technical support for establishing mathematical model of C-leg wheel-soil interaction.

2. Situation of test

2.1 The loop experiment platform

The overall structure of the loop experiment platform is shown in Figure. 1[5]. The mechanical and electrical parts are composed of central column, the supporting connecting rod and the three C-legs on one side with its support structure. The three C-legs on one side are respectively driven by AC servo motor to revolve around the center column. C-leg 3 is used for testing, which installs a 16-way sensor and a digital encoder in total. There are 4 ways for three-dimensional force sensor (with torque measurement) installed on the drive axle of the C-leg to test ground support force, drawbar pull, axial force and resistance torque on axle of C-leg separately. The rest 12 ways are used for 6 sets of sensors installed on the wheel flange, each of which includes a normal stress sensor and a shear stress sensor.

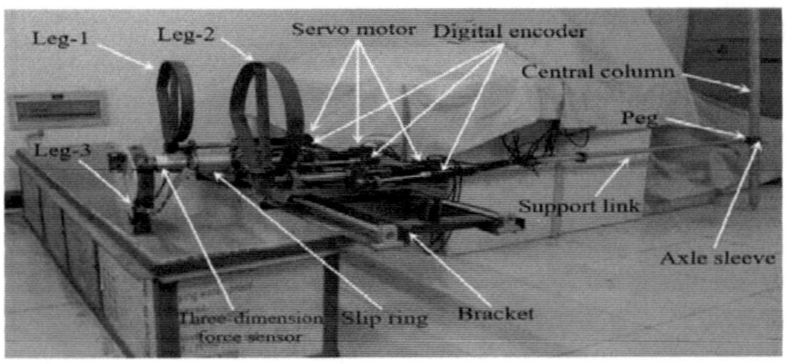

Figure.1 The loop experiment

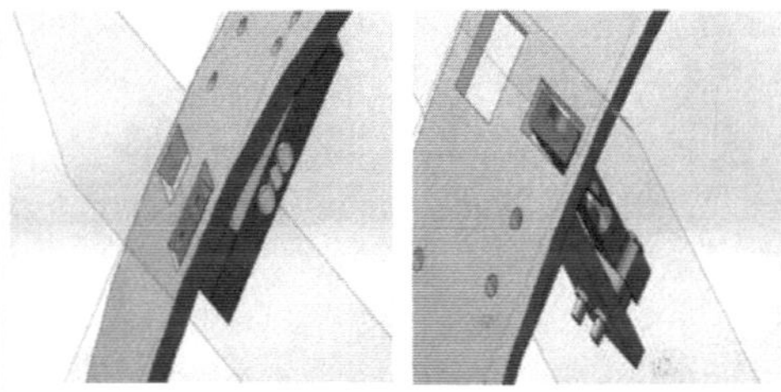

(a) (b)

Figure.2 Installation of normal ang shear stress sensors

The normal stress and the shear stress sensor's installation are shown on Figure. 2. The Figure. 2(a) describes the normal stress sensor installation. Figure. 2(b) describes the shear stress sensor installation.

The radius of C-leg, R= 165mm, and the width of C-leg wheel, b= 40mm.The weight of C-leg for testing is 2.47 kg, and the weight endurance of the tested wheel is 14kg in static state. The material of C-leg is steel plate and the outer wheel is smooth without wheel thorn.

The hardware of data acquisition system is the NI PCI-6143S multi-function data acquisition (DAQ) board card[6]. Two types of tests can be carried out on the platform. One is test experiment of the dynamic characteristics,The other is the test experiment in sand tank: The C-leg is placed in the sand tank, only to limit its advancing, which can complete the test of slip rate of 1. This experiment mainly tests the relationship between the drawbar pull and its relative velocity that taking sand medium as the reference point of C-leg.

2.2. The Sand tank

The overall dimension of the sand tank is 2000mm × 400mm × 400 mm. The sand on the sand tank is taken from the natural aeolian sand of the Tengger desert (dry-sand) , and the sampling site is zhongwei shapo head.

2.3. Testing principle

Figure. 3 portrays the C-leg and sand interaction analysis of forces. The vertical baseline, BD, and the reference line of C-leg wheel, OA, are defined. The rotation angle,θ, displays the displacement of OA relative to BD during the movement of C-leg. Φ denotes the deflection angle of the observation point on the wheel flange relative to the reference line OA(Clockwise rotation angle is positive, the counter clock wise deflection is negative). Φ1 and Φ2 are the entry and departure angle respectively. z1 is the front settlement and z2 is for settlement. R, b, and αdenote the wheel radius width, and round corner of the C-leg respectively.

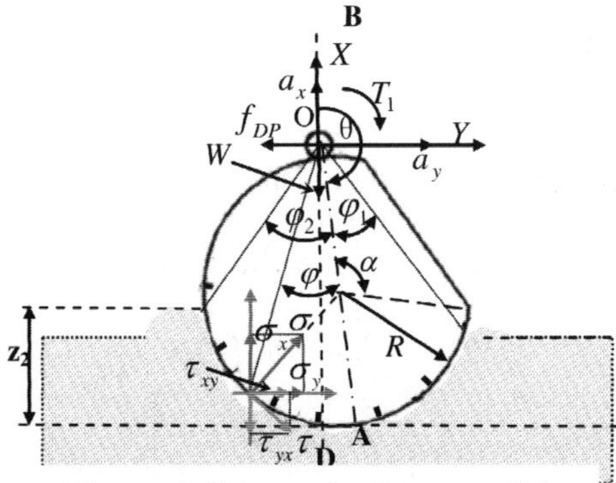

Figure.3 Schematic diagram of the sensor group installation of the C-leg

C-leg movement equations are expressed as

$$F_N(\theta) - W - Mg = Ma_X \tag{1}$$

$$F_{DP}(\theta) - f_{DP} = Ma_Y \tag{2}$$

$$T_1 - T_2(\theta) = J\frac{d^2\theta}{dt^2} \tag{3}$$

Here:M and J denote quality and rotary inertia of C-leg, respectively. $F_N(\theta)$ and $F_{DP}(\theta)$ are ground support force and drawbar pull, respectively. X-direction and Y-direction acceleration are a_X and a_Y , respectively, and g is acceleration of gravity. T1 is motor driving torque. $T_2(\theta)$ is resisting torque of wheel-soil interaction.

The ground support force and drawbar pull caused by C-legs' wheel and soil interaction:

$$F_N(\theta) = F_{N\sigma}(\theta) + F_{N\tau}(\theta) \tag{4}$$

$$F_{DP}(\theta) = F_{DP\sigma}(\theta) + F_{DP\tau}(\theta) \tag{5}$$

Here: $F_{N\sigma}(\theta)$ and $F_{N\tau}(\theta)$ are respectively normal stress component and sheer stress component of the ground support force. $F_{DP\sigma}(\theta)$ and $F_{DP\tau}(\theta)$ are respectively normal stress component and sheer stress component of the drawbar pull.

The integral model of the above components expressed as Eqs. (6) to (9):

$$F_{N\sigma}(\theta) = -2bR\int_0^{\varphi_2}\sigma(\varphi+\theta)\cos(2\varphi+\theta)d\varphi - 2bR\int_{\varphi 1}^{0}\sigma(\varphi+\theta)\cos(2\varphi-\theta)d\varphi \tag{6}$$

$$F_{N\tau}(\theta) = 2bR\int_0^{\varphi_2}\tau(\varphi+\theta)\sin(2\varphi+\theta)d\varphi - 2bR\int_{\varphi 1}^{0}\tau(\varphi+\theta)\sin(2\varphi-\theta)d\varphi \tag{7}$$

$$F_{DP\sigma}(\theta) = -2bR\int_0^{\varphi_2}\sigma(\varphi+\theta)\sin(2\varphi+\theta)d\varphi + 2bR\int_{\varphi 1}^{0}\sigma(\varphi+\theta)\sin(2\varphi-\theta)d\varphi \tag{8}$$

$$F_{DP\tau}(\theta) = -2bR\int_0^{\varphi_2} \tau(\varphi+\theta)\cos(2\varphi+\theta)d\varphi - 2bR\int_{\varphi 1}^{0} \tau(\varphi+\theta)\cos(2\varphi-\theta)d\varphi \tag{9}$$

Here: $\sigma(\varphi+\theta)$ is the normal stress while C-leg rotation angle of the C-leg is θ and the deflection angle on the flange is φ. $\tau(\varphi+\theta)$ the sheer stress on the above corresponding position.

From Eq. (6) - Eq. (9), it can be seen that for the above formulas, the expression of normal $\sigma(\varphi+\theta)$ and sheer stress $\tau(\varphi+\theta)$ on the wheel flange should be derived first.

2.4. The test process

The sand tank test was studied in this paper. The tested wheel was placed in the middle of the sand tank, and the rope is used to prevent platform from advancing. It is not allowed to affect the up-and-down movement of the test wheel. In the test, the weight of the three legs on one side and its bracket is shared by the tested wheel and the pin of center column. When the test wheel leaves the sand tank, the platform is balanced by wheel 1, wheel 2, and pin. While studying the relationship between the ground support force, drawbar pull of the C-leg and the normal force, shear force on the wheel flange, it is needed to make the tested wheel rotate at a certain speed, repeating 20 times. The test result is stored in the computer directly by the data acquisition system. The ultimate test data is from the average 20-time test results.

3. Results and analysis

3.1. The result of the experience

In tank test, the change law of the sheer and normal stress in wheel flange with rotation angle θ can be tested. In the same time, the change law of the support ground and drawbar pull on the axle with rotation angle also can be obtained, as shown in Figure. 4. In Figure. 4(b), the ground support force and drawbar pull are measured by the three-dimensional force sensor on the C-leg shaft. GSFfit is a fitting curve of the ground support force, and DPfit is a fitting curve of drawbar pull.

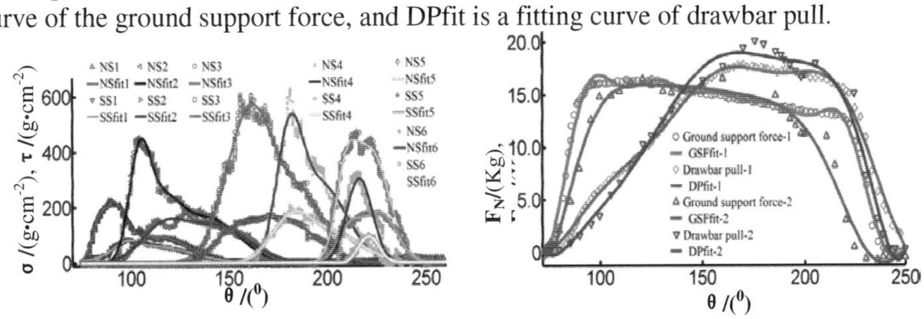

(a) Test data of normal stress and shear stress of wheel flange (b) Ground support force and Drawbar pull

Figure.4 Test data of C-shaped leg wheel

3.2. Test data processing

3.2.1. Numeric insertion. It can be seen from the test data Figure. 5(a) that six sets of sensors are installed on the C-leg wheel, but in most cases the normal stress sensor (or shear stress sensor) has only 2 to 3 readings. It is clear that the measurement data is small if using it to analyze the stress of C-leg distribution rule. Therefore, numerical insertion method is used to insert some values between the sensors. The normal stress (or shear stress) on the wheel flange is continuously changed. The closer the point is to the sensor, the more similar normal stress (or shear stress) changes with the sensors. So the position weighted average numerical interpolation method used. On the basis of the measurement value, it will be inserted between the two sensors to make up for the data shortage. The detailed operation of numerical insertion in this paper is as follows: Firstly, aligning the data columns measured by the two sensors during the process of sand entry referring to the peak. Secondly, the weight of the two sensor measurements in the insertion value is determined by the number of values to be insertedThen the insertion value is obtained according to the weight of the average. The data column of the insertion value can be obtained by means of average of the one by one correspondence. Finally it will be inserted into the corresponding position.Whether the inserted value is accurate or not, the ground support force and the hook traction force obtained by the C-shaped leg and sand action can be directly measured by installing a three-dimensional force sensor on the C-shaped leg shaft, and the calculated values of the Eq. (4) and (5) can be obtained. In comparison, if the ground support force and hook traction force obtained by the two methods are the same, it indicates that the method of inserting values is feasible.

3.2.2 Sensor measurements and insertion values and regression equations. The normal stress and shear stress data columns can be obtained from measurement and inserted values. The curve fitting tool of MATLAB software (Curve Fitting Tool) can be used. The regression equations are obtained by Gaussian function fitting. They can be express as:

$$\sigma(\theta)_{jks} = \sum_{i=1}^{n} a_{jk_i} \exp(-((\theta - b_{jki}) / c_{jki})^2) \quad (10)$$

$$\tau(\theta)_{jks} = \sum_{i=1}^{n} u_{jki} \exp(-((\theta - v_{jki}) / w_{jki})^2) \quad (11)$$

Here, fitting parameters include a_{jki}, b_{jki}, c_{jki}, u_{jki}, v_{jki} and w_{jki}. j and k are the serial number of the sensor group, taking six numbers 1-6. s is the serial number of the inserted value. Figure. 5 shows the fitting curve for the measurement and insertion values. Figure. 5 (a) shows the 32 normal stress fitting curve. The red are the fitting curves of the normal stress measurement(Fitting curve of test values of normal stress), and the blue are the fitting curves of the normal stress insertion value (Fitting curve of inserted values of normal stress). Figure. 5 (b) shows the 32 shear stress fitting curve. The red are the fitting curves of the shear stress measurement(Fitting curve of test values of shearing stress), and the blue curve is the fitting curve of the shear stress insertion value(Fitting curve of insert values of shearing stress) .

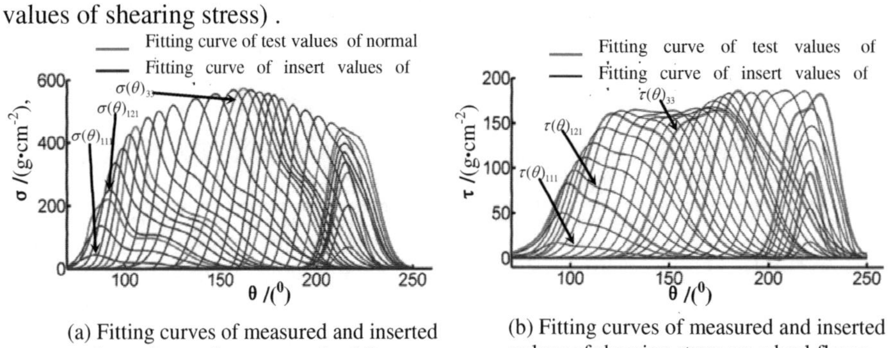

(a) Fitting curves of measured and inserted values of normal stress on wheel flange

(b) Fitting curves of measured and inserted values of shearing stress on wheel flange

Figure.5 Fitting curves of measurement and insertion values on the C leg flange

3.2.3 The normal and shear stress on the flange and its regression equation While C-leg enters the sand and revolves to arbitrary angle θ, the sheer and normal stress value of the sensors and insertion points on the flange would be derived by the above 64 regression equations. Then, it can be obtained through the means of curve fitting that the regression equations of sheer stress $\sigma(\varphi+\theta)$ and normal stress $\tau(\varphi+\theta)$ in θ changed with φ.The equations are as follows

$$\sigma(\varphi+\theta) = \sum_{i=1}^{n} a_i \exp(-(((\varphi+\theta) - b_i) / c_i)^2) \quad (12)$$

$$\tau(\varphi+\theta) = \sum_{i=1}^{n} u_i \exp(-(((\varphi+\theta) - v_i) / w_i)^2) \quad (13)$$

Here: a_i, b_i, c_i, u_i, v_i and w_i are fitting parameters.

3.3. Results analysis

3.3.1. The ground support force and drawbar pull. The ground support force and drawbar pull can be obtained when the C-leg rotates to θ angle by substituting Eq. (12) and Eq. (13) into formula (4) to (9). In this paper, the values of the ground support force and drawbar pull can be calculated when θ angle is set in the range of 75° to 250° (C-leg enters into the sand at the angle of 78° and goes out the sand at the angle of 240°) at 5° intervals as shown in Figure.5(b).The regression equation of each fitted curve can be obtained by software MATLAB. This paper uses sine function fitting.

From the fitting parameters, the average value of the ground support force and the hook traction force calculated from the measured values of the rim normal stress and the shear stress sensor are calculated within the range of the C-leg sand entering angle. The average ground support force and hook traction force obtained were 14.21 Kg and 13.28 Kg, respectively.

Similarly, in the range of $\theta = 85^o$ to $\theta = 210^o$, the average values of the ground support force and the drawbar pull are calculated by three-dimension force sensor are respectively 14.91kg and13.12kg.

3.3.2. Analysis of experimental results. Through Figure.5(b), there are the consistent changing rule of the ground support force and drawbar pull, which are measured directly by the three-dimensional force sensor on the C-leg's axle and calculated by the given integral model for the normal stress and the shear stress (measured and inserted value) on the wheel flange, respectively, as shown in Figure.5(b).Contrasting with the average value from $\theta = 85^o$ to $\theta = 210^o$, it can be seen that the relative error, $\gamma_{NSJ} = (\overline{F}_{NS} - \overline{F}_{NJ}) / \overline{F}_{NS}$ between $\overline{F}_{NS}$ and $\overline{F}_{NJ}$, $\gamma_{DPSJ} = (\overline{F}_{DPS} - \overline{F}_{DPJ}) / \overline{F}_{DPS}$ between $\overline{F}_{DPS}$ and $\overline{F}_{DPJ}$, are respectively 4.69% and 1.16%. In other word, the relative error are both less than 5%.

4. Conclusion

This paper addressed the integral model of the interaction of C-leg wheel and soil, studied numerical interpolation between stress sensors on the C-leg flange, proposed a "positional weighted average interpolation method", and given the concrete method to calculate the expression $\sigma(\varphi+\theta)$ of normal stress, and the expression of shear stress $\tau(\varphi+\theta)$ on the C-leg flange. The change rules of the ground support force and drawbar pull obtained by the integral model are consistent with that measured by three-dimensional force sensor. In the θ_1=85° to θ_2=210° range. The relative error of the ground support force and the drawbar pull calculated from the integral model and measured by the three-dimensional force sensor, respectively, is less than 5%.This shows that the integral model presented is correct, the proposed "positional weighted average interpolation method" is feasible. The research results could provide theoretical and technical support for establishing the mathematical model of wheel-soil interaction.

Acknowledgement

This research is supported by Chinese National Natural Science Foundation (61463043). The authors would like to thank Zhuangzhuang Ruan, and Meng Liu at Ningxia University for their contributions to this experiment.

References

[1] Shibly H, Iagnemma K, Dubowsky S. An equivalent soil mechanics formulation for rigid wheels in deformable terrain, with application to planetary exploration rovers[J]. *Journal of Terramechanics*, 2005, 42(1): 1 - 13.

[2] Meirion-Griffith G, Spenko M. A modified pressure–sinkage model for small, rigid wheels

IOP Conf. Series: Materials Science and Engineering **428** (2018) 012010 doi:10.1088/1757-899X/428/1/012010

on deformable terrains[J]. *Journal of Terramechanics*, 2011, 48(2): 149 - 155.

[3] Moreland S, Skonieczny K, Inotsume H, et al. Soil behavior of wheels with grousers for planetary rovers[C]// *Aerospace Conference. IEEE*, 2012: 1-8.

[4] DENG Zong-quan,DING Liang,GAO Hai-bo,TAO Jian-guo,WANG Shao-qian. Influence of soil properties on lunar rover's wheel-soil interaction mechanics[J].*JOURNAL OF HARBIN INSTITUTE OF TECHNOLOGY.* 2010,42(11): 1724 - 1728.

[5] Chen Li, Tingnan Zhang, Daniel I. Goldman. A Terradynamics of Legged Locomotion on Granular Media[J]. Science,2013, 339(6126): 1408 - 1412.

[6] HU Gang-dun, Guo hui, LEI Ning.Dedign of C-leg Experiment Platform Based on LabVIEW Date Acquisition System. *Measurement & Control Technology*,2017,(09):127-130.

IOP Conf. Series: Materials Science and Engineering **428** (2018) 012011 doi:10.1088/1757-899X/428/1/012011

An Investment Decision Model of Distribution Network Planning Based on Correlation Mining of Reconstruction Measures and Loss Load Index

Jun Xiong[1], Zhixuan Liu[2], Yue Xiang[3], Yanxin Chai[3], Junyong Liu[3] and Youbo Liu[3]

[1.] State Grid Xiamen electric power supply company, Xiamen, 361000, China
[2.] State Grid Fujian Electric Power Research Institute, Fuzhou, 350007, China
[3.] Sichuan University, Chengdu 610065 China

The corresponding author's e-mail address: ziqi_chai@163.com

Abstract. In recent years, how to build a reliable, high-efficient, high-tech and environmental-friendly distribution network infrastructure and service system, especially to accurately assess the reliability impact of various technical measures and establish optimal investment decision model for the future distribution network planning, has great theoretical and practical significance. However, the traditional analysis of investment decision model of distribution network is based on complex power flow calculation, which is not suitable for current construction of large and smart distribution network with multi-agent interaction under electrical market environment. Therefore, in this paper, based on the sample data, the correlation mining methods of the reconstruction measures and loss load index, such as neural network method, are applied to analyse the relationship between different types of reconstruction measures and loss load index. Based on the correlation mining methods, the relationship between the reconstruction measures and loss load index can be obtained and the value of loss load can be fast predicted. Through the comparison the relationship between reconstruction measures and loss load index, the optimal reconstruction measures can be chosen, and the data relationship based accurate investment decision model can be built. Experimental result shows the accuracy and effectiveness of the presented methodology.

1. Introduction

With the increasing permeability of clean energy, the transformation and investment decision of distribution network is the need of the development of smart distribution network in the future. [1-2]. However, traditional way of extensive investment projects has been unable to meet the accurate investment demand of the future active distribution network.

To establish the precise investment decision model, the relationship between the transformation reconstruction measures and performance indexes is crucial. However, the traditional correlation analysis involves complicated power flow calculation. Therefore, common data correlation mining methods [3-8] seem to be better choices. Based on the artificial neural network prediction method, the reliability index of the operation reliability of a certain time scale could be predicted [9].

At present, the investment decision has evolved into a large scale dynamic combinatorial optimization problem. Taking the maximization of the project portfolio as the target, establish the investment decision model [10-11]. Besides, the analytic hierarchy process and the principal component and set pair analysis were also used to analyze the comprehensive investment decision based on the evaluation index system [12-14]. Moreover, an investment evaluation model based on triangular fuzzy number is put forward [15]. Correlation analysis is also widely used. The investment evaluation model based on the improved grey relational degree and TOPSIS was constructed, and the investment benefit is evaluated and compared according to the degree of relevance of the sample and the ideal solution [16-17].

In this paper, the artificial neural network algorithm is applied to obtain the association expressions between the reconstruction measures and loss load index. And then the investment decision model of distribution network planning based on correlation mining is established, taking the optimal performance indexes as the target. The IEEE 33 node distribution network is used as an example to verify the effectiveness of the proposed investment decision model in the paper.

2. Distribution network index correlation mining based in BP neural network

In this paper, considering the complex relationship between the loss load index and reconstruction measures, BP Neural Network is used to achieve correlation mining.

2.1. BP neural network principle

BP Neural Network (Back Propagation Neural Network, the BPNN) is a kind of multilayer feedforward Neural Network based on error Back Propagation algorithm. The training process of BP neural network includes two parts. The structure is shown in the Figure 1 below.

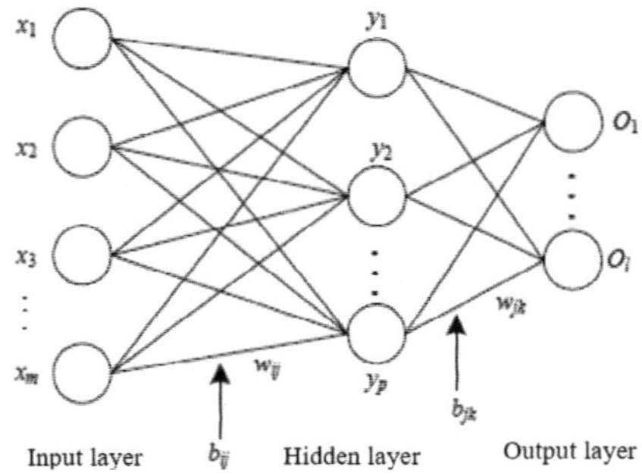

Figure 1. The structure of BP Neural Network

2.2. BP neural network training

According to the structure of BPNN in Figure. 1, signal forward propagation can be described as formula (1-2). x_i represents the input sample. y_j represents the hidden layer output. o_k represents the output layer output. w represents the weight. b represents the threshold.

$$y_j = f[\sum_{i=1}^{m} w_{ij}x_i + b_{ij}], j = 1,2,..., p \tag{1}$$

$$o_k = f[\sum_{j=1}^{p} w_{jk}y_j + b_{jk}], k = 1,2,...,l \tag{2}$$

Error back propagation can be described as formula (3-5). The gradient descent method is applied. e represents the output error. d_k represents the expected output. Δw and Δb represent the weight and threshold of the adjustment. η represents the learning rate, which is an important factor impacting the convergence rate of the algorithm.

$$e = \frac{1}{2}(D-O)^2 = \frac{1}{2}\sum_{k=1}^{l}(d_k - o_k)^2$$
$$= \frac{1}{2}\sum_{k=1}^{l}\{d_k - f[\sum_{j=1}^{p} w_{jk}y_j + b_{jk}]\}^2$$

$$=\frac{1}{2}\sum_{k=1}^{l}\{d_k - f[\sum_{j=1}^{p} w_{jk} f[\sum_{i=1}^{m} w_{ij}x_i + b_{ij}] + b_{jk}]\}^2 \tag{3}$$

$$\nabla w = -\eta \partial e / \partial w \tag{4}$$

$$\nabla b = -\eta \partial e / \partial b \tag{5}$$

3. The investment decision model based on the correlation mining

3.1. Reconstruction measure set of distribution network

Generally speaking, the optional plan of reconstruction measure can be expanded from the following three aspects:
1) Reconstruction of infrastructure;
2) Deployment of modern facilities;
3) Improvement of technology and management level.

3.2. Loss load index

Loss load index can be described as formula (6-7).

$$I_D = \lambda_F(\beta_1 T_1 + \beta_2 T_2 + \beta_3 T_3) \tag{6}$$

I_D and λ_F represent the SAIDI (System Average Interruption Duration Index) and the total failure rate of the feeder. T_1, T_2 and T_3 respectively represent the power failure users type, then β_1, β_2 and β_3 are their distribution coefficient.

$$I_E = I_D P_L \tag{7}$$

The loss of load index can be represented by EENS (expected energy not supplied) index I_E. And P_L is the total quantity of load of the feeders.

3.3. The investment decision model of distribution network

The traditional correlation analysis of the reconstruction measures and performance indexes involves the constraints of the complex power flow, which can be described as follow:

$$\max \bar{I} - I \tag{8}$$

$$s.t. \ \sum_{i=1}^{n} K(X_i) \le K_{\max} \tag{9}$$

$$X_i \in \{0,1\} \tag{10}$$

$$\varphi_i(I_i, X_i, y) = 0 \tag{11}$$

The objective of the model is maximizing the index as formula (11). Investment cost constraint, reconstruction measures inseparable constraint and correlation constraint are respectively shown in formula (9-11). $\bar{I}$ and I represent the index after and before the implementation of the reconstruction measures. X_i is the reconstruction measure type and $K(X_i)$ is the investment cost. Besides, φ_i expresses the relationship between X_i and I_i. In particular, y in formula (11) represent the network and power flow condition.

Compared with the traditional and complicated power flow calculation shown as formula (11), the investment decision model proposed in this paper introduces the BPNN described as formula (12). w and b represents the weight and threshold of the applied BP neural network.

$$\max \bar{I} - I$$

$$s.t. \ \sum_{i=1}^{n} K(X_i) \le K_{\max}$$

IOP Conf. Series: Materials Science and Engineering **428** (2018) 012011 doi:10.1088/1757-899X/428/1/012011

$$X_i \in \{0,1\}$$
$$\varphi_i(I_i, X_i, w, b) = 0 \tag{12}$$

4. Case study

In this paper, the offline BP neural network is trained based on a large number of reconstruction measures and performance indexes data of distribution network, and then the sensitivity of each reconstruction measure to each performance index can be obtained, based on which the optimal combination of reconstruction scheme can also be gained. The initial distribution network applied in the paper is shown in Figure 2, and the reconstruction measures implemented are listed in Table 1.

Table 1. Reconstruction measures of distribution network

Serial Number	Measure	Cost(￥)
P1	additional remote terminals at node 9	100,000
P2	additional remote terminals at node 26	100,000
P3	new construction of DG at node 14: Photovoltaic 0.5MW	1000,000
P4	existing DG expansion at node 14: Photovoltaic 1MW	1500,000

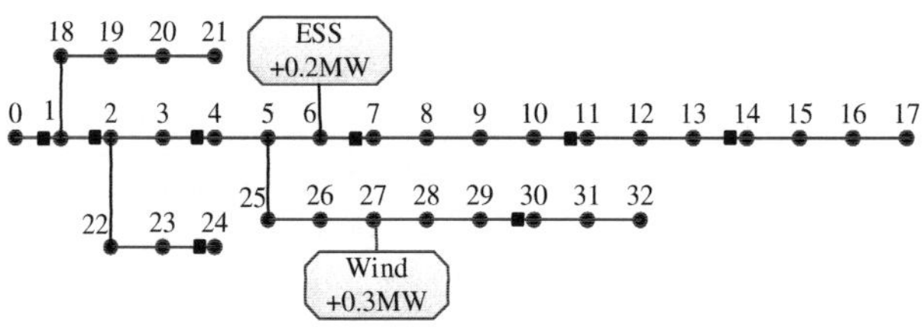

Figure 2. The initial distribution network with DGs and switches

At the same load level, as the result is shown in Table 2, all the four measures have influence on the loss load index. For example, the quantity of lost load of initial network reaches 1.4194MW, while the quantity of lost load of the network with 1.5MW additional photovoltaic at node 14 and remote terminals at node 9 and 26 is 1.2725MW, which has reduced 10.35%.

Table 2. The EENS index with reconstruction measures

Terminal / Lost load(MW) / PV	P1	P2	P1	P2	P1	P2	P1	P2
	0	0	0	1	1	0	1	1
0MW	1.4194		1.3569		1.3674		1.3141	
0.5MW(P3)	1.3965		1.3390		1.3324		1.2864	
1.5MW(P4)	1.3288		1.3127		1.3118		1.2725	

In order to make optimal investment decision, the cost-effectiveness ratio which can be described in formula (13) is introduced to measure the various reconstruction measures. The cost-index ratio of measures P1, P2, P3 and P4 are shown in Figure 3.

$$BCR = \frac{|\bar{I} - I|}{\cos t} \times 100\% \tag{13}$$

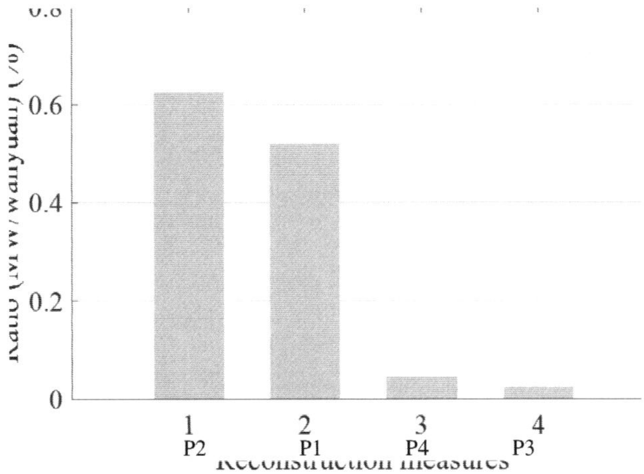

Figure 3. The cost-EENS index ratio of the four reconstruction measures

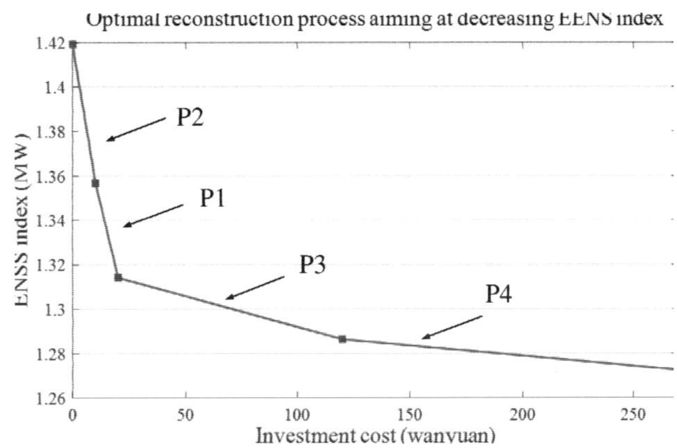

Figure 4. The optimal reconstruction process aiming at decreasing EENS index

In this paper, based on the calculation of the cost-effectiveness ratio above, the optimal reconstruction process can be presented as Figure 4 with the decrease of EENS index of the distribution network as the target.

Table 3. The optimal investment decision with two targets

Target	Constraint	Optimal investment
Max ΔI	$K_{max} = 200$	$P2 \rightarrow P1 \rightarrow P3$

As shown in Table 3, when setting the reduction of lost load as the target, the optimal investment process of distribution network is adding the remote terminals at node 9 and 26 in turn, and finally constructing new photovoltaic at node 14 with the maximum investment cost of ￥2000,000. At this time, the quantity of lost load is 1.2864MW and the reduction is 9.37%.

5. Conclusion

In this paper, BP neural network is utilized to realize the correlation mining of the reconstruction measures and loss load index of distribution network, which replaces the complex power flow calculation in distribution network planning. And the optimal investment process and scheme can be obtained. Therefore, the planning schemes decided by this method have more practical guidance.

Acknowledgement

Project Supported by Science and Technology Foundation of State Grid Corporation of China (Research and Application of Technical Path Development and Its Precise Investment Decision in Active Distribution Network) <52130417002R>.

Reference

[1] ZHENG Gang, XU Hongbing, WANG Xinheng, et al, "Distributed control strategy for variable speed constant frequency wind power unit based on intelligent multi-agents," Automation of Electric Power Systems, vol. 34, no. 6, 2010, pp. 94-97.
[2] Yunche S U, Liu J, Liu Y, et al, "Optimization Model of Selecting Power Supply Reliability Reconstruction Measures in Large-Scale MV Distribution Network and Its Solution Method," Power System Technology, vol. 41, no. 1, 2017, pp. 201-210.
[3] Heinemann F, Moessinger M D, Yeter M, "Do fiscal rules constrain fiscal policy? A meta-regression-analysis," European Journal of Political Economy, vol. 51, 2017, pp. 69-92.
[4] Velden T, Yan S, Lagoze C, "Mapping the cognitive structure of astrophysics by infomap clustering of the citation network and topic affinity analysis," Scientometrics, vol. 111, 2017, pp. 1-19.
[5] Kalyani G, Rao M V P C S, Janakiramaiah B, "Privacy-Preserving Association Rule Mining Using Binary TLBO for Data Sharing in Retail Business Collaboration," Proceedings of the 5th International Conference on Frontiers in Intelligent Computing: Theory and Applications, 2017, pp. 189-197.
[6] Martsenyuk V, "Indirect method of exponential convergence estimation for neural network with discrete and distributed delays," Electronic Journal of Differential Equations, vol. 2017, no. 246, 2017, pp. 1-12.
[7] Wu W, Chen Y, Seng D, "Implementation of Web Mining Algorithm Based on Cloud Computing," Intelligent Automation & Soft Computing, vol. 12, 2017, pp. 1-6.
[8] Tian H, Kong L A, Huang Z, "Based on Fuzzy Grey Correlation Method in the Relation Analysis between Safety Investment and Benefit," International Conference on Bioinformatics and Biomedical Engineering. IEEE, 2010, pp. 1-4.
[9] HU Lijuan, DIAO Yinglong, LIU Keyan, et al, "Operational Reliability Analysis of Distribution Network Based on Big Data Technology," *Power System Technology,* vol. 41, no. 1, 2017, pp. 265-271.
[10] Tian K, Zheng L, Zeng M, et al, "Optimal models of grid project portfolio," East China Electric Power, vol. 36, No. 10, 2008, pp. 5-8.
[11] Tang Y, "The Research of an Optimization Model of Distribution System Annual Investment Decision Based on the Annual Maximum Expected Profit," Power System & Clean Energy, vol. 79, no. 4, 2015, pp:133 - 139.
[12] Tang Y, "The evaluation methodology of investment decision-making for distribution network construction project based on analytical hierarchy process," Modern Machinery, vol. 6, 2014, pp. 75-79.
[13] Qi L I, Haojin Q I, Yichen H U, "Investment Decision-making for Power Distribution Network Project Group Based on Principal Component and Set Pair Analysis," Guangdong Electric Power, vol. 30, no. 4, 2017, pp. 15-20.
[14] Liu F, "Study on Investment Efficiency Evaluation System for Distribution Networks Based on Investment Correlation Analysis," Zhejiang Electric Power, vol. 35, no. 3, 2016, pp. 68-71.
[15] Xiaojun L I, Tengkai Y U, Liu X, "An Investment Benefit Assessment Method Based on Triangular Fuzzy Numbers for Distribution Network," Distribution & Utilization, vol. 3, 2015, pp. 62-66.
[16] Zhao S Q, Cheng D C, Yan-Feng M A, "Analysis of correlative factors about reliability of distribution system based on improved synthetic incidence model," Electric Power, vol. 41, no. 10, 2008, pp. 20-23.
[17] Xiaojun L, Jifeng L, Xiangyu, L, "Investment Assessment on Distribution Network Projects Based on Improved Grey Correlation Degree and TOPSIS Method," Northeast Electric Power Technology, vol. 38, no. 8, 2017, pp. 56-59.

IOP Conf. Series: Materials Science and Engineering **428** (2018) 012012 doi:10.1088/1757-899X/428/1/012012

Reduce Energy Consumption in SWIPT System by Discontinuous Service Base Stations

Wenjie Dong*, He Zhang and Xiang Liao

School of Mechanical Engineering, Nanjing University of Science and Technology, Nanjing, China

*E-mail: 2578726158@qq.com

Abstract. In this paper, we investigate energy consumption reduction problem of simultaneous wireless information and power transfer (SWIPT) systems by establishing a discontinuous service base station (DSBS) scheme for SWIPT. Service time of the DSBS is minimized by optimizing the time allocation between energy harvesting (EH) and information transfer (IT), and the power consumption allocation between intermittent IT and continuous target detect. To solve the optimization problem, A method to reduce the dimensions of optimization subject is put forward. Out of service probability of the system is analysed, where the failure event is defined as the terminal required service time is lager then the maximum allowed DSBS service time. Analytical and simulation results reveal that, the system we presented possesses effective energy consumption ability with little reliability lost.

1. Introduction

Energy harvesting (EH) is a promising method to expand the life length of energy limitation devices, which are always constrained by energy storage capacity, such as wireless sensor networks (WSNs) and cellular networks. Among the EH methods being studied, one of the method where wireless devices harvest energy from dedicated RF power is wireless powered communication (WPC), as introduced in [1]. With consideration of transmit power limitation, the tradeoffs between information transfer rate and power harvest efficiency were analyzed in [2], [3], where simultaneous wireless information and power transfer (SWIPT) was first put forward. Authors in [4], [5] designed dynamic power splitting (DPS) schemes in receiver to split SWIPT signal into EH steam and information transfer (IT) steam and proposed optimal PS rules to achieve maximum ergodic capacity and average harvested energy. Two special case of DPS schemes, time switching (TS) and power splitting (PS) were widely discussed in later researches, hybrid PS receiver architectures which combined TS and PS scheme were designed in [6], [7]. Optimization analyses in [4], [5] only considered simplex information and power transfer. [6], [8], [9], [10] studied duplex SWIPT systems with downlink energy transfer and duplex information transfer. A half-duplex SWIPT protocol for an amplify-and-forward relaying system was designed in [6], where the system receives power and information from BS first, then closes the receiver and switches to the information forward mode. [8], [9], [10] analyzed full-duplex relay systems with power recycle, where the relay node receives and forward information simultaneously, and their antennas would recycle the relay transmission energy. Information self-interference effect of the energy recycle was analyzed by treating the self-interference as a noise signal at antenna. outage probability of SWIPT relaying system was analyzed in [11], [12], where channels fading and noise cause the outage. To the best of our knowledge, most researches about SWIPT systems assumed that the BS in SWIPT system is stationary and has continuous power supply.

However, for green energy powered BS like the solar powered BS (SPBS) in [13], energy consumption reduction must be considered, where an optimal energy allocation policy for maximizing total utility and reducing energy consumption of the SPBS was proposed. Authors in [14] introduced a kind of sleep-mod BS to increase energy efficiency of cellular networks, where the BSs switch between low and high energy consumption modes based on their loads. These BS are named as discontinuous service BS (DSBS). In this paper, in order to reduce energy consumption of the whole SWIPT system, we apply DSBS to SWIPT systems. To keep the SWIPT terminal continuous working, based on hybrid PS receivers, we propose an adaptive SWIPT terminal scheme, where the terminal has an intermittent SWIPT module, and a continuous module for target detection in WSNs, which makes our scheme different from the sleep-wake policy of sensor nodes in [15]. Service time of the DSBS is minimized by optimizing the time allocation between energy harvesting (EH) and information transfer (IT), and power allocation between IT and target detect consumption.

2. System model

Figure1 shows a working period of the system with a fixed length T, which is divided into two phases. In the first phase with dynamic length T_1, DSBS powers the terminal up and communicates with the terminal. In the second phase with length T-T_1, DSBS shutdown the SWIPT power. The target detector in terminal works continuously in the whole T. The first phase is divided into three time slots. In the first time slot of t_1 length, the DSBS powers the terminal up and estimates the channel; In the second time slot of t_2 length, the terminal feedback power up and channel estimation information. In the third time slot of t_3 length, the terminal upload target information to the DSBS and download updated setting information from the DSBS. The terminal harvests energy in all the three time slots, upload information in the second and third time slots, and download information in the third time slot.

The terminal antenna received signal in first phase is

$$y_{Ti}(t) = \sqrt{P_B} h_{BT} s_i(t) + w_i \sqrt{P_T} h_{TS} s_U(t) + n_a(t) \tag{1}$$

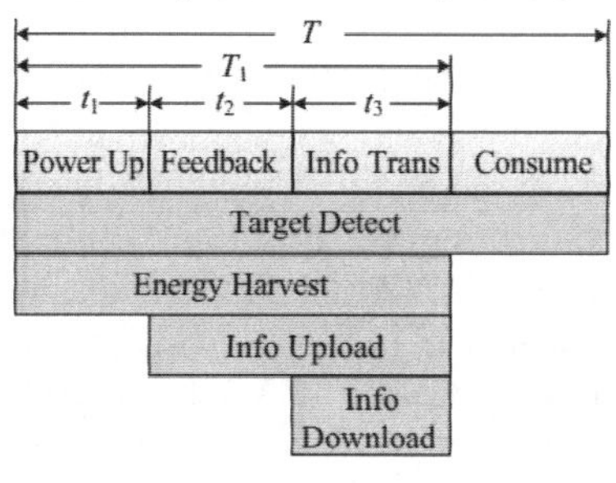

Figure1. A working period of the system

where $y_{Ti}(t)$ represents the received signal in time slot i, $i \in [1,2,3]$, P_B is the BS source power, $s_i(t)$ is the normalized source signal in time slot i, w_i is full-duplex self-interference weights of the terminal, P_T is uplink drive power of the terminal, $S_U(t)$ denotes the normalized sensor uplink signal, $\mathbb{E}(S_U(t)) = 1$, and $n_a(t) \sim \mathcal{CN}(0, \sigma_a^2)$ represents the AWSN noise introduced by antenna at the terminal with variance σ_a^2. Since there is no downlink information transmission in time slot 1 and 2, $s_1(t) = s_2(t) = 1$ is the unmodulated carrier signal, and $s_3(t)$ is the normalized downlink signal, $\mathbb{E}(s_3(t)) = 1$. We assume that self-interference energy of the terminal will be completely harvested by terminal antenna, w_2=w_3=1 and w_1=0.

The base band signal at the terminal is

$$y_{T3}(k) = \sqrt{1 - w_{\rho i}} \left(\sqrt{P_B} h_{BT} s_i(k) + w_i \sqrt{P_T} h_{TS} s_U(k) + n_a(k) \right) + n_d(k) \tag{2}$$

where $n_d(t) \sim \mathcal{CN}(0,\sigma_d^2)$ represents the AWSN noise introduced by base band conversion at the terminal with variance σ_d^2, $w_{\rho i}$ are the power splitting weights, $w_{\rho 1} = w_{\rho 2} = 1$ and $w_{\rho 3}$ equals to power storage factor ρ, which means the terminal harvests the whole received power in time slot 1 and 2.
The DSBS antenna received signal in first phase is

$$y_{Bi}(t) = w_i \sqrt{P_T} h_{TB} s_U(t) + n_b(t) \tag{3}$$

where $n_b(t) \sim \mathcal{CN}(0,\sigma_b^2)$ represents the AWSN noise introduced by antenna at the DSBS with variance σ_b^2. Since the noise power cannot be harvested. The harvested power in the first phase equals to

$$P_i = \eta w_{\rho i}\left(P_B |h_{BT}|^2 + w_i^2 P_T |h_{TS}|^2\right) \tag{4}$$

where η denotes the energy harvest efficiency. For the sake of simplicity of hardware implementation, the terminal uplink drive power P_T is static and proportional to P_3 in time slot 2 and 3, which is

$$P_T = \frac{\xi\eta\rho P_B |h_{BT}|^2}{1\text{-}\xi\eta |h_{TS}|^2} \tag{5}$$

where ξ is the uplink power consumption factor.

3. Service time optimization

3.1. Optimization Problem

In this subsection, we analysis the optimal problem of this system. As we focus on reducing energy consumption of the whole system, the optimization goal is minimizing the DSBS service phase T_1.

Allocation policy between the time slot length t_1, t_2, t_3, the power storage factor ρ and the power consumption factor ξ has a significant effect on T_1. Thus, we formulate the optimization problem as follows:

$$\min_{t_1,t_2,t_3,\rho,\xi} T_1 \tag{6a}$$

$$\text{s.t.}\ W_U \log_2(1+\lambda_U)t_3 \ge I_i \tag{6b}$$

$$W_U \log_2(1+\lambda_U)(t_2+t_3) \ge I_u \tag{6c}$$

$$W_D \log_2(1+\lambda_D)t_3 \ge I_d \tag{6d}$$

$$P_1t_1 + P_2t_2 + P_3t_3 \ge P_T(t_2+t_3) + P_C T \tag{6e}$$

$$P_1t_1 \ge E_p \tag{6f}$$

$$0 < \rho < 1 \tag{6g}$$

$$0 < \xi \tag{6h}$$

where I_i in (6b) specifies the minimum required power up feedback information, W_U is the uplink bandwidth, and $\lambda_U = \frac{P_T |h_{TB}|^2}{\sigma_b^2}$ is the uplink SNR. I_u in (6c) specifies the total need uploading information. In (6d), I_d limits the minimum terminal downlink information, W_D is the downlink bandwidth, $\lambda_D = \frac{(1-\rho)P_B |h_{BT}|^2}{(1-\rho)(P_T |h_{TS}|^2 + \sigma_a^2) + \sigma_d^2}$ is the downlink SNR. Inequality (6e) represents the total energy balance, where left of it is the harvested energy and right of it is the energy consumption, where P_C is the terminal continuous power consumption, and it is assumed that information processing power consumption is negligible. Ep in (6f) is the power up threshold limit. (6g) is the limit that both downlink information demodulation and energy harvesting power must be larger than zero. (6h) is the limit of minimum uplink transmission power.

3.2. Solve the Problem

In this subsection, we seek the optimal value of ρ, ξ, and the optimal allocation between t_1, t_2 and t_3 to minimize T_1. In order to solve the five-dimensional non-convex optimization problem in (6), We present a method to reduce the dimensions. Since $T_1 = t_1 + t_2 + t_3$, (6) is a linear problem with non-linear boundaries when ρ and ξ are fixed, where the optimal value is obtained at the boundaries. We replace ξ with P_T by

$$\xi = \frac{P_T}{\eta\rho\left(P_B\left|h_{BT}\right|^2 + P_T\left|h_{TS}\right|^2\right)} \tag{7}$$

where the ξ is completely determined by P_T and ρ. Then we give out the following proposition to find which boundaries the optimal value is obtained on.

Proposition 1: For any fixed value of P_T and ρ, T_1 gets its optimal value when (6b) is tight and ether (6c) or (6d) is tight. Moreover, for any fixed value of ρ, T_1 gets its optimal value when (6c) and (6d) are both tight.

Proof of the proposition 1 appears in appendix A.

According to the proposition 1, by combining tight (6c) and tight (6d), the relationship between P_T and ρ is

$$\rho = 1 - \frac{\sigma_d^2(F-1)}{(P_B\left|h_{BT}\right|^2 + P_T\left|h_{TS}\right|^2(1-F) + \sigma_a^2(1-F))} \tag{8}$$

Where, $F = \left(\frac{P_T\left|h_{TB}\right|^2 + \sigma_b^2}{\sigma_b^2}\right)^{\frac{I_d W_R}{I_u W_T}}$.

The multi-dimensional problem (6) is transformed to a one-dimensional optimization Problem

$$\min_{P_T} T_1 \quad \text{s.t. (6e),(6f),(6g),(6h)} \tag{9}$$

which can be solved by using line search method. The DSBS energy consumption reduction algorithm is:

Algorithm: DSBS energy consumption reduction

1: **Initial** $P_T^{(0)} > 0$, the iteration index $\mu = 0$, the searching distance $d = 1$, and tolerance $\varepsilon > 0$;
2: **while** μ ≤max-index **or** $\left|T_1^{(\mu)} - T_1^{(\mu-1)}\right| < \varepsilon$ **do**
3: $\mu = \mu + 1$, $P_T^{(\mu)} = P_T^{(\mu-1)} + d$;
4: Calculate $P^{(\mu)}$ by equation (8), Solve $T_1^{(\mu)}$ through (6) with $\rho^{(\mu)}$ and $P_T^{(\mu)}$;
5: **if** $(T_1^{(\mu)} - T_1^{(\mu-1)}) \geq \varepsilon$ **then**
6: $d = -d/2$;
7: **end if**
8: **end while**
9: Output current $\mu, P_T^{(\mu)}, T_1^{(\mu)}, \rho^{(\mu)}, \xi^{(\mu)}$.

3.3. Out of Service Probability Analysis

In this subsection, we analyze the out of service probability of the system. It is assumed that all channels in this system are independent quasi-static flat Rayleigh fading channels in a working period, $h_{BT} \sim \mathcal{CN}(0,|h_1|^2)$, $h_{TS} \sim \mathcal{CN}(0,|h_2|^2)$, $h_{TB} \sim \mathcal{CN}(0,|h_3|^2)$. We define random variables $x \triangleq |h_{BS}|^2$, and $x \triangleq |h_{TB}|^2$, whose probability density functions are $f_x(x) = \frac{1}{|h_1|^2} e^{\frac{x}{|h_1|^2}}$, $f_y(y) = \frac{1}{|h_2|^2} e^{\frac{y}{|h_2|^2}}$, and

$f_z(z)=\frac{1}{|h_3|^2}e^{\frac{z}{|h_3|^2}}$. Out of service situation includes leakage of energy and failure of information transfer, whose probability is

$$P_o = P\{T_{1\min} > T_{1M}\} = \iiint T_{1\min}(x,y,z) > T_{1M} f_x(x) f_y(y) f_z(z) dxdydz \tag{10}$$

where $T_{1M} \leq T$ is the maximum allowed service time. We assume that fading of the self-interference channel is negligible, $y=|h_2|^2$. An effective approximation of optimization problem (9) is tightening (6e), (6f) and solving the combined equation. The approximation may cause the second type of error in low SNR situation, where some of the result may outside the boundaries in (6). The function of T_1 is

$$T_1 = \frac{(A-D\rho)P_T - B(\rho I_d + I_i)x}{Bx\log_2(1+\frac{P_T z}{\sigma_b})} + \frac{E_C}{Bz} \tag{11}$$

where $A = I_d + I_i(1-\eta|h_2|^2 W_R)$, $B=\eta P_B W_R$, $D = I_d \eta |h_{TS}|^2 W_R$. P_T is solved by equation

$$P_T = \frac{(W_U E_C - W_U E_P)\log_2(1+\frac{P_T z}{\sigma_b}) - \eta P_B I_u x}{(\eta|h_2|^2-1)I_u} \tag{12}$$

Numerical result of integration (10) will be given out in the next section.

4. Numerical results

Numerical simulation results are presented by assuming that the P_B is normalized to 1W, h_{BT}=-20dB, $\sigma_a = -70$dBm, $\sigma_d = -65$dBm, $\sigma_b = -70$dBm, $h_{TS} = -15$dB, $h_{TB} = -20$dB, W_U = 1Hz, W_D = 1Hz, η = 0:9, P_C = 0dBm, T = 10s, I_i = 1bit, I_u = 2bit, I_d = 1bit, E_p = 1mJ.

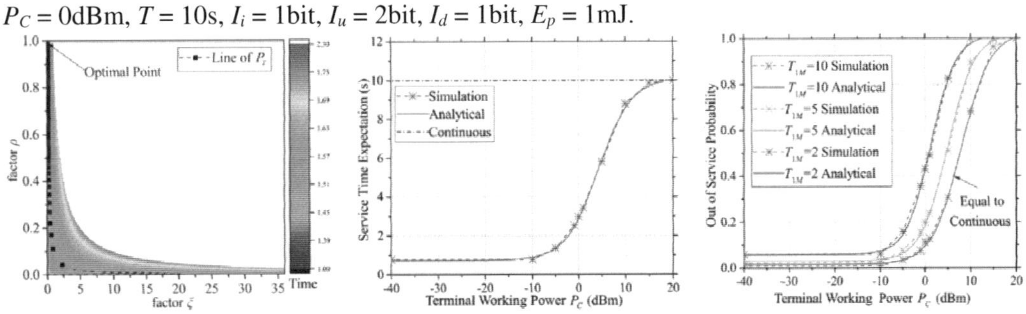

(a) T_1 vs. ρ and ξ (b) $\mathbb{E}[T_1]$ vs. P_C (c) P_o vs. P_C

Figure2. Numerical Results

Figure. 2a shows the service time optimization result by using our algorithm vs. direct searching the ρ-ξ plane, where the color represents optimized service time. In Figure2.a, the optimal result lies in the flat green zone. Searching line P_T is a curve line in the ρ-ξ plane. Searching points on the searching line are equal spacing P_T values, which become dense on the ρ-ξ plane when approaching the optimal value. Thus, the algorithm reduces the computation complexity and improves the computation accuracy. The optimal value is ρ = 0:9997, ξ = 0:0111, $P_T = -23$dBm and T_1 = 1:4s.

Then, we focus on the system reliability performance versus P_C, which represents how much energy the terminal could use besides the SWIPT consume. The solid line in Figure2.b is the analytical expectation of service time E(T_1) versus P_C calculated by (10), and the dash line with star points represents its simulative result. Figure2.b shows that $\mathbb{E}[T_1]$ increases with P_C increasing when $P_C > -10$dBm. When $P_C \leq -10$dBm, $\mathbb{E}[T_1]$ approaches to a non-zero service time constant, which represents the SWIPT energy consumption of the system. It is worth noting that, as the P_B is nominalized, $\mathbb{E}[T_1]$ can be treated as the expectation of DSBS energy consumption. The dash-dot line represents the energy consumption of continuously full duplex PS, which are constant under any P_C value. Therefore, the DSBS effectively reduce the system energy consumption. Figure2.c shows the analytical and simulative out of service probability versus P_C with different maximum service time T_{1M} values. In Figure2.c, under the small power storage condition, where the terminal power storage capacity equals to P_CT, continuous BS out of service probability equals to the out of service probability DSBS at $T_{1M} = T$. The DSBS can save energy without causing performance degradation. In low P_C condition, better energy saving effect can be gotten with a small performance degradation.

5. Conclusion

In this paper, energy consumption reduction problem of the SWIPT system has been studied, where the DSBS are applied to the SWIPT system. To minimize the service time of DSBS, A dynamic hybrid PS scheme is designed, and the optimal time and power consumption allocation are solved. The out of service probability of the system is analyzed, where the failure event is defined as the terminal required service time is lager then the maximum allowed DSBS service time. Analytical and simulation results reveal that, the system we presented possesses effective energy consumption reduction ability with little reliability lost under small power storage condition.

6. Appendix A: proof of the proposition

Proof: For any fixed value of P_T and ρ which satisfies (6g) and (6h), we assume a set of $[t_1, t_2, t_3]$ which satisfy all constraints in (6), and transform t_2 and t_3 to

$$t_2 = \Delta t_2 + \frac{I_i}{I_1} = \Delta t_2 + t_2^*, t_3 = \Delta t_3 + \max(\frac{I_u - I_i}{I_2}, \frac{I_d}{I_D}) = \Delta t_3 + t_3^* \tag{13}$$

where $\Delta t_2 \geq 0$, $\Delta t_3 \geq 0$. The energy balance constraint (6e) is transformed to

$$P_1 t_1 + P_2(\Delta t_2 + t_2^*) + (P_3 - P_T)(\Delta t_3 + t_3^*) \geq E_C \tag{14}$$

Since $|h_{TS}|^2 \leq 1$ and $\eta < 1$, $\eta P_T |h_{TS}|^2 - P_T < 0$. According to equation (4), we can get inequity $P_1 > (P_2 - P_T) > (P_3 - P_T)$. Thus, (6e) is transformed to

$$P_1(t_1 + \Delta t_2 + \Delta t_3) + (P_2 - P_T)t_2^* + (P_3 - P_T)t_3^* \geq E_C \tag{15}$$

where the equal achieves only when $\Delta t_2 = \Delta t_3 = 0$. When $\Delta t_2 > 0$ or $\Delta t_3 > 0$, a Δt_1 can always be found, which satisfies $0 \leq \Delta t_1 \leq \Delta t_2 + \Delta t_3$ and inequality

$$P_1(t_1 + \Delta t_1) + (P_2 - P_T)t_2^* + (P_3 - P_T)t_3^* > E_C \tag{16}$$

We define a new $T_1^* = t_1 + \Delta t_1 + t_2^* + t_3^*$. According to the analyses above, $T_1^* \leq T_1$, where the equal achieves only when $\Delta t_2 = \Delta t_3 = 0$. Thus, the first part of the proposition that For any fixed value of P_T and ρ, T_1 gets its optimal value when (6b) is tight and ether (6c) or (6d) is tight is proved.

Then, for any fixed value of ρ, we assume a P_T to make (6c) tight and (6d) loose, which is

$$P_T = \frac{(2^{\frac{I_u - I_i}{W_R t_3^*}} - 1)\sigma_b^2}{|h_{TB}|^2} < \frac{P_B |h_{BT}|^2}{(2^{\frac{I_d}{W_T t_3^*}} - 1)|h_{TS}|^2} - \frac{\sigma_d^2}{(1-\rho)|h_{TS}|^2} - \frac{\sigma_a^2}{|h_{TS}|^2} \tag{17}$$

With the value of ρ and P_T we assumed, the optimal value of the first phase length is $T_1^* = t_1^* + t_2^* + t_3^*$. We can find a $\Delta P_T > 0$ which satisfies inequality

$$\frac{(2^{\frac{I_u - I_i}{W_R t_3^*}} - 1)\sigma_b^2}{|h_{TB}|^2} < P_T + \Delta P_T \leq \frac{P_B |h_{BT}|^2}{(2^{\frac{I_d}{W_T t_3^*}} - 1)|h_{TS}|^2} - \frac{\sigma_d^2}{(1-\rho)|h_{TS}|^2} - \frac{\sigma_a^2}{|h_{TS}|^2} \tag{18}$$

As $P_T + \Delta P_T > P_T$, $[t_1^*, t_2^*, t_3^*]$ still satisfy the constraints in (6). By using the first part of the proposition with ρ and $P_T + \Delta P_T$, a new value $T_1^{**} \le T_1^*$ can be found, thus, the second part of the proposition is proved.

Acknowledgements
This work was supported by the Key Basic Research Projects of Basic Strengthening Plan (No.2017-JCJQ-ZD-004), Postgraduate Research & Practice Innovation Program of Jiangsu Province (KYCX17_0334). The authors would like to express their appreciation to the reviewers for their valuable suggestions which have helped to greatly improve the manuscript.

References
[1] S. Bi, C. K. Ho, and R. Zhang, "Wireless powered communication: Opportunities and challenges," IEEE Commun. Mag., vol. 53, no. 4, pp. 117–125, Apr. 2015.
[2] L. R. Varshney, "Transporting information and energy simultaneously," in Information Theory, 2008. ISIT 2008. IEEE International Symposium On. IEEE, 2008, pp. 1612–1616.
[3] P. Grover and A. Sahai, "Shannon meets Tesla: Wireless information and power transfer," in Information Theory Proceedings (ISIT), 2010 IEEE International Symposium On. IEEE, 2010, pp. 2363–2367.
[4] X. Zhou, R. Zhang, and C. K. Ho, "Wireless information and power transfer: Architecture design and rate-energy tradeoff," IEEE Trans. Commun., vol. 61, no. 11, pp. 4754–4767, 2013.
[5] L. Liu, R. Zhang, and K.-C. Chua, "Wireless information and power transfer: A dynamic power splitting approach," IEEE Trans. Commun., vol. 61, no. 9, pp. 3990–4001, 2013.
[6] R. Tao, A. Salem, and K. A. Hamdi, "Adaptive Relaying Protocol for Wireless Power Transfer and Information Processing," IEEE Commun. Lett., vol. 20, no. 10, pp. 2027–2030, 2016.
[7] B. He, N. Yang, S. Yan, and X. Zhou, "Regularized Channel Inversion for Simultaneous Confidential Broadcasting and Power Transfer: A Large System Analysis," IEEE J. Sel. Top. Signal Process., vol. 10, no. 8, pp. 1404–1416, 2016.
[8] D. Wang, R. Zhang, X. Cheng, and L. Yang, "Capacity-Enhancing FullDuplex Relay Networks based on Power Splitting (PS-) SWIPT," IEEE Trans. Veh. Technol., vol. PP, no. 99, pp. 1 – 1, 2016.
[9] Y. Zeng and R. Zhang, "Full-duplex wireless-powered relay with selfenergy recycling," IEEE Wirel. Commun. Lett., vol. 4, no. 2, pp. 201– 204, 2015.
[10] Z. Wen, X. Liu, N. C. Beaulieu, R. Wang, and S. Wang, "Joint source and relay beamforming design for full-duplex MIMO AF relay SWIPT systems," IEEE Commun. Lett., vol. 20, no. 2, pp. 320–323, 2016.
[11] G. Pan and C. Tang, "Outage performance on threshold AF and DF relaying schemes in simultaneous wireless information and power transfer systems," AEU-Int. J. Electron. Commun., vol. 71, pp. 175–180, 2017.
[12] H. Lee, C. Song, S.-H. Choi, and I. Lee, "Outage Probability Analysis and Power Splitter Designs for SWIPT Relaying Systems With Direct Link," IEEE Commun. Lett., vol. 21, no. 3, pp. 648–651, 2017.
[13] E. T. Ceran, T. Erkilic, E. Uysal-Biyikoglu, T. Girici, and K. Leblebicioglu, "Optimal energy allocation policies for a high altitude flying wireless access point," Trans. Emerg. Telecommun. Technol., 2016.
[14] J. Wu, Y. Zhang, M. Zukerman, and E. K.-N. Yung, "Energy-efficient base-stations sleep-mode techniques in green cellular networks: A survey," IEEE Commun. Surv. Tutor., vol. 17, no. 2, pp. 803–826, 2015.
[15] P. Du, Q. Yang, Z. Shen, and K.-S. Kwak, "Distortion Minimization in Wireless Sensor Networks with Energy Harvesting," IEEE Commun. Lett., vol. PP, no. 99, pp. 1–1, 2017.

IOP Conf. Series: Materials Science and Engineering **428** (2018) 012013 doi:10.1088/1757-899X/428/1/012013

Optical Circumferential Scanning Azimuth Measurement Method

Lin Gan*and He Zhang

School of Mechanical Engineering, Nanjing University of Science and Technology, Nanjing 210094, China

*E-mail: kg568605@163.com

Abstract. End defence is the last barrier to active protection system, which is a hot topic in recent years. As the multi-radiation program, partition program and partition scanning program need a plurality of lasers and laser receivers, and the synchronous scanning program has a complex structure, these four schemes are all hard to apply directly in compact system. A new type of laser detection system has been developed for the position measuring (≤15m). Unlike traditional static gaze detection manner, we use biaxial micro-motor to drive laser dynamic scanning, and the solver of the distance and azimuth was achieved by laser and magnetic. Based on the static detection, we conduct the experiment of dynamic rotation scanning detection. In each scan cycle, the system can effectively detect the target echo and accurately solver target azimuth, which verifies the feasibility of laser proximity circumferential detection methods. The system contains only one laser and laser receiver, which is small enough to measure proximity target distance and azimuth.

1. Introduction

End defence is the last barrier to active protection system, which is a hot topic in recent years. Quickly and accurately positioning technology of proximity (≤15m) targets is the key technology [1]. Laser has directionality, monochrome, high brightness and coherence characteristics, so it can be used in detecting targets by echo analysis [2]. Laser detection system has been used wildly in weapons, such as American AGM-88 high-speed anti-radar missiles, Sweden RDS-70 missile and so on [3, 4]. In order to increase laser detection coverage, laser detection system requires omnidirectional detection, rapid identification and accurate positioning features [5, 6]. Laser orientation detection system is mainly involved in four common beam layout programs [7, 8]. As the multi-radiation program, partition program and partition scanning program need a plurality of lasers and laser receivers, and the synchronous scanning program has a complex structure, these four schemes are all hard to apply directly in compact system [9, 10].

In our work, we propose a new type of laser proximity detection method by using biaxial micro-motor to drive laser dynamic scanning to improve the traditional static gaze[11], only one laser and laser receiver are used to achieve a 360°-circumferential proximity detection.

2. Description of the system

We have designed the detection system shown in Figure 1.

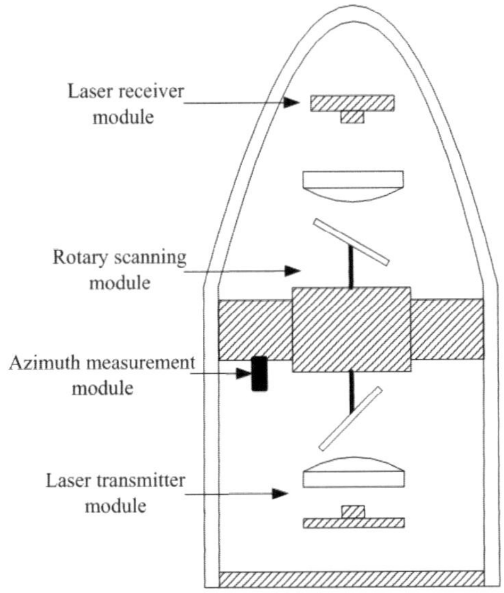

Figure 1. The detailed setup of the Laser proximity azimuth measurement system

The rotary scanning module includes biaxial micro-motor and all-trans plane mirror. The upper and lower total reflection plane mirror is driven by biaxial micro-motor to do the scanning, so that the laser proximity circumferential detection comes true.

The laser transmitter module includes laser drive power, semiconductor laser and collimating lens. Laser drive power makes semiconductor laser send out the initial laser beam, the original pulsed laser beam is shaped by collimating lens and then the beam is exited.

The laser receiver module includes focusing lens, photoelectric sensor and signal processing circuit. Target echo signal reaches photoelectric sensor after it through the focusing lens, then, optical signal is converted to electric signal and processed by signal processing circuit.

The magneto resistive sensor is the core device of azimuth measurement module, this module includes magneto resistive sensor matching circuit and induction magnetic core, and periodic scanning magnetic signals are recorded by this module.

3. Experimental results

The shaft length of biaxial micro-motor is S, the angles between the upper and lower total reflection plane mirror are σ_L and σ_r

$$P_R = \frac{\rho P_L T_E T_R A_R \cos\beta}{\pi R^2} \cdot e^{-2\sigma R} \tag{1}$$

Hypothesis: Target surface area is greater than the laser spot size. For extended target, For extended target, laser echo power equation is represented by formula (1).

ρ is target reflectivity, P_L is peak optical power of the laser, T_E is transmittance of emission optical system, T_R is transmittance of receive optical system, σ is atmospheric attenuation coefficient, β is the angle between the target reflective surface normal and the optical axis, A_R is aperture area of receive optical system, R is received power for system, R is the target distance. According to this formula, the target distance is solved.

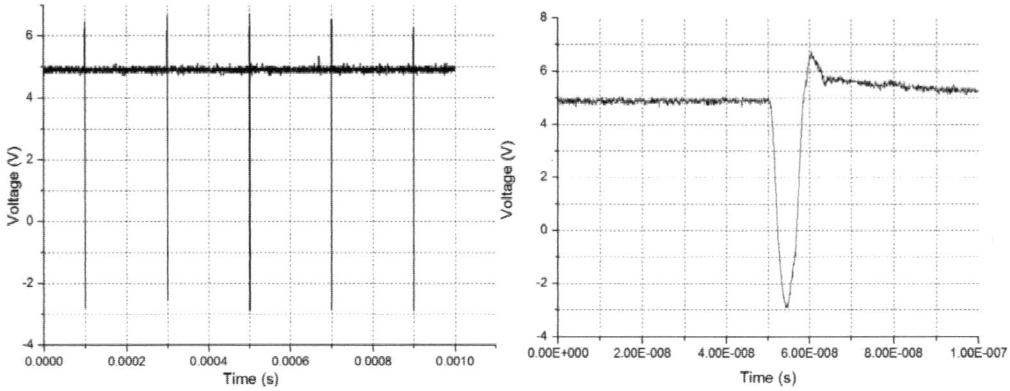

Figure2. The curve of echo pulse

The voltage output curve of APD detection system shown in Figure 2, when the detection distance is 10m, the repetition rate of echo is 5KHz, the pulse width is nearly 5ns, the amplitude is about 8V, in addition, echo signal amplitude and circuit noise satisfy processing requirements of the subsequent circuit.

Assuming that the target azimuth angle is the angle which is between x axis positive and the line from the target location to system axis. Magnetic induction signal is at peak when induction diskette and magnetic sensor are alignment at Zero time.

Signal processing system does A/D converter to the analog output signal of the sensor; using software comparison to find out peak time (T_1(s) and T_3(s)) of the goal appears period in the A/D conversion data, and temporarily stored in the microprocessor.

When the output signal of the laser receiver module is judged as the target echo signal, extract T_1 and T_3, the motor speed in real time can be calculated as:

$$n = \frac{2\pi}{T_3 - T_1} \tag{2}$$

The azimuth angle of the target is:

$$\alpha = \frac{2\pi\left(T_2 - T_1\right)}{T_3 - T_1} \tag{3}$$

T_2(s) is the time when the target is detected.

If receiver module has no signal output when a peak occurs in the next time, then the microprocessor discards T_1, T_3 and records the next peak time, prepare for the target which appears at any time.

The azimuth detection methods used in this paper is based on the motor speed of the target appears cycle, it only be affected by the target appears cycle speed variation, A/D conversion frequency and system fuzzy delay. The circuit has the characteristics of simple structure, strong real-time and small interfering by external electromagnetic environment.

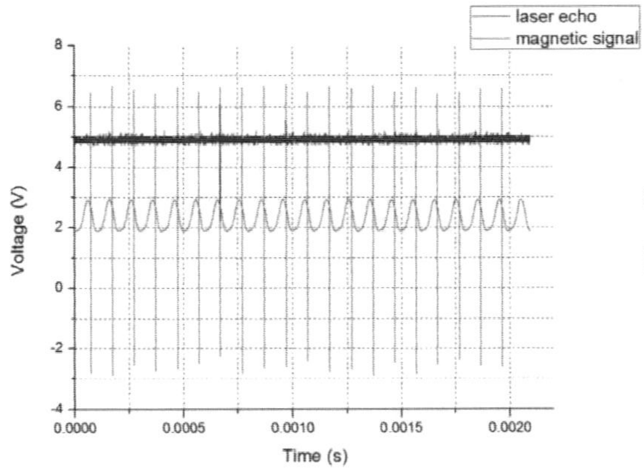

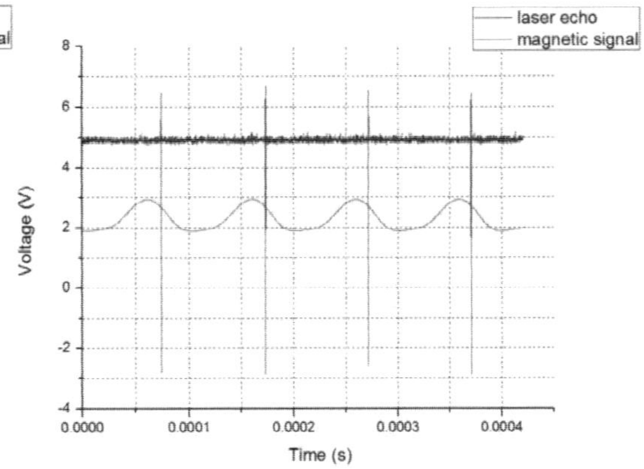

Figure3. The composite waveform of laser and magnetic **Figure4.** Expanded view of the composite waveform

Based on the static detection, we conduct the experiment of dynamic rotation scanning detection. Run the rotating scanning system so that the laser is driven to realize the proximity circumferential detection. Placed the target in different directions, rotate scan signals and the target echo signals were recorded. When the target located at a distance of 10m and an azimuth of 45°, Figure 3 and Figure 4 show the composite waveform of laser and magnetic signal while the target's distance is 10m and azimuthal angle is 45°. According to the algorithm to calculate the first test results, $T_1=6.1019\times10^{-5}$s, $T_2=7.3503\times10^{-5}$s, $T_3=16.1018\times10^{-5}$s,$\alpha=\frac{2\pi(T_2-T_1)}{T_3-T_1}$ =44.94°, Figure 5 shows 20 test results, the maximum and the minimum error were -2.48° and -0.06°, the average of test results is 44.39°. In each scan cycle, the system can effectively detect the target echo and accurately solver target azimuth, which verifies the feasibility of laser proximity circumferential detection methods.

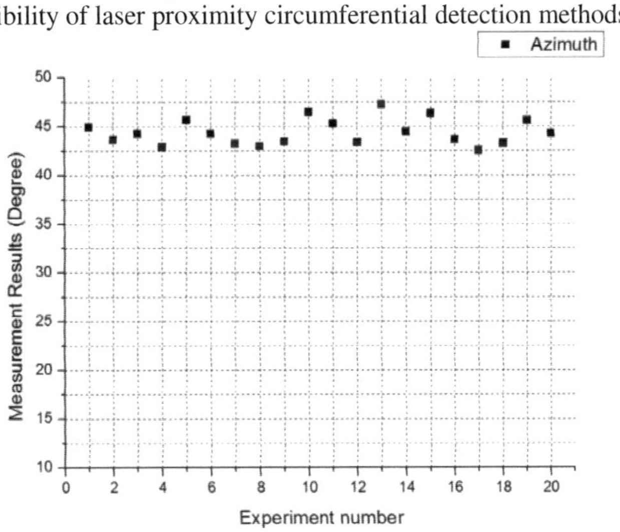

Figure5. The distribution of the experiment results

4. Conclusion

We propose a detection method which compound laser and magnetic. To compensate for the lack of traditional proximity detection methods, we complement advantages of laser and magnetic detection, use laser and magnetic measure distance and angle information of short-range target. This method is suitable for fast and accurate measurements of proximity target position in different environments.

Acknowledgements
This research is supported by National Natural Science Foundation of China (51605227) and the Fundamental Research Funds for the central universities (NUST30915011303).

References

[1] United States Department of Defense. (2000) Joint Vision 2020. *USA: United States Department of Defense*.
[2] Guo, J., ZHANG, H., ZHANG, X.J., WANG, X.F., (2010). Avlanche Photodiode Detecting Technology for Laser Fuze. *Journal of Detection & Control, 32(1),* 77-79.
[3] Ding, Z.Z., Wang, K.Y., Zhan, Z.X., (2010). Application of aspherical liquid lens in laser diode beam shaping. *Infrared and Laser Engineering, 39(4),* 623-626.
[4] Wang, D., LI, X.Q., (2001).New progress in semiconductor lasers and their applications. *Optics and Precision Engineering, 9(3),* 279-283.
[5] Amann, M.C., Bosch, T., Myllyla, R., (2001). Laser ranging: a critical review of usual techniques or distance measurement. *Optical Engineering,* 10-19.
[6] Steinvall, O., Chevlier, T., (2005). Range accuracy and resolution for laser radars. *SPIE Electro-Optical Remote Sensing,* 598801-5988018.
[7] Zhang, Z.H., Yang, M., Xu, S.W., (2006). Analysis of FOV configuration in laser proximity fuze. *Infrared and Laser engineering, 35(6),* 700-704.
[8] Tao, M., Wang, Z., BAI, X.Y., (2007). The choice and analysis of laser fuze beam position. *Journal of Projectiles; Rockets; Missiles and Guidance, 27(2),* 62-65.
[9] G. Buzzard. (2010). "Modeling the Interaction of a laser target detection device with the sea surface", *in Proceedings of 54th Annual Fuze Conference 67.*
[10] Gan, L., Zhang, H., Zhang X.J., (2013). Research on Large FOV Single Transceiver Bidirectional-driving Detection Technology for Laser Fuze. *Acta Armamentarii, 34(8),* 942-947.
[11] Barna, A., Földes, I.B., Bohus, J., et al. (2015). Active Stabilization of the Beam Pointing of a High- Power KrF Laser System. *Metrology & Measurement Systems, 22,* 165-172.

IOP Conf. Series: Materials Science and Engineering **428** (2018) 012014 doi:10.1088/1757-899X/428/1/012014

Demagnetization Process Analysis and Numerical Simulation of Magnet with Shock Wave from Compact Explosively Driven Ferromagnetic Generator

Jian Wei[1], Shaojie Ma[1]*, Peng Liu[1]and Yunlei Shi[2]

[1]Ministerial Key Laboratory of ZNDY, Nanjing University of Science and Technology, Nanjing, China
[2]Quality Inspection and Testing Cneter, China Electronic Product Reliability and Environmental Testing Research Institute, Guangzhou, China

*Corresponding author: 438729185@qq.com

Abstract. For compact explosively driven ferromagnetic impulse source used in magnetic flux compression generators, the relations between the magnet's demagnetization rate and the shock wave pressure are theoretically analyzed. A nonlinear finite element method is adopted to conduct numerical simulation on the magnet's impact loading process. The propagation rules and pressure distributions of shock wave within hollow cylindrical NdFeB magnet caused by lateral impact of single-ended and double-ended initiating explosives are explored. The numerical simulation results show that, the pressures at both ends resulting from lateral impact of double-ended initiating explosives on hollow cylindrical magnet are low; double-ended initiating has faster demagnetization rate, swifter variation of flux and higher electromotive force of inductive output than single-ended initiating.

1. Introduction

Compact explosively driven ferromagnetic generator is a device that makes use of the impact of ferromagnetics to achieve the transduction from magnetostatic energy to electric pulse energy. As shown in Figure 1, the shock wave driven by the explosion of explosives works on ferromagnet wrapped by conductor coils (magnetization direction M) for rapid demagnetization of magnet. In accordance with Faraday's law of electromagnetic induction, the rapidly changing flux within the conductor coils will produce electromotive force in the conductor, thereby forming pulse current i in the discharge circuit composed by loads and providing pulse energy for post-stage circuits.

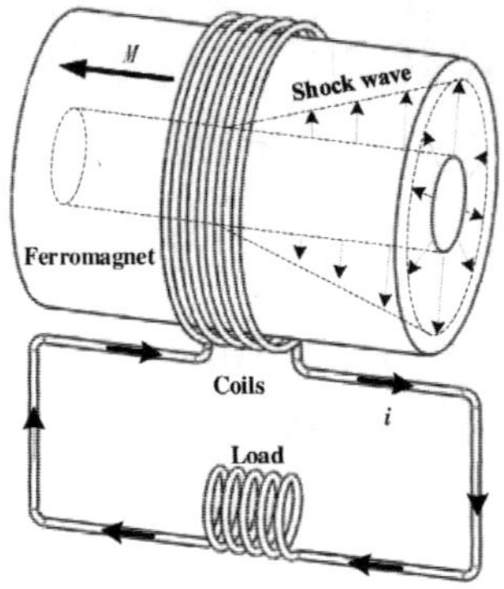

Figure 1. Schematic diagram of explosively driven ferromagnetic generator

Domestic and foreign scholars have explored explosively driven ferromagnetic generator from the theoretical, numerical simulation and experimental perspectvies. Authors in [1], [2], [3], [4] theoretically studied the principle of explosive demagnetization of ferromagnet, and carried out experiments on lateral and longitudinal shock waves of explosively driven ferromagnetic generators. Authors in [5], [6] investigated the dynamic features of NdFeB under the shove wave, and analyzed and computed the magnetic-electric conversion princple of explosively driven ferromagnetic generators. Authors in [7] theoretically and empirically studied explosively driven ferromagnetic generators. For compact explosively driven ferromagnetic impulse source used in magnetic flux compression generators, the relations between the magnet's demagnetization rate and the shock wave pressure, as well as the propagation rules and pressure distributions of shock wave within hollow cylindrical NdFeB magnet caused by lateral impact of single-ended and double-ended initiating explosives, are dissected.

2. System model releations between demagnetization rate and shock wave pressure

The magnet's demagnetization rate reflects the demagnetization degree of a magnet under the action of shock wave. To explore the relations between demagnetization rate and shock wave pressure is of great significance to the design of explosively driven ferromagnetic seed source. A magnet's demagnetization rate can be expressed as the ratio of total flux variation to initial flux on a cross-sectional unit of the magnet, namely:

$$\eta^x = \frac{\Delta\Phi_f^x}{\Phi_0^x} \times 100\% \tag{1}$$

The superscript x represents a certain cross-sectional unit of the magnet. According to the law of electromagnetic induction, the variation in flux will bring about induced electromotive force within the coils around the cross-section:

$$E_g^x(t) = -d\Phi^x(t)/dt \tag{2}$$

By calculating the integral of Eq. (2) within the time range of shock wave, we can get:

$$\Delta\Phi_f^x = -\int_0^{+\infty} E_g^x(t)\,dt \tag{3}$$

where the variation in $Ex\ g(t)$ is aquired through experiments; the total flux variation $\Delta\Phi x\ f$ of the cross-section x is the area formed by the waveform and time axis of induced electromotive force; the initial flux $\Phi x\ 0$ can be obtained through magnetostatic field computation of open-circuit magnet. Substitude $\Delta\Phi x\ f$ and $\Phi x\ 0$ into Eq. (1) to get the demagnetization rate of the cross-section x under certain shock wave.

Next, the calculation of shock wave pressure resulting from the shock wave on the cross-section x is introduced. According to the momentum conservation expression during the impact loading process, we can get:

$$P_m = \rho_{m0} D_m u_m \tag{4}$$

P_m is the shove wave pressure of the magnet; ρ_{m0} is the density of the magnet without impact; D_m is the shove wave speed of the magnet; u_m is the speed of particle in the magnet. The magnet's Hugoniot equation can be expressed as:

$$D_m = c_0 + \lambda u_m \tag{5}$$

where c_0 and λ are Hugoniot parameters of the magnet material; for pure iron material, c_0=3.78km/s, λ=1.652; for ferroalloy magnet, Hugoniot parameters can be obtained by adding up the volumes of mixtures or by data acquisition and linear fitting via comparative experimental method; D_m can be obtained by calculating the electric pulse signal generated at the electirc probes or parallel coils along the propogation direction of shock wave within the magnet. Assume that ΔL is the distance between two electric probes or coils; t_{x1} and t_{x2} are the moments when pulse signals are generated; then, the shove wave speed within the magnet can be expressed as:

$$D_m = \Delta L / (t_{x1} - t_{x2}) \tag{6}$$

After getting the values of c_0, λ and D_m, in combination of Eq. (4) and (5), P_m and u_m can be obtained. In [8], for explosion products of explosives, Hugoniot equation can be expressed as:

$$P_0 = 2.412 p_{CJ} - \left(1.7315 \frac{p_{CJ}}{u_{CJ}}\right) u_m + \left(0.3195 \frac{p_{CJ}}{u_{CJ}^2}\right) u_m^2 \tag{7}$$

P_0 is the shock wave pressure of the contact surface between explosive and magnet; p_{CJ} and u_{CJ} are the pressure and particle speed of CJ (Chapman-Jouguet). By substituting the particle speed u_m in Eq. (7), we can obtain P_0 value; the pressure range within the magnet under the loading of explosive shock wave is $P_m \sim P_0$. Sergey I. Shkuratov concluded that $Nd_2Fe_{14}B$ rare earth permanent magnet is nearly fully demagnetized along the magnetization direction under the shock wave pressure of 28.2GPa-38.7Gpa in [9]; Wu Junying et al. found that when the incident shock wave pressure of the magnet is 45GPa, 13.87GPa and 8.52GPa, the maximum demagnetization rate is 87.94%, 44.07% and 35.60%, respectively in [10].

3. Numerical simulation of demagnetization process under the loading of explosively driven shock wave

Whether the shock wave pressure in the magnet satisfies the demagnetization conditions of magnet is a precondition for the design of compact explosively driven ferromagnetic generator. A nonlinear finite element method (ANSYS/LS-DYNA) is adopted to conduct numerical simulation of impact loading process of the magnet and to analyze the propoganda rules and pressure distribution of shock wave within the magnet.

The physical model is a hollow cylindrical magnet loaded with explosives with single-ended or double-ended initiating for lateral loading of shock wave. In the model, NdFeB is chosen as the material for the magnet, and Explosive 8701 is used. NULL model and Gruneisen state equation are used to describe the impact loading process of the magnet. In NULL model, deviatoric stress calculation can be avoided by calling state equation, and the lower limit of pressure is determined by cutoff pressure. NdFeB material has a density of 7.5g/cm^3 and an elasticity modulus of 160Gpa. The parameters of Gruneisen state equation can be obtained from Hugoniot curve data of NdFeB material. As shown in Table 1, c_0, S_1, S_2 and S_3 are the Hugoniot curve parameters of the magnet; γ_0 is Gruneisen coefficient; a is the first-order volume correction of γ_0; E_0 is the internal energy density. Explosive 8701 has a density of 1.72g/cm^3, an explosive velocity of 8.43km/s and an explosive pressure of 30.56Gpa. High-energy explosive model and JWL state equation are adopted. Table 2 presents the parameters of the state equation for explosive product JWL of Explosive 8701. The magnet in the model has an outer diameter of Φ50mm, an internal diamter of Φ15mm and a height of 30mm; the explosive has a diameter of Φ15mm and a height of 30mm.

Table 1 Parameters for Gruneisen state equation of NdFeB material [11]

c_0(cm/μs)	S_1	S_2	S_3	γ_0	a	E_0(J/m^3)
0.368	1.059	0	0	0.25	0.48	0

Table 2 Parameters for JML state equation of Explosive 8701

A_1(100GPa)	A_2(100GPa)	R_1	R_2	ω	E_0(J/m^3)
8.802	0.174	4.600	1.200	0.300	0.104

3.1. Single-ended initiating

Figure 2 presents the isobaric lines of explosive and magent at different moments when the single-ended initiating explosive applies load to hollow cylindrical NdFeB magnet. With the initiating moment of the explosive as the zero time, the detonation wave spreads downwardly as a spherical wave. At t=0.89μs, the spherical detonation wavefront reaches the contact surface of explosive and NdFeB, the shock wave enters NdFeB magnet, and the explosive continues to explode along the axial direction. At t=2.01μs, isobaric lines suggest that the detonation wave reaches the center section of NdFeB magnet, and it can be seen that the propagation speed of shock wave within NdFeB magnet is significantly slower than that of detonation wave. At this time, only about one fifths of the volume of the magnet is under impact. At t=3.27μs, the explosive wavefront has a maximum pressure value of 37.4GPa, and the explosive cylinder is nearly fully exploded, but the shock wave in the magnet still does not reach the side wall, and only about a half of the volume is under impact. At t=3.67μs, the explosive is fully exploded, and the detonation wavefront reaches the lower end, the shock wavefront in the magnet is significantly backward than detonation wavefront; next, the shock wave in the magnet continues to spread and reflect at the lower end. At t=4.48μs, the shock wave reaches the side wall of the magnet. It can be clearly seen that, there is a low pressure zone at the edge of the upper end of the magnet for single-ended initiating, and that this part of magnet cannot be demagnetized. In order to futher analyze the variations in pressure, it is needed to monitor the variations in pressure of center section of NdFeB magnet at different heights.

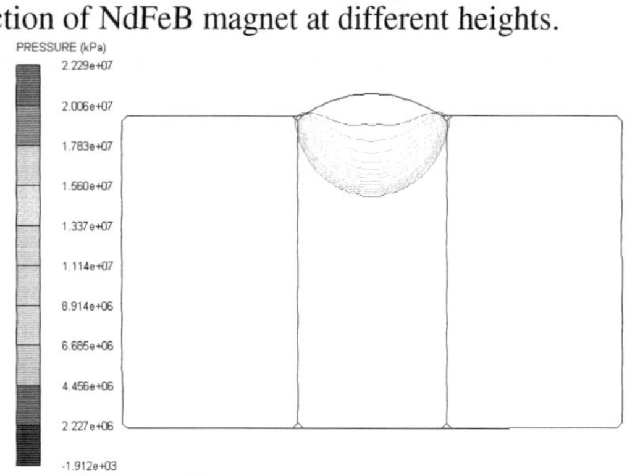

(a) t=0.89μs

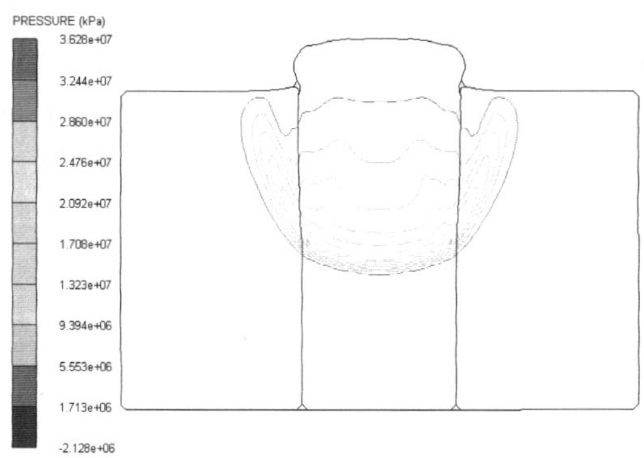

(b) t=2.01μs

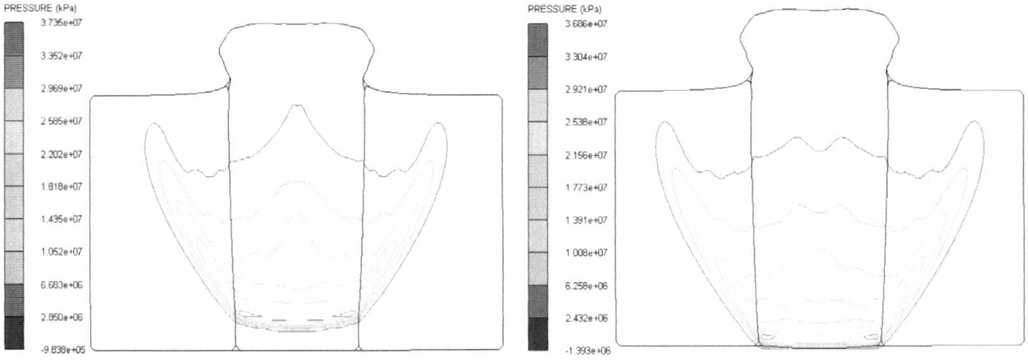

(c) t=3.27μs (d) t=3.67μs

Figure 2. Isobaric lines in the explosive and magnet for single-ended initiating at different moments

Fig.3 delivers the pressure variation curve at different locations within NdFeB magnet in the case of single-ended initiating. In Fig.3(a), at the center section of the magnet, the pressure variation curves at seven locations (with a radius of 7.5mm, 10mm, 15mm, 18mm, 21mm, 23mm and 25mm, respectively) along the radial direction are given. At r=7.5mm, the peak pressure reaches 36.2GPa; at r=10mm, the peak pressure quickly attnuates to about 19GPa; at r=15mm, 18mm, 21mm and 23mm, the peak pressure is 12.3GPa, 8.5GPa, 7.2GPa and 5.5GPa, respectively, and the attenuation degree of shock wave pressure gradually decreases. In Fig.3(b), at the location with 1/2 radius to the magnet, the pressure variation curves at seven locations (at 0mm, 3mm, 5mm, 9mm, 11mm, 13mm and 15mm, respectively) along the depth direction are presented, and the peak shock wave pressures at these seven locations are 0.9GPa, 8.3GPa, 11.1GPa, 12.5GPa, 12.9GPa, 13.1GPa and 13.2GPa, respectively. Thus, it can be seen that the pressure increases slightly along the axial direction by following the propagation path of detonation wave, which also reflects the superiority of lateral impact over the longitudinal impact.

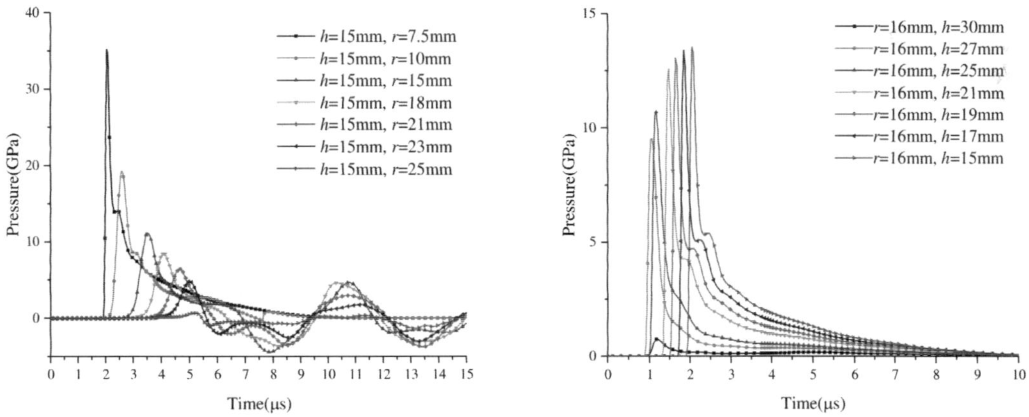

(a) At different locations on center section along the radial direction

(b) At different locations at 1/2 radius along the depth direction

Figure 3. Pressure variation curves at different locations within the magent in the case of single-ended initiating

3.2. *Double-ended initiating*

Figure 4 presents the isobaric lines of explosive and magent at different moments when the double-ended initiating explosive applies load to hollow cylindrical NdFeB magnet. With the initiating moment of the explosive as the zero time, the detonation waves spread from two ends to the center as two spherical waves. At t=0.96μs, the spherical detonation wavefronts reache the contact surface of explosive and NdFeB, and the shock waves enter NdFeB magnet. At *t*=1.74μs, the two spherical waves meey at the center of the explosive, and the detonation waves overlay and enhance each other; at *t*=2.32μs, the overlapped shock waves in the magnet have a lateral impact on NdFeB magnet; at *t*=4.18μs, the shock wave reaches the edge of the magnet. It takes about 3.22μs for the shock wave to fully penetrate through the magnet from entering the magnet. Since the two shock waves overlay at the center, the pressure value at the center is high, while the pressures at both ends are low. Yet, the shock wave pressures at the edge for double-ended initiating are significantly higher than those at the edge for single-ended initiating, and all magnet experiences the process of demagnetization.

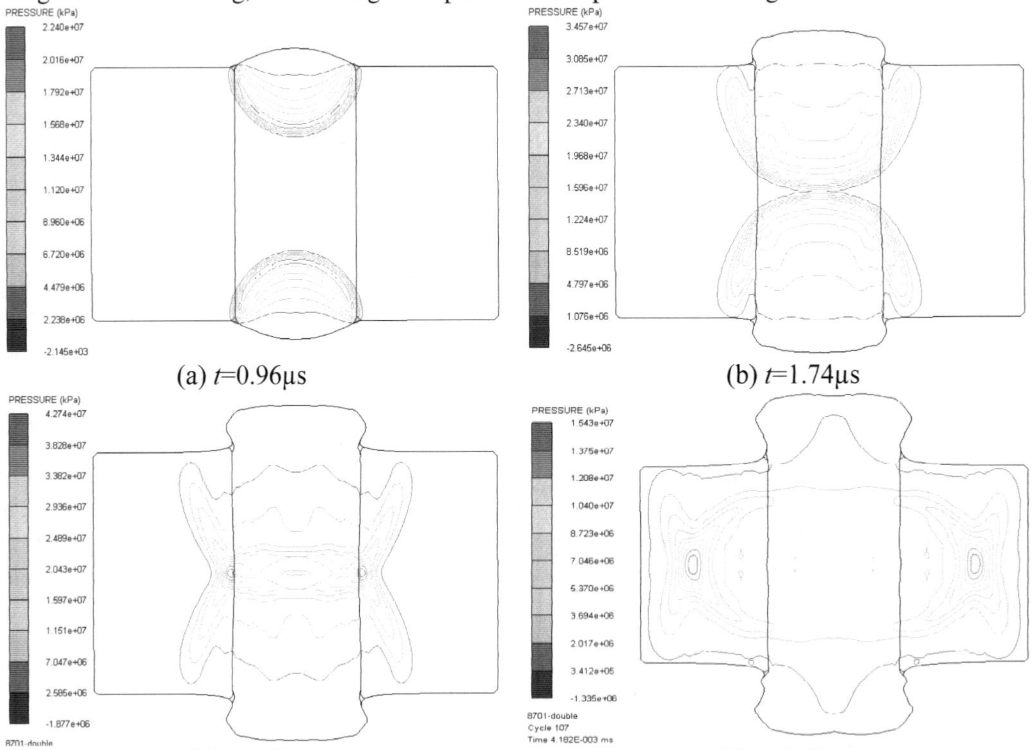

(a) *t*=0.96μs (b) *t*=1.74μs

(c) *t*=2.32μs (d) *t*=3.67μs

Figure 4. Isobaric lines in the explosive and magnet for double-ended initiating at different moments

In order to futher analyze the variations in pressure, it is needed to monitor the variations in pressure of center section of NdFeB magnet at different heights. Fig.5 presents the pressure variation curve at different locations within NdFeB magnet in the case of double-ended initiating.

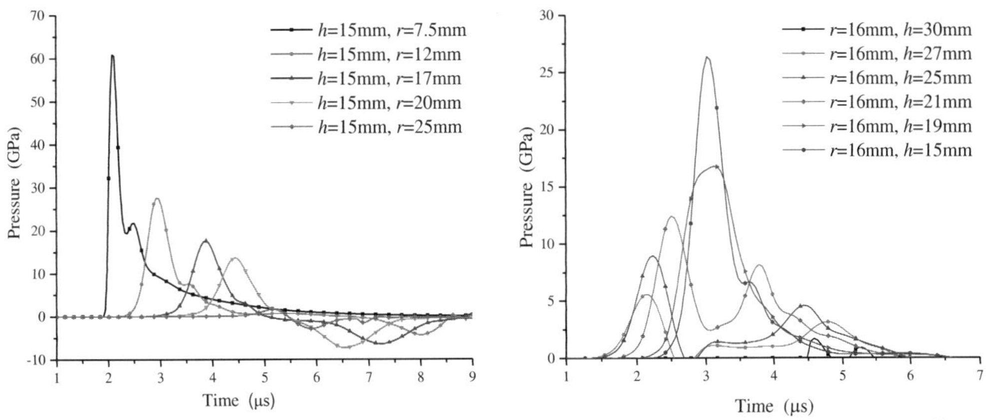

Figure 5. Pressure variation curves at different locations within the magent in the case of double-ended initiating

In Figure 5 (a), at the center section of the magnet, the pressure variation curves at five locations (with a radius of 7.5mm, 12mm, 17mm, 20mm and 25mm, respectively) along the radial direction are given. At *r*=7.5mm, the peak pressure reaches up to 60.90GPa, far exceeding the critical demagnetizing pressure of the magnet; at *r*=12mm, the peak pressure rapidly attenuates to 27.62GPa; at *r*=17mm and *r*=20mm, the peak pressures are 18.5GPa and 14.8GPa, respectively, and the attenuation degree of shock wave pressure gradually decreases. In Figure 5 (b), at the location with 1/2 radius to the magnet, the pressure variation curves at five locations (at 3mm, 5mm, 9mm, 11mm and 15mm, respectively) along the depth direction are presented, and the peak shock wave pressures at these five locations are 5.9GPa, 8.5GPa, 12.8GPa, 17.3GPa and 26.5GPa, respectively.

4. Conclusions

According to the theoretical analysis and numerical simulation results, it can be concluded that when the single-ended initiating explosive has a lateral impact on hollow cylinderical magnet, there is a low-pressure zone at the edge of the initiating end, and this part of magnet cannot be demagnetized; there is a huge difference in the propagation speeds of shock wave and denotation wave in NdFeB magnet; when the impact wave enters NdFeB magnet, the pressure will soon attenuate; with the increase of propagation distance, the attenuation degree of the pressure gradually decreases; the center section of the magnet has the most flux where the incident shock wave pressure basically meets the demagnetization requirements. When the double-ended initiating explosive has a lateral impact on hollow cylinderical magnet, since the shock waves overlay at the center, the pressure at the magnet center is very high, while the pressures at both ends are low; double-ended initiating has faster demagnetization rate, swifter variation in flux and higher induced electromotive force compared to single-ended initiating; the incident shock wave pressure at the center section of the magnet reaches more than 60GPa, far exceeding the critical demagnetization pressure of the magnet.

Acknowledgements

This work was supported by the Key Basic Research Projects of Basic Strengthening Plan (No.2017-JCJQ-ZD-004), Postgraduate Research & Practice Innovation Program of Jiangsu Province (KYCX18_0390). The authors would like to express their appreciation to the reviewers for their valuable suggestions which have helped to greatly improve the manuscript.

References

[1] Shkuratov, S. I., Talantsev, E. F., Baird, J. and L. L. Altgilbers, 2010. “Miniature explosively

driven high-current transverse-shock-wave ferromagnetic generators", *IEEE Transactions on Plasma Science*, Vol.38 (8), pp.1784-1793.

[2] Shkuratov, S. I., Talantsev, E. F., Baird, J., Rose, M. F., Shotts, Z. and Z. Roberts, et al, 2006, "Completely explosive autonomous high-voltage pulsed-power system based on shockwave ferromagnetic primary power source and spiral vector inversion generator", *IEEE Transactions on Plasma Science*, Vol.34 (5), pp.1866-1872.

[3] Shkuratov, S. I., Talantsev, E. F., Dickens, J. C. and M. Kristiansen, 2002, "Transverse shock wave demagnetization of nd2fe14b high-energy hard ferromagnetics", *Journal of Applied Physics*, Vol.92 (1), pp.159-162.

[4] EVGUENI F. TALANTSEV, SERGEY I. SHKURATOV, JAMES C. DICKENS and MAGNE KRISTIANSEN, 2008, "The conductivity of a longitudinal-shock-wave-compressed, nd2fe14b hard ferromagnetics", *Modern Physics Letters B*, Vol.16 (15n16), pp.545-554.

[5] F. Lu, L. Chen, C. G. Feng, J. Y. Wu, Y. F. Li and M. G. Zhu, 2014, "The dynamic mechanical properties of ndfeb ferromagnet under shock wave compression", *Binggong Xuebao/acta Armamentarii*,Vol.35, pp.1-7.

[6] F. Lu, L. Chen, C. G. Feng and J. Y. Wu, 2015, "The theory of magnetoelectric conversion in explosive driven shockwave demagnetization generator", *Journal of Beijing Institute of Technology*, Vol.35 (S2), pp.102-106.

[7] J. Y. Wu, L. Chen and C. G. Feng, 2007, "Experiments and theoretical calculation of expl0sive_driven shock wave ferromagnetic generators", *Explosion and Shock Waves*, Vol.27 (5), pp.398-404.

[8] P. W. Cooper, 1996, Explosives engineering. Wiley & Sons.

[9] Shkuratov, S. I., Talantsev, E. F., Dickens, J. C., Kristiansen, M. and J. Baird, 2003, "Longitudinal-shock-wave compression of nd2fe14b high-energy hard ferromagnet: the pressure-induced magnetic phase transition", *Applied Physics Letters*, Vol.82 (8), pp.1248-1250.

[10] J. Y. Wu, L. Chen W. Li and C. G. Feng, 2011, "Experimental study on shock wave demagnetization of nd2fe14b hard ferromagnetic", *Chinese Journal of High Pressure Physics*, Vol.25 (1), pp.29-35.

[11] J. Y. Wu, L. Chen W. Li and C. G. Feng, 2010, "Nd_2fe_(14)b hard ferromagnetic under shock wave loading: experiments and numerical simulation", *Transactions of Beijing Institute of Technology*, Vol.30 (12), pp.1391-1394.

Simulation and Experimental Research of Continuous High Impact Test Equipment

Feiyin Li and Shaojie Ma*

School of Mechanical Engineering, Nanjing University of Science and Technology, Nanjing, China

*E-mail: 438729185@qq.com

Abstract. In order to realize the simulation test of key components under continuous high impact environment, a continuous high impact test equipment based on direct impact loading of multiple impact heads is designed. In view of the collision contact problem between the impact head of semi-cylindrical structure and the tested object, a mathematical model of collision contact force is established based on Hertz line contact theory and hysteretic damping model, the change of the contact force and the energy loss are described in the form of nonlinear elastic force and equivalent damping, provide theoretical support for impact parameter analysis of the test equipment. On the basis of theoretical analysis, the overload acceleration of continuous high impact test equipment is verified combined with dynamic simulation and experiment, demonstrate the rationality and feasibility of the test equipment design.

1. Introduction

Many of the motion components, control components and other key components in industrial production, such as engine cylinder exhaust valve, concave-convex mold, rock drill piston, cutters, long-term working under high amplitude and repeated impact loading, even though the subjected contact stress is far less than the yield limit of the material, however, the plastic deformation, wear and various types of fatigue which lead to failure will still be produced under alternating impact[1]. The existing research shows that, the changes in structure and performance of material under repeated impact loading are quite different from that of general static low cycle, there may be some unique changes in structure and performance, such as yield strength, elastic hysteresis loop, strain hardening exponent and so on[2]. Therefore, during the design and production of such components, is there an effective method to carry out continuous high impact test, plays a vital role in stability and reliability of the above components and the whole industrial equipment.

In order to realize continuous impact test, many experts and scholars at home and abroad have carried out active research and exploration, a variety impact loading methods of different principle are introduced. Wellinger[3] had introduced an impact loading method of multiple free falling balls, Li Xuemin[4] had developed a multiple impact equipment of cantilever beam structure, Xu Mei[5] had developed a hydraulic multiple impact equipment on the control of solenoid valve, He Ling[6] had researched a multiple impact loading method based on straight roller cam mechanism, Hua Minglei[7] had developed a multiple impact system based on electro-hydraulic servo actuator for the impact test requirement of precision optical component. The above researches have greatly enriched the existing continuous impact test methods, but there are still limitations in impact frequency and impact amplitude. Therefore, in order to meet the actual engineering test requirement better, researching a test method of short interval and high amplitude continuous impact loading has become an urgent technical problem in scientific research and production.

In view of this problem, Nanjing University of Science and Technology had designed a rotary continuous impact test equipment based on direct impact loading of multiple impact heads. Comparing with the existing indirect continuous impact loading method of cam structure or cantilever beam structure, the impact loading is realized directly based on the collision contact between the impact head and the tested object, which can significant increase the amplitude of impact force and overload acceleration of the tested object; the impact loading method of multiple impact heads which driven by a high speed rotating turntable, which can meet the requirement of short interval continuous impact, the impact frequency is also can be adjusted freely, provides a new method for continuous high impact test, the research has good advancement and practical engineering value.

2. Principle of test equipment

The principle prototype of continuous high impact test equipment as shown in Figure 1, the structure diagram as shown in Figure 2, the test equipment is mainly composed of base, turntable, turntable bracket, work-holding device, stand bracket, electric motor, hydraulic station and so on. A series of impact heads are installed on the turntable and the impact loading kinetic energy is provided by the rotating turntable; the work-holding device is installed on the stand bracket and connected with the motion platform through a spring damping mechanism, the retreat and return after each collision can be realized through the compression and restoration of the spring damping mechanism; the motion platform is directly connected to the piston of the hydraulic system, the occurrence and stop of the impact test can be controlled by the retreat and return controlling of the whole impacted mechanism. During the impact test, the turntable is accelerated to the rated speed by the driven motor through the inverter, then feeding the specimen to the motion trajectory of the impact heads along the guide rail of the stand bracket by controlling the hydraulic system, a number of impact heads collide with the tested object in turn, produce continuous high impact loading on the tested object.

Figure 1. Principle prototype

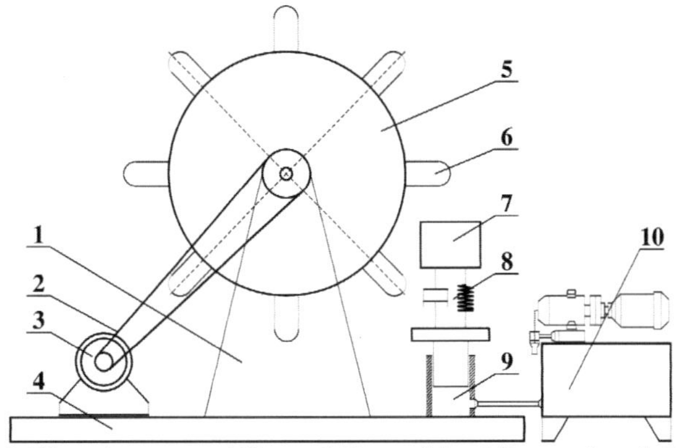

1 Stand bracket, 2 Belt, 3 Electric motor, 4 Base, 5 Turntable, 6 Impact head, 7 Tested object, 8 Spring damping system, 9 Piston, 10 Hydraulic station

Figure 2. Structure diagram

3. Dynamic model

The impact performance parameter is an important indicator of the continuous high test equipment for testing the tested object, in order to simulate the impact test for the tested object of actual working condition effectively, it is necessary to analyze and establish the collision contact force model of the test equipment, so the impact parameters can be adjusted according to test requirement.

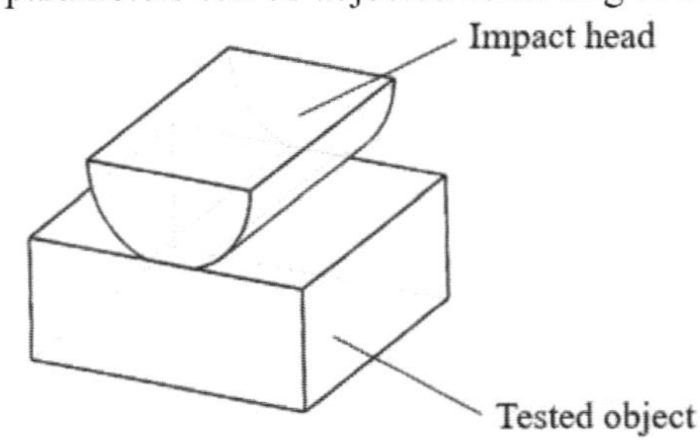

Figure 3. Collision contact diagram

The impact loading of the continuous high impact test equipment is generated by the collision contact between the impact head and the tested object, as the structure of the impact head is semi-cylindrical, therefore, the collision contact between the impact head and the tested object is usually a line contact problem, as shown in Figure 3, the contact force can be obtained according to the Hertz line contact theory, but the Hertz theory is established on complete elastic contact, for the modeling of collision contact problem, the energy conversion process is the most complex part, the Hertz elastic model cannot reflect the physical characteristics of energy conversion during the collision[8], therefore, we introduce a hysteretic damping model based on Hertz model to analyze the collision contact force between the impacted head and the tested object; at the same time, according to the research of Hutchings and Wu, the proportion of stress wave propagation energy is small during the impact process, which can be approximately ignored[9], at this time, the collision force between the impact head and the tested object can be expressed as:

$$F_n = K\delta^n + C\dot{\delta} \tag{1}$$

Where, K represents the equivalent Hertz contact stiffness, C represents the equivalent hysteretic damping coefficient, δ represents the embedded depth of collision contact, $\dot{\delta}$ represents the relative velocity of collision contact, n is the exponential coefficient related to the shape of the collision contact object, for the line contact between the cylinder and the plane, n takes 10/9.

The effective contact length between the impact head and the tested object is l, the contact loading is Q, the loading line density is q, according to the Hertz theory, the relationship of them satisfies:

$$q = \frac{Q}{l} \tag{2}$$

The contact loading forms a rectangular contact surface with a length of l and a width of $2b$, in the range of $-b \leq x \leq b$, the contact stress is elliptically distributed along the contact area, which can be expressed as[10]:

$$p = \frac{2Q}{\pi l b}\sqrt{1-(\frac{x}{b})^2} \tag{3}$$

The half contact width is:

$$b = \sqrt{\frac{4Q}{\pi l \sum \rho}(\frac{(1-\upsilon_1^2)}{E_1} + \frac{(1-\upsilon_2^2)}{E_2})} \tag{4}$$

For the embedded depth of collision contact δ, usually lack of a theoretical solution, but according to the empirical formula which given by Plamgren[11]:

$$\delta = 3.81 \times (\frac{1-v_1^2}{\pi E_1} + \frac{1-v_2^2}{\pi E_2})^{0.9} \frac{Q^{0.9}}{l^{0.8}} \tag{5}$$

The above formula can be rewritten as:

$$Q = K \cdot \delta^{\frac{10}{9}} \tag{6}$$

The equivalent Hertz contact stiffness:

$$K = [\frac{l^{0.8}}{3.81 \times (\frac{1-\upsilon_1^2}{\pi E_1} + \frac{1-\upsilon_2^2}{\pi E_2})^{0.9}}]^{\frac{10}{9}} \tag{7}$$

Where E_1 and E_2 are respectively the elastic modulus of material, υ_1 and υ_2 are respectively the Poisson ratio of the material.

Furthermore, according to the nonlinear damping model of Lankarani and Nikravesh[8], the equivalent hysteretic damping coefficient C can be expressed as:

$$C = \frac{3K(1-e^2)\delta^{\frac{10}{9}}}{4\dot{\delta}_0} \tag{8}$$

The final expression of collision contact force can be expressed as:

$$F_n = K\delta^{\frac{10}{9}} + C\dot{\delta} = [\frac{l^{0.8}}{3.81 \times (\frac{1-\upsilon_1^2}{\pi E_1} + \frac{1-\upsilon_2^2}{\pi E_2})}]^{\frac{10}{9}} \delta^{\frac{10}{9}} [1 + \frac{3(1-e^2)}{4\dot{\delta}_0}\dot{\delta}] \tag{9}$$

It can be seen from the above formula that, the collision contact force of the continuous test equipment is related to the material, shape and collision speed of the collision pair, by changing the material, shape and impact loading speed of the collision pair, can realize the simulation test requirements of different impact parameters. it also can be concluded that, in order to realize effective comparison verification, the simulation model and experiment should be established of the same conditions.

4. Simulation analysis

For the actual continuous impact process, because of the high amplitude and high dynamic characteristics of collision contact force, it is usually difficult to be measured directly, but for the collision contact process of rigid body, the overload acceleration is mainly produced by the collision contact force, therefore, on the basis of theoretical analysis, we can take the overload acceleration as the parameter and combine with rigid dynamic simulation and experimental research to analyze the

IOP Conf. Series: Materials Science and Engineering **428** (2018) 012015 doi:10.1088/1757-899X/428/1/012015

continuous impact process. ADAMS is a widely used simulation software in mechanical analysis of interactive image environment and part library, constraint library and force library, a completely parametric dynamic model of mechanism system can be created for static, dynamic and kinematic analysis[12]. In the research of this paper, because of the interested object is the collision contact process of continuous impact, in order to reduce the interference of noncritical components to simulation process and simulation result, adjust the full-scale 3D model of the test equipment in SolidWorks, remove the unnecessary components, only reserve the key structures which produce continuous impact, at the same time, according to the experimental model, the tested object is modeled as a frame with a horizontal boss on the top and a cylindrical specimen with an acceleration sensor inside, import the above model into ADAMS simulation environment, as shown in Figure 4. The material of each component in simulation model is defined according to the actual principle prototype, the driver and the constrain is also created based on actual assembly relationship and working principle. Defining the collision contact type of 8 impact heads and tested objects of solid to solid, and calculate the collision contact force based on impact function method, due to the friction force is less than normal collision contact force, so the friction force can be ignored in the simulation.

Figure 4. Dynamic simulation model

In order to make the simulation conditions consistent with the experiment, set the rotary speed of the turntable as 720r/min, then can be obtained that 1 turn of the turntable is about 0.083s, thus the simulation time can be configured as 0.1s, the simulation step can be configured as 2000, the obtained continuous impact overload acceleration of the tested object as shown in Figure 5.

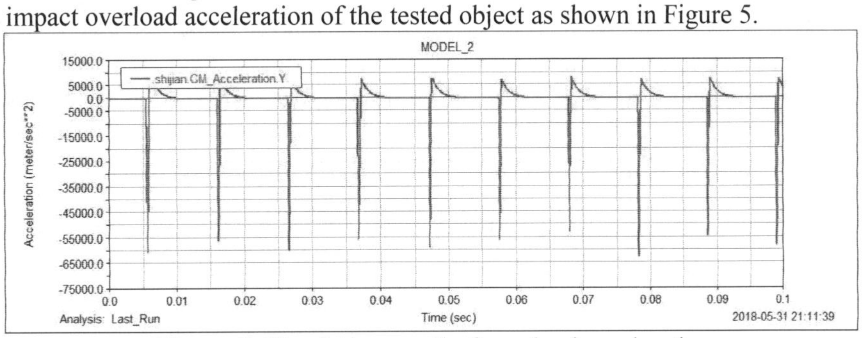

Figure 5. Simulation result of overload acceleration

It can be seen from the figure that, the peak acceleration of continuous impact is usually about 5500g, affected by the natural motion period of spring mass system of the retreat and return mechanism, it is difficult to ensure the tested object of the same motion state and motion position as the original after each impact, result in inconsistency of continuous impact overload acceleration, but the peak acceleration does not change significantly, also presents the periodic change rule.

5. Experimental verification

Carry out experimental research based on continuous high impact test equipment, the tested object is a cylinder specimen of an acceleration sensor inside, as shown in Figure 6, the specimen is installed in work-holding device of the continuous high impact test equipment, ensure that the cylinder specimen will not oscillate during continuous impact process, the overload acceleration of the test equipment can be obtained by the output of the acceleration sensor.

Figure 6. Cylindrical specimen and installation

The obtained overload acceleration curve as shown in Figure 7, it can be seen from the figure that, the overload acceleration of continuous 10 impact is well achieved by the acceleration sensor, affected by the natural motion period of the spring mass system, the measured overload acceleration presents a similar change rule to the simulation result, but the fluctuation is obviously stronger than the simulation, and the change rule is also more complex, this is because the simulation is based on ideal parameter conditions, but in actual continuous impact process there will be more reasons that can affect the motion rule and the overload acceleration of tested object, the reason is also more complex. The actual measured overload acceleration peak can reach as 7000g, higher than simulation result, this is because the dynamic simulation is established on the ideal rigid body, the overload acceleration and collision contact force satisfy Newton's second law, but in actual impact process, the tested object cannot be an ideal rigid body, both structural deformation and stress-strain under high impact loading will result in the measured overload acceleration is higher than that of ideal condition, moreover, affect by the structural response vibration under high impact loading, there will be multiple accompanying sub-pulses after each main pulse of the impact overload acceleration, the peak value of the sub-pulse is obviously lower than that of main pulse, and also change irregularly, it reflects the complexity of actual impact process, especially for these continuous impact process, however, because there is no other external reinforcement after each impact, the structural response vibration of the tested object will gradually attenuate, therefore, the acceleration sub-pulse of each impact also presents an attenuation trend.

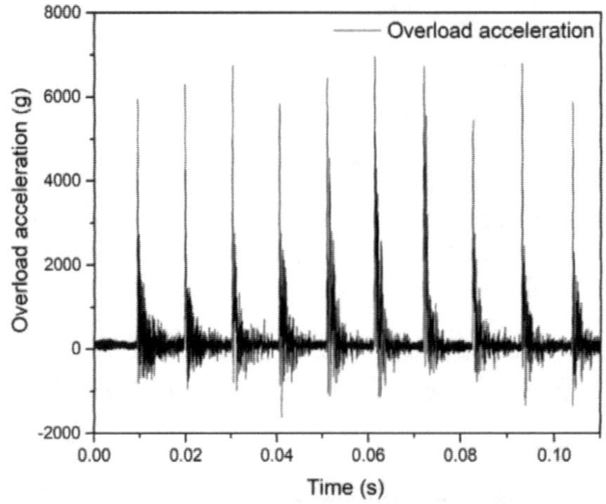

Figure 7. Measured overload acceleration

6. Conclusion

Aiming at the designed continuous high impact test equipment of direct collision contact method, a collision contact model is established based on Hertz line contact theory and hysteretic damping model, analysis the reasons that affect the collision contact force, provide theoretical support for simulation test of specified parameters. The dynamic simulation of continuous high impact test equipment is established and the experimental research is also carried out, considering the feasibility and convenience of measurement, taking the overload acceleration as the parameter, comparative analysis and discussion of the simulation and experimental results. The simulation result and experimental result both reflect the actual performance of the continuous high impact test equipment, verify the rationality and feasibility of the designed impact test equipment, provide a new method for continuous high impact testing, the research is of good advancement and practical engineering value.

Acknowledgement

This research is supported by Pre Research Project of General Armaments Department (41419070201), Postgraduate Research & Practice Innovation Program of Jiangsu Province (KYLX16_0420, KYCX18_0388), The Key Basic Research Projects of Basic Strengthening Plan (2017-JCJQ-ZD-004).

References

[1] T. Xu (2011). *Analysis the T10 carbon tool steel accumulative macrography plastometric set and mechanism at the multi-shock compression collision burden* (Master dissertation, Soochow university).

[2] Y. Wang (2011). *Study on the impact wear behaviors of Ti6A14V titanium alloy and surface modification layers* (Doctoral dissertation, Southwest jiaotong university).

[3] Wellinger K, Breckel H. "Kenngröessen und verschleiss beim stoss metallischer", *Wear*, Vol.1969(13), pp.257-281.

[4] X.M. Li, "Development of a new test rig for impact wear", *Journal of Gansu University of Technology*, Vol.23(1), pp.45-52.

[5] M. Xu, G.Y. Fu, L. Liu, J.G. He. "Amelioration of the new type of repeated impact testing machine", *Journal of Soochow University (Engineering Science Edition)*, Vol.29(1), pp.12-15.

[6] L. He, C. Xu. "Design Optimization and Performance Analysis of Loading Mechanism of Impact Fatigue Test Machine for Firearm Automaton", *Acta Armamentarii*,Vol.32(7), pp.805-811.

[7] M.L. Hua (2011). *Impact properties research on intensity simulation system of some precision optical devices* (Master dissertation, Changchun University of Science and Technology).

[8] Y. Zhang (2012). *Dynamic modeling and analysis of crank-slider mechanism with kinematic pair clearance* (Master dissertation, Harbin Institute of Technology).

[9] F.Q. Shi (2011). *Dynamic Analysis and Modeling to elastic-plastic collision between two spheres* (Master dissertation, Lanzhou University).

[10] Y.X Chen (2005). *Research on vibrations of cyclindrical roller bearings based on contact mechanics* (Doctoral dissertation, Huazhong University of Science and Technology).

[11] D.S. Huang (2013). *Complex system finite element modeling technology and its application in light artillery* (Master dissertation, Nanjing University of Science and Technology).

[12] A.F. Yuan, L.F. Xiang. "Modeling and simulation of cam drive system of rotary press based on adams and solidworks", *Manufacturing Automation*, Vol.32(11), pp.125-128.

Finite Element Analysis of New Downhole Throttle Based on Orifice Valve

G Y HUO, S Y DIAN* and X XU

School of Electrical Engineering and Information, Sichuan University, Chengdu Sichuan, 612265, China

*Email: scudiansy@scu.edu.cn

Abstract. In order to realize the intelligentization of downhole natural gas intelligent gas production technology, a set of downhole intelligent throttles suitable for remote control was designed. This article describes the mechanical structure design and remote-control mechanism of smart throttle. Using ANSYS finite element analysis software, the mechanical structure of underground intelligent throttle is analyzed and the fluid field of key parts of throttle is analyzed. According to the finite element analysis results, the mechanical design of the throttle satisfies the requirements and can provide equipment support for the implementation of downhole natural gas intelligent gas recovery technology. Finally, through the simulated well test, the feasibility of the throttle design was verified.

1 Introduction

In recent years, in order to meet the intelligent development of underground natural gas intelligent talents, researchers have been eager to carry out research on the throttling process and the intelligentization of throttling devices [1]. Advanced downhole process technology has become more and more automated, and the downhole throttle control tool can be remotely controlled through the operation on the well. However, the development of domestic underground intelligent throttling devices is relatively slow, and the finished product of new-type downhole intelligent tools is not high. Therefore, it is increasingly important to research and develop autonomous intelligent throttling tools. The main application of throttling in the domestic natural gas production industry is active and fixed pure mechanical downhole throttles with throttling nozzles [2]-[3]. The biggest drawback of this type of restrictor is that it cannot adjust the size of the nozzle diameter according to the real-time demand, and it must be salvaged through the salvage tool. Manual adjustment of the tool is then placed in the downhole, and the time-consuming adjustment process is severe and economical. not tall. Therefore, the development of intelligent throttling tools to control the production of borehole diameter through remote operation will be the trend of the development of natural gas technology in the future.

2 New Intelligent Throttle Mechanical Structure

2.1 Mechanical design

The downhole intelligent throttle aims at realizing the remote control of the communication signal and adjusting the aperture size of the throttle using computer control technology, automatic control theory and modern signal processing technology [4]. However, the mechanical structure is the carrier of the intelligent control of the throttle. Therefore, it is necessary to analyze the mechanical structure of the

IOP Conf. Series: Materials Science and Engineering **428** (2018) 012016 doi:10.1088/1757-899X/428/1/012016

throttle. The mechanical structure of the throttle as shown in Figure 1.

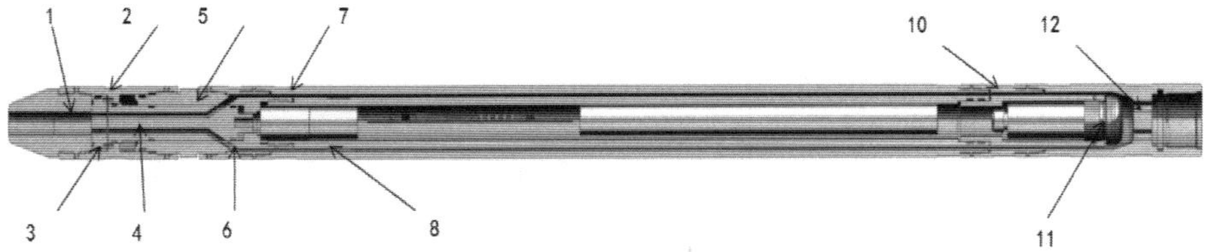

1. Air inlet; 2. Valve seat; 3. Static valve; 4. Flap; 5. Transfer sleeve one; 6. Transfer sleeve two; 7. Drive motor; 8. Air guide tube; ;10. Retainers; 11. Heads; 12. Female sets.

Figure 1. Throttle mechanical structure

Throttle structure is mainly composed of motor transmission mechanism, electrical structure, orifice valve body, pressure wave acquisition system and sealing structure. The transmission mechanism formed by the motor and the speed reducer controls the supporting valve structure of the intermediate transmission mechanism to control the rotation of the valve flap. After the natural gas passes through the air intake nozzle, it passes through the air-guiding channel between the electric sealing cylinder and the air-guiding tube after throttling and flows out from the end of the female sheath. There is a uniform flow hood in the female sleeve, and the gas flows out through the common flow hood after passing through the end of the throttle valve.

2.2 Design calculation methods

The orifice is designed with an orifice plate valve. The valve flap is divided into two parts, the flap and the static valve. The schematic diagram of the orifice valve structure is shown in Figure 2.

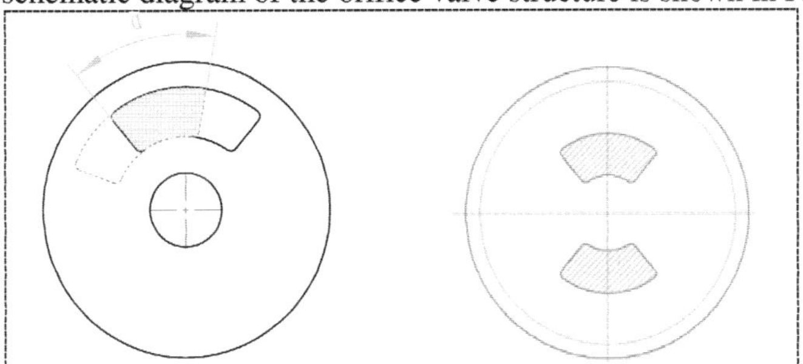

Figure 2. Orifice valve structure schematic

According to the existing method for calculating the throttle diameter of the throttle [5], the diameter of the nozzle of the throttle can be obtained through the judgment formula of the gas frontage state and the mouth flow equation:

$$\frac{P_1}{P_2}=\left(\frac{2}{k+1}\right)^{\frac{k}{k-1}} \tag{1}$$

When $\frac{P_1}{P_2}<\left(\frac{2}{k+1}\right)^{\frac{k}{k-1}}$ is satisfied, the current state is the critical flow state, the gas flow through the throttling nozzle reaches a maximum, and the production is:

$$q_{\max}=\frac{0.408p_1d^2}{\sqrt{r_gT_1Z_1}}\sqrt{\frac{k}{k-1}\left[\left(\frac{2}{k+1}\right)^{\frac{2}{k-1}}-\left(\frac{2}{k+1}\right)^{\frac{k+1}{k-1}}\right]} \tag{2}$$

When $\frac{P_1}{P_2}>\left(\frac{2}{k+1}\right)^{\frac{k}{k-1}}$ is satisfied, the current state is non-critical flow state, the gas flow through the throttling nozzle reaches a maximum, and the production is:

$$q_{sc} = \frac{0.408 p_1 d^2}{\sqrt{r_g T_1 Z_1}} \sqrt{\frac{k}{k-1}\left[\left(\frac{P_1}{P_2}\right)^{\frac{2}{k}} - \left(\frac{P_1}{P_2}\right)^{\frac{k+1}{k}}\right]} \tag{3}$$

Where $q_{\max}$ and q_{sc} represent the natural gas volume flow through the throttling nozzle in critical and non-critical states, respectively $\left(10^4 m^3 / d\right)$; p_1 indicate the pressure at the entrance of the nozzle before throttling and p_2 indicate the outlet pressure at the throttle after throttling (MPa); d is the nozzle aperture diameter (mm); T_1 is the temperature at the inlet of the nozzle before throttling (K); Z_1 is the gas deviation coefficient of throttling nozzle at T_1 and P_1; r_g is the relative density of natural gas ; The natural gas adiabatic coefficient is k , General value $1.25-1.30$.

Considering the actual situation, the throttling nozzle is set to operate in a critical state. According to (2), the diameter of the throttling nozzle can be determined as

$$d = \frac{\sqrt{q_{\max}}}{\sqrt{0.408 p_1}} \sqrt{\frac{\sqrt{r_g T_1 Z_1}}{\sqrt{\left(\frac{2}{k+1}\right)^{\frac{2}{k-1}} - \left(\frac{2}{k+1}\right)^{\frac{k+1}{k-1}}}}} \tag{4}$$

As shown in Figure 2. above, the nozzle area is calculated as $\frac{\pi d^2}{4} = \frac{2\pi\alpha}{360^o}\left(r_1^2 - r_2^2\right)$, Where r_1 is the outer radius of the sector area, r_2 is the sector area inner radius, and α is the angle between the moving valve flap and the static valve flap.

3 Analysis of the main results

This paper uses ANSYS finite element analysis software to perform mechanical analysis and flow field analysis of the downhole intelligent throttle, and to simulate the working environment of downhole tools. Through the simulation analysis results, improve the mechanical structure, so that the structure can work better in the harsh environment downhole [6]-[7].

3.1 Static analysis

The downhole working environment is rather harsh, so it must be ensured that the restrictor can work in the high temperature and pressure environment downhole. Due to the limitations of downhole working space and the fragility of electrical structures, this paper uses ANSYS finite element analysis software to perform static analysis of downhole intelligent throttle [8]. Mainly on the electric seal tube and air tube in the high temperature and high pressure static analysis and modal analysis. The static analysis includes the stress and strain data, as shown in Figure 3 and 4. The modal analysis refers to the natural frequency and vibration shape of the mechanical structure, as shown in Figure 5.

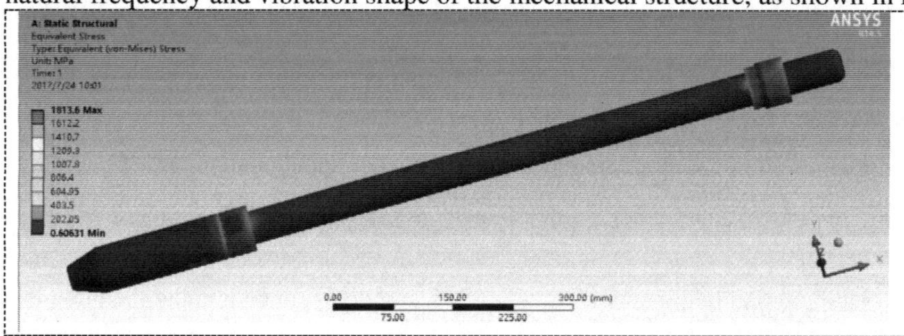

Figure 3. Strain and stress cloud diagram of electric seal barrel in static state

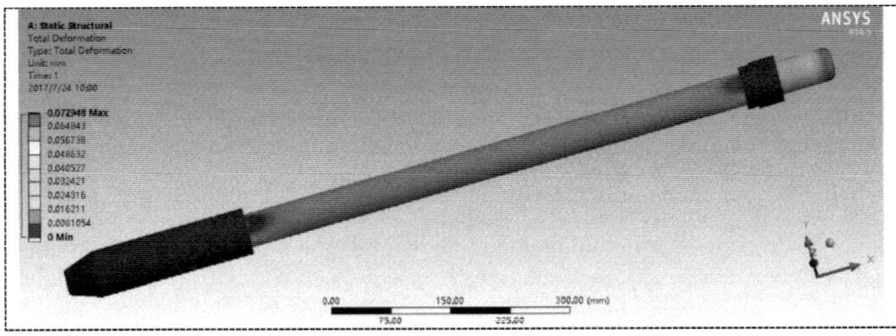

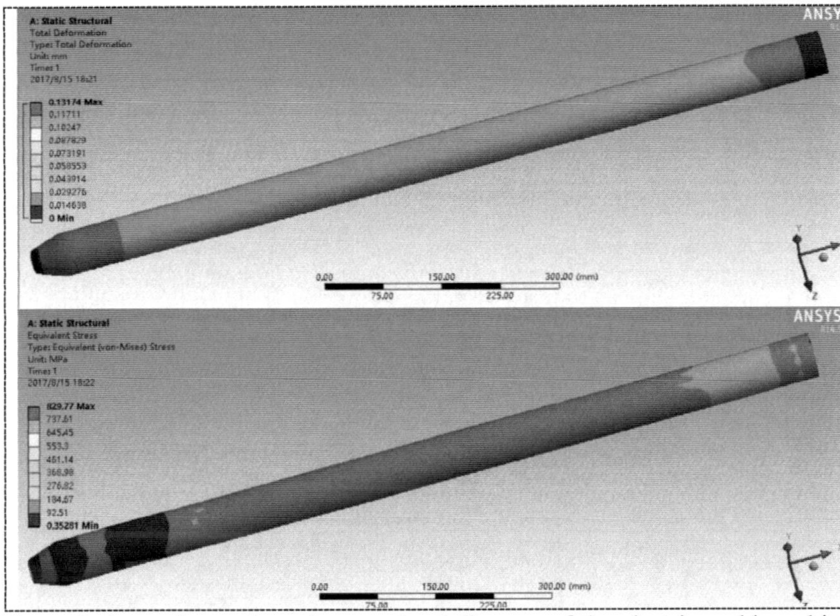

Figure 4. Strain and stress cloud diagram of air-guiding tube in static state

Figure 3 shows the strain and stress cloud diagram of the electric seal barrel in static state. It can be seen from the figure that the maximum stress of the electric seal barrel is 1813.6MPa. The maximum deformation is 0.072947mm, which occurs at the end of the electric seal barrel. In addition, the deformation of the outer wall of the seal barrel is also different. Figure 4 shows the static state strain and stress cloud diagram of the air-conductor. It can be seen from the figure that the maximum stress of the air-guiding tube is 829.77 MPa. The maximum deformation is 0.13174mm.

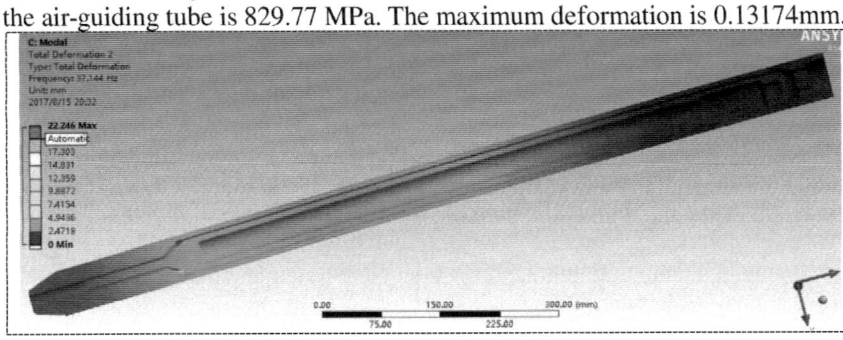

Figure 5. Throttle section profile modal analysis

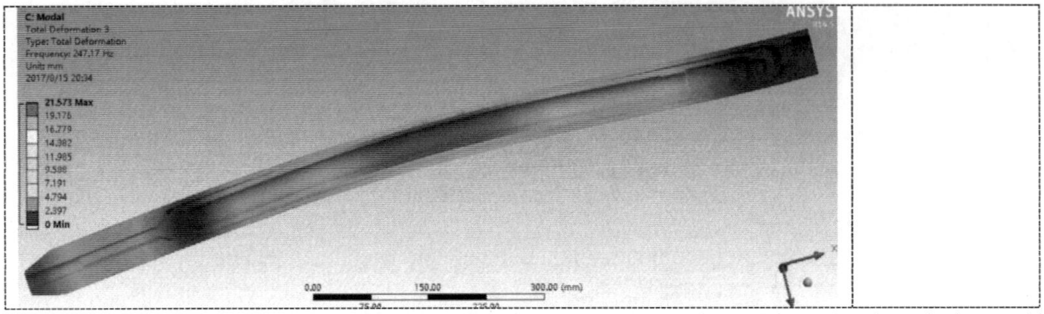

The above figure shows the modal analysis of the throttle section. When the throttle is working downhole, high-speed eroding of the air flow will generate vibration. When the frequency is too high and reaches the natural frequency, resonance will occur. The modal analysis can obtain different natural frequencies. In the case of vibration, the figure above shows two natural frequencies. The first natural frequency is 37.144 Hz and the second natural frequency is 247.17 Hz. The structure oscillates only slightly at the first natural frequency; the second natural frequency swings significantly, mainly due to the distortion of the electrically sealed bucket. Combined with downhole vibration, it can be analyzed that the throttle can work normally downhole.

3.2 Flow field analysis

According to the actual parameters and technical requirements of gas wells in the Sichuan-Yunnan region [9], the internal diameter of the natural gas pipeline is 62mm, the pressure at the inlet of the throttle is 35MPa, the outlet after the throttle is 7MPa, and the throttle is placed in the ground at a distance of 2000m below the ground. The formation temperature is 90°C; the natural gas relative density is 0.85, the gas compression coefficient is 0.7, and the adiabatic coefficient is 1.25.

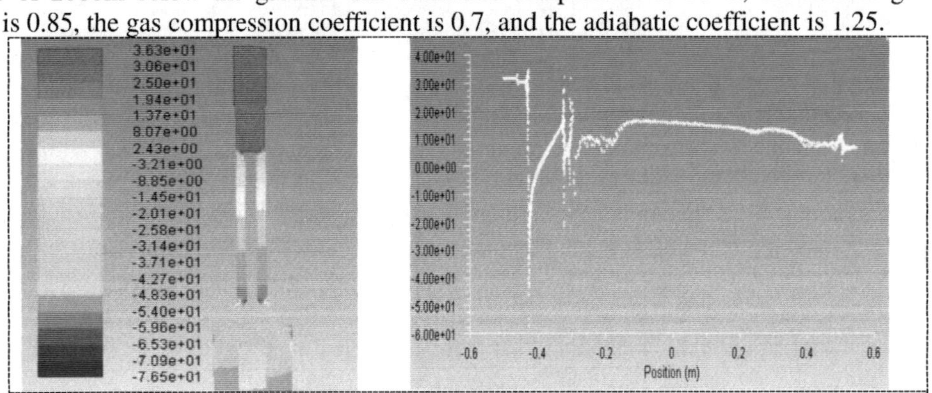

Figure 6. Throttle mouth pressure change cloud

As shown in Figure 6, the set value of the pressure at the inlet of the throttle valve is 35 MPa. Through the throttling effect of the throttling nozzle, the flow rate of the gas in the channel increases rapidly and the pressure drops rapidly as the area inside the pipeline decreases rapidly. The pressure area in the channel rises, and the pressure change tends to be stable [10]-[11]. After the throttling is over, the pressure value is stabilized at about 7MPa, and the throttling process is completed.

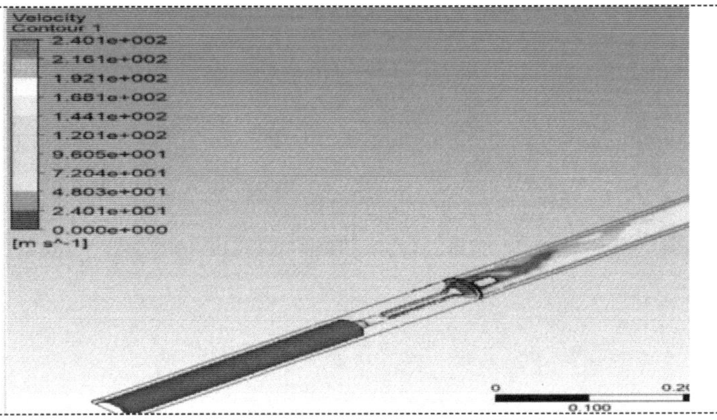

Figure 7. Throttle front speed change cloud

As shown in Figure 7, the figure shows the speed change cloud diagram of the front end of the throttle. It can be seen from the figure that the speed change of the gas in the channel before and after throttling occurs, and the gas undergoes sudden expansion and sudden flow during the flow. During the shrinking process, the speed suddenly increases near the orifice, and the maximum speed reaches about 210~240 m/s. After the throttling effect of the throttling nozzle, the area inside the throttle pipe suddenly increases and the speed decreases. Finally stabilized at around 20 m/s.

4 Conclusion

The new type of orifice plate valve wireless smart festival was tested in a simulated well by testing the overall tightness of a new type of orifice valve wireless smart throttle, the stability of electrical components operating at high temperatures, and the stability of pressure waves transmitted under high temperature and high pressure. The flowmeter has achieved good stability under various experiments. At the same time, the downhole intelligent throttle can provide equipment support for the implementation of downhole intelligent throttling technology and has reference significance for the development of relevant intelligent downhole tools.

References

[1] Yan D, Qing D, Li L, et al. Numerical Simulation of Throttle Valves with Different Spool and Valve Body by Fluent 2011 *J Mechanical Engineering & Automation,* **1** 11.
[2] HAN, Dan-xiu, LI, et al. Adaptability Study of Intelligent Well Systems in East China Sea Oil Field 2008 *J International Journal of Plant Engineering & Management* **13(4)** 205-213.
[3] Hu G, Zhang P, Wang G, et al. Performance study of erosion resistance on throttle valve of managed pressure drilling 2017 *J Journal of Petroleum Science and Engineering* **156** 29-40.
[4] Duan B, Liu L, Xu X, et al. Research and application of steel wire free dropping and fishing downhole chock 2011 *J Oil Drilling & Production Technology* **1** 046.
[5] Wang X, Economides M. Advanced natural gas engineering 2013 *M Elsevier.*
[6] Zhang Z, Hao Y. Research and application of downhole throttling technology in Yulin gas field 2009 *J Oil Drilling & Production Technology* **1** 038.
[7] Scheaua F D. Functional description of a hydraulic throttle valve operating inside a hydraulic circuit 2016 *J Hidraulica* **1** 47.
[8] He M G, Ma F M et al. Design and Flow Field Analysis of Downhole Intelligent Throttle 2013 *J China Petroleum Machinery* **6** 20.
[9] Yong X, Qian Y M, Yacong Y, et al. Development mode and surface supporting technology in the Changqing gas zone 2010 *J Natural Gas Industry* **2** 029.
[10] Hou C, Qian J, Chen F, et al. Parametric analysis on throttling components of multi-stage high pressure reducing valve 2018 *J Applied Thermal Engineering* **128** 1238-1248.
[11] Ashraf, Walid, et al. Effect of Replacement of Butterfly Throttle Body by Barrel Throttle Body on Mass Flow Rate using CFD 2017 *R SAE Technical Paper* **1** 1078.

IOP Conf. Series: Materials Science and Engineering **428** (2018) 012017 doi:10.1088/1757-899X/428/1/012017

Sleep and Wake Classification Based on Heart Rate and Respiration Rate

W Li[1,a], X D Yang[1,b], A N Dai[1,c] and K Chen[1,d]

[1]Department of mechanical engineering, Tsinghua University, Beijing 100084, China

[a]weili16@mails.tsinghua.edu.cn
[b]yangxd@tsinghua.edu.cn
[c]daian15@mails.tsinghua.edu.cn
[d]kenchen@tsinghua.edu.cn

Abstract. Sleep plays a significant role in human health. Along with the emergency of diseases related to sleep, sleep monitoring is becoming a hotspot. Traditionally, Polysomnography (PSG) is considered as the gold standard for sleep monitoring, but it is expensive, time-consuming, and uncomfortable, especially not ideal for long-term sleep studies. Thus, exploring alternative sleep monitoring method has drawn much attention of researchers. Currently most of the sleep monitoring methods require complicated signals and feature extraction process. However, this paper innovatively proposes a method using simple signal of heart rate and respiration rate, and integrating the prior knowledge of experts to simplify the feature extraction process, which can make the subsequent classifier distinguish wake state and sleep state easily. In this paper, we used multi-layer neural network as a classifier, getting a Cohen's kappa value of 0.739, and the overall accuracy of classifying sleep and wake state reached to 88%. Furthermore, we improved the ability of detecting wake states greatly by oversampling method, and the sensitivity and specificity reached to 91.3% and 82.3% respectively. Nowadays, the development of various bed sensors makes it easy to obtain the heart rate and respiration rate signals of human beings. Therefore, the method proposed in this paper is conducive to people's long-term sleep monitoring more conveniently and has great significance to people's health.

1. Introduction

Sleep is one of the most important physiological activities of human body. Recently, some studies have found that lack of sleep may cause many diseases such as hypertension and cardiovascular diseases [1-2], which brings sleep monitoring more and more attention. Currently the most standard sleep monitoring method is PSG, but the recording process is costly, and many electrodes can interfere sleep process as well. So researchers have shown great interest in finding other effective ways to replace PSG.

At first, people use signals similar to those in PSG such as EEG and ECG signals, which get a classification accuracy of 90% [3-4]. But these signals are also collected from the electrodes attached to the human body, thus will interfere people's sleep, which makes these sleep monitoring methods not ideal for long-term practice. Actigraphy, on the other hand, is able to measure the body's motion through acceleration sensors. The wristband-style Actigraphy is popular with researchers because it greatly reduces the disturbance of sleep when collecting information [5]. However, motion signal alone often fails to recognize sleep states correctly, so additional signals need to be acquired to get more accurate results [6]. And then researchers, for instance Xi Long, start to use cardiorespiratory signals to classify sleep states. He extracted a large number of features from body motion and respiration, which made the accuracy improved a lot [7]. Nevertheless, most of these methods introduced above require sophisticated instruments to get signals and complex feature extraction processes such as time-frequency analysis [8], wavelet transformation and dynamic time wrapping [7], and those will cost tons of money and time. Besides, the accuracy of detecting wake state is lower compared with sleep state. Therefore, a simpler and more practical method to classify the state of sleep needs to be studied urgently.

With the development of various bed sensors, it becomes easy to extract the physiological signals of people in bed, including heartbeat, respiratory, etc. These signals are quite accurate and contain information of sleep, but unfortunately they have not been further utilized so far. As we all know, heartbeat and respiratory will both slow down and become more stable after people fall asleep, which means that heartbeat and respiration are related to sleep states. Therefore, this paper proposes a method of recognizing the human sleep states through heart rate and respiration rate based on the above background information. This method has a simple and fast feature extraction process, and uses neural networks as the final classifier. Besides, this method also balances the number of sleep and wake state in the training process, improving the ability of distinguishing wake state.

2. Methodology

The sleep monitoring method introduced in this paper has three steps, including data preprocessing, feature extraction and neural network construction. Data showed in this paper was collected from 10 healthy experiment volunteers using Ann Pollen PSG. Acquisition frequency of heart rate and respiration rate is 1 Hz, and the images of corresponding sleep states are saved in PDF. After processing the image, we can get the state of a certain heart rate and respiration rate. Due to the image resolution problems, a state label will correspond to about 20 seconds' heart rate and respiration rate, which will lead to some errors in the place of state transition. Fortunately, there is few transitions between wake state and sleep state under normal circumstance, so it can be considered that we get accurate heart rate, respiration rate and the corresponding state of the original data.

2.1 Data Prepocessing

Some raw data of heart rate and respiration rate from Ambler PSG can be abnormal sometimes, which should be modified to avoid affecting subsequent classification. This paper set a threshold 40 to 110 for heart rate and 10 to 30 for respiration rate to make sure the data applied in analysis is within a reasonable range. The raw data will be considered as the boundary value if it falls out of the range. With consideration of reducing jitter of raw data, data of heart rate and respiration rate are smoothed by the method of moving average. At each time point, a rectangular window of 60 seconds with the time point as midpoint is applied, and the calculation formula shows below:

$$\overline{x_i} = \frac{1}{60}\sum_{j=i-30}^{i+30} x_j \tag{1}$$

In addition, the original label of polysomnography is divided into five states. But the goal of this paper is to distinguish sleep and wake state, which means that we also need to change the labels to two states. After preprocessing, the dataset of October 20th, 2016 shows below in figure 1.

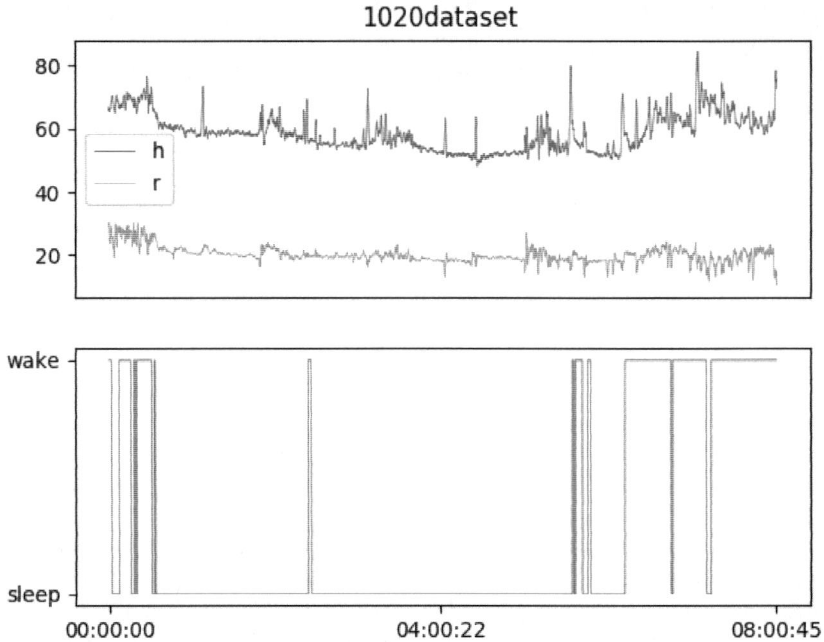

Figure 1. October 20th, 2016 dataset's heart rate, respiration rate and corresponding sleep states

2.2 Feature extraction

To some extent, the instantaneous value of heart rate and respiration rate are random, but after statistical process, they become useful to classify sleep states. What is known to people is that when the human body fall in asleep, the heartbeat and breathing will become slow and steady. According to above prior knowledge, we can use mean to indicate the trend of slowing down, and use variance to indicate the trend of becoming steady. Further, this paper proposes a simple but effective feature extraction method that can be seen in table 1. These features are based on statistical characteristics of heart rate and respiration rate for a period of time, including mean, variance, skewness and kurtosis. And the length of time used in this paper is 400 seconds (t = 400). Higher-order statistics skewness and kurtosis represent more detailed information about data distribution. Skewness shows the deviation of data distribution from normal distribution, and kurtosis shows the sharpness of data distribution relative to the normal distribution. They are calculated as: ($\bar{x}$, σ represent the mean and variance of corresponding period, and n is the length of one period)

$$\text{skewness} = \frac{1}{n-1}\sum_{i=1}^{n}\left(\frac{x_i - \bar{x}}{\sigma}\right)^3 \tag{2}$$

$$\text{kurtosis} = \frac{1}{n-1}\sum_{i=1}^{n}\left(\frac{x_i-\bar{x}}{\sigma}\right)^4 - 3 \tag{3}$$

However, these calculation is a little complicated which will slow down the feature extraction process. Fortunately, experiments show that the accuracy improved by these two features is less than 1%, so it is acceptable to abandon these two features. The last feature of heart rate and respiration rate measures the similarity of two adjacent periods. It is more likely to be classified as sleep state if the two adjacent periods are similar, so we choose this as an important feature, which is calculated as: (x, y represent the value of adjacent period.)

$$\text{feature} = \frac{1}{n}\sum_{i=1}^{n}(x_i - \bar{x})(y_i - \bar{y}) \tag{4}$$

On the other hand, different people's heartbeat and breathing will have different basic values, and the features extracted from data will have different scales. In order to make classification better, the features of data must be normalized. In this paper, the data normalization method is as follows,

$$x_{new} = \frac{x-\mu}{\sqrt{\sigma+\varepsilon}} \tag{5}$$

where μ and σ represent the mean and variance, and ε is generally a small number to prevent the occurrence of the divided-by-zero error. And such a normalization method changes the data into a zero-mean, unit variance distribution. Figure 2 shows all features extracted from dataset of October 20, 2016.

Table 1. Description of all kinds of features

Feature number	Calculation method (t = 400)
1	The average of heart rate for a length of t
2	The variance of heart rate for a length of t
3	The skewness of heart rate for a length of t
4	The kurtosis of heart rate for a length of t
5	The average difference of heart rate between two adjacent periods
6	The variance difference of heart rate between two adjacent periods
7	The similarity of heart rate between two adjacent periods
8-14	Above process operated on respiration rate

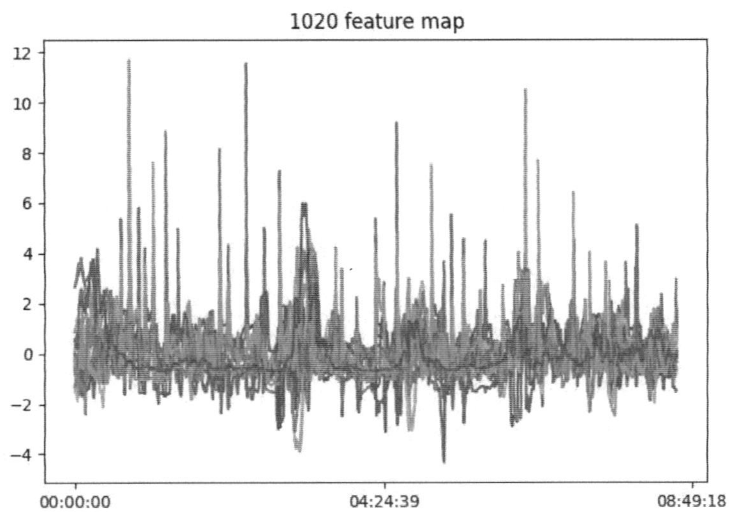

Figure 2. October 20th, 2016 dataset's feature map including 14 features

2.3 Build neural network

In this paper, the structure of neural network is the 14-32-16-4-1, with the hidden layers taking Relu as activation function while the last layer taking sigmoid as activation function. Due to the depth of the neural network, this paper uses Batch Normalization method to solve the problem of training deep network. Because there is a large amount of data (more than one million) that can be utilized, this paper does not use any regularization method. What's more, in order to speed up the training process, this paper adopts the gradient optimization method with momentum. The momentum factor is 0.9, the initial length of step is 0.01, and the decay factor is 1e-6. It is found that the loss drops slowly after 50 epochs and then we can choose to stop the optimization.

The amount of sleep state and wake state in the dataset acquired from the original data is quite different, so the previous classifier [9] tends to be more sensitive and less specific. In this paper, we use an oversampling method to double the wake state, and then merge them with sleep state to form the final dataset. This dataset will be shuffled and divided into three parts, including training set, validation set and test set, which account for 64%, 16%, 20% respectively. The training set will be trained in neural networks to solve the problem of low specificity. Although oversampling makes accuracy drop down slightly, the model is even more powerful. And figure 3 shows the change of loss and accuracy in training process.

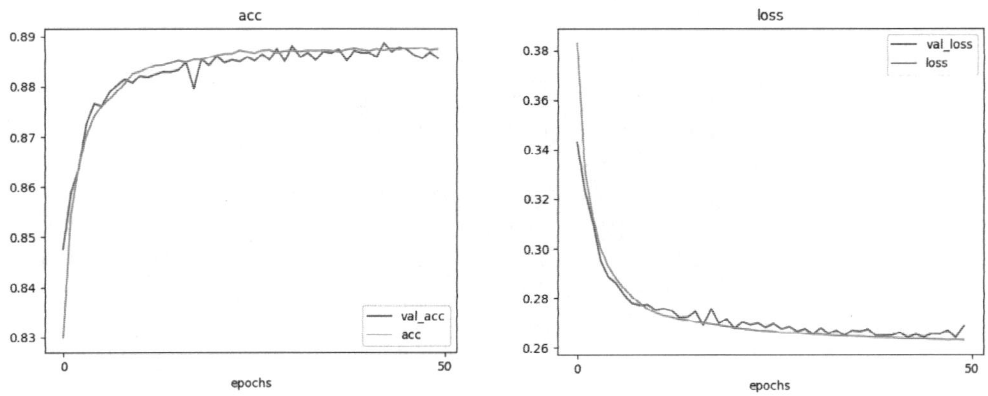

Figure 3. Training process on data including training set and validation set. Left is loss curve, right is accuracy curve.

3. Results and discussion

Following the above steps, we can get the final neural network classification model. This model achieved an overall accuracy of 88% in training set, as well as in validation set and test set. By balancing the number of sleep states and wake states, we got the confusion matrix as table 2, and the sensitivity and specificity reached to 91.3% and 82.3% respectively. And this classifier got a Cohen's kappa value of 0.739, showing the high coherence of the ground truth. Compared with the similar method for classifying the states of sleep in which pressure sensors are installed on a bed [10], the accuracy is improved by about 9% and the specificity is improved by nearly 20%. In table 3, we show the results of 20 subjects randomly selected from all data, which is also better than the classification results of Bayesian approach using heart rate and surplus pulse O2 signals in [9]. And there is comparison of these methods mentioned above in table 4.

Table 2. Confusion matrix on all data

Prediction \ ground truth	sleep states	wake states
sleep states	567588 (TP)	57080 (FP)
wake states	54176 (FN)	268161 (TN)

$$\text{sensitivity} = \frac{TP}{TP + FN} = 0.913 \quad (6)$$

$$\text{specificity} = \frac{TN}{TN + FP} = 0.823 \quad (7)$$

$$\text{kappa} = \frac{p_o - p_e}{1 - p_e} = 0.739 \quad (8)$$

This paper proposes a new model using heart rate and respiration rate signals that people can easily and inexpensively acquire to classify sleep states. What's more, the statistical characteristics of heart rate and respiration rate are used as features, which simplifies the feature extraction process greatly and reduces the time cost at the same time. Because we have found the hidden relation between heart rate, respiration rate and sleep states by neural network, people can use bed sensors to collect the data of heart rate and respiration rate, and then get the state of sleep. Therefore, sleep monitoring can be used more widely.

Table 3. Sensitivity, specificity and accuracy of 20 nights' data randomly selected from data

ID	sensitive	specificity	accuracy
1	0.9550	0.8727	0.9430
2	0.9520	0.8020	0.9195
3	0.9363	0.9134	0.9349
4	0.9376	0.7903	0.8902
5	0.8824	0.6321	0.8413
6	0.9085	0.9127	0.9097
7	0.9234	0.9263	0.9239
8	0.8961	0.8352	0.8887
9	0.9091	0.8544	0.8952
10	0.9292	0.8639	0.9111
11	0.9212	0.7262	0.8761
12	0.9435	0.8841	0.9321
13	0.9255	0.9248	0.9254
14	0.8328	0.7998	0.8252
15	0.8844	0.9078	0.8896
16	0.9653	0.7721	0.8805
17	0.8559	0.7238	0.8199
18	0.9322	0.8757	0.9245
19	0.9002	0.8980	0.8994
20	0.8660	0.7309	0.8368
mean	0.9128	0.8323	0.8933

Table 4. A comparison of methods

method	sensitive	specificity	accuracy
This paper	0.9128	0.8323	0.8933
Bayesian approach [9]	0.9274	0.5917	0.8269
Pressure-Based [10]	0.955	0.532	0.743

4. Conclusion

Nowadays, sleep has drawn great attention due to its close relation with health. At the same time, the monitoring of sleep states has naturally become a hot topic. In this paper, a method of using heart rate and respiration rate to classify sleep states is proposed. Through simple feature extraction process based on prior knowledge and multi-layer neural network, we got an overall accuracy of 88%. Moreover, by the way of oversampling, we enhanced the ability to detect wake state and improved the specificity obviously. With the trend of the development of various kinds of bed sensors, this method makes it possible to monitor sleep at low cost, as well as to prevent various sleep-related diseases. On the basis of this method, people can improve the accuracy by adding more useful features, and we will search for these features to make the classification better in the future.

Acknowledgment

Thank my tutor Yang for his patient guidance and help, and also thank Softide for its support and thanks for the experimenters who provided the data.

Reference

[1] Yan C. Sleep monitoring for prevention of cardiovascular and cerebrovascular diseases[J]. Medical information management, 2011, 24(6):2587

[2] Lei W. Research on relation between sleep quality and blood pressure rhythm in hypertensive patients[D]. Beijing University of Chinese Medicine, 2014.

[3] Fraiwan L, Lweesy K, Khasawneh N, et al. Automated sleep stage identification system based on time-frequency analysis of a single EEG channel and random forest classifier[J]. Computer Methods & Programs in Biomedicine, 2012, 108(1):10-19.

[4] Da S T, Kozakevicius A J, Rodrigues C R. Single-channel EEG sleep stage classification based on a streamlined set of statistical features in wavelet domain[J]. Medical & Biological Engineering & Computing, 2017, 55(2):343.

[5] Morgenthaler T, Alessi C, Friedman L, et al. Practice parameters for the use of actigraphy in the assessment of sleep and sleep disorders: an update for 2007.[J]. Sleep, 2007, 30(4):519-29.

[6] Paquet J, Kawinska A, Carrier J. Wake Detection Capacity of Actigraphy During Sleep[J]. Sleep, 2007, 30(10):1362-9.

[7] Long X, Fonseca P, Foussier J, et al. Sleep and Wake Classification With Actigraphy and Respiratory Effort Using Dynamic Warping[J]. IEEE Journal of Biomedical & Health Informatics, 2014, 18(4):1272-1284.

[8] Xiao M, Yan H, Song J, et al. Sleep stages classification based on heart rate variability and random forest[J]. Biomedical Signal Processing & Control, 2013, 8(6):624-633.

[9] Fu C, Zhang P, Jiang J, et al. A Bayesian approach for sleep and wake classification based on dynamic time warping method[J]. Multimedia Tools & Applications, 2015, 76(17):1-20.

[10] Walsh L, Mcloone S, Ronda J, et al. Noncontact Pressure-Based Sleep/Wake Discrimination[J]. IEEE Transactions on Biomedical Engineering, 2016, 64(8):1750-1760.

Swing Process Model Design of a Cutter Suction Dredger Based on RBF-ARX Model

Shuang Jiang[1], Fusheng Ni[1] and Lina Qian[1]

[1]Engineering Research Center of the Education Ministry of Dredging Technology, Hohai University, 200 jinling north road, changzhou city, jiangsu province, PRC.

Jiangshuang01@163.com; Fushengni@163.com; 20071760@hhu.edu.cn

Abstract. The swing process of a cutter suction dredger are affected by many factors, which changed time and nonlinear. So it is difficult to describe the change law accurately. The swing process model based on RBF-ARX model is established after detailed analysis of the formation process of dredger production. The shortcomings of the traditional neural network could be overcome by this method, such as slow convergence rate, more hidden layer. After modeling, the linear and nonlinear parameters of the RBF-ARX model are identified off-line by using SNPOM (Structured Nonlinear Parameter Optimization Method). Then the errors of the output of the model and the real data are compared and analyzed by simulation. The result show that the swing process model can accurately describe the dynamic characteristics of the system in the global range, and the model output is well fitted with the real data of a cutter suction dredger.

1. Introduction

Due to advantages such as low engineering cost, high production efficiency, wide scope of application, etc., cutter suction dredger is widely used in such dredging engineering as earth reclamation, channel construction and maintenance [1]. But because its complex and variable dynamic characteristics varies with soil quality, landform and operating condition of ship, dredging operation process cannot be summarized and expressed with an accurate equation. For a long time, construction efficiency of dredger mainly depends on experience of operators, with problems such as large yield fluctuation, high energy consumption and experience difficult to copy, enormous difference between individuals, etc. In recent years, rapid development of computer technology and modern control theory creates necessary conditions for modelling and automatic control of dredger construction process which is a nonlinear system with obvious time variant characteristics.

Overseas, the research on automation of dredging operation began in 1990s, and a dredging comprehensive detection system developed by Dutch L.H.C can express dredging process information by combining text and graph [2]. Jelmer Braaksma of DELFT conducted deep research on modelling and control of trailing suction hopper dredger [3]. Domestically, related research on dredger has been conducted for nearly 10 years, but those on swing process control are few. Tang Jianzhong of Zhejiang University proposed swing speed control aiming at a steady working point [4]. Bi Yuezhi proposed inline optimization of soil and sand delivery system operating point based on the concentration process model proposed by Tang JianZhong [5]. Ni Fusheng of Hohai University conducted much research on dynamic characteristics of soil and sand delivery system, laying a foundation for research on swing process [6]. Li Zhigang proposed a swing control mathematical model of current protection under ideal

IOP Conf. Series: Materials Science and Engineering **428** (2018) 012018 doi:10.1088/1757-899X/428/1/012018

condition [7]. Zhu Wenliang used modern control theory to resolve the modelling problem of complex parameter of swing process, proposing a spatial model of swing process state based on measured data [8].

RBF-ARX model is an emerging offline identification model proposed for dealing with nonlinear control system with a complex object [9]. It is based on linear ARX model, but the difference between them is that the model has functional factors based on RBF neural network. Based on full understanding of RBF-ARX model, by combining dynamic characteristics of swing process, directly starting from measured input/output data of the dredger, this paper identifies linear and nonlinear parameters in the model using SNPOM method to establish dredger swing process model, compares and analyses the error result of established model and real data using simulation.

2. RBF-ARX model

2.1. Basic structure of RBF-ARX model

RBF-ARX model is a composite structure of RBF neural network and linear ARX model [10], with equation as shown in (1).

$$\begin{aligned} y(t) = \phi_0(X(t-1)) + \sum_{i=1}^{n_y}\phi_i^y(X(t-1))y(t-i) + \sum_{i=1}^{n_u}\phi_i^u(X(t-1))u(t-i) \\ + \sum_{i=1}^{n_v}\phi_i^v(X(t-1))v(t-i) + \xi(t) \end{aligned} \quad (1)$$

Where, y(t) is output; u(t) is input; v(t) is external measurable interference quantity; $\xi(t)$ is white noise; X(t-1) is state vector which can have direct or indirect relation with the system; ny, nu, nv is the order of model; $\phi_0(X(t-1))$ and $\phi_i^j(X(t-1))$, $i=1,2,...n_j$; $j=y,u,v$ are state dependent functional coefficients which rely on state vector X(t-1). When state vector X(t-1) at moment t-1 is decided, functional coefficient can be calculated through RBF neural network.

2.2. Identification method for RBF-ARX

The functional coefficient in RBF-ARX model is obtained by RBF neural network training and calculating offline data, its structural diagram is shown in Figure1.

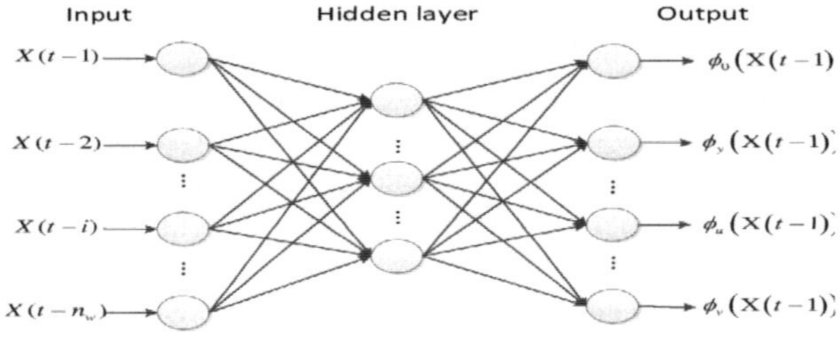

Figure1. Structural diagram of RBF identification of RBF-ARX model functional coefficient

SNPOM parameter identification method was used to solve parameters of functional coefficient, the core of the method is to construct nonlinear parameter subspace and liner parameter subspace by dividing model parameters into nonlinear and linear parameters [11]. Nonlinear parameters are optimized using LMM method, linear parameter is optimized using LSM method.

First write equation (1) as:

$$y(t) = f(\theta_L, \theta_N, X(t-1)) + \xi(t); \quad y(t) = \varphi(\theta_N, X(t-1))\theta_L^T + \xi(t) \quad (2)$$

Where, θ_N is nonlinear parameter space, $\theta_N @ \{\lambda_k^j, Z_k^j | k=1,2,...,m; j=y,u,v\}$; θ_L is linear parameter subspace. $\theta_L @ \{c_k^0, c_{i,k}^j | k=1,2,...,m; i=1,2,...,n_j; j=y,u,v\}$. The initial value of RBF network centre Z_k^j in nonlinear parameter subspace θ_N can be given at random, to ensure steady boundedness of linear weight space, the initial value of proportion coefficient of RBF network also needs to be decided:

$$\lambda_k^0 = \frac{-log\varepsilon}{\max_{t-1}\left\{\left\|X(t-1)-Z_k^j\right\|_2^2\right\}} \qquad \varepsilon\in[0.1\sim0.00001] \tag{3}$$

The initial value of linear parameter θ_L is calculated using LSM algorithm:

$$\theta_L^0 = \left[R(\theta_N^0)^T R(\theta_N^0)\right]^{-1} R(\theta_N^0)^T \bar{Y} \tag{4}$$

Where, $R(\theta_N^0)=\begin{Bmatrix}\varphi(\theta_N^0,\bar{X}(\tau))\\ \varphi(\theta_N^0,\bar{X}(\tau+1))\\ \text{M}\\ \varphi(\theta_N^0,\bar{X}(M-1))\end{Bmatrix}$, $\bar{Y}=(\bar{y}(\tau+1),\bar{y}(\tau+2),...,\bar{y}(M))$, here $\left\{\bar{y}(i),\bar{X}(i-1)\middle|i=\tau+1,\tau+2,...,M\right\}$ is the set of observed data, τ is the time delay of the system, M is length of observed data.

After being initialized, parameters need to be optimized further. Define target function:

$$V(\theta_L,\theta_N)=\frac{1}{2}\left\|F(\theta_L,\theta_N)\right\|_2^2 \tag{5}$$

Where, $F(\theta_L,\theta_N)=\begin{bmatrix}f(\theta_L,\theta_N,\bar{X}(\tau))-\bar{y}(\tau+1)\\ f(\theta_L,\theta_N,\bar{X}(\tau+1))-\bar{y}(\tau+2)\\ \text{M}\\ f(\theta_L,\theta_N,\bar{X}(M-1))-\bar{y}(M)\end{bmatrix}$.

So, parameter optimization is transformed into the following equation:

$$(\hat{\theta}_L,\hat{\theta}_N)=\arg\min_{\theta_L,\theta_N}(V(\theta_L,\theta_N)) \tag{6}$$

Updating strategy of nonlinear parameter θ_N^{k+1} is:

$$\theta_N^{k+1}=\theta_N^k+\beta_k d_k \tag{7}$$

Where, d_k is search direction, β_k is step size. Every iteration d_k is solved by the following equation:

$$\begin{cases}\left[J(\theta_N^k)^T J(\theta_N^k)+\gamma_k I\right]d_k=-J(\theta_N^k)^T F(\theta_L^k,\theta_N^k)\\ J(\theta_N^k)=\left(\dfrac{\partial F(\theta_L^k,\theta_N^k)}{\partial\theta_N^k}\right)^T\end{cases} \tag{8}$$

Where, $J(\theta_N^k)$ is the Jacobean matrix of $F(\theta_L^k,\theta_N^k)$ with respect to θ_N^k ; where, γ_k controls the amplitude and search direction of d_k .when γ_k approaches 0, d_k will be in Gauss-Newton direction; γ_k approaches infinite, d_k will be in max decline direction. After γ_k is selected d_k can be obtained. When next nonlinear parameter θ_N^{k+1} is determined, next linear parameter θ_L^{k+1} needs to be optimized using LSM again. During parameter optimization, $V(\theta_L^{k+1},\theta_N^{k+1})<V(\theta_L^k,\theta_N^k)$ needs to be ensured, otherwise optimization is over, parameter identification end.

3. Model design of cutter suction dredger swing process

3.1. Variable analysis of swing process

Production of cutter suction dredger refers to the earthwork volume delivered in pipe in unit time. When flow rate of slurry is stable, concentration variation of slurry can reflect variation of output to certain extent, so expression of slurry concentration is:

$$C_w=f(\lambda,\eta,A_c,v)V_s \tag{9}$$

Where, C_w is concentration of slurry(%), λ is the rate of sedimentation loss, $\eta(0 \le \eta \le 1)$ is slurry leakage rate, A_c is actual cutting area of cutter(m^2), v is velocity of slurry(m/h), V_s is swing speed(m/min). During dredging construction output is expected as large as possible, but limited by the cutter power, A_c cannot be too large, nor adjustable. Slurry flow rate depends on characteristics of pump, unable to be adjusted frequently, either. So, the swing process model established uses concentration reflecting production as output, swing speed regulated most frequently as control input.

3.2. Basic structure of swing process model

Without considering external measurable interference, RBF-ARX based cutter suction dredger swing process model is as follows:

$$\begin{cases} C_w(t) = \phi_0(\mathrm{X}(t-1)) + \sum_{i=1}^{n_c} \phi_i^c(\mathrm{X}(t-1)) C_w(t-i) + \sum_{i=1}^{n_v} \phi_i^v(\mathrm{X}(t-1)) V_s(t-i) + \xi(t) \\ \phi_0(\mathrm{X}(t-1)) = c_0^0 + \sum_{k=1}^{m} c_k^0 \exp\left\{-\lambda_k^c \left\|\mathrm{X}(t-1) - Z_k^c\right\|_2^2\right\} \\ \phi_i^j(\mathrm{X}(t-1)) = c_{i,0}^j + \sum_{k=1}^{m} c_{i,k}^j \exp\left\{-\lambda_k^j \left\|\mathrm{X}(t-1) - Z_k^j\right\|_2^2\right\} \\ \mathrm{X}(t-1) = [x(t-1), x(t-2), \ldots, x(t-n_w)]^T \\ x(t-l) = [c_w(t+\tau-l), v_s(t-l)],\ l = 1,2,\ldots,n_w \\ Z_k^j = [z_{k,1}^j, z_{k,2}^j, \ldots, z_{k,n_x}^j]^T, \quad j = c,v \end{cases} \tag{10}$$

Where, $C_w(t)$ is concentration of slurry, $V_s(t)$ is swing speed, $\xi(t)$ is white noise array, X(t-1) is state vector which contains combination of input variable of slurry concentration and output variable of swing speed; n_c, n_v, m, n_w are the orders of model; $\phi_0(\mathrm{X}(t-1))$ and $\phi_i^j(\mathrm{X}(t-1))$, $i = 1,2,\ldots n_j$; $j = c,v$ are state dependent functional coefficients which are approximated by a group of RBF network respectively; $Z_k^j, (k = 1,2,\ldots,m; j = c,v)$ is RBF network center; $\lambda_k^j (k = 1,2,\ldots,m; j = c,v)$ is proportion coefficient of RBF network; $c_k^0 (k = 1,2,\ldots,m)$, $c_{i,k}^j (k = 1,2,\ldots,m; i = 1,2,\ldots,n_j; j = c,v)$ are linear weights of RBF network; $\|\bullet\|$ represents vector quadratic norm. τ is lagging time of slurry concentration.

3.3. Parameter identification of swing process model

Basic procedure of parameter identification of swing process model is shown in the following figure 2.

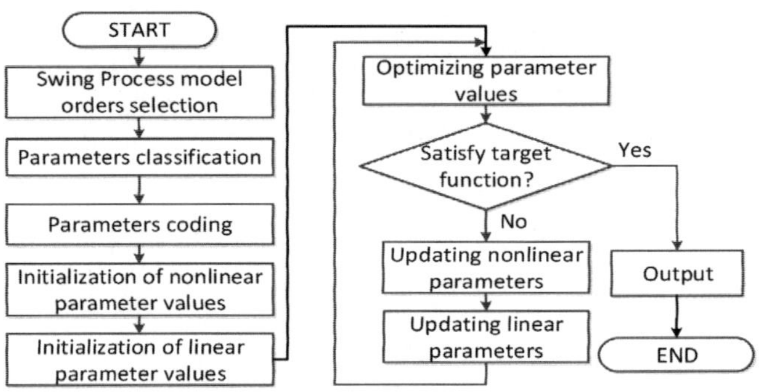

Figure2. Flow chart of parameter identification of swing process model

3.3.1. Determination of orders of swing process model m and n_w are selected based on experience, typically m is 1~2, in this paper $m = 1$, n_w is typically selected as 2. Then use Akaike Information Criterion, AIC:

$$AIC = N\log V + 2d \tag{11}$$

Where, V is the covariance of model, N is the length of observed data. d is the sum of numbers of all linear parameters and nonlinear parameters to be identified. Because the smaller the AIC value, the better swing process model output and actual value fit, it can be found from Figure3 that by comparing AIC value, combining model characteristics, they can be selected $n_c = 3, n_v = 2$.

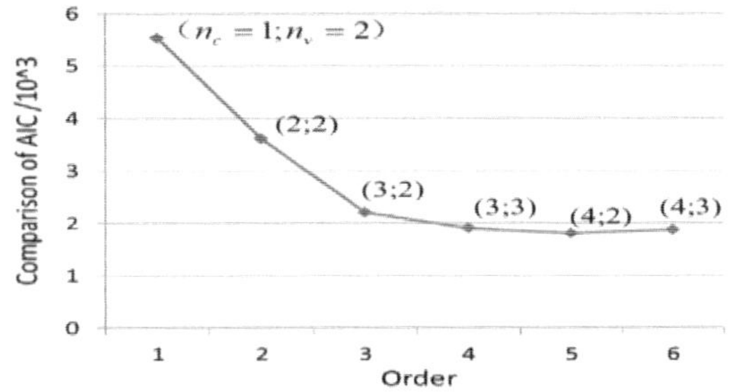

Figure3. Comparison curve of AIC value of different order

3.3.2. Determination of linear parameter and nonlinear parameter Next model parameters will be identified using measured data of system global input/output, the result of linear parameter and nonlinear parameter of parameter identification is shown in Table 1.

Table 1. Result of parameter identification of swing process model

Parameter class	Symbol	Recognition values	
Linear parameters		-3.36E-09	5.62E-06
$\theta_L @\{c_k^0, c_{i,k}^j\}$		0.003327283	0.013614834
$k = 1,2,...,m$;	θ_L	1.000000344	1.20E-09
		120.8401331	-178.809188
$i = 1,2,...,n_j$;		8.72E-06	2.18E-10
$j = y,u,v$		246.9656647	19.50627259
Nonlinear parameters			
$\theta_N @\{\lambda_k^j, Z_k^j\}$	λ_k^j	0.048	0.11
$k = 1,2,...,m$;	Z_k^j	1.13E-45	0.186836255
$j = y,u,v$		21.875732	6.68505241

4. Test of swing process model

To test prediction capacity of established swing process model, select the data of the same cutter suction dredger measured at another moment, with length of 4000s. Comparison diagram of swing process model prediction output and measured data is shown in figure 4.

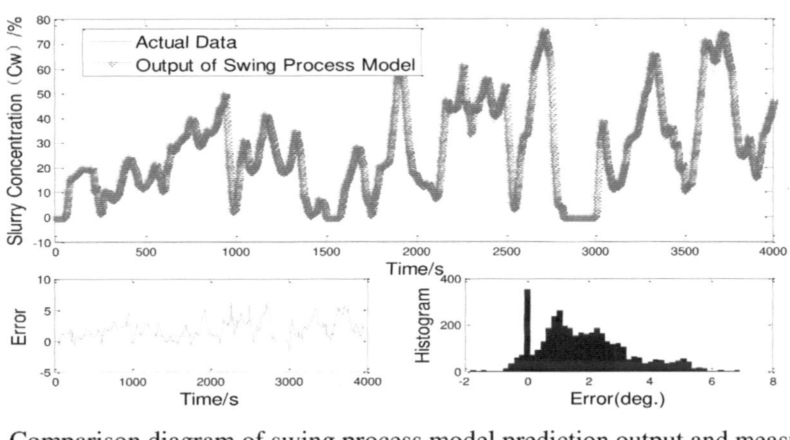

Figure4. Comparison diagram of swing process model prediction output and measured data

It can be found from figure4 that variation of predicted value and measured value of slurry concentration have the same overall trend, prediction result is relatively accurate. But comparing with measured data, model output result of slurry concentration has a larger error at higher and lower concentration, because actual dredging process is very complex, those factors such as flow rate in pipe, cutter power and vacuum can influence slurry concentration to different extent. In this paper only swing speed is used as control input, and state variable only includes swing speed and slurry concentration itself, later such factors as flow rate, cutter power and vacuum will be included in the process of model establishment.

5. Conclusion

In the paper swing process model is established using RBF-ARX model without needing accurate parameter relation. By directly starting from the data measured by the dredger, through RBF network and SNPOM combined parameter identify algorithm, swing process modelling is transformed into solving linear ARX model of the operating point, and these models are integrated into a global model issue through functional coefficient. Through simulation the model is tested to clearly describe swing process dynamic characteristics, also laying a foundation for next design of swing controller using prediction control theory.

Acknowledgments

This research was supported by the special fund of Jiangsu Province for the transformation of scientific and technological achievements (NO.BA2016143), and also supported by the Fundamental Research Funds for the Central Universities (NO.2017B21014). The master student of Wei Dai assist to complete the simulation

References

[1] NI Fu-sheng. *Development of International Dredging Equipment* [J]. Journal of Hehai University Changzhou, 2004(1):1-9.
[2] WANG Yan, XIE Hai-ying, WU Ying. *Present State and Development of Domestic and International Dredging Markets* [J]. JOURNAL OF SSSRI, 2004, 27(2):109-113.
[3] Jelmer Braaksma. *Model-Based Control of Hopper Dredgers* [D]. TU Delft, 2008:5-6.
[4] Tang Jianzhong, Wang Qing feng, Zhang Qi, Bi Zhiyue. *Theoretical and Experimental Research on Cutter Suction Dredging Expert System*[J].China Mechanical Engineering, 2007,9:2218-2212.
[5] Bi Zhiyue. *Reaserch on Efficiency Optimization and Control for Dredging Slurry Pipeline Transport System*[D].Zhejiang Hangzhou, Zhejiang University, 2008:16-18.
[6] Li Zhigang. *Research of Cutter Suction Dredger Swing Automation Based on Swing Resistance* [D]. Hohai University, Changzhou Jiangsu, 2015:7-8.
[7] NI Fu-sheng, ZHAO Li-juan, GU Lei, et al.. *Simulation of Dredging of Cutter Suction Dredger*[J]. Journal of System Simulation, 2012,24(11):2271-2274.
[8] ZHU Wen-liang, NI Fu-sheng, WEI Chang-yun, et al. *Linear Quadratic Optimal Control of Cutter Suction Dredger during Swing Process*[J]. Journal of System Simulation Ship Engineering, 2016,38(2):24-32.
[9] Peng Hui, Tohru Ozaki, Valerie Haggan-Ozaki, et al. *A parameter optimization method for Radial Basis Function type models*[J]. IEEE Transactions on Neutral Networks, 2003,14(2): 432-438.
[10] Hui Peng, Zi-Jiang Yang, Weihua Gui, et al. *Nonlinear system modeling and robust predictive control based on RBF-ARX model*[J]. Engineering Applications of Artificial Intelligence, 2006,20(1):1-9.
[11] GAN Min, PENG Hui. *Effect of different basis functions on RBF-ARX model*[J] Journal of Central South University (Science and Technology) , 2010,41(6): 2231-2235.

IOP Conf. Series: Materials Science and Engineering **428** (2018) 012019 doi:10.1088/1757-899X/428/1/012019

The Study of Ocean Ambient Noise Reconstruction in Low and Medium Frequency Bands

Zhang Liang[1,a], Wen Fengdan[2,b], Meng Chunxia[3,c] and Han Jing[4]

[1,2,3,4] Science and Technology on Underwater Test and Control Laboratory Dalian, 116013, China

[a]53359229@qq.com, [b,c] dlinstitute@163.vip

Abstract. Ocean ambient noise is the general background of the sea, the sources of which is highly variable, the generated noise signal is random, it changes with time and space. It is usually considered that ocean ambient noise is a special random process with certain power spectrum, so the broadband stationary random process can be used for fitting the continuous spectral noise. This paper carried out statistical analysis on sample data obtained by continuous observation in one day, and then the statistic characteristics of ambient noise frequency spectrum in low and medium frequency bands can be calculated. Generating filter with specific frequency response based on the probability density function of sound pressure level in each frequency randomly, and ocean ambient noise is reconstructed effectively by Gaussian white noise passing the filter above. The simulation results have important implication for low and medium frequency underwater acoustic signal analysis in lab conditions, and it is helpful to test the performance of platform in the oceanic background.

1. Introduction

Ocean ambient noise may be said to be the noise of the sea itself, the sources of which is highly variable, it is a mixture of several different types of noise: ocean dynamic noise, biological noise, shipping and industrial noise, thermal noise[1] and so on. The characteristics of ambient noise varies with different frequency bands[2], its spectrum is composed of segments of different slope having a different behavior under different condition such as wind speed or ocean depth[3].

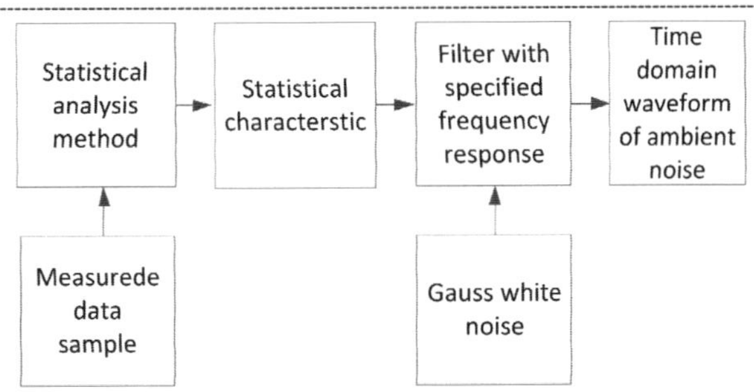

Figure 1. Flowchart of ocean ambient noise reconstruction

Ocean ambient noise is the background of the sea [4], the properties of oceanographic equipment are different at the different marine environment [5]. Chinese scholars have conducted a significant amount of researches on ambient noise in shallow water over the last decades [6], less research has been done in deep-sea environment owing to the lack of measured data [7], the properties of ambient noise in deep-sea still need to be studied [8]. In this paper, base on lots of ambient noise data measured in deep-water, the frequency spectrum characteristic of ocean ambient noise is analyzed statistically, and the time domain waveform of ambient noise is reconstructed, the entire flow is illustrated in Figure 1.

2. Statistical analysis of ocean ambient noise in low and medium frequency

2.1. Description of ocean ambient noise data
In the summer of 2015, ocean ambient noise data of different time period in one day was obtained during an underwater acoustic experiment in the South China sea. Hydrophone is located about in 10 meters underwater, the sensitivity of which is -172dB, it's data sampling rate is 12kHz. In the course of that day, oceanic condition was 2 grades, the weather was clear, and the number of boat around the experimental sea area is smaller.

The effective ocean ambient noise was selected to perform time frequency analysis, Figure 2 shows the time-domain signal of ocean ambient noise, which was measured in a certain period of time.

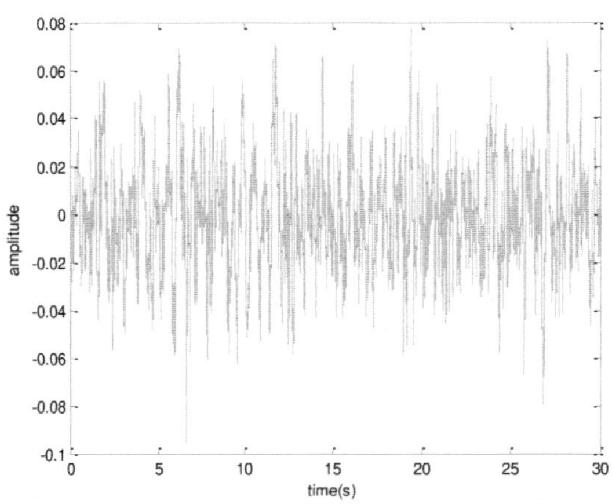

Figure 2. Time-domain waveform of ocean ambient noise

2.2. Statistical analysis method of ambient noise
Ocean ambient noise signal is random, and the energy is discontinuous distributed. The random signal cannot be expressed by a certain time function, for this reason, ocean ambient noise's level can only be expressed by statistical characters for a certain frequency.

- Mathematic expectation, describe the average value of a random signal. Supposing the experiments times is n, the times with measurement results of x_1 is m_1,and with results of x_2 is m_2, and so on, the times with measurement results of x_k is m_k, the average is shown below.
- Variance yields, describe the variation intensity of random signal amplitude, it reflects the fluctuation of random variable's value. Supposing X is a random variable, and variance of X exists.
- Probability density function.The probability density function of one-dimensional real random variable describes the possibility of random variable's output quantity near a certain value point.

2.3. Statistical properties of ocean ambient noise
Selecting the ambient noise data obtained by continuous observation in one day, which length in each time period is 10 seconds, and then frequency spectrum analysis is adopted for single sample data with

frequency resolution less than 1Hz. The change rule of ambient noise spectrum in different time period of one day is shown in figure 3.

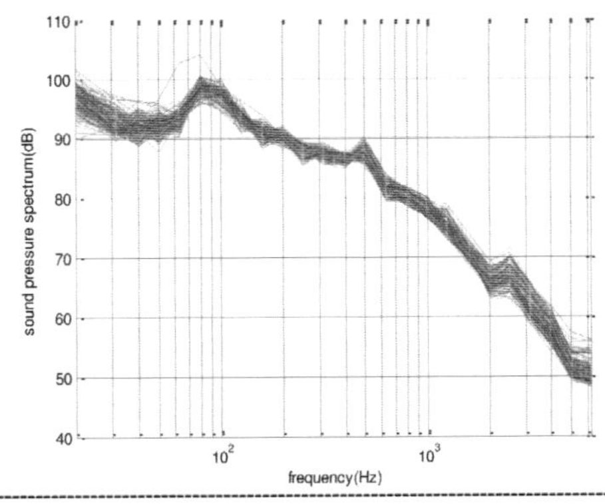

Figure 3. The statistic result of sound pressure level in different time period

As can be seen from the figure, the distribution of ambient noise in different frequency remains consistent roughly in different time period, the maximal value of spectra level in 100Hz may be caused by ship traffic. The result shows that ambient noise level changes over time in one day, the variation reach up to 10dB approximately in low frequency band, which may be caused by the changes of surface wave and current all the time.

Some statistical feature can be obtained by performing statistical analysis on sound pressure level of all the data samples, such as average, variance, probability density function and so on. Figure 4 shows the average and variance of sound pressure level at different frequency point.

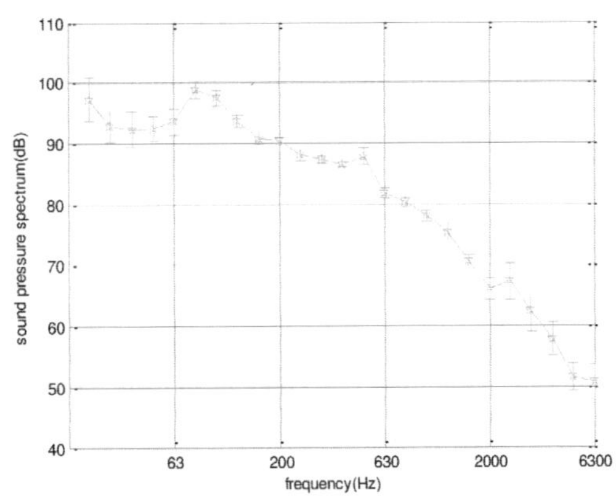

Figure 4. Sample mean and variance

As is shown in figure 4, the variation tendency of ocean ambient noise measured in South China sea is roughly same with Wens-noise-level in summer: the noise level is high in low-frequency band, and is low in mid-frequency band, the maximal value is about 100Hz. The analysis results in this paper show that ocean ambient noise level fluctuate obviously in frequency less than 100Hz and more than 2kHz, the variation of which in frequency from100Hz to 2kHz is small.

Acoustic pressure spectrum of each data sample at each frequency point is calculated, and the probability density function is obtained, as an example, figure 5 shows the probability distribution of ocean ambient noise at frequency of 200Hz.

As can be seen from the analysis results in figure 5, the probability distribution at 200Hz is similar to Gaussian distribution, which is related to the randomness of ocean ambient noise.

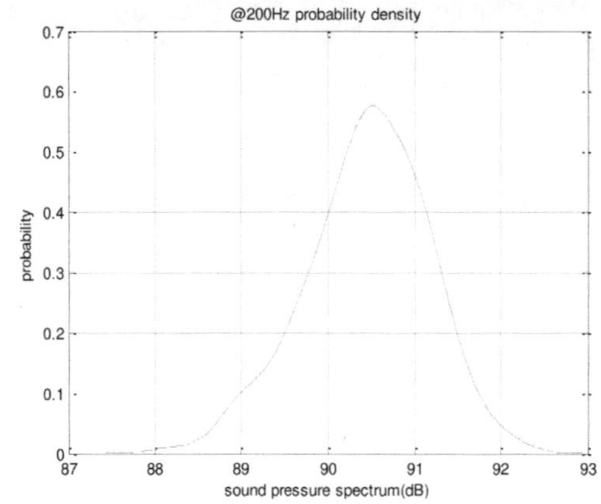

Figure 5. The probability density of ocean ambient noise in 200Hz

Basing on these analyses, the probability density distribution can be calculated, as is shown in Figure 6.

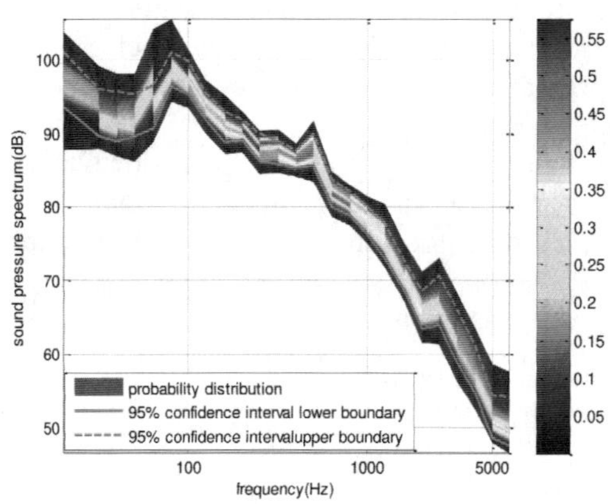

Figure 6. Sample probability density

Figure 6 shows the probability distribution of ambient noise pressure level at different frequency point, and the changing curve of 95% confidence intervals can also be found in the above picture. As is shown, the sample is dispersive in frequency less than 100Hz and more than 1 kHz; this situation is more obvious in low frequency band.

3. Algorithmic generation of Gaussian white noise random sequence

There are many methods of generating Gaussian white noise random sequence, Box-Muller method is a typical in signal processing at present, the long period Gaussian white noise random sequence obtained by this method is authentic enough.

4. Generating filter with specific frequency response based on measured data

By using the statistical analysis algorithm, the random variable in each frequency can be generated randomly, which satisfies the probability density function, the random variable can be treated as the amplitude-frequency value of the filter with specific frequency response. Figure 7 shows the filter with specific frequency response generated randomly. It is worth reminding that ocean ambient noise spectrum in the acoustic field is dynamic, each of the reconstructed frequency spectrum satisfy statistical regularities.

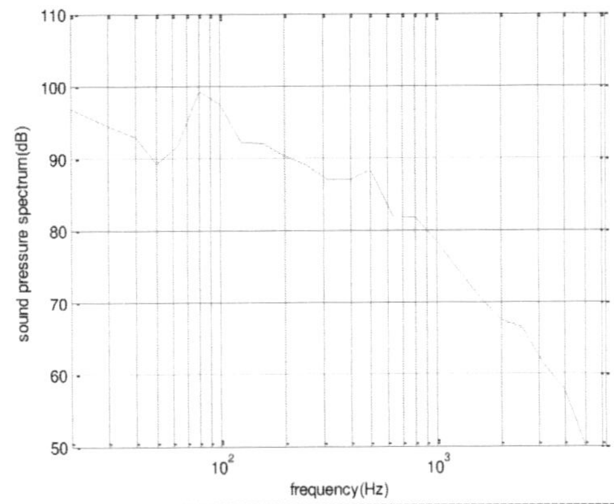

Figure 7. Spectrum of ocean ambient noise reconstructed by measured data

5. Signal reconstruction of ocean noise background in time domain

In order to get the signal of ocean noise background in time domain, filter with specific frequency response in part 3 is adopted to filter the white noise in part 2. The generated time-domain signal meet the requirement of ocean noise background, which amplitude satisfies the statistical regularity, and the phase angles are randomly distributed.

6. Conclusions

In this paper, basing on statistical analysis, lots of measured data were used to study the reconstruction method of middle-frequency and low-frequency ambient noise in deep water. Filter with specific frequency response is adopted to filter the Gaussian white noise, and the time-domain waveform of ocean ambient noise was obtained by simulation, the waveform meets the requirement of spectrum statistical characteristic.

Ocean ambient noise has a variety of different sources, in the deep water, sound speed profile together with the sea and its boundary, forming a remarkably complex medium for the sound propagation; all of them have great influence on acoustic energy emitted from the sources. The sound vertical velocity varies with the seasons, and the surface roughness is different under different sea conditions. From the above mentioned, therefore, the method proposed in this article can be used to analyze the ocean ambient noise in different seasons and different sea conditions, so that the statistical property has a universal meaning. Further researches are still needed.

Acknowledgments

This study was funded by the Defense Advanced Research project (No. 414160603) and foundation research project (No. JCKY2016207A037). The authors would like to thank colleagues for technical assistance with acoustic data collection, and the organizers are thanked very much by the author for providing the international academic exchange opportunity.

References

[1] W.A.Kuperman, F.Ingenito and Laurie T.Fialkowski. Modeling Ambient Noise in Three-Dimensional Ocean Environments . *J.Acoust.Soc.Am.* 1993,vol 93, pp739–52

[2] YANG Shi-E, YANG YIXIN and MA YUANLIANG, Investigation of the Horizontal Directivity of Underwater Ambient Noise with the circular Hydrophone Array, *ACOUSTICS 2012* .HONGKONG, p20

[3] R.L.Weaver and O.I.Lobkis. Elastic wave thermal fluctuations, ultrasonic waveforms by correlation of thermal phnon. *J.Acoust. Soc.Am* . 2003,vol 113, pp2611–21

[4] G.M.Wenz. Acoustic ambient noise in the ocean spectra and sources. *J.Acoust. Soc.Am.* 1962, vol 34, pp1936–56
[5] R.M.Hamson. The modelling of ambient noise due to shipping and wind sources in complex environments. *Applied Acoustics* .1997, vol 51, pp251–87
[6] SHI Jie, ZHANG Xiaomin, HOU Tieshuang, et al. Noise level evolution of low-frequency ocean ambient noise caused by ship sources. *TORPEDO TECHNOLOGY* .2010,vol 18, pp112–6
[7] Stephanie E.Fried, Shane C. Walke,William S.Hodgkiss, and W.A.Kuperman. Measuring the effect of ambient noise directionality and split-beam processing on the convergence of the cross-correlation function. *J.Acoustic. Soc.Am.* 2013, vol 134, pp 1824–32
[8] O.I.Lobkis and R.L.Weaver. On the emergence of the Green's function in the correlations of a diffuse field. *J.Acoust. Soc.Am* . 2001,vol 110, pp3011–7

IOP Conf. Series: Materials Science and Engineering **428** (2018) 012020 doi:10.1088/1757-899X/428/1/012020

Analysis on Mars Atmospheric Composition Probing Technology in the Future

Man Li*, Yuming Liu and Dongbo Tian

Beijing Institute of Spacecraft Environment Engineering, Beijing, China

*li-man8319@163.com

Abstract. Mars atmospheric composition probing is one of critical tasks of mars exploration. The mars atmosphere may have the existence of water, methane, organic matter, conducting mars atmospheric composition probing is very critical to studying whether the mars has, or once had the existence of living matter, analyzing the forming, evolution of mars atmosphere as well as its change with season. Restricted by launching technology, platform technology, mars exploration task has very strict requirement to weight, volume and power consumption of probing load. At present, the mars atmospheric composition in-situ probing load in abroad usually adopts mass spectrometry with load weight from several kilogram to tens kilogram, and power consumption from several watts to tens watts. Carbon Nanotubes (CNTs) has large surface area with sensitivity to gas absorption reaction, and can be used to prepare the high performance gas sensor with high sensitivity, quick response, light weight, small volume, and low power consumption, and it is possible to realize atmospheric composition probing load with 100g weight and milliwatt power consumption. CNTs gas sensor technology shall be the critical development trend of mars atmospheric exploration technology in the future.

1. Introduction

Mars exploration is an important field of global spaceflight activity at present. For a period of time in the future, the trend of mars exploration shall be inclined to probing of living matter. According to near term launching and future planned launching of deep space probes in abroad, its main objective is to study whether the mars has the existence, or once had the existence of living matter, while mars atmospheric composition in-situ probing load is one of main means to realize this objective.

The mass spectrometer is an important gas analytical instrument and has played a critical role in mars exploration task already conducted. Nevertheless, with gradual development of scientific understanding and development of basic technology, like micro-electronics and new type material, the probing load gradually developes towards light and small type, long service life, high reliability, and high integration. The load function is gradually increasing, while the weight and power consumption is gradually decreasing, and this is an important direction of mars probing load technical development in the future. Therefore, the atmospheric composition probing load on the basis of a new type technology may distinguish itself in mars atmospheric in-situ probing in the future.

On the basis of analyzing Mars atmospheric composition probing load in abroad, this paper focuses on development and application of new technology, researches the application of CNTs gas sensor in Mars atmospheric composition probing, and discusses the technical bottleneck of carbon nano tube gas transducer.

2. Analysis on mass spectrometer for superior planet atmospheric composition probing

According to difference of scientific probing goals and probing objects, and impact of load volume, weight, power consumption, sensitivity, resolution, at present the mass spectrometer for probing planetary atmospheric composition mainly includes magnetic deflection mass spectrometer, quadrupole mass spectrometer, time of flight mass spectrometer, ion trap mass spectrometer, and they were all once used for planetary atmospheric composition probing.

2.1. Magnetic deflection mass spectrometer

Viking 1 and Viking 2 launched in 1976 adopted the uniform structure magnetic deflection spectrometer to analyze surface atmosphere of Mars[1]. The mass spectrometer was of Nier-Johnson structure, with one electron impact ion source of sensitivity 10^{-8}/pa, and one double focusing mass analyzer. The instrument weight was 10kg, ion mass number probing range was 1~230amu.

The Phoenix Mars Lander, which arrived at Mars in 2008, carried a thermal evolved gas analyzer TEGA, the core of lander was a magnetic deflection mass spectrometer for analyzing volatile substance content after heating the soil of Mars, and determining main composition of Mars atmosphere[2]. The structure of mass spectrometer was shown in Fig.1. The weight of mass spectrometer was 4.3kg, mean power consumption was 13W, the magnetic analyzer included four channels with related mass number range 0~14aum, 7～35amu, 14～70amu, 28～140amu.

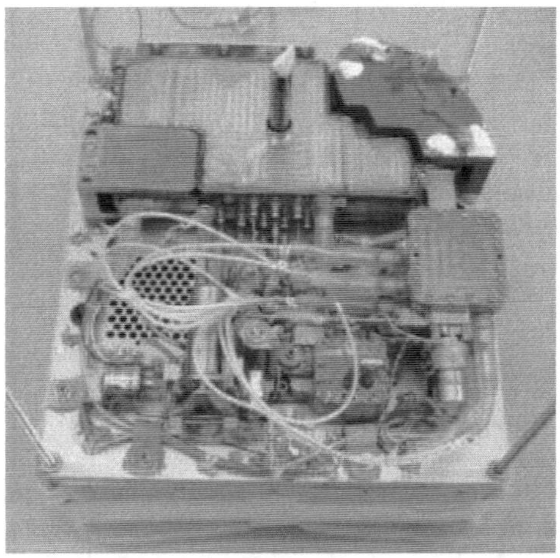

Figure.1 The structure of Phoenix mass spectrometer

2.2. Quadrupole mass spectrometer

Galileo Jupiter Probe, which was injected into Jupiter orbit in 1989, carried the quadrupole mass spectrometer[3]. The weight of mass spectrometer was 13.2kg, mean power consumption was 13W. The mass number measurement range of mass spectrometer was 2~150amu, and the resolution was higher than 1ppb. Accuracy of He partial pressure proportional measurement was 1%, accuracy of H_2O, CH_4, NH_3 partial pressure probing was 5%.

The Huygens Probe of Saturn also carried quadrupole mass spectrometer to measure the Titan atmospheric composition[4]. The mass spectrometer of Huygens Probe adopted the aspirator pump and an ion sputtering pump to maintain cleanness of vacuum chamber. The weight of mass spectrometer was 17.3kg, mean power consumption was 28W, mass number measurement range was 2~150amu, mass resolution was 10^{-6} for 60amu.

2.3. Time of flight mass spectrometer

The Rosetta Spacecraft, which was launched by ESA in 2004, carried a test pack ROSINA[4]. The test pack was fitted with a time of flight mass spectrometer to detect the organic matter as polycyclic aromatic hydrocarbon carbohydrate. The structure of mass spectrometer was shown in Fig. 2. The weight of mass spectrometer was 15kg, mean power consumption was 30W, mass number measurement range was 1~300amu.

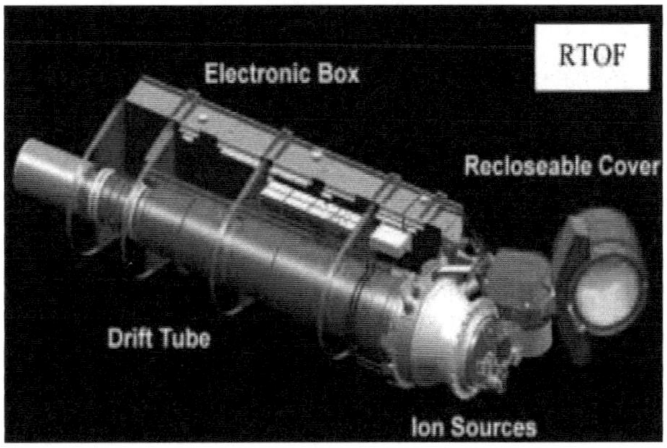

Figure.2 The structure of Time of flight mass spectrometer

2.4. Ion trap mass spectrometer

The Rosetta Spacecraft of ESA was also fitted with an ion trap mass spectrometer to probe the gas with mass number 1~150amu.

In the design of Extraterrestrial Mars Program of ESA, the Mars organic mass analyzer (MOMA) shall be carried, its core is an ion trap mass spectrometer. The mass number probing range can be extended to 2000amu, and it can be used to probe the organic matter with high molecular weight.

At present the weight of various mass spectrometers ranges from several kg to tens kg, power consumption from several watts to tens watts, and they are mainly used to probe the gas molecule with small mass number. The mass spectrometer is also confronted with some technical bottlenecks for Mars atmospheric probing. The development and application of some latest technologies is hopeful to surpass mass spectrometer to greatly improve the technical level of Mars atmospheric probing.

3. Introduction of CNTs gas sensor

In 1991, Iijima the scientist of NEC Company of Japan found a new carbon structure [5], as shown in Fig.3.

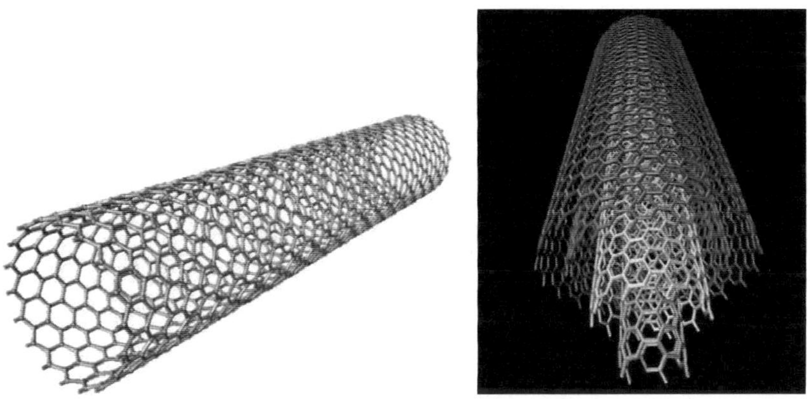

Figure.3 The structure of CNTs

The electrical property of CNTs seriously depends on outermost layer carbon atom, after combination of external layer carbon atom with gas molecule, electrical property of CNTs may change, therefore CNTs can be used as gas sensor to probe ppm ($1/10^6$) and even ppt ($1/10^{12}$) concentration gas[6,7]. CNTs sensor can be sorted as semiconductor sensor for spatial atomic oxygen detection, in contrast with existing atomic oxygen detector. It is featured by quick response speed, high sensitivity, small volume, low power consumption, high accuracy, and has a huge technical advantage. At present AMES research center of NASA of USA has manufactured 32 channel CNTs sensor unit for some satellite of USA navy, for detecting NO_2 concentration[8] of orbit, and realizing spatial application of CNTs sensor.

3.1. Working principles of CNTs gas sensor

The traditional gas sensor is a gas sensor that causes a change in the physical and chemical properties of the sensor unit after contact with the sensitive unit of sensor, and reacts to the output of the electrical signal to be determined by the change in the size of the electrical signal.

CNTs gas sensors can be divided into resistor /conductivity type and capacitor type according to the different electrical signals detected. At present, the most majority of CNTs gas sensors analyze gas concentrations by measuring changes in resistance / conductivity of CNTs [9].

As shown in Figure 4, when the gas molecules under test are adsorbed on the surface of the CNTs, the resistive/conductivity regularity of the CNTs will be changed, thereby detecting the gas composition and content.

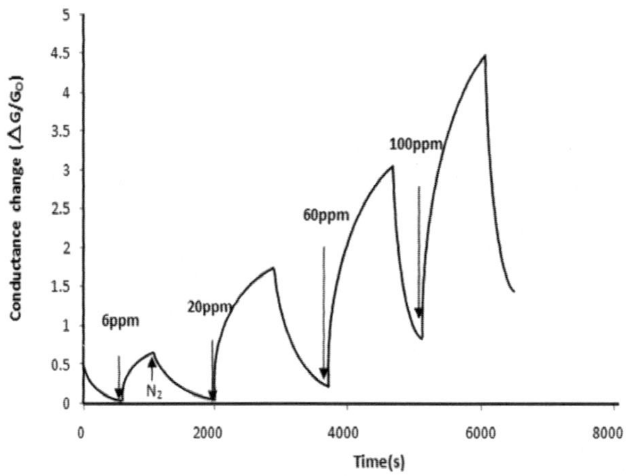

Figure.4 Representative sensor response for NO_2[9]

3.2. Performance of CNTs gas sensor

The main performance parameters of CNTs gas sensor include:

- Sensitivity-the minimum content of gas that can be detected;
- Response time-the time from start of gas environment change to stabilization of electrical signal change of sensor;
- Recovery time- the time from eliminating environment of gas under detection, to the time when the electrical performance of sensor is recovered to original value;
- Selectivity- the sensor only has response to certain category gas without response to other gases. This property determines whether the sensor can be used for probing the mixed gas consisting of multiple gases.

Carbon nanotubes used in CNTs gas sensor can be divided into two major categories, they are pristine carbon nanotubes and functionalized carbon nanotubes. Pristine carbon nanotubes sense gases by the properties of pristine carbon atoms on the surface of carbon nanotubes Functionalized carbon nanotubes are prepared by chemical or physical modification. Their surface carbon atoms are combined with other atoms or molecules to function.

Pristine CNTs gas sensor has the following disadvantages: poor selectivity, low sensitivity and the response time and recovery time are long.

IOP Conf. Series: Materials Science and Engineering **428** (2018) 012020 doi:10.1088/1757-899X/428/1/012020

Many gases can influent the electrical properties of pristine carbon nanotubes. Therefore, when the CNTs sensor is exposed to a mixture of unknown components, the sensor will respond, but it can not distinguish these gases and can not determine the concentration of the gas. Even under the known main gas composition conditions, the gas concentration can not be accurately judged. The sensitivity of carbon nanotubes to different gas varies greatly, ranging from ppb level to more than 100 ppm level. Even if the same gas is detected, the results obtained by different research teams are also different. It shows that the sensitivity of the sensor has a great relationship with the preparation process of the carbon nanotube sensor and is not easy to control. In practical applications, ppt level gas concentrations need to be detected. The performance of sensors should be relatively stable. Pristine CNTs sensor is not enough to meet such requirements.

For most gases, the response time and recovery time of the carbon nanotube sensor are relatively long, taking several minutes or even ten minutes. For certain gases, the performance of the pristine carbon nanotube should not be fully restored, and the sensor should not be reused.

Functionalized CNTs gas sensor can detect more kinds of gases. The performance indexes such as sensitivity, response time, recovery time, and selectivity are superior to those of pristine carbon nanotubes. Therefore, this kind of sensor has more practical application value. A more effective method at present is to functionalize the surface of the CNTs with atoms, molecules, functional groups. The functional combination of CNTs is divided into two types: covalent bonding and non-covalent bonding.

4. Analysis on feasibility of applying CNTs gas sensor to mars atmospheric composition probing

4.1. Feasibility analysis

For mars atmospheric probing, the main probing objective gas includes CO_2, H_2O, CH_4 and organic molecule, etc. While CNTs sensor has realized detection to such gas, e.g., CH4 with available detection concentration 6ppm, ethyl alcohol, glucose, etc. Furthermore, according to requirement of objective gas under probing, more gas probers can be developed. Therefore, CNTs sensor is hopeful to realize probing of objective gas of mars.

4.2. Analysis on challenge of technology

The mars atmosphere has multiple gases which may all influence electrical signal of sensor, therefore the sensor must have fairly strong selectivity, during design and manufacturing of CNTs sensor, CNTs shall be specially treated so that electrical property of CNTs shall only respond to certain gas composition instead of other gases. The CNTs sensor with different gas selectivity shall be integrated into a chip to recognize the gas category and content in the mars atmosphere. Performance index of CNTs gas sensor shall be further improved in accord with mars atmospheric probing requirement, although CNTs gas sensor has very high sensitivity to some gas perception and detection, sensitivity of CNTs gas sensor is not yet ideal for most gases. The mars atmosphere has a variety of gas contents, wherein, some gas has extremely low content and shall not be detected until the sensor has fairly high sensitivity. Therefore, according to mars atmospheric probing objective, high performance gas sensor shall be researched and manufactured, to improve perception capability of sensor to gas via certain CNTs functional method.

Researching CNTs gas sensor performance in the lab is usually conducted at room temperature. The environment condition is fairly simple, while mars atmospheric environment differs greatly from lab environment, the atmosphere mainly includes CO_2 with high, low temperature environment and radiation environment, etc. The influence of these environmental factors to CNTs sensor is still unclear. Environment effect research of sensor is almost a blank, therefore sensor environmental adaptability research and protection design research shall be enhanced to resolve the environment problem of CNTs gas sensor engineering application.

The current created of CNTs gas sensor is only microamp level and power consumption microwatt level, but the electrical signal measurement, treatment system as well as data transmission system may greatly increase the weight and power consumption of mars atmospheric probing load system. Therefore, the system design shall be optimized in accord with electrical signal characteristic of CNTs sensor to realize light and small type orientation of the whole system.

5. Conclusion

The mars atmosphere in-situ probing is one of important constituted parts of mars exploration in the future. The deep space probing has strict requirement to dimension, weight and power of probing load. Although various mass spectrometers have plenteous performances for planetary atmospheric probing, the miniaturization, low power consumption and high performance of mass spectrometer shall be further improved to satisfy the requirement of deep space exploration. CNTs gas sensor has the advantages of high probing sensitivity, quick response speed, a variety of gases under probing, small volume, low power consumption, simple structure, batch production, low cost. Therefore CNTs gas sensor shall be the ideal probing load for mars atmosphere in-situ probing in the future.

References

[1] Biemern K, Oro J, Toutrin P. J. Geophys. Res, 1977, 82:4641-4658
[2] John H. Hoffman, Roy C. Chaney, Hilton Hammack. Phoenix mars mission-The thermal evolved gas analyzer. J. Am. Soc. Mass Spectrom, 2008, 07:15
[3] Theresa Evans-Nguyen, Luann Becher, Vladimir Doroshenko, Robert J. Cotter. Development of a low power, high mass range mass spectrometer for mars surface analysis. International journal of mass spectrometry. 2008, 278:170-177
[4] Chicarro A, Clavel J, Escoubet C P, Favata F, Foing B, Fridlund M, Gimenez A. ESA's Report to the 35th COSPAR Meeting. Paris, France. July, 2004
[5] S.Iijima. Helical microtubules of graphitic carbon[J]. Nature,1991,354 pp. 56-58.
[6] P.Qi, O.Vermesh, M.Grecu, et al. Toward large arrays of multiplex functionalized carbon nanotube sensors for highly sensitive and selective molecular detection[J]. Nano Lett.,2003, vol. 3, pp. 347–351.
[7] Snow E R, Perkins F K, Robinson J A. Chemical vapour detection using single-walled carbon nanotubes. Chemical Society Reviews, 2006, vol. 35, pp.790-798.
[8] http://www.nasa.gov/centers/ames/home/index.html.
[9] Li P F, Martin C M, Yeung K K et al. Dielectrophoresis aligned single-walled carbon nanotubes as pH sensors. Biosensors, 2011 vol.1, pp.23-35.

IOP Conf. Series: Materials Science and Engineering **428** (2018) 012021 doi:10.1088/1757-899X/428/1/012021

Harnessing Frequency and Language Features for Keyword Extraction on E-commerce Platforms

Chit Khine[1] and K Nongpong[1]

[1] Intelligent Systems Laboratory (ISL), Vincent Mary School of Science and Technology, Assumption University of Thailand, Bangkok 10240, Thailand

E-mail: ck.light@gmail.com; kwan@scitech.au.edu

Abstract. Automatic keyword extraction has become essential with the growing number of internet resources. This method aims to extract quality keywords that are relevant to products in e-commerce platforms. The problem with e-commerce products is that there are wide ranges of categories in which a product can be in it. We propose a technique to improve the extraction of keywords from products information for the e-commerce systems by applying both frequency-based and language-based features. For frequency-based features, we consider the fact that products in the same category may have popular keywords which are different from other frequency-based features. For language-based features, different types of noun phrases are extracted and ranked accordingly. In this work, the proposed category-based document frequency is combined with the traditional TFIDF and noun phrases ranking. The approach is evaluated using product descriptions from Amazon. The results show that our approach performs better than the traditional TFIDF and RAKE by at least 10 percent on various categories of e-commerce products.

1. Introduction

The requirement for retrieving the precise information of text has become the crucial task in information retrieval. Instead of reading a long paragraph or a document, concise keywords can give us the meaning of the whole information. This statement is also true for products where we need to provide the important information for both customers and providers. Many e-commerce businesses need to associate products with keywords to facilitate product search or for product recommendation. However, manual keyword labelling is a tedious and troublesome process. The main challenge of this problem is how to extract quality keywords that best represent and are most relevant to the products themselves.

Regardless of the domain, the automatic keyword extraction can be divided into three important processes: (1) preprocessing, (2) keyword or keyphrase extraction and (3) keyword evaluation. Pre-processing techniques are usually applied to the documents before performing the actual extraction of the keywords to remove the noises from the documents *i.e.* product descriptions in this case. Pre-processing techniques include stop word removal [1-3], stemming or lemmatization [4-6], and n-gram creation techniques [7-8]. Stop word removal is a technique that filtered out the words that are common in the language and provides very little meaning to the document. Stemming or lemmatization generates base form the words using either the linguistical or morphological analysis of words. "N-gram is a set of n following characters extracted from a word. The main idea behind this approach is that similar words will have a high quantity of *n*-grams in common. For *n* equals to 2 or 3, the words extracted are called digrams (bigrams) or trigrams, respectively" [2-3].

Extraction techniques are usually categorized into supervised [9] and unsupervised techniques [10]. Supervised techniques can be defined as choosing the best keywords from the prepared set of keywords. Unsupervised keywords are the keywords that are extracted from the structure of the document without requiring the training data. Some popular techniques for keyword extraction techniques are TFIDF [11], Part-of-Speech (POS) keyword extraction [12], RAKE [13] and KEA [13], Noun phrases and named entity recognition [14-15]. The details of these extraction algorithms will be explained Section 2.

Evaluation techniques [16-17] are divided into two categories: *manual evaluation* in which the evaluation is performed by human judges whether the keyphrases represent the document's content and *automatic evaluation* which can be categorized into exact matching, morph matching, part of/include matching). These techniques are used to evaluate the performance of extracted keywords.

2. Related works

The extraction of keywords can usually be categorized into supervised and unsupervised keywords where keywords are domain independent or linguistic approach while others can be statistically extracted. There are also approaches which combine the two approaches mentioned before.

Statistical methods are useful for extracting both single or multiple documents. The techniques for extracting keywords in statistical methods include TFIDF [11] and word-occurrence statistical methods TF [17], which stands for term frequency of words that exist in a single document. These term frequencies can be used to determine the popularity of a term in a document. However, popular words such as articles, common nouns can be seen multiple times in many documents. Therefore, the inverse document frequency is used to rationalize the frequency of the term appearing in whole list of documents. TFIDF combines these two values, tf and idf, to generalize the best key terms in documents. Word-occurrence approach is similar to TF-IDF apart from the fact the keywords are made to check the linking connections with other keywords. The number of times a keyword is occurred next to a certain keyword is used as a factor for calculating the relevancy scores of the keywords.

Linguistics approaches consider the linguistic features of the words, sentences and document [18]. Some popular techniques are using part-of-speech hierarchy [12] and noun phrases [14]. Part of speeches are commonly used in linguistical methods and rules are defined to extract different keywords based on part-of-speech hierarchy [12]. Again, 51 of the 56 part-of-speech patterns in English contains noun tags and most popular patterns from documents are made up of noun-phrases [19]. Therefore, many researchers emphasize on using noun phrases for keyword extraction.

Some popular automatic keyword extraction techniques are RAKE and MAUI [13]. These techniques provide extraction frameworks for evaluating keywords. For instance, RAKE extracts keywords by cutting out the stop words (a, an, the, about, etc) and use the remaining keywords. The rest of the extracted keywords are manually assigned to different groups; hence, different values for the keywords' relevancy scores. These can be created by corresponding algorithms using linguistic features, word co-occurrences, statistically assigned values and normally defining the group types for the extracted keywords. MAUI, an extension of KEA is a supervised machine learning method. This technique includes the candidate selection and machine learning based filtering [1]. For candidate selection, they remove stop words, punctuations and create the n-gram keywords which is followed up by defining candidate features with TFIDF, first occurrence and key-phrase-frequency.

3. Framework

Our proposed framework, as shown in Figure 1, can be divided into two main parts which is keyword segmentation where we extract the keywords from the document and keyword analysis where we provide the relevancy scores for the extracted keywords.

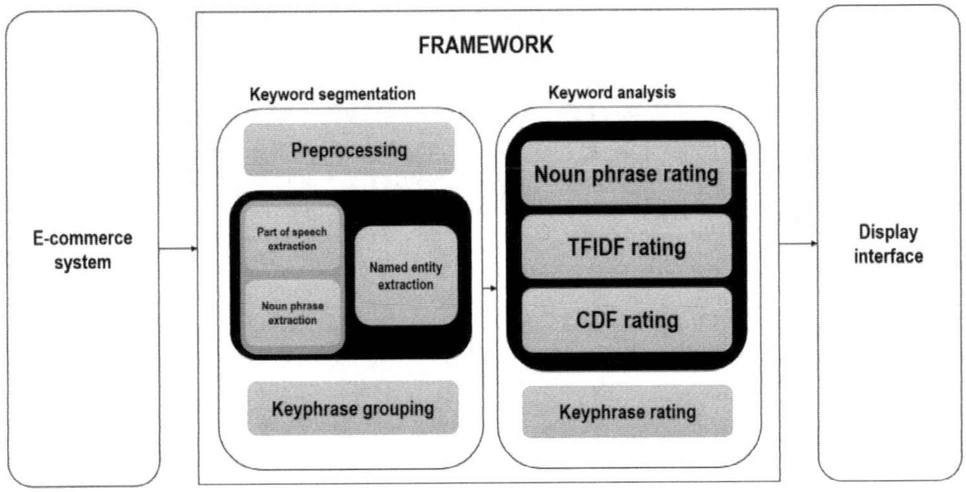

Figure 1. Framework for extracting relevant keywords and keyphrases using noun phrase weighting and frequency based techniques

3.1 Pre-processing
It is preferable to have each document as one long text of one or more sentences. Hence, we perform the formatting of document, line breaks removal and removing noises from the documents such as unnecessary white spaces. Unlike other approaches, standard preprocessing techniques are not performed in this step since we do not want to destroy the structure of sentences before extracting noun phrases. Such operations will be performed after extracting out the keywords. In this work, we assume that the input document must at least contain the product title, product description and the category that such a product belongs to.

3.2 Keyword segmentation
Steps in keyword segmentation includes part-of-speech (POS) extraction, noun phrase extraction, named entity extraction and keyword grouping.

3.2.1 Extraction of part-of-speeches. For part-of-speech extraction, we use the Stanford NLP Parser and POS Tagger to assign part of speech values to each token in the document.

3.2.2 Extraction of noun phrases. Only the noun phrases whose size are not greater than five terms are extracted as keywords. After extracting the noun phrases, we perform the following tasks.

1. Stop word removal – remove stop words which are the same as noun phrases
2. Pronoun removal (e.g. I, we, he, she, they, it)
3. Lemmatization – lemmatize keywords while retaining the part-of-speech of such words.
4. Article removal – remove articles since they have little meaning to the keywords.

3.2.3 Defining named entities. We also consider the fact that noun phrases which include named entities are generally unique in each document; hence, they should be differentiated from non-named entity noun phrases. In this step, Stanford Named Entity Recognizer is used to extract named entities.

3.2.4 Finalize keywords. After performing those three steps above, we attach the properties to the keywords. Such properties are the part-of-speech and a flag for the named entities.

3.3 Keyword analysis
We define two important factors in defining relevance values for our keywords: frequency-based relevancy and language-based relevancy. For keyword analysis, we include three major factors which

are the TFIDF of the keyword, the category-based document frequency of the keyword and the noun phrase rating of the keyword.

3.3.1 Term frequency – inverse document frequency of the keyword. We define two important factors in defining relevance values for our keywords: frequency-based relevancy and language-based relevancy. We calculated the TFIDF of the keyword by using the standard TFIDF formula as shown in equation (1) to get the standard TFIDF score which will get the keywords.

$$tfidf(\mathrm{t, d, D}) = tf(\mathrm{t, d}) \,.\, idf(\mathrm{t, D}) \quad (1)$$

where *tf* = term frequency, *idf* = inverse document frequency, t = term, d = document that the term is inside and D = all documents.

3.3.2 Category-based document frequency of the keyword. We would like to argue that document frequency of a category plays an important role in defining the relevancy values for each keyword. In the normal TFIDF, high document frequency indicates that the token is relatively general in all the documents. However, in this framework, we include the document frequency of a keyword in which the category exists in the document.

$$CDF_{i=0}^{n} = \sum (df(i) \,/\, n) \quad (2)$$

Where *df* = document frequency of the category and *n* = the total number of categories

3.3.3 Noun phrase rating of the keyword. Finally, the scores are assigned based on the nature of noun phrases as different types of noun phrases gives different information and impact on the document. For instance, the noun phrase "black-laced Adidas shoe" provides more impact meaning than "black shoe" or "shoe". Table 1 shows the rating assigned to each noun phrase class.

Table 1. Noun phrase ratings

Classes of Noun Phrase	Rating
Noun phrases containing named entities	2
Noun phrases containing adjectives or participles	1.5
Multiple word nouns	1
Single word nouns without proper nouns	0.5

3.3.4 Keyphrase scoring. Keyphrase score is computed by combining all features with the adjustable weights as shown in equation (3).

$$\mathrm{KPR} = \beta\,(\mathrm{TFIDF}) + \alpha\,(\mathrm{CDF} / \mathrm{TKC}) + \mu\,(\mathrm{NPR} / 2) \quad (3)$$

where KPR = key phrase rating, TFIDF = term frequency – inverse document frequency, CDF = category-based document frequency, TKC = total number of times a keyword is in each category, NPR = noun phrase rating and β, α, μ = weights for each rating (where $\beta + \alpha + \mu = 1$)

The weight used in this work are 0.4 for NPR and 0.3 for both TFIDC and CDF. Though many existing works extracted noun phrases as keywords, this work uses a different approach in keyword rating assignment by taking into consideration of category-based document frequency and noun phrase classes.

4. Evaluation

Keywords extracted from the documents are evaluated using the techniques mentioned in [16]. We have fetched documents from the Amazon dataset [20] which contains product information collected over 11 years. Domain experts are then asked to define golden keywords for each product based on its production description.

4.1 Matching Techniques
We evaluate how effective our proposed technique is by comparing the assigned golden keywords with the keywords extracted by our algorithm using three different matching techniques.

4.1.1 Exact Matching. The extracted keyphrase is considered relevant to the document if it is identical to the golden keyword as shown in equation (4). In other words, "lemon drink" is considered relevant according to exact match if the abstracted keyword is also "lemon drink" without requiring to be case sensitive. Given that

$$\text{Extracted Keyphrase} = e_1\, e_2\, \dots\, e_n \tag{4}$$
$$\text{Golden Keyphrase} = g_1\, g_2\, \dots\, g_n,$$

the extracted keyphrase is considered a match, if $\forall_{1\le i\le n}, e_i = g_i$.

4.1.2 Morph Matching. The extracted keyphrase is considered relevant to the document if its morphological structure is identical to the golden keyword as shown in equation (5). Both keywords are transformed to their basic forms using lemmatization to check the morph matching. The order of the keywords is important to satisfy this matching. Given that

$$\text{Extracted Keyphrase} = e_1\, e_2\, \dots\, e_n \tag{5}$$
$$\text{Golden Keyphrase} = g_1\, g_2\, \dots\, g_n,$$

the extracted keyphrase is considered a match, if $\forall_{1\le i\le n}, stemmed(e_i) = stemmed(g_i)$.

4.1.3 Partial Matching. The extracted keyphrase is considered relevant to the document if it is part of the golden keywords as shown in equation (6). Note that stop words are disregarded in partial matching. If a lemmatized form of either golden keyword or extracted keyword is in the keyphrase of the other one, we consider it as partially matched. Given that

$$\text{Extracted Keyphrase} = e_1\, e_2\, \dots\, e_n \tag{6}$$
$$\text{Golden Keyphrase} = g_1\, g_2\, \dots\, g_m,$$

the extracted keyphrase is considered a match, if $\exists_{1\le i\le n}\exists_{1\le j\le m}, stemmed(e_i) = stemmed(g_j)$.

From these frequency matching counts, we calculate the precision, recall and f-measures. If the extracted keyword matches with the rule, we consider such a keyword relevant to the document. The precision can be determined by dividing the relevant score by the total extracted keywords while recall can be calculated by dividing the relevant keywords by the total number of keywords.

The evaluation is performed using three different product categories from Amazon products which are Baby, Beauty and Health because each category has different and unique characteristics in the description. This way will show the proposed algorithm's ability to handle various kinds of documents and noises. We decided not to include books in this work because most product information only contain the ISBN instead of a long book description which is hard for the experts to provide golden keywords without going through the actual book contents.

Products in baby and beauty categories have distinct features in which they contain the size/volume/quantity, the year that the product has been created, the materials produced for creating that item, the country that is made from, safety and occasions to be used at. Sometimes, they also include the instructions on how to use the product which contains a lot of unnecessary information for information retrieval. Health products' description contains different kind of information which are symptoms, diagnosis, diseases, experiences, precautions, vitamins and side effects.

4.2 Comparison over baby category
Tables 2, 3 and 4 show the accuracy of the proposed algorithm on the baby category. In this category, our algorithm proves significant advantage in both precision and f-measure in all matching evaluations due to our candidate selection nature between noun phrases and category document frequency. However, the recall was stronger for TFIDF which contains multiple keywords than the other two algorithms.

IOP Conf. Series: Materials Science and Engineering **428** (2018) 012021 doi:10.1088/1757-899X/428/1/012021

Table 2. Exact matching in baby category

Algorithm	Precision	Recall	F-measure
Our Algorithm	0.220666	0.323482	0.251575
RAKE	0.127057	0.26284	0.127366
TFIDF	0.033644	0.592813	0.035345

Table 3. Morph matching in baby category

Algorithm	Precision	Recall	F-measure
Our Algorithm	0.232078	0.340088	0.264261
RAKE	0.127366	0.272578	0.165953
TFIDF	0.035345	0.621687	0.065649

Table 4. Partial matching in baby category

Algorithm	Precision	Recall	F-measure
Our Algorithm	0.739721	0.945714	0.814089
RAKE	0.641971	0.986783	0.758592
TFIDF	0.652437	1.000000	0.77257

4.3 Comparison over health category

In Tables 5, 6 and 7, we calculate the results for the health category. In the health category, RAKE gives the best result in exact matching since the RAKE can totally filter out symptoms, diseases without involving the stop words which gives them slight advantage over items which is described using many medical terms. However, in partial matching, our algorithm gives better results than both RAKE and TFIDF since we can generate closer results due to the categorization and noun phrase scoring. Partial matching includes most of the keywords which makes it harder for algorithm to differentiate from one another, but our algorithm shows a slight advantage on precision and f-measure over other algorithms.

Table 5. Exact matching in health category

Algorithm	Precision	Recall	F-measure
Our Algorithm	0.318005	0.3213463	0.319901
RAKE	0.355151	0.344492	0.341751
TFIDF	0.124516	0.33435	0.17625

Table 6. Morph matching in health category

Algorithm	Precision	Recall	F-measure
Our Algorithm	0.831113	0.847931	0.837333
RAKE	0.349222	0.347355	0.341396
TFIDF	0.12242	0.336222	0.174886

Table 7. Exact matching in health category

Algorithm	Precision	Recall	F-measure
Our Algorithm	0.924465	0.938984	0.930288
RAKE	0.88968	0.842659	0.850262
TFIDF	0.86069	0.998042	0.919349

4.4 Comparison over beauty category

The precision, recall and f-measure of beauty category are compared and shown in Tables 8, 9 and 10. Beauty category shows the very similar result as the baby category since the features of the two categories are very similar. In this category, our algorithm outperforms other techniques in both precision and f-measure.

Table 8. Exact matching in beauty category

Algorithm	Precision	Recall	F-measure
Our Algorithm	0.028766	0.059414	0.037048
RAKE	0.027095	0.058404	0.035414
TFIDF	0.01334	0.087718	0.022605

Table 9. Morph matching in beauty category

Algorithm	Precision	Recall	F-measure
Our Algorithm	0.029197	0.060121	0.037584
RAKE	0.027322	0.059515	0.035792
TFIDF	0.013423	0.088818	0.022758

Table 10. Partial matching in beauty category

Algorithm	Precision	Recall	F-measure
Our Algorithm	0.638428	0.986715	0.754965
RAKE	0.614579	0.985892	0.735427
TFIDF	0.332015	0.99009	0.478469

4.5 Comparison across all categories

The overall performance of automatic keyword extractions for all product categories are compared in Tables 11, 12 and 13. The results show that our algorithm yields better precision and f-measure than other approaches on the exact and partial matchings while outperforming other existing approaches on the morph matching. The evaluation results show that our automatic keyword extraction technique that utilize both frequency-based and language-based features is very promising as it outperforms other techniques in many aspects.

Table 11. Exact matching in all categories

Algorithm	Precision	Recall	F-measure
Our Algorithm	0.189146	0.234747	0.202841
RAKE	0.169768	0.221912	0.168177
TFIDF	0.057167	0.338294	0.078067

Table 12. Morph matching in all categories

Algorithm	Precision	Recall	F-measure
Our Algorithm	0.364129	0.416047	0.379726
RAKE	0.16797	0.226483	0.181047
TFIDF	0.057063	0.348909	0.087764

Table 13. Partial matching in all categories

Algorithm	Precision	Recall	F-measure
Our Algorithm	0.767538	0.957138	0.833114
RAKE	0.71541	0.938445	0.781427
TFIDF	0.615047	0.996044	0.723463

References

[1] Leung A 2016 Evaluating automatic keyword extraction for internet reviews, Master Report, University of Lorraine

[2] Medelyan A 2016 NLP keyword extraction tutorial with RAKE and Maui, [Online]. Available: https://www.airpair.com/nlp/keyword-extraction-tutorial. [Accessed 10 September 2017]
[3] Buckley C 1985 *Implementation of The SMART Information Retrieval System*, Cornell University, Ithaca
[4] Vijayarani S, Ilamathi J, Nithya, Phil M 2015 Preprocessing techniques for text mining - an overview, *Int. J. of Computer Science & Communication Networks*, *5*(3), 7-16
[5] Ramasubramanian C and Ramya R 2013 Effective pre-processing activities in text mining using improved Porter's stemming algorithm, *Int. J. of Advanced Research in Computer and Communication Engineering, 2*(12), ISSN (Online): 2278-1021
[6] Manning C D, Surdeanu M, Bauer J, Finkel J, Bethard S J and McClosky D 2014 The Stanford CoreNLP natural language processing toolkit, *Proc. of the 52nd Annual Meeting of the Association for Computational Linguistics: System Demonstrations*, 55-60
[7] Jivani A G et al., A comparative study of stemming algorithms, *Int. J. Comp. Tech. Appl.*, 2(6), 1930-1938, ISSN:2229-6093.
[8] Sharma D 2012 Stemming algorithms, a comparative study and their analysis, *Int. J. of Applied Information Systems (IJAIS)* – ISSN: 2249-0868, Foundation of Computer Science FCS, New York, USA, *4*(3), September
[9] Mestrovic A, Beliga, Sanda Martincic-Ipsic, Slobodan 2015 An overview of graph-based keyword extraction methods and approaches, *J. of Information and Organizational Sciences 39*
[10] Sharan A, Siddiqi, Sifatullah 2015 Keyword and key-phrase extraction techniques: A literature review, *Int. J. of Computer Applications 109*
[11] Manning C D, Raghavan P and Schütze H 2009 *An Introduction to Information Retrieval*, Cambridge University, England
[12] Khoury R, Karray F and Kamel M S 2008 Keyword extraction rules based on a part-of-speech hierarchy, *Int. J. of Advanced Media and Communication, 2*(2), ISSN: 1462-4613
[13] Rose S, Engel D, Cramer N and Cowley W 2010 Automatic keyword extraction from individual documents, in Text Mining: Applications and Theory, West Sussex, Wiley, 1-20
[14] Singh Kathait S, Varshney A, Tiwari S, Sharma A 2017 Unsupervised key-phrase extraction using noun phrases, *Int. J. of Computer Applications, 162*(1), 1-5
[15] Lahriri S, Mihalcea R and Lai P H 2015 Keyword extraction from emails, *Int. J. of Natural Language Engineering, 23*(02), 1-23
[16] Zesch T and Gurevych I 2009 Approximate matching for evaluating keyphrase extraction, *Proc. of Recent Advances in Natural Language Processing*, 484-489
[17] Matsuo Y and Ishizuka M 2004 Keyword extraction from a single document using word co-occurrence statistical information, *Int. J. on Artificial Intelligence Tools, 13*(1), 157–169
[18] Dutta A 2016 A novel extension for automatic keyword extraction, *Int. J. of Advanced Research in Computer Science and Software Engineering, 6*(5), ISSN: 2277-128X, 160-163
[19] Hulth A 2003 Improved automatic keyword extraction given more linguistic knowledge, *Proc. of the 2003 conference on Empirical methods in natural language processing*, 216-223
[20] He R and McAuley J 2016 Modeling the visual evolution of fashion trends with one-class collaborative filtering, *Proc. of the 25th Int. Conference on World Wide Web,* 507-517

Weighted Focal Loss: An Effective Loss Function to Overcome Unbalance Problem of Chest X-ray14

Ruoxi Qin, Kai Qiao, Linyuan Wang, Lei Zeng, Jian Chen and Bin Yan*

National Digital Switching System Engineering and Technological Research Center, Zhengzhou, China, 450001

ybspace@hotmail.com

Abstract. X-ray film has been applied widely in the elementary diagnosis and screening of thorax diseases. The rapid development of deep learning has improved automatic diagnosis of thorax disease. But different thorax disease in database inevitably exist the problem of unbalanced quantity because of the different morbidity in real life. So, the restricted factor is not just the architecture of the models but the unbalanced problem of database. What's more, different from the same problem of natural images, this unbalanced problem is not just the unbalanced quantity but also the unbalanced complexity of diagnosis between samples. In this study, we introduced the Focal Loss and proposed the W-FL (weight Focal Loss) function which makes the model pay more attention to the difficult samples and those kinds of diseases which have small quantity and protect the loss and computed gradients from overwhelming by easy negatives. We evaluate the efficiency of W-FL function on the public Chest X-ray14 database using prevailing per-trained models including AlexNet, Inception-v3 and ResNet. We exceed at least 2% average AUC value on 14 different thorax diseases than the W-CEL function which is commonly proposed to solve the unbalanced problem in this database.

1. Introduction

X-rays film is the most common type of radiology examination in the world and a particularly challenging example of multi-label classification in medical diagnosis. Making up nearly 45% of all radiological studies, the x-ray achieves global ubiquity as a screening tool for pathologies including thorax diseases. Automatically detecting the abnormalities with high accuracy system could greatly enhance the diagnosis's process. But the real challenge is the difficulty for classification on some diseases which even the expert radiologist may misdiagnose, let alone the multifarious diseases and independence between them in each film. Especially for the thorax diseases whose fundamental diagnosis and screening mostly depend on the x-ray films, the diagnosis drives the demand of accurate and quick diagnosis through a multitude of pathological outcomes.

With the development of deep learning, more and more people want to use this method to build high precision computer-aided diagnosis (CAD) systems, and more and more experiments proved the effectiveness of deep learning to achieve this goal [1-5]. The appearance of hospital-scale Chest X-ray14 database, which 112,120 Chest X-ray images are individually labelled with up to 14 different thoracic diseases makes it possible for deep learning combining with the diagnosis of thorax diseases closely.

Mined from the text radiological reports via NLP techniques [6], the Chest X-ray14 database inevitably suffer from the unbalance-class problem because of the different morbidity in real life.

Table 1 shows the proportion of abnormalities. Among abnormalities, the unbalanced problem is also serious while the "Infiltration" makes up most quantity of the data (14.05%) and the "Hernia" occupies the least part of database (0.16%). On the contrast, half proportion of this database is "No Finding". There are many methods to deal with unbalance-class problem of database [7]. Some previous studies showed results on cost sensitive learning of deep neural networks [8-10] and more work is about data level method such as oversampling and undersampling [11-12]. There also many study about this problem of medical decision [13]. Although there are lots of methods, we can't ignore the unbalanced complexity of diagnosis – ideally with the use of Focal loss function to overcome the complicated unbalanced problem between samples.

Some notable work about this database includes a baseline work about Chest X-ray14 using transfer learning model [6]. A partial solution using LSTMs to extract interdependencies among target labels in predicting 14 pathologic patterns from chest x-rays [14]. A new network named CheXNet with 121-layers trained on Chest X-ray14 achieves state of the art on all 14 diseases [15]. Compared with baseline work, all progresses on this database pay attention to the improvement of network architecture but ignore the unbalanced problem. W-CEL function proposed in [6] is the only method which is also used in other network focusing on the unbalanced problem. The definition of W-CEL is shown as below:

$$\beta = \begin{cases} \beta_P = \dfrac{|P|+|N|}{|P|} & y=1 \\ \beta_N = \dfrac{|P|+|N|}{|N|} & y=0 \end{cases} \tag{1}$$

In this function, the P defined as the total numbers of "1" in a batch of labels while the N is the total numbers of "0". y means each element of labels. Through the statistic quantity of "0" and "1"in each label, the W-CEL function only makes the model focus on the unbalanced problem between abnormalities and normal samples.

According to the above analysis, in order to achieve accurate diagnosis on Chest X-ray14, it is necessary for us to overcome the unbalanced problem in this database. In this paper, we excavate the complexity of the Chest X-ray14 data by contrasting the different results using different loss function and improve the FL loss functions [16] to accommodate the multi-labels classification. We apply and improve the FL loss function to achieve a quick and reliable transfer learning on this database. Through the experiments, we find that the mutual influence of unbalance-class and complicated diagnosis make it more difficult for training. Our experiments show a simple extension of loss function to classify multiple diseases outperforms previous transfer learning on Chest X-ray14.

Table 1. Proportion of different diseases in Chest X-ray14.

Disease	Atelectasis	Effusion	Infiltration	Pneumonia	Pneumothorax	Pleural_Thickening	Edema
Proportion	8.16%	9.41%	14.05%	0.957%	3.75%	2.39%	1.63%
Disease	Cardiomegaly	Emphysema	Nodule	Fibrosis	Mass	Consolidation	Hernia
Proportion	1.96%	1.78%	4.47%	1.19%	4.06%	3.30%	0.161%

2. Method

2.1 Pre-trained model

Our goal is finding the mutual influence of unbalanced problem in Chest X-ray14 and the effectiveness of Focal Loss function. We tackle this problem by using a transfer learning model trained on ImageNet and comparing the result with other loss function (CEL, W-CEL) to verify our ideas. For the quick transfer learning on Chest X-ray14 dataset, we defined the unified transfer learning framework as mentioned in [6]. We use the pre-trained models as the instrument of extracting high-dimensional features. To roundly demonstrate the complex interactions between this database and the universality of Focal Loss function, we use the per-trained AlexNet [17] and Inception-v3 [18] and ResNet [19] as our transfer learning models.

We firstly use the per-trained AlexNet model which has 650 thousand neurons and 60 million parameters and reduce the top-5 error rate to 16.4% as one of our baseline networks [17]. As one of the representative Convolutional Neural Networks, AlexNet promotes the development of deep learning in computer vision. To completely verify our efficiency of the Focal loss function is better than W-CEL function, we also make experiment on the ResNet [19], which has the best performance in [6] using transfer learning model.

As a creative architecture recently, Inception-v3 promotes the performance in ways that utilize the added computation as efficiently as possible by suitably factorizing convolutions and aggressive regularization [18]. The architecture of Inception-v3 implements the idea of "factorization into small convolutions", which can make the network handle more plentiful features.

2.2 Fine-tune network

2.2.1 Transition layer connected to bottleneck layer. We train our fully-connected layers as the decoder of the high-dimensional features. The transition layers help pass down the weights from pre-trained DCNN models and transform the activations into a uniform dimension of output. Through the decoder of high-dimensional features, we can use activations to represent the possibility of each disease. The transition layers have the most of parameters for us to train and the nonlinear expression of that decide the performance of our model.

2.2.2 Loss layer. To make the output more reasonable for multi-label problem, we replace the softmax nonlinearity which is often used in traditional loss layers with a sigmoid nonlinearity, outputting the probability that the image contains each disease in each dimensionality (14 in this database). Without the normalization of softmax, that is the way to furthest address the multi-label issue. In this way we can make each dimensionality of the output represent the probability of each disease among "0" to "1", which are more reasonable when we calculate the loss with multi-labels. This definition of output transits the multi-label problem into a regression-like loss setting. What's more, as note in [16], the implementation of the loss layer combines with the sigmoid operation for computing the loss computation, resulting in greater numerical stability. That's another reason for us to use sigmoid nonlinearity as our activated function in loss layer.

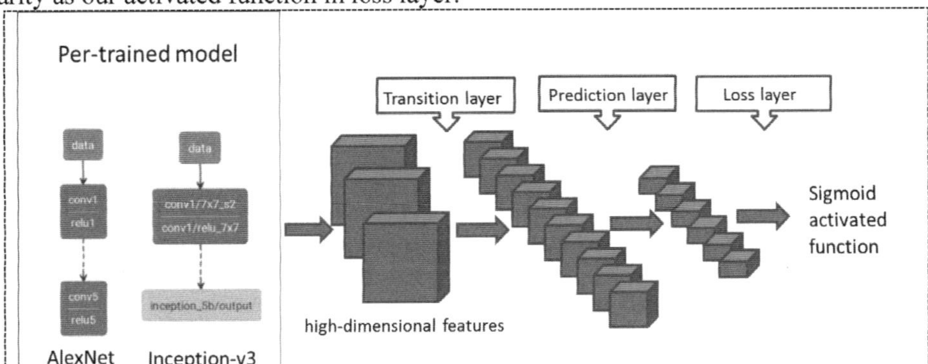

Figure 1. The flow-chart of our transfer learning model

Figure 1 illustrates the DCNN (Deep Convolutional Neural Network) architecture we adapted. As shown in Figure 1, we perform the network surgery on the pre-trained models (using ImageNet [20]), e.g., AlexNet [17], Inception-v3 [18] and ResNet [19], by leaving out the transition layers and the final prediction layers. Among the aforementioned architecture, we set the amount of the neurons in transition layers as 4096 in Inception-v3 and 2048 in AlexNet.

2.3 Improved Focal Loss for Classification

Recently more and more new kinds of loss function for neural networks training were also developed [21]. About Chest-Xray14 database, they tacitly approve the W-CEL loss function [6] as their method to solve the unbalanced problem about the database. The original intention of W-CEL loss function is solving the unbalanced problem caused by different quantity of pathologies and normality. Inspired by the W-CEL loss function, we creatively use the Focal Loss function [16] which is proposed to promote dense detectors. Furthermore, we modify the Focal loss function to pay attention to the mutual influence between unbalance-class and complex diagnosis. The focal loss function defined as below:

$$p_t = \begin{cases} p & \text{if } y=1 \\ 1-p & \text{otherwise} \end{cases} \tag{2}$$

$$FL(p_t) = -(1-p_t)^{\gamma} \log(p_t) \tag{3}$$

In the above y ∈ {0,1} specifies the ground-truth class; γ ∈ {0,5} aims at adjusting the rate by down-weight the loss of easy examples [16]. In our experiment, we set the γ as 2, which is mentioned in [16] that has the best performance. The experiment shows some disadvantage about FL function, to overcome that, we improve the definition about the parameter β. Different from the originally definition about α in two-dimension. We defined β as the corresponding weight similar as the W-CEL function in formula (1). So our modified FL Loss function defined as below:

$$\mathrm{FL}(P_t) = -\beta(1-p_t)^{\gamma} \log(p_t) \tag{4}$$

Through the modified Focal Loss function, the parameter β and γ solve both sides of unbalanced problem and achieve a good performance with transfer learning models. What's most important is that we propose a new application of Focal Loss function which we think will be more suitable for the unbalanced problem of Chest X-ray14 database.

3. Experiments and results

3.1 Experiment design

3.1.1 Processing Chest X-ray14 Data. To reduce the operating pressure of computer and make our input image suitable for our transfer learning model, we transform the X-rays images' size to 299×299 for Inception-v3 and 227×227 for AlexNet and unify them to grayscale images (some image of this data is not grayscale). For the pathology classification, we randomly shuffled the entire dataset into two subgroups: i.e. training (80%) and validation (20%). We make sure that the data of training and validation are strictly separated so that we can have an appraisal about our classification model. About the multi-label setup, we do the same way as is mentioned in [6].We define a 14-dimensional label vector for each image and the all-zero vector represents the "No Finding" while the factor "1" in labels means the existent of corresponding disease.

3.1.2 Metrics. Considering the relevance between different diseases, pathological diagnosis is not like traditional image classification. Instead of using the common top accuracy judging criteria to define the loss, we set the corresponding Area-Under-Curve (AUC) values as the judging criteria adjusted to multi-class tasks since accuracy metric is associated with notable difficulties in the context of imbalanced data. The corresponding Area-Under-Curve (AUC) values is only calculated on validation to demonstrate the performance of model. We select 4800 images randomly in validation data to

IOP Conf. Series: Materials Science and Engineering **428** (2018) 012022 doi:10.1088/1757-899X/428/1/012022

calculate the AUC value. The set of metrics provides a meaningful quantification of this performance to verify the effectiveness of our loss function.

3.2 Classification Result

Comparing the result using CEL with that using W-CEL, Figure 2 shows the interaction between proportion and performance. With W-CEL function, the AlexNet's performance of the AUC values is connected to the unbalance-class of different diseases. The W-CEL function has already balance the loss between normal examples and abnormal examples and has integrally promotes the performance of examples which has poor proportion (except "Hernia"). But the rising value of AUC connected to the proportion is not unconditional: the performance of "Hernia" which has the poorest proportion even decline in AlexNet and compared with the result of ResNet the performance of the AUC values is not connected to proportion. Those performances verify two aspect disadvantage of W-CEL function: a) the W-CEL function doesn't pay attention on the different proportion of abnormalities, which will let the examples with poor proportion has bad performance; b) the performance of W-CEL varies greatly. In network with simple architecture, the performance is connected to the proportion while it is unconnected with deeper network and the average value of AUC even decline in Inception-v3. It will get worse performance with the deeper architecture. The poor performance of "Hernia" compared with that using CEL function also proves the unbalance-class problem between abnormalities and prove the inefficiency of W-CEL function. We must address this problem not only in statistics but in pathology, which shows the complicated unbalanced problem in this database. This problem immensely hinders the application of high precision computer-aided diagnosis (CAD) systems on large-scale disease diagnosis.

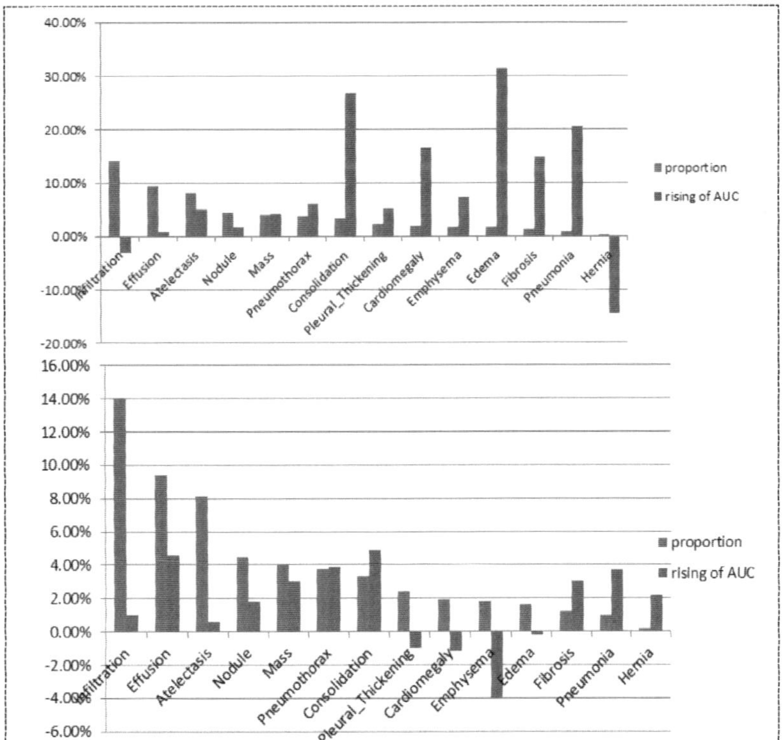

Figure 2. Interaction between proportion and performance.
Top: Alexnet with W-CEL *Bottom*: ResNet with W-CEL

Table 2. Classification result using different pre-trained model with different loss function

	Pre-trained AlexNet				Pre-trained Inception-v3				Pre-trained ResNet-50			
	CEL	W-CEL	FL	W-FL	CEL	W-CEL	FL	W-FL	CEL	W-CEL	FL	W-FL
Atelectasis	0.665	0.715	0.721	0.709	0.683	0.666	0.679	0.697	0.620	0.626	0.702	0.678
Infiltration	0.611	0.58	0.623	0.597	0.558	0.583	0.598	0.567	0.567	0.577	0.615	0.605
Pneumothorax	0.663	0.724	0.722	0.713	0.703	0.782	0.766	0.695	0.615	0.654	0.762	0.721
Edema	0.538	0.852	0.835	0.826	0.706	0.616	0.785	0.792	0.826	0.824	0.830	0.829
Emphysema	0.610	0.683	0.714	0.745	0.643	0.610	0.763	0.754	0.581	0.541	0.781	0.708
Fibrosis	0.576	0.723	0.760	0.739	0.675	0.586	0.745	0.736	0.659	0.689	0.688	0.734
Consolidation	0.453	0.722	0.706	0.729	0.598	0.667	0.776	0.753	0.687	0.736	0.725	0.762
Effusion	0.760	0.786	0.783	0.790	0.803	0.691	0.810	0.778	0.709	0.755	0.820	0.800
Pneumonia	0.345	0.550	0.666	0.580	0.337	0.566	0.639	0.597	0.616	0.653	0.710	0.679
Pleural_Thickening	0.587	0.64	0.696	0.723	0.548	0.524	0.670	0.682	0.650	0.606	0.679	0.712
Cardiomegaly	0.623	0.79	0.848	0.865	0.584	0.668	0.758	0.813	0.618	0.636	0.847	0.757
Nodule	0.529	0.547	0.591	0.596	0.588	0.543	0.631	0.585	0.504	0.522	0.603	0.544
Mass	0.611	0.652	0.688	0.649	0.612	0.527	0.684	0.679	0.54	0.570	0711	0.640
Hernia	0.714	0.569	0.516	0.551	0.711	0.561	0.489	0.526	0.532	0.554	0.472	0.504
Average	0.591	0.680	0.705	0.701	0.625	0.610	0.705	0.689	0.623	0.638	0.712	0.692

The results of all those four loss functions with three different models are demonstrated in Table 2. Compared with the result with W-CEL, we can find some correlation between the pathologies' proportion and the change of different pathologies' AUC value. With the less proportion of the data, Focal Loss universally works better, which solve a degree of problem about unbalance-class. But we can't ignore the phenomenon that the AUC value of "Hernia" which has the least proportion and is easy to diagnose (having good performance with CEL function) decreases by 19.5 instead of increasing. That may because the γ make the model give more weight on the complexity example without a parameter to balance the complexity and unbalance-class problem making even the easy examples are overwhelming because of the poor proportion. That also shows the unbalanced problem in statistics and pathology. The decline of the AUC values is not only because the unbalance-class of different diseases, but also because the complexity of diagnosis. The combine of those problems makes the classification of diseases more difficult than that of nature images.

To use the FL Loss function more effective on this problem, we improve the definition about the parameter β . We use the loss function in formula (4). With the modified FL loss function, our performance on "Hernia" outperforms that without parameter α (increase by 3.5% on AlexNet and increase by 8.7% on Inception-v3). The introduction of parameter β makes the performance of model more balanced without enhances the performance entirely. But we can't ignore the advantage it takes. We believe that through the painstaking work on selecting parameter, the modified FL loss function can further improve the performance when we train the model from scratch.

4. Conclusion

In this paper, we firstly demonstrate the influence of the unbalanced problem in two aspects. According to the result, we find that the influence of unbalanced problem will amplify when more samples with small quantity are introduced to the database and the different result with different loss function shows the interaction of unbalance-class and complexity of diagnosis, which also verify the common loss function's limitation. To eliminate the negative effects from FL function and solve the problem mentioned above, we creatively combine the FL function with W-CEL function and propose loss function named W-FL. With that Weighted Focal Loss function our transfer learning models achieve a good performance on Chest X-ray14.

Acknowledgments
This work was supported by the National Key R&D Program of China under grant 2017YFB1002502 and National Natural Science Foundation of China (No. 61701089, No.61601518).

References

[1] Lakhani, Paras and Sundaram, Baskaran. Deep learning at chest radiography: Automated classification of pulmonary tuberculosis by using convolutional neural networks. Radiology, pp. 162326, 2017.

[2] Esteva, Andre, Kuprel, Brett, Novoa, Roberto A, Ko, Justin, Swetter, Susan M, Blau, Helen M, and Thrun, Sebastian. Dermatologist-level classification of skin cancer with deep neural networks. Nature, 542(7639):115-118, 2017.

[3] Islam, Mohammad Tariqul, Aowal, Md Abdul, Minhaz, Ahmed Tahseen, and Ashraf, Khalid. Abnormality detection and localization in chest x-rays using deep convolutional neural networks. arXiv preprint arXiv:1705.09850, 2017.

[4] Rajpurkar, Pranav, Hannun, Awni Y, Haghpanahi, Masoumeh, Bourn, Codie, and Ng, Andrew Y. Cardiologist-level arrhythmia detection with convolutional neural networks. arXiv preprint arXiv:1707.01836, 2017.

[5] Grewal, Monika, Srivastava, Muktabh Mayank, Kumar, Pulkit, and Varadarajan, Srikrishna. Radnet: Radiologist level accuracy using deep learning for hemorrhage detection in ct scans. arXiv preprint arXiv:1710.04934, 2017.

[6] Wang, Xiaosong, Peng, Yifan, Lu, Le, Lu, Zhiyong,Bagheri, Mohammadhadi, and Summers, Ronald M.Chestx-ray8: Hospital-scale chest x-ray database and benchmarks on weakly-supervised classification and localization of common thorax diseases. arXiv preprint arXiv:1705.02315, 2017

[7] Buda M, Maki A, Mazurowski M A. A systematic study of the class imbalance problem in convolutional neural networks[J]. 2017.

[8] Salman H Khan, Mohammed Bennamoun, Ferdous Sohel, and Roberto Togneri. Cost sensitive learning of deep feature representations from imbalanced data. arXiv preprint arXiv:1508.03422, 2015.

[9] Vidwath Raj, Sven Magg, and Stefan Wermter. Towards effective classification of imbalanced data with convolutional neural networks. In IAPR Workshop on Artificial Neural Networks in Pattern Recognition, pages 150-162. Springer, 2016.

[10] Yu-An Chung, Hsuan-Tien Lin, and Shao-Wen Yang. Cost-aware pre-training for multiclass cost-sensitive deep learning. arXiv preprint arXiv:1511.09337, 2015.

[11] Gil Levi and Tal Hassner. Age and gender classification using convolutional neural networks. In Proceedings of the IEEE Conference on Computer Vision and Pattern Recognition Workshops, pages 34-42, 2015.

[12] Andrew Janowczyk and Anant Madabhushi. Deep learning for digital pathology image analysis: A comprehensive tutorial with selected use cases. Journal of pathology informatics, 7, 2016.

[13] Maciej A Mazurowski, Piotr A Habas, Jacek M Zurada, Joseph Y Lo, Jay A Baker, and Georgia D Tourassi. Training neural network classifiers for medical decision making: The effects of imbalanced datasets on classification performance. Neural networks, 21(2):427-436, 2008.

[14] Yao, Li, Poblenz, Eric, Dagunts, Dmitry, Covington,Ben, Bernard, Devon, and Lyman, Kevin. Learning to diagnose from scratch by exploiting dependencies among labels. arXiv preprintarXiv:1710.10501,2017.]

[15] Rajpurkar P, Irvin J, Zhu K, et al. CheXNet: Radiologist-Level Pneumonia Detection on Chest X-Rays with Deep Learning[J]. 2017.

[16] Lin T Y, Goyal P, Girshick R, et al. Focal Loss for Dense Object Detection[J]. 2017:2999-3007.

[17] A. Krizhevsky, I. Sutskever, and G. E. Hinton. Imagenet classification with deep convolutional neural networks. In Advances in neural information processing systems, pages 1097–1105, 2012.
[18] Szegedy C, Vanhoucke V, Ioffe S, et al. Rethinking the Inception Architecture for Computer Vision[C] IEEE Conference on Computer Vision and Pattern Recognition. IEEE Computer Society, 2016:2818-2826.
[19] He K, Zhang X, Ren S, et al. Deep Residual Learning for Image Recognition[C] IEEE Conference on Computer Vision and Pattern Recognition. IEEE Computer Society, 2016:770-778.
[20] J. Deng, W. Dong, R. Socher, L.-J. Li, K. Li, and L. FeiFei. Imagenet: A large-scale hierarchical image database. In Computer Vision and Pattern Recognition, pages 248–255. IEEE, 2009. 5
[21] Shoujin Wang, Wei Liu, Jia Wu, Longbing Cao, Qinxue Meng, and Paul J Kennedy. Training deep neural networks on imbalanced data sets. In Neural Networks (IJCNN), 2016 International Joint Conference on, pages 4368-4374. IEEE, 2016.

Deep Learning Based Semantic Labelling of 3D Point Cloud in Visual SLAM

Xuxiang Qi[1, 2], Shaowu Yang[1, 2, *] and Yuejin Yan[2]

[1]State Key Laboratory of High Performance Computing, National University of Defense Technology, 410073, Changsha, China
[2]College of Computer, National University of Defense Technology, 410073, Changsha, China

* Corresponding author: shaowu.yang@nudt.edu.cn

Abstract. Three-dimensional (3D) point cloud understanding is important for autonomous robots. However, point clouds are normally irregular and discrete. It is challenging to obtain semantic information from them. In this paper, we present a method to build a dense semantic map, which utilizes both two-dimensional (2D) image labels and 3D geometric information. The dense point cloud is built by using a state-of-the-art RGB-D SLAM system. It is further segmented into meaningful clusters using a graph-based method. Then, image keyframes during the SLAM process are used to extract semantic image labels by a convolution neural network (CNN). Finally, these semantic labels are projected to the point cloud clusters to achieve a 3D dense semantic map. The effectiveness of our method is validated on a popular public dataset.

1. Introduction

Scene understanding is crucial to autonomous driving and mobile robotics. Recent research focuses on scene reconstruction to build a 3D sparse or dense map, such as KinectFusion [1], ElasticFusion [2] and DynamicFusion [3]. However, there is no semantic information in those maps, from which robots cannot obtain a semantic-level understanding of surroundings. In fact, recent years have witnessed great progress in 2D image semantic segmentation. With the help of CNN, we can analysis 2D level semantic information in images, such as the work in FCN [4], U-Net [5], SegNet [6], RefineNet [7], PSPNet [8] and DeepLab [9-11]. Meanwhile, 3D point cloud semantic segmentation is a hot issue in computer vision, and it has made great progress in the recent years, including PointNet [12], PointNet++ [13] and PointCNN [14]. However, these approaches merely make use of 3D information to analyse point cloud.

Actually, point cloud can be generated by RGB-D SLAM, e.g. ORB-SLAM2 [15], using cheap RGB-D sensors like Asus Xtion. In RGB-D SLAM, RGB images have rich texture information, and point clouds have geometrical information. In 3D semantic segmentation and mapping, most of existing methods use either RGB images or point clouds as input. However, few methods make use of both 2D and 3D information. In this paper, we introduce a system that fuses 2D and 3D information to build a dense semantic map. Our main contributions of this paper can be summarized as follows:

- An efficient segmentation method for 3D point cloud
- A semantic labelling method for 3D point cloud that fuses both 2D image information and 3D geometric information

IOP Conf. Series: Materials Science and Engineering **428** (2018) 012023 doi:10.1088/1757-899X/428/1/012023

- A 3D dense semantic mapping system

2. Related Work

2.1. Semantic SLAM

Traditional simultaneous localization and mapping (SLAM) systems mainly focus on using low-level geometric features, such as points, lines, and planes, which cannot provide semantic information. Semantic SLAM can give semantic information of environments. It can help robots to understand surrounding scenes in both geometrical and content level. Salas-Moreno et al. proposed the SLAM++ [16], which can perform object detection in RGB-D tracking and mapping. The study by John McCormac et al. on semantic SLAM was SemanticFusion [17]. The method uses a convolution neural network to produce class probability maps, and fuses these predictions into the 3D map. Keisuke Tateno et al. presented a real-time dense monocular CNN-SLAM [18]. With the aid of CNN, CNN-SLAM can perform not only depth prediction, but also semantic segmentation. DA-RNN [19] introduces a new recurrent neural network (RNN) architecture for semantic labelling on RGB-D videos, which utilizes information in multiple viewpoints to improve segmentation performance. Tong [20] combines different SLAM systems facilitated by a scene detection method.

2.2. 2D Object Detection and Semantic Segmentation

An essential component to get semantic information is object detection, which can localize object instances in images. Girshick et al. [21] presented R-CNN, which proposed to apply CNN to object detection. Other similar methods have been proposed in recent years, like Fast R-CNN [22], Faster R-CNN [23], Mask-RCNN [24] and YOLO [25-26]. R-CNN uses selective search algorithm for generating region proposals, which runs very slow. Faster R-CNN replaces the slow selective search algorithm with a fast neural net. Mask R-CNN improves the region of interest (ROI) pooling layer and extends Faster R-CNN to pixel-level image segmentation.

Semantic segmentation is to understand an image at a pixel level, which can label each pixel with a class identity. Similar to object detection, state-of-the-art semantic segmentation approaches also rely on CNN. FCN [4] by Long et al. is the first end-to-end system, which popularizes CNN architecture for semantic segmentation. U-Net [5] is a popular encoder-decoder architecture which can make use of annotated samples more efficiently and have a higher accuracy. SegNet [6] is a similar encoder-decoder architecture. SegNet copies indices from max-pooling for up-sampling, which makes it more memory efficient. RefineNet [7] proposes a method called RefineNet block which fuses both high resolution and low resolution features. It solves the problem of significant decrease in image resolution when we repeat the sub-sampling operation. PSPNet [8] introduces a pyramid pooling method to aggregate the context. DeepLab [9-11] utilizes dilated convolutions to increase the field of view.

2.3. 3D Point Cloud Segmentation and Semantic Analysis

Point cloud segmentation is the process of dividing point clouds into different regions, each of which has similar properties. It is an essential step towards scene understanding from point clouds. Driven by specific applications, like environment modelling in robotics, 3D point cloud segmentation becomes a very active research topic. Point Cloud Library (PCL) [27] is a popular library which provides open-source segmentation algorithms. Early approach [28] uses RANSAC to detect planes from the point clouds, and then it divides objects with Euclidean separation. Region growing [29] algorithm was proposed in 2D image processing work. Later, it was used in the work related to 3D point cloud. Rabbani et al. [30] presented a method for segmentation of point clouds using smoothness constraint, which finds smoothly connected areas in point clouds. Vo et al. [31] introduced a novel octree-based region-growing algorithm for the fast surface patch segmentation of 3D point clouds in urban environments. Stein et al. [32] divided the point cloud into some individual segments using Locally Convex Connected Patches (LCCP) algorithm, which uses normal vector to judge local convexity.

Golovinskiy et al. [33] proposed a min-cut based method of segmenting objects in point clouds, which can be adapted for both automatic and interactive segmentation.

Different from 2D images, 3D point clouds are irregular and unordered. Thus, the common way in 2D image processing like convolution is ill-suited for them. Recently, deep neural network based methods are proposed for 3D point cloud classification and segmentation, such as PointNet [12] and PointNet++ [13]. The PointNet is able to learn directly from unordered point clouds, which combines local point features and global information to perform 3D segmentation. Based on PointNet, PointNet++ introduces a hierarchical neural network to learn local features with increasing contextual scales, which can learn deep point set features efficiently and robustly. PointCNN [14] presents a novel approach named X-transformation, which can take advantage of CNNs for point cloud processing. However, these methods use only point cloud information, which is difficult to be extended to semantic labelling.

3. Deep Learning Based Semantic labelling in 3D Point Cloud

Our system is based on ORB-SLAM2 [15] and consists of 3 modules: visual SLAM module, point cloud segmentation module and semantic labelling module. We use a keyframe-based update strategy to generate point cloud data. When a keyframe is inserted, we use a modify deep learning framework to process it to obtain object labels. Furthermore, we segment the point cloud with a graph-based method. Finally, we project image labels to the point cloud segments to achieve semantic labelling of the point cloud. The above process is shown in figure 1.

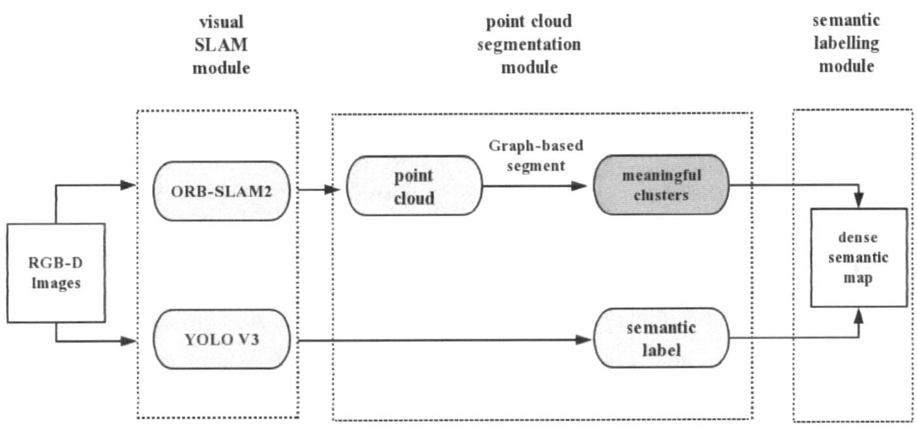

Figure 1. Overview of our semantic 3D mapping system

3.1. The visual SLAM module

The visual SLAM module generates the point cloud data from the RGB-D dataset, which will be further processed by the point cloud segmentation module. The visual SLAM module works in three threads: A tracking thread, a local mapping thread and a loop closing thread.

The tracking thread takes charge of extracting and matching ORB features in gray-scale images converted from raw RGB images for localizing the camera. Besides, it decides when to insert a new keyframe.

The local mapping thread focuses on building the local 3D sparse map. It optimizes both local map and the keyframe poses by performing local bundle adjustment (BA). To generate point cloud data, only keyframes are utilized, while other frames are used to compute camera poses. Point clouds are generated by transforming the 3D points in the depth images from the camera coordinate system to the world coordinate system. We also obtain a rough semantic map in this thread, which will be described in Sect. 3.3.

The loop closing thread detects appearance loops and then corrects the accumulated drifts by pose graph optimization. It is accomplished by using a bag-of-words method among the keyframes.

3.2. The point cloud segmentation module

In the point cloud segmentation module, we only rely on geometric information to segment the point cloud, while the image texture corresponding to the point cloud has no effect on the segmentation result. First, we use the supervoxel method [34] to segment the original point cloud as shown in figure 2. In this way, we can not only reduce the computational cost but also convert point cloud into surface patches according to the similarities of the points. The result of supervoxel can be represented by an adjacency graph $G=\{\nu,\varepsilon\}$ where $\nu_i \in \nu$ are patches and ε connect adjacent patches (ν_i, ν_j).

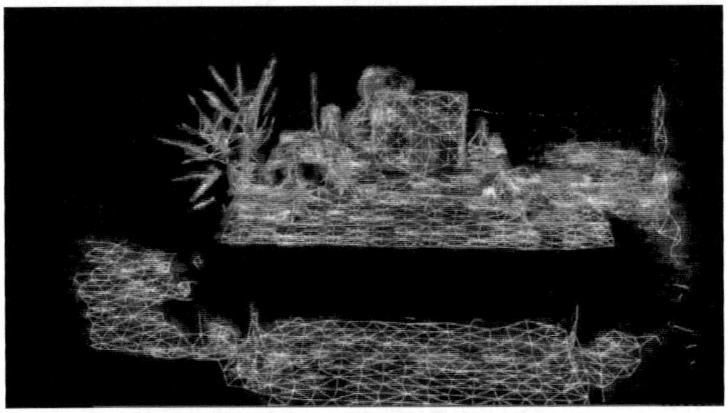

Figure 2. supervoxels of point clouds

After the process of supervoxel, there is a centroid c_i, and a normal vector n_i in each surface patch. The scene segmentation can be framed as a graph partitioning problem. Make it clear in figure 3, nodes are the surface patches, each of node belongs to an object. We are supposed to determine whether edges are ON or OFF.

It is a common way to compute the similarity of nodes by using Euclidean distance [35] with mean shift algorithm [36]. However, the method has a high computational complexity. Normal vector [32] is a reflect of local convexity information, which can be used for clustering. However, it becomes unreliable when there is a higher percentage of noise in point clouds. Thus, we propose to fuse support plane to suppress this negative effect. Supposing that we have K support planes $\{s_1, s_2, ..., s_k\}$ in a point cloud, and surface patches have been obtained in these planes. We define a variable $\{b_i\}_1^N, b_i \in [0, K], b_i = K$ indicates that surface patch belongs to surface plane s_k. Then, we extract planes of all objects and distribute surface patches to them.

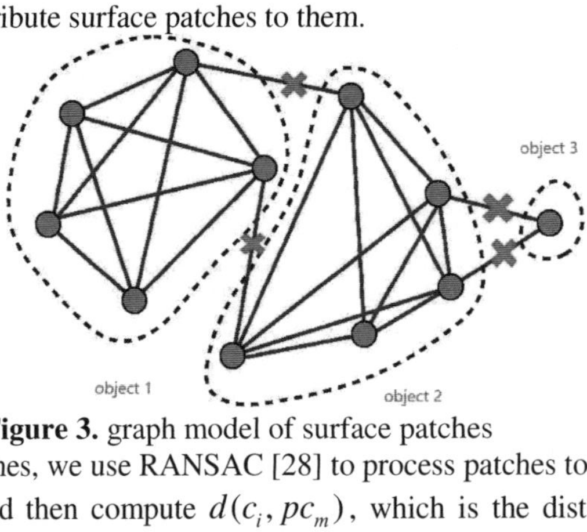

Figure 3. graph model of surface patches

After we get surface patches, we use RANSAC [28] to process patches to generate plane candidates $PC=\{pc_1, pc_2, ...pc_m\}$, and then compute $d(c_i, pc_m)$, which is the distance of the surface patch centroid c_i to plane pc_m. With a threshold δ, we can get all patches within the distance δ to plane pc_m, named $\Pi=\{v_i \in V \mid d(c_i, pc_m) < \delta\}$. Then, we define:

$$D(p_{cm}) = \begin{cases} 1 - \dfrac{\Pi}{\eta} & \Pi < \eta \\ \exp(1 - \dfrac{\eta - \Pi}{\eta}) & \Pi \geq \eta \end{cases} \tag{1}$$

$D(p_{cm})$ are possible planes for objects in the point cloud. In the experiments, we set $\eta = 45$ and $\delta = 4.5cm$. Then the plane extraction problem becomes to minimize the energy formulation:

$$P^* = \arg\min E(P), P \subset PC. \tag{2}$$

And our fitting energy formulation is:

$$E(P) = \sum_{pc_m \in E} D(pc_m). \tag{3}$$

After we finish plane extraction, we get planes L and surface patches K .We can directly use the fast graph-based [37] method to perform segmentation.

3.3. The semantic labelling module

Original map generated by ORB-SLAM2 has no semantic information (see figure 4(a)), which is merely a set of irregular and unordered points. We perform semantic labelling in point cloud with the aid of a deep learning framework.

We use a modify YOLO v3 [26] framework to detect objects in keyframes extracted by the SLAM module. Image boundaries of the labels are not accurate enough in 2D images (see figure 4(b)), but we can project 2D semantic labels to 3D point cloud to achieved reasonable 3D semantic mapping, as shown in figure 4(c). However, it is a rough semantic map and not accurate enough to use. As mentioned before, we have already performed point cloud segmentation, then we can fuse segmentation and rough semantic map to improve mapping performance. Original labels come from RGB images, here we make use of both 2D label and 3D segment result to perform semantic segment. Finally, each object region in the point cloud gets a specific semantic label. More details of the process will be further described in Sect. 4.3.

(a) original point cloud

(b) object detection

(c) point cloud with semantic labels, using specific colors

Figure 4. Semantic labelling from 2D image to 3D point cloud

IOP Conf. Series: Materials Science and Engineering **428** (2018) 012023 doi:10.1088/1757-899X/428/1/012023

4. Experiments and Results
We evaluate our system with the popular TUM dataset [38]. Without training the neural network for semantic labelling in TUM dataset, we only focus on validating our 3D semantic mapping method with a few kinds of known objects.

4.1. Experimental setup
The system runs on a laptop with Ubuntu 14.04 64-bit operating system, an Intel Core i5-6300HQ (4 cores @2.3GHZ) CPU, 16G DDR4 RAM and a GTX690M GPU. Our visual SLAM module is able to reach real-time performance. With the help of GPU, YOLO v3 is able to detect images at 8 frames per second.

4.2. Point cloud segment results
In this experiment, we evaluate our point cloud segmentation method qualitatively by comparing the segmentation results with other existing methods. Locally Convex Connected Patches (LCCP) method only uses local convexity or concavity for segmentation, which ignores global geometrical information.

Lack of global information may lead to wrong segment result, as we can see in figure 5(a), LCCP method is unable to segment small objects such as the keyboard from indoor scene.

Region Growing method initializes the seed points by their curvature value, and then each region from these points will grow by adding neighbour points. However, the method is sensitive to noise data. In the segment result, we can see that background object wall and foreground object desktop are mixed (see figure 5(b)).

Thus, we use supervoxel method to reduce the time costing, and then make use of both global information such as supporting planes and local information to get a better segmentation performance. We can check whether different methods can segment out objects in the point cloud. As it shows in figure 5(c), we can see that our method is able to segment the majority objects in this indoor scene.

Furthermore, based on an efficient graph-based approach, our segmentation method is faster than LCCP and Region Growing method. In this experiment, our SLAM system uses 82060 RGB images and 82060 depth images to build a 3D point cloud map, which contains 952563 points. Our system is able to finish segmentation within 5 seconds. Time costs of these methods are shown in table 1.

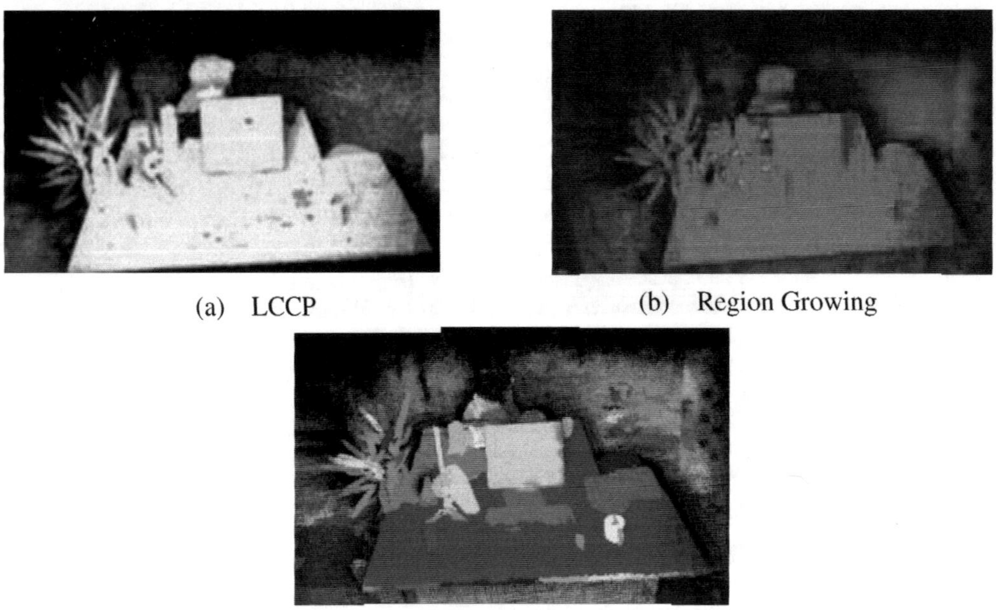

(a) LCCP (b) Region Growing

(c) Ours

Figure 5. Segment results

Table 1. Time cost comparison, in milliseconds.

Method	Time-costing
LCCP	7138
Region Growing	13214
Ours	4638

4.3. Semantic segmentation results in the point cloud

Our visual SLAM module uses YOLO v3 to obtain object labels and project them to point cloud. The point cloud has XYZ and RGB information, which represents Euclidean coordinates and the color data for each point (see figure 6(a)). Besides, the point cloud has semantic information from 2D images. After the point cloud segmentation, we also obtain a segment label for each point (see figure 6(b)). However, this segment label is only a random label without semantic meaning. Then we can fuse two point clouds to combine semantic information with the segment labels, in this way we obtain the final semantic point cloud as shown in figure 6(c).

In fact, two point clouds have the same number of points, and we can use PCL library to fuse them. After fusion, we can not only correct the segment result, but also perform semantic labelling. In this way, we convert 2D image labels to 3D point cloud labels. By the way, our semantic labelling module cannot reach a real-time performance, but it can be a helpful module for a visual SLAM system.

(a) point cloud with semantic label

(b) segment result

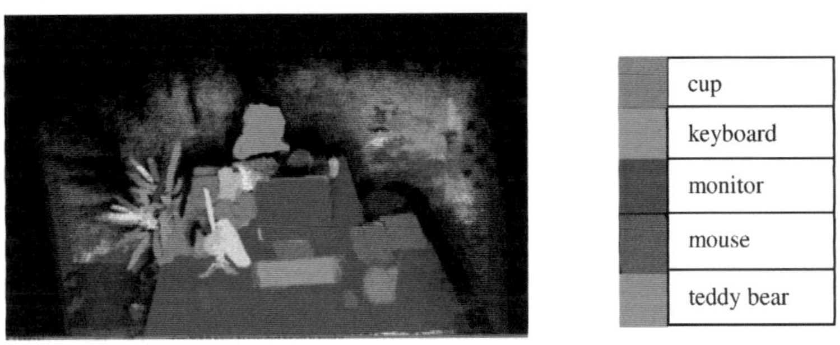

(c) semantic map

Figure 6. fuse label and segmentation

YOLO v3 was trained on *coco*[1] dataset which contains 80 classes. However, we only detect several classes in our indoor scene (cup, keyboard, monitor, mouse, teddy bear) as objects of interest. Most of the classes in *coco* dataset are outdoor objects classes, which do not exist in our experiment.

5. Conclusion

In this work, we present a novel approach to combine 2D object labels and 3D point cloud segmentation to achieve 3D semantic mapping in visual SLAM. We propose an effective method of graph-based 3D point cloud segmentation. Another contribution of this paper is that we fuse 2D and 3D information to build a dense semantic map. We share the source code on github[2]. Besides, a video demonstration can be found online[3].

References

[1] Newcombe, Richard A., et al. "KinectFusion: Real-time dense surface mapping and tracking." Mixed and augmented reality (ISMAR), 2011 10th IEEE international symposium on. IEEE, 2011.

[2] Whelan, Thomas, et al. "ElasticFusion: Dense SLAM without a pose graph." Robotics: Science and Systems, 2015.

[3] Newcombe, Richard A., Dieter Fox, and Steven M. Seitz. "Dynamicfusion: Reconstruction and tracking of non-rigid scenes in real-time." Proceedings of the IEEE conference on computer vision and pattern recognition. 2015.

[4] Long, Jonathan, Evan Shelhamer, and Trevor Darrell. "Fully convolutional networks for semantic segmentation." Proceedings of the IEEE conference on computer vision and pattern recognition. 2015.

[5] Ronneberger, Olaf, Philipp Fischer, and Thomas Brox. "U-net: Convolutional networks for biomedical image segmentation." International Conference on Medical image computing and computer-assisted intervention. Springer, Cham, 2015.

[6] Badrinarayanan, Vijay, Alex Kendall, and Roberto Cipolla. "Segnet: A deep convolutional encoder-decoder architecture for image segmentation." IEEE transactions on pattern analysis and machine intelligence 39.12 (2017): 2481-2495.

[7] Lin, Guosheng, et al. "Refinenet: Multi-path refinement networks for high-resolution semantic segmentation." IEEE Conference on Computer Vision and Pattern Recognition (CVPR). 2017.

[8] Zhao, Hengshuang, et al. "Pyramid scene parsing network." IEEE Conf. on Computer Vision and Pattern Recognition (CVPR). 2017.

[9] Chen, Liang-Chieh, et al. "Deeplab: Semantic image segmentation with deep convolutional nets, atrous convolution, and fully connected crfs." IEEE transactions on pattern analysis and machine intelligence 40.4 (2018): 834-848.

[10] Chen, Liang-Chieh, et al. "Rethinking atrous convolution for semantic image segmentation." arXiv preprint arXiv:1706.05587 (2017).

[11] Chen, Liang Chieh, et al. "Encoder-Decoder with Atrous Separable Convolution for Semantic Image Segmentation." (2018).

[12] Qi, Charles R., et al. "Pointnet: Deep learning on point sets for 3d classification and segmentation." Proc. Computer Vision and Pattern Recognition (CVPR), IEEE 1.2 (2017): 4.

[13] Qi, Charles Ruizhongtai, et al. "Pointnet++: Deep hierarchical feature learning on point sets in a metric space." Advances in Neural Information Processing Systems. 2017.

[14] Li, Yangyan, et al. "PointCNN." arXiv preprint arXiv:1801.07791 (2018).

[15] Mur-Artal, Raul, and Juan D. Tardós. "Orb-slam2: An open-source slam system for monocular, stereo, and rgb-d cameras." IEEE Transactions on Robotics 33.5 (2017): 1255-1262.

[1] *http://cocodataset.org*

[2] *https://github.com/qixuxiang/orb-slam2_with_semantic_label*

[3] *http://v.youku.com/v_show/id_XMzYyOTMyODM2OA*

[16] Salas-Moreno, Renato F., et al. "Slam++: Simultaneous localisation and mapping at the level of objects." Computer Vision and Pattern Recognition (CVPR), 2013 IEEE Conference on. IEEE, 2013.
[17] McCormac, John, et al. "SemanticFusion: Dense 3D semantic mapping with convolutional neural networks." Robotics and Automation (ICRA), 2017 IEEE International Conference on. IEEE, 2017.
[18] Tateno, Keisuke, et al. "CNN-SLAM: Real-time dense monocular SLAM with learned depth prediction." arXiv preprint arXiv:1704.03489 (2017).
[19] Xiang, Yu, and Dieter Fox. "DA-RNN: Semantic mapping with data associated recurrent neural networks." arXiv preprint arXiv:1703.03098 (2017).
[20] Tong, Zhehang, Dianxi Shi, and Shaowu Yang. "SceneSLAM: A SLAM framework combined with scene detection." Robotics and Biomimetics (ROBIO), 2017 IEEE International Conference on. IEEE, 2017.
[21] Girshick, Ross, et al. "Rich feature hierarchies for accurate object detection and semantic segmentation." Proceedings of the IEEE conference on computer vision and pattern recognition. 2014.
[22] Girshick, Ross. "Fast r-cnn." arXiv preprint arXiv:1504.08083 (2015).
[23] Ren, Shaoqing, et al. "Faster r-cnn: Towards real-time object detection with region proposal networks." Advances in neural information processing systems. 2015.
[24] He K, Gkioxari G, Dollár P, et al. Mask R-CNN[J]. 2017.
[25] Redmon, Joseph, et al. "You only look once: Unified, real-time object detection." Proceedings of the IEEE conference on computer vision and pattern recognition. 2016.
[26] Redmon, Joseph, and Ali Farhadi. "YOLOv3: An incremental improvement." arXiv preprint arXiv:1804.02767 (2018).
[27] Rusu, Radu Bogdan, and Steve Cousins. "3d is here: Point cloud library (pcl)." Robotics and automation (ICRA), 2011 IEEE International Conference on. IEEE, 2011.
[28] Schnabel, Ruwen, Roland Wahl, and Reinhard Klein. "Efficient RANSAC for point - cloud shape detection." Computer graphics forum. Vol. 26. No. 2. Blackwell Publishing Ltd, 2007.
[29] Adams, Rolf, and Leanne Bischof. "Seeded region growing." IEEE Transactions on pattern analysis and machine intelligence 16.6 (1994): 641-647.
[30] Rabbani, Tahir, Frank Van Den Heuvel, and George Vosselmann. "Segmentation of point clouds using smoothness constraint." International archives of photogrammetry, remote sensing and spatial information sciences 36.5 (2006): 248-253.
[31] Vo, Anh-Vu, et al. "Octree-based region growing for point cloud segmentation." ISPRS Journal of Photogrammetry and Remote Sensing 104 (2015): 88-100.
[32] Stein, Simon Christoph, et al. "Object Partitioning Using Local Convexity." CVPR. 2014.
[33] Golovinskiy, Aleksey, and Thomas Funkhouser. "Min-cut based segmentation of point clouds." Computer Vision Workshops (ICCV Workshops), 2009 IEEE 12th International Conference on. IEEE, 2009.
[34] Papon, Jeremie, et al. "Voxel cloud connectivity segmentation-supervoxels for point clouds." Computer Vision and Pattern Recognition (CVPR), 2013 IEEE Conference on. IEEE, 2013.
[35] Aldoma, Aitor, et al. "Tutorial: Point cloud library: Three-dimensional object recognition and 6 dof pose estimation." IEEE Robotics & Automation Magazine 19.3 (2012): 80-91.
[36] Comaniciu, Dorin, and Peter Meer. "Mean shift: A robust approach toward feature space analysis." IEEE Transactions on pattern analysis and machine intelligence 24.5 (2002): 603-619.
[37] Boykov, Yuri, Olga Veksler, and Ramin Zabih. "Fast approximate energy minimization via graph cuts." IEEE Transactions on pattern analysis and machine intelligence 23.11 (2001): 1222-1239.
[38] Sturm, Jürgen, et al. "A benchmark for the evaluation of RGB-D SLAM systems." Intelligent Robots and Systems (IROS), 2012 IEEE/RSJ International Conference on. IEEE, 2012.

IOP Conf. Series: Materials Science and Engineering **428** (2018) 012024 doi:10.1088/1757-899X/428/1/012024

An Auxiliary Notification System for Amnesiac Patients Based on Face Recognition and Voice Synthesis Using Cloud Computing

Qiming Sun[1], Ping Guo[2] and Changjiang Zhang[1,*]

[1] Department of Computer Science, Wenzhou-Kean University, Wenzhou, Zhejiang, China
[2] Department of Computer Science, University of Illinois at Springfield, Springfield, Illinois, USA

*czhang@kean.edu

Abstract. Many patients with amnesiac symptoms have difficulties to recognize people, even relatives' names. This paper aims at building an auxiliary system for patients with memory obstacles, especially for those who have difficulty in recognizing their relatives and close friends. The proposed system can detect the patient's surroundings in real time and use face recognition and voice synthesis techniques to recognize someone's approach to the patient. If the face recognition is achieved, it will notify the patient by text and voice messages, which is saved on cloud server during the training. Based on a novel framework, the system has a good usability and does not require complex human operations. It is friendly to patients. All its functionalities are designed to be automatic and efficient.

1. Introduction

Many patients with amnesiac symptoms suffer from recognizing people's names. For example, people with Alzheimer's Disease (AD) are very likely to get cognitive dysfunction in few years [1]. According to Laffan and Bier's research, as AD patients suffer from memory impairment, they may get trouble to make face-name associations. The re-learning of the associations and training of basic cognitive process can be applied to moderate the disease [2, 3] and improve memory for elders [4].

In this paper, we proposed a system to help amnesiac patients address face-name association problems by utilizing the combination of cloud server and smartphone application. The use of cloud computing could guarantee the efficiency and performance of face identification [5]. Backed by cloud server, smartphones with limited power can also accomplish complex computation. In addition, our system adopts the techniques of face recognition and voice synthesis, which can equip our system with automated and intelligent functions for users.

Our system allows users to register faces to the database on the cloud, the system can train a model online in a short time. After the registration, users can open the real-time function for patients. The application is able to automatically monitor the surroundings and detect the person who is approaching the patient. The system will then use voice synthesis technique to notify the patient.

Our proposed system could be a good auxiliary tool for people with memory obstacle. Each module is automated enough and does not require complex human operations. The system can automatically scan the camera's view and notify the patients in voice if someone else is coming. It runs in a good performance without interruption unless the application is closed manually, or the smartphone runs out of power.

2. Related Work

Face recognition and voice synthesis, which mean identify specific faces and converting text to speech, could be a helper for patients with amnesiac symptoms. Currently, researchers have proposed some successful examples to setup face recognition service on cloud service [6]. Many models and frameworks have combined mobile processing and cloud computing together with a good performance [7].

Researchers have shown that building an auxiliary system on a mobile phone for health-care has a very good usability [8-10]. Wade's team suggests that the mobile phone has a potential to enhance independence, reduce stress and could be a good memory aid tool [11]. Migo et al conclude that smartphone applications have the potential to help patients with memory problems [12]. In addition, Evald's research proved that smartphones are useful compensatory tools in memory recovery process [13].

3. Implementation Details

The proposed system uses offline face detection to check if there is any face information inside the camera view. Once a face is detected, the system can upload the images to the cloud server to do online face registration or recognition. The server maintains a database that organizes different devices and users, and the cloud services, including face recognition and voice synthesis, are under the support of Baidu AI Platform Java SDK. The entire system is built on Android and requires a smartphone with at least one camera and Internet connection. Figure 1 describes the general logic of the system.

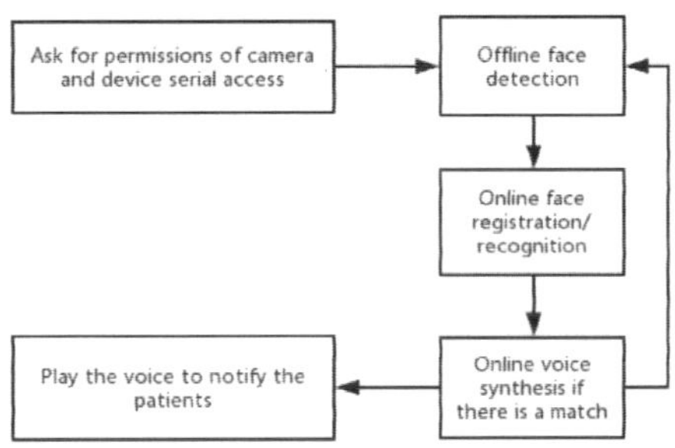

Figure 1. The overall logic of the proposed system.

3.1. Offline Face Detection

The face detection module aims at checking each frame from the camera's stream and decides whether a face exists or not. The detection process is used to make sure that the current frame is available to do further computation on the cloud, and since its operations are done offline, it significantly reduces the Internet usage and improves the system performance.

OpenCV Library, which is designed for real-time applications and multi-core processing, is used here to do the image process as well as the detection. For each frame captured by the camera, the image will be transformed to grayscale. Then a built-in cascade classifier will be loaded to try to generate the size and features of a face from the grayscale image.

Since the images taken by a cell phone camera are sometimes fuzzy because of slight hands shaking, in order to improve the image quality, an anti-shaking algorithm is introduced. The algorithm selects the sixth image with detected face after a continuous five images with faces, which means the camera should be stable in at least six frames, and it can avoid shakings in most cases.

Once an image in a frame passed the face detection with an available face as well as the anti-shaking check, it will be converted to a byte array and sent to the cloud server to do more computation for face recognition or face registration.

3.2. Online Face Registration and Recognition

Two core cloud-based modules, face registration and face recognition, are demonstrated in this section. Face registration is used to train the system to distinguish a certain face. Users should use this function at first to let the system build a specific model in order to recognize their faces. After a face is registered, the user can be identified by the face recognition module once he/she appears in the camera view.

In face registration, the module accepts three images from the detection module, and the interval between two captures is 5 seconds. In the whole process, instructions will be shown on the screen to suggest users moving their head. In this way, three images of a face in different angles are retrieved, then uploaded and registered in the cloud server. Besides, the system requires users to provide their names, which will be used to synthesize a fragment of voice to notify the patients in recognition module.

Face recognition module can keep monitoring the surrounding environment and notify users in voice once a person appears and the face has a similar matching in the current database. By uploading an image contains face information through the SDK, a JSON string with the similarity score of the best match and its description string will be returned. If the score is greater than a threshold value, such as 80, meaning 'very similar', the system will consider the person as a registered user, and the description string will be sent to voice synthesis module to generate and play a speech.

3.3. Voice Synthesis Module

Voice synthesis is an auxiliary function that helps the patients know who is coming or talking to him/her. Based on the Baidu's platform, the module can get a byte array with the synthetic voice fragment by sending a text to the cloud server.

The module is integrated with a converter and a media player. The converter uses an output file stream to save the voice fragment to a cache directory in MP3 format. The media player then uses an input stream to load and play the MP3 file. The temporary files will be deleted automatically after the user terminates the application.

3.4. Control of Users

The proposed system supports multi-device as well as multi-users. Figure 2 shows the framework of the user repository.

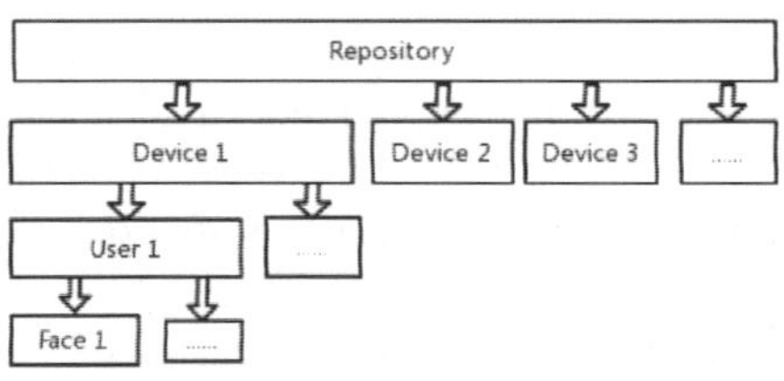

Figure 2. Organization framework of users.

The system uses a cell phone's IMEI as a unique key to distinguish different devices. When a user installs and starts the application for the first time, it will automatically register an account on the cloud server using IMEI number of the cell phone. Each device supports multiple users and each user supports multiple face images.

3.5. Multi-thread Framework

To optimize the user experience and reduce the negative influence caused by Internet latency. A novel multi-thread framework is designed to build the system. Figure 3 demonstrates the details of the framework.

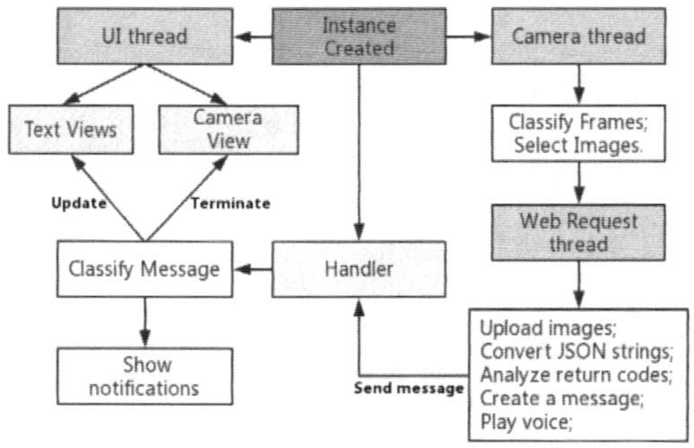

Figure 3. The multi-thread framework inside the system.

Once an instance is created, the system will create a UI thread, a camera thread, as well as a handler for cross-thread communications. The UI thread organizes instructions and elements on the screen. The camera thread is used to classify the frames and select suitable images with face information. After a suitable image is selected, a web request thread will be created to communicate with the server and play the synthetic voice. The message returned from the cloud server will then be passed to the handler. The handler can classify the messages, determine the next steps, and update the other threads.

4. Results

A smartphone with Android system is required to run the application. Figure 4 is a flowchart of the detailed usage of the proposed system.

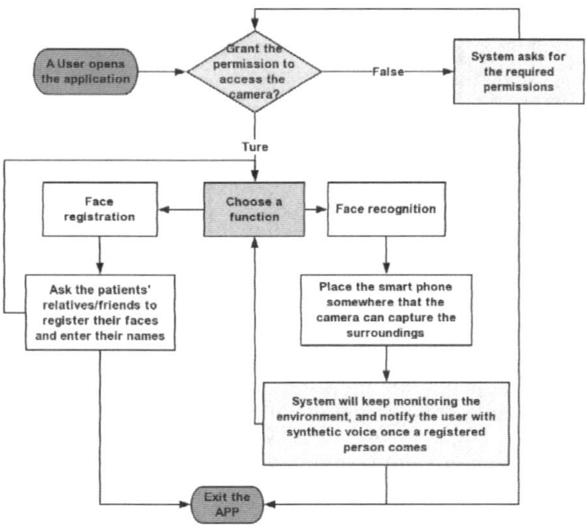

Figure 4. The flow chart to use the system.

Initially, when users open the application, they are required to grant the permissions to allow the program to invoke the cameras and get the phone's IMEI code. Then people should register their faces before they can be recognized by the system. Users can ask the relatives or friends of the patients with memory problems to register their faces. Before the registration, the system asks the user to enter a text description, such as a name, as the label to the registered face. The text will be used to generate a voice fragment to notify the patients later. Then the system will open the camera and start to collect the facial information. This process could last for 10-15 seconds. Figure 5 shows a screenshot of the registration module. 'David' is registering his face and the system is training a model online.

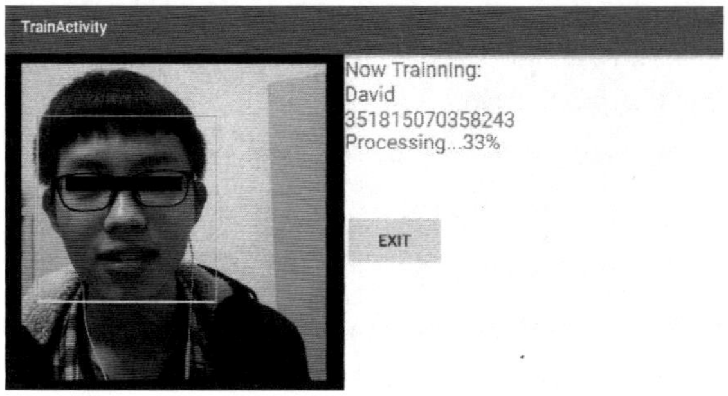

Figure 5. The user interface of face registration.

After the registration, the system will be able to recognize the specific person. When the user chooses the recognition function, simply place the smartphone somewhere that can capture the faces is enough. For example, fix the smartphone on the patients' wheelchair. The system can keep monitoring the surroundings. For instance, an amnesiac patient cannot recognize the relatives in a short time, when a relative, who has registered his/her face in the system, comes to talk to the patient, the system can capture the person's face and tell the patient who is coming. Figure 6 shows the face identification interface; when a registered person comes inside the camera view, the system can recognize him and display his name. At the same time, His name 'David', which has been registered as the label text, will be spoken out to notify the patient.

Figure 6. The user interface of face recognition.

5. Conclusions

The proposed system has the ability to recognize registered people's face in real time and notify the person in voice when the person is approaching him/her. It is easy to learn and operate with a simple user interface and could be a good auxiliary tool for elder people with amnesiac symptoms. Also, it does not need the patients themselves to be able to use the smartphone. Instead, simply install the phone near the patient is enough, the application can keep notifying the patient's surroundings for someone's approaching without human intervention until it runs out of power or the application is closed.

In the future, we will ask more patients to try the system. Based on their feedbacks, some new features or functions might be added. We will try to make the application more helpful to the people in need.

We may also consider using an embedded system to build a special face recognition device considering the limited power of a smartphone.

IOP Conf. Series: Materials Science and Engineering **428** (2018) 012024 doi:10.1088/1757-899X/428/1/012024

References

[1] Amieva, Hélène, et al. "Prodromal Alzheimer's disease: successive emergence of the clinical symptoms." Annals of neurology 64(5), 492-498 (2008).

[2] Laffan, Amanda J., et al. "Making errorless learning more active: Self-generation in an error free learning context is superior to standard errorless learning of face–name associations in people with Alzheimer's disease." Neuropsychological rehabilitation 20(2), 197-211. (2010).

[3] Bier, Nathalie, et al. "Face–name association learning in early Alzheimer's disease: A com-parison of learning methods and their underlying mechanisms." Neuropsychological reha-bilitation 18(3), 343-371. (2008).

[4] Hering, Alexandra, et al. "Prospective memory training in older adults and its relevance for successful aging." Psychological research 78(6), 892-904. (2014).

[5] Yuan, Jiawei, and Shucheng Yu. "Efficient privacy-preserving biometric identification in cloud computing." INFOCOM, 2013 Proceedings IEEE, pp. 2652-2660. IEEE (2013).

[6] Soyata, Tolga, et al. "Cloud-vision: Real-time face recognition using a mobile-cloudlet-cloud acceleration architecture." Computers and Communications (ISCC), 2012 IEEE Symposium on, pp. 000059-000066. IEEE (2012).

[7] Shiraz, Muhammad, et al. "A review on distributed application processing frameworks in smart mobile devices for mobile cloud computing." IEEE Communications Surveys & Tuto-rials 15(3), 1294-1313. (2013).

[8] Yamagata, Christina, et al. "Mobile app development and usability research to help dementia and Alzheimer patients." Systems, Applications and Technology Conference (LISAT), 2013 IEEE Long Island, pp. 1-6. IEEE (2013).

[9] Ventola, C. Lee. "Mobile devices and apps for health care professionals: uses and benefits." Pharmacy and Therapeutics 39(5), 356 (2014).

[10] Zorluoglu, Gokhan, et al. "A mobile application for cognitive screening of dementia." Com-puter methods and programs in biomedicine 118(2), 252-262. (2015).

[11] Wade, Tracy K., and Juliette C. Troy. "Mobile phones as a new memory aid: a preliminary investigation using case studies." Brain injury 15(4), 305-320. (2001).

[12] Migo, Ellen M., et al. "mHealth and memory aids: levels of smartphone ownership in pa-tients." Journal of Mental Health 24(5), 266-270. (2015).

[13] Evald, Lars. "Prospective memory rehabilitation using smartphones in patients with TBI." Disability and rehabilitation 25(2), 1-10. (2017).

IOP Conf. Series: Materials Science and Engineering **428** (2018) 012025 doi:10.1088/1757-899X/428/1/012025

A High-precision Pressure Controller Based on Zigbee Transmission

Wang Chao[1] and Pu Han[2]

[1]Changji university, Chang Ji 831100, China
[2]Nanjing tech university ,Nanjing 211800, China

[a]Corresponding Email:wang0001chao@163.com

Abstract: Nowadays, the pressure controller is one of the most commonly used sensors in the market, it has advantages of good real-time performance, good efficiency, but also exist precision is poor, low resolution and volume weight, inconvenient to carry shortcomings. Therefore, this article designed a high accuracy pressure controller based on zigbee transmission, and gives the design diagram, hardware composition and software system, the system USES 51 single chip microcomputer, ten high precision A/D conversion chip HX711, cantilever type resistance strain sensor and zigbee wireless transmission module, system can be collected in real-time and high precision gravity through to the digital tube display; , at the same time, through the zigbee wireless transmission way of transmitting current data terminal equipment (such as: PC, mobile phone, etc.), convenient for the user in the long-distance real-time view the latest data, in improving the efficiency of using the system at the same time, also ensure the data real-time and efficiency.

1. Introduction

With the continuous improvement of people's living standard and the increasing modernization of commercial behaviors, people also put forward new requirements on the speed and precision of commodity measurement [1-3].

Currently, the pressure controller is one of the most commonly used sensors in the market, and all have one thing in common: on the scene, the pressure sensor displays the value on the current screen. However, in the field of precision control such as current industry, gravity controller has the disadvantages of low precision and difficult to view data in different places.

For using these two shortcomings, therefore, to design a set of high precision pressure controller based on zigbee transmission, this paper discusses the design idea, method and design of the system implementation process, a detailed analysis on the selection of various modules, functions and implementation method, including the system hardware structure, the selection of sensor, the flow chart of system operation, etc.

Control core unit using STC89C52 as CPU control module, sensor using A sensor with high precision, stability, good resistance bridge - 8 HL type sensor, A/D conversion module USES HX711 of high resolution and high accuracy.

2. Overall design of digital electronic scale

This digital electronic scale system can be divided into single chip computer control circuit, zigbee wireless transmission circuit, sensor sensor circuit and digital tube display circuit, etc, as shown in figure 1.

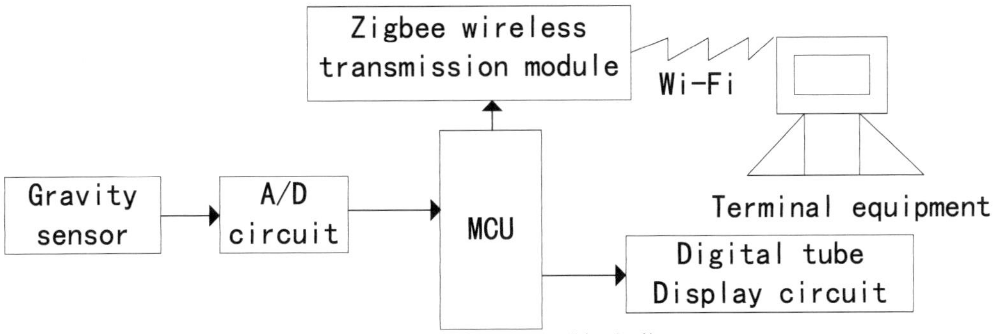

Figure 1. System block diagram

3.System hardware design

3.1. Selection of piezoelectric sensors

The strain gauge resistance sensor is a sensor that USES the strain gauge as the identification element. It has many advantages, such as: 1. High precision and wide measurement range; 2. Good frequency response; 3. Variety of strain gauge and low price; 4. High sensitivity of system contact points, high integration degree and simple structure [4-6].

The sensor used in this design is a full bridge measuring circuit. In the absence of pressure, the bridge is in equilibrium and the output voltage is 0V; When there is pressure, the bridge arm resistance of the bridge is similar to the linear change, and the bridge gradually loses its balance. In the whole bridge measuring circuit, the two strain plates with the same force nature are connected to the opposite side of the bridge. The output sensitivity of the half-bridge is approximately doubled, and the nonlinear error and temperature error [7] are also improved.

The circuit diagram of the full bridge measurement circuit is shown in figure 2.

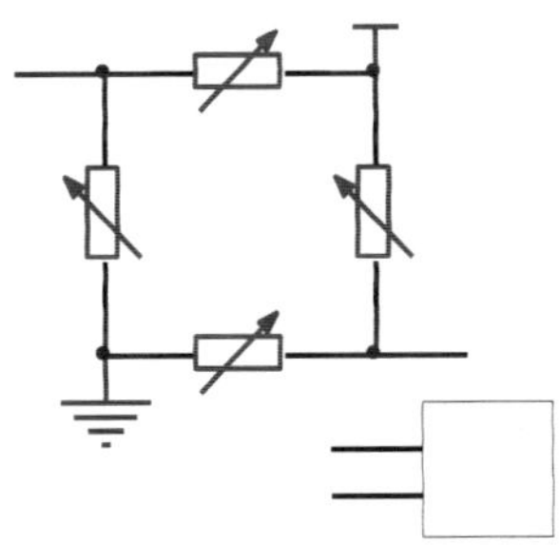

Figure 2. Full bridge measuring circuit

3.2. A/D conversion chip HX711 interface circuit design

According to the requirements of the system to use, must use reach 10 conversion precision A/D conversion equipment, however, considering the STC89C52 SCM I/O port resource factors such as tension, quantitative accuracy of final use HX711, it can reach 1/4096 < 1/1000, comply with the design requirements.

The HX711 module diagram is shown in figure 3.

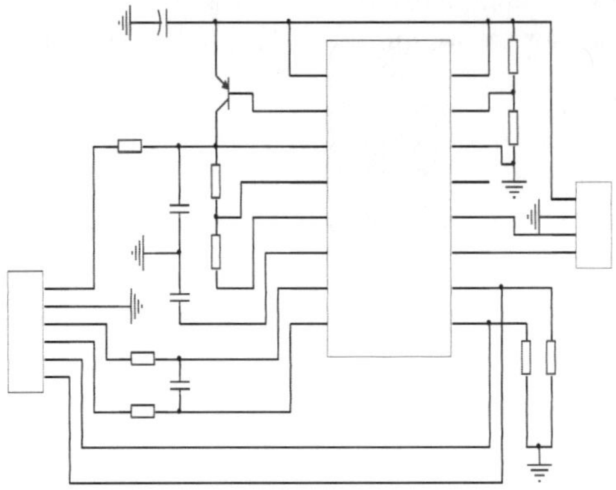

Figure 3. HX711 module chart

3.3. Digital tube display circuit design

The system USES four segments of digital tubes as peripheral typical devices. At the same time, in order to enhance the capacity of MCU para choose signal driver, the P2 port (a) in the 9012 transistors (PNP) base, 9012 triode act as the role of the switch, and at the same time, greatly improving the digital tube display brightness, for observation of the output data.

The digital tube display circuit of this design is shown in figure 4.

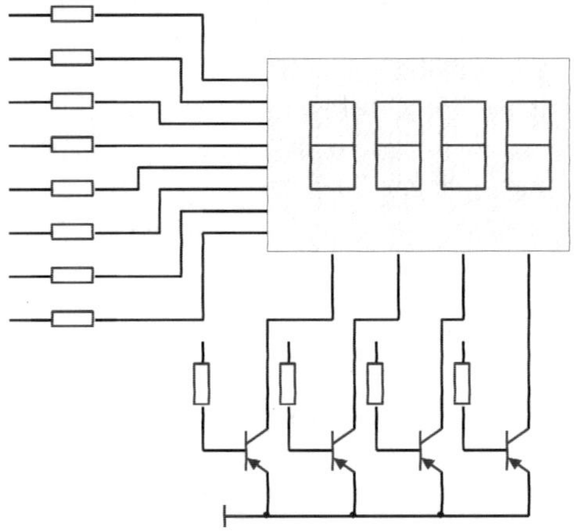

Figure 4. Digital tube display circuit diagram

3.4. Zigbee wireless transmission circuit design

Zigbee wireless transmission technology, is now one of the most popular wireless control mode, has a good real-time performance and long transmission distance, and the advantages of good stability, the system USES zigbee technology, makes the current collected gravity value can be displayed on the terminal equipment, greatly improve the utilization rate of the data and convenient for the user in real time, the accuracy and stability of different view the data.

4. System test and calibration

4.1. System hardware test

In the process of system hardware debugging, first check whether the components on the PCB welding correctly, observe carefully if there is any leakage welding phenomenon, with a multimeter test PCB whether there is the phenomenon of short circuit or open circuit. After the test, the power supply is connected. Meanwhile, the voltage value of the power part is measured with the multimeter, and the error is obtained. First press microcontroller reset button reset the system, digital tube display subroutine call 1234, after determine the digital tube display correctly, connected to sensors and composed of 24 HX711 serial A/D converting circuit and A/D conversion circuit, the burning of all of the application to the MCU processor, observe whether digital tube display, 0000.

If it is not, the software will be adjusted to zero. After the zero is completed, the pressure will be exerted on the sensor by hand, and the display value on the digital tube will increase with the increase of pressure. When you let go, observe whether the number of digital tubes is returned to around 0000.

4.2. The determination of linearity

In the process of selection, in turn increase the quality of the weights, try to put weight on sensor test side edge, Jordan farmar's centre of gravity in the center of sensor test area, so it is concluded that the value of the more accurate. After the display value of the digital tube is stable, record the display value and use MATLAB to draw the curve.

4.2.1 Display value of digital tube when unmarked. In the software part of the design, if no clear program is added, when no weight is placed on the sensor, the digital tube displays the initial value of 70 (g). Then we put 10 (g), 20 (g), 40 (g), 50 (g)... , write down the number of digital tubes. Through the MATLAB simulation, it is concluded that the display value of the digital tube is linear with the weight value. FIG. 5 is a linear test curve when uncalibrated;

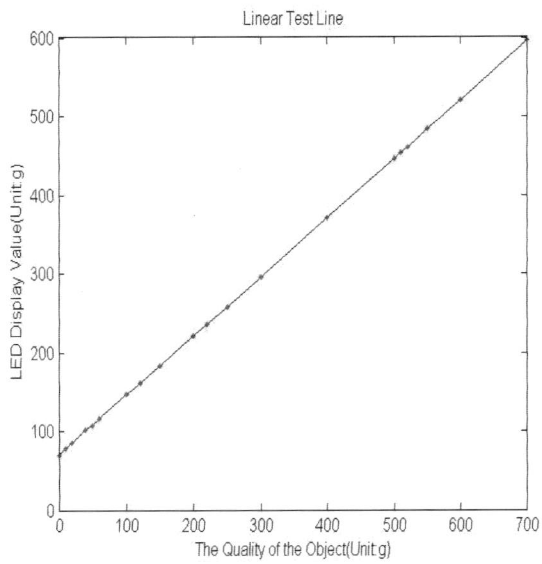

Figure 5. Linear test curve

4.2.2 Digital tube display value when unmarked. The display value of the digital tube when unmarked.

The least square method has strict mathematical basis. Although the calculation is complicated, the fitting linear precision is high, namely, the error is small. Then, in the code design, add the digital tube to display the zero - zero procedure, the calibration process is completed. The calibration curve after calibration is shown in figure 6.

IOP Conf. Series: Materials Science and Engineering **428** (2018) 012025 doi:10.1088/1757-899X/428/1/012025

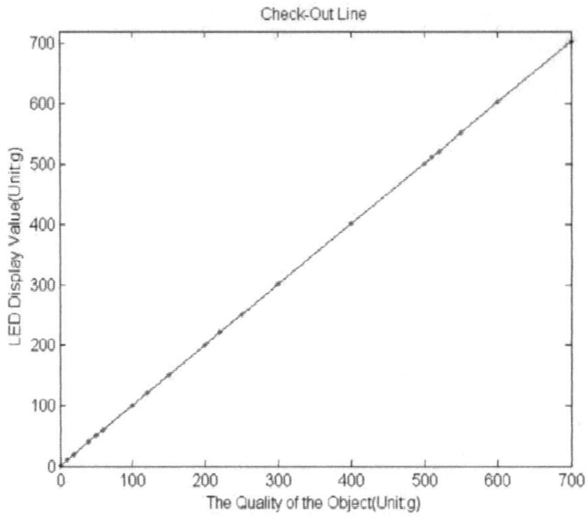

Figure 6. Linear test curve

4.2.3 The error analysis. The experimental results show that HX711 itself is highly sensitive, and if there is no fixed position of the sensor, there will be some errors in the data obtained.

In addition, the lead of HX711 is also very sensitive, even if it is slightly touched, it will produce mechanical errors.

5. The conclusion

Based on zigbee transmission, the high-precision pressure controller is characterized by high real-time performance, high accuracy, and remote viewing of real-time data, portability and high cost performance.

This system can get high accuracy in the field of gravity, also can by connecting LAN, through the terminal device to view real-time gravity, bringing great data repeated utilization ratio and improve the limited accuracy.

Especially in the process of determination of linearity, the experimental test data using MATLAB graphics, digital tube display value and weight value can be drawn from the relationship between close to a straight line, therefore, it is concluded that the linear fitting relationship between them, finally complete the calibration.

References

[1] JI Zhong-yong. Design of electronic scale based on single-chip microcomputer [J]. Technology information,2013(18):267-268.
[2] GUO Zhi-yong. Design and implementation of high-precision electronic scale based on ADS1100 [J]. Journal of An Hui university of technology (natural science),2010,27(4):385-392.
[3] Sun Na. Design of portable electronic scale based on single-chip microcomputer [J]. China science and technology information,2012(1):99-100.
[4] YANG Cheng-hui, WANG Shu-zhi, YANG Wei-xin. Research and design of a digital electronic scale [J]. Journal of northwestern university of nationalities (natural science edition),2012,33(3):22-27.
[5] HU Mei-jiao, GAO Mei-chun. Design of high-precision electronic scale based on 24 bit A/D conversion [J]. Modern computer (professional edition),2013(32):61-63.
[6] Wang Feng, XU Hong-yao. Design of electronic scale based on single-chip microcomputer [J]. Technology innovation and application,2013(15):26-26.
[7] Cheng Zhi. Design of simple electronic scale [J]. Electronic world,2013(7):133-133.

IOP Conf. Series: Materials Science and Engineering **428** (2018) 012026 doi:10.1088/1757-899X/428/1/012026

Design and Implementation of Inertial Measurement Information Acquisition System Based on XML

Pei Yu[1,a], Ting Wang[1], Jinyun Zhang[1] and Jing Li[2,b]

[1]Beijing Aerospace Control Instrument Research Institute, Beijing 100854
[2]Information Engineering College, Beijing Institute of Petrochemical Technology, Beijing 102617

[a]biacd_yupei@163.com
[b]lijing0408220@163.com

Abstract. Platform inertial measurement information acquisition system is to acquire, store and display inertial measurement information of inertial platform navigation system by means of data communication. Due to the different types of platform inertial navigation system, the communication protocols are diverse. In this paper conducts the repetitive coding problem caused by different formats of communication protocols and put forward an improved method based on XML template applied to format of data unit in protocols. This method greatly improved the generality of acquisition software. In the premise that the acquisition board is not replaced, this method according to the type of platform to change the XML template, without compiled twice time. The inertial information acquisition system has been applied in practice. Practice has proved that this method can significantly reduce the workload of code maintenance and cut down the cost of personnel.

1. Introduction

Platform inertial measurement information acquisition System acquire, store and display the inertial measurement information of the platform inertial navigation through data communication. The system needs to communicate with the underlying equipment according to the relevant protocol format. Due to different types of inertial navigation platforms different types of protocol frames and different communication protocols exist. As a result, the data processing complexity increases when data packet analysis is performed, and the communication format needs to be analyzed according to the corresponding protocol type.

XML is a structural markup language used to mark electronic documents, a language that can be used to mark data, define data types, and allow users to define their own markup language. Hao Chao[1] using C # language through Microsoft.NET FrameWork developed XML-based monitoring software rapid design system, using object-oriented and modular ideas to achieve the software scalability. Xianyu Long[2] adopted XML standard model file and put forward a framework for data exchange of device model. The model was actually used in SCADA system software developed independently by the company.

In order to reduce the waste of resources caused by various protocol types and the resulting data interface design in the common platform inertial measurement information acquisition system. Based on the design idea of scalable low coupling, the protocol of message format with descriptive language XML is adopted. Through the XML protocol to achieve the protocol format and content description.

By using XML file creation and parsing, a small number of code modifications are made to resolve the communication protocol of inertial platform inertial navigation system. The system function module is independent and scalable, which solves the problem of data analysis and interaction of heterogeneous protocol in data communication.

2. Protocol Description Based on XML Protocol

2.1. Introduction to XML technology

Extensible Markup Language is a subset of SGML that provides a uniform way to describe and exchange structured data. XML is scalable, self-describing, structured, content and presentation apart, giving it a natural advantage in data description and information sharing. XML technology provides a standard markup language that can be used for data exchange, initially used to describe the document elements on the WEB page [3-5]. Compared with other existing data exchange technologies, XML technology has the following advantages:
1) Good readability;2) Convenient and scalable3) Data content and form of separation;4) Facilitate the retrieval of information;5) Easily cross-platform applications;

2.2. The sample code to parse XML configuration files

Platform inertial measurement information acquisition system using C + + Builder 6.0 environment for software development. The handling of XML is mainly encapsulated in the TXMLDocument component, which is part of the Internet Components component. This component has the following property methods:
1) Use the control method to load the XML file;2) Use of control methods to save the XML file;
3) Read the node, attribute data;4) Add child nodes and set properties5) Formatting XML
This article only uses the first method and third method of the above attribute method, the following code parsing XXXXML intermediate file protocol parsing code example:

```
1.Form1->XMLDocument1->LoadFromFile("XXX.xml");//load XML document
2.IXMLNode *my=Form1-> XMLDocument1->DocumentElement->ChildNodes->GetNode(0);
//get the first node
3._di_IXMLNodeList nodes=my->ChildNodes;4.int cc= nodes->Count;int *TaskID ; //create a task
5.ID dynamically String *TaskIDName; // create a task identifier name dynamically 6.TaskID=new
int[cc] ; // create a task ID 7.TaskIDName=new String[cc]; // create a task identifier name
 for (int i=0;i<cc-1;i++)// Circulate the task ID name and ID of the xml intermediate file to a dynamic
array through the loop
 {
TaskID[i]= StrToInt(nodes->Nodes[i]->ChildNodes->FindNode("ID")->Text);// get task ID
TaskIDName[i]= nodes->Nodes[i]->ChildNodes->FindNode("IDname")->Text ; } //get task identifier
name
```

3. Design and Analysis of Common Platform Inertial Measurement Information Acquisition System

3.1. Composition of platform inertial measurement information

In the platform inertial navigation system, the accelerometer used to measure the acceleration of the carrier and the gyroscope to measure the angular velocity of the carrier are mounted on the inertial platform.

The system uses the precession of the gyroscope and applies torque control to the gyroscope to offset the rotation angle of the gyroscope in order to keep relative stability with the inertial space so that the gyroscope always tracks the specified navigation coordinate system. The inertial platform avoids the influence of carrier motion on the acceleration measurement and provides the measurement basis for the whole system[6].The inertial platform of the platform can avoid the influence of the carrier motion on the inertial element, and the angle sensor on the frame corner can directly measure the attitude angle for navigation estimation.

Platform-based inertial navigation system mostly multi-axis multi-frame system, which is designed to accurately control all aspects of space and design, in the structure of the various vertical to each other, the framework without mutual coupling[7]. The inertial navigation system needs to measure the following information:

1) Frame Angle information

The three-axis platform inertial navigation system is used to measure the frame angle, inner ring shaft frame angle and outer ring axis frame angle information. The four-axis platform inertial navigation system is used to measure the frame angle, inner ring shaft frame angle, outer ring shaft frame angle, and the frame angle of the axis. The frame angle sensor is mounted on the frame axis of the platform and the output circuit of the frame is completed by the frame angle sensor and the output of digital output. The data acquisition software collects the output of the frame angle digital output after the output circuit is converted.

2) Acceleration information

The accelerometer is mounted on the triaxial and four-axis platform inertial navigation system. The acceleration measured is the component of the axial acceleration along the platform axis. In addition to the accelerometer for acceleration measurement, the sensor is also used as the sensor of the platform system to adjust the loop[8]. The accelerometer analog quantity is transformed by the I/F conversion circuit (range conversion circuit) into the pulse number and the counting circuit is transformed into a digital quantity. This data acquisition software collects the pulse output after conversion.

3) Gyro torque current

The gyro plus moment current is the gyroscopic precession driven by the instruction current generated by the inertial navigation computer. Through the platform stable loop, the platform realizes the rotation of relative inertial space, or counterbalances the gyroscope drift and ground speed, so that the platform coordinate system is followed by a designated geographic coordinate system[9]. The data acquisition software collects the digital quantity of the torque current output of the gyro plus moment circuit.

4) Platform task status and location number

Platform main task status including flight navigation, lock to zero, end of zero lock , self-calibration, end of self-calibration, self - targeting, end of self – targeting etc. of twenty kinds of state. Each state corresponds to the position of the platform body, so the location number is different. If the task identity of the platform is self-targeting, the position number is from 1 to 2 if the XX platform adopts the two-position self-aim method. If three or four positions are used, the position number is 1~4.

3.2. Design of XML file format based on inertial measurement information

The general platform inertial measurement information acquisition system collects, displays and saves the above data, which is convenient for subsequent data processing and analysis. Due to the diversity of platform inertial navigation system, the use of different types of platform inertial navigation system with separate types of platform inertial navigation inertial measurement information acquisition system, need a number of personnel for maintenance, when agreement to change or increase the control command platform, need a number of researchers to inertial measurement information acquisition system for the corresponding code modifications, maintenance difficulties and high cost. Based on this paper, we propose a protocol based on the XML technology in the first section of this paper. The XML file format is as follows:

1) Platform task status and location number

The platform task identity XML file is used to specify the ID number and task ID name corresponding to the platform task identity. Different types of frame type platform system task identifier with ID number and the task of Chinese name is different, there is the same ID number corresponding to different task identifier name different phenomena, so to solve by means of as shown in figure 1:

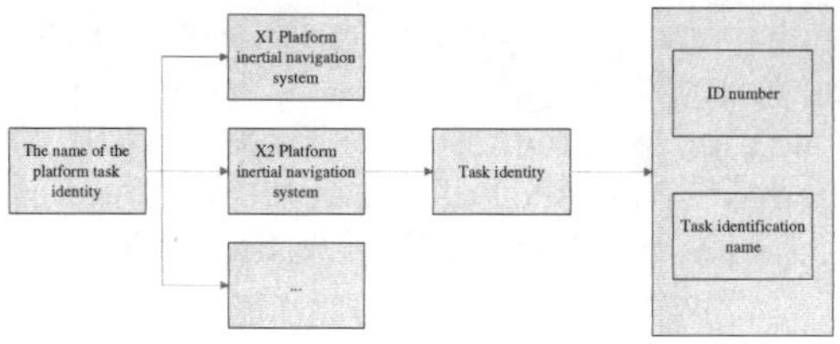

Figure 1: The XML file format of the platform task identity name.

2) XML file format of the protocol format

The protocol message format XML file is used to specify the parsing rules for valid data areas in different platform communication protocols. Due to the different types of frame type platform system corresponding to the instrument configuration and frame angle coordinate system, framework agreement from the angle of information data, accelerometer pulse output data, the position of the platform, mission of the position of bytes identify different platform. The solution is shown in figure 2.

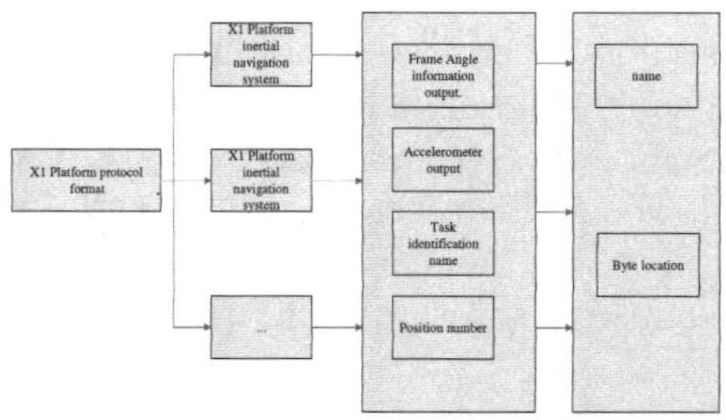

Figure 2: The XML file format of the platform protocol

3.3. The parsing process for the XML protocol specification template

Two structure types are defined according to the specification template that has been established for XML. The first structure is the task identity structure, which contains two attributes. When parsing XML template file as shown in figure 3: first of all, according to the corresponding platform type read the corresponding XML documents, access to the root of the XML document tag in the frame format of the number, for a total of several child nodes. Then the information of the frame subtag of the sub-node is resolved. Through the above process, the XML formalized description of the entire protocol structure is identified.

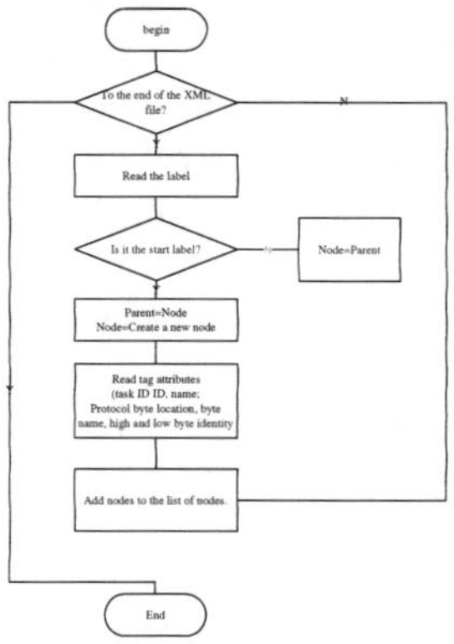

Figure 3: The parsing process for the XML file

4. Design and implementation of inertial measurement information acquisition system for platform inertial measurement

4.1. Description of system functions

Inertial navigation information acquisition software platform to achieve the following six functions: the card interface initialization, according to the corresponding platform type selection data storage paths, receiving navigation data, analytical processing navigation data, save the navigation data, show the navigation data. Among them, the board card interface is initialized and the data storage path is selected according to the type of the platform, which requires user participation, and the remaining functions are automatically realized by the software.

4.2. The description of the structure system

Platform inertial navigation information collection software consists of data acquisition module, data analytical processing module, configuration management module, data display module, data storage module and user interface.

1) Data acquisition module

The data acquisition module used API functions that call the underlying driver software to collect data. The underlying driver software is provided by the data acquisition board vendor, and the manufacturer will package the required functional functions into DLL and header files. Therefore, the collection module can only call the initialization and interrupt function of the corresponding plate card, and the data can be collected.

2) Configuration management module

The configuration management module implementation is resolved according to the platform type to use the configuration template in the valid data resolution area of the data receiving area, and the configuration template includes two parts. Through configuration templates to different types of platform framework from the Angle of the data, the accelerometer and gyro torque current task identification data, platform byte position, overcomes the problem of poor universality of software.

3) Data analysis module

The data analysis module analyzes the public data analysis area and the valid data analysis area respectively. The framework Angle data, accelerometer data, gyro plus moment current data, platform task identification and location number are analyzed in the valid data analysis area. The inertial information of different types of platforms is analyzed and stored in the corresponding variables to carry out subsequent data display and storage module.

4) Data display module

Depending on the platform type, the user's graphical interface is displayed according to the type of the platform selected. The user selects the corresponding type of platform inertial navigation and displays the inertial measurement information of the corresponding platform inertial navigation.

5) Data storage module

The data of inertial measurement information of different types of platform inertial measurement is stored according to the unified storage type. The system switches the data files according to the task handover and file size, ensuring the continuous unification of the files and the readability of the data files.

4.3. The workflow of the system

The workflow of the system is shown in figure 4. The first step is to initialize the data, which is to configure the receiving register before receiving it. If the receiving state is valid, the data is received, the read application control register is configured, the cache is applied, and the data can be read until the data is received. The receive cache is released, and the receiving state is checked if it continues to read. Repeat the above process, otherwise the receiving end.

IOP Conf. Series: Materials Science and Engineering **428** (2018) 012026 doi:10.1088/1757-899X/428/1/012026

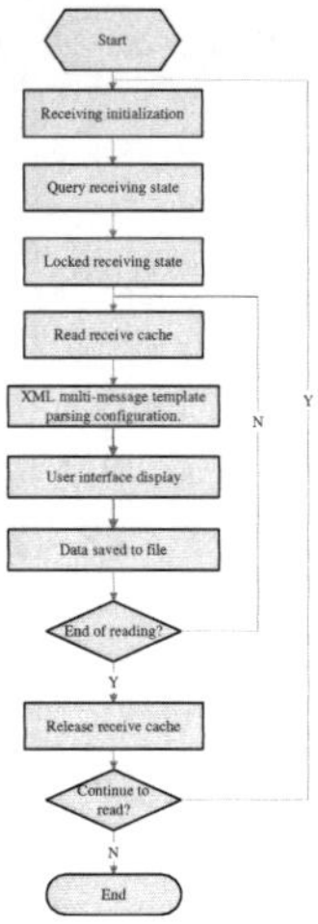

Figure 4: The workflow of the system

5. Conclusion

Due to the types of platform system framework system and instrument configuration caused by different communication protocols different kinds of difficulties, need many people maintain many sets of data acquisition software and repeat coding issues. This paper presents an xml-based message format specifications, greatly improving the inertial platform inertial navigation system information acquisition software versatility, ensures that different types of derivative according to the consistency and security of the exchange platform used to. The software has been used in the actual work. The practice proved: the method is significantly reduced code maintenance workload, the need to people before maintenance reduced set of software to one more person to maintain a set of software, reduce staff costs.

References

[1] Chao Hao, Zhe Yang, Rapid design technology of monitoring software based on XML[J], China Science and Technology Information, 2017(06): 36–37.

[2] Yulong Xian, Jingyang Zhang, Peng Chen, etc. An interactive standardization method for oil and gas pipeline equipment model based on XML[J], Electronic Science& Technology, 2017(04):28-32.

[3] Liansheng Yang, Aiping Li,Jiawei Li, etc. Research and implementation of data exchange interface for ship production design based on XML[J], Shipbuilding of China, 2016(04):164-174.

[4] Zhenjie Zhao, Yuehui Yan, Hao Wang, etc. Data exchange technology for heterogeneous PDM platform based on XML[J], Intelligent Manufacturing, 2017(04): 35–38.

[5] Jinyu Gao, Ping Cao, Run Chen, etc. Research on structured method based on XML[J]. Aeronautic Standardization & Quality, 2017(01): 45–47.

[6] Yuanjiu Lu, Inertial components[M]. Beijing: China Aerospace Press, 2009:Chapter 16 p169

[7] Yiyuan Deng, Hydrostatic floating gyro platform system[M].Beijing: China Aerospace Press,2012:Chapter 8. p121

[8] Ying Gu, The inertial guidance accelerometer technology overview[J].Winged Missiles Journal, 2001(06): 78–85.

[9] ErWei Wang, Haifeng Jiang, Gang Huang, etc, Based on the LabView gyro moment circuit testing method[J].Navigation and Control, 2016(05): 108–112.

IOP Conf. Series: Materials Science and Engineering **428** (2018) 012027 doi:10.1088/1757-899X/428/1/012027

The Research on the Ice Thickness Detection System Based On Capacitance Sensor

Du Xiaoting

Anhui sanlian college No. 47, He'an Road, Economic and Technological Development Zone, Hefei City, Anhui Province

xt198200@163.com

Abstract. A transmission line remote monitoring system combined with wireless sensor technology was designed, through the single-chip computer control to complete the system data from the collection to send all aspects of data processing work.The icing monitoring system uses capacitive thickness detection sensors to collect data from multiple capacitance values and send the collected data to the control center processor through the serial port. The processor then processes the data and stores it on the SD card. in. Through the single-chip computer control to complete the system data from collection to send all aspects of data processing. The processor finishes the data transmission and connects with the GPRS communication module through the RS232 serial port. The automatic detection of ice thickness using the difference in resistance characteristics of air and ice, it is a new type of object surface ice coating thickness detection method. The utility model can be applied to real-time continuous automatic monitoring engineering application fields of high-pressure transmission lines and fixed towers, buildings or various equipment surfaces, suspension brackets, forest branches, and the like.

1. Introduction

With the rapid development of wireless communication technology and low power consumption embedded computing technology, Wireless Sensor network (WSN) has emerged as a combination of various modern information technologies. The WSN has the characteristics of fast deployment, low cost and strong adaptability. It is widely used in environmental monitoring, disaster prediction and other fields, it is an indispensable key technology in new interactive systems such as the Internet of things and information physical fusion system[1-3]. The high voltage overhead transmission line's safe operation is the key to the reliable transmission of electric energy and the stable operation of the whole network. overhead lines icing situation is usually occurs in areas with complex terrain and harsh environment. Therefore, traditional manual inspection methods cannot be solved(traditional manual inspection methods cannot solve it). A variety of overhead line icing monitoring system has emerged , at present, the prediction model detection system has been widely applied. However, there are a number of factors affecting the ice formation of transmission lines. The prediction model may produce an obvious deviation, it's not very reliable [4-7] .

2. Principle of line icing

Transmission line icing is related to the geography of the wire, if the transmission lines are erected in the mountains, the direction of the mountain range and the slope of the slope will have a great impact on the icing. In addition, the relative position of transmission lines and water sources will directly affect the thickness of the ice. The common types of ice cover are showed in table 1[8-9].

Table 1. The type of ice cover and the formation reason

NO.	type	reason	characteristic
1	glaze	Because the near-earth temperature is low,as the snowflakes fall, they turn into drops of water., the water droplets fall to the surface of the object in the form of cooling water droplets.	The adhesion is very strong and the density is very high
2	Mixed battle	The ambient temperature is below freezing point and the wind is strong.	The adhesion force is stronger, the density is larger, and it's application grows faster
3	Soft rime	The lower clouds in the mountain range have lower temperature and less wind speed.	Weak adhesion, small density, less threat to wire
4	white frost	Moisture in the air comes in contact with objects below the freezing point, and the moisture condenses on the surface.	The cohesive force is very weak, and it is hardly a serious hazard to the conductor
5	snow	The snow in the air is high humidity, and the wind is insufficient.	Wind - driven, plain or low-lying areas are more common.

3. Ice monitoring system scheme

The block diagram of icing monitoring system is shown in figure 1:

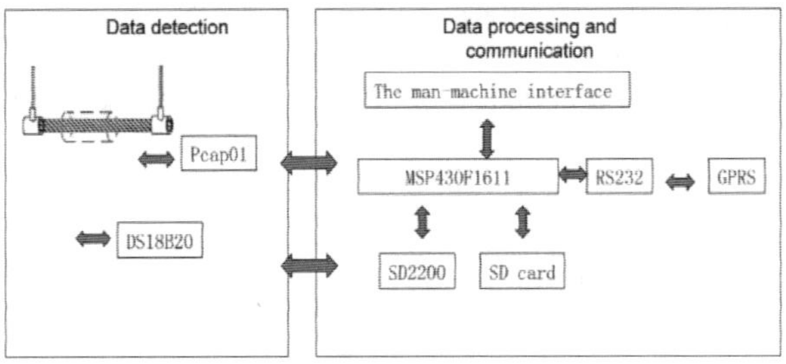

Figure 1. Block diagram of ice thickness detection system.

The measuring module of ice monitoring system includes two parts: capacitance measuring circuit and temperature sensor. We used a capacitive sensor that reflects the thickness of the ice coating, an overhead transmission line is used as one of the plates of this capacitor, and the other plate is a parallel induction line erected along the overhead line, therefore, if there is ice covering between the plates, the medium between the plates will change, and the capacitance value of the sensor will change accordingly, moreover, the variation value varies with the thickness of the ice cover. The temperature of the transmission line will increase with the increase of the temperature current Because the current power line carrier is larger when the temperature increases, and that will affect the measurement results. A temperature compensation circuit is added to the monitoring system to ensure that the temperature of the two plates is consistent. The temperature compensation device detects the temperature difference between the two plates by the temperature sensor installed on the sensor line, and the temperature of the plate is kept in accordance with the control temperature of the heating device. The power module of the monitoring system adopts solar power supply mode to charge the system battery.

4. Ice Thickness Detection Algorithm

The structure diagram of ice thickness detection circuit is shown in figure 2, transmission line parallels 1 to induction line, the cross-sectional area of the two lines is denoted as R1, R2, respectively, the distance between the transmission line and the sensing line is d. Assuming that the thickness of the ice overlay on the two lines is D at a certain moment, the dielectric constant of air is expressed, and the dielectric constant of the ice overlay is ε_0, The dielectric constant of ice coating is ε_1.

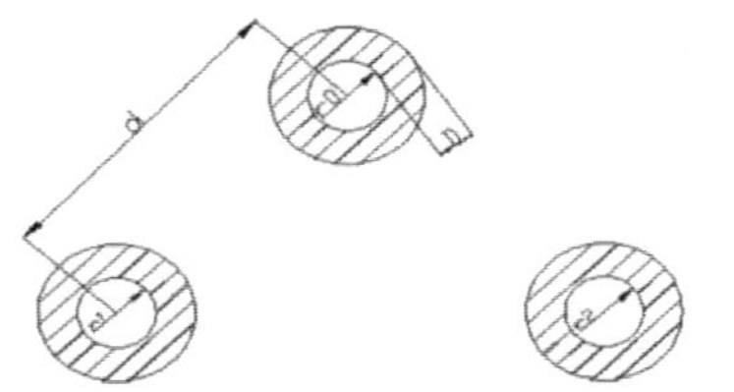

Figure 2. The structure diagram of ice thickness detection circuit

According to Chang Ching, etc [8], The capacitance between two bare wires in the air with a radius of R and a distance of d is shown in Equation 4-1:

$$C = \frac{Q}{U} = \frac{2\pi\varepsilon_r}{\ln\left[\frac{d^2 - R_1^2 - R_2^2}{2R_1R_2} + \sqrt{\left(\frac{d^2 - R_1^2 - R_2^2}{2R_1R_2}\right) - 1}\right]} \tag{4-1}$$

5. Ice thickness detection sensor coverage model

The ice coverage detection system sensor network probability coverage model is shown in Equation 5-1:

$$P_p(S) = \begin{cases} 1, D(S,P) < R \\ \lambda e^{-a(D-RS)}, D(S,P) \geq R \end{cases} \tag{5-1}$$

The distribution of ice coating directly affects the distribution of wireless sensor networks. The geographical factors in the area are directly related to the distribution of ice coating. For example, ice thickness increases with altitude. Taking into account the impact of terrain and other factors on the coverage of wireless sensor networks, a new variable λ is introduced, and $0 \leqslant \lambda \leqslant 1$, when the height difference between the target position and the sensor is greater, λ is smaller, and when The greater the change in climate conditions between the target location and the sensor, the smaller the value of λ .

In the formula:

D(s.p) refers to the distance between the sensor and the target location

R: refers to the vertical span of the sensor; a: indicates the sensor's detection capability coefficient

6. Ice coating sensor detection system designing

6.1. Controller selection

Based on the requirements of low power consumption and data processing capability required for actual projects, the MS16430 processor F1611 from Texas Instruments was selected.As shown in table 2:

Table 2. MSP430 I/O ports and function allocation

I/O port	Features
P1	Clock reset

P2	Power module control port
P3	LCD control port
P4	Sensor connection port
P5	Data access port
P6	Alternate port

6.2. Temperature sensor selection

According to the requirements of ambient temperature and measurement accuracy of transmission lines, DS18B20 digital temperature sensor was selected as the temperature measurement device, Temperature measurement from minus 55 degrees Celsius to zero over 125 degrees Celsius can be achieved, measurement accuracy can reach 0.5 degrees Celsius, and multi-point networking capability.

6.3. A/D conversion module selection

We selected TI's ADS1100, with a 6-pin DIP package, it can sample 8, 16, 32, or 128 samples per second for conversion.On-Chip PGA Provides 8x Gain.it has a high resolution, smaller signal detection can be achieved.As shown in Figure 3:

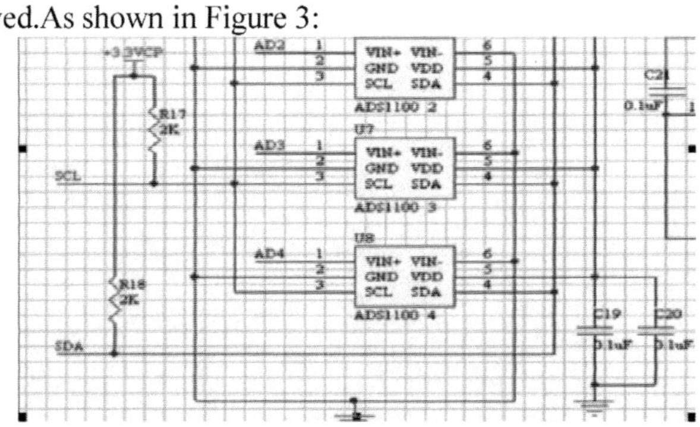

Figure3. ADS1100 single-ended input typical connection circuit

6.4. Data transmission module selection

According to the transmission line detection position is mostly away from the detection center characteristics,we selected GPRS mobile communication system.As shown in Figure 4:

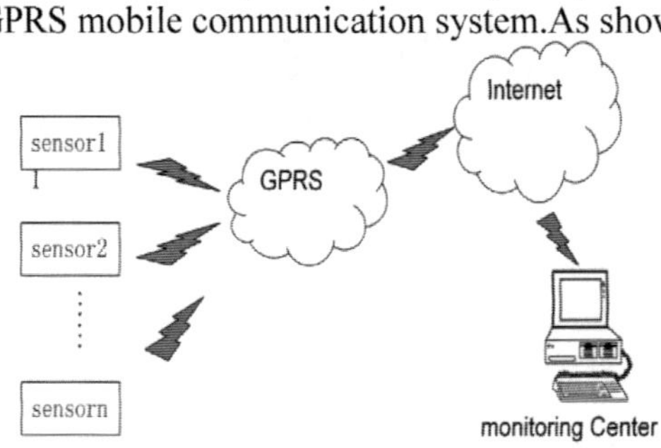

Figure4. Wireless data network diagram

6.5. SD card data storage module design

We chose 1-bit SD mode, namely SPI mode:It has independent instructions and data channels, unique transmission format; SCM and SD card interface circuit shown in Figure 5:

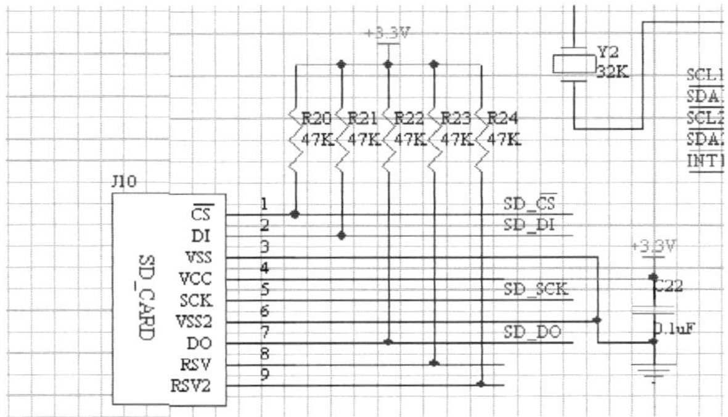

Figure 5. Connection circuit of SD card and one-chip computer

6.6. System power design

Because the transmission line environment is relatively complicated,,solar panel battery supply side is used in this system.The power supply used is the LM2575-3.3. The circuit is shown in Figure 6:

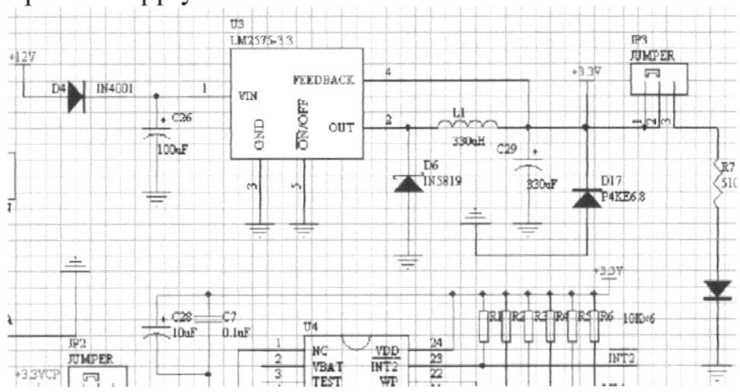

Figure6. Circuit Diagram of Power System

6.7. Serial communication module design

We selected MAX3232 and MSP430F1611, interface circuit shown in Figure 7:

6.8. The system flow chart

Monitoring system flow chart shown in Figure 8:

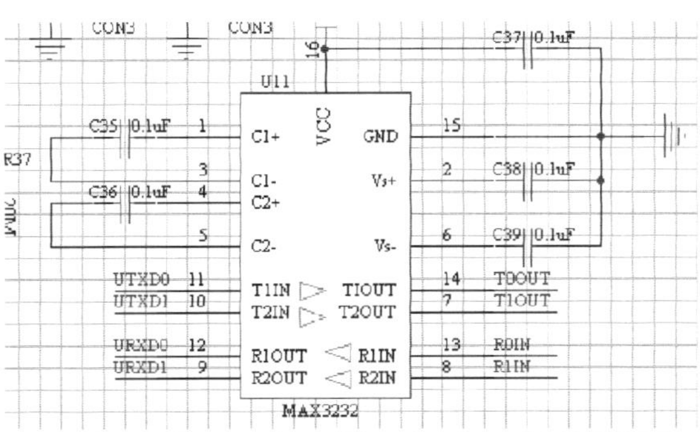

Figure7. Interface Circuit of the MAX3232

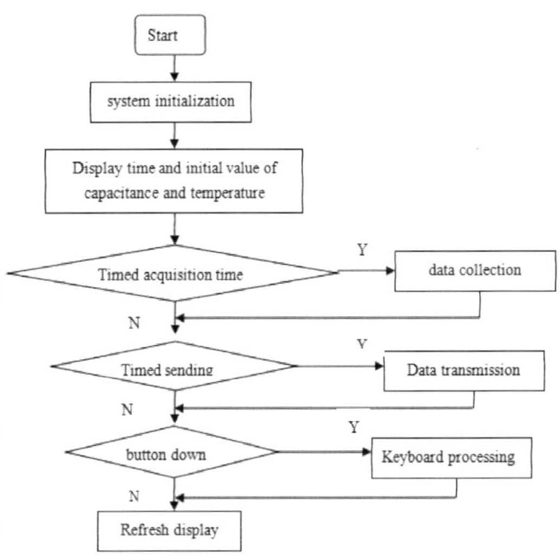

Figure8. System Flow Chart

7. Conclusion

The basic module of the ice-covering monitoring system is a capacitive thickness detection sensor. It can perform data acquisition on multiple capacitance values, send the collected data to the control center processor through the I2C serial port, Then the processor processes the data and stores it in the SD card. MSP430 MCU is the core component of the system, Through the single-chip computer control to complete the system data from the collection to send all aspects of data processing work. Processor SCM completes data transmission through RS232 serial port and GPRS communication module connection.

Acknowledgements

This topic is based on the outstanding young talents support program for colleges and universities in Anhui Province . Number:gxyq2017136;Anhui Sanlian University Research Fund Project. Number: KJZD2017001;Anhui Provincial Natural Science Research Project. Number: KJ2017A522.

References

[1] Wu Libing et al2015 *Selection method of dynamic K value cluster heads for wireless sensor networks*.(wuhan:Journal of Huazhong University of Science and Technology-Nature Science) , pp37-41
[2] Hu X et al2013 *A wireless sensor network-based structural health monitoring system for highway bridges*(Compute-Aided Civil and Infrastructure Engineering). pp193-209.
[3] Kulakowski Pet al 2013. *Performance study of wireless sensor and actuator networks scenarios* (International journal of communication systems) .pp 515-529.
[4] Zhou Shaoyi.2010 .*Study on the division of ice-covered areas of transmission lines in Guangxi based on ice model and gis* .(Guangxi Electric Power) pp11-13.
[5] Lu Jiazheng 2009. *Results and analysis of ice disaster monitoring in Hunan power system*(Power System Protection and Control) pp99-105.
[6] Huang Xinbo2007 *Mechanics analysis of wire icing and on-line monitoring system for ice coating* (Automation of Electric Power Systems) pp98-101,
[7] Zhang Yu 2008. *On-line monitoring system for wire icing on overhead transmission lines* (High Voltage Technology).pp 1992-1995.
[8] Zhang Yuyu 2008.*Analysis of Ice Coating Mechanism on Transmission Lines*(Jiangxi Electric Power) pp46-47
[9] Zhang Qing 2003.*An exact solution of the capacitance between two infinitely long parallel straight wire*(Journal of Anhui Shangye University).pp84-86.

MCU System-based Intelligent High-speed Elevator Door Operator Fault Analysis and Research

Wang Qibing[1, 2,*], Leng Yonggang[1], Li Dongliu[2], Zhang Xiaoqiang[2], Li Ren[2], Zhu Hongmei[2] and Zhang Hao[2]

[1]Tianjin University Tianjin 300072
[2]Sicher Elevator Co., Ltd. Zhejiang 313009

*Wang Qibing: 43410697@qq.com

Abstract: The elevator hall door lock MCU system is the important safety core component installed on the elevator. It can monitor the fault information of the elevator operation in real time. At present, MCU system can not adapt to the changes in the application of the intelligent high-speed elevator. This paper first analyzes the reasons of MCU system fault on the intelligent high-speed elevator based on the actual case analysis, and summarizes the optimization measures for the safety monitoring, and provides reference for the further research on the performance and the system intelligent technology of the intelligent high-speed elevator door operator system.
Key words: MCU system; high-speed elevator; elevator door operator; fault analysis

1. Introduction

With the rapid development of the modern cities in China, the elevators are widely applied in multi-floor, high-rise and super high-rise buildings. Featuring with convenient, fast and labor-saving performance, the elevators improve people's life quality. According to the statistics of China Elevator Association, 679 thousand and 779 thousand elevators were produced in China respectively in 2017 and 2016. The output and sales volume of elevators in China ranked No. 1 in the world. The intelligent technology of elevator keeps innovating, with higher automation degree. However, the casualties caused by elevator accidents always exist in both developing countries and developed countries. According to the statistics of elevator accidents at home and abroad, the elevator accidents caused by the door system fault and maintenance operation error accounted for more than 80% of all the elevator accidents. The high proportion shows that the elevator door system is the main cause for serious accidents. Improving the reliability of the door control system is one of the key methods to eliminate the elevator accidents [1-2].

2. Basic structure and working principle of elevator door system

The door system, as shown in Fig. 1, is mainly consisted of landing door (hall door) and car door, door operator, door lock device, driving mechanism, protection system and accessory parts. The hall door is located at the landing entrance, the car door is set on the elevator car, the door that the passenger sees after he/she enters into the elevator. When the elevator stops at the landing, the car door is the drive door, and the hall door is the driven door. The door vane is embedded on both sides of the door lock roller. When the car door opens, the door vane presses the roller to one side to unlock the hall door lock and the car door will open or close the hall door. When the hall door and car door are completely closed, the elevator can run normally [3-6]. From the structural function, the elevator door system is consisted of driving motor, door controller, motion actuator and related driving device.

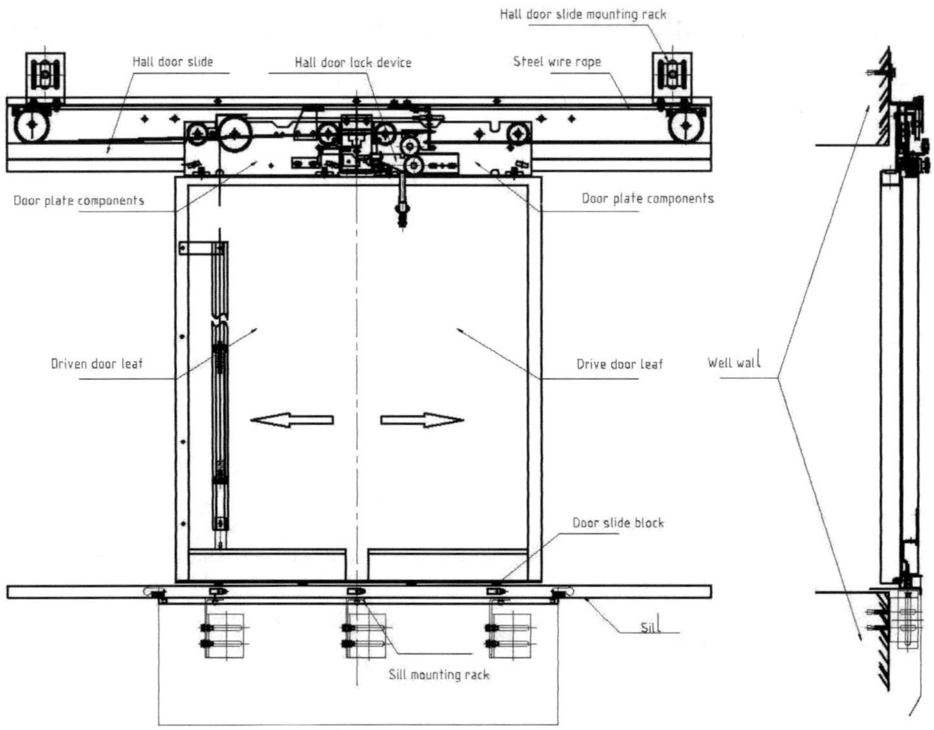

Figure 1. Elevator door structure diagram

3. Main fault and reason of elevator hall door lock MCU system

3.1. Shearing and collision accidents

The accident is caused by the case that the door lock switch is short circuited, the door lock switch contacts can not be reliably opened, and the door lock relay device can not be open or delayed to open. Such problems are mainly caused by the case that the elevator linkage door interlocking is invalid due to human factors or non-human factors. The elevator can operate normally when the hall door opens or is ajar. But if there is a person between the hall door and the car door, it may cause shearing or collision, even cause casualties if the case is serious.

3.2. Well dropping accident

Such accidents are caused by the case that the door lock is short circuited, in that case, the emergency unlocking device opens the hall door when the elevator car stays at other floors when the hall door and car door open; the hall door is forced open when the door lock is damaged or the door lock has small meshing length; the damaged hall door is also the reason causing the accident. The short circuit of the door lock may also cause the accidents. Meanwhile, the hall door positive closing device fails or is damaged [7-9]. Such case will cause that the elevator hall door and car door open while the elevator car does not stop in the floor due to human factors or non-human factors. It will easily cause the well dropping accidents and then cause injury or even death.

4. Fault and safety hazard analysis of elevator door MCU system

IOP Conf. Series: Materials Science and Engineering **428** (2018) 012028 doi:10.1088/1757-899X/428/1/012028

The elevators work frequently in places with high visitors flow rate, such as station, airport and office buildings, the elevator door system keeps working at high frequency. The failure rate of the electromechanical equipment is a function of time, and the elevator is also consistent with the fault bathtub curve distribution rules [10]. Continuous working at high frequency and increasing time also increase the elevator door system fault rate.

4.1.Mechanical fault summary

When the clearance between the elevator door vane and the roller is less than 5mm, when the elevator passes through the floor, the door vane will collide with the roller, which causes the roller bumped or the hall door unlocking, so the electrical interlocking contact is open and the elevator is suddenly stopped; if the clearance is too large, and the door vane and the roller will be misplaced. During elevator leveling, it will cause that the door vane can't drive the roller to move horizontally, causing the hall door cannot be open. This will cause the hazards for persons trapped in the elevator.

For example, when the lock hook position of the elevator door is offset, the meshing depth of the lock hook and the door lock block is too small, less than 7mm, the door lock contact is not completely pressed, which leads to the poor contact of the electrical safety loop, increasing the fault rate of the abrupt stopping in the normal operation. The hall door will suddenly open when the passenger pulls down the door due to too small meshing depth. It provides the probability of well dropping accident with potential hazard.

As shown in Fig. 2, the clearance between the lock hook and the door lock block is too large. If the passengers pull down the door outside the hall door, the electrical contact of the main lock or the auxiliary lock will be open and the elevator in the normal operation will stop suddenly, causing the accident.

Because (1) the safety shoe microswitch is pressed without action (2) the safety shoe is blocked (3) the safety shoe contacts the casing, the elevator door will be closed, the safety shoe is clamped, and can not automatically rebound to open, causing the potential hazard.

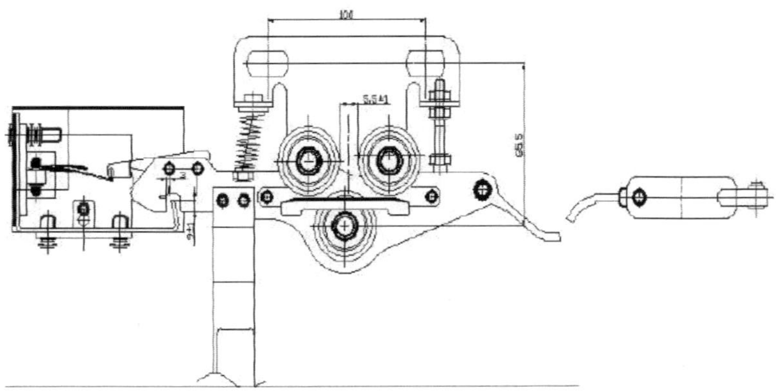

Figure 2. Elevator door lock structure

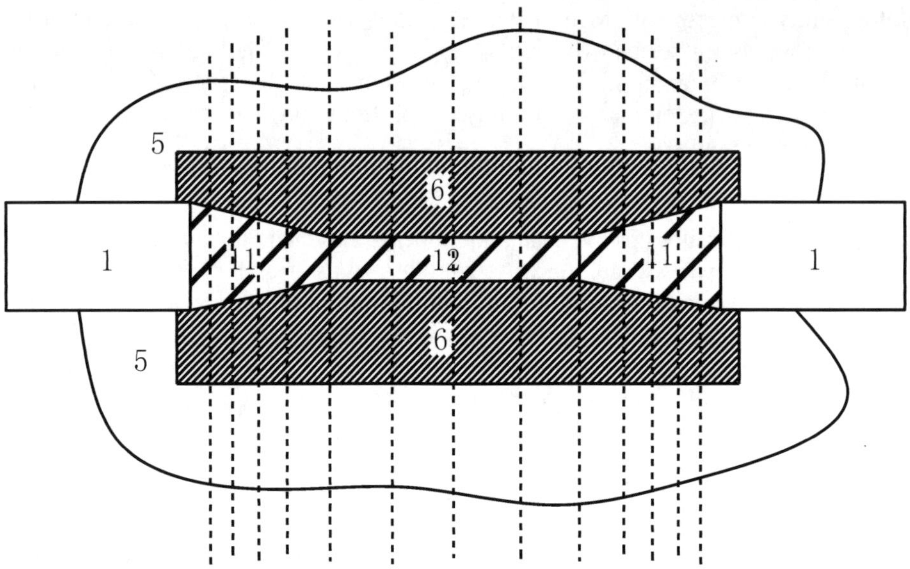

1- elevator door, 2- HD camera, 3- image processing device, 4- elevator control system, 5- ground, 6- high reflective surface, 11- door edge, 12- door sill.

Figure 3. Elevator door anti-clamp device

From the analysis above, we realize that the locking of the drive door and the driven door doesn't has a necessary relationship, but associated relationship. The driving parts such as the belt and the wire rope between the drive door and the driven door may fail, no matter from the manufacturing process, or material or installation process. Taking the common steel wire rope linkage door as example, the linkage steel wire rope will easily rust, fracture or even fail because of the influence of the frequent bending and other factors in the working process [11-12]. As shown in Fig. 3, we need to design a new elevator door anti-clamp device to ensure the safe and reliable operation of the elevator door operator to solve the problems above.

4.2.Elevator door system electrical fault and potential hazard analysis

The door lock safety contact is not connected through due to pollution and poor contact caused by the human factor, the safety loop is not connected, the signal feedback is blocked, causing the elevator stopping. It will cause persons trapped in the elevator. Improper human operation: the door lock switch is short circuited; the emergency button is short circuited, and the door lock is short circuited. Non-human factor: door lock switch contacts are not open; door lock relay is delayed to open or not open; door lock is short circuited due to fault. [13-14] The elevator door lock electrical loop fails. As shown in Fig. 4, when the electrical safety contact of the door lock is open, the elevator works with the door open, and the passengers will have the risks of being sheared, crushed and dropping into the well. The safety shoe and the light curtain wiring are short circuited, the automatic closing device fails.

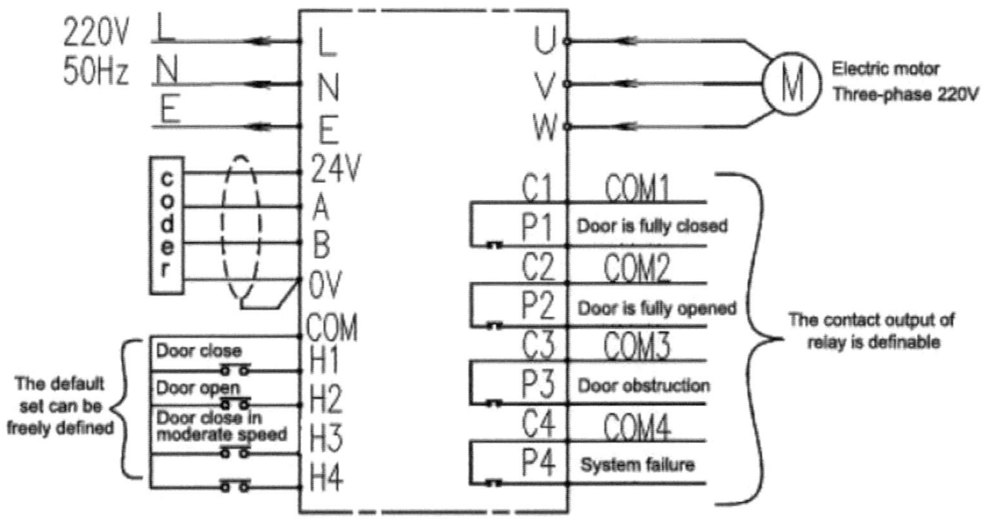

Figure 4. Electrical control system and working principle of elevator door operator

4.3. Other possible situations

In addition to the problems mentioned above, we analyze the steel wire rope breaking in the elevator door system. In the actual operation, the steel wire rope will be broken, as shown in Fig. 5, but the researchers still don't focus on this possibility at present. If the steel wire rope is broken, and the elevator is not in the unlocking area, the drive door is locked effectively. Restrained by the steel wire rope, the driven door can not be opened, and it is also locked effectively.

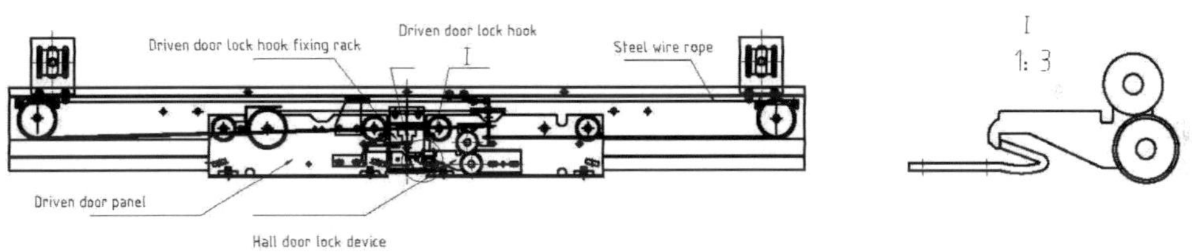

Figure 5. Elevator door operator steel wire rope action

If the elevator stops to the locking area during leveling, the car door is opened because the door operator starts, meanwhile the drive door is open being driven by the door vane, and the driven door is closed with the effect of the automatic closing device since the driven door is not driven by the door vane. The door is likely to be ajar. However, since the steel wire rope is in a relaxed state when the drive door opens, it is easy to get out of the pulley groove. It is easy to be stuck by the related hall door parts. If the steel wire rope is tore during the closing, the steel wire rope will be broken and eventually lead to the condition that the driven door can not be effectively locked [15].

5. Preventive measures for elevator accidents

5.1. Precaution in type selection

The elevator with national license shall be selected. The reliability is not guaranteed if the elevators are produced without national license. Do not select the elevators at lower price that are produced without the national license. The elevator reliability and technology shall be taken into consideration when purchasing. Especially, we must pay attention not to purchase the elevator with variable frequency and variable voltage technology [16]. Usually, if the installation technology is guaranteed, the fault rate of the elevator is very low, so we must pay attention to the reliability and technology. We must purchase the qualified products and strictly control the installation, to avoid potential hazard during installation.

5.2. Precautions in design

In the elevator design, we shall design two safety contacts in the dual-lock hook via the single door lock and dual-lock hook structure, called as single door lock dual-hook dual-contact type. The two safety contacts verify the locking positions of the drive door and the driven door. The elevator can operate normally only when two safety contacts are connected and the two lock hooks are locked on the drive door and the driven door. This will effectively prevent the elevator from operating normally when the hall door is not locked.

5.3. Precautions in maintenance

In the elevator maintenance, the relevant staff in the maintenance unit shall improve the elevator quality effectively. In the actual situation, pay attention to check the steel wire ropes and the joints. Once the potential hazards are found, it shall be eliminated within the shortest time. The elevator with fault must not be used.

5.4.Precautions in testing

The inspectors shall inspect the elevator regularly. At the same time, the inspectors shall further enhance their sense of responsibility and improve the inspection. It is noteworthy that this problem is also involved in the Elevator Supervision Inspection and Regular Inspection Rules - Traction and Forced Driving Elevators, but it emphasizes the electrical verification of the hall door closing in place only. The inspectors shall fully understand the corresponding provisions of the rules, and learn from others in actual working, so as to effectively avoid missing and incorrect judgment.

5.5. Remote monitoring and fault warning system for elevator door system

With the development of electronic information technology and data processing theory, the maintenance of mechanical equipment is developing from traditional fault diagnosis to intelligent maintenance and early warning. The key point of elevator door system maintenance is gradually changing from fault diagnosis to the research in door operator intelligent maintenance area mainly including state monitoring, predictive maintenance and performance degradation analysis. With artificial intelligence theory, advanced information technology and network monitoring technology, the elevator door system is modeled and feature signals of the system are extracted, processed and analyzed to judge which parts of the elevator door system may have faults in the future [17-20]. The famous elevator manufacturers both at home and abroad are able to provide remote monitoring and management system matching with the elevators with improved functions. Its principle is shown in Fig. 6:

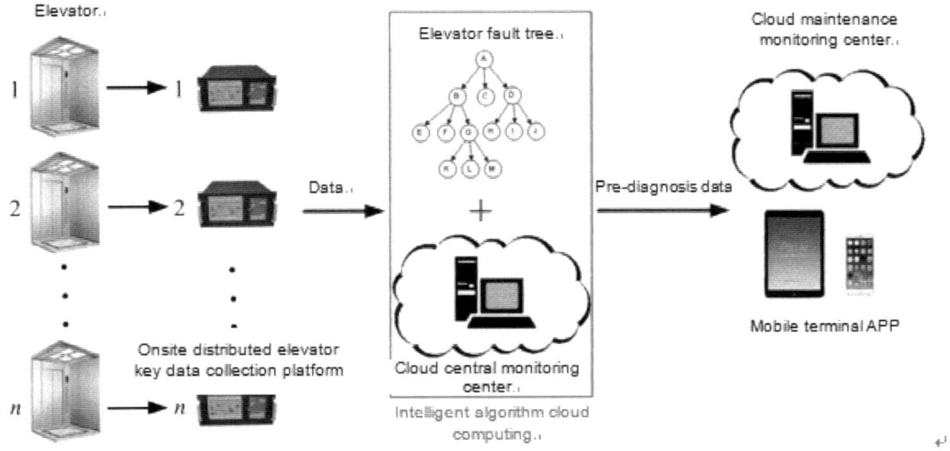

Figure 6. Remote monitoring and fault warning system of elevator door system state

The state requires the supervision inspection and the annual regular inspection for special elevator equipment. The equipment shall only be used after passing the annual inspection. This artificial regular inspection method plays a role in ensuring the safe operation of the equipment, but it can only detect the status of the monitoring points of the elevator at that time, and can not know the daily operation state in the whole year, as well as the real-time operation status view. It has defect in fault warning and troubleshooting. The elevator door system intelligent monitoring system is designed to continuously monitor and make fault warning of the daily operation state of the elevator, and provide the basis for the elevator inspection and management department to learn the elevator operation condition.

6. Conclusion

To sum up, the elevator door MCU system has a direct impact on the safety and comfort of the elevator operation. In this case, the elevator manufacturing, design, installation, inspection, maintenance shall be improved. At the same time, there are defects in fault warning and troubleshooting. The elevator door system intelligent monitoring system is designed to continuously monitor and make fault warning of the daily operation state of the elevator, and provide the basis for the elevator inspection and management department to learn the elevator operation condition.

References

[1] Special equipment safety technical regulation TSG T5001-2009 Elevator service management and daily maintenance regulation [J].Modern Property Management, 2009, (12).
[2] Zhou Jie, Wen Zengyu. Technical retrofit for eliminating the self-closing fault of elevator hall door [J]. China Plant Engineering,2011, (3).
[3] Zheng Weifeng. Discussion on fault diagnosis and repair of elevator electrical control system [J].China Homes (last third publication), 2014, (4).
[4] Wang Qibing, Li Dongliu, Zhu Liqiang etc. Research on application of ship elevator in ship engineering [J]. Ship Engineering, 2012, Ship Engineering, 2012, (supplement 2): (S2):145-147.
[5] Wang Qibing, Hu Jie, Ge Qing. Analysis of traction explosion-proof elevator design and installation [J]. Lifting and Transportation Machinery, 2014 (4): 34-35.
[6] Wang Qibing, Li Dongliu, Liang Deliang etc. Research on Design an Elevator to be Used in Clean Room and Controlled Environment [J]. Machinery, 2014(S1):11-12.

[7] Q.B.WANG,D.L.LI,L.Q.ZHU,Q.M.LI,Z.B.CHEN.B.Z,Z.H.LI : Research on Design of an Elevator in the Controlled Environment of Clean Room. International Asia Conference on Industrial Engineering and Management Innovation. (IEMI2012) Proceedings.(2013).
[8] F.J.MA,G.J.ZHAO,H.T.ZHANG:Malfunction Detection of Elevator Based on Expert System.Computer Automated Measurement & Control, (2002).
[9] Zhu Changming, Hong Zhiyu, Zhang Huiqiao. Principle, structure, installation and testing of elevator and escalator [J] Shanghai: Shanghai Jiao Tong University Press. 1995
[10] Wang Baoqiang, Yang Chunfan, Jiang Xuesong. Latest elevator principle, use and maintenance [J] Beijing: China Machine Press. 2006
[11] Jiang Chunyu, Zhang Yuanpei, Chen Jiafang. Elevator installation, use and maintenance manual [M] Beijing: China Machine Press. 2002.1
[12] Wang Qibing. Analysis on safety protection function of mine lift auxiliary elevator [A]. Zhejiang Association for Science and Technology, Shanghai Association for Science and Technology, Jiangsu Association for Science and Technology. Marine Economy and Power Development – Symposium of the 10th Yangtze Delta Motor, Power Technology Sub Forum [C].Zhejiang Association for Science and Technology, Shanghai Association for Science and Technology, Jiangsu Association for Science and Technology, 2013:3.
[13] Wang Xin. Application of fuzzy PI control in permanent magnet synchronous motor elevator door operator system [D]. Tianjin University, 2012.
[14] Weng Dan, Zhu Changming, Hu Hui, etc. Experimental study and analysis of electromechanical performance of elevator door operator system [J]. Hoisting and Conveying Machinery, 2008, 2008(8):93-96.
[15] Luo Dan, Xiao Huaming, Xie Bohuai. Discussion on common problems in inspection and use of elevator door operator system [J]. China Elevator, 2009(23):58-60.
[16] Ren Y, Xue-Juan M O. Simulation and analysis of harmonic suppression strategy in elevator door-machine servo system [J]. Journal of Mechanical & Electrical Engineering, 2012.
[17] Wan J R. Soft-drive elevator door machine without mechanical link[J]. Hoisting & Conveying Machinery, 2003.
[18] Hong Y H, Zi-Min Y U, Luo H C, et al. Remote Fault Monitoring and Early Warning System for Industrial Robots [J]. Science & Technology Vision, 2017.
[19] Tian L Y, Xiao-Zhu L I, Wang H J. Mine Air Compressors' Remote Monitoring Early Warning and Fault Diagnosis Expert System [J]. Instrument Technique & Sensor, 2012.
[20] Tao Z, Zhang B, Zhang P, et al. Fusion design between beidou satellite communication technology and sliding force remote monitoring warning system [C]// International Conference on Information Science and Engineering. IEEE, 2011:4542-4545.

IOP Conf. Series: Materials Science and Engineering **428** (2018) 012029 doi:10.1088/1757-899X/428/1/012029

Research on Gear-Shift Strategy without Disengaging the Clutch for Automated Manual Transmission in Hybrid Electric Vehicle

Shaohua SUN, Baogui WU, Yupeng ZOU and Xinqing WANG

College of Mechanical and Electronic Engineering,China University of Petroleum

Email:sunshaohua518@163.com

Abstract. In order to realize a quick, smooth and precise shifting process of hybrid electric vehicle (HEV) equipped automated manual transmission (AMT), a research platform for HEV driven by electrically controlled engine and drive motor was established. The key technologies including its system components, working principle, control strategy of engine and motor during shifting process were analyzed. A technology of AMT shift without disengaging the clutch was proposed. According to the dynamic characteristics of the HEV's engine, motor, and transmission, the dynamic model of transmission system was established. The various stages of shifting process were emphatically analyzed. Its shift characteristics and the key factors affecting the shift quality were grasped. On the basis of the above researches, the shift control strategy without disengaging the clutch was acquired. It was verified on a demo HEV platform. The results showed that the technology of AMT shift without disengaging the clutch for HEV not only ensured smooth shifting through the coordinated control of engine, motor and actuator, but also shortened the shift time and improved the shift quality significantly.

1. Introduction

HEV with AMT can achieve three different driving mode, which are pure electric, pure engine and hybrid, through clutch engagement and disengagement [1]. Shifting process of hybrid mode is most complex, so this research focuses on the shifting of hybrid mode.

Shifting jerks of AMT are mostly caused by the clutch's engagement and disengagement in shifting process [2]. Shifting without clutch disengagement technology can not only eliminate shifting jerks, simplify control logic and reduce difficulty of controlling, but also effectively shorten shifting time as well as improving shifting quality. This technology of heavy-duty truck is put forward in paper [3], and the key point is how to realize precise torque and speed control of diesel engine. Because of the existence of auxiliary motor, it is easier for HEV to achieve than conventional engine vehicles.

HEV's shifting strategy is different from traditional vehicles. Thus the shifting process of HEV is analyzed and its shifting control is simulated. In hybrid mode, in order to realize shift without clutch disengagement, it is needed to coordinated control the two power sources, the engine and the motor [4]. At the same time, the control effects will directly affect shift quality and driving performance.

2. System Composition and Working Principle

This research is based on a HEV with 4-speed AMT. 4 forward gears are driven by constant mesh gear sets through shift sleeves without synchronizers, and motor rotates in reverse to achieve reverse gear.

2.1. System Composition

Figure 1 shows the system composition of a HEV. Power provided by engine and motor is transmitted to wheels through gearbox, drive shaft, and final drive. AMT system consists of gear sets, shift actuator, clutch, clutch actuator, TCU and corresponding position sensors, etc. TCU communicates with VCU, ECU, MCU and BMS through CAN bus.

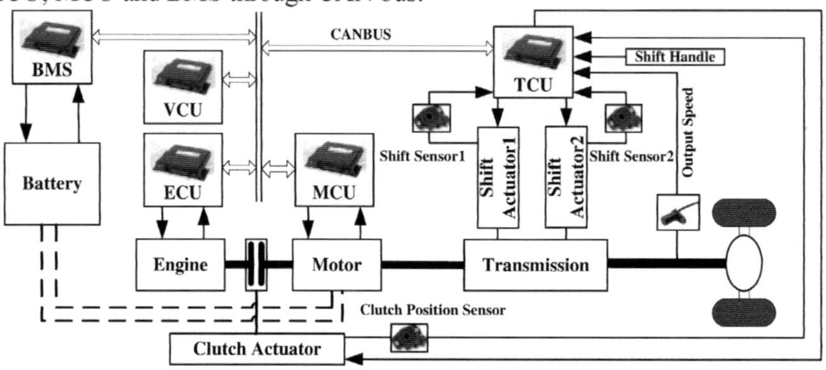

Figure 1. AMT System Composed of Hybrid Vehicle

In order to guarantee the vehicle starting with pure electric mode and mode switches of pure electric and hybrid, engine transmits power to motor through clutch, and motor output shaft is connected to the input shaft of the gear box by splines.

When the vehicle starts, clutch is disengaged, and motor is used to start vehicle. When switching from pure electric to hybrid, the speed differences between clutch driving plate and driven plate is eliminated by closed-loop speed regulation function of the engine, then the clutch is engaged, and finally hybrid driving is achieved.

2.2. Working Principle

The key point of AMT shift without clutch disengagement is the coordination control of engine and motor in different phases of shifting process. The control process is as follows:

1. Before shift to neutral, transmission input torque is reduced closely to zero through the coordination control of the engine and motor torque to ensure shifting easily and swift.
2. Eliminate speed difference between the two shift elements through the engine and motor speed coordination control.
3. When the speed difference is in acceptable range, the engine and motor is controlled in again to reduce transmission input torque closely to zero and then shift sleeve is engaged with the target gear.
4. The engine and motor torque is synergistically controlled to satisfy the power distribution requirements between engine and motor set in the control strategy.

Advantages of AMT shift without clutch disengagement are as follows:

1. Control of clutch in shift process is cancelled, so the control logic is simplified and the control difficulty is reduced.
2. A range of issues such as clutch wear due to long time slipping, shift shock and engine extinguishment due to the driver's improper manipulation are eliminated, which extend clutch service life of and reduce driver's burden .
3. The power interruption time during shifting process is shortened and the power and economy of the vehicle is effectively improved.

3. Dynamics Analysis of Shifting Process

When driving in hybrid mode, the power required for driving is provided by engine and motor according to the power distribution set. AMT shift process based on the rev-synchronization is mainly composed of shifting to neutral phase, speed adjustment phase, gear engaging phase and torque recovery phase. Not only the gearbox, but also the engine and motor are involved in any phase. In order to establish an effective transmission model, following assumptions are first made:

1. The system is composed of inelastic inertia elements.
2. Each rotary member has only one degree of freedom.

Based on above assumptions, the system can be simplified as a discrete equivalent system as shown in figure 2. Parameters or variables before and after the sleeve are respectively converted into the gearbox input shaft and output shaft, and parameters or variables after the final drive are converted into the vehicle. Kinematics and dynamics relationships of powertrain in hybrid mode are as follows:

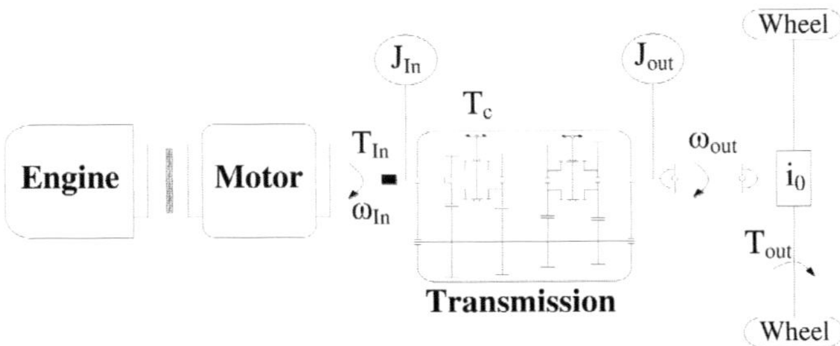

Figure 2. Model of Powertrain System for Hybrid Vehicle

$$\frac{T_c}{i_g} = T_{In} - J_{In} \cdot \dot{\omega}_{In} \tag{1}$$

$$\frac{T_{out}}{i_o} = T_c - J_{out} \cdot \dot{\omega}_{out} \tag{2}$$

Where, T_c is the torque transmitted by sleeve, T_{In} is the sum of engine and motor output torque, T_{out} is the vehicle driving resistance torque, J_{In} and J_{out} are respectively the inertias converted to the transmission input and output shafts, ω_{In} and ω_{out} are respectively the speed of transmission input and output shafts, and i_g and i_o are respectively the gearbox ratio and final drive ratio.

3.1. Shifting to Neutral Phase

Figure 3 shows the forces applied on the sleeve in this phase. F is shifting force, F_c is tangential force between the sleeve and gear, F_f is friction force between mesh gears, and f is friction coefficient.

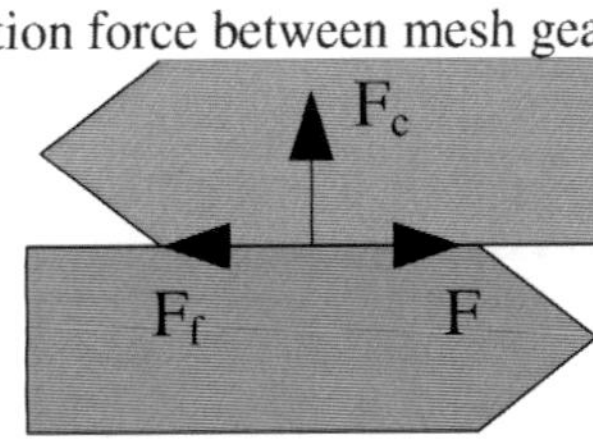

Figure 3. Force Applied on Gears in Shifting to Neutral

R is referred as the radius of the point at which the forces apply on meshed gears.

$$F_f = F_c * f = T_c f / \int dR \tag{3}$$

As shown in equation (3), in order to reduce the friction force to the lowest level, T_c should be minimized as much as possible. T_c shown in equation (4) can be derived from equation (1) and (2).

$$T_c = \frac{J_{In} \cdot T_{out} \cdot \frac{i_g}{i_0} + J_{out} \cdot T_{In}}{J_{In} \cdot i_g + \frac{J_v}{i_g}} \tag{4}$$

Equation (4) shows that the magnitude of T_c is determined by T_{In} and T_{out}. Assuming T_{out} is a constant when shifting, so if T_{In} is reduced to zero, the torque transmitted by the sleeve will reach minimum. Therefore, the engine and motor torque are co-ordinately controlled to make T_{In} to zero.

3.2. Speed Adjustment Phase

After shifting to neutral, the engine and motor are unloaded. In order to reach the input speed needed for target gear, the engine and motor speed should be swiftly adjusted. The target speed and the speed adjustment range can be respectively expressed as:

$$\omega_{In} = \omega_{out} * i_g \tag{5}$$

$$\Delta\omega = (i_{g2} - i_{g1})\frac{u_a i_0}{0.377r} \tag{6}$$

Where, $\Delta\omega$ is speed adjustment range, u_α is current vehicle speed, r is wheel rolling radius, and i_{g1} and i_{g2} are respectively current and the target gear ratios.

The duration of power interruption determined by the time of speed adjustment affects shifting performance, especially when the vehicle is going uphill, because a long-time power interruption will cause vehicle speed drops too much, not only influencing the driving experience, but also leading to security threat. Therefore, the engine and motor response speed and speed adjustment precision are essential to determine shift time as well as shift quality.

3.3. Gear Engaging Phase

The gear engaging phase is divided into "chamfer touching" and "chamfer crossed"[5]. Since the absolute rev-synchronization cannot be guaranteed during speed adjustment phase, a certain adjustment of speed synchronization is still needed at the beginning of this phase using gear chamfer. The force applied on the gears in this phase is showed in figure 4.

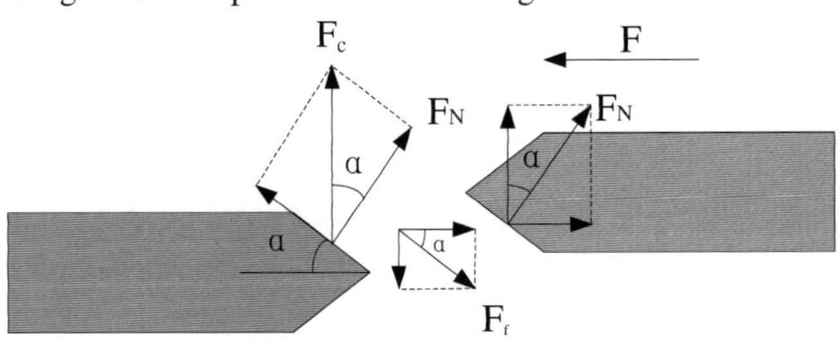

Figure 4. Force Applied on the Gears in Gear Engaging Phase

$$F_f = F_N \cdot f = F_c \cdot \cos\alpha \cdot f \tag{7}$$

Where F_N is the normal pressure on the contact surface and *a* is the cone angle of the contact surfaces. Shifting force F should satisfies equation (8).

$$F > F_N \sin\alpha + F_f \cos\alpha \tag{8}$$

Equation (9) can be derived from equation (3), (6), (7) and (8),

$$F > \frac{T_c}{\int dR}\cos\alpha\sin\alpha + \frac{T_c}{\int dR}\cos\alpha \cdot f \cdot \cos\alpha = kT_c \tag{9}$$

Substitute equation (1) to equation (9), equation (10) is obtained.

$$F > ki_g(T_{In} - J_{In} \cdot \dot{\omega}_{In}) \tag{10}$$

Equation (10) suggests that shift success or not is determined by T_{In} and $\dot{\omega}_{In}$. Thus in order to shift smoothly, on the one hand, T_{In} during this phase should be reduced closely to zero; on the other hand, to decrease $\dot{\omega}_{In}$, the speed difference of actual speed and target speed should be as small as possible.

After sleeve crosses the chamfer, the speed difference is fully eliminated, and cone angle α becomes 0. Resistance of gear engaging is mainly friction force, which is much smaller than shifting force. Then the engagement of sleeve to target gear can be quickly completed.

3.4. Torque Recovery Phase

After gear engaging phase finishes, torque should be recovered as soon as possible according to the vehicle requirement by coordination control of engine and motor to ensure shifting smoothly without shocks. Jerk, referenced as j, which is the change rate of vehicle longitudinal acceleration during shifting, is usually used to evaluate shift smoothness [6]. Since the torque recovery phase is very short, the driving force is assumed constant in this phase. Jerk can be calculated in equation (11).

$$j = \frac{da}{dt} = \frac{d^2u_a}{dt^2} = \frac{i_g i_0 \eta_T}{\delta m r}\frac{dT_{In}}{dt} \tag{11}$$

Where η_T is the efficiency of transmission, δ is the correction coefficient of rotating mass, m is the vehicle mass and r is the wheel rolling radius.

Equation (11) shows that the jerk is proportional to the change rate of torque. So the maximum jerk that passengers can withstand should be as a boundary condition and the torque recovery rate of each gear is calculated according to vehicle power distribution requirement. Another conclusion that can be drawn from equation (11) is the torque recovery of low gear is slower than that of high gear.

4. Shift Control Strategy without Disengaging Clutch

4.1. Coordination Control of Multi Controllers

Shift of HEV is a coordination control process between multi controllers, which communicate with each other through CAN bus network [7]. VCU is used to collect vehicle state, forward CAN messages as well as arbitrating vehicle control right; ECU and MCU are respectively used to collect engine and motor states, speed, torque, etc. and respond the VCU torque and speed request; TCU is responsible for collecting handle information and sending speed and torque requirements to engine and motor.

4.2. Shift Process Control Strategy

From the view of powertrain structure, the engine and motor can be considered as a whole. The shift control process of hybrid vehicle without clutch disengagement is shown in figure 5[8].

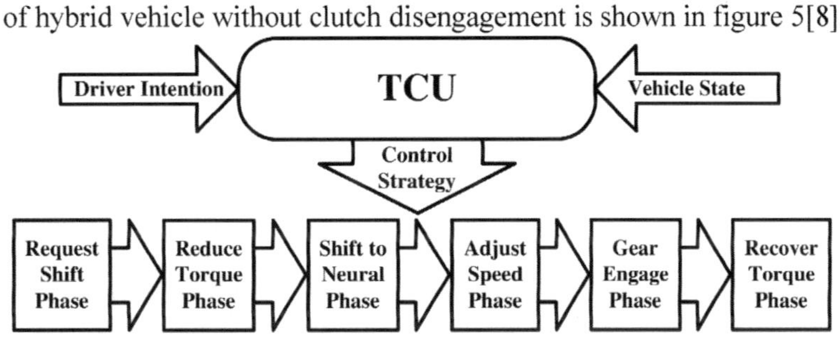

Figure 5. Shift Process Control of AMT for Hybrid Vehicle

When driving normally, driver's intends are determined by TCU according to vehicle driving state and then the shift decision is made.

1. Requesting shift phase. When the driving states satisfy shift condition, shift request is sent to VCU. After permission of VCU, the highest control rights of engine and motor are obtained by TCU.

2. Reducing torque phase. The engine and motor are co-ordinately controlled by TCU to make the transmission input torque close to zero as soon as possible [9].

3. Shifting to neutral phase. The shift actuators are controlled by TCU to pick neutral.

4. Adjusting speed phase. The target speed is sent to ECU and MCU, and the engine and motor are controlled co-ordinately to adjust speed. When the speed difference is smaller than setting range, this phase is finished.

5. Gear engaging phase. In order to prevent the output torque of the engine and motor after rev-synchronization and protect against shift shock and noise, engine and motor torque are co-ordinately controlled again. So the AMT input torque is close to zero and then shift to target gear.

6. Recovering torque phase. Engine and motor are co-ordinately controlled to recover torque and drive normally according to the power distribution relationship between engine and motor.

Shift without clutch disengagement is simpler, because clutch control is avoided, and a range of issues such as clutch wear due to long time slipping, shift shock and engine extinguishment due to the driver's improper manipulation are eliminated, which extend the service life of clutch[10].

5. Experiment Verification and Analysis

A hybrid demo vehicle is selected for shifting experiment. Parameters of vehicle are shown in Table 1.

Table1. Vehicle main parameters

Number	Parameters	Value	Number	Parameters	Value
1	Curb Weight	13500kg	5	Engine Type	NQ210KN5
2	Final Drive Ratio	4.875	6	Engine Max Power	140KW
3	Tire Radius	0.465m	7	Motor Voltage	600V
4	Gear Ratio	4.43/2.41/1.46/1	8	Motor Max Power	80KW

Figure 6 shows the curves of shifting from 1 gear to 2. It can be seen that the variation of the current gear, the position of shift fork 1 and 2, the engine/motor speed, the output speed of gearbox, the engine throttle, the motor torque and the jerk. The time of six phases is shown in table 2.

Table2. Time of Six Shift Phases

Number	Phases	Time (ms)	Number	Phases	Time (ms)
1	Requesting Shift	30	4	Adjusting Speed	541
2	Reducing Torque	208	5	Gear Engaging	133
3	Shifting to Neutral	118	6	Recovering Torque	241

Shift time and power interruption time are comprehensive indexes to evaluate shift quality [11]. In this study, the shift time is defined as the duration from the moment TCU sends out shift request to the end of recovering torque phase. The power interruption time is defined as the duration from the beginning of shifting to neutral phase to the end of gear engaging phase. The shifting time from 1 gear to2 is1271ms, and the power interruption time is 792ms, which satisfy the requirements of the shift time of the hybrid vehicle. And there is no significant power interruption sense during driving process.

In shift process, the time of adjusting speed is 541ms, which accounts for about 42.6% of the total time. It shows that the coordinate control of engine and motor speed has great affect on shift quality. It can be seen from the figure 6 that the engine throttle needn't fully close and only maintain an appropriate degree of opening in the adjusting speed phase. At this time, the motor torque is negative, which indicates that the motor is generating. The time of reducing torque is 208ms and accounts for about 16.4%. Without prejudice to shift to neutral, it is allowed to properly amplify the residual torque range of engine and motor before shifting to neutral. The time of shifting to neutral and engaging gear is 251ms, which accounts for about 19.7% and mainly depends on factors such as how big the synchronous speed difference is, the efficiency of shift actuators as well as the response of shift motors. The time of recovering torque is 241ms and accounts for about 19%. The rate of recovering torque is dependent on the vehicle power distribution requirement and the acceptable jerk.

The degree of jerk is an important index to evaluate shift quality. As can be seen from figure 6, the curve of jerk shows that there is great degree of jerk during reducing torque phase and recovering torque phase. Jerk is mainly caused by torque mutations of the gearbox output torque. jerk is proportional to gear ratio. The bigger the speed ratio is, the greater the jerk is. In the shift process, the maximum of jerk is $14.62 m/s^3$ which is smaller than industry-recommended $17.64 m/s^3$.

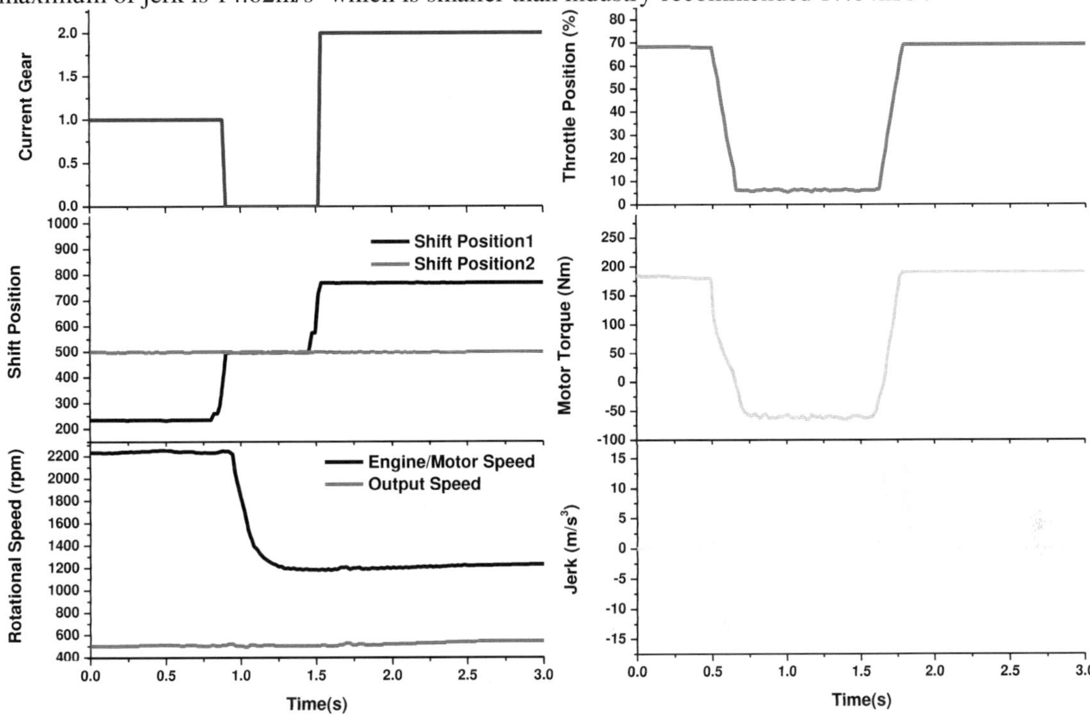

Figure 6. Shift Curves of 1-2 Process

In addition, speed difference directly affects the vehicle shift quality: the smaller the speed difference is, the shorter the shift time and the smaller the impact will be. Therefore, controlling speed difference effectively is very important to improve shift quality. The speed difference range of each gear permitted can be obtained by method of experimental calibration. Taking hybrid AMT demo-car as an example, the speed difference range of the first gear is (-35, 40) rpm, the second gear is (-25, 30) rpm, the third and forth gear is (-10, 15) rpm. In this speed range, there is no shock and jerk when shifting, the shift time is short, and the quality of shift is good.

6. Conclusions

For the structural features and technical requirements of HEV with AMT, the key technologies such as system composition, working principle, shift process and control of engine and motor are analyzed deeply. A control technology that can realize shift without disengaging clutch is proposed, and corresponding shift control strategy is developed. The strategy is verified and optimized through the experiment based on a hybrid demo vehicle. Road test of 5000km proved that HEV equipped with AMT controlled by the strategy could effectively extend the service life of clutch, shorten the shifting time and reduce shifting impact as well as improving the driving performance and comfort.

Acknowledgments

This work was supported by Shandong province natural science foundation of China (No. ZR2016EL11).

References

[1] Ryan, A. and Mc Gee., "Model Based Control System Design and Verification for a Hybrid Electric Vehicle. Future Transportation Technology," Future Transportation Technology, Conference Costa Mesa, California, June 23-25, 2013.

[2] Lishu, Guo., Anlin, Ge., Tai, Zhang., and Yingjie, Yue., " AMT Shift Process Control," Transactions of the Chinese Society for Agricultural Machinery, 2003,34(2): 1-4.

[3] Yuncheng, Wang., Qingnian, Wang., Fei, Xie., and Shunli, Wang., " Non-release-clutch Technology of AMT for Heavy Duty Truck," Vehicle Technology, 2015,5: 14-18. doi: 1000-3703(2015)05-0014-04.

[4] Haotao, Zhu., Junzhi, Zhang., Chenglin, Liao., and Zhaolin, Hu., "Simulation of Integrated Gear Shifting Control for Hybrid Electric Vehicle," Journal of Highway and Transportation Research and Development, 2005,22(8): 131-135.doi: 1002-0268 (2005) 08-0131-04.

[5] Zhisheng, Yu., "Automobile Theory," China Machine Press, 2011,6.

[6] Masaru, Yamasaki., Hitoshi, Konno., Hiroshi, Kuroiwa., and Naoyuki, Ozaki., "Automated Manual Transmission with Torque Assist Mechanism for Reducing Shift Shock," SAE Technical Paper 2005-01-1783, 2005.

[7] Yingji, Liu., Tianxia, Zhang., Bangchun, Wen., and Wanke, Cao., "Online Fault Diagnosis System of AMT Based on CAN Bus and PSA Model," Transactions of the Chinese Society for Agricultural Machinery, 2008,39(6): 132-136.

[8] Junqiang, Xi., Lei, Wang., Wenqing, Fu., and Wanwu, Liang., "Shifting Control Technology on Automated Mechanical Transmission of Pure Electric Buses," Transactions of Beijing Institute of Technology, 2010,30(1):42-45. doi:1001-0645(2010)01-0042-04.

[9] Yong, Li., Siqin, Chang., and Yingjun, Wei., "Shift Strategy of AMT Vehicle Based on Speed and Torque Control," Chinese Journal of Mechanical Engineering, 2011,22(15):1880-1885. doi: 1004-132X(2011)15-1880-06.

[10] Dong, Pei., "Research on the Integrated Control of No_separated_clutch AMT," Changchun: College of Automotive Engineering, Jilin University, 2007.

[11] Yulong, Lei., and Anlin, Ge., "Testing Study of Improving Shift Quality of Automated Mechanical Transmission," China Journal of Highway and Transport,1996,12(2):95-99. doi: 1001-7372(1999)02-0095-05.

IOP Conf. Series: Materials Science and Engineering **428** (2018) 012030 doi:10.1088/1757-899X/428/1/012030

RIHOG-BoVWs for Rotation-invariant Human Detection

Baozhen Liu[1*], Hang Wu[2], Yu Zhang[1], Wenlong Xu[1], Kaiyu Mu[1] and Yuhua Yao[1]

[1] China Astronaut Research and Training Center, Beijing, 100094, P.R.China
[2] Institute of Medical Support Technology of Academy of System Engineering, Academy of Military Sciences, Tianjin, 300161, China

*liubaozhen91@126.com

Abstract: Rotation-invariant human detection is of vital significance to many applications, including victim detection via a daylight camera on an unmanned aerial vehicle (UAV). With this problem in mind, we propose a powerful rotation-invariant object-level description method, i.e., RIHOG-BoVWs, which is achieved by local-level feature calculation and high-level semantic feature extraction. In local-level feature calculation, we consider Rotation-invariant Histograms of Oriented Gradients (RIHOG) as the local descriptor and project gradient information into the frequency domain, where the rotation-invariant representation vector is constructed. In high-level semantic feature extraction, Bag of Visual Words (BoVWs) model is employed to achieve global description from local features without considering their spatial structures. Experimentally, we first confirm the sufficient discrimination power of the RIHOG-BoVWs on the public *Freestyle Motocross* dataset, and then demonstrate the high performance of RIHOG-BoVWs on a victim dataset which has varied backgrounds and body postures.

1. Introduction

After disasters, finding human victims is one of the primary goals of search and rescue efforts. It is possible to quickly survey a disaster site from the air to identify humans in need of help. In this paper, we focus on the human detection problem in disaster scenes from a UAV perspective using a daylight camera.

Figure 1. Posture difference between pedestrians (a) and victims (b). The pedestrian images are from INRIA dataset [1] and the victim images are from [2].

As an important branch of computer vision, the human detection problem has attracted a lot of attention and most research focuses on pedestrian detection [1]. As shown in figure 1, the human body seen from aerial images has more varied postures and is very distinct from pedestrians appearing in upright poses such as standing or walking. The most fundamental limitation of imaging equipment is the varied relative orientations of human bodies in the image. In this paper, we concentrate on the modification of Histograms of Oriented Gradients (HOG) [1], which is the most successful feature in pedestrian detection, and integrate Rotation-invariant HOG (RIHOG) [3] with Bag of Visual Words (BoVWs) [4] model to achieve rotation-invariant semantic description for human bodies.

The rest of the paper contains an overview of related work in Section 2, a description of HOG in Section 3 and the semantic description method we propose in Section 4. A thorough set of experiments is then presented in Section 5, where the performance of RIHOG-BoVWs is proven, before we finally conclude our work in Section 6.

2. Related work

In computer vision, many researchers seek to find solutions to rotation-invariant object detection. The simplest approach is detecting an image from several discretized orientations using a learning classifier trained by calibrated samples [2, 5]. In addition to this method, rotated object detection is also addressed as a multi-class detection task, where the orientation are discretized artificially and every direction of objects is regarded as one category [6, 7]. Another approach to in-plane rotation is the development of learning procedures. [8] decouples the orientation estimation from object classification and [9] incorporates the rotation-invariance into a discriminatively trained structured Support Vector Machine (SVM). Besides, many researchers pay attention to the rotation-invariant semantic description.

Orientation normalization in Scale-Invariant Feature Transformation (SIFT) [10] is one of the common ways to construct the rotation-invariant feature, which aligns the local coordinate system to the dominant direction at each detected point. Learning from SIFT, [11] first predicts the dominant direction in the histogram bins of HOG and then rotates the image to the estimated orientation to calculate the final HOG feature. The method of orientation normalization relies on the assumption that such a dominant direction is available and robust enough, which restricts its application to arbitrary positions or dense feature computation.

Compared with the Cartesian coordinate system, the polar representation has a unique advantage for achieving rotation-invariance. In polar coordinate systems, [12, 13] propose a rotation-invariant HOG-like feature, which takes the tangent direction as the reference of gradient orientation and then accomplishes the information aggregation on multiple concentric spatial bins. The new feature, i.e., RIFF, is originally proposed for the purpose of keypoint matching and has no sufficient discrimination in object detection. [14] improves RIFF and propose Sector-ring Histograms of Oriented Gradient (SRHOG) method to obtain sufficient discriminations of human bodies. However, RIFF and SRHOG adopt the local coordinate system to determine pixel gradients and have no translation invariance, which limits their detecting efficiency.

Another way to realize the feature's rotation-invariance is via the Fourier transform, which extracts the rotation-invariant information of an image in frequency domain. [15] represents Circular Fourier-HOG (CHOG) and [16, 17] propose Local Binary Pattern Histogram Fourier Features (LBP-HF). [3] considers the gradient orientation as continuous, and redefines Histograms of Oriented Gradient (HOG) in the frequency space to obtain the new descriptor, i.e., RIHOG. In contrast to the common Fourier-based features, RIHOG focuses not only on the frequency magnitude in the spectrum of the image but also the phase information and the global dependency of those frequency terms, improving the method's descriptive power in a significant manner.

Although there are a multitude of rotation-invariant features, the unavoidable information loss of their construction weakens their discriminations, and they are not powerful enough to discriminate human bodies. In this paper, we consider RIHOG as the low-level descriptor to catch object's specialties and generate a set of local features, upon which the Bag of Visual Words (BoVWs) model is then applied to extract the global description. The method is called RIHOG-BoVWs and has stronger discrimination for human detection with rotation-invariance.

3. HOG descriptor

As a gradient based feature, HOG has been paid great attention and applied to human detection extensively [18, 19] since it was proposed in 2005. HOG concentrates on the gradient orientations of local image patches and employs a histogram binning, spatial aggregation and overlapping local contrast normalization to strengthen the robustness to illumination and shadow.

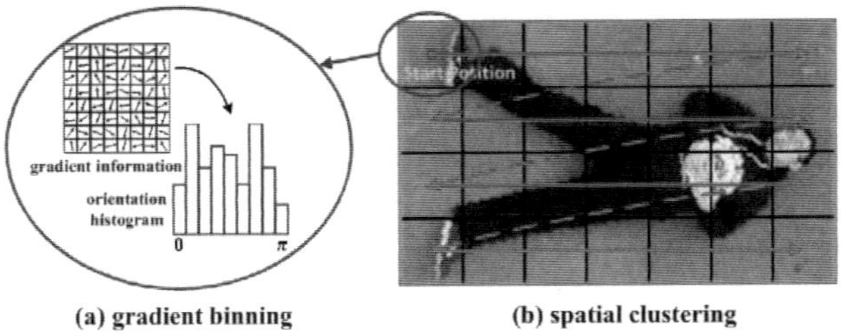

Figure 2. Illustration of HOG's calculation. HOG extracts the object's local features by the orientation histogram and clusters them to the global descriptor in the form of series. Both of two steps are contrary to rotation-invariance.

As shown in figure 2, the calculation of HOG can be summed up into two parts: the low-level feature calculation; and the high-level semantic feature extraction. The former transforms gradient information to orientation histograms, using local image patches and the latter accumulates local histograms to the final descriptor. Apparently, both of these two steps are not rotation-invariant.

4. The proposed method

In this section, we put forward a method to achieve the rotation-invariant global description for human detection. The method is completed from two aspects of low-level feature calculation and high-level semantic feature extraction based on the analysis of HOG in Section 3.

4.1. *Rotation-invariant low-level feature*

Following the idea of HOG, RIHOG [3] encodes the gradient information in a rotation invariant manner. The calculation of RIHOG can be divided into the following three steps.

4.1.1. Pixel-level features In HOG, a pixel's gradient orientation is discretized into the histogram, which changes in a complex manner when the image rotates. In contrast, RIHOG employs the continuous representation of gradient orientation, which means creating an orientation distribution function at each pixel. Taking the pixel with gradient $\boldsymbol{g}$ as an example, the function can be defined as:

$$\rho(\varphi) = \|\boldsymbol{g}\|\delta(\varphi - \psi(\boldsymbol{g})) \quad (1)$$

where $\|\boldsymbol{g}\|$ is the gradient magnitude and $\psi(\boldsymbol{g})$ denotes the gradient orientation that can be any value in $[0, 2\pi)$. Then, project the function into the Fourier space and the Fourier coefficients read

$$f_m = \frac{1}{2\pi}\int_0^{2\pi} \rho(\varphi)e^{-im\varphi}d\varphi = \frac{1}{2\pi}\|\boldsymbol{g}\|e^{-im\psi(\boldsymbol{g})} \quad (2)$$

where $m \in ¢$ denotes the order of Fourier basis. Equation 2 means that the pixel's gradient information can be represented by a series of Fourier coefficients. Since gradient magnitude and orientation are real-valued, the terms of $m < 0$ are redundant and can be discarded. When implementing, m has to be limited to $0 \le m \le M$, only getting the approximate representation of gradient information. Controlling the value of M is equivalent to low-pass filtering in the frequency domain, which can weaken the aliasing effect in the orientation quantization because that higher frequency terms are more sensitive to noise.

Extend equation 2 to the image scale and get the Fourier coefficients matrices that encode the image's gradient information:

$$F_m = \frac{1}{2\pi}\|\boldsymbol{G}\|e^{-im\psi(\boldsymbol{G})} \quad (3)$$

where $\boldsymbol{G} = \nabla I$ denotes the gradient field of the image I.

To further improve the robustness to local deformation, soft binning, like linear interpolation in HOG, is necessary and can be realized here by spatial convolutions upon the Fourier coefficients matrices. The convolution kernel must be isotropic and triangle or Gaussian kernel is optional.

Additionally, the local contrast-normalization can also be implemented based on convolutions. Therefore, the Fourier coefficients of the matrices are updated to

$$P_m = \frac{F_m * K_1}{\sqrt{\|G\|^2 * K_2}} \tag{4}$$

where K_1 and K_2 is the convolution kernel for the soft binning and the local contrast normalization, respectively. $P_m \in ¡^2$ is the final pixel-level feature matrix, and its size is as same as the raw image. Figure 3 shows an example.

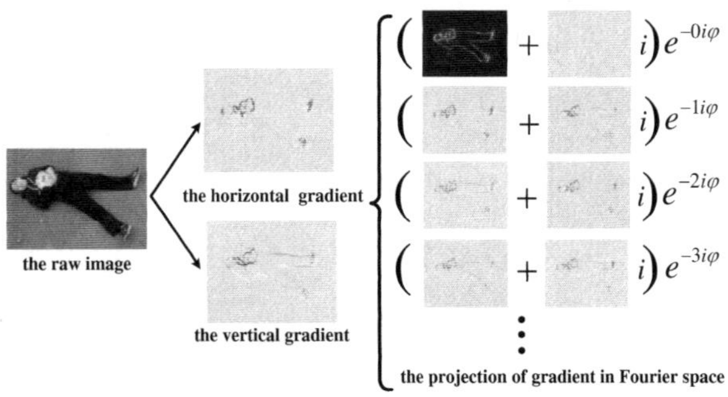

Figure 3. The example of gradient projection in Fourier space. For one image, its gradient information is represented by $(2M-1)$ coefficient matrices if the scope of Fourier order is $0 \le m \le M$. And every matrix has the same size as the raw image.

4.1.2. Local aggregation Local aggregation is important to the feature's property. According to [20], the optimal choice in the angular part of the circular ring is the Fourier basis and the function for computing local features are ready:

$$S_{j,k}(r,\varphi) = \Lambda(r - r_j, \sigma)e^{ik\varphi} \tag{5}$$

where k is the degree of Fourier basis, r_j denotes the middle radius of the j th circular ring and $\Lambda(x,\sigma) = \max(\frac{\sigma - |x|}{\sigma}, 0)$ defines a triangular function of width 2σ . Figure 4 presents an example of the local region partition, from which we find that the Fourier basis $e^{ik\varphi}$ can be considered as the weight distribution term upon pixels within the circular ring.

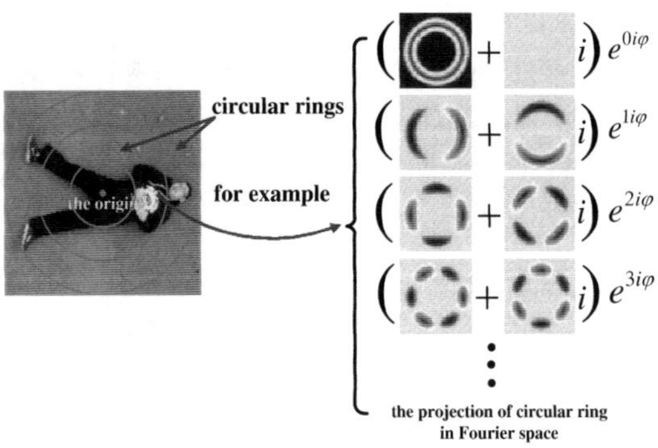

Figure 4. The example of local region partitions. When aggregating, the local patch is divided into several concentric circular rings, each of which can be represented by $(2K-1)$ matrices in Fourier space if the scope of Fourier degree is $0 \le k \le K$. Every matrix can be viewed as the weight distribution of pixels in the ring.

Subsequently, local aggregation can be completed via the convolution operation between P_m and $S_{j,k}$, i.e.,

$$H_{j,k,m} = S_{j,k} * P_m \tag{6}$$

$H_{j,k,m}$ denotes the regional features in one image patch. When implementing the process, k and m can be chosen and combined freely for the independence between gradient information and spatial region.

4.1.3. The selection of rotation-invariant terms According to the aforementioned derivations, every feature from equation 6 can be described as a separable form:

$$h = Ae^{-im\alpha+ik\varphi} \tag{7}$$

where A denotes the magnitude of the spectrum while $e^{-im\alpha+jk\varphi}$ represents the frequency and phase. It is well-known that the power spectrum, i.e., A, will not change when the image rotates, but it is far from enough to describe the characteristics of one object category. So, we need analyze the h and create more rotation-invariant features.

Taking the pixel p as an example, when the image rotates by angle θ, its gradient orientation α will change to $\alpha' = \alpha + \theta$. Similarly, all pixels have the same transformation law. Moreover, the absolute positions of pixels are also changed, while the relative positions between pixels remain invariant, which is performed as the translation of phase in the Fourier space. Consequently, the regional feature after rotation can be represented as:

$$h_r = Ae^{-im\alpha'+ik(\varphi+\theta)} = Ae^{-im(\alpha+\theta)+ik(\varphi+\theta)}. \tag{8}$$

Compared with the feature before rotation, we discover that $h_r = he^{i(k-m)\theta}$ and the image rotation's influence on the regional feature depends on the values of k and m. When $k = m$, h_r remains the same as h, which can be explained by the impact of image rotation compensated by the change of the pixels' weights. Both real and imaginary parts of h are put into the final descriptor.

There are also some dependencies between the image information of different circular rings. Consider the following examples, $h_1 = A_1 e^{-im_1\alpha_1+ik_1\varphi}$ and $h_2 = A_2 e^{-im_2\alpha_2+ik_2\varphi}$. When the proper values of Fourier order are chosen to satisfy $k_1 - m_1 = k_2 - m_2$, the two features are subject to the same impacts from image rotation. Under this condition, we can construct the union feature $h_{1,2} = h_1 * \overline{h_2}$ to preserve the relative phase information, which is rotation invariant, too.

In conclusion, to achieve the rotation-invariant description, RIHOG consists of three types of image formations, which include the power spectrum of gradients in frequency domain, the relative spatial structure of pixels in the same annulus, and the relative spatial structure of pixels in different annuli.

4.2. Rotation-invariant high-level semantic feature

In view of the discussions, RIHOG is more suitable to be the local descriptor and it is necessary to seek a rotation-invariant spatial clustering method. BoVWs model provides a powerful tool. In this paper, we focus primarily on the feasibility of the BoVWs model in rotation-invariant human detection, and we pay no attention on its optimization. Thus, we adopt the classical framework, under which the rotation-invariant global description can be divided into the following three steps:

Step 1: Sample the image densely using RIHOG as the local descriptor. Let every RIHOG be a *D*-dimensional vector, so the local feature set can be described as $X = [x_1, x_2, ..., x_K]$ $X \in \mathbb{R}^{D\times K}$, where K is the number of local features in one image.

Step 2: Mix the RIHOG features of training images and generate the codebook via *K*-means clustering algorithm. Consider N training images, and the $X_{train} = [x_1, x_2, ..., x_{NK}] \in \mathbb{R}^{D\times NK}$ denotes the RIHOG feature pool. Firstly, M basis vectors are selected randomly in the pool as the original clustering centers, each of which represents the different feature space. Then, assign each RIHOG feature to one feature space according the minimum Euclidean distance principle and re-determine the clustering centers of every space. Repeat the steps until the maximum iterations or the demanded cumulative approximation error $\sum_{i=1}^{T} \left\| x_i - b_{q_i} \right\|^2$. The final clustering centers are visual words, which compose the codebook $B = [b_1, b_2, ..., b_M] \in \mathbb{R}^{D\times M}$ and are stored offline.

IOP Conf. Series: Materials Science and Engineering **428** (2018) 012030 doi:10.1088/1757-899X/428/1/012030

Step 3: Redescribe the image with the codebook. Firstly, for each local feature contained in a image, select the most similar visual word $d_i = \arg\min_j \|x_i - b_j\|_2$. Then, create a sparse histogram $H \in \mathbb{R}^{M\times 1}$ over the visual words set $D=[d_1,d_2,...,d_K]$ and consider it as the final coding result after normalization.

After completing the feature description of all training samples, the classification model is trained via the selected method. When detecting, the test image is also described by the above processes and then decided by the classifier.

5. Experiments

To facilitate the comparison with other research and prove the effectiveness of the proposed method, two datasets are used in our experiments: one is the *Freestyle motocross* public dataset; and the other is our victim dataset.

The *Freestyle motocross* dataset was collected by Villamizar et al [8] and has been used in [21]. In *Freestyle motocross* dataset, there are two image sets for testing: one containing motorbikes without rotations; and another one corresponding to motorbikes with various in-plane rotations. Our victim dataset is established by simulating the perspective of a camera mounted on a UAV during the victim detection task. All images are collected with a quadrotor, in the genuine environment, under uncontrolled illuminations and shadows. We simulate the typical realities and difficulties of the rescue scenario as much as possible, forcing the dataset to have a large diversity of body postures, which include multiple in-plane rotations and various out-of-plane rotations.

5.1. Experiment setup

Our experiments contains four parts. First, using *Freestyle motocross* public dataset, the feasibility of RIHOG-BoVWs for complicated object detection is confirmed through an orientation-specific motorbike detection. Second, while still using *Freestyle motocross* public dataset, we conduct rotated motorbike detection tests to demonstrate the effectiveness of our method for rotated object detection. Third, using our victim dataset, we compare the RIHOG-BoVWs with other features and prove its superiority for rotation-invariant human detection. Finally, the processing time of different methods for victim detection is analyzed.

In the evaluation, whether a prediction is consider true or false depends on if the ratio of the intersection over the union of the bounding box and the ground truth exceeds 0.5. Multiple predictions of the same ground truth object are considered to be false positive except one as true. For all of our experiments, we report equal error rate (EER) and show precision-recall (PR) curves.

In addition, for all detection tests, we employ linear SVM as the classification method with its value optimized by 5-fold cross-validation.

5.2. Experiment 1: orientation-specific object detection on Freestyle motocross dataset

The first test dataset in *Freestyle motocross* has 69 images containing 78 motorbikes without explicit planar rotations. This dataset is used to verify the feasibility of RIHOG-BoVWs for object detection. In addition to our proposed method, we apply RIFF, RIHOG and HOG as the global descriptors to make comparisons. Additionally, the popular classification strategy, Dense SIFT plus BoVWs, is being tested in the detection task too. Relative to the standard SIFT, the Dense SIFT adopts a dense sample and cancels the dominant gradient direction, which make it very similar to HOG and is not rotation-invariant. So the method of the Dense SIFT plus BoVWs can also be viewed as the application of BoVWs model on HOG. The learning is done using 500 positive samples and 2000 negative samples selected randomly from the Caltech motorbike dataset [22].

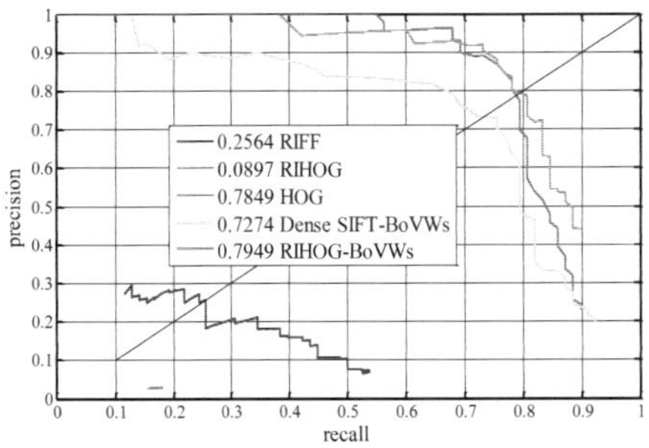

Figure 5. PR curves and EER values for orientation-specific motorbike detection.

Results of the first experiment are shown in figure 5. RIHOG-BoVWs achieves 0.7949 EER, which is comparable with HOG and is much better than other methods. Comparing to [8], their best result is 0.9103 EER, which is much higher than our results. In addition to the non-linear classification method, their results are achieved after training with 800 images that are generated by adding affine transformations from the most closely related images to the target motorbike model according to the author's doctoral thesis [23].

5.3. Experiment 2: rotation-invariant object detection on Freestyle motocross dataset

Regarding the second motorbike dataset of *Freestyle motocross*, there are 100 images of 128 motorbikes with challenging rotations. Like [21], in our experiment, the training set is created by taking the first 40 images and the remaining 60 images are considered as the test dataset

In rotation-invariant motorbike detection, we also use RIFF, RIHOG and HOG as the global descriptors to compare with. For HOG, when detecting, the image detection is recalled 15 times with a step of 24 degrees to respond to objects with different orientations.

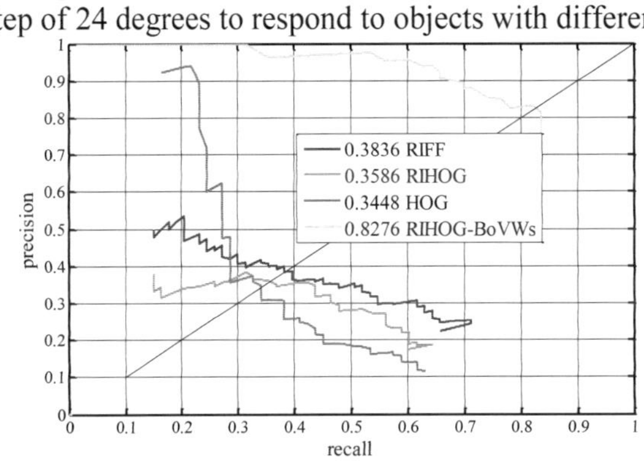

Figure 6. PR curves and EER values for rotation-invariant motorbike detection.

Detection performance of the second experiment is shown in figure 6. Among all the methods, RIHOG-BoVWs achieves the best performance, which demonstrates its superiority for rotation-invariant object detection. In [21], the result is 0.90 EER, which is better than our method. However, their approach is based on complex harmonic basis and equivariant filters. On the contrary, we focus on the rotation-invariant description and employ a linear classification method, which is more intuitive and concise.

5.4. Experiment 3: rotation-invariant human detection on our victim dataset

Finally, we conduct the experiment on our victim dataset. Apart from the methods in Section 5.3, we also compare the HogLbp[24], DPM[25], and ACF (Aggregate Channel Features) [26], which have revealed high-performance results in pedestrian detection. As a note, we recall the image detection 15 times with a step of 24 degrees to realize the desired victim detection when using HOG or other pedestrian detectors. Our victim dataset consists of 158 images with 58 training images and 100 testing images, where bounding boxes have been manually labeled.

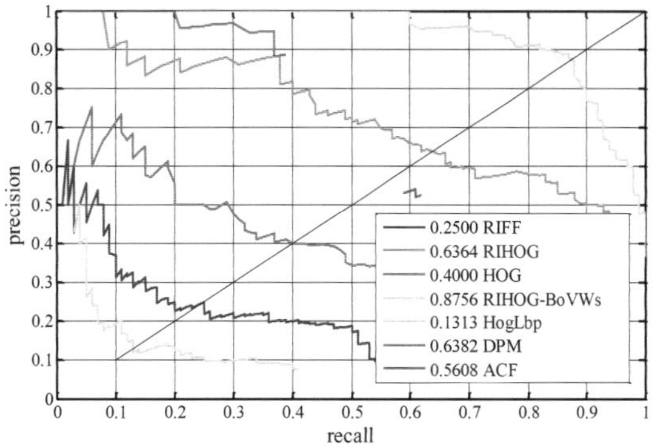

Figure 7. PR curves and EER values for victim detection.

Detection results are shown in figure 7. From the comparisons, we find that RIHOG-BoVWs is more suitable for victim detection from areal views Figure 8 shows instances of detection results and presents the robustness of the RIHOG-BoVWs method to various rotations and complex postures of human bodies.

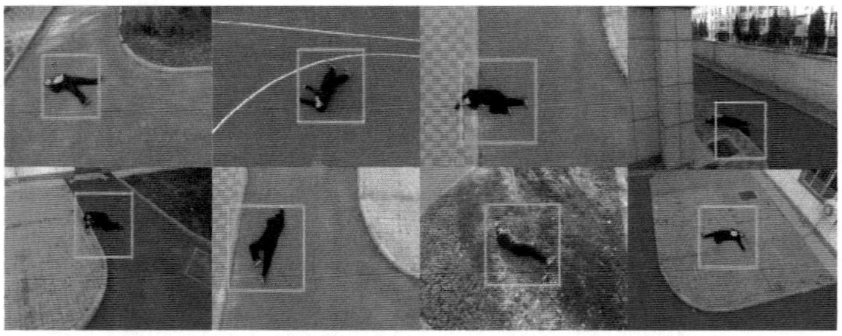

Figure 8. Examples of detection results using RIHOG-BoVWs on our victim dataset.

5.5. *Experiment 4: computational time analysis*

We test the time associated with different methods when working on victim detection and present them in figure 9 with HOG as the bench mark. Among all methods, RIHOG-BoVWs has the similar computational complexity with ACF and has huge efficiency in victim detection task.

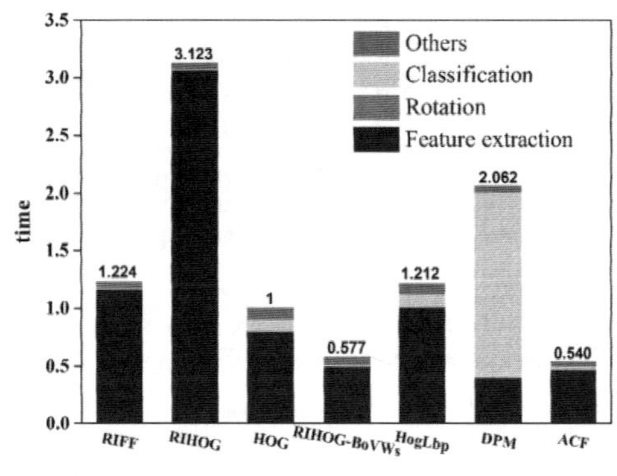

Figure 9. Processing time analysis of different detectors in victim detection. For each method, we present the processing time in four parts: feature extraction, image rotation, classification and others.

6. Conclusion

In this paper, we presented a rotation-invariant semantic description method, called RIHOG-BoVWs, to investigate rotation-invariant human detection, which is very significant in the victim detection case from a UAV. Learning from HOG, we concentrate on the object's local characteristics, which are robust against the geometric deformation and illumination variation. In the proposed method, RIHOG is employed upon image patches to get a local feature set and BoVWs model is then used to extract the semantic object-level description. Both steps guarantee the rotation-invariance of the final descriptor. Experimentally, we tested the sufficient discrimination of RIHOG-BoVWs for complicated objects and evaluated its ability in addressing planar rotations. Finally, we have established a victim dataset and compared RIHOG-BoVWs with other pedestrian detection methods. Detection results show that RIHOG-BoVWs has the high performance for victim detection with the advantage of being computationally more efficient.

Acknowledgments
Supported by the Key Program of Experimental Technique of General Armament Department (2014SY54A1102).

References

[1] N. Dalal and B. Triggs, "Histograms of Oriented Gradients for Human Detection," in IEEE Computer Society Conference on Computer Vision & Pattern Recognition, 2005, pp. 886-893.
[2] M. Andriluka, P. Schnitzspan, J. Meyer, and S. Kohlbrecher, "Vision based victim detection from unmanned aerial vehicles," in Ieee/rsj International Conference on Intelligent Robots and Systems, 2010, pp. 1740-1747.
[3] K. Liu, H. Skibbe, T. Schmidt, T. Blein, K. Palme, T. Brox, et al., "Rotation-Invariant HOG Descriptors Using Fourier Analysis in Polar and Spherical Coordinates," International Journal of Computer Vision, vol. 106, pp. 342-364, 2014.
[4] L. Fei-Fei and P. Perona, "A bayesian hierarchical model for learning natural scene categories," in 2005 IEEE Computer Society Conference on Computer Vision and Pattern Recognition (CVPR'05), 2005, pp. 524-531.
[5] T. Moranduzzo and F. Melgani, "Detecting Cars in UAV Images With a Catalog-Based Approach," IEEE Transactions on Geoscience & Remote Sensing, vol. 52, pp. 6356-6367, 2014.
[6] D. P. Casasent, "Real-time people and vehicle detection from UAV imagery," Proceedings of SPIE - The International Society for Optical Engineering, vol. 7878, pp. 78780B-78780B-13, 2011.
[7] S. Tuermer, F. Kurz, P. Reinartz, and U. Stilla, "Airborne Vehicle Detection in Dense Urban Areas Using HoG Features and Disparity Maps," IEEE Journal of Selected Topics in Applied Earth Observations & Remote Sensing, vol. 6, pp. 2327-2337, 2013.
[8] M. Villamizar, F. Moreno-Noguer, J. Andrade-Cetto, and A. Sanfeliu, "Efficient rotation invariant object detection using boosted random ferns," in Computer Vision and Pattern Recognition (CVPR), 2010 IEEE Conference on, 2010, pp. 1038-1045.
[9] A. Vedaldi, M. Blaschko, and A. Zisserman, "Learning equivariant structured output SVM regressors," in Computer Vision (ICCV), 2011 IEEE International Conference on, 2011, pp. 959-966.
[10] D. G. Lowe, "Distinctive Image Features from Scale-Invariant Keypoints," International Journal of Computer Vision, vol. 60, pp. 91-110, 2004.
[11] W. Zhang, X. Sun, K. Fu, C. Wang, and H. Wang, "Object detection in high-resolution remote sensing images using rotation invariant parts based model," Geoscience and Remote Sensing Letters, IEEE, vol. 11, pp. 74-78, 2014.
[12] G. Takacs, V. Chandrasekhar, S. S. Tsai, D. Chen, R. Grzeszczuk, and B. Girod, "Fast computation of rotation-invariant image features by an approximate radial gradient transform," IEEE Transactions on Image Processing, vol. 22, pp. 2970-2982, 2013.
[13] G. Takacs, V. Chandrasekhar, S. Tsai, D. Chen, R. Grzeszczuk, and B. Girod, "Unified real-time tracking and recognition with rotation-invariant fast features," in Computer Vision and Pattern Recognition (CVPR), 2010 IEEE Conference on, 2010, pp. 934-941.
[14] B. Liu, H. Wu, W. Su, and J. Sun, "Sector-ring HOG for Rotation-Invariant Human Detection," Signal Processing: Image Communication.
[15] H. Skibbe and M. Reisert, "Circular Fourier-HOG features for rotation invariant object detection in biomedical images," Proceedings, vol. 88, pp. 450-453, 2012.
[16] T. Ahonen, J. Matas, C. He, and M. Pietikäinen, "Rotation Invariant Image Description with Local Binary Pattern Histogram Fourier Features," Image Analysis, vol. 5575, pp. 61-70, 2009.

[17] G. Zhao, T. Ahonen, J. Matas, and M. Pietikainen, "Rotation-Invariant Image and Video Description With Local Binary Pattern Features," IEEE Transactions on Image Processing A Publication of the IEEE Signal Processing Society, vol. 21, pp. 1465-1477, 2012.
[18] S. Tang, M. Andriluka, and B. Schiele, "Detection and tracking of occluded people," International Journal of Computer Vision, vol. 110, pp. 58-69, 2014.
[19] P. F. Felzenszwalb, R. B. Girshick, D. McAllester, and D. Ramanan, "Object detection with discriminatively trained part-based models," Pattern Analysis and Machine Intelligence, IEEE Transactions on, vol. 32, pp. 1627-1645, 2010.
[20] Q. Wang, O. Ronneberger, and H. Burkhardt, "Rotational invariance based on Fourier analysis in polar and spherical coordinates," IEEE Transactions on Pattern Analysis and Machine Intelligence, vol. 31, pp. 1715-1722, 2009.
[21] K. Liu, Q. Wang, W. Driever, and O. Ronneberger, "2d/3d rotation-invariant detection using equivariant filters and kernel weighted mapping," in Computer Vision and Pattern Recognition (CVPR), 2012 IEEE Conference on, 2012, pp. 917-924.
[22] R. Fergus, P. Perona, and A. Zisserman, "Object class recognition by unsupervised scale-invariant learning," in Computer Vision and Pattern Recognition, 2003. Proceedings. 2003 IEEE Computer Society Conference on, 2003, pp. II-264-II-271 vol. 2.
[23] M. A. Villamizar Vergel, "Efficient approaches for object class detection," 2012.
[24] X. Wang, T. X. Han, and S. Yan, "An HOG-LBP human detector with partial occlusion handling," in IEEE International Conference on Computer Vision, 2009, pp. 32-39.
[25] P. Felzenszwalb, D. McAllester, and D. Ramanan, "A discriminatively trained, multiscale, deformable part model," in Computer Vision and Pattern Recognition, 2008. CVPR 2008. IEEE Conference on, 2008, pp. 1-8.
[26] P. Dollar, R. Appel, S. Belongie, and P. Perona, "Fast Feature Pyramids for Object Detection," IEEE Transactions on Pattern Analysis & Machine Intelligence, vol. 36, pp. 1532-45, 2014.

Target Tactical Clustering Algorithm Based on Grouping Field

Na Hao[a], Kaixuan Chu[b], Tianqing Chang[c], Bo Zhang and Lei Zhang

Army Academy of Armored Forces, 100072, Beijing

[a]lzygdshn@163.com, [b]1016938865@qq.com, [c]changtianqing@263.net

Abstract. This paper proposes a target-based tactical clustering algorithm based on the grouping field. Inspired by the midfield theory of physics, the concept of gravitational force is used in the target tactical grouping of the armored unit level, and the concept of grouping field and grouping force is proposed. After acquiring the target identity, the state of motion parameters and the trend of the movement, the information of the identity attributes and state parameters of each target are fused and analyzed, the group field strength function is constructed, and a single target with a certain cooperation relationship in tactics is classified, forming tactical groups. Simulation results show the effectiveness of the proposed algorithm.

1. Introduction

The target tactical grouping is to cluster the targets with similar combat functions, similar intentions and close coordinate positions, and then conduct research on them. The purpose is to make inferences about their operational intentions based on clustering characteristics and cluster parameters. The most widely used is MacQuccn in 1967 proposed K-Means clustering algorithm. The algorithm finds out k clustering centers, divides the data in the dataset into k classes according to the distance relation, and adjusts the clustering center to minimize the sum of the square distances of the corresponding clustering centers in the dataset. The algorithm is simple and easy to implement, but it is sensitive to the initial value, and the number k must be specified manually[1] .In order to overcome the shortcomings of algorithms, scholars have studied the optimization of K-means clustering based on swarm intelligence. Literature [2] and [3] use artificial bee colony algorithm and particle swarm algorithm respectively to improve them and overcome the sensitivity of initial clustering center and easy to fall into the local shortcomings. Spectral Analysis[4] is widely used in clustering, pattern recognition and meshing. The method mainly uses the eigenvalues and eigenvectors to cluster the data in the dataset. Complex network clustering[5] has a wide range of applications in social networks, biometrics and the World Wide Web. In practice, it often happens that data points in a dataset belong to more than one class at a time. After the researchers' research, a fuzzy clustering method[6-7] has been proposed, such as FCM algorithm, FBSA algorithm, Gustafson-kessel algorithm, Gath-Geva algorithm. Hierarchical clustering is also a commonly used clustering method.Inspired by the midfield theory of physics, literature [8] proposes a hierarchical clustering method.

At present, there are few researches on armored tactical tactical clustering algorithm. On the one hand, the traditional combat mode is based on commander's command, and there is less demand for this; on the other hand, the armored detachment target clustering needs to consider not only the target geographical position attribute but also the Consider the target feature attributes. Currently, the algorithm only considers the geographical attributes of the target itself, and the effect of direct use for

IOP Conf. Series: Materials Science and Engineering **428** (2018) 012031 doi:10.1088/1757-899X/428/1/012031

combat target grouping is not satisfactory enough. This paper combines the Voronoi diagram theory in computational geometry with the field theory in physics. Based on the FTASC algorithm[9], according to the characteristics of terrestrial armored targets, the radiation parameters are set according to the threat degree of different armored targets, and a clustering-based approach is proposed. Target tactical clustering algorithm to achieve armored team-level target tactical grouping.

2. The principle of grouping

Due to the tactical cooperation between multiple targets in the tactical group, the target position characteristics must have certain constraints in the spatial region, and there are some similarities between the target speed features. Based on this, the tactical clustering feature space can be constructed based on the obtained target state parameters, and the similarity measure function of the target in the tactical clustering feature space can be established to measure the characteristics of the target group. The standard of the metric function construction is to make the difference between the targets belonging to the same tactical group within a certain threshold range, while the targets belonging to different tactical groups have big differences.

A number of goals for each tactical group will inevitably have a core, that is, the group's command node. The synergy between command nodes and other targets in the tactical group can be vividly described as the attraction between objects with different masses in the gravitational field. Therefore, the concept of gravitation can be used in the objective tactical clustering of armored sub-team level. Make the following convention before grouping:

(1) In battlefield space, our armor units will face multiple enemy targets that have certain tactical values.

(2) The target will have some restraining effect on the other targets around it, and the scope of this restraining effect is limited. In the limited battlefield space around the target, the role of restraint is obvious. The space region is defined as the tactical grouping field. The restraining effect is defined as the tactical grouping force, referred to as the grouping force for short.

From the above agreement can be determined within the battlefield armored sub-team level target tactical clustering principles:

(1) For a certain tactical group, there must exist a target that has obvious restraining effect on the other targets in the group, that is, it produces the largest grouping force.

(2) Among the targets in the tactical group, some tactical coordinated actions will be adopted, and the targets are closely linked and the force of grouping is strong. Due to the different tactics between different tactical groups, the linkages will inevitably loose and will be weaker in the grouping forces.

(3) The goal that does not belong to any tactical group belongs to a single goal, independence of action, the grouping force will inevitably be smaller.

3. Grouping field and grouping force

This paper proposes the concept of 'grouping field' and 'grouping force' for land armored targets, defines a function that characterizes the mechanism of action of the flocking field, and combines the flocking field with the grouping force. For armored team-level targets and tactical groups, due to the limitation of the number of team-level targets, the team tactical group has limited space within the battlefield. Therefore, the scope of the clustering field should be compatible with the scale of the battlefield control area at the tactical group. The range of influence produced by the divergence field, that is, the scope of the grouping force should be limited to the battlefield area of the tactical group at the detachment level. The force beyond this area should be significantly reduced. The improved Voronoi diagram is used to divide the tactical clustering feature space, and then the function mechanism function of the target tactical group clustering field is constructed.

Whether the multiple targets in the battlefield space can be divided into one tactical group is closely related to the degree of similarity of the groups in the target. Some features in the same tactical group among the various objectives must have the characteristics of high similarity. Because the targets in the same tactical group will maintain a relatively stable trend in the spatial range, the position coordinate difference is within a certain range, and there is a high degree of similarity between the average velocity, the heading angle and the direction of the weapon barrel. Therefore, The target position coordinates x_i, y_i, z_i, the target speed V_{ix}, V_{iy}, V_{iz}, the target heading angle α_i, and the target weapon pointing angle β_i as the eigenvectors. Considering that the state parameter of a single target in a tactical group is a function of a discrete time series, in order to ensure the stability of the target tactical grouping, the average value of the parameters over a period of time is taken as the value of the eigenvector. The target tactical clustering eigenvector is constructed as:

$$Oj_i = [\bar{x}_i, \bar{y}_i, \bar{z}_i, \bar{V}_{ix}, \bar{V}_{iy}, \bar{V}_{iz}, \bar{\alpha}_i, \bar{\beta}_i]$$

The target tactical clustering feature space is a 7-dimensional space containing 7 characteristic indexes, denoted as **QT**.

Definition 1: Feature Voronoi space area

For a certain battlefield space target set O, $O = \{O_1, O_2, O_3, \cdots, O_n\}$, any goal O_i of O ,its corresponding characteristic Voronoi space area O_i^V can be defined as

$$O_i^V = \{x \mid d(q, O_i) \le d(q, O_j), O_i, O_j \in O, i \ne j, q \in \mathbf{QT}\}$$

All the corresponding Voronoi spatial regions of O_i constitute the Voronoi field of the target set O .That is to say

$$O^V = \{O_1^V, O_2^V, O_3^V, \cdots, O_n^V\}$$

Definition 2: Recent Voronoi space domain target

For a battlefield space goal set O, where $O_i, O_j \in O$ .If O_i^V and O_j^V have adjacent spatial surfaces ,their relationship can be described as :they act as each other's closest Voronoi space target. Delaunay set is a collection of all its nearest Voronoi space domain targets, denoted as $\mathrm{NSD}(O_i)$.

Definition 3: Recent Voronoi space domain

At a certain target O_i in the optional battlefield space, the characteristic V-space region to which the target of the nearest V-space domain owned by O_i belongs forms a spatial region that forms the nearest V-space domain of O_i with O_i^V, denoted as $\mathrm{DNV}(O_i)$.

Definition 4: Field strength function of the grouping field

An important principle to be followed in constructing the field strength function of a grouping field is to consider the area of the target feature Voronoi space as the applicable range of the target's restraining effect. The restraining effect of each target spread outwards around its own position until Encouragement of other goals.

According to the above characteristics, we can establish the field strength of the sub-field as

$$E_{O_i} = k \frac{1}{d(O_i, q)^{2\sigma}} e_{O_i q}$$
$$\sigma = \begin{cases} 1, & q \in \mathrm{DNV}(O_i) \\ +\infty, & q \notin \mathrm{DNV}(O_i) \end{cases} \tag{1}$$

where, E_{O_i} is the target field strength of the subfields generated in the feature space, k is grouping field radiation parameters, q is any point in the feature space, $d(O_i, q)$ is the distance between target O_i and point q, σ is field strength parameters , $e_{O_i q}$ is the unit vector of O_i to q.

The selection of the group field radiation parameters k needs to be adapted to the target type. For different types of equipment selection of radiation parameters, using expert scoring method. The construction of the target type T_i and k corresponding relationship shown in Table 1.

Table 1. Radiation factor k value table

T_i	Main battle tank	Infantry chariot	Transport car	Self-propelled artillery	Light vehicle
k	1	0.8	0.5	0.4	0.2

According to formula (1), we can express the grouping power between two targets

$$F_Q(O_i, O_j) = E_{O_i} m_{O_j} = k \frac{m_{O_j}}{d(O_i, O_j)^{2\sigma}} e_{O_i O_j}$$
$$\sigma = \begin{cases} 1, & O_j \in \text{NSD}(O_i) \\ +\infty, & O_j \notin \text{NSD}(O_i) \end{cases} \tag{2}$$

where, m_{O_j} is the quality of O_j ,supposing the target is a unit of particle, so $m_{O_j} = 1$, $d(O_i, O_j)$ is distance metric function of target O_i and O_j in feature space, σ is field strength parameters.

4. Grouping algorithm

4.1. Principle of algorithm

According to the hypothesis, armored unit-level tactical clustering algorithm consists of the following two basic steps:

(1) For each target in the battlefield space, obtain the target of the most recent space domain;

(2) Choose a target that has greater containment power for other targets, that is, generate the target with the largest grouping field strength, and divide the targets of its recent space domain into a category to form a tactical group. The basic principle of the algorithm is as follows.

During the tactical grouping of armored unit-level targets, when a group of objects that interact with each other is a vector, and a target has a restraining effect on its surrounding targets, it will inevitably be adversely checked by other target directions, that is, There is a reverse group of forces acting on this goal. All the opposite direction restraints inevitably form a combined force in the space, and the target force generally appears in the opposite direction of the combinatorial force. The target can be considered as a tactical cluster center. The combined force at the tactical grouping center O_i in the grouping field is

$$F_C = \sum F_Q(O_i, O_j) \tag{3}$$

The angle between the combined force and the component force will have a certain impact on the tactical grouping. If the angle is smaller than 90°, the component force can enhance the effect of the combined force. If the angle is greater than or equal to 90°, then the component force will have a weaker effect on the combined force, which will cause grouping errors when the target is at multiple tactical group boundaries. Certain constraints must be set on the grouping algorithm.

4.2. Algorithm constraints

For the problem of wrong grouping caused by the target at the boundary of tactical group, we set the constraints on the size of grouping when grouping, as follows:

(1) Select a target within the battlefield space as the center of the grouping, count the target of the nearest Voronoi space domain, and select all the targets whose angles of component force and combination force are less than 90° are expressed as $\text{CNS}(O_i)$, which is:

$$\text{CNS}(O_i) = \{O_j \mid \theta(F_Q(O_i, O_j), F_C) < 90^\circ, O_j \in \text{NSD}(O_i)\}$$

Where, $\theta(F_Q(O_i, O_j), F_C)$ is the angle of the forces.

(2) Summarize the reverse groupings suffered by the cluster centers and calculate their average

Sort all the target groups in the $\text{CNS}(O_i)$ of the grouping center by the grouping power of the grouping centers, remove the minimum $[N/2]$ targets, sum the grouping forces of the remaining targets, and average them, expressed as $E(F)$, N is the nearest target number of Voronoi space domain, [ ] means rounding.

(3) Determine the target of tactical grouping objects

The target of the tactical grouping target must meet: The target grouping group has a grouping force greater than $E(F)/5$. The set of goals that satisfy the condition is denoted as $\mathrm{NS}(O_i)$, which is

$$\mathrm{NS}(O_i) = \{O_j \mid |F_Q(O_i, O_j)| > E(F)/5, O_j \in \mathrm{CNS}(O_i)\}$$

4.3. Algorithm Description

According to the above basic principle, given a battlefield space target set O, armored unit-level tactical clustering algorithm can be described as follows:

Step A: For each target O_i in the grouping target set, calculate its characteristic Voronoi space area and generate $\mathrm{NSD}(O_i)$; and obtain the combined force of the grouping forces O_i received.

Step B: Generate $\mathrm{NS}(O_i)$ for target O_i.

Step C: Calculate the combined force of the reaction force of the grouping force of each target in the target set. Select the target of the maximum combined force O_x as the tactical grouping center, and classify all the targets in the $\mathrm{NS}(O_x)$ as O_x which is marked as "grouped."

Step D: Select the non-clustering target O_j in C_i and assign $\mathrm{NS}(O_j)$ to C_i until all the targets are marked as "grouped" to generate a target tactical group.

Step E: Calculate the number of targets contained in the set C_i. When $Count=1$, indicating that only one target is included and the target is an independent target.

Step F: If there are no targets in the statistical target set O that are included in the tactical group, look for new tactical grouping centers in these targets and perform *Step* C ~ *Step* E again until all targets are clustered or become independent targets.

5. Simulation example

In order to validate the tactical clustering algorithm and tactics recognition method proposed in this paper, we use Matlab2014a software to write the simulation program and simulate the computer with the frequency of 1.6GHz CPU and memory 2G.

(1) Battlefield Environment

In the $OXYZ$ coordinate system, set the simulated area size of 1km × 1km × 800m undulating flat terrain environment, as the target sports venue.

(2) Scale of Troops

Set the target number of enemy targets by unit size. At a certain moment, the target location of the target units in the battlefield space is shown in Figure.1, and each target is displayed by a diamond logo.

(3) Target Tactical Clustering Algorithm Simulation

We observe that the enemy's target cluster eigenvectors are shown in Table 2. Among them, the heading angle and the pointing angle of the weapon are all based on the positive direction of the Y-axis and rotate clockwise to be positive.

Table 2. Target cluster eigenvector

Target	$\bar{x}$	$\bar{y}$	$\bar{z}$	$\bar{V}_x$	$\bar{V}_y$	$\bar{V}_z$	$\bar{\alpha}$	$\bar{\beta}$
1	198	385	20	-5	15	0.1	-18	-10
2	245	510	22	-2	18	0.2	-6	-12
3	400	410	23	-2	15	0.2	-8	-5
4	490	620	21	0	21	-0.1	0	10
5	590	780	25	2	22	0	5	15
6	703	620	20	1	30	0.1	2	10
7	600	520	25	2	24	0	5	10

8	790	400	22	3	18	0.2	9	0
9	900	505	25	5	20	0.2	14	5
10	990	390	22	4	17	0.1	13	8

According to the target cluster eigenvector in Table 2, the characteristic distances among the targets are solved as shown in Table 3.

Table 3.The distances among target features

Target	1	2	3	4	5	6	7	8	9	10
1	0	1.06	1.04	2.24	2.76	2.25	2.44	2.42	3.00	2.88
2	1.06	0	0.53	2.18	2.35	1.83	2.09	1.63	2.22	2.27
3	1.04	0.53	0	1.93	2.06	1.53	1.82	1.47	2.05	2.04
4	2.24	2.18	1.93	0	0.82	1.09	0.74	1.87	2.05	1.59
5	2.67	2.35	2.06	0.82	0	0.78	0.48	1.55	1.51	1.13
6	2.25	1.83	1.53	1.09	0.77	0	0.65	1.12	1.29	1.06
7	2.44	2.08	1.82	0.74	0.48	0.66	0	1.29	1.35	0.94
8	2.42	1.63	1.47	1.87	1.55	1.12	1.28	0	0.63	0.82
9	3.00	2.22	2.05	2.05	1.51	1.29	1.35	0.63	0	0.62
10	2.88	2.27	2.04	1.59	1.13	1.06	0.94	0.81	0.62	0

Enemy target is the main battle tank, then k = 1. For NSD(O_i), which generates the target O_i, the following processing is performed in the simulation. The k targets whose distance metric function $d(O_i, O_j)$ in the feature space between the target O_i and the O_j is the smallest form the NSD(O_i) of the target, and the value of k is the same as that of the local team In the simulation example, the enemy unit contains 10 targets, and the k value of 3 is more appropriate, corresponding to the rank tactical units. The field strength variation σ is 1. $e_{O_iO_j}$ is the unit vector, and the value is 1. The force of grouping among targets can be calculated as shown in Table 4.

Table 4. Grouping force among targets

Target	1	2	3	4	5	6	7	8	9	10
1	0	0.90	0.92	0.20	0	0	0	0	0	0
2	0.90	3.54	0	0	0	0	0	0.38	0	0
3	0	0	0	0	0	0	0	0.46	0	0
4	0	0	0	0	1.48	0.84	1.81	0	0	0
5	0	0	0	1.48	0	1.68	4.41	0	0	0
6	0	0	0	0	1.68	0	2.35	0	0	0.90
7	0	0	0	1.81	4.41	2.35	0	0	0	0
8	0	0	0	0	0	0.79	0	0	2.52	1.50
9	0	0	0	0	0	0.60	0	2.52	0	2.58
10	0	0	0	0	0	0	1.14	1.50	2.58	0

Based on the above calculation results, a tactical clustering algorithm is run and tactical clustering is carried out on the target of the contingent tank. The result is shown in Figure 2.

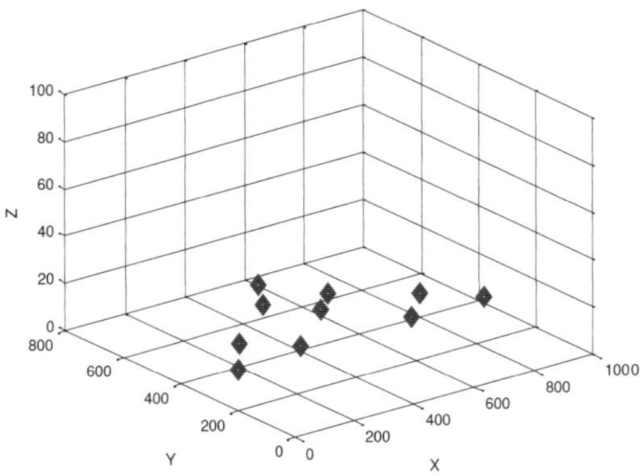

Figure 1. Unit target geographic location map.

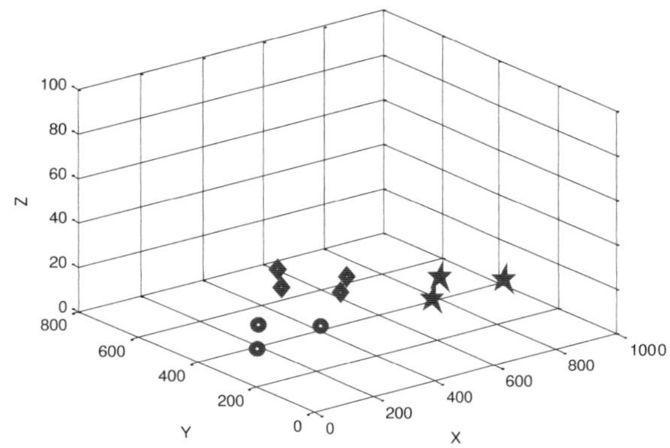

Figure 2. Unit target grouping map.

6. Conclusion

The simulation results show the effectiveness of the target-tactical clustering algorithm based on the clustered field proposed in this paper, which can quickly and accurately achieve target tactical clustering at the armored unit level and provide an important basis for further obtaining the tactical intent of the enemy and assessing the threat level of the enemy target.

References

[1] Wazavkar S V, Manjrekar A A. Text Clustering Using HFRECCA and Rough K-Means Cluster Algorithm[J]. Discovery, 2014, 15(40):44-47.
[2] Yu Jinping, Zheng jie, Mei Hongbiao. K-means clustering algorithm based on improved artificial bee colony algorithm [J]. Journal of Computer Applications, 2014, 34(4):1065-1069.
[3] Yang Zhi, Luo Ke. Improved clustering algorithm based on particle swarm optimization [J]. Application Research of Computers, 2014, 31(9):2597-2599.
[4] Liu R, Zhang H. Segmentation of 3D Meshes through Spectral Clustering[C]// Computer Graphics and Applications, 2004. PG 2004. Proceedings. 12th Pacific Conference on. IEEE, 2004:298-305.
[5] Yang Bo, Liu Dayou, Liu Ji-ming. Application Research of Computers [J]. Journal of Software, 2009, 20(1): 54-66.
[6] Basu B, Srinivas V V. Regional flood frequency analysis using kernel-based fuzzy clustering approach[J]. Water Resources Research, 2014, 50(4):3295-3316.
[7] Li Xiang, Wang Hau-san, Wu Si. A fuzzy minimax clustering model and its application[J]. Information science, 2012, 186(1):114-125.
[8] GAN W,LI De. An Hierarch ical Clustering Method Based on Data Fields [J]. ACTA ELECTRONICA SINICA, 2006, 34(2):258-262.
[9] DENG Min, LIU Qiliang, LI Guangqiang, CHENG Tao .Field-theory based spatial clustering method[J]. Journal of Remote Sensing,2010,14(4):694-709

IOP Conf. Series: Materials Science and Engineering **428** (2018) 012032 doi:10.1088/1757-899X/428/1/012032

Dynamic Differential Evolution Algorithm Applied in Point Cloud Registration

C L Li and S Y Dian*

School of Electrical Engineering and Information, Sichuan University, Chengdu, 610065, China;

*Corresponding e-mail: scudiansy@scu.edu.cn

Abstract. Aiming at the problem of point cloud registration, a registration method based on dynamic differential evolution algorithm (DDE) is proposed. The voxel grid method is used to uniformly sample the point cloud to reduce the time complexity of DDE. The individuals of DDE are coded by rotation angle and the translation distance. The kd-tree is used to search for corresponding point pairs, and the root mean square error is defined as the objective function of DDE. Experiments show that the proposed method can register point clouds with a large difference in position, and has better registration accuracy than iterative closest point (ICP).

1. Introduction

With the rapid development of 3D scanning technology, 3D reconstruction of objects based on point cloud has been widely used in agriculture, medicine, and reverse engineering [1-3], which has important research value. Due to the object's size and the linear propagation characteristics of light, to obtain the complete point cloud on the object surface, point cloud collection of multiple views is required. And the point cloud registration refers to transform the data collected in multiple views into a common coordinate system by seeking an appropriate coordinate transformation. The accuracy of point cloud registration will directly affect the accuracy of subsequent 3D reconstruction.

At present, the ICP registration algorithm proposed by Besl [4] is the most influential, and many improved algorithms have emerged [5-9]. However, the algorithm needs a good initial position, or it may fall into a local optimum, which result in registration failure. Therefore, it is very necessary to study the coarse registration algorithm to transform the position of point cloud to convergence domain of ICP. The common method is to use point cloud's normal vector [10-13] and curvature [14] and other feature information to establish the correspondence between point cloud data to achieve coarse registration, or use PCA [15-16] method and genetic algorithm [17]. For the above methods, feature histograms and curvature maps generally need to be established, which are time-consuming. The application of PCA needs to meet the extremely high similarity between point clouds to be registered. Therefore, it is of great significance to continue the research on the coarse registration of point clouds.

Because the differential evolution (DE) algorithm has powerful global search capability and fast convergence performance [18-19], this paper applies a dynamic differential evolution algorithm to point cloud coarse registration. Then, the ICP algorithm is used to the fine registration of the point cloud.

2. Dynamic differential evolution algorithm

Some typical test functions have shown that DDE has a faster convergence rate than original DE [20]. So, DDE is used to the coarse registration rather than DE.

2.1. The principle of dynamic differential evolution algorithm

The main steps of the DDE algorithm based on optimal ordering mutation operation [20] are shown as follows:

a) Set the population size NP, evolutionary generation D, scaling factor F and crossover rate CR.

b) Randomly initialize the population X in the search space. The i-th individual is shown as:

$$X_i = \{x_{i1}, \cdots, x_{ij}\} \tag{1}$$

Where $i \in [1, \text{NP}]$. j represents the dimension of the individual population.

c) Perform mutation operation to generate mutant individuals:

$$V_i = X_1 + F \cdot (X_2 - X_3) \tag{2}$$

Where X_1, X_2 and X_3 are three different individuals randomly selected from the current generation population, but they satisfy the following relationship of objective function value:

$$f(X_1) \le f(X_2) \le f(X_3) \tag{3}$$

That is, the best individual X_1 among the three individuals is taken as the base vector, and the search is performed along the direction $(X_2 - X_3)$ in which the individual objective function value is better. In this way, the random selection of three individuals can guarantee the global search performance, and the optimal sorting mutation strategy can improve the convergence rate of the algorithm.

d) Perform crossover operation to generate trial individuals.

$$Q_i = \{q_{i1}, \cdots, q_{ij}\} \tag{4}$$

$$q_{ij} = \begin{cases} v_{ij}, & rand_{ij} \le CR \;\; or \; j = j_{rand} \\ x_{ij}, & otherwise \end{cases} \tag{5}$$

Where, $rand_{ij}$ represents a random number between [0,1]. j_{rand} represents a random integer between $[0, j]$.

e) Individuals with better objective function values are selected to enter current population.

$$X_i = \begin{cases} Q_i, & if \;\; f(Q_i) \le f(X_i) \\ X_i, & otherwise \end{cases} \tag{6}$$

2.2. Registration parameter calculation

Given a source point cloud S and a target point cloud M, the goal of 3D registration is to find the six parameters for point cloud position transformations that can minimize the distance between two point clouds. These six parameters are rotation angle $\alpha_x, \alpha_y, \alpha_z$ around xyz axis and translation distance t_x, t_y, t_z along xyz axis, which are defined as the individual X of DDE expressed as follows:

$$X = [\alpha_x \;\; \alpha_y \;\; \alpha_z \;\; t_x \;\; t_y \;\; t_z] \tag{7}$$

Then, rotation matrix R and translation vector T can be calculated based on X:

$$R = \begin{bmatrix} \cos\alpha_z & -\sin\alpha_z & 0 \\ \sin\alpha_z & \cos\alpha_z & 0 \\ 0 & 0 & 1 \end{bmatrix} \begin{bmatrix} \cos\alpha_y & 0 & \sin\alpha_y \\ 0 & 1 & 0 \\ -\sin\alpha_y & 0 & \cos\alpha_y \end{bmatrix} \begin{bmatrix} 1 & 0 & 0 \\ 0 & \cos\alpha_x & -\sin\alpha_x \\ 0 & \sin\alpha_x & \cos\alpha_x \end{bmatrix} \tag{8}$$

$$T = \begin{bmatrix} t_x & t_y & t_z \end{bmatrix}^{\mathrm{T}} \tag{9}$$

The root mean squared error (RMSE) of the registration is used as the objective function of DDE:

$$f_{\text{RMSE}}(R,T)=\sqrt{\frac{1}{n}\sum_{i=1}^{n}\| m_i - s_i' \|^2}=\sqrt{\frac{1}{n}\sum_{i=1}^{n}\| m_i-(T+Rs_i)\|^2} \tag{10}$$

Where n indicates the number of corresponding points. s_i represents the point in the source cloud and s_i' represents the point after coordinate transformation of s_i. In order to speed up the search, the kd-tree [21] algorithm is used, and m_i, searched from M, represents the closest point to s_i'. R and T can be calculated by DDE minimizing $f_{\text{RMSE}}(R,T)$, and finally coarse registration result can be obtained based on R and T.

3. Experimental results

The experimental dataset contains "bunny" and "dragon" obtained from Stanford University Graphics Lab [22]. Our algorithm is implemented in MATLAB 2015a, while all the experiments are conducted on a 3.2GHz Intel(R) core i5-4460 processor with 8GB RAM.

Firstly, the voxel grid method is used to down sample the point cloud. Figure 1 (a) and Figure 2 (a) show original point clouds, where blue and red represent two perspectives. After sampling, bunny data was reduced from 40256 to 2067 for blue and from 40097 to 2004 for red, while dragon data was reduced from 41841 to 2156 for blue and from 22092 to 1499 for red. In order to verify the validity of DDE for point cloud registration with a large difference in position, the down sampling point cloud is arbitrarily transformed as (11) to increase its position distance. The point clouds after sampling and transformation are shown in Figure 1 (b) and Figure 2 (b).

$$[\alpha_x,\alpha_y,\alpha_z,t_x,t_y,t_z]=[50,-20,30,0.5,-0.2,0.1] \tag{11}$$

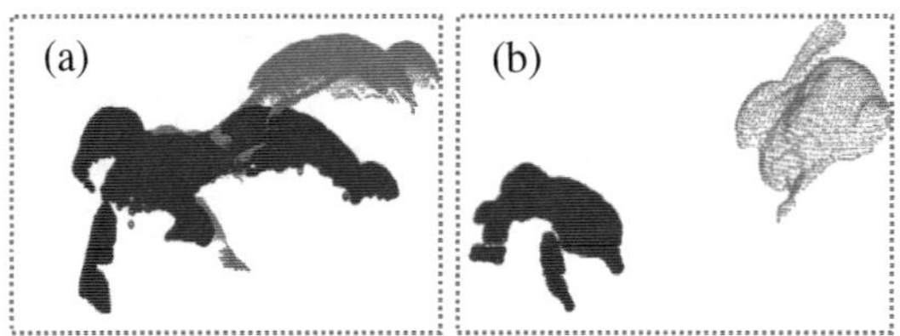

Figure 1. Point cloud sampling and transformation of bunny. (a)Original point clouds; (b) The point clouds after sampling and transformation

Figure 2. Point cloud sampling and transformation of dragon. (a)Original point clouds; (b) The point clouds after sampling and transformation

Considering that if the search space of DDE is too large, it will affect the registration speed, so it is necessary to choose a suitable search space. According to the position of sampled and transformed point cloud in Figure 1 (b), the search space roughly is set as shown in (12). The parameters of DDE are set to: NP=30, D=80, F=0.4 and CR=0.8.

$$\left(\alpha_x,\alpha_y,\alpha_z,t_x,t_y,t_z \mid \alpha_x,\alpha_y \in[-\frac{\pi}{2},\frac{\pi}{2}],\alpha_z \in[-\frac{\pi}{4},\frac{\pi}{4}],t_x,t_y,t_z \in[-0.6,0.6]\right) \tag{12}$$

Considering that for a point cloud that does not have any predictable information, it is difficult to determine the registration error because a strict corresponding point pair cannot be determined. To solve this problem, this paper uses the method of [9] to define the registration error parameter e as follows:

$$\begin{cases} e = \frac{1}{N}\sum_{i=1}^{N} J(M_i, S_i) \\ J(M_i, S_i) = \begin{cases} 1 & Dist(M_i, S_i) > \delta \\ 0 & otherwise \end{cases} \end{cases} \tag{13}$$

Where $Dist(M_i, S_i)$ indicates euclidean distance of point pair (M_i, S_i). δ represents registration accuracy and N is the number of point. $J(M_i, S_i)$ is used to measure whether the current point pair meets the registration accuracy. Obviously, e represents the proportion of data point pairs that do not meet the accuracy in the point cloud.

Figure 3 shows the registration results for the bunny and Figure 4 shows the registration results for the dragon. In the two figures, (a) is ICP registration. (b) is DDE registration, and (c) is DDE-ICP registration which represents a fine registration using ICP [5] based on the DDE coarse registration. Table 1 shows the registration error comparison for the above registration algorithms.

Figure 3. Registration results of bunny. (a) ICP registration; (b) DDE registration; (c) DDE-ICP registration

Figure 4. Registration results of dragon. (a) ICP registration; (b) DDE registration; (c) DDE-ICP registration

Table 1. Registration error of different algorithms

Model	ICP	DDE	DDE-ICP
	e	e	e
Bunny	0.9384	0.6777	0.3677
Dragon	0.9316	0.6785	0.6707

From the above experimental results, we can see:

For point cloud registration, the original ICP is likely to fall into a local extreme for point cloud with a large distance in position, which causes registration failure. DDE has strong global search performance, and successfully register the point cloud. Based on the coarse registration of DDE, the ICP algorithm is used again to perform fine registration of point clouds. For the bunny model, the DDE-ICP registration error is significantly smaller than that of the DDE, but for the dragon model, the change is not obvious, indicating that the DDE registration accuracy can basically reach the ICP registration accuracy for some models, which verifies that the DDE coarse registration has a good registration effect.

4. Conclusions

Although ICP is widely used in point cloud registration, it requires a proper initial position and it is easy to fall into a local extreme. Because differential evolution algorithm has strong global search performance, so in order to solve this problem, a point cloud registration algorithm based on DDE was proposed. By using the voxel grid method to down sample the point cloud, the time complexity of the DDE is greatly reduced. Compared with the traditional ICP algorithm, DDE can effectively deal with the problem of point cloud registration with large distance, and it can be used as an effective coarse registration algorithm.

Acknowledgments

This work is supported by Fundamental Research Special Funds for Central Universities (20826041A4133).

References

[1] Sun Zhihui, Lu Shenglian, Guo Xinyu, et al. Surfaces reconstruction of plant leaves based on point cloud data[J]. Transactions of the CSAE, 2012, 28(3): 184－190.

[2] Wang Junchen, Wang Tianmiao, Xu Yuan, et al. Registration method based on ICP algorithm for 3D surgical navigation[J]. Journal of Beijing University of Aeronautics and Astronautics, 2009, 35(4): 434-438.

[3] Tu Zhiqiang, Zhang Ke, Yang Chenglong, et al. Improved ICP registration in 3-D model reconstruction[J]. Transactions of the China Welding Institution, 2013, 34(1): 97-100

[4] Besl P J, Mckay H D. A method for registration of 3-D shapes[J]. IEEE Transactions on Pattern Analysis and Machine Intelligence, 1992, 14(2): 239-256.

[5] Li S, Wang J, Liang Z, et al. Tree point clouds registration using an improved ICP algorithm based on kd-tree[C]. Geoscience and Remote Sensing Symposium. IEEE, 2016:4545-4548.

[6] Rusinkiewicz S, Levoy M. Efficient variants of the ICP algorithm[C]. International Conference on 3-D Digital Imaging and Modeling, 2001. Proceedings. IEEE, 2002:145-152.

[7] He Y, Liang B, Yang J, et al. An iterative closest points algorithm for registration of 3D laser scanner point clouds with geometric features[J]. Sensors, 2017, 17(8):1862.

[8] Wei Shengbin, Wang Shaoqing, Zhou Changhe, et al. An iterative closest point algorithm based on biunique correspondence of point clouds for 3D reconstruction[J]. Acta Optica Sinica, 2015, 35(3): 0515003.

[9] Wang Xin, Zhang Mingming, Yu Xiao, et al. Point cloud registration based on improved iterative closest point method[J]. Optics and Precision Engineering, 2012, 20(9): 2068-2077.

[10] Jiaqi Yang, Zhiguo Cao, Qian Zhang. A fast and robust local descriptor for 3D point cloud registration[J]. Information Sciences, 2016, s346–347: 163-179.

[11] Rusu R B, Blodow N, Beetz M. Fast point feature histograms (FPFH) for 3D registration[C]. IEEE International Conference on Robotics and Automation. IEEE Press, 2009: 1848-1853.

[12] Tao Haiji, Da Feipeng. Automatic registration algorithm for the point clouds based on the normal vector[J]. Chinese Journal of Lasers, 2013(8): 0809001.

[13] Shen Yinghua, Li Zhuojia, Yang Cheng, et al. Point cloud registration with normal feature histogram[J]. Optics and Precision Engineering, 2015, 23(10): 591-598.

[14] Ge Baozhen, Peng Bo, Tianqingguo. Registration of three-dimensional point-cloud data based on curvature map[J]. Journal of Tianjin University(Science and Technology), 2013, 46(2): 174-180.

[15] Qin Nannan, Lai Xudong, Min Xiaofeng. Registration of spatial data based on principal component analysis[J]. Engineering Journal of Wuhan University, 2012, 45(4): 529-531

[16] Zhong Ying, Zhang Meng. Automatic registration technology of point cloud based on improved ICP algorithm[J]. Control Engineering of China, 2014, 21(1):37-40.

[17] Yan L, Tan J, Liu H, et al. Automatic registration of TLS-TLS and TLS-MLS point clouds using a genetic algorithm[J]. Sensors, 2017, 17(9).

[18] Wang Shenwen, Ding Lixin, Zhang Wensheng, et al. Advances in Differential Evolutionary Algorithms [J]. Journal of Wuhan University Science Edition, 2014, 60(4): 283-292.

[19] Qin, A. K, Huang, et al. Differential evolution algorithm with strategy adaptation for global numerical optimization[J]. IEEE Transactions on Evolutionary Computation, 2009, 13(2):398-417.

[20] WU Liang-hong, WANG Yao-nan. Dynamic differential evolution algorithm and its application [M]. Beijing: Science Press, 2014: 9-20.

[21] Kjer H, Wilm J. Evaluation of surface registration algorithms for PET motion correction[D]. Technical University of Denmark Kgs Lyngby Denmark, 2010: 12-24.

[22] Stanford University. The Stanford 3D Scanning Repository. http://graphics.stanford.edu/data/

IOP Conf. Series: Materials Science and Engineering **428** (2018) 012033 doi:10.1088/1757-899X/428/1/012033

The Application of PID-DMC Algorithm with Feedforward Compensation in the PTZ Platform

Y H Zhao*, L Chai and L Z Jin
Four Pai Lou campus of Southeast University, Nanjing, China

*Zhaoyahan0926@163.com

Abstract. According to the characteristics of PTZ camera itself pure delay of measurement and control, discontinuity of control system and unknown model, a new PID-DMC control algorithm with feedforward compensation is proposed. When using the conventional PID algorithm to control, due to the discontinuity of the control amount of the PTZ camera, the serious control jitter when the closed-loop system operating, the stability of the system is affected critically, and even the equipment will be damaged. For this reason, feedforward compensation is added on the basis of conventional PID control to alleviate the problem of increasing of the overshoot caused by the integral action. Meanwhile, DMC control is introduced to effectively reduce the fluctuation of the system, which makes the follow-up tracking of the system more stable and effective.

1. Introduction

In the application of intelligent video surveillance, PTZ (Pan/Tilt/Zoom) cameras can realize full-range (horizontal, vertical) rotation of intelligent cloud platform and zoom control, and are considered to be one of the most useful sensors[1]. In the monitoring system, the main task of the PTZ camera is to perform target detection and target tracking [2].

In the moving target tracking system, the rotation of the camera will cause changes in the background of the monitor screen, which brings great challenge to the recognition and tracking of the target in the image processing[3]. Therefore, this paper uses joint tracking scheme of PTZ camera platform and static camera. Under this scheme, the main task of the PTZ camera rotation control is to control the rotation of the camera in the both horizontal and the vertical direction, so that the given target angle position is tracked by the horizontal and vertical angle of the camera cloud.

This paper first discusses the characteristics of PTZ camera control, and then designs the PID-DMC predictive controller with feedforward compensation to control the horizontal and vertical axis of the PTZ camera separately according to the characteristic of the pure lag of the camera. Compared with the traditional PID controller from simulation results, the PID-DMC predictive controller with feedforward compensation can effectively prevent the large fluctuation of the system control and make the tracking of the control system more stable.

2. The features of the PTZ camera

2.1. Pure lag of measurement and control

The PTZ camera used in this paper is BRC-Z330 of Sony Corp, and its control communication adopts VISCA protocol based on RS-232C/RS-422. The command sequence of the VISCA protocol is shown in Figure 1. V represents the command processing cycle of the VISCA protocol and under the PAL system 1V represents 20ms[4]. As shown in Figure 1, the VISCA protocol command sequence specifies that the two adjacent command sending intervals must be at least 1V, and the response packet of the command should be return within the 4V interval. According to the actual measurement data, the command of the angle measurement generally returns the measured data 16ms or 32ms after command, while the speed control command will need 47ms or even longer time to return the response packet and the completion package, thus bringing the pure lag and uncertainty of the angle measurement and control of the PTZ camera.

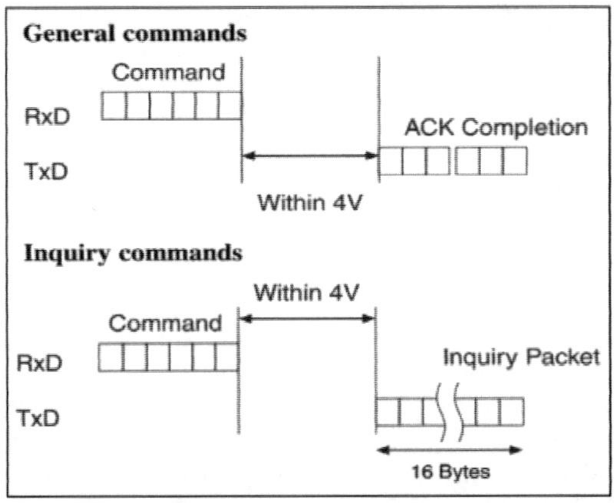

Figure 1. Sequence diagram of VISCA protocol command

2.2. Discontinuity of control system

The command group that controls the rotation of the PTZ camera in the VISCA protocol is Pan-tilt Drive, which is divided into the speed rotation control and angle rotation control. But the execution of angle control command is not suitable for the follow-up tracking system where the target position rapidly changes. In applications, because of this command, the next instruction won't be read until the systems reaches specified position. Therefore, the rotation control of the PTZ camera is achieved through its speed control command here. The speed control command is divided into the direction domain and the speed domain. The speed word domain is composed of two 4 bits. The allowable transmission speed is divided into 24 levels of 01H-18H from low to high. The PTZ camera has a minimum panning speed of $0.25°/s$ and a maximum panning speed of $60°/s$. The actual running speed corresponding to the 24 speed levels of horizontal axis is shown in Figure 2.

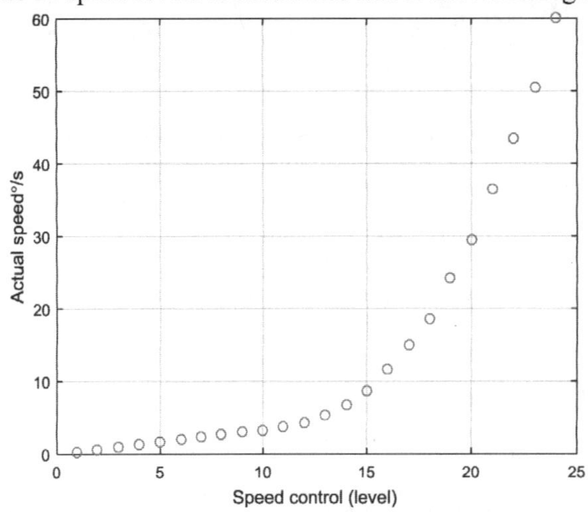

Figure 2. Actual speed curve of horizontal axis

As shown in Figure 2, the effective speed control of the PTZ camera is only a limited number of discrete point, and it is unevenly distributed throughout the entire speed range, which will inevitably cause a great difficulty for the continuous smooth operation of the system.

2.3. Unknown model properties

BRC-Z330 as a mature commercial product, the internal implementation details are well packaged, only leaving user with the necessary interface, which is convenient for installation, connection and use of the system. But that also brings difficulties for the real-time control of the PTZ camera.

Because it is impossible to understand the specific implementation details such as motor type, driving method and speed control algorithm, the establishment of accurate dynamic model of the PTZ camera becomes very difficult or even impossible, which brings great difficulties for not only the design of control algorithm but also the tuning of the controller parameters.

In summary, PTZ camera has the characteristics of unfavourable control, such as pure delay in measurement and control, discontinuous control and unknown model characteristics. In order to analyze and attempt to overcome the influence of the above characteristics on control during the design of the controller, the following discussion is made on how to carry out the controller design of PID-DMC predictive control with feedforward compensation. Through its result of tracking control, it can be seen that compared with the classic PID controller, the PID-DMC controller with feedforward compensation has smaller operating error and runs smoothly.

3. Design of PTZ camera controller

3.1. Feedforward control

The PID control algorithm is composed of three parts including Proportion, Integration and Differentiation[5]. Because the cloud platform controller generally uses the discrete control system implemented by the embedded computer or the industrial control computer, we use its discrete form for computer implementation[6]. However, integral action must be introduced in the control of PID, and the integral effect makes the overshoot and adjusting time of the system increase, which is not conducive to the rapid response of the system. Therefore, in order to ensure that the system can still track without static error, the feedforward compensation of target speed is introduced to the control system, as shown in Figure 3.

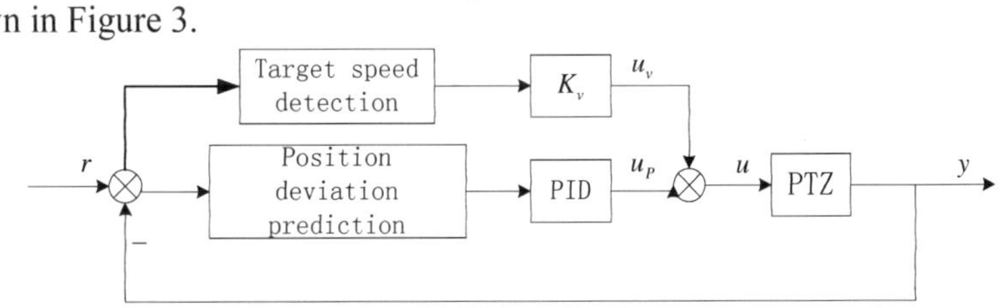

Figure 3. Forward compensation PID controller structure diagram

The control algorithm of PID controller with feedforward compensation is shown in the following expressions:

$$\begin{cases} u(k) = u_p(k) + u_v(k) \\ u_p(k) = K_p e(k) + K_i \sum_{i=0}^{k} e(i) + K_d[e(k) - e(k-1)] \\ u_v(k) = K_v \times [r(k) - r(k-1)]/T \\ e(k) = r(k) - (y(k) + u(k-1) \times D) \end{cases} \quad (1)$$

K_p, K_i, K_d, K_v represent proportional coefficient, integral time constant, differential time parameter and feedforward coefficient respectively.

3.2. DMC control

IOP Conf. Series: Materials Science and Engineering **428** (2018) 012033 doi:10.1088/1757-899X/428/1/012033

DMC control is a kind of advanced control algorithm which is first developed in the field of process industry in the 1970s and 1980s. It is based on the "three major principles" such as prediction model, rolling optimization and feedback correction, which overcomes the shortcomings of modern control theory which relies on the precise model of the system, and has been successfully applied [7] in the practical engineering.

1) The predictive model DMC control only focuses on the predictive function of the model rather than the specific form of the model [8]. The system uses a step response model as a prediction model for DMC controller. The unit step response curve of the PTZ camera measured by experiment is shown in Figure 4:

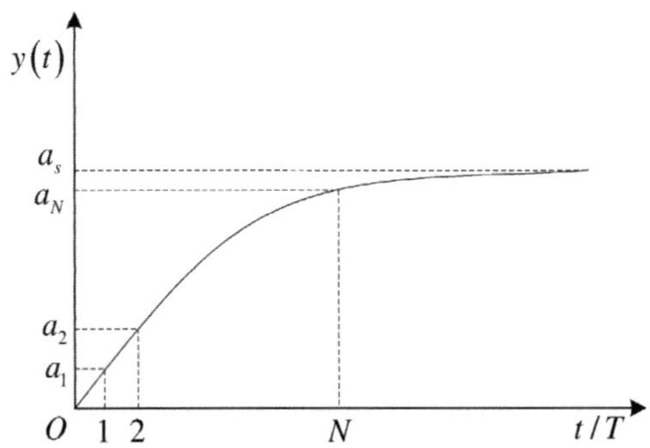

Figure 4. Unit step curve

a_s is steady state value; N is truncated step length. Once a constant control increment $\Delta u(k)$ is applied at k time, the system step response model can be used to predict the output value of the system for the next N time in the future, as shown by formula 2.

$$y_m(k+i) = y_0(k+i) + a_i \Delta u(k), i = 1,2 \cdots N \tag{2}$$

$y_0(k+i)$ is the output of the system without$\Delta u(k)$ at k time.

According to formula 2, the relationship between predictive output vector and control increment vector can be expressed as:

$$Y_m(k+1) = Y_0(k+1) + A\Delta U(k) \tag{3}$$

$Y_m, Y_0, \Delta U$ are respectively the vector of time sequence of future sampling, and A is the dimensional matrix $P \times M$ composed of step response coefficients.

$$\begin{cases} Y_m(k+1) = [y_m(k+1) \quad \cdots \quad y_m(k+P)]^T \\ Y_0(k+1) = [y_0(k+1) \quad \cdots \quad y_0(k+P)]^T \\ \Delta U(k) = [\Delta u(k) \quad \cdots \quad \Delta u(k+M-1)]^T \end{cases} \tag{4}$$

$$A = \begin{bmatrix} a_1 & \cdots & 0 \\ \vdots & \ddots & \vdots \\ a_M & \cdots & a_1 \\ \vdots & \vdots & \vdots \\ a_P & \cdots & a_{P-M+1} \end{bmatrix}_{P \times M} \tag{5}$$

2) Predictive error feedback correction

Predictive error detection is as follows:

$$e(k) = y(k) - \hat{y}(k|k-1) = y(k) - y_0(k) \tag{6}$$

After introducing the prediction error feedback, the recursive update Y_0 is changed into formula 7.

$$\begin{cases} Y_0(0) = [0 \ \cdots \ 0]^T \\ e(k) = y(k) - y_0(k) \\ Y_0(k) = Y_0(k) + He(k) \\ Y_0(k) = [y_0(k+1) \ \cdots \ y_0(k+P-1)]^T \\ Y_0(k+1) = Y_0(k) + \sum_{i=1}^{P} a_i \Delta u(k) \end{cases} \tag{7}$$

H is an error feedback gain matrix of P-dimensional vectors. Due to the lack of sufficient understanding of the inherent mechanism of model mismatch and external disturbance, the dynamic characteristic of prediction error is generally described by step model that the prediction error remains unchanged at the future P time, and the typical H values of each component are taken as formula 8.

$$H = [h \ \cdots \ h]^T \tag{8}$$

3) The rolling optimization and control quantity solving

At the time k, the optimal performance index of DMC is:

$$\min J(k) = \sum_{i=1}^{P} q_i [y_r(k+i) - y_m(k+i)]^2 + \sum_{j=1}^{M} r_j \Delta u^2(k+j-1) \tag{9}$$

y_r is the target trajectory; y_m is the p step prediction output under the M step control increment Δu; q_i represents the tracking error weighting coefficient. And r_j represents the control function weighting coefficient.

After the upper expression is transformed into vector form, get the following formula:

$$\min J(k) = \|Y_r(k+1) - Y_0(k+1) - A\Delta U(k)\|_Q^2 + \|\Delta U(k)\|_R^2 \tag{10}$$

Q, R are respectively weights of the weights of the sampling points in the optimal time domain.

The optimal performance index can be obtained when $\partial J(k)/\partial \Delta U(k) = 0$ and it shows as following:

$$\Delta U(k) = (A^T Q A + R)^{-1} A^T Q [Y_r(k+1) - Y_0(k+1)] \tag{11}$$

In this experiment, the flow chart of DMC controller method is shown in Figure 5.

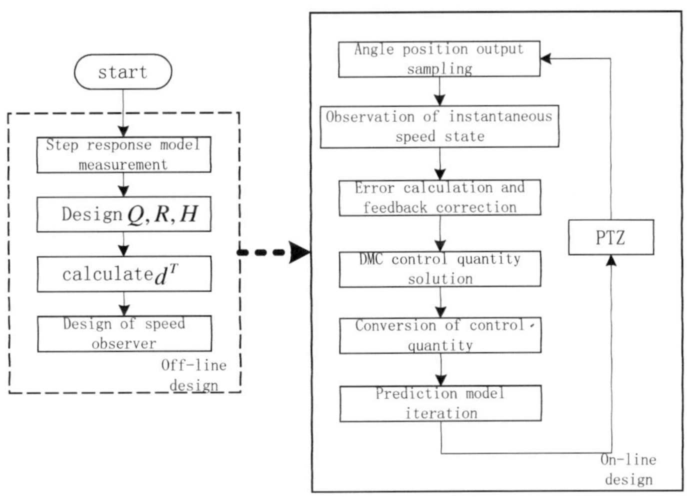

Figure 5. Flow chart of DMC speed loop correction control algorithm.

In summary, the structure of PID-DMC control algorithm with feedforward compensation is shown in Figure 6.

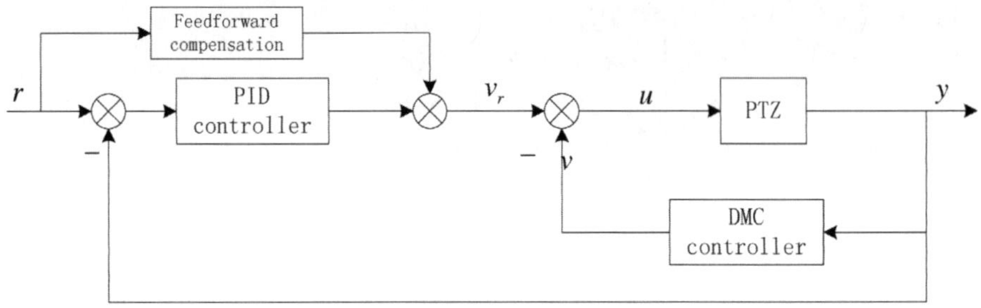

Figure 6. Flow chart of PTZ platform control algorithm.

4. Simulation experiment

Taking sampling period $T = 62ms$, the PID controller and PID-DMC controller with feedforward compensation are respectively used to track the superimposed signals of step signal and ramp signal. By simulating the read-back operation data, target trajectory and control quantity curve as shown in figure 7, 8, 9, 10. The compared result of statistics of the IAE (Integrated Absolute Error) and PAE (Peak of Absolute Error) of the operation data is shown in Table 1.

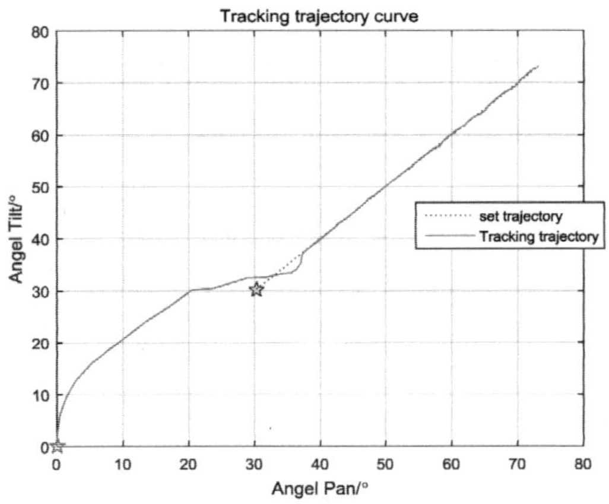

Figure 7. Tracking trajectory curve of PID

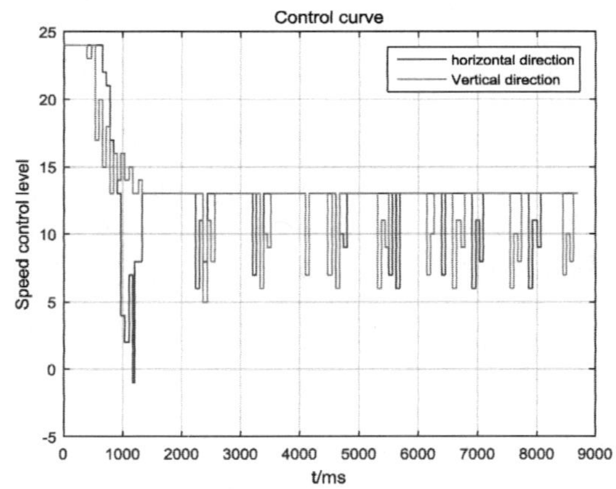

Figure 8. Control curve of PID

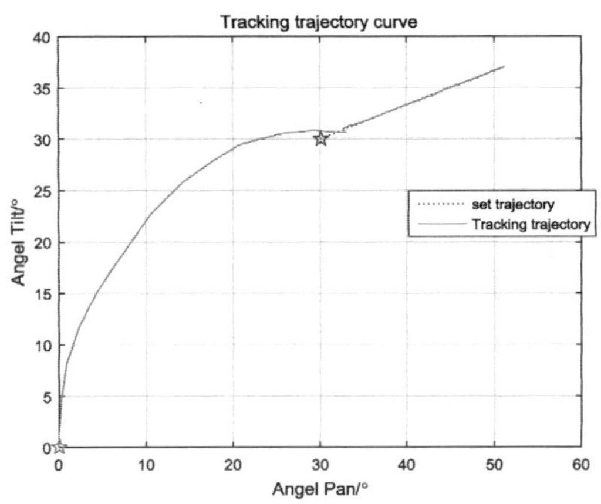

Figure 9. Tracking trajectory curve of PID-DMC

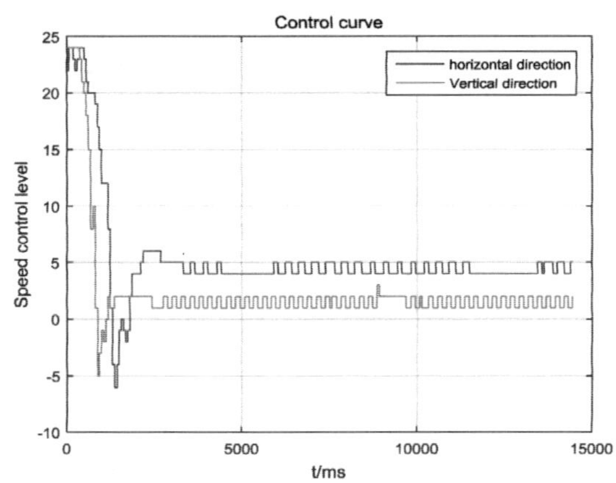

Figure 10. Control curve of PID-DMC

Table 1. Comparison of simulation experiment results of pan tilt servo control algorithm

Algorithm	IAE/°	PAE/°
PID controller	283.2	35.7
PID-DMC controller	112.3	18.5

From the Figure 7 and Figure 8, it can be seen that the speed control amount under the PID controller fluctuates greatly, and there is a clear overshoot during the step motion, and the system operation is unstable. While in the Figure 9 and Figure 10, under the control of the PID-DMC controller with feedforward compensation, the fluctuation of speed control is significantly reduced and the overshoot was further reduced. The operation was also very smooth.

5. Conclusion

In this paper, a PID-DMC controller with feedforward compensation is designed in the case that the PTZ camera has the characteristics of the pure time delay of measurement and controll, the discontinuous control quantity and the unknown model characteristics. The experimental results show that the algorithm prevents the large fluctuation of the system and improves the stability of the system effectively.

References

[1] Chen Chunghao, Yao Yi, Page D, et a1. Heterogeneous fusion of omnidirectional and PTZ cameras for multiple object tracking [J]. *IEEE Transactions on Circuits and Systems for Video Technology,* 2008, **18(8):1052--1063**.

[2] Bevilacqua A, Azzari P. A Fast and Reliable Image Mosaicing Technique with Application to Wide Area Motion Detection[C]// *Proceedings of the 4th international conference on Image Analysis and Recognition. Springer-Verlag,* 2007:**501--512**.

[3] Xiang Jinhai, Liao Honghong, Pei Renwang, et al. Real-time video target tracking [J]. *Camera application and software under the PTZ camera,* 2012 **(11): 275-280**.

[4] Liu Jianchang, Guan Shouping, Zhou Wei. Computer control system [M]. *Science Press,* 2009.

[5] Wu Qi, Wang Shi Si. Automatic control principle (Second Edition) - General Higher Education "11th Five-Year" national planning textbook [M]. *Tsinghua University press,* 2006.

[6] Ding Baocang. Theory and method of predictive control [M]. *Machinery industry*, 2008.

[7] Qian Jixin Zhao Jun Xu Zuhua. Predictive control (Graduate Teaching Book) [M]. *Chemical industry,* 2007.

[8] Wang Wenbo. Disturbance rejection analysis and improvement of model predictive control [D]. *Zhejiang University,* 2012.

IOP Conf. Series: Materials Science and Engineering **428** (2018) 012034 doi:10.1088/1757-899X/428/1/012034

Research on Intelligent Detection Technology for Bundle Branch Conduction Block

Hongwei Fang and Songyi Dian*

College of Electrical Engineering and Information Technology, Sichuan University, Chengdu, China

*scudiansy(at)mail.scu.edu.cn

Abstract. To help doctors diagnose heat disease effectively, an Intelligent detection method of bundle branch conduction block(BBB) heart rhythm is proposed in this paper. Firstly, continuous wavelet transform is used to preprocess experimental data obtained from MIT/BIH arrhythmia database, and then, some features of ECG signals are extracted from the time domain, frequency domain, and wavelet domain. After normalization, these features are fed into the BP neural network optimization algorithm based on genetic algorithm(GABPNN) to train the network. The training error, testing error, and F_1 score are evaluation criterions for classification in this paper. Compared with BP neural network algorithm(BPNN), GABPNN algorithm behaves better in accuracy. The accuracy of 99.502% was achieved, which is better than the results of some literature.

1. Introduction

Electrocardiogram (ECG) is critical for doctors to diagnose heart disease in patients. And a complete heart beat in ECG includes P, Q, R, S, T, U, and so on several parts, each part reflects the physiological information. Doctors can diagnose whether the patient is ill or what the disease is according to the abnormal information in each part. Doctors' interpretation of ECG is often based on their experience, and their diagnosis rely on a description of some causes in ECG signal based on the standard of time, and such a standard is derived from the cardiovascular system of physiological, doctors can have a good understanding of this standard[1]. In order to help doctors to make quick judgment and improve the efficiency of patients' treatment, this paper has done some research on the intelligent identification of BBB in abnormal ECG.

2. Data acquisition and preprocessing

ECG signals in this paper are obtained from the MIT-BIH arrhythmia database in *physionet*. In order to faciliate extraction of features, those signals should be well preprocessed.

2.1. Data acquisition

MIT-BIH arrhythmia database contains discontinuous numbers from 100 to 124 and from 200 to 234 of 48 group of ECG signals, that includes more than 10 types of ECG, the sampling frequency is 360 Hz, each set of data is obtained from two lead channels. In this paper, four kinds of MLII lead ECG signals are selected, the normal (N), left bundle branch block (L), right bundle branch block (R), and pacemaker heartbeat (P), pacemaker heartbeat exists as interference category. The four ECG signals are labeled from 1 to 4 in turn, then the total number is divided by 30% into the training set and the rest as a test set. The obtained data is shown in Table 1.

Table 1. Selected ECG records from the MIT/BIH arrhythmia database

ECG Type	MIT-BIH data file	Training Set	Test Set	Total
N	100-101-103-112-115-117-122-205-213-219-234	218	94	312
L	109-111-214	97	41	138
R	118-124-212	71	31	102
P	102-107-217	101	43	144
Total	20 files	487	209	696

2.2. Preprocessing

There are some interference factors with raw data, power frequency interference, baseline drift, and myoelectricity interference. Studies have shown that the interpretation of ECG features in the computer-assisted manner has more than 90% accuracy in the underlying current and hemodynamic causes, so appropriate treatment can be provided to patients[1]. Signal processing technology is such a manner that can detect the small changes human cannot notice. In this paper, the original ECG signals is processed by wavelet analysis[2]. The Mexican hat function is similar to ECG signals, so it is chosen as the mother wavelet in the wavelet transform. And an attempt is made to obtain a better result when the transformation scale is 6.

This processing works well, before preprocessing, serious baseline drift and other interference can be seen in left image of Figure 1. But after preprocessing, ECG signal's baseline alignment to the location of 0 on the longitudinal axis, and ECG becomes smoother.

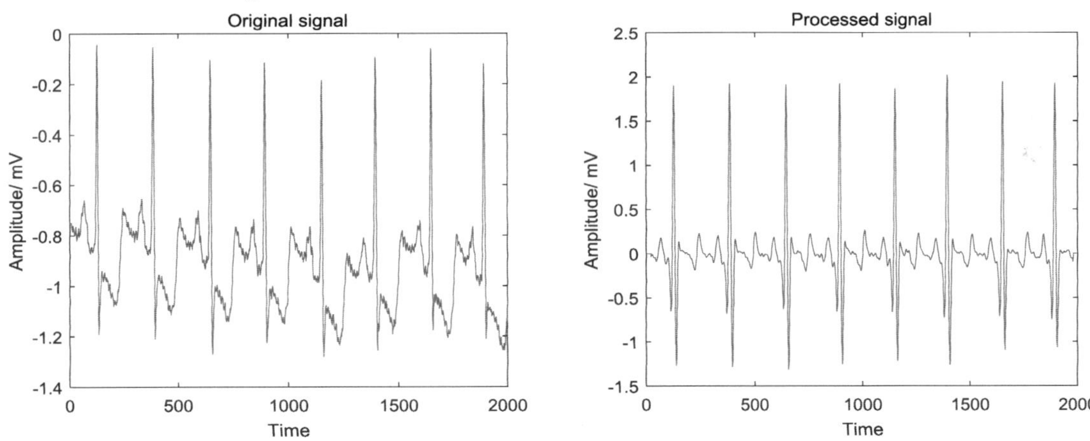

Figure 1. ECG signal preprocessing image

3. Feature extraction

The features of ECG signals are extracted from time domain, frequency domain and wavelet domain, so that data characteristics can be detected from various aspects so as to accurately classify.

3.1. Feature extraction in time domain

In time domain, one of the most important features of ECG is the cardiac cycle, it is defined as the time span from one R wave to the next, the corresponding amplitude of R wave is another feature. Heart rate can be computed with cardiac cycle, it refers to the number of heart beats per minute in the quiet state of a normal person, is usually 60 to 100 times per minute. It can be calculated as Heart rate = 60/R_R, where R_R is cardiac cycle.

Different types of ECG have a certain difference in the visual waveform, so some morphological features can be extracted according to their waveform. The main energy in a heartbeat cycle concentrated in QRS complex. Before extracting the morphological features, each heartbeat cycle is intercepted by capturing the 26 points before R wave and 44 points after R wave, such a total of 71 data points will represent a cardiac cycle. Then the standard deviation, the peak factor, the waveform factor, and the impulse factor of a heartbeat cycle are calculated as the morphological features of ECG.

3.2. Feature extraction in frequency domain

ECG morphological features and time-domain features associated with interphase cannot reflect the characteristics of the frequency domain, while the information in frequency domain can sometimes reflect the differences in signals that cannot be expressed in the time domain. In order to extract ECG features in frequency domain, the FFT transform of ECG signal is necessary. In frequency domain, the relation between frequency and amplitude can be obtained. The QRS waves in the frequency domain are located in the high frequency region, while P wave and T wave are located in the low frequency region. Normal ECG and other abnormal ECG have great difference in the frequency components. In this paper, the spectral mean, standard deviation, spectrum deviation and frequency kurtosis are mainly four features selected in the frequency amplitude spectrum[3].

3.3. Feature extraction in wavelet domain

In the time-frequency domain analysis of ECG signal, the one-dimensional signal can be expressed as a 2-d joint function of time and frequency, and the spectral composition of each moment can be described and analysed in the time frequency, which can easily reflect the change rule of features in frequency domain over time. In this paper, the wavelet packet decomposition method is used to analyse ECG signal, and the wavelet domain features are extracted. The wavelet packet can analyse the ECG signal in fine detail. It divides the frequency band into multiple layers, and then decomposes the high frequency parts again that are not decomposed in the general wavelet analysis. After a series of experiments, this paper selects the "Shannon" entropy type, takes "db6" as the mother wavelet, the ECG signal is decomposed by 4 layers decomposition, and the fourth layer wavelet packet coefficient is used to extract the features. Singular value, standard deviation, maximum value[4] of the fourth layer of the wavelet packet coefficient are calculated as three kinds of feature, and due to there are 16 band in the fourth layer, a total of 48 features can be obtained.

3.4. Feature processing and analysis

A total of 59 features are obtained in those three domains. If the data set is directly fed to the learning algorithm, the weight value will oscillate in the process of convergence, which is because the orders of magnitude vary greatly from feature to feature, and it is easy to converge to the local optimal result, thus resulting in poor classification results. To avoid such a result, the features should be scaled and normalized. Suppose that the acquired feature values are $Feature_i$, where $i = 1,2,\dots,M$, M is the total number of features, and the processing process is as follows:

$$\begin{cases} \mu_i = \frac{\sum_{i=1}^{M} Feature_i}{M} \\ S_i = Max(Feature_i) - Min(Feature_i) \\ Feature_{Normal_i} = \frac{Feature_i - \mu_i}{S_i} \end{cases} \tag{1}$$

Where $Feature_{Normal_i}$ is the processed $Feature_i$.

After being processed, each sample of the four ECG signals is taken at random, and as is shown in Figure 2, 59 features in the 4 samples are obviously different in some places, and 4 samples can be easily distinguished from each other. Therefore, the features extracted in this paper are reasonable.

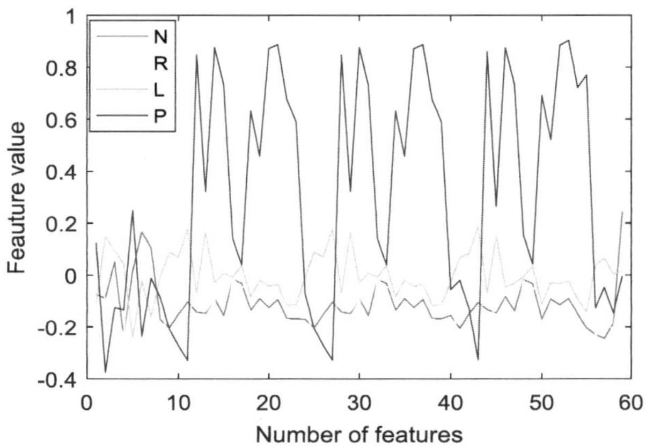

Figure 2. Feature images of 4 ECG samples

4. BP neural network optimization algorithm based on genetic algorithm

BP neural network (BPNN) algorithm is a multi-layer feedforward network based on error inverse propagation algorithm, which is one of the most widely used neural network models. BPNN has the ability to learn and store a large number of input-output mode mapping relationships without having to disclose the mathematical equations describing the mapping relationship in advance. Its learning rule is to use the gradient descent method to continuously adjust the weights and thresholds of the network through the reverse propagation, so as to minimize the error sum of the network.

In the initial iteration of BP neural network algorithm, the weight and threshold value of each layer are randomly generated, and the weight and deviation are constantly updated in the direction of decreasing the cost function gradient in the subsequent iteration process. But there exists a problem, the updation of weight and threshold may be trapped in local optimum. If the weight and threshold of the first generation are optimized so that the error at this time is minimized, then the weight and threshold values will be updated in a better direction, and then a smaller error and better classification results will be obtained. Genetic algorithm is chosen to optimize the initial rights and threshold of BP neural networks. The optimized BP neural network is a better predictor, and the flowchart of GABPNN algorithm is shown in Figure 3.

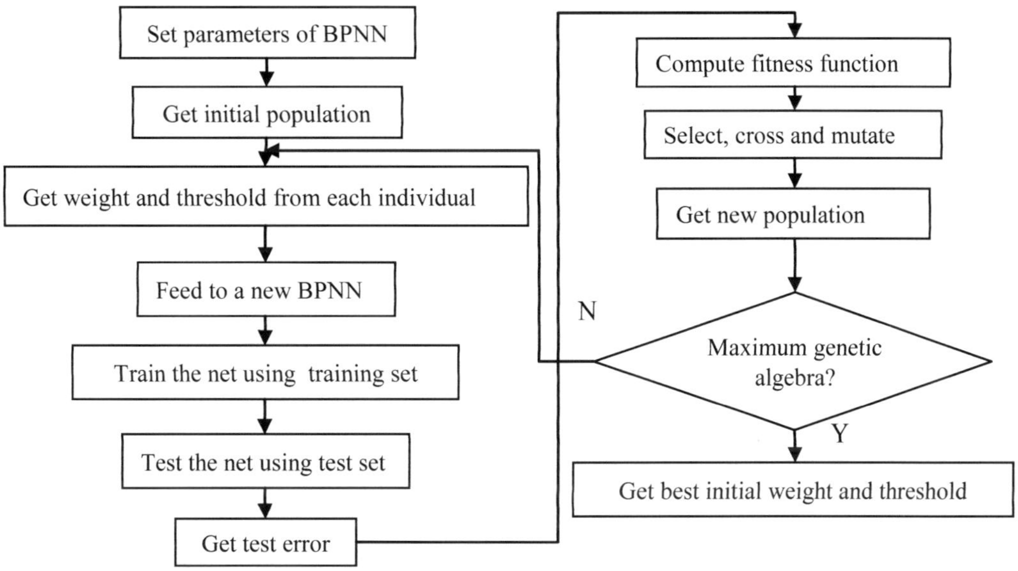

Figure 3. Flowchart of GABPNN algorithm

5. Simulation results

The BP neural network adapted in this paper with 59 neurons in the input layer, 8 neurons in hidden layer, 1 neuron in output layer. The training error, test error, and F_1 score are the evaluation criteria of algorithm. These two errors are calculated as two-norm, the smaller test error and train error, the better the prediction of the algorithm and the stronger the generalization. F_1 score is defined as the weighted average of precision and recall. The higher it is, the more effective the method is, and the performance of the algorithm is better.

In GABPNN algorithm, the number of individuals in a population is 40, the largest genetic algebra is 8, in the process of every genetic, each individual in the population as the weights and thresholds are fed into BPNN to train the network and calculate the test error as a fitness value.

Compared to other works in the field, the work in this paper shows better results, which is shown in Table 2.

Table 2. Performance of several methods for ECG arrhythmias classification in comparison with the proposed method

Method	Arrhythmia types	Training error	Test error	F_1 score	Accuracy %
Wavelet/neural network[5]	4 types of Arrhythmia	-	-	-	97
DFT, DCT and wavelet / neural network[6]	4 types of Arrhythmia	-	-	-	95
Maximum margin clustering/ Immun evolutionary algorithm[7]	5 types of Arrhythmia	-	-	0.9245	95.9
PCA and SVM[8]	4 types of Arrhythmia	-	-	-	99.17
BPNN in this paper	4 types of Arrhythmia	2.119	2.1959	0.9755	98.01
GABPNN in this paper	4 types of Arrhythmia	2.2674	1.4953	0.9956	99.502

6. Conclusion and prospect

As shown in Table 2, based on the features extracted from the method herein, a better result is achieved in the BPNN algorithm and the GABPNN algorithm on the recognition of the BBB rhythm. Compared with BPNN, GABPNN optimizes the weight and threshold of the initial training neural network, so that it can get a better network, thus increasing the correct recognition rate of ECG signals. This algorithm recognizes that the improvement of accuracy is at the cost of training time. In the GA part, every individual in the population will be used as weight and threshold to be trained in neural network during each genetic process, so the algorithm time is greatly increased.

This paper is the author's first exploration of the ECG signal recognition, and the way that the ECG signal features are extracted is not likely to have a good effect on all the ECG signals, and it's also possible that there are some features that are not found more easily to recognize. In future work, other cardiac arrhythmical features will be tried, more research will also be done on more categories of ECG signal intelligence.

Acknowledgments

This work is supported by Fundamental Research Special Funds for Central Universities (20826041A4133).

References

[1] Najarian K, Splinter R. Biomedical Signal and Image Processing. [M]. Boca Raton: CRC press.2012:190-191.
[2] Zhang F D. Wavelet Analysis in Matlab. [M]. Beijing: Mechanical industry press,2012:53-58.
[3] Chen Y. Study on Abnormal Heart Rate Classification Algorithm for ECG Signals. [D]. College of Mechanical Engineering of Chongqing University,2016:38-40.
[4] Yuan D Y. Study on The Classification of ECG Signals Based on Wavelet Packet and Neural Network. [D]. Tianjin Polytechnic University, 2017:13-19.
[5] Kianisarkaleh A, Fereshteh P, Ali K, et al. ECG Arrhythmias Detection Using a New Intelligent System Based on Neural Networks and Wavelet Transform. [J]. 2012, 9:33-39.
[6] Belkheiri M, Douidi Z, Belkheiri A. ECG Beats Extraction and Classification Using Radial Basis Function Neural Networks. [M]. Proceedings of the Fourth International Conference on Signal and Image Processing 2012 (ICSIP 2012). Springer India, 2013:127-136.
[7] Zhu B, Ding Y, Hao K. A novel automatic detection system for ECG arrhythmias using maximum margin clustering with immune evolutionary algorithm. [J]. Computational and Mathematical Methods in Medicine,2013, (2013-4-18), 2013, 2013(2):453402.
[8] Zhang H, Zhang L Q. ECG analysis based on PCA and Support Vector Machines. [C]. International Conference on Neural Networks and Brain, 2005. Icnn&b. IEEE, 2005:743-747.

IOP Conf. Series: Materials Science and Engineering **428** (2018) 012035 doi:10.1088/1757-899X/428/1/012035

The State-of-the-art of Model Predictive Control in Recent Years

Jixia Han, Yi Hu and Songyi Dian*

College of Electrical Engineering and Information Technology, Sichuan University, Chengdu, China

*scudiansy(at)mail.scu.edu.cn

Abstract. Model predictive control is a control algorithm based on model and online application optimization performance. In the past 40 years, the feedback control strategy has been widely studied. However, with the rapid development of the economy, the requirements for online optimization and constrained performance have been improved, and the current model predictive control theory can not meet the demand any more. This paper first briefly describes the current situation of model prediction, industrial development, and application areas, and then analyzes the limitations of theory and technology at the current stage, then proposes the significance of the study of predictive control of large-scale systems, fast dynamic systems, and nonlinear systems for the development of model predictions.

1. Introduction

Model predictive control (MPC) has mainly focused on online optimization since it is studied. After a long period of development, it now has a rich theoretical and practical foundation in the academic community [1]. The view "process control: theory and benefits" pointed in the 12th IFAC World Conference on model predictions can be simplified into the following features:

- The main feature of model predictive control is online constraint optimization.
- Model predictive control is mainly used in industrial process control.
- There is a disconnect between the practical application of model predictive control and theory

In recent years, through reading a large number of documents, it can be understood that the field of using model predictive control to solve constrained optimization problems is being expanding, such as urban life, medical care, power grids, energy, development, aerospace and so on. For example, microgrid operation management [2], the strategy of drinking water network management [3], Type I diabetes insulin delivery control [4]. System for building heating [5], vehicle traffic simulation autopilot [6], spacecraft intersection [7], etc. This is in contrast to the model prediction that mainly application in industrial process in the early stage. This directly shows people's increasing demand for model forecasting technology and the expectation of rapid development. This paper analyzes the current domestic and foreign predictive control technologies, points out the problems, and puts forward several opinions on the development prospects of model predictive control.

2. The Model Predictive Control Theory and Research Status

2.1. The Model Predictive Control Theory

In recent years, traditional monotonous predictive control algorithms can't meet increasingly complex industrial process requirements any longer. It is continuously combining with other algorithms to develop more advanced and accurate predictive control strategies. For example. combine engineering economics with predictive control to economic nonlinear model predictive control [8],composite neural network predictive optimal anti-interference control algorithm combined with neural network [9], model predictive control algorithm based on T-S fuzzy modeling [10], model predictive control based on reinforcement learning [11], etc.

Since the development of predictive control, it dealt with the complex constraints mainly through optimization. A large number of papers point that the advantage of predictive control being widely used and developed is that its algorithm has two important features. The first is the capability of display the constraint processing [12], the second is that it easily extend the algorithm to a multiple-input multiple-output system. At present, the successful application of thousands of industrial equipment to this technology shows that predictive control can be used as a mature control algorithm and is recognized by industrial control. In China, predictive control has been incorporated into national key scientific research projects, and universities and companies have successively developed predictive control software with independent intellectual property rights.

2.2. The Research Status

With the development of science and technology and the increased industrial demand, people are more clearly aware of the deficiencies both predictive control theory and industrial applications that mentioned above, thus promoted in-depth research in this field. According to incomplete statistics, we counted the number of papers published in several well-known journals in the control field. The statistics are shown in Figure 1.

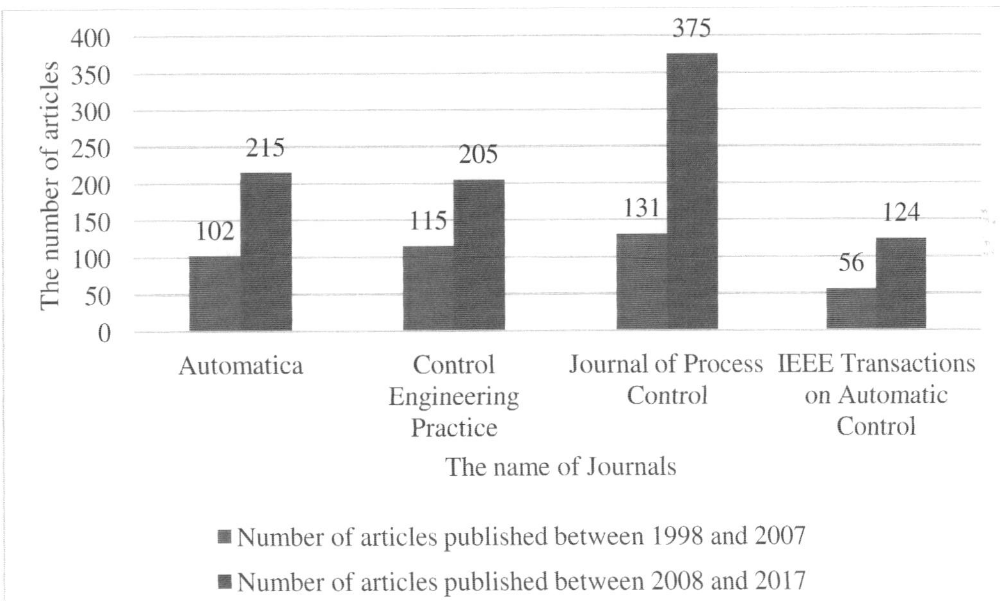

Figure 1. The statistics of related papers for Model Predictive Control

According to the comparison between the number of articles written in the past decade and the number of articles written in the past two decades, we can see that there has been a large increase in research on Model Predictive Control in recent years. At the same time, in China, the scope of industrial applications for predictive control is also expanding, from the distillation, petrochemical, and other industries to energy saving [13], irrigation [14], power generation [15], medical [16], shipping [17] and urban transportation [18], various new algorithms and control strategies were gradually proposed. The research conducted at home and abroad at this stage can be mainly divided into the following aspects:

2.2.1 Non-linear Predictive Control. a) Linearized nonlinear predictive control: linearized the nonlinear model firstly, then rolling optimizations for linearized models, while a nonlinear model is used in the model design and feedback part.

b) Non-linear predictive control of hybrid systems: This type of system is well suited for describing discrete dynamic processes with logical switching. By adding logic variables, the segmented system with the logic switching characteristics is converted into a hybrid logic system, and then the optimal control is designed.

c) Multi-model nonlinear predictive control: This type of system introduces multi-model methods into predictive control and uses several different linear models to approximate nonlinearity.

2.2.2 Random Predictive Control Model. The stochastic system uses probabilistic uncertainty to describe the system, but the uncertainty meets certain probability statistics. According to the nature of the research object, it can be divided into two categories: random uncertainty model and random disturbance model. For the stochastic uncertain model with normal distribution, Tube technology can be used for further research [19], random perturbation model can refer to Wang [20] for affine perturbation feedback.

2.2.3 Predictive Control Applied to More Fields. a) New prediction algorithm combined with other control algorithms.

In recent years, predictive control has broken the original monotony of the algorithm research model, and began to combine with other algorithms to develop a class of advanced predictive control strategies. The mainly aspects is as shown in table 1:

Table 1. New prediction algorithm and Applications.

Control Strategy	Application Examples
PID Predictive Control	Using predictive function control optimization and designing of a new PID controller for chamber pressure in a coke furnace [21]
Fuzzy Predictive Control	Using the Fuzzy Predictive Control Strategy to control the heating and cooling of the buildings [22]
Adaptive Predictive Control	Using the Adaptive Model Predictive Control to achieve quadrotor altitude efficiently [23]
Neural Network Predictive Control	Using the Neural Network Predictive Control structure for control of thermal processes to save energy [24]

b) Industrial Application Extensions

In recent years, industrial applications for predictive control have focused more on the integration of economic indicators. On one hand, economic indicators are directly added to the dynamic control of economic predictive control. On the other hand, there are many changes and uncertainties in the production process. As a result, real-time optimization technologies are increasingly used and replacing traditional ones gradually. At the same time, in the face of large-scale and complex industrial production processes, how to implement real-time optimization to achieve global optimization effectively is still an important issue.

3. Problems With Model Predictive At This Stage

Although predictive control has been successfully applied both at home and abroad, it still has the following limitations to effectively solve the complex constrained optimization problems

3.1. From the algorithm point of view

At this stage, the predictive control algorithm is mainly adapted to work in environments with high-performance computers or slow dynamic processes, it limits its broader development space and applications in a large extent. According to the statistics of Qin in the literature [14], the trend can be reflected as follows: Model prediction technology is still only scale application in industrial processes. The limitations of fast dynamic systems such as aeronautics and manufacturing are mainly due to the requirements for high-performance computer equipment. It is difficult for ordinary equipment to perform accurate real-time calculations in small sampling cycles, so it has not been possible to use a large number of applications in these fields.

3.2. From the perspective of the application object: mainly limited to linear or quasi-linear process

The main reason for the limitation is that nonlinear modeling of industrial processes is generally difficult to find precise models, and there is a lack of effective algorithms for nonlinear constraint optimization. At this stage, nonlinear predictive control has become a research hotspot, but industrial practice applications are still in the initial stage.

3.3. A Large number of online computing requirements, it is difficult to meet the application needs

The main research of predictive control is to guarantee the stability and robustness of the closed-loop system in rolling optimization. Most of the prediction algorithms proposed for specific problems are theoretically guaranteed performance, and do not pay attention to the cost of practicing online applications. In recent years, this type of problem has received more attention, but due to the large amount of online calculations, a great deal of attention in the application field can not be obtained

Due to the brief description of the application of the prediction field and the theoretical analysis, although the application in the industrial process is very successful and the theoretical basis system is also very complete, there is a problem of disconnection between theory and practice, and it can not meet the needs of economic and social development.

4. Model Predictive Outlook

From a large number of predictive control papers, it can be seen that although the industrial development and application of predictive control is relatively successful, the theoretical research of predictive control is relatively complete, however, there is a great degree of disconnection between them, which cannot meet the needs of current social development. Therefore, the research of predictive control theory and efficient algorithms is imminent, gradually reduce the disconnect between theory and practice. Taking this as a goal, we can study from several different aspects, such as large-scale systems, rapid systems, and nonlinear systems, starting from the limitations of existing predictive control technologies, conducting theoretical and algorithmic research, and applying pilot projects in combination with typical cases, improve the practicality and scientific of predictive control algorithms.

4.1. Large-scale System

For large systems with distributed features such as urban traffic and sewage discharge, the existing industrial predictive control algorithms can not be applied directly at this stage. It is necessary to study the convergence, global stability and disturbance robustness of the online information interaction iteration of the *predictive* control algorithm under different control structures. So far, the development of practical and effective predictive control strategies remains a challenge.

4.2. Fast dynamic system

Equations From the development of predictive control, the industrial process of slow dynamic systems has been *successfully* applied. However, in the field of fast dynamic systems such as aerospace, electromechanical, etc. It is considered unsuitable for predictive control because the industrial process prediction algorithm requires a large number of iterations to solve the constraint optimization problem online. At present, it is necessary to study algorithms that can guarantee the optimization of the constraints and satisfy the real-time requirements.

4.3. Nonlinear systems

The predictive control of nonlinear systems is not mature either from theoretical research or industrial process practice. Although the in-depth development of predictive control theory has brought new directions to nonlinear systems, it still has no mature conditions for practical application. The problem of nonlinear systems is mainly to reduce the amount of online calculations and nonlinear processing.

5. Conclusion

The core of predictive control theory is online constraint optimization. Confronted with the higher demand for processing optimization in the development of various fields, predictive control also faces more challenges. The study of predictive control should be aimed at reducing the direction of disconnection between theory and practice, which not only has theoretical support but also meet real-time, stability, and rapidity requirements for each application area. This is not only the goal of the relentless pursuit of control theory research, but also the research direction for the future development of predictive control.

Acknowledgments

This work is supported by Fundamental Research Special Funds for Central Universities (20826041A4133).

References

[1] Qin S J, Badgwell T A. A survey of industrial model predictive control technology. Control Engineering Practice, 2003, 11(7): 733-764.
[2] Parisio A, Rikos E, Glielmo L. Stochastic model predictive control for economic/environmental operation management of microgrids: An experimental case study [J]. Journal of Process Control, 2016, 43:24-37.
[3] Grosso J M, Ocampo-Martínez C, Puig V, et al. Chance-constrained model predictive control for drinking water networks[J]. Journal of Process Control, 2014, 24(5):504-516.
[4] Percival M W, Wang Y, Grosman B, et al. Development of a multi-parametric model predictive control algorithm for insulin delivery in type 1 diabetes mellitus using clinical parameters.[J]. Journal of Process Control, 2011, 21(3):391-404.
[5] ODwyer E, Tommasi L D, Kouramas K, et al. Prioritised objectives for model predictive control of building heating systems[J]. Control Engineering Practice, 2017, 63.
[6] Formentin S, Guanetti J, Savaresi S M. Least costly energy management for series hybrid electric vehicles[J]. Control Engineering Practice, 2016, 48:37-51.
[7] Hartley E N, Trodden P A, Richards A G, et al. Model predictive control system design and implementation for spacecraft rendezvous[J]. Control Engineering Practice, 2012, 20(7):695-713.
[8] Faulwasser T, Grüne L, Müller M A. Economic Nonlinear Model Predictive Control[J]. 2017, 5(1):224-409.
[9] Li J, Wu C, Li S, et al. Optimal disturbance rejection control approach based on a compound neural network prediction method[J]. Journal of Process Control, 2014, 24(10):1516-1526.
[10] Feng X, Patton R. Active fault tolerant control of a wind turbine via fuzzy MPC and moving horizon estimation[J]. IFAC Proceedings Volumes, 2014, 47(3):3633-3638.

[11] Gorges D. Relations between Model Predictive Control and Reinforcement Learning[J]. 2017, 50(1):4920-4928.
[12] Henson M A. Nonlinear model predictive control: current status and future directions. Computers and Chemical Engineering, 1998, 23(2): 187-202.
[13] Kennedy J L, Travis R C. Nonlinear model reduction and model predictive control of residential buildings with energy recovery [J]. Journal of Process Control, 2014, 24(6):723-739.
[14] Fele F, Maestre J M, Hashemy M, et al. Coalitional model predictive control of an irrigation canal[J]. Journal of Process Control, 2014, 24(4):314-325.
[15] Wu X, Shen J, Li Y, et al. Fuzzy modeling and stable model predictive tracking control of large-scale power plants[J]. Journal of Process Control, 2014, 24(10):1609-1626.
[16] Mythreyi K, Subramanian S C, Kumar R K. Nonlinear glucose–insulin control considering delays—Part II: Control algorithm[J]. Control Engineering Practice, 2014, 28(1):26-33.
[17] Chen Hong-Li, Lan Hai, Shen Yi. Study on generalized predictive control for ¼ type rudder pitching stabilization. Journal of System Simulation, 2006, 18(12): 3508-3511.
[18] Sun Guang, Huo Wei. Direct-adaptive fuzzy predictive control of satellite attitude. Acta Automatica Sinica, 2010, 36(8): 1151-1159.
[19] Kouvaritakis B, Cannon M, Rakovillc S V, Cheng Q F. Explicit use of probabilistic distributions in linear predictive control. Automatica, 2010, 46(10): 1719-1724.
[20] Wang C, Ong C J, Sim M. Convergence properties of constrained linear system under MPC control law using affine disturbance feedback. Automatica, 2009, 45(7): 1715-1720.
[21] Zhang J. Design of a new PID controller using predictive functional control optimization for chamber pressure in a coke furnace[J]. Isa Transactions, 2016, 67:208.
[22] Killian M, Mayer B, Kozek M. Cooperative fuzzy model predictive control for heating and cooling of buildings[J]. Energy & Buildings, 2016, 112:130-140.
[23] Chikasha P N, Dube C. Adaptive Model Predictive Control of a Quadrotor[J]. IFAC-PapersOnLine, 2017, 50(2):157–162.
[24] Vasičkaninova A, Bakosova M, Mészáros A, et al. Neural network predictive control of a heat exchanger[J]. Applied Thermal Engineering, 2011, 31(13):2094-2100.

IOP Conf. Series: Materials Science and Engineering **428** (2018) 012036 doi:10.1088/1757-899X/428/1/012036

Controller Design for Compound Pendulum with PID and MRAC Switch Control

Shuang Zhang and Songyi Dian*

Department of Automation, College of Electrical Engineering and Information Technology, Sichuan University

*scudiansy@mail.scu.edu.cn

Abstract. Compound pendulum swings with the reference trajectory accurately so that we could get the motion states of it easily. To ensure the compound pendulum tracking performance, we propose a switch controller of PID and model reference adaptive controller (PID+ MRAC). According to its equations of motion, we can establish a suitable model and obtain correspond adaptive reference rules. We determine a suitable threshold, then compare the error and threshold to choose PID control or MRAC control. The same output results could be obtained from reference model and plant model by adjusting the gain. Similarly, we compare the preference of PID controller, MRAC and PID+MRAC switch controller to evaluate which one is better. The results show that the switch control is the best one and the MRAC control is better than the PID control.

1. Introduction

As a simple physical model, compound pendulum system has compound nonlinear characteristics. Compound pendulum can be seen as an approximation of many complex systems, such as the crane operation process, aircrafts carry items with the rope and so on [1]. All of them can be simplified as the pendulum system for analysis their motion states [2], and we can obtain the control law for specific problems and control well. Naturally, we must consider that the motion of the compound pendulum has the damping and driving force [3]. Otherwise, we will get the wrong model and wrong approximate results.

In many industrial fields, proportional-differential-integral (PID) control is widely used for its simple structure and reliable operation. Desired performance would be got by adjusting the PID gain. With the development of control theory [4], many new control methods have been proposed, including model reference adaptive control (MRAC) control [5], [6]. Compared to PID control, MRAC has better performance on overcoming disturbances (including disturbances to system parameters and disturbances to control system performance [7]). Although MRAC can overcome disturbances, static errors often exist in practical applications which results poor control effect.

Therefore, a PID+ MRAC switch controller is used to solve the above problem. For the compound pendulum, the results show that PID+ MRAC has better control performance than the others. The MRAC control is better than PID control.

2. Mathematical model of compound pendulum system

For the circular compound pendulum shown in Figure 1, we can assume it meets the following conditions:

- The circular compound pendulum is subject to a certain amount of damping torque
- The driving torque acts on the circular complex compound pendulum.
- Compound pendulum movements follow the law of rigid body rotation strictly.

The equation of the compound pendulum system movements can be expressed as (1):

$$I\ddot{\theta} + c\dot{\theta} + mdg\sin\theta = F(0.5L + d) \tag{1}$$

Where m is the compound pendulum quality; I is the moment of inertia of the axis o; $c\dot{\theta}$ indicates the damping torque; $F(0.5L+d)$ is the driving torque. For a specific compound pendulum system, the parameters are set as (2).

$$m = 0.43kg, L = 0.495m, d{=}0.023m, c = 0.00035N\cdot m\cdot s, I_0 = 9.008\times 10^{-3}kg\cdot m^2 \tag{2}$$

Then (1) can be rewritten as (3):

$$\ddot{\theta} + 0.04\dot{\theta} + 10.78\sin\theta = 30.03F \tag{3}$$

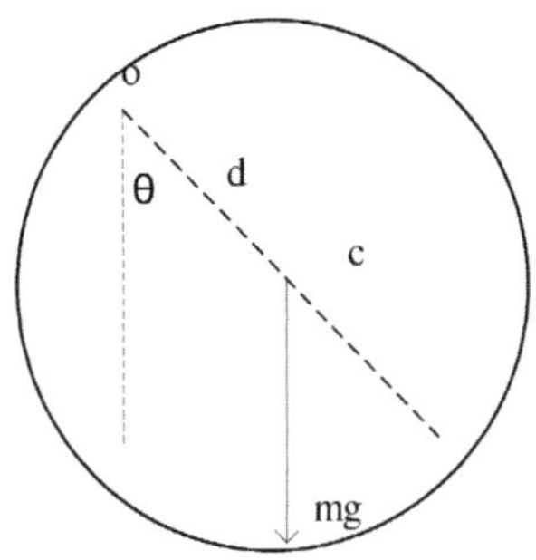

Figure 1. Circular compound pendulum model

3. Controller design

3.1. PID controller

PID controller design rules can be described as: the input of PID controller is error signals $e(t)$ and the output of PID controller is $u(t)$. Therefore, we can get (4) as follows [7]:

$$u(t) = K_p[e(t) + \frac{1}{T_i}\int_0^t e(t)dt + T_d\frac{de(t)}{dt}] \tag{4}$$

Where K_p, T_i, T_d are the scale factor, integral time constant and differential time constant, respectively. Using difference equations to discretize (4), we can get (5), where k_p, k_i, k_d are scale factor, integral coefficient, differential coefficient.

$$u(k) = k_p e(k) + k_i\sum_{i=1}^{k} e(i) + k_d de(k) \tag{5}$$

3.2. MRAC controller

In order to simplify the pendulum system and allow it to be handled under a certain degree of accuracy, a relative simple method is used to deal with this problem. However, some assumptions listed below should be taken into account [8].

- The reference model is a linear time-invariant system and is fully controllable and fully observable.
- The dimensions of the reference model are the same as the adjustable system model.
- In the parameter adjustment adaptive system, all parameters of the adjustable system can be adjusted.
- During the adaptive adjustment process, the parameters of the adjustable system are only affected by the adaptive mechanism.

- The initial deviation between the parameters of the reference model and the parameters of the adjustable system is unknown.
- Generalized error vectors and output error vectors are measurable

In the model reference adaptive control system, the desired dynamic characteristics are represented by the ideal model, so that the characteristics of the controlled system are consistent with the ideal model [9]. It is assumed that there are several parameters that can be adjusted in the adjustable system, for example, the parameters in the gain or feedback compensation network. When the external conditions change or disturbances occur, the characteristics of the controlled object also make corresponding changes. By detecting the error between the actual system and the ideal model, an adaptive mechanism adjusts the parameters of the adjustable system to compensate for the influence of the external environment or other disturbances on the system, and gradually makes the performance index reach to a minimum value.

For the complex pendulum system, we adopt an adaptive control law based on the local parameter optimization method (MIT control law). The system structure of model reference adaptive control with adjustable gain is shown in Figure 2, where the K_c is the gain which can be changed during the adjustment process; u is the input of the system; ym is the output of the reference model; yp is the output of the plant model; e is the error which is the difference between ym and yp.

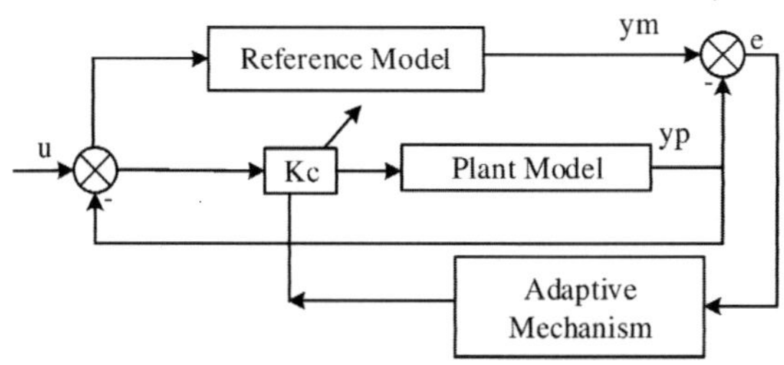

Figure 2. System structure of model reference adaptive control with adjustable gain

For compound pendulum system described above, we can use the (6) to show.

$$u = k_1 u_c - k_2 \dot{\theta} - k_3 \theta \tag{6}$$

Where k_1 is the adaptive control gain; k_2, k_3 is the feedback gain; u_c is the given input; θ is the controlled output. Then we take the (6) into the (3), we can get the (7):

$$\ddot{\theta} + 0.04\dot{\theta} + 10.78\sin\theta = k_1 u_c - k_2 \dot{\theta} - k_3 \theta \tag{7}$$

The adaptive control law of each control parameter is shown as (8):

$$k_i = -\gamma \cdot e \cdot \frac{\partial e}{\partial k_i}, i = 1,2,3 \tag{8}$$

Where γ is the design parameter; e is the difference between the output of the controlled object and the output of reference model; $\frac{\partial e}{\partial k_i}$ is the sensitivity. Our object is to predict the corresponding control parameters by the error of plant model and reference model. The reference model of compound pendulum system is written as (9):

$$\ddot{\theta} + 20\dot{\theta} + 100\sin\theta = 100u \tag{9}$$

Combined with the above formulas, the sensitivity of each parameter is (10):

$$\begin{aligned}
&(\frac{\partial e}{\partial k_1})'' + 20(\frac{\partial e}{\partial k_1})' + 100\frac{\partial e}{\partial k_1} = u_c \\
&(\frac{\partial e}{\partial k_2})'' + 20(\frac{\partial e}{\partial k_2})' + 100\frac{\partial e}{\partial k_2} = -\frac{\dot{\theta}_p}{k_1} \\
&(\frac{\partial e}{\partial k_3})'' + 20(\frac{\partial e}{\partial k_3})' + 100\frac{\partial e}{\partial k_3} = -\frac{\theta_p}{k_1}
\end{aligned} \tag{10}$$

IOP Conf. Series: Materials Science and Engineering **428** (2018) 012036 doi:10.1088/1757-899X/428/1/012036

3.3. PID+ MRAC switch controller

The switch controller structure of PID+ MRAC is shown in Figure 3, where the u, ym, yp, e are the same as described above. The difference from the MRAC control structure and PID+ MRAC control structure is the switch of the system feedback, which select the corresponding control method to control [10]. The switch will use the magnitude of the error to choose control this system via PID or via MRAC. Therefore, we can get better dynamic performance and steady state performance.

$$u(t)=\begin{cases} k_1u_c-k_2\dot{\theta}(t)-k_3\theta(t) & e\geq\alpha \\ k_p[e(t)+k_i\int_0^t e(t)dt+k_d\dfrac{de(t)}{dt}] & else \end{cases} \tag{11}$$

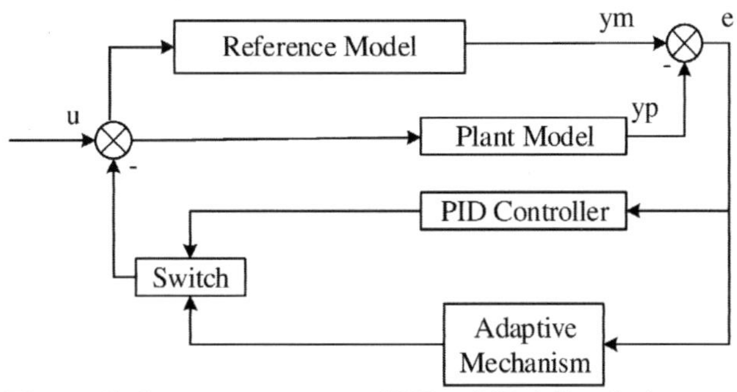

Figure 3. System structure of PID+ MRAC switch control

Similarly, we can get the control law (11), where ∝ is the threshold which set to different values for different systems; the remaining parameters have the same meaning as the above parameters.

4. Simulation analysis

In this section, we use the above three controllers to simulate the compound pendulum system for simulation. We set the angle as 60° and angle conversion is necessary for the final display. The threshold ∝ is set as 1.2 to meet the control request. And the input u is step signal is the sample time 1. In order to check the anti-interference performance of the system, a sine signal with a duration of 1s was added to interfere at t=7 s. The results are shown in figure 4.

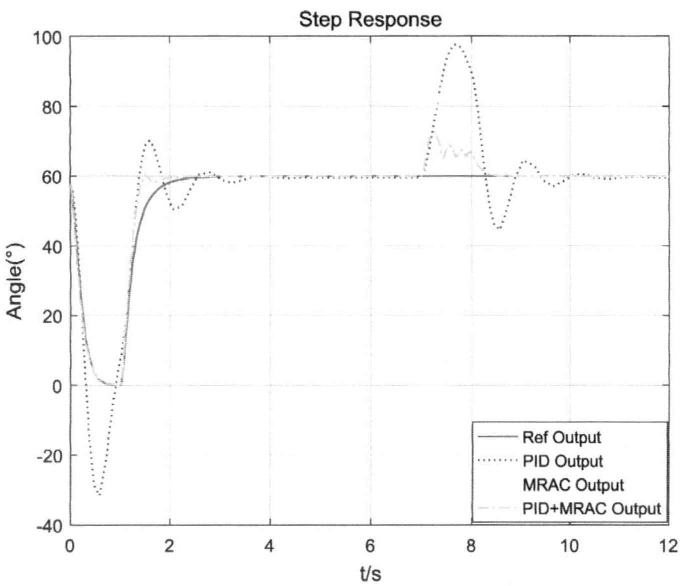

Figure 4. The corresponding step response of three methods

From figure 4, we can get the messages as follow:

- PID+ MRAC control is the best and MRAC control is better than PID control
- The threshold ∝ makes the PID+ MRAC curve coincides the MRAC curve before 1s, then the PID+ MRAC curve has a smaller overshoot than the MRAC curve and faster response.
- The adjustment of the parameters has not yet reached an optimal level, which results a very long adjustment time when the disturbances produce.

5. Conclusion

This paper has proposed a PID+ MRAC controller to control the compound pendulum. Then, three control method methods are used to compare the effect of the control. Compared PID control and MRAC control, the results show that the PID+ MRAC has the best performance of the control. But it should be note that the adjustment of parameters is not good. This study would give readers more ideas to design excellent controllers for different systems.

Acknowledgments

This work is supported by Fundamental Research Special Funds for Central Universities (20826041A4133).

References

[1] Li Q R, Tao W H, Sun N, et al. Stabilization Control of Double Inverted Pendulum System[C] International Conference on Innovative Computing Information and Control. IEEE Computer Society, 2008:417.

[2] Rsetam K, Cao Z, Man Z, et al. Optimal second order integral sliding mode control for a flexible joint robot manipulator[C]// IECON 2017 -, Conference of the IEEE Industrial Electronics Society. IEEE, 2017:3069-3074.

[3] Sun H, Zhao Y, Liu G, et al. Simulation of the nonlinear behavior of compound pendulum's oscillation based on MATLAB[J]. Journal of Shangqiu Normal University, 2012.

[4] Zuo X, Liu J W, Wang X, et al. Adaptive PID and Model Reference Adaptive Control Switch Controller for Nonlinear Hydraulic Actuator[J]. Mathematical Problems in Engineering, 2017, 2017(4):1-15.

[5] Zhou X, Yang C, Cai T. A Model Reference Adaptive Control/PID Compound Scheme on Disturbance Rejection for an Aerial Inertially Stabilized Platform[J]. 2016, 2016(5):1-11.

[6] Peng Y Q, Luo J, Zhuang J F, et al. Model reference fuzzy adaptive PID control and its applications in typical industrial processes[C]// IEEE International Conference on Automation and Logistics. IEEE, 2008:896-901.

[7] Zhen-Shun W U, Yao J J, Yue D H. A self-tuning fuzzy PID controller and its application[J]. Journal of Harbin Institute of Technology, 2004, 36(11):1578-1580.

[8] Zinober A. Adaptive Control: The Model Reference Approach[J]. Electronics & Power, 1980, 26(3):263.

[9] Zhang D, Wei B. Design of a joint control system for serial mechanical arms based on PID and MRAC control[C]// Intelligent Robot Systems. IEEE, 2016:91-96.

[10] Lei Y, Ze-Tao L I. Adaptive fuzzy PID control of temperature system of greenhouse[J]. Control Engineering of China, 2014.

Robust Predictive Control Based on LMI Optimization for Active Suspension System

Hao Peng, Hao Jiang and Songyi Dian*

College of Electrical Engineering and Information Technology, Sichuan University, Chengdu, 610065, China

*scudiansy@mail.scu.edu.cn

Abstract. The vehicle will inevitably be vibrated in the process of running. The suspension is an important device to ensure the comfort and safety of the vehicle. In order to study the control method of the suspension, we simulate a quarter of the active suspension device which uses longitudinal acceleration as a measurement index. we compare the control effect of traditional PID control with robust predictive control. Through simulation, we find that the traditional PID control can realize the stability control of the active suspension without constraints. However, on the one hand, it brings serious jitter, on the other hand, it is not suitable for the actual constrained situation. On this basis, we consider the robust predictive control with constraints. By means of simulation, we obtain that the robust predictive control can produce less jitter under the premise of achieving the stable control of the constrained system, which is more suitable for practical engineering requirements.

1. Introduction

Vehicles will cause vibrations when it is running, which is not conducive to safety and comfort. In order to improve the stability and comfort of the vehicle, vehicles usually use a suspension system to suppress the vibration. The traditional vehicle suspension system consists of springs and a damping system. It does not require external control, but only dissipates or temporarily stores energy. It is also called a passive suspension system [1]. This system has poor adaptability, and it limits the improvement of vehicle vibration damping performance. So, it is a deadlock for traditional passive suspension system because of its fixed characteristics of springs and dampers. Good tune and design of a passive suspension can, to some extends, optimize and trade off riding comfort quality and stability, but cannot eliminate these conflicts completely [2]. The proposal of the active suspension provides a new method for vehicle vibration damping. In the 20th century, the idea of active suspension was put forward. Later, it was widely used in trains and automobile vehicles [3-7].

In recent years, controllable suspension systems have attracted considerable attentions by researchers and automobile industries, because it can overcome some limitations of traditional passive suspension systems. The controllable suspension system is always divided into active and semi-active system. Active suspension systems have greater advantages than passive suspension systems. Mainly reflected in that the vehicle system can be adjusted with changes in speed, road surface by self-regulation. The active suspension system is more suitable to use the control algorithm to achieve stability and simulation verification than the semi-active suspension system. Therefore, we mainly discuss the control design of the active suspension system here.

PID control methods are widely used in industry [8]. As the earliest practical algorithm, it has been widely applicated. The modernization control designs PID as a module, this module not only can greatly save the time, but also can facilitate the adjustment. The PID controller consists of a proportional unit (P), an integral unit (I), and a derivative unit (D), which can meet the control requirements of most linear systems, but PID is not suitable for some lag systems and high-order systems. Here we use PID control algorithm to design active suspension control, aiming to test the active suspension performance under the control of PID algorithm.

Linear H_∞control theory is a very important method to solve the robust control of linear systems. When the traditional constraint H_∞ control method is used to attain controllers that satisfy both time-domain hard constraints and robust performance [9], the upper bound of the external disturbance energy must be assumed as a priori, so it is not possible to coordinate constraints and satisfy system performance. Considering this question, we use LMI-based state feedback control strategy to ensure the efficiency of the algorithm. By ensuring the system's H_∞ performance under the conditions of system dissipation, the rolling time-domain closed-loop system can coordinate control performance and system constraints real-time.

2. Active suspension system modelling

This design considers a two-degree-of-freedom 1/4 car model. First, based on the assumption that the mass distribution coefficient of the car body design tends to be between 0.8 and 1.2, it can be considered that the body vibrations are independent from each other before and after the vibration, and the complete decoupling of the front and rear wheels can be achieved. With a symmetric excitation input, we can build a two-degree-of-freedom 1/4 suspension model. The physical model shown in Figure 1 was constructed from the load changes of the active suspension and the physical structural features of the suspension system. The model has the following features:

- Under the premise of correctness, the description parameters of the system are simplified.
- It is convenient for us to model.
- The model can basically represent the physical structure of the active suspension.

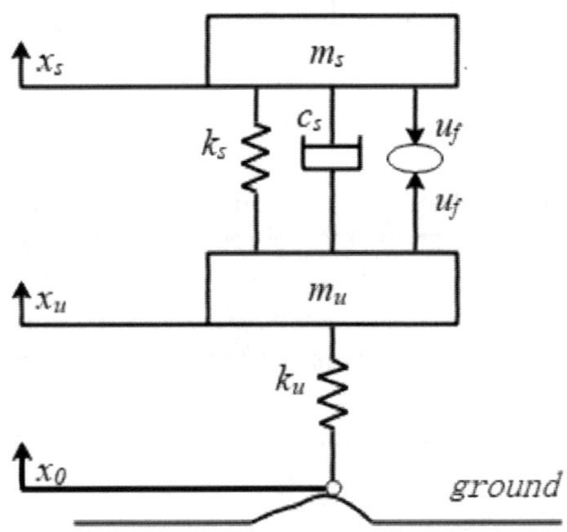

Figure 1. Two-degree-of-freedom 1/4 car model

Where m_s and m_u represent the sprung and unsprung mass. $x_s - x_u$ is the suspension travel stroke, $x_u - x_0$ is the tire movement displacement, x_0 is the road surface longitudinal displacement caused by the road surface roughness; u_f is the main power, provided by the vehicle's internal hydraulic transmission device, and can be regarded as the control input.

According to Newton's second law, we can establish the dynamic equation:

$$m_s\ddot{x}_s + c_s(\dot{x}_s - \dot{x}_u) + k_s(x_s - x_u) - u_f = 0 \tag{1}$$

$$m_u\ddot{x}_u - c_s(\dot{x}_s - \dot{x}_u) - k_s(x_s - x_u) + k_u(x_u - x_o) + u_f = 0 \tag{2}$$

Simplify the formulae of (1) and (2) to obtain the following formula:

$$\ddot{x}_s = \frac{1}{m_s}\left[-c_s\left(\dot{x}_s - \dot{x}_u\right) - k_s\left(x_s - x_u\right) + u_f\right] \tag{3}$$

$$\ddot{x}_u = \frac{1}{m_u}\left[c_s\left(\dot{x}_s - \dot{x}_u\right) + k_s\left(x_s - x_u\right) - k_u\left(x_u - x_o\right) - u_f\right] \tag{4}$$

In the above equation, when $u_f = 0$, the active suspension model is actually called the passive suspension model [10]. At the same time, the control input is also limited:

$$\left|u_f(t)\right| \le u_{\max}, \quad \forall t \ge 0. \tag{5}$$

Combining (3) and (4) with the constraint problem, the state space model of the active suspension can be constructed. In addition, we define the state variables: $x = (x_s - x_u, \dot{x}_s, x_u - x_0, \dot{x}_u)$. If the ground unevenness causes the vertical velocity to be regarded as interference, the state space model can be described as:

$$\dot{x}(t) = \begin{pmatrix} 0 & 1 & 0 & -1 \\ -\frac{k_s}{m_s} & -\frac{c_s}{m_s} & 0 & \frac{c_s}{m_s} \\ 0 & 0 & 0 & 1 \\ \frac{k_s}{m_u} & \frac{c_s}{m_u} & -\frac{k_u}{m_u} & -\frac{c_s}{m_u} \end{pmatrix} x(t) + \begin{pmatrix} 0 \\ 0 \\ -1 \\ 0 \end{pmatrix} \omega(t) + \begin{pmatrix} 0 \\ \frac{u_{\max}}{m_s} \\ 0 \\ -\frac{u_{\max}}{m_u} \end{pmatrix} u(t) \tag{6}$$

According to the vehicle model, the nominal values of the physical parameters of equation (6) are as follows:

$$m_s = 320kg,\ m_u = 40kg,\ k_s = 18kN/m,\ c_s = 1kN \cdot s/m,$$
$$k_u = 200kN/m,\ x_{\max} = 0.08m,\ u_{\max} = 1.5kN.$$

In order to simulate vehicles passing through pits or bags, we assume that the system's disturbance input is:

$$\omega(k) = \begin{cases} \frac{\pi VA}{L}\sin\frac{2\pi V}{L}kT_s, & 0 \le k \le \frac{L}{VT_s}, \\ 0, & k > \frac{L}{VT_s}, \end{cases} \tag{7}$$

The corresponding parameter attributes in the formula are shown in Table 1:

Table 1. Parameter attributes

parameter	unit	value
V	km/h	60
A	m	0.1
L	m	5
T_S	s	0.02

Combining people's feelings of comfort, we chose longitudinal acceleration of the body to describe the ride comfort [11]:

$$y(t) = \begin{pmatrix} -\frac{k_s}{m_s} & -\frac{c_s}{m_s} & 0 & \frac{c_s}{m_s} \end{pmatrix} x(t) + \frac{u_{\max}}{m_s} u(t) \tag{8}$$

3. H_∞ robust predictive control

This chapter first introduces the unconstrained H_∞ control, then introduces the time-domain constrained H_∞ control, and finally derives the rolling time domain constrained H_∞ control. Robust predictive control after linear matrix inequality optimization can coordinate constraints and system performance [12]. The LMI optimizes the state feedback control strategy, and then refreshes the system state at the sampling time, so that the system reaches the rolling optimization control.

3.1. Unconstrained H_∞ control

First, regardless of the time domain constraints, the state feedback is $u = Kx$. The closed-loop system that has been discretized is described as:

$$\begin{aligned} x(k+1) &= A_{cl}x(k) + B_1\omega(k) \\ z_1(k) &= C_{1,cl}x(k) + D_1\omega(k) \end{aligned} \tag{9}$$

The matrix $P(Q = P^{-1})$ is introduced according to the Schur complement formula. Combined with the system dissipation conditions, we transform the matrix. Finally, we make it satisfy the following inequality:

$$\begin{pmatrix} Q & O & * & * \\ 0 & \gamma I & * & * \\ AQ + B_u Y & B_1 & Q & 0 \\ C_1 Q + D_{1u} Y & D_1 & 0 & \gamma I \end{pmatrix} > 0 \tag{10}$$

Where the discrete closed-loop H_∞ norm from ω to z_1 is smaller than γ, $Y = KQ$

By solving the above matrix inequality, the system can realize unconstrained H_∞ control.

3.2. Time domain constraint conversion and constrained H_∞ control

The parameter e_υ is the standard base vector of $\mathbb{R}^{p2}$. Considering the time-domain constraints, the system satisfies the terminal domain of the standard MPC stability, so Q and Y satisfy (11) while also satisfying inequality:

$$\begin{pmatrix} z_{2\upsilon,\max}^2 & e_\upsilon^T\left(C_2Q + D_{2u}Y\right) \\ * & Q \end{pmatrix} > 0, \quad \upsilon = 1, 2, \cdots, p_2 \tag{11}$$

3.3. Rolling time domain H_∞ control algorithm

Applying the rolling optimization principle, we will solve the LMI optimization problem updated by the latest measurement state $x(k)$ at each sampling time k, and add a constraint to re-dissipate the closed-loop system. Namely:

$$\begin{pmatrix} p_0 - p_{k-1} + x(k)^T P_{k-1} x(k) & x(k)^T \\ x(k) & Q \end{pmatrix} \geq 0 \tag{12}$$

Combining the unconstrained H_∞control algorithm with time domain constraints, and we add the rolling optimization, robust predictive control can be achieved out and the system can have self-optimizing performance.

4. Experimental evaluation

We hope that the active suspension system can achieve the function of tracking disturbances stably and being able to predict road surface disturbances at time to achieve a function of noise reduction. Therefore, we start from the unconstrained control algorithm and gradually consider adding time domain constraints and rolling time domain optimization, eventually, the system can achieve a certain degree of robustness.

By solving the linear matrix inequality of the above optimization problem, an optimal algorithm for robust predictive control can be realized. The simulation data uses the data given in the previous chapter. The disturbance input is added at 20s~35s. Figure 2 shows the input as external disturbance, the simulation results are in Figure 3. In order to reflect the superiority of robust predictive control in active suspension systems, we added PID control for comparison, the parameters are set through multiple tests to obtain the optimal tuning result: $K_p = 3000, K_i = 1200, K_d = 3$. (Both simulations are in the same disturbance input in Figure 2):

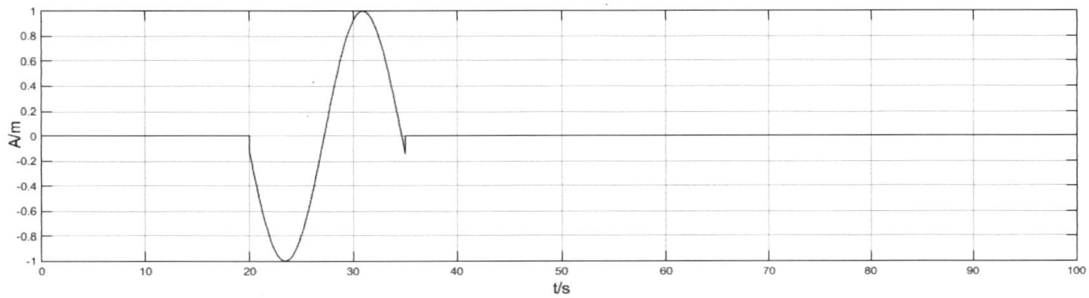

Figure 2. Disturbance input

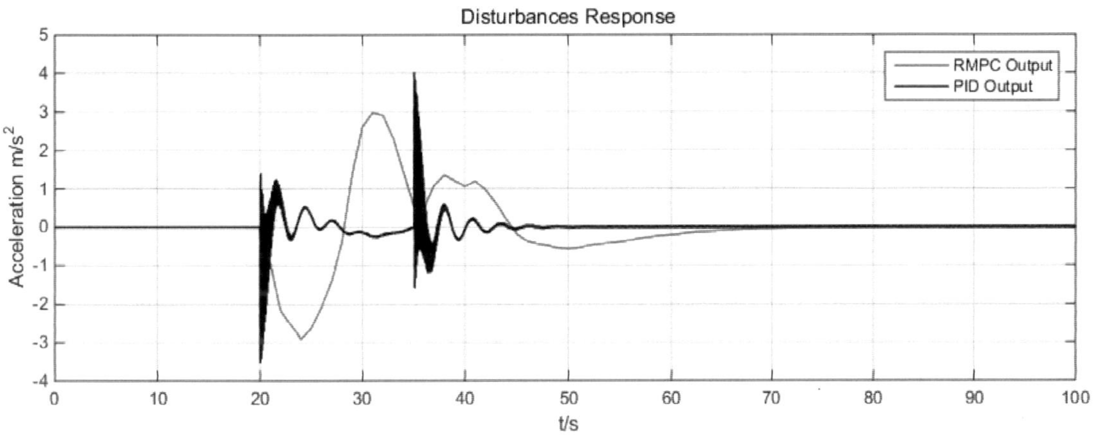

Figure 3. Robust predictive active suspension control and PID output after optimization

Analysis of the results, using PID regulation can adjust the longitudinal acceleration and be stability finally. However, during the adjustment process, the PID adjustment will cause the waveform to vibrate and the acceleration to change drastically. During normal PID adjustment, this process can be ignored. However, in the active suspension system, passengers will feel extremely uncomfortable if this happens. As a result, PID regulation cannot solve active suspension control well.

It can be seen from the simulation figure that the optimized rolling time domain H_∞ control method can effectively solve the control problem of the active suspension system. There is no sudden change in acceleration and system stability is better than PID regulation. Since the LMI optimization is performed online, the closed-loop system can coordinate the contradictory relationship between the control performance and the system constraints in real time and obtain the best possible control quality under the premise of ensuring the control constraints.

5. Conclusion

First, we construct a state space model for active suspension. Then we used PID control and H_∞ robust predictive control to adjust. Among them, the traditional PID adjustment results show that the system can achieve stability, and the adjustment time is within the acceptable range. However, in the disturbance rejection process, the system output is vibrated violently, which is unacceptable for a suspension system. Through the robust predictive control after optimization, we find that the system stability is outstanding, and it can realize online optimization to feedback the system state in real time.

More importantly, the accuracy of robust predictive control is also better than PID control. From the simulation image, it can solve the active suspension problem very well.

Similarly, in the high-precision control cases such as active suspension, we cannot rely on simple control methods such as PID, so we need introduce many advanced and stable control strategies to solve such problems.

References

[1] Ding Wensi, Bu Jiling, Liu Youmei. Strategy and method of high-speed train suspension's lateral semi-active control in China [J]. China Railway Science, 2002, 23(4):1~7.

[2] AHMADIAN M E H D I, DAVID E, SIMON. An analytical and experimental evaluation of magneto rheological suspensions for heavy trucks [J]. Vehicle System Dynamics, 2003, 37(S): 38-49

[3] Alleyne A, Hedrick J K. Nonlinear adaptive control of active suspensions [J]. IEEE Transactions on Control Systems Technology, 1995,3(1):94~101.

[4] Zhang Yuchun, Wang Liangxi, Cong Hua. Present situation and future challenges for automobile active suspension control [J]. Control Theory & Applications, 2004, 21(1):140~144.

[5] UTKIN V I. Variable structure systems with sliding mode [J]. IEEE Trans on Automatic Control, 1977, 22(2): 212~222.

[6] Guan Jifu, Gu Liang, Hou Chaozhen, Adaptive control for semi-active suspension tracked vehicle [J]. Acta Armamentaria, 2004, 25(4):390~393.

[7] DO YLEJC, STEIN G. Multivariable feedback design concepts for a classic/modern synthesis [J]. IEEE Trans on Automatic Control, 1981, 26 (1):4~16.

[8] Fu Yongfa. Several new PID regulator parameter setting method [J]. Chemical Automation and Instrumentation, 2004(11):39~41.

[9] Chen H, Guo K H. Constrained H_∞ control of active suspensions: An LMI approach. IEEE *Trans. Contr. Syst. Technology*, 2005,13(3): 412~421.

[10] Gordon T J, Marsh C, Milsted M G. A comparison of adaptive LQC and nonlinear controllers for vehicle suspension systems. *Vehicle System Dynamics*, 1991, 20:321~340.

[11] Hrovat D. Survey of advanced suspension developments and related optimal control applications. *Automatica*, 1997, 33(10):1781~1817.

[12] Scherer C W, Weiland S. Linear Matrix Inequalities in Control. Delft Center for Systems and Control, 2000, DISC Lecture note, Dutch Institute of Systems and Control.

IOP Conf. Series: Materials Science and Engineering **428** (2018) 012038 doi:10.1088/1757-899X/428/1/012038

Based On K-Means and Nearest Neighbor Algorithm for Fuzzy System Used for Data Fitting

Yi Hu, Jixia Han and Songyi Dian*

College of Electrical Engineering and Information technology, Sichuan University, Chengdu, China

*scudiansy(at)mail.scu.edu.cn

Abstract. This paper proposes using classical clustering algorithm in fuzzy system to fit experimental data. The main purpose is to reduce the number of fuzzy rules by clustering. K-means algorithm and nearest neighbor algorithm clusters classify data into classes so that each cluster corresponds to a fuzzy rule, the number of fuzzy rules is also determined by the number of clusters. We compare and discuss the results of data fitting between K-means algorithm and nearest neighbor algorithm. It find that two algorithm both can achieve better data fitting performance.

1. Introduction

Nowadays, many things cannot be simply divided into yes or no. The fuzzy system provides new ideas and tools for people to analyze and solve this problem. The theory of fuzzy systems was founded by Zadeh in 1965. It is a system based on knowledge or rule. With the continuous development of fuzzy theory, fuzzy systems began to be combined with various algorithms. In paper [1] improved the T-S fuzzy system on stabilization. The author [2] did research on a on smooth compactly supported radial fuzzy system. In paper [3] used participatory search algorithm in fuzzy systems modeling. In [4] proposed a rainfall forecasting using fuzzy system based on genetic algorithm.

Clustering is an unsupervised identification and classification process that divides the data into different groups. Clustering has been widely used in many fields, including signal processing, pattern recognition, bio-engineering, and image segmentation. Several clustering methods have been proposed. Classical clustering algorithms include K-means algorithm and nearest neighbor algorithm. In order to improve the classification speed and effectiveness, people have made the following efforts. The paper [5-7] proposed a K-means clustering algorithm based on feature weight. In paper [8-9] enhanced K-means by using PSO algorithm. The research [10] used nearest neighbor algorithm in human activity recognition.

This paper applies the K-means algorithm and nearest neighbor algorithm to the fuzzy system. The basic idea is to reduce the fuzzy rules by clustering, and one set corresponds to a fuzzy rule. Experiments show that the fitting error of data pairs can be reduced.

2. Fuzzy system structure

The fuzzy system is composed of four parts: fuzzy rule base, fuzzy inference machine, fuzzifier, and defuzzifier. We often use product inference engine, single-valued fuzzifiers and center average defuzzifier.

The fuzzy rule base is composed of the following fuzzy "If-Then" rules:

$$R_u^{(l)}: if\ x_1\ is\ A_1^l\ and \cdots and\ x_m\ is\ A_m^l,\ then\ y\ is\ B^l \tag{1}$$

Where A_i^l and B^l are fuzzy set in $U^i \subset R$ and $V \subset R$, $x = (x_1, x_2, \cdots, x_n)^T \in U$ and $y \in V$ are respectively input variables and output variables of fuzzy system.

It is given a finite set of input-output data couples $(x_0^l; y_0^l)(l = 1, 2, \cdots, N)$. Fuzzy system consists of N rules like (1), it described as follows [11]:

$$f(x) = \sum_{l=1}^{N} y_0^l \exp\left(-\frac{|x - x_0^l|^2}{\sigma^2}\right) / \sum_{l=1}^{N} \exp\left(-\frac{|x - x_0^l|^2}{\sigma^2}\right) \tag{2}$$

Where y_0^l is equal to the center of B^l and the fuzzy set $\mu_{A_i^l}$ uses a Gaussian membership function given by:

$$\mu_{A_i^l}(x_i) = \exp\left(-\frac{|x_i - x_{0i}^l|}{\sigma}\right) \tag{3}$$

The appropriate value σ can make the fuzzy system to fit all input-output data couple within any error. We have $|f(x_0^l) - y_0^l| < \varepsilon\ (l = 1, 2, \cdots, N)$ when every $\varepsilon > 0$.

3. Clustering algorithm

3.1. K-means algorithm

The K-means algorithm proposed by MacQueen is a classic algorithm to solve the cluster analysis problem [12]. The main principle of k-means is to divide a group of data into k clusters based on the distance between the data and have a higher similarity within the cluster and a lower degree of similarity between clusters. Algorithm works by minimizing the cost function using an iterative optimization technique as follows:

$$J = \sum_{i=1}^{K} J_i = \sum_{i=1}^{K} \sum_{j=1}^{N} w_{ji} \| X_j - C_i \|^2 \tag{4}$$

Where K and N are the number of clusters and samples in data set, respectively. $\| X_j - C_i \|$ is the Euclidean distance between the i th cluster center C_i and j th data X_j. w_{ji} presents the weight of the data X_j in the i th cluster. It can be calculated by following equation:

$$w_{ji} = \begin{cases} 1, & if\ \| X_j - C_i \| \leq \| X_j - C_m \|,\ \forall m \neq j \\ 0, & otherwise \end{cases}$$
$$s.t.\ \sum_{i}^{K} w_{ji} = 1, \forall j = 1, 2, \cdots, N;\ \sum_{i}^{K} \sum_{j}^{N} w_{ji} = N \tag{5}$$

The following data clustering techniques using the K-means algorithm as follows.

Step1: Initialize algorithm parameters, including initial cluster random centers, cluster number, iteration number and termination criterion;

Step2: Calculate the euclidian distance of the remaining data points from the means of all the clusters. If the data point X_j is determined to belong to the i th cluster, the weight value $w_{ji} = 1$ otherwise it is 0.

Step3: Calculate the cost function. If $\| J^{k+1} - J^k \| < \varepsilon$ then stop; else continue.

Step4: Update the cluster centers using the following formula(6) and go to step2.

$$C_i = \sum_{j=1}^{N} w_{ji} X_i / \sum_{j=1}^{N} w_{ji} \tag{6}$$

3.2. Nearest neighbor algorithm

Nearest neighbor algorithm is highly popular and effective in the field of pattern recognition. It is different from K-means algorithm that it does not need to specify the cluster number. Similarly, it divides a group of data into k clusters based on the distance.

The algorithm of nearest neighbor is as follows:

Step1: Set up radius r and select the first data as initial cluster centers.

Step2: Calculate the euclidian distance of the remaining data points from the centers of all clusters. $\| X_j - C_i \|$ is the shortest Euclidean distance between the i th cluster center C_i and j th data X_j.

(a) If $\| X_j - C_i \| > r$, X_j becomes a new cluster centers.

(b) If $\| X_j - C_i \| \le r$, X_j is assigned to C_i.

Step3: Until all point is assigned to the nearest cluster based on its distance from the center of each cluster and stop.

4. Experiment and discussion

It is given N input-output data couples. We separately use K-means algorithm and nearest neighbor algorithm in fuzzy system.

Step1: Using K-means to cluster data couples and until the cost function stabilize or using nearest neighbor to cluster data couples and until all point is assigned to the nearest cluster.

Step2: If the cluster C_i $(i = 1, 2, \cdots, K)$ has m data couples, $A^i = y_i^1 + y_i^2 + \ldots + y_i^m$, and $B^i = m$.

Step3: The fuzzy system is as follows:

$$f(x) = \sum_{i=1}^{K} A^i \exp\left(-\frac{|x - C_i|^2}{\sigma}\right) / \sum_{i=1}^{K} B^i \exp\left(-\frac{|x - C_i|^2}{\sigma}\right) \tag{7}$$

For K-meas algorithm, we initialize initial random cluster centers, cluster number, iteration number and termination criterion. The result is shown below.

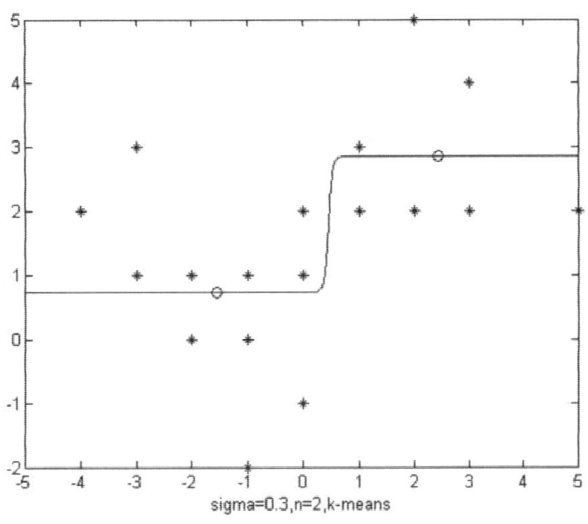

Figure 1. Data fitting of K-means algorithm using 2 clusters

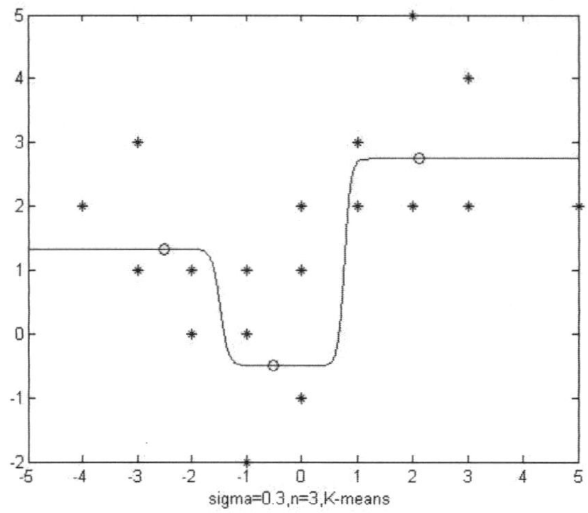

Figure 2. Data fitting of K-means algorithm using 3 clusters

From the above figure 1 and figure 2, we can find that it have a poor performance on fitting data when cluster number is 2. It shows that the fitting effect is related to the number of clusters. We can have better data fitting effects by appropriately increasing the number of clusters.

For nearest neighbor algorithm, we set up radius and initial cluster centers. The result is shown below.

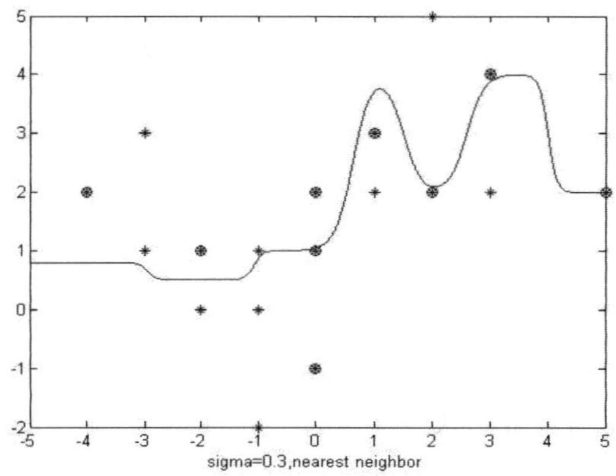

Figure 3. Data fitting of nearest neighbor algorithm

Nearest neighbor algorithm does not need to specify the cluster number. From the above figure 3, we find that it is over-fitting because of having too much cluster. When there are too many data couples, it is easy to over fitting.

Mean Squared Error(MSE) can evaluate the effect of data fitting. The smaller the value of MSE, the better the accuracy of the prediction model to describe the experimental data.

$$MSE = 1/n * \sum_{i=1}^{n}(X_n - X_n^*)^2 \tag{8}$$

Where X_n and X_n^* is the Experimental data and Model data, n is the number of data.

We use MSE to measure the data fitting results and calculate MSE values for K-means algorithms and nearest neighbor algorithm separately. As shown in Table 1 below.

Table 1. The MSE of K-means and nearest neighbor algorithm.

	K-means algorithm (cluster number=2)	K-means algorithm (cluster number=3)	nearest neighbor algorithm
MSE	1.6133	1.2684	2.0301

From Table 1, we find that the MSE of K-means algorithm (cluster number=3) is the smallest. It has a better behavior on data fitting. But using nearest neighbor algorithm can't fit data well. If we increase the cluster number of K-means algorithm, we can get better data fitting results.

5. Conclusion

This paper reduces the number of fuzzy rules by combining clustering algorithms and fuzzy systems. Classic clustering algorithms include K-means and nearest neighbor. The performance on fitting data of the fuzzy system is related to the number of clusters. The disadvantage of the K-means algorithm is that the number of clusters needs to be set in advance, but the nearest neighbor algorithm does not need to. In the case of a large number of data, the nearest neighbor algorithm tends to generate too many clusters leading to over fitting. We can improve the nearest neighbor algorithm that if the membership of a cluster has not changed for a long time, cancel this cluster.

Acknowledgment

This work is supported by Fundamental Research Special Funds for Central Universities (20826041A4133).

6. Reference

[1] Lian, Z., He, Y., Zhang, C., & Wu, M. (2017). Improved stabilization conditions for T-S fuzzy systems with interval time-varying delay. Chinese Control Conference (pp.384-389).

[2] David Coufal M.(2017). On Smooth Compactly Supported Radial Fuzzy System (pp.39-43). IEEE

[3] Liu, Y. L., & Gomide, F. (2017). Fuzzy systems modeling with participatory search algorithm. Fuzzy Systems Association and, International Conference on Soft Computing and Intelligent Systems(pp.1-6). IEEE.

[4] Nhita, F., & Adiwijaya. (2013). A rainfall forecasting using fuzzy system based on genetic algorithm. Information and Communication Technology(pp.111-115). IEEE.

[5] Wang, S., Fan, Y., Zhang, C., Xu, H. X., Hao, X., & Hu, Y. (2008). Subspace Clustering of High Dimensional Data Streams. Ieee/acis International Conference on Computer and Information Science (pp.165-170). IEEE Computer Society.

[6] Ren, J. T., Shi, X. X., Sun, J. H., Huang, H. Y., Yin, J., M.(2006). An improved k-means clustering algorithm based on feature weighting. computer science,33(7), 186-187.

[7] Zeng, B., Zhao, W., Luo, C., & Chen, B. (2010). The Optimization Arithmetic of K-means Clustering Based on Indirect Feature Weight Learning. International Conference on Computer and Communication Technologies in Agriculture Engineering(cctae 2010) (volume (Vol.2, pp.243-246).

[8] Gupta, A., Pattanaik, V., & Singh, M. (2017). Enhancing K means by unsupervised learning using PSO algorithm. International Conference on Computing, Communication and Automation (pp.228-233).

[9] Wang, X., & Sun, Q. (2017). The Study of K-Means Based on Hybrid SA-PSO Algorithm. International Symposium on Computational Intelligence and Design (pp.211-214). IEEE.

[10] Zul, M. I., Muslim, I., & Hakim, L. (2017). Human Activity Recognition by Using Nearest Neighbor Algorithm from Digital Image. International Conference on Soft Computing, Intelligent System and Information Technology (pp.58-61). IEEE Computer Society.

[11] Wang, L. X. (1996). A course in fuzzy systems and control. Prentice-Hall, Inc.

[12] Tou, J. T., & Gonzalez, R. C. (1974). Pattern recognition principles., 17(10), 274-282.

IOP Conf. Series: Materials Science and Engineering **428** (2018) 012039 doi:10.1088/1757-899X/428/1/012039

Comparative Study of PID, Type-1 and Type-2 Fuzzy Control for Nonlinear Systems

Yao Chen[1 a], Jiahao Liu[2 b] and Tao Zhao[3 c]

[1.] School of Electrical Engineering and Information Technology, Sichuan University, Chengdu 610065, China
[2.] School of Electrical Engineering and Information Technology, Sichuan University, Chengdu 610065, China
[3.] School of Electrical Engineering and Information Technology, Sichuan University, Chengdu 610065, China

[a] 2017223035189@stu.scu.edu.cn; [b]2017223030076@stu.scu.edu.cn
[c]zhaotaozhaogang@126.com

Abstract. Traditional PID control is widely used in industry due to its advantages of stability. The traditional PID controller is very difficult to obtain satisfied control performance for complex nonlinear systems with uncertainties. Fuzzy control approach has better anti-interference ability than traditional PID control method, and thus it can be used to deal with many complex nonlinear systems. In this paper, a comparative study for the control performance and anti-interference ability of traditional PID, type-1 and type-2 fuzzy control approaches are considered. By numerical simulations, the performance of fuzzy control is better than that of traditional PID in some specific situations, and the anti-interference performance of type-2 fuzzy control is better than type-1 fuzzy control and traditional PID control for a certain nonlinear problem.

1. Introduction

The use of classical controllers such as PID controller can be a good choice for controlling some systems. However, the higher and the more accuracy requirement in system control make it difficult to adjust the parameters of the controller, which is quite sensitive to noise signals [1].
Fuzzy logic controllers are used successfully in many application areas, and these include control, classification, etc [2-6].

The traditional fuzzy controller refers to type-1 proposed by Zadeh in 1965[7],can take advantage of the human experience and knowledge to deal with many control problems. As a kind of nonlinear controller, type-1 fuzzy logic controller provides a kind of language information by experts and convert them to control method, the control strategy of the system that can get the accurate model of the complex and difficult system. Therefore, it is an effective mean to deal with uncertainty and inaccuracy. Type-1 fuzzy system (Type-1 FLS), however, it will also face many uncertainties, and the limitation of the type-1 fuzzy logic controller is that it uses the membership function by accurate representation of a type-1 fuzzy sets. Therefore, type-1 fuzzy logic controller does not directly deal with the uncertainty of fuzzy rules. As an extension of type-1fuzzy logic control(T1-FLC) Zadeh proposed the type-2 fuzzy set(T2-FS) in 1975. The characteristic of T2-FS is the fuzzy representation of the membership value in the fuzzy set, that is, its membership is T2-FS [8-9].

This paper presents a comparative study for the control performance and anti-interference ability of traditional PID, type-1 and type-2 fuzzy control approaches. Numerical simulation is provided to show the advantages of type-2 fuzzy control in handling uncertainties. The paper is organized as follows. In section 2, some preliminaries are introduced. The PID, type-1 and type-2 fuzzy controller design is presented in section 3. The simulation comparisons of three kinds of control methods are provided in section 4.The last section draws the conclusion and puts forward the prospect.

2. Preliminaries

2.1 Controlled object analysis

In this paper, we use the nonlinear system shown in formula (1) to make a comparison among the control performance of PID, type-1 and type-2 fuzzy controllers, the control object adopted is the nonlinear model as shown in formula (1) [10].

$$\begin{aligned} Y(i) = 0.2 * y(i-3) * 0.07 * y(i-2) + 0.9 * y(i-1) \\ + 0.05 * u(i-1) + 0.5 * u(i-2) \end{aligned} \quad (1)$$

It can be seen from the formula (1) that the output of the nonlinear plant is related to the previous three outputs and the previous two control inputs. PID, as a classic and stable controller,has been widely applied in industry. Firstly, we explore whether the PID controller can meet the control requirements. Especially, some stochastic disturbances exist in the feedback channel, and thus the anti-interference ability of controller is required. Comparing with PID control method, fuzzy control method has stronger robustness. Secondly, type-1 and type-2 fuzzy controller is designed to enhance the ability to deal with nonlinear effects and uncertainties.

2.2 Performance criteria

To more clearly compare the control performance and anti-interference ability of PID, type-1and type-2 fuzzy control approaches, three functions are used to evaluate the control results quantitatively. These evaluation functions have been widely used in literature0.

1) Integral of Square Error(ISE)

$$ISE = \int_0^{\infty} [e(t)]^2 \, dt \quad (2)$$

2) Integral of the Absolute value of the Error(IAE)

$$IAE = \int_0^{\infty} |e(t)| dt \quad (3)$$

3) Integral of the Time multiplied by the Absolute value of the Error(ITAE)

$$ITAE = \int_0^{\infty} t\,|e(t)| dt \quad (4)$$

3. Controller design

In this section, PID, type-1 and type-2 fuzzy controllers are designed for the nonlinear system provided in formula (1)respectively.

3.1 PID controller design

The control law of PID control is as follows:

$$u(t) = K_p \left[e(t) + \frac{1}{T_i} \int_0^t e(t) dt + T_d \frac{de(t)}{dt} \right] \quad (5)$$

where e(t) is error, T_i is the integral coefficient, and $T_d$is the differential coefficient. The sampling time T is 0.1, and the sampling total number N=150. According to the responses of nonlinear system in (1), PID parameter is selected as K_p=1, T_i=-1/6, T_d=1.

3.2 Type-1 fuzzy controller design

IOP Conf. Series: Materials Science and Engineering **428** (2018) 012039 doi:10.1088/1757-899X/428/1/012039

Firstly, the tracking error e(t) and the error variation Δe(t) are selected as the inputs of type-1 fuzzy controller and then act on the plant. Secondly, for each of the inputs of the type-1 fuzzy logic system, e(t)and Δe(t), three type-1 fuzzy Gaussian MFs were defined as: negative, zero, positive. The universe of discourse of these membership functions is in the range [-10 10]; their centers are -10,0 and 10 respectively, and their standard deviations is 4.2466 as is illustrated in Figure.1[10].

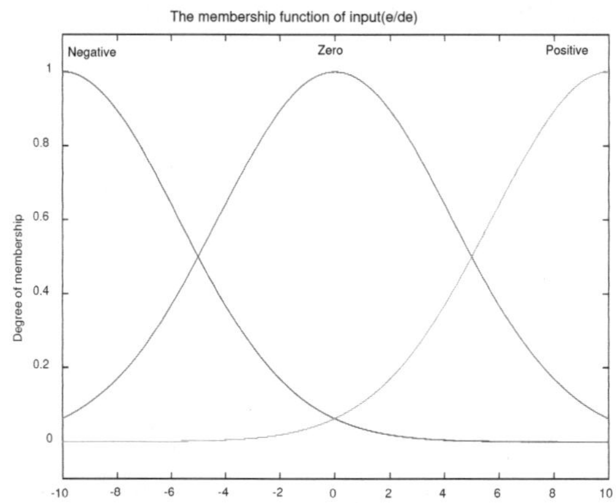

Figure 1. The membership function diagram of inputs

For the output of the type-1 fuzzy logic system, we have five type-1fuzzy Gaussian MFs: NG, N, Z, P and PG. They are in the interval [-10,10], their centers are -10, -.5, 0, 5 and 10 respectively; and their standard deviation is 2.1233 as can be seen in Figure.2[10].

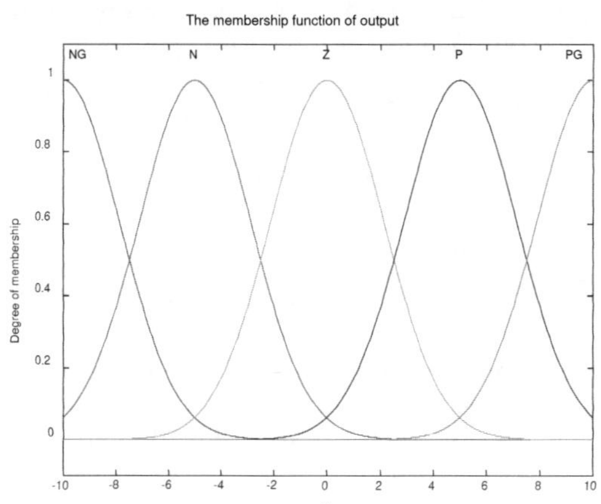

Figure 2. The membership function diagram of output

Table 1. Rules table of type-1 fuzzy logic system

e(t) \ Δe(t)	N	Z	P
N	NG	N	Z
Z	NG	Z	P
P	Z	P	PG

The rules of type-1 fuzzy controller is shown in Table.1 and centroid defuzzification are used to obtain a crisp output.

3.3 Type-2 fuzzy controller design

Firstly, the type-1 fuzzy logic system principle diagram is presented by Figure.3

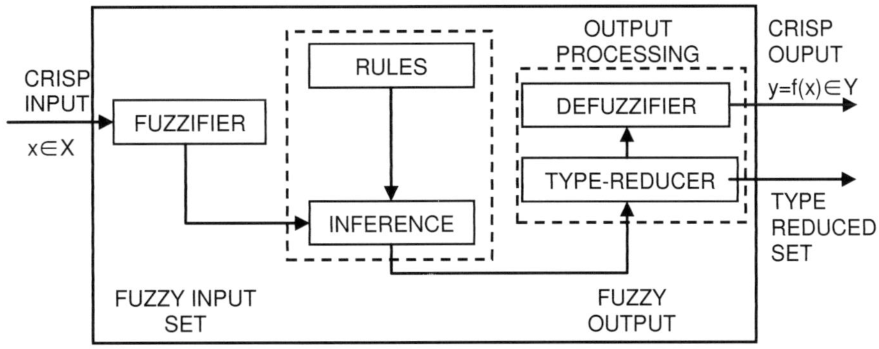

Figure 3. Type-2 fuzzy logic system

Secondly, the type-2 fuzzy controller also selects the error e(t) and the error variation as input, and the membership function of type-2 fuzzy controller is shown in Figure.4;The type-2 fuzzy controller can deal with uncertainties directly because its' membership function is also fuzzy. In other words, atype-1fuzzy set is a special case of a type-2 fuzzy set; the type-2 fuzzy system's secondary membership function is a subset with only one element [10].

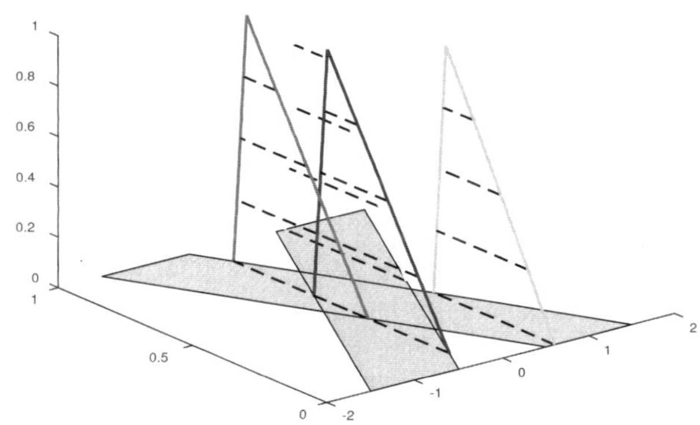

Figure 4. The membership function diagram of type-2 fuzzy controller

Thirdly, the rule base has the following four rules:

$$R^1: IF x_1 is \widetilde{X_{11}} and x_2 is \widetilde{X_{21}}, THEN y is Y^1.$$
$$R^2: IF x_1 is \widetilde{X_{11}} and x_2 is \widetilde{X_{22}}, THEN y is Y^2.$$
$$R^3: IF x_1 is \widetilde{X_{12}} and x_2 is \widetilde{X_{21}}, THEN y is Y^3.$$
$$R^4: IF x_1 is \widetilde{X_{12}} and x_2 is \widetilde{X_{22}}, THEN y is Y^4.$$

Where $x_i$and y are the input and output of the type-2 FLS, respectively; $\widetilde{X}_\imath$and$Y^i$are the type-2 antecedent and the consequent sets, respectively. And COS type-reduction + average of end-points defuzzification are used to obtain a crisp output.

4. Simulation

4.1 Simulation of an Ideal System

The simulation results of the tradition PID, type-1 and type-2 fuzzy controllers are shown in Figure.5.

IOP Conf. Series: Materials Science and Engineering **428** (2018) 012039 doi:10.1088/1757-899X/428/1/012039

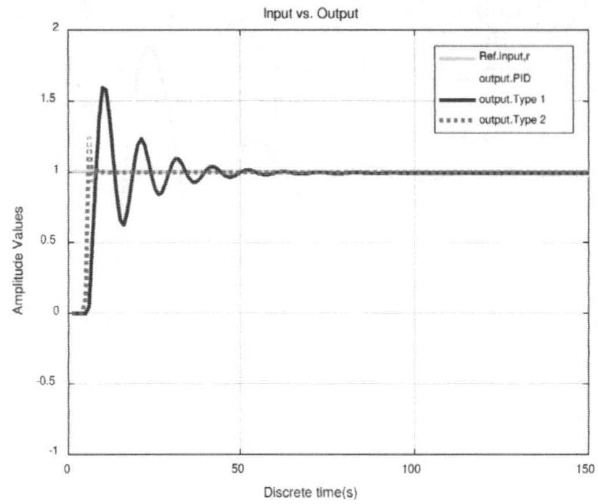

Figure 5. Simulation results of the tradition PID controller, type-1 and type-2 FLC

Table 2. Comparison of performance criteria for the traditional PID controller, type-1 and type-2 FLC

Performance Criteria	PID	Type-1 FLC	Type-2 FLC
ISE	24.97	4.79	1.87
IAE	33.90	9.70	2.72
ITAE	109.50	20.10	5.05

As shown in Figure.5 and Table.2, The highest point of the traditional PID response curve is 1.73, which means the overshoot is 73%. The adjust speed is slow and the final tracking input signal has a static error. Obviously, the performance of PID controller is not ideal, we consider to use the fuzzy logic controller to get a better control performance.

Firstly, we can see the performance of type-1 and type-2 FLC are faster and more accurate than traditional PID controller, especially the type-2 fuzzy controller is much better. The type-2 fuzzy controller's response speed is quick and the overshoot is small. Therefore, we can conclude that the fuzzy controller has better control performance.

4.2 Simulation result of a System with Uncertainty

To observe anti-interference ability of the fuzzy controllers, the feedback channels of the three controllers were added to the random disturbance of 0.05*random(1), and then the simulation results are shown in Figure.6

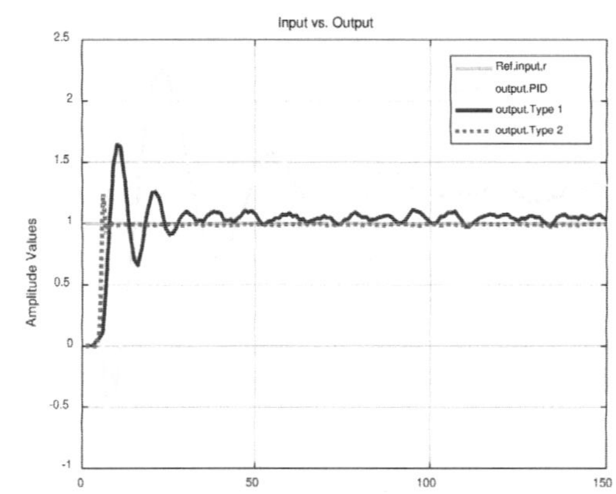

Figure 6. Simulation results of the traditional PID controller, type-1 and type-2 FLS with uncertainties

Table3. Comparison of performance criteria for PID controller, type-1 and type-2 FLC with uncertainties

Performance Criteria	PID	Type-1 FLC	Type-2 FLC
ISE	34.20	4.86	1.87
IAE	53.97	13.73	3.60
ITAE	306.99	61.37	11.73

As can be seen from Figure.6 and Table.3, the anti-interference ability ranges from high to low: type-2 fuzzy controller, type-1 fuzzy controller and the traditional PID controller.Thetype-2 fuzzy controller has strong anti-interference ability, which also demonstrates that type-2 fuzzy controller does have a strong ability to deal with uncertainties.

5. Conclusions and prospects

The simulation results indicate that the fuzzy control's response speed is quick, overshoot is small, the ability of resistance is better. Thetype-2 fuzzy control shows good control performance and anti-interference ability. However, the traditional PID controller also has some advantages that fuzzy controller does not have. Therefore, it is a wise choice to combine fuzzy control with PID control to achieve better control performance and anti-interference ability in a subsequent study.

Acknowledgments

This work is supported by the National Natural Science Foundation of China (61703291) and the Applied Basic Research Program of Science and Technology Department of Sichuan Province, China (2016JY0085).

References

[1] Castillo, O., Melin, P.: Soft Computing for Control of Non-Linear Dynamical Systems. Springer, Heidelberg (2001)
[2] Al-Jaafreh, M.O., Al-Jumaily, A.A.: Training type-2 fuzzy system by particle swarm optimization. In: IEEE Congress on Evolutionary Computation, CEC 2007, Singapore, pp. 3442–3446 (2007)
[3] Cai, A., Quek, C., Maskell, D.L.: Type-2 GA-TSK fuzzy neural network. In: Proceedings of IEEE Congress on Evolutionary Computation, CEC 2007, pp. 1578–1585(2007)
[4] Hidalgo, D., Melin, P., Mendoza, O.: Evolutionary optimization of type-2 fuzzy systems based on the level of uncertainty. In: Proceedings of the IEEE World Congress on Computational Intelligence, WCCI 2010, Barcelona (July 2010)
[5] Juang, C.-F., Hsu, C.-H.: Reinforcement ant optimized fuzzy controller for mobile robot wall-following control. IEEE Transactions on Industrial Electronics 56, 3931–3940 (2009)
[6] Castro, J.R., Castillo, O., Melin, P., Martinez, L.G., Escobar, S., Camacho, I.: Building fuzzy inference systems with the interval type-2 fuzzy logic toolbox. Advances in Soft Computing, vol. 41, pp. 53–62 (2007)
[7] L. A. Zadeh, "Fuzzy sets," Inf. Control, vol. 8, no. 3, pp. 338–353, 1965
[8] Mendel J M. Type-2 Fuzzy Sets and Systems: an overview[J]IEEE Computational Intelligence Magazine,2007,2(1):30-43
[9] Mendel J M. Uncertain Rule-Based Fuzzy Systems: Introduction and New Directions, 2nd Edition[M]. Springer Publishing Company, Incorporated, 2017.
[10] Janusz Kacprzyk. Type-2 Fuzzy Logic in Intelligent Control Applications. Studies in Fuzziness and Soft Computing, Volume 272

IOP Conf. Series: Materials Science and Engineering **428** (2018) 012040 doi:10.1088/1757-899X/428/1/012040

Optimization of Interval Type-2 Fuzzy Logic Controller for Uncertain Inverted Pendulum System

Qing Wu[1a], Jiahao Liu[2b] and Tao Zhao[3b]

1. College of Electrical Engineering and Information Technology, Sichuan University, Chengdu 610065, China
2. College of Electrical Engineering and Information Technology, Sichuan University, Chengdu 610065, China
3 College of Electrical Engineering and Information Technology, Sichuan University, Chengdu 610065, China

[a]15882181576@163.com; [b]2017223030076@stu.scu.edu.cn; [c]zhaotaozhaogang@126.com

Abstract. In this paper, an uncertain inverted pendulum system is controlled by an IT2 fuzzy logic controller (IT2 FLC) which was optimized by the particle swarm optimization (PSO) algorithm. The IT2 FLC, as an extension of the type-1 fuzzy logic controller (T1 FLC), has better control performance in dealing with uncertainties than T1 FLC. However, IT2 FLC has more complexity because the parameters of the IT2 FLC are more than the parameters of T1 FLC, so it is difficult to select appropriate parameters to get better performance. The PSO algorithm is used to optimize several parameters of IT2 FLC via two cost functions. According to simulations, the performance of the IT2 FLC is optimized by the PSO algorithm is better than the IT2 FLC without optimization, and the IT2 FLC has better performance in dealing with uncertainties than T1 FLC.

1. Introduction

The type-1 fuzzy logic controller (T1 FLC), presented by Prof. Lotfi Zadeh in 1965, has been studied for decades by many experts, and has many applications in the fields of industry. T1 FLC has the ability to deal with the uncertainties and nonlinearity. With the increase of uncertainty of the object, the interval type-2 fuzzy logic controller (IT2 FLC) is presented on the foundation of the T1 FLC. Compared with the T1 FLC, the T2 FLC has better ability to handle uncertainties, but the IT2 FLC is much more complex than the T1 FLC. The membership function is a mathematical tool for characterizing fuzzy sets and also is one of the most important parts of the IT2 FLC, whose parameters will have a great effect on the performance of the system. To get the better performance, we need to tune the parameters of membership function. However, the number of membership function is too many, so it is difficult to tune it with general methods.

The particle swarm optimization (PSO) algorithm is a kind of evolutionary algorithm developed by J. Kennedy and R.C. Eberhart in 1995, which starts from the random solution to optimal solution by iterative computation. The PSO algorithm seeks the global optimum by following the current searched optimal value, so we can use PSO algorithm to seek the optimal parameters of membership function in a certain range of space. Furthermore, the PSO algorithm has attracted much attention from the academic circles because of its advantage of easy to achieve, high precision and fast convergence.

This paper studies the optimal design of IT2 FLC for the uncertain inverted pendulum system by PSO algorithm. Simulation results show the effectiveness of the proposed method. The paper is organized as follows. Section 2 introduces preliminaries. In Section 3, the optimal design of IT2 FLC is presented. Section 4 provides numerical simulation to demonstrate the effectiveness of the proposed approach. The conclusion is drawn in the last section.

2. Preliminaries

2.1. Interval type-2 fuzzy sets

An interval type-2 fuzzy set (IT2 FS) $\tilde{A}$ can be expressed as

$$\tilde{X} = \int_{x\in D_{\tilde{X}}} \int_{u\in J_x \subseteq [0,1]} u_{\tilde{A}}(x,\mu)/(x,u), \tag{1}$$

Where x and u are the primary variable and secondary variable respectively; and J_x is called the support of the secondary membership function; and the amplitude of $u_{\tilde{X}}(x,u)$, called a secondary grade of $\tilde{X}$, equals 1 for $\forall x \in D_{\tilde{X}}$ and $\forall u \in J_x \subseteq [0,1]$ [1].

2.2. IT2 FLC

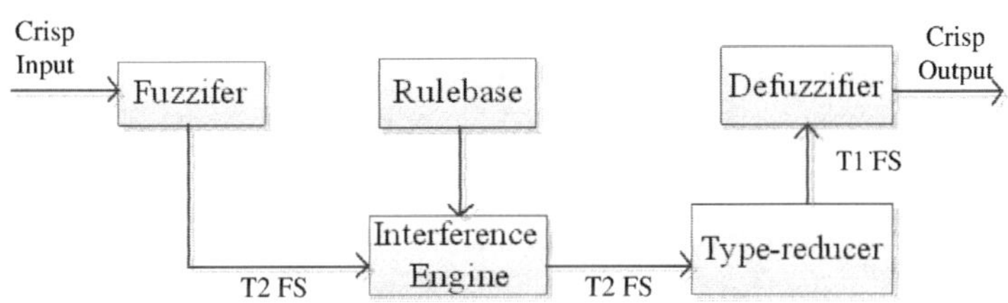

Figure 1. Schematic diagram of IT2 FLC

As the Figure.1 showed, an IT2 FLC contains five parts, called fuzzifier, interference engine, rulebase, type-reducer and defuzzifier respectively [2]. The fuzzifier can covert a crisp input to an IT2 FS by the membership function. In general, the shape of membership function is Triangle, Gauss and trapezoid. The defuzzifier can covert an IT2 FS to a crisp output. The rulebase is the heart of the IT2 FLC, which has fuzzy if-then rules of the form:

$$\text{IF } x_1 \text{ is } \tilde{X}_1^n \text{ and } L \text{ and } x_I \text{ is } \tilde{X}_I^n \text{, THEN } y \text{ is } Y^n \qquad n=1,2,\cdots,N$$

where $\tilde{X}_i^n (i=1,\ldots, I)$ are IT2 FSs, and Y^n is an interval which can be the centroid of an IT2 membership function or the simplest TSK model [3]. The choice of rulebase is depend on the experience of experts or the input-output data with intelligence algorithms. The inference engine uses minimum or product t-norm generally. To simplify the calculation, we use product t-norm in this paper [3]. The type-reducer is unique for type-2 FLC, converting IT2 FSs into T1 FSs[4].

2.3. PSO algorithm

In this paper, we use PSO algorithm to optimize the IT2 FLC for the inverted pendulum system. In the PSO algorithm, each particle can be viewed as a point and all the particles make up a 'swarm'. The particle has two information, called weight and velocities. Each individual seek the most optimal relay on the information of its own and other individuals. At each iteration, the variables of ith particle update by

$$\begin{cases} v_i = \omega v_i + c_1 r_1 (pbest_i - w_i) + c_2 r_2 (gbest - w_i) \\ w_i = w_i + v_i \end{cases} \tag{2}$$

where the w_i and v_i represents the position and velocity respectively; $pbest_i$ is the optimal weight in all iteration for a particle and $gbest$ is the optimal weight in all iteration for all particles; and the c_1 and c_2 are positive constant learning rates; and the ω is called inertia weight; the r_1 and r_2 are random numbers between 0 and 1[5].

2.4. Dynamic model of the inverted pendulum system

The inverted pendulum system is a typical multi-variable, nonlinear and fast-moving system that can effectively reflect many keys issues in control, whose state space model can be represented as:

$$\begin{cases} \dot{x}_1 = x_2 \\ \dot{x}_2 = \dfrac{g\sin x_1 - \dfrac{mlx_2^2 \cos x_1 \sin x_1}{m_c + m}}{l(\dfrac{4}{3} - \dfrac{m\cos^2 x_1}{m_c + m})} + \dfrac{\dfrac{\cos x_1}{m_c + m}}{l(\dfrac{4}{3} - \dfrac{m\cos^2 x_1}{m_c + m})} \end{cases} \tag{3}$$

where g is the gravity acceleration; and m_c is the mass of the car; and m is the mass of the inverted pendulum; and l is the half of length of the inverted pendulum; and u is the control force; and x_1 is the angle of the inverted pendulum, and x_2 is the derivative of x_1 [3].

3. Optimization of the FLC for the inverted pendulum system

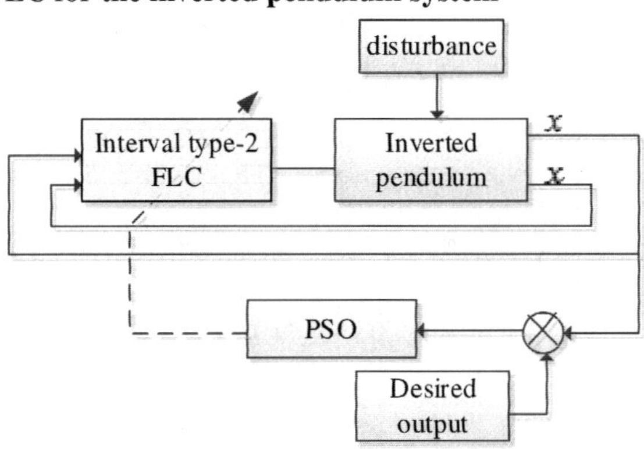

Figure 2. Overall control diagram for the inverted pendulum system

Table 1. The rulebase of the inverted pendulum system

x_2 / x_1	NB1	NS1	ZE1	PS1	PB1
NB2	[-10.1 -9.9]	[-7.1 -6.9]	[-5.1 -4.9]	[-5.1 -4.9]	[-0.1 0.1]
NS2	[-7.1 -6.9]	[-5.1 -4.9]	[-3.1 -2.9]	[-0.1 0.1]	[2.9 3.1]
ZE2	[-5.1 -4.9]	[-3.1 -2.9]	[-0.1 0.1]	[2.9 3.1]	[4.0 5.1]
PS2	[-3.1 2.9]	[-0.1 0.1]	[2.9 3.1]	[4.0 5.1]	[6.9 7.1]
PB2	[-0.1 0.1]	[2.9 3.1]	[4.0 5.1]	[6.9 7.1]	[9.9 10.1]

IOP Conf. Series: Materials Science and Engineering **428** (2018) 012040 doi:10.1088/1757-899X/428/1/012040

The overall control diagram for the invertd pendulum system shown in Figure.2 can be divided into two parts. The first part is FLC system and the second part is about optimization of PSO algorithm. The IT2 FLC has two inputs and one output, which represents two states of the inverted pendulum system and control force of the inverted pendulum system. The two inputs of FLC are represented by five linguistic terms NB, NS, ZE, PS, and PB, as shown in Figure.3. The fuzzy rulebase is showed in Table.1, which is determined by the experience of experts. The second part is about the optimization of IT2 FLC by PSO algorithm. The flow diagram of PSO algorithm is shown in Figure.4. In each iteration, the fitness value was computed by cost function, which was used to evaluate the weight of particles. The cost functions of PSO used as mean of root of squared error (MSE) and mean of absolute magnitude of the error (MAE) are represented as follow:

$$\mathrm{MSE} = \frac{1}{N}\sum_{i=1}^{N}(x_1 - x_d)^2 \tag{4}$$

$$\mathrm{MAE} = \frac{1}{N}\sum_{i=1}^{N}\left|x_1 - x_d\right| \tag{5}$$

where x_d is the desired output of the inverted pendulum system. The selecting of *pbest* and *gbest* is relay on the fitness value [6, 7, 8].

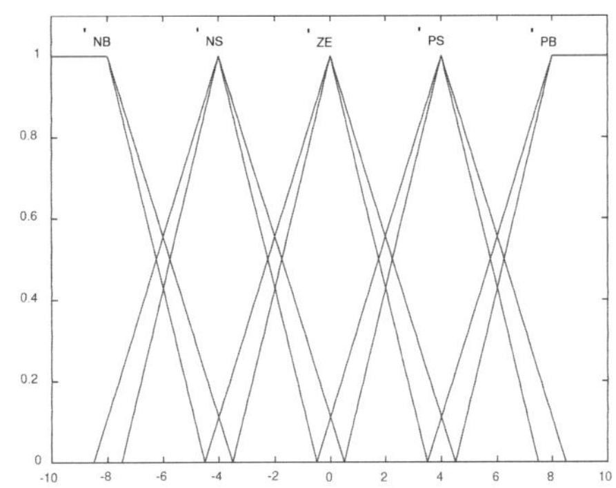

Figure 3. The membership function of IT2 FLC

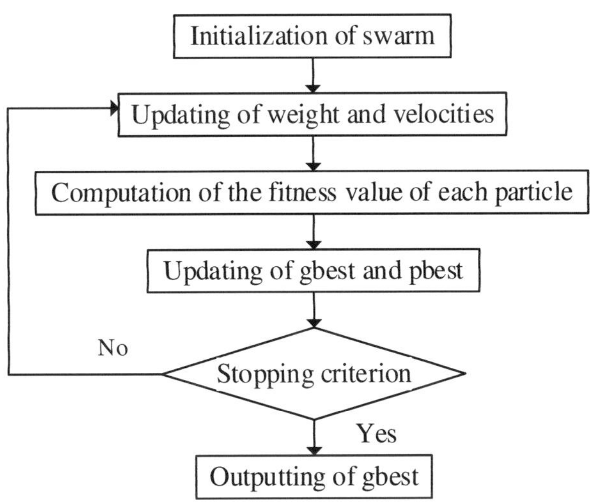

Figure 4. The flow diagram of PSO algorithm

4. Simulation results

Let the initial value of x_1 and x_2 be equal to -0.6 and 0.25 respectively. The desired balance position is $x_d = 0$.

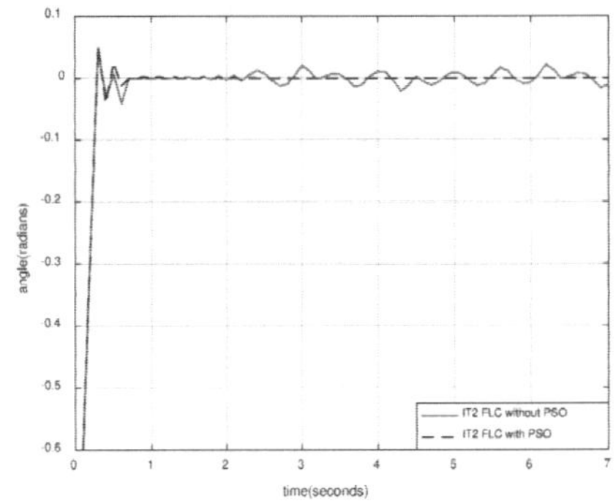

Figure 5. Simulation result of the angle of the inverted pendulum

The simulation result is showed as Figure.5. Because the initial membership functions is not suitable for the controlled system, the angle of the inverted pendulum shake around the balanced position. After optimization, the angle of the inverted pendulum is stable at the balanced position. The different cost functions for the IT2 FLC can be viewed in Table.2.

Table 2. The cost function value of MAE and MSE

Cost functions	IT2 FLC without PSO	IT2 FLC with PSO
MSE	0.0059	0.0058
MAE	0.0197	0.0139

In the following analysis, a disturbance with a sine signal whose amplitude is 0.8, is increased to the inverted pendulum system. The T1 FLC and IT2 FLC are designed for the uncertain inverted pendulum system, respectively. The corresponding simulation result is shown in Figure.6. It can be seen from Figure.6 that the IT2 FLC has better performance in dealing with uncertainties than T1 FLC.

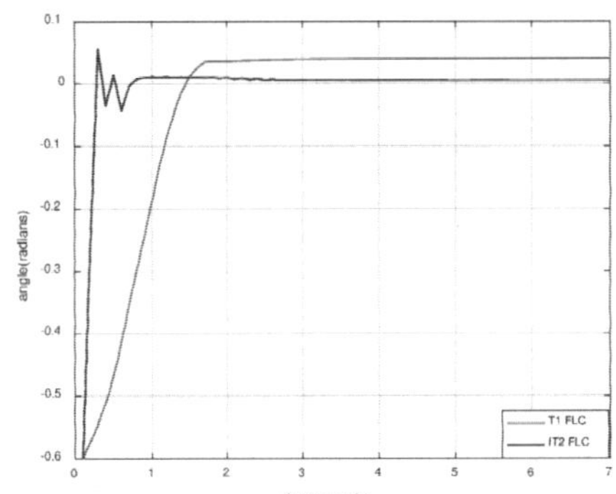

Figure 6. The control for the angle of the inverted pendulum

5. Conclusions

In this paper, an IT2 FLC for the uncertain inverted pendulum system is designed. The proposed IT2 FLC is optimized by PSO algorithm. The simulation results illustrate that the performance of system after being optimized is better than the system without optimization. When disturbance exists in system, the IT2 FLC has more robustness than T1 FLC.

Acknowledgments

IOP Conf. Series: Materials Science and Engineering **428** (2018) 012040 doi:10.1088/1757-899X/428/1/012040

This work is supported by the National Natural Science Foundation of China (61703291) and the Applied Basic Research Program of Science and Technology Department of Sichuan Province, China (2016JY0085).

References

[1] Atanassov K, Gargov G. Interval valued intuitionistic fuzzy sets[M]. Elsevier North-Holland, Inc. 1989.
[2] Mendel J M. Uncertain Rule-Based Fuzzy Systems: Introduction and New Directions, 2nd Edition[M]. Springer Publishing Company, Incorporated, 2017.
[3] Wang L X. A course in fuzzy systems and control[M]. Prentice-Hall, Inc. 1996.
[4] Wu D. Approaches for Reducing the Computational Cost of IT2 Fuzzy Logic Systems: Overview and Comparisons[J]. IEEE Transactions on Fuzzy Systems, 2013, 21(1):80-99.
[5] Eberhart R, Kennedy J. A new optimizer using particle swarm theory[C]// International Symposium on MICRO Machine and Human Science. IEEE, 2002:39-43.
[6] Castillo O, Melin P, Kacprzyk J, et al. Type-2 Fuzzy Logic: Theory and Applications[M]. Springer Publishing Company, Incorporated, 2008.
[7] Melin, Patricia, et al. "Optimal design of type-2 and type-1 fuzzy tracking controllers for autonomous mobile robots under perturbed torques using a new chemical optimization paradigm." *Expert Systems with Applications*40.8(2013):3185-3195.
[8] Bingül Z, Karahan O. A Fuzzy Logic Controller tuned with PSO for 2 DOF robot trajectory control[J]. Expert Systems with Applications, 2011, 38(1):1017-1031.

IOP Conf. Series: Materials Science and Engineering **428** (2018) 012041 doi:10.1088/1757-899X/428/1/012041

Research on Nonlinear Decoupling Method of Piezoelectric Six-Dimensional Force Sensor Based on BP Neural Network

Yingjun Li[1,*], Guicong Wang[1], Binbin Han[1], Xue Yang[1] and Zhiquan Feng[2]

[1]School of Mechanical Engineering, University of Jinan, Jinan 250022, PR China
[2]School of Information Science and Engineering, University of Jinan, Jinan 250022, PR China

[*]me_liyj@ujn.edu.cn

Abstract: The six-dimensional force sensor has become one of the major bottlenecks restricting the development of robots in China. In this paper, the problem of the decoupling of the piezoelectric six-dimensional force sensor with four-point support structure is studied, and the static decoupling method is studied. Firstly, the principle of nonlinear decoupling algorithm for six-dimensional force sensor is analyzed, and experimental data obtained by decoupling are acquired through calibration experiments, and sample selection and normalization processing are performed. After that, the BP forward feedback neural network was used to optimize the multi-dimensional nonlinear characteristics of the sensor output system, and the input and output mapping relationship of the sensor was determined, and the decoupled sensor output data was obtained. The determinant sensor's measurement accuracy evaluation index is compared with linearity error and coupling rate error.

1. Introduction

The development characteristics and trends of robotics technology are intelligence and the interaction between humans and robots. Intelligence is the third stage of robot development. Robots with sensing and autonomous behavior control functions have become a trend [1-2]. The six-dimensional force sensor can sense all information of the space six-dimensional external forces and moments of the joints of the robot wrist, etc., and is an important part of the intelligent robot. The piezoelectric sensor is a non-elastic body type power sensor, which is one of the most promising sensors in the development of smart structures and robotics [3-4].

The accuracy and reliability of the six-dimensional force sensor's technical parameters are directly related to the quality of the robot, and thus relate to the accuracy, safety, and reliability of the robot's work. In the signal output process, coupling between the dimensions of the signal will reduce the sensor's measurement accuracy, restricting the application of the sensor in the high, fine, and sharp areas [5]. Although the linear decoupling algorithm can reduce the influence of inter-dimensional coupling to a certain extent, the decoupling principle model is established under the assumption that the sensor system is linear. So the decoupling effect is not ideal. In order to further improve the measurement accuracy, research on nonlinear decoupling algorithms must be carried out. Many scholars have proposed various decoupling algorithms for strained six-dimensional force sensors with different configurations [6-7]. For piezoelectric six-dimensional force sensors, there are relatively few reports on nonlinear decoupling algorithms [8-10], and the coupling error is large.

At present, neural networks play an extremely important role in the fields of computer science, artificial intelligence, engineering, and robotics, and their application scope is also expanding. It is an important branch of intelligent development. The neural network can be subjected to a simple non-linear mapping and further stronger non-linear processing capabilities. It can build a coupling relationship between multidimensional data based on experimental data and achieve nonlinear decoupling [11]. In this paper, the decoupling algorithm of BP neural network is mainly used to study the nonlinear decoupling of piezoelectric six-dimensional force sensor.

2. Nonlinear decoupling principle

The principle of the nonlinear decoupling algorithm for piezoelectric six-dimensional force sensor based on neural network is shown in Figure 1.

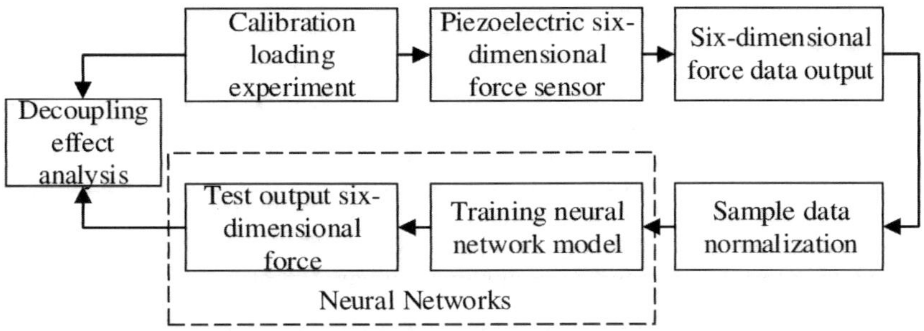

Figure 1. Nonlinear decoupling principle of six-dimensional force sensor based on neural network

The experimental object of this experiment is the four-point support structure piezoelectric six-dimensional force sensor developed by the research group [9]. Through the reasonable arrangement of the four quartz crystal groups, the generalized six-dimensional force is detected. The piezoelectric six-dimensional force sensor was calibrated to obtain the calibration experimental sample data, and then the nonlinear decoupling algorithm model for the piezoelectric six-dimensional force sensor based on BP neural network was designed.

3. Static Decoupling of Piezoelectric Six-Dimensional Force Sensor Based on BP Neural Network

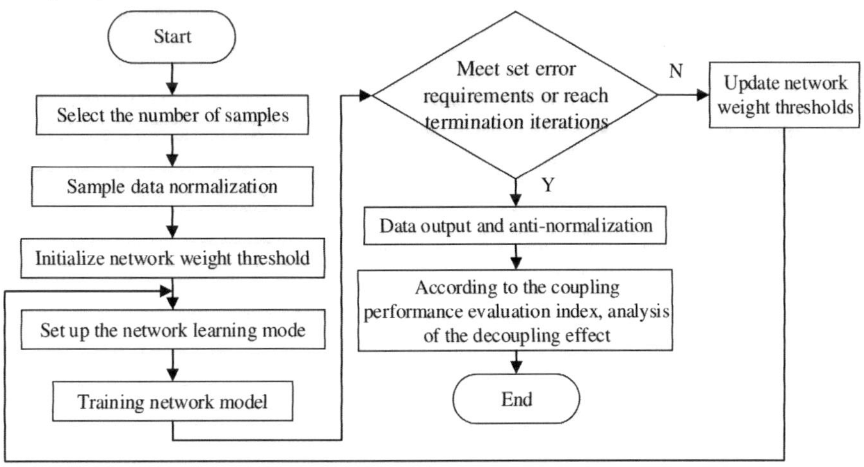

Figure 2. Flowchart of nonlinear decoupling algorithm based on BP neural network

In this paper, a three-layer BP neural network is used to construct the model of decoupling algorithm for six-dimensional force sensor. Each neuron in the next layer is connected to each neuron in the previous layer with a certain weight, and the connection weight between the neurons in the same layer is zero. Connection weight determines the connection strength between neurons. Changing the weight and threshold can change the mapping relationship between input and output of neural network. The

BP neural network model of piezoelectric six-dimensional force sensor is shown in Figure 2.

The S-type (Tansig) function can be used as an activation function for hidden layer cell nodes, and the Purelin function can be used as an activation function for the output layer cell nodes. The six-dimensional output signal sample collected in the loading experiment is used as the input data of the input layer unit node. The actual loading force vector in the loading experiment is used as the output data of the output layer unit node.

When decoupling the forces/torques of each dimension, five sets of experimental data are selected to be normalized and treated as network training samples. A group of data that has not participated in training is normalized and used as a network test sample. The structural model is trained to obtain the right corresponding layer weights and thresholds, so that the mean square error between the predicted value and the expected value meets the set conditions, and the corresponding parameters of the sensor decoupling network model are obtained.

4. Training, Testing and Experimental Data Analysis of BP Neural Network

Using the BP algorithm toolbox that comes with MATLAB software, the network structure model of the BP algorithm that meets the decoupling requirements can be easily constructed, and the weight threshold initialization, data training, and test output work can be quickly completed [11]. In the decoupling model training process, because there is no uniform method to select the number of hidden layer element nodes, it needs to be preset based on experience. The number of hidden layers is initially set to 4-13, and then the network is trained separately. The number of hidden layers is determined by the results of multiple experiments. In the BP neural network model parameter setting, set the number of termination iterations epochs=10000, the momentum term coefficient to mc=0.9, the training speed set to lr=0.05, and the mean square error set to 10^{-5}. Other parameters are default. When the number of hidden layer cell nodes is different, the results of the decoupling effects are listed in Table 1. It can be seen from Table 1that when the number of nodes in the hidden layer is 9, it is appropriate to analyze from the perspective of mean square error, convergence speed, or network complexity. And the training effect is optimal. Figure 3 shows the minimum square root error performance drop curve and error gradient drop curve for the BP feedforward neural network when the hidden layer is 9.At this point, the number of network training steps is only 41 steps, and the mean square error can meet the target requirement of less than 10^{-5}, and the network training ends.

Table 1. Comparison of decoupling performance with different number of hidden layer cell nodes

Hidden node	4	5	6	7	8	9	10	11
Training step length	10000	949	10000	55	70	41	41	74
Training time	124.9	32.8	37.0	37.8	32.1	33.4	37.4	21.5
Detection error	0.0556	0.0139	1.1e-5	9.9e-6	9.8e-6	9.9e-6	9.8e-6	9.9e-6

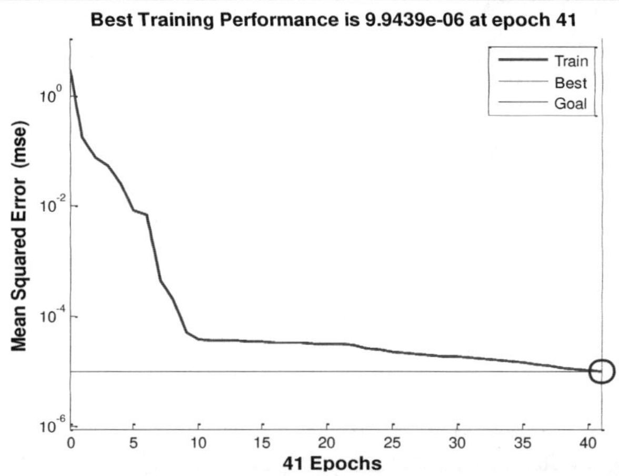

(a) Minimum square root error drop curve

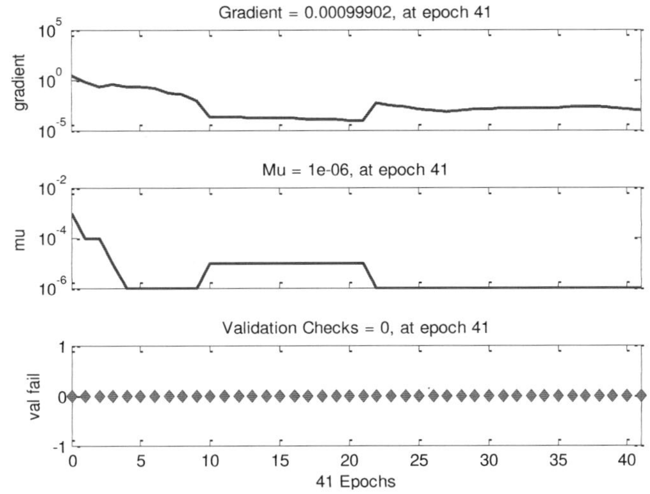

(b) Error gradient descent curve

Figure 3. BP neural network training error performance curve (hidden layer is 9)

5. Decoupling effect analysis of BP neural network decoupling algorithm

Using the neural network structure obtained by training, the test sample is input into the network structure, and partial data of the decoupled input force value F_{BP}' is obtained. According to the data of the applied load value F and the input force value F_{BP}', the performance curve of the six-dimensional force measurement force is shown in Figure 4. It reflects the correspondence between the decoupled applied load value and the input force value F_{BP}'.

When the various force/moment loadings are applied, it can be seen from the figure 4 that the decoupled input force value and the actual expected value curve are nearly completely coincident, and the linearity is good. When a three-dimensional force load is applied, the influence on the coupling output of the three-dimensional torque is small; when three-dimensional torque is applied, the three-dimensional force is coupled and output, but the influence on the three-dimensional force full-scale value is small.

The percentage static coupling rate matrix algorithm is used to verify the decoupling effect of the BP neural network decoupling algorithm. The decoupled input force value F_{BP}' and the percentage static coupling rate matrix E_{BP} are as shown in Equation 1 and 2.

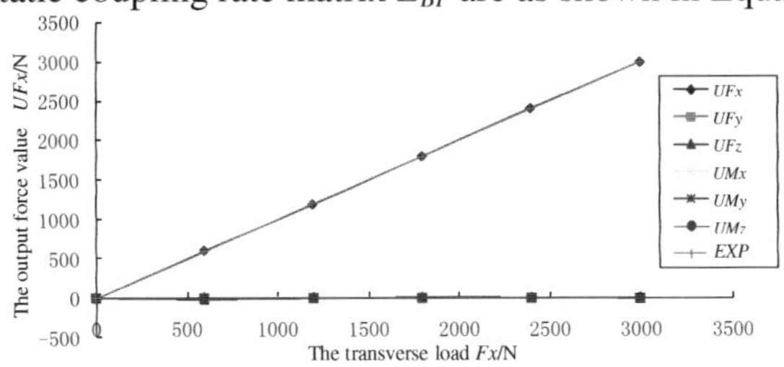

(a) in F_x direction

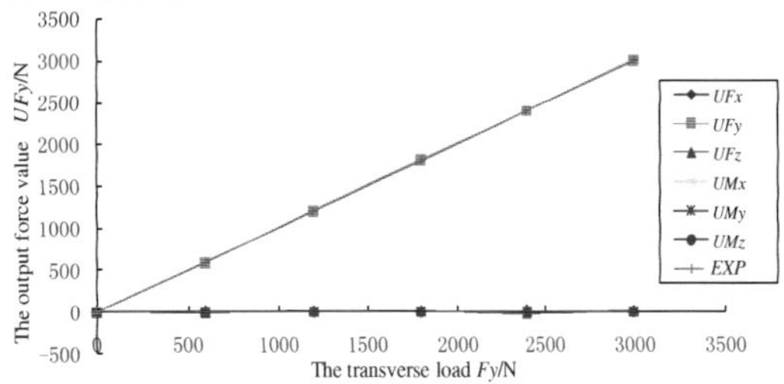

(b) in F_y direction

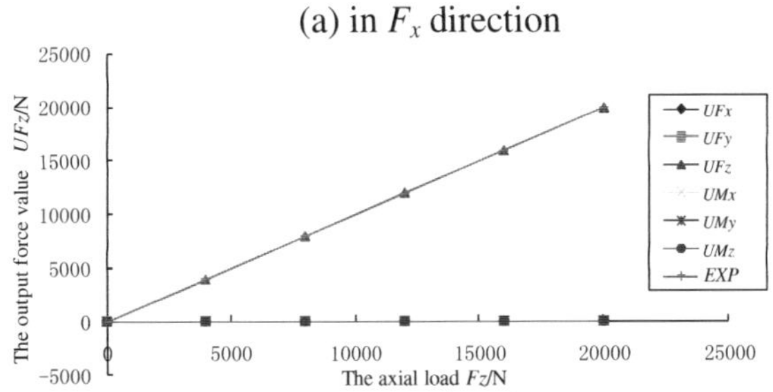

(c) in F_z direction

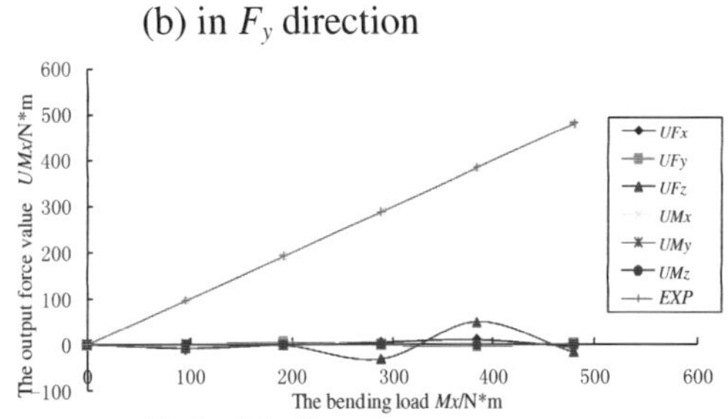

(d) in M_x direction

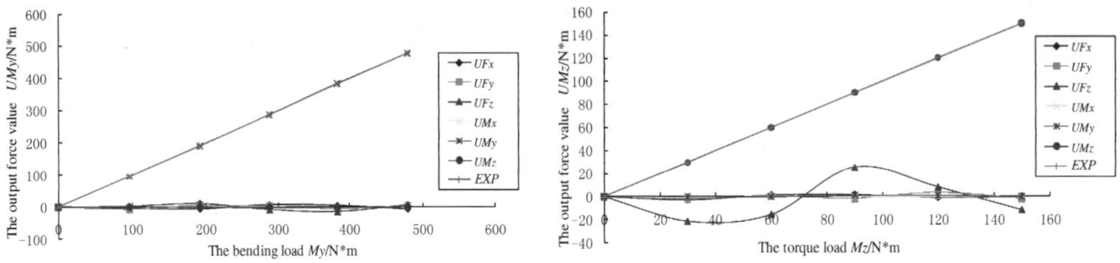

(c) in M_y direction (d) in M_z direction

Figure 4. Curve of the relationship between input loads and output loads after decoupling

$$F_{BP}' = \begin{bmatrix} 3002.49 & -66.84 & -29.38 & -21.46 & -13.31 & -23.03 \\ 20.99 & 2984.98 & 43.45 & 16.20 & 23.08 & -87.75 \\ 115.28 & 64.43 & 20027.1 & 124.00 & -6.61 & 150.15 \\ 4.50 & -2.14 & 5.34 & 479.02 & 2.98 & -12.05 \\ 0.01 & 9.12 & 5.54 & 5.02 & 479.82 & 5.15 \\ -1.65 & -3.27 & -2.13 & -4.14 & -1.86 & 149.35 \end{bmatrix} \quad (1)$$

$$E_{BP} = \begin{bmatrix} 0.0829 & 2.2278 & 0.9792 & 0.7153 & 0.4437 & 0.7677 \\ 0.6998 & 0.5007 & 1.4483 & 0.5399 & 0.7693 & 2.9251 \\ 0.5764 & 0.3222 & 0.1355 & 0.6200 & 0.0331 & 0.7507 \\ 0.9382 & 0.4463 & 1.1133 & 0.2040 & 0.6205 & 2.5102 \\ 0.0023 & 1.8998 & 1.1555 & 1.0461 & 0.0368 & 1.0735 \\ 1.0967 & 2.1815 & 1.4210 & 2.7605 & 1.2369 & 0.4331 \end{bmatrix} \quad (2)$$

From the static coupling rate matrix, after the decoupling of the BP forward feedback neural network, the I-type errors of the six-dimensional force sensor are: F_x is 0.08%, F_y is 0.50%, F_z is 0.14%, and M_x is 0.20%. M_y is 0.04% M_z is 0.43%, That is, the maximum Class I error is 0.50%. When force M_z is loaded, F_y is strongly coupled by M_z, and the inter-dimensional coupling strength is maximum, That is, the maximum class II error is 2.93%. After decoupling by the BP decoupling algorithm, the linearity error and the coupling rate error of the sensor are reduced, and the measurement accuracy of the sensor can be more effectively improved.

6. Conclusion

In this paper, a static calibration experiment was conducted for a four-point support structure piezoelectric six-dimensional force/torque sensor prototype developed by the laboratory, and the reasons for the coupling of the sensors were briefly analysed. A mathematical model based on BP neural network decoupling algorithm was established and decoupled calculations were performed. After compensating the experimental data obtained by the calibration experiment of the sensor, it is verified that the decoupling algorithm based on BP neural network is adopted. Can more accurately reflect the nonlinearity of the sensor system, and to a greater extent improve the measurement accuracy of the sensor. Effectively solve the problem of sensor decoupling difficulties, and provide a basis for the engineering application research of the decoupling method of piezoelectric six-dimensional force sensor.

Acknowledgments

This work is supported by a grant from National Natural Science Foundation of China (GrantNo.51875250), Shandong Provincial Natural Science Foundation, China (Grant No. ZR2017QEE012), Key research and development plan of Shandong Province (Grant No. 2018GGX103028) and the National Natural Science Foundation of China (Grant No.51705200).

IOP Conf. Series: Materials Science and Engineering **428** (2018) 012041 doi:10.1088/1757-899X/428/1/012041

References

[1] Song A. Research on human-robot interaction telerobot with force telepresence. Science & Technology Review, 2015, 33(23):100-109.
[2] Palli G, Moriello L, Scarcia U, et al. Development of an optoelectronic 6-axis force/torque sensor for robotic applications. Sensors & Actuators A Physical, 2014, 220:333-346.
[3] Li D. Research on the Measuring Principle and Performance of Integrated Piezoelectric 6-Axis Force Sensor. Dalian: Dalian University of Technology, 2015.
[4] Li Y J, Wang G C, Zhao D, et al. Research on a novel parallel spoke piezoelectric 6-DOF heavy force/torque sensor. Mechanical Systems & Signal Processing, 2013, 36(1):152-167.
[5] Zhang J, Dezhang X U. Study on Decoupling Algorithms for Six-Axis Force Sensor Based on Ant Algorithm and BP Neural Network. Light Industry Machinery, 2016.34(1):1-4.
[6] Zhao Y, Zhang C, Zhang D, et al. Mathematical Model and Calibration Experiment of a Large Measurement Range Flexible Joints 6-UPUR Six-Axis Force Sensor. Sensors, 2016, 16(8):1271.
[7] Shi Z, Zhao T, Li M, et al. Research on static decoupling of large range flexible joint six-axis force sensor. Chinese Journal of Scientific Instrument, 2012, 33(5):1062-1069.
[8] Zhang J, Wang K, Qian M, et al. Component Decoupling of Piezoelectric Vector Force Sensor and Experiment Research. Instrument Technique & Sensor, 2016, 1:1-3.
[9] Li Y J, Yang C, Wang G C, et al. Research on the parallel load sharing principle of a novel self-decoupled piezoelectric six-dimensional force sensor. Isa Trans, 2017, 70:447-457.
[10] LI Y J, HAN B B, WANG G C, et al. Decoupling algorithms for piezoelectric six-dimensional force sensor based on RBF neural network. Optics and Precision Engineering, 2017, 25(5):1266-1271.
[11] L I H B, Duan Z X, Gao L F, et al. An approach to static calibration for 6-dimensional force sensor based on neural network. Journal of Inner Mongolia University of Technology, 2006, 25(2):85-89.

IOP Conf. Series: Materials Science and Engineering **428** (2018) 012042 doi:10.1088/1757-899X/428/1/012042

Research on An Adaptive Optimization Method of Anti-tank Missile in Plateau Environment

H L Liu[1,4], S Y Dian[1], L Wang[2] and Y H Wu[3]

[1]School of Electrical Engineering and Information, Sichuan University, Chengdu, 610065, China
[2]N0. 78167 Unit of PLA, Chengdu, 610065, China
[3]Sichuan Electric Power Company, Chengdu, 610065, China

[4]809942814@qq.com

Abstract. By analyzing the atmospheric elements of the plateau environment, the main factors affecting the dynamic characteristics of the missile are identified. From the analysis of the frequency domain characteristics of guided missiles and the guidance system parameters, we study the problem of the altitude adaptation of missiles. Comprehensive economy, adaptability, reliability and other factors were proposed to improve the dynamic characteristics of the missile's monomer under high altitude environment. The simulation results show that the method used in this paper can effectively improve the stability of the guidance system at the plateau, and the improved missile performance can basically meet the requirements of the plateau.

1. Introduction

During the trial test of a certain type of anti-tank missile in the plateau, some missiles dropped after launch. This type of anti-tank missile was mainly used in low-altitude plain environmental conditions at the beginning of design. In order to study the mechanism of the occurrence of land-drop and instability of this type of missile under the condition of high altitude, the problem of missile adaptability in high altitude environment was solved. The author considers this type of missile to be adapted to the plateau in terms of reliability, economy, and applicability.

Looking at the literature, we can see that some scholars have proposed a nonlinear method to solve missile ballistic problems[1]. Some researchers started from the conditions of the plateau environment and analyzed in detail the mechanism of the impact of the high-cold environment on missiles, and proposed a simple solution[2][3]. Several of these researchers started with correcting the network to solve the problem better, but the ability to solve complex situations is not enough[4][5]. Based on the above research, this paper starts with the dynamic characteristics of the plateau environment missiles, and analyzes the impact of altitude changes on the dynamic characteristics of missiles to design an improved solution with better reliability, economy, and applicability. Through simulation tests, the improved program meets operational requirements at different altitudes.

2. Plateau environment analysis

The main influencing factors of missile performance on the plateau environment are sound velocity, air density, gravitational acceleration, atmospheric pressure, oxygen content, temperature, and sunshine intensity[6]. The quantitative analysis of the impact of the air density on the dynamic characteristics of the missile is used to obtain the variation of each factor with altitude.

At sea level, temperature T_0, static pressure P_0 are $T_0 = 288.1667(K)$, $P_0 = 101314.628(N/m^2)$.

At an altitude, according to the standard atmospheric model, the highland area of China is located in the first interval, that is, $z \leq 11000m$.

$$T_a = T_0 - (0.006499708)z(K) \tag{1}$$

$$P_a = P_0(1 - 2.255692257 \times 10^{-5} z)^{5.2561}(N/m^2) \tag{2}$$

In this interval, the density and sound velocity of the ambient atmosphere are given by the following formula:

$$\rho_a = P_a / RT_a(kg/m^3) \tag{3}$$

$$V_a = 20.037673\sqrt{T_a}(m/s) \tag{4}$$

R is the gas constant $(286.992360) \times [(N-m)/(K_p - K)]$.It should be noted that the speed of sound V_a can also be obtained from the following relationship:

$$V_a = kRT \tag{5}$$

k is the specific heat ratio of gas(The specific heat ratio of air is 1:4),R is the gas constant, T is the absolute temperature of the standard atmosphere, T_a is the ambient atmospheric temperature at the altitude z, P_a is the atmospheric pressure at altitude z, ρ_a is the ambient atmospheric density at the altitude z, V_a is the speed of sound at altitude z. Through simulation, we can get the curve of each element with altitude, as shown in Figure 1.

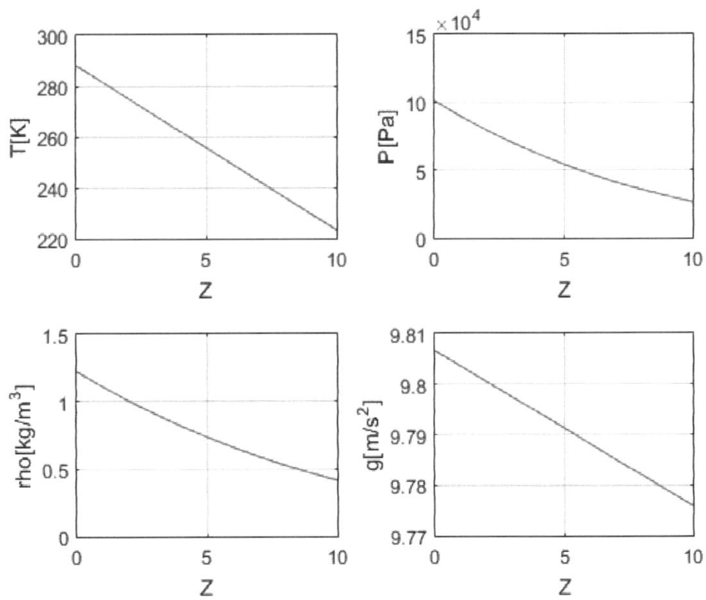

Figure1. Changes in atmospheric characteristics at different altitudes

As can be seen from Figure 1, the greatest differences between atmospheric characteristics of the plateau and the plains are temperature, air density, and static pressure. However, due to the hardware characteristics of the missile, atmospheric pressure and temperature have little impact on the dynamic performance of the missile. Therefore, we mainly analyze the impact of air density changes on the dynamic performance of the missile.

3. Control system principle

The guided missile equipment of the modified missile is guided by the three-point guidance law. In the three-point guidance, it is necessary to constantly manipulate the missile so that it is located between the target tracker and the target. The three-point guidance method involves ground trackers, missiles, and targets. The combination of various functional modules shown in Figure 2 Style and spacing.

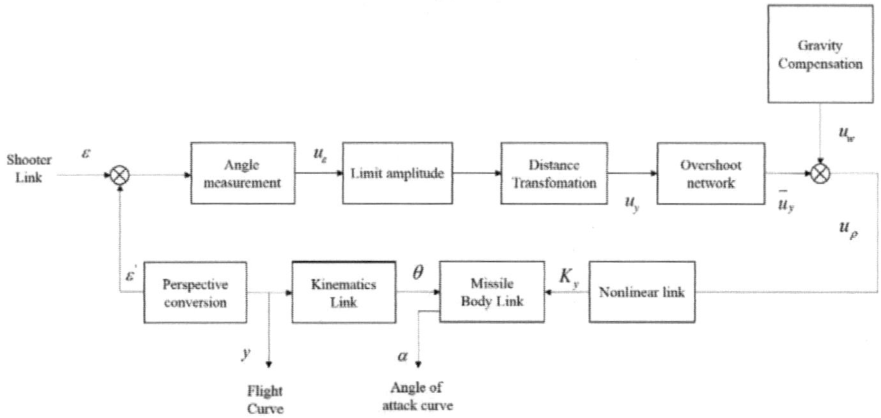

Figure 2. Diagram of guidance link

According to the composition shown in Figure 2, missile-controlled flight trajectories can be constructed. In order to analyze the dynamic characteristics of missiles, a missile mathematical model was established. Due to the missile's axisymmetric layout, only the pitch direction is modeled and analyzed. After simplification and introduction of the kinetic coefficients, while ignoring some minor conditions, such as wash delay, etc., a simplified form of the longitudinal channel equations can be obtained[7].

$$\begin{cases} \ddot{\vartheta} + a_{22}\dot{\vartheta} + a_{24}\alpha = -a_{25}\delta_z \\ \dot{\theta} - a_{34}\alpha = a_{35}\delta_z \\ \vartheta - \theta - \alpha = 0 \end{cases} \tag{6}$$

a_{ij} is the Pitch motion coefficient, $\vartheta, \theta, \alpha$ are respectively the pitch angle, the ball inclination angle and the angle of attack, ,in rad. Pulling the above equation can get the transfer function:

$$\begin{aligned} W_K^{\dot{\vartheta}}(s) &= \frac{\dot{\vartheta}(s)}{\delta_z(s)} = \frac{-a_{25}s - a_{25}a_{34}}{s^2 + (a_{22} + a_{34})s + a_{22}a_{34} + a_{24}} \\ W_K^{\dot{\theta}}(s) &= \frac{\dot{\theta}(s)}{\delta_z(s)} = \frac{-a_{25}a_{34}}{s^2 + (a_{22} + a_{34})s + a_{22}a_{34} + a_{24}} \\ W_K^{\alpha}(s) &= \frac{\alpha(s)}{\delta_z(s)} = \frac{-a_{25}}{s^2 + (a_{22} + a_{34})s + a_{22}a_{34} + a_{24}} \end{aligned} \tag{7}$$

$$s^2 + (a_{22} + a_{34})s + a_{22}a_{34} + a_{24} = 0 \tag{8}$$

From Equation (7), the characteristic equation of each transfer function is Equation (8), with similar system characteristics. Among them, the coefficient a_{22} indicates the aerodynamic damping of the missile, a_{24} indicates the steady state of the missile, a_{25} indicates the efficiency of the elevator, and a_{34} indicates the angular deviation of the tangential rotation of the ballistic tangent caused by the angular deviation of the unit. Their expressions are:

$$a_{22} = \frac{m_z^{\bar{\omega}_z} qSb_A}{J_z} \frac{b_A}{V} \tag{9}$$

$$a_{24} = \frac{m_z^{\alpha} qSb_A}{J_z} \tag{10}$$

$$a_{25} = \frac{m_z^{\delta_z} qSb_A}{J_z} \tag{11}$$

$$a_{34} = \frac{c_y^{\alpha} qs + P}{mV} \tag{12}$$

q is the flow pressure($q = \frac{1}{2}\rho V^2$) and positive correlation with air density. Therefore, dynamic pressure determines the change law of the dynamic coefficient. It can be known from the characteristic equation that the missile's transfer coefficient decreases as the altitude increases. It shows that the control efficiency of the steering gear to the missile decreases with the elevation, and the maneuverability of the missile decreases. The transfer function of the system control object changes, resulting in instability of the control system.

4. Control system altitude adaptability improvement

The addition of autopilot to the original control system is an effective way to improve the dynamic performance of the system, but it is not feasible to consider issues such as cost and redesign. Without changing the hardware, changing the control commands in the control box has low cost and high reliability. According to the components of the control system, at an altitude of 4,000 meters, the open loop gain of the pilot loop is:

$$G_k(s) = [W_\varepsilon(s)W_D(s) + G(s)]W_K^{\dot{\vartheta}}(s)W_Y(s) \tag{13}$$

Bring in related parameters:

$$G_k(s) = \frac{0.0037s^3 + 1430.4s^2 + 3152s + 76.49}{s^3(0.05s+1)(0.045s+1)(0.022s+1)(s^2+1.93s+121.13)} \tag{14}$$

From this we can get the open-loop Bode diagram of the original system, as shown in Figure 3.

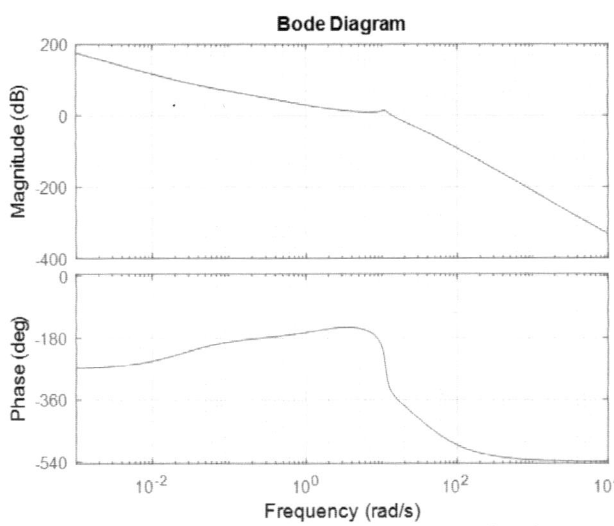

Figure 3. Control System Open Loop Bode Diagram

From figure 3, the open-loop system p=0, the phase-frequency curve crosses the -180° line. According to the stability criterion of the logarithmic phase frequency characteristics, the closed-loop instability of the original control system in the plateau environment can be known. At the same time, during the flight of the missile, the turbulence is larger, the number of times is greater, the stability is not good, and the system adjustment time is too long and the response is slow. Therefore, corrections must be made through the corrective network to ensure that the dynamic characteristics of the system and the attitude of the plain remain stable at high altitudes. The design steps are as follows:

a) Open-loop frequency domain indicators such as slope and height of the lowest frequency band, open-loop cut-off frequency, and stability margin are converted from the required performance indicators (stabilization accuracy, adjustment time, and overshoot) of the missile.

b) Use the tool Sisotool of Matlab to adjust the open loop type and gain. Constantly try to make adjustments to change the shape of the high-frequency band in the open-loop Bird map to meet the requirements for stability, rapidity, and others (such as resistance to high-frequency interference).

c) Integrate correction network parameters for various time periods to obtain a suitable correction network function.

According to the flight characteristics of the missile, adjustments need to be made in two stages. The first stage is the limitation of the additional angle of attack on the adjustment instructions in the early period of the ballistics, reducing the system oscillation during the disturbance stage and ensuring a stable flight path. The second paragraph is the stability trajectory, the normal rotation speed does not need to carry on the big adjustment, the overload has reduced, returned to normal, therefore no longer limits the additional angle of attack. Through Sisotool try to figure out the best PD correction network parameters, get the time 3s when the PD correction network is shown by equation (15), and the open-loop system correction is shown in figure 4. Compared with figure 3, it can be seen that the system tends to be stable and the adjustment time and response speed are improved.

$$C=0.00012975\times\frac{1+51s}{1+0.66s} \tag{15}$$

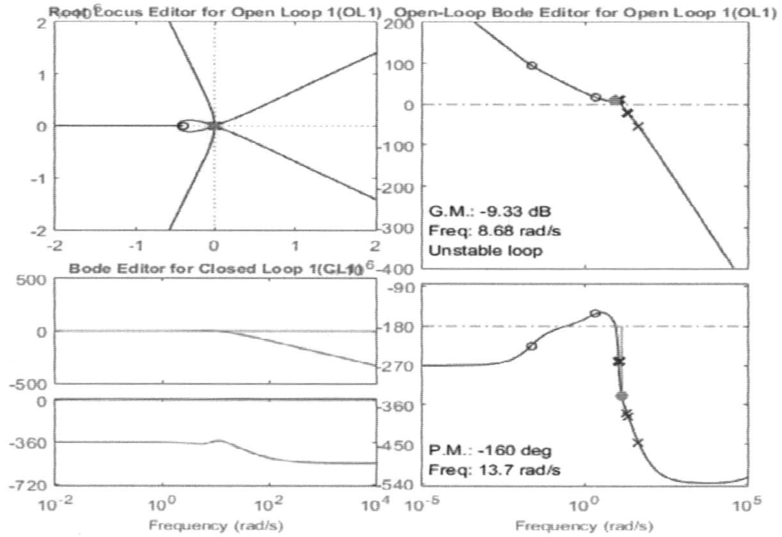

Figure 4. Corrected system dynamic characteristic diagram

In the same way, the corrected network parameters of each time period during which the missile is launched to hit the target can be found. After the integration, it can be found that the correction network of the entire process is a time-varying variable coefficient lead correction network.

$$G_c = J_t \frac{is+1}{js+1} \tag{16}$$

Where J_t is the gain function changes with time. The i and j satisfy equation (17)

$$j = \frac{1}{\omega_n \sqrt{i/j}} \tag{17}$$

Through this network ballistic simulation, as shown in figure 5, when the missile is flying at high altitude, the closed loop system re-adjusts the re-filling instructions and adjustment instructions in the control law, and the flight trajectory tends to be flat to the line of sight and does not diverge.

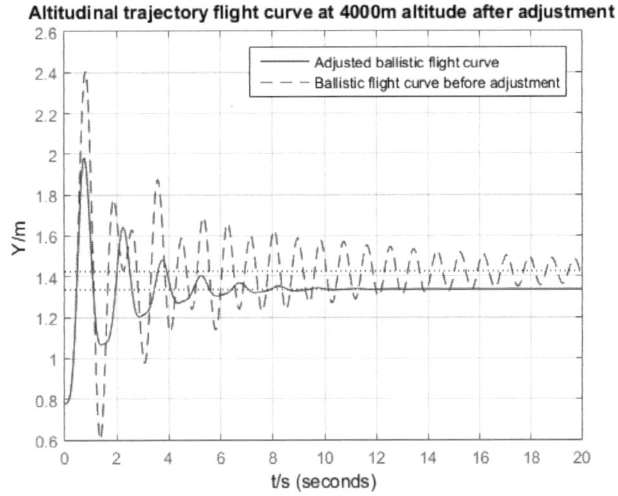

Figure 5. Adjust the vertical surface ballistic flight curve at 4000m above sea level

5. Conclusion

Through the analysis of the flight characteristics of the plateau missiles, taking the plateau 4000m as an example, the original control system was adapted to the plateau. The simulation results prove the feasibility of the improved flight control law of the missile plateau, which makes the high-altitude combat performance of this type of missile more significant.

References

[1] Zhu Boli, Yang Shuxing, Chang Hao. A non-linear control method to improve the adaptability of anti-tank missiles at high altitude [J]. Journal of Beijing Institute of Technology, 2007, Vol.27, No.3, pp.235-239.

[2] Guo Xiwei, Zhao Shen, He Peng. Control System Simulation Of an Anti-tank Missile ［J］.Tactical Missile Technology, 2015, No.3, pp.101-110.

[3] Zhang Yue, Tang Sheng-jing. Flight Characteristics Analyses and Adjustment of the Thrust Vector Control Missile at an Elevated Plain[J]. Journal of Projectiles, Rockets, Missiles and Guidance.2005, Vol.26, No.1, pp.711-716.

[4] Feng Tianyu, Zhang Jing, Shi Bin. Analysis and Improvement of Dynamic Characteristics of Anti-tank Missile in Plateau Environment[J].Tactical Missile Technology, 2014,No.1,pp.27-31.

[5] Cai Pu, Xu Hualong, Xia Kehan. Modification of Corrective Network Used in Attitude Control System for Missile[J].Missile and Space Vehcile, 2005,No.1,pp.10-15.

[6] Wu Xun. Inference of High Altitude Environment to Solid Missile[J].Equipment Environmental Engineering,2006,Vol.3,No.3,pp.102-105.

[7] Lu Xiaodong, Guo Jianguo, Lin Peng, Liu Yingying. Design of Missile Guidance and Control System and Matlab Simulation[M]. Northwestern Polytechnical University.

IOP Conf. Series: Materials Science and Engineering **428** (2018) 012043 doi:10.1088/1757-899X/428/1/012043

Age Classification Using Convolutional Neural Networks with the Multi-class Focal Loss

Wei Liu[1], Lin Chen[2], and Yajun Chen[3,*]

[1]School of Computer Science & Technology, China West Normal University, No.1, Shida Road, Shunqing District, Nanchong, China
[2]Chongqing Institute of Green and Intelligent Technology, Chinese Academy of Sciences, No.266 Fangzheng Avenue, Beibei District, Chongqing, China
[3]Education and Information Technology Center, China West Normal University, No.1, Shida Road, Shunqing District, Nanchong, China

[*]Corresponding author: scnccyj@cwnu.edu.cn

Abstract. Automatic age classification has drawn significant interest in plenty of applications such as access control, human-computer interaction, law enforcement and surveillance. Automatic age classification is a challenging task due to the complexity of facial images. A large number of approaches have been investigated on unconstrained datasets. However, most of these approaches have focused on the network architecture rather than the distribution of data, i.e., the extreme class imbalance existing among different age groups as the difficulty of data collection. In this paper, we propose a convolutional neural networks model based on the multi-class focal loss function. Specifically, our approach is designed to address the class imbalance via reshaping the standard cross entropy loss that it down-weights the loss assigned to well-classified examples. We validate our approach on well-known Adience benchmark. Finally, the experimental analysis shows that the proposed model achieves a significant improvement in performance for age classification.

1. Introduction

Human face conveys many kinds of important information, such as identity, expression, emotion, gender, age, etc. With the rapid emergence of the intelligent applications, there is an increasing demand for automatic extraction of facial attributes. As one of the key facial attributes, age plays a fundamental role in social interactions. Therefore, automatic age estimation from facial images is an important and challenging task investigated in numerous applications, including access control, human-computer interaction (HCI), law enforcement and surveillance [1].

Automatic age estimation aims to assign a label to a face image regarding the exact age or the class of age it belongs. Most of the previous approaches for age estimation are intended for accurate estimation of the actual age. However, as a result of the speciality of aging effects on the face, it is difficult to estimate an actual age. There are different aging patterns among different people, which are influenced by many internal and external factors, including genes, ethnicity, health condition, lifestyle and environment [2]. Therefore, the aging progress is uncontrollable, making age estimation challenging for humans in some cases, and even greater challenging for computer vision systems [3]. Furthermore, the ability of automatic age estimation is influenced by additional factors, such as lighting conditions, facial expressions and pose variations.

In recent years, the automatic age classification of facial images has been investigated extensively[4]. In particular, convolutional neural networks (CNNs) have attracted substantial research attention [5, 6], which are capable of learning a compact and discriminative feature representation with large-scale data. As a consequence, CNNs achieved a great success in the vision community, significantly improving the performance in classification problems.

In this study, we concentrate on the issue of age group classification rather than that of exact age estimation. Though extensive approaches for age classification have been proposed, most of them focused on constrained images such as FG-NET [7] and MORPH [8]. Unlike these methods discussed above, we devote to unconstrained images and study the issue of extreme class imbalance among different age categories: there are a great number of face images in the age group of 25-32, while a few face images belong to the older groups. In this paper, we propose a CNN model based on the multi-class focal loss function, to efficiently improve the performance of age classification. Our main contributions are summarized as follows:

- We propose a CNN model based on the multi-class focal loss function to address the imbalance of different classes for age classification.
- Experimental results show that the proposed approach could benefit the class imbalance problem in the CNN based classification. Despite of the very challenging nature of the images in the Adience dataset, our approach achieves significant improvements in age classification accuracy over the state-of-the-art methods.

The remainder of this paper is organized as follows: Section 2 briefly reviews the related work. Section 3 illustrates the proposed method and network architecture in detail. The experimental results are given in Section 4. Finally, conclusions are drawn in Section 5.

2. Related work

A comprehensive review of age representation methods from facial images was presented in [9]. Kwon et al. [10] proposed a method to classify images into different age categories by calculating ratios of different measurements based on facial features. However, this method may not be suitable for in-the-wild images which have a large of variations in pose, illumination, expression and occlusion. Geng et al. [11] proposed an automatic age estimation method named AGES, which constructs a representative subspace and models the face pattern in time order as a sequence of the individual face images. However, this method demands for face images of a person at different ages. In order to address this problem, Fu et al. [12] proposed a novel manifold space that analyzes and learns the face manifold for different age stages to find a low-dimensional embedding space.

Feature extraction is a crucial step in automatic age estimation. Several methods for feature extraction have been proposed, for instance, Active Appearance Model (AAM) [13], Local Binary Patterns (LBP) [14-16], Anthropometric Features [17] and Biologically-Inspired Features (BIF) [18]. In addition, regression and classification methods are utilized to estimate the accurate age or the category of age from facial images. SVM was utilized for age classification in [19]. Some methods for regression were utilized to predict the accurate age, such as SVR [20], linear regression [21] and CCA [22]. Even though all of these methods have proven effective on constrained benchmarks, they could fail to tackle large variations in unconstrained images.

Recently, increasing attention has been investigated on CNN for age estimation from facial images. Yi et al. [23] proposed an age estimation method based on Multi-Scale CNN to learn parameters instead of hand-crafted. Levi et al. [5] proposed a simple CNN for age classification on unconstrained Adience benchmark, to effectively avoid the overfitting problem. Ranjan et al. [24] proposed an approach based on deep convolutional neural networks (DCNNs) for age estimation on unconstrained challenge dataset, and outperformed conventional methods. Residual Networks of Residual Networks (RoR) [25] was capable of tackling large variation in the wild for age classification. Chen et al. [26] proposed ranking-CNN via taking the ordinal relation between ages into consideration for age estimation.

To summarize, CNNs have achieved great success on age classification task, significantly improving the performance of classification. Most of the existing work improved classification accuracy via modifying the network architecture. However, the age estimation with imbalanced distribution has not been investigated in depth so far. In this study, we propose a CNN model based on the multi-class focal loss function, which achieves better results than state-of-the-art methods.

3. Proposed method

In this section, we elaborate our multi-class focal loss and describe the details of the network architecture.

3.1. Multi-class focal loss

The original focal loss [27] was designed to address the class imbalance for binary classification but it can be extended to address the multi-class classification problems. For multi-class issue, the Cross Entropy (CE) loss for an example is given in equation (1):

$$L_{CE} = -\sum_{i=1}^{C} t_i \log(y_i) \quad (1)$$

where C denotes the number of categories, t_i denotes a real probability distribution, y_i denotes a probability distribution of the prediction. As shown in equation (2), where $t_i = 1$ if i belongs to the true label, else it is 0.

$$t_i = \begin{cases} 1 & (i = true\ label) \\ 0 & (i \neq true\ label) \end{cases} \quad (2)$$

The Adience dataset is split into eight age categories. Moreover, there is a great difference in the number of different categories. It is difficult to classify the age groups of 15-20, 38-43, and 48-53. In contrast, the age groups of 0-2 and 25-32 are more easily classified.

Our multi-class focal loss is designed to address the class imbalance via down-weighting easy examples such that their contribution to the total loss is small even if their number is large. In other words, it focuses training on minority examples. A method for addressing class imbalance is to add a modulating factor $(1-y_i)^\gamma$ to the cross entropy loss, with adjustable focusing parameter $\gamma \geq 0$. We define the multi-class focal loss, as formulated in equation (3):

$$L_{fl} = -\sum_{i=1}^{C} (1-y_i)^\gamma t_i \log(y_i) \quad (3)$$

There are two parameters related to the multi-class focal loss. In the first place, a focusing parameter γ is utilized to control the rate at which easy examples are down-weighted. Especially, when $\gamma = 0$, L_{fl} is equivalent to L_{CE}. The effect of the modulating factor $(1-y_i)^\gamma$ is increased with γ. Secondly, when an example is misclassified and y_i is small, the modulating factor nearly tends to 1, and thus the loss is unaffected. As y_i increases to 1, the modulating factor nearly tends to 0 and the loss for well-classified examples is down-weighted. We adopt the multi-class focal loss in our experiments with $\gamma = 1.5$, which achieve significant improvements in age classification accuracy.

3.2. Network architecture

Our experiments for age classification are implemented using CNN with the multi-class focal loss. As depicted in Figure.1, the network contains three convolutional layers and two fully connected layers. The choice of the simple model aims to reduce the risk of overfitting [28]. Unlike the network architecture proposed in [5], we adopt the multi-class focal loss function instead of softmax loss function to effectively address class imbalance. During training, the input image is resized to 256×256 pixels, and then a random crop of 227×227 pixels is fed to the network. Moreover, the image is passed through three convolutional layers as follows:

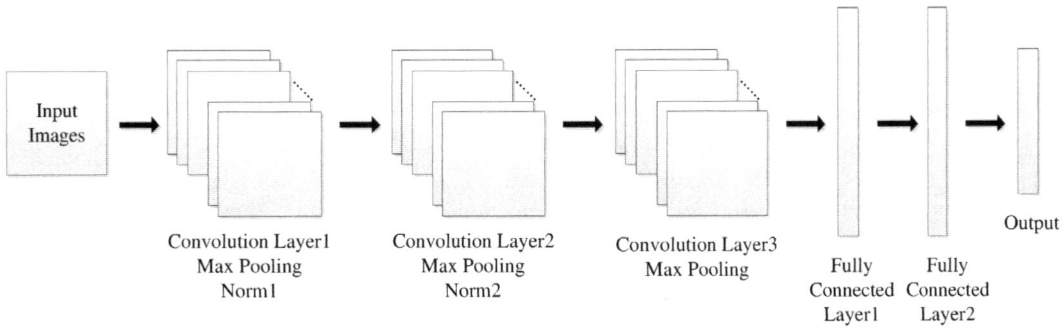

Figure 1. The CNN architecture for age classification.

The first two convolutional layers are followed by a rectified linear unit (ReLU) as the activation layer, a max pooling layer and a local response normalization layer [6] with the same hyper parameters, respectively. Then, the third convolutional layer is followed by only a ReLU and a max pooling layer. Moreover, all max pooling layers are taken the maximal value of 3x3 regions with two-pixel strides. Furthermore, spatial padding is utilized in the second and the third convolutional layers, to preserve the resolution after the convolution operation.

To reduce the risk of overfitting, dropout learning [29] is utilized after two fully connected layers. The remaining architecture of the network is as follows:

The configuration of two fully connected layers is consistent, which contains 512 neurons, followed by a ReLU and a dropout layer at a dropout ratio of 0.5. In the final layer, the output from the previous layer of size 512 features is densely mapped to 8 neurons for age classification. A soft-max with the multi-class focal loss layer is adopted to obtain a probability for each class. More details of the network architecture are given in Table 1.

Table 1. Detailed architecture of the CNN for age classification.

	Kernel size	Stride	Pad	Output size
Input	-	-	-	227×227×3
Conv1	7×7	4	-	56×56×96
Pool1	3×3	2	-	28×28×96
Conv2	5×5	1	2	28×28×256
Pool2	3×3	2	-	14×14×256
Conv3	3×3	1	1	14×14×384
Pool3	3×3	2	-	7×7×384
FC1	-	-	-	512
FC2	-	-	-	512
Output	-	-	-	8

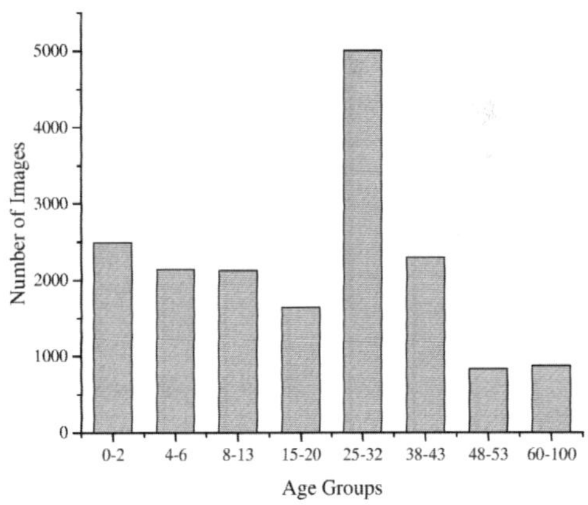

Figure 2. Images distribution of age groups.

4. Experiments

Our approach is implemented using Caffe [30], an open-source framework. Our implementations are based on AMD Ryzen 7 1800X Eight-Core 3.6GHz processor and Nvidia GeForce GTX 1080 Ti GPU.

4.1. Data preprocessing and Training

In order to decrease the influence of the background of images, the aligned images with 816x816 pixels are cropped to 256x256 pixels around the face center. The weights are initialized using random values from a zero mean Gaussian with standard deviation of 0.01. It must be noted that none of pre-trained models are utilized to initialize the network. Moreover, the network is trained without using any additional data. We adopt stochastic gradient descendent (SGD) with a mini-batch size of 64 for training. The learning rate starts from 0.001, reduced to 0.0001 after 10000 iterations. Furthermore, the maximum number of iterations is 50000.

4.2. The Adience Dataset

We conduct experiments using the unconstrained Adience benchmark [15] for age classification. The Adience dataset consists of images automatically uploaded to Flickr from iPhone 5 or later smart-phones without prior manual filtering. Compared with other unconstrained datasets, the Adience dataset is greater variable in terms of pose, lighting conditions and quality. The entire Adience contains over 26,000 images of 2,284 subjects.

In this study, we utilize the in-plane aligned version of Adience. Figure.2 shows the distribution of images from different age groups. And eight groups are defined with the age ranges as same as [5]. Training and testing are carried out using a standard 5-fold, subject-exclusive cross validation protocol, defined in [15]. Besides, experiments are implemented using the aligned and cropped faces, respectively.

4.3. Results

To evaluate the performance of the proposed approach in age classification, we utilize two different criterions, including exact accuracy and 1-off accuracy. Exact accuracy rate is the exact age-group classification result, while 1-off accuracy is off by one adjacent age-group i.e., the subject belongs to the group immediately older or immediately younger than the predicted group. To our best knowledge, facial features may change very little between oldest faces in one age class and the youngest faces of the subsequent class. We compare our approach against some state-of-the-art methods for age classification on the Adience dataset: 1) [15] proposed dropout-SVM to avoid overfitting for age estimation. 2) [5] adopted the same CNN model but it was based on the softmax loss function.

As can be seen from Table 2, the results listed are the mean accuracy ± standard error all age categories. The reported exact and 1-off accuracies in [15] are 45.1±2.6 and 79.5±1.4, respectively. The reported exact and 1-off accuracies in [5] are 50.7±5.1 and 84.7±2.2, respectively. These low accuracies of facial images demonstrate the difficulty involved with the Adience dataset. Finally, with the multi-class focal loss, our approach achieves the best results, not only improving the accuracy of aligned faces, but also improving the accuracy of cropped faces. It is evident that the performance of all models has been improved via preprocessing data.

Figure.3 provides a few examples of the age classification errors made by our approach on the unconstrained benchmark. It is clear that most of the mistakes are caused by extremely challenging viewing conditions of the Adience benchmark, such as low resolution, lighting conditions, and pose variations.

Table 2. Age classification results on the Adience benchmark.

Method	Exact (%)	1-off (%)
Best from[15]	45.1 ± 2.6	79.5 ± 1.4
Best from[5]	50.7 ± 5.1	84.7 ± 2.2
Our approach with aligned faces	53.0 ± 6.5	87.1 ± 2.7
Our approach with cropped faces	54.0 ± 6.4	88.2 ± 2.3

Figure 3. Age misclassifications. The younger is mistakenly classified as the older in the top row, the older is mistakenly classified as the younger in the bottom row.

5. Conclusions

In this paper, we consider class imbalance as the primary issue on the Adience benchmark for age classification. To address this issue, we have proposed a new CNN model with the multi-class focal loss function to improve the performance of age classification, which applies a modulating term to the cross entropy loss to focus learning on minority examples. Extensive experiments on the Adience benchmark have demonstrated the effectiveness of the proposed approach. The absence of large scale training data could be an issue in age classification task. In future work, we plan to investigate the performance of more complex network architecture using more sufficient data.

Acknowledgments

This work is supported by the National Natural Science Foundation of China (Grant Nos. 61602434 and 91646114), Youth Innovation Promotion Association CAS, No. 2017393, Talent Foundation of China West Normal University (No.463177), Chongqing key standard technologies innovation program of key industries under grant cstc2017zdcy-zdyfX0076 and Social Livelihood Foundation of Chongqing (No. cstc2017shmsA120010).

References

[1] Rothe R, Timofte R and Gool L V 2016 Deep Expectation of Real and Apparent Age from a Single Image Without Facial Landmarks *International Journal of Computer Vision* 1-14

[2] Geng X, Zhou Z H, Zhang Y, Li G and Dai H 2006 Learning from facial aging patterns for automatic age estimation. In: *ACM International Conference on Multimedia, Santa Barbara, Ca, Usa, October,* pp 307-16

[3] Tistarelli M, Yadav D, Vatsa M and Singh R 2013 *Short- and long-time ageing effects in face recognition*: IET Digital Library)

[4] Ramanathan N, Chellappa R and Biswas S 2009 Age progression in Human Faces : A Survey *Journal of Visual Languages & Computing* **15**

[5] Levi G and Hassncer T 2015 Age and gender classification using convolutional neural networks. In: *Computer Vision and Pattern Recognition Workshops,* pp 34-42

[6] Krizhevsky A, Sutskever I and Hinton G E 2012 ImageNet classification with deep convolutional neural networks. In: *International Conference on Neural Information Processing Systems,* pp 1097-105

[7] Panis G and Lanitis A 2015 An Overview of Research Activities in Facial Age Estimation Using the FG-NET Aging Database *Journal of American History* **84** 455-62

[8] Ricanek K and Tesafaye T 2006 MORPH: a longitudinal image database of normal adult age-progression. In: *International Conference on Automatic Face and Gesture Recognition,* pp 341-5

[9] Fu Y, Guo G and Huang T S 2010 Age Synthesis and Estimation via Faces: A Survey *IEEE Transactions on Pattern Analysis & Machine Intelligence* **32** 1955-76

IOP Conf. Series: Materials Science and Engineering **428** (2018) 012043 doi:10.1088/1757-899X/428/1/012043

[10] Kwon Y H and Lobo N D V 1999 Age Classification from Facial Images. In: *Computer Vision and Pattern Recognition, 1994. Proceedings CVPR '94., 1994 IEEE Computer Society Conference on,* pp 762-7
[11] Geng X, Zhou Z H and Smithmiles K 2007 Automatic age estimation based on facial aging patterns *IEEE Transactions on Pattern Analysis & Machine Intelligence* **29** 2234-40
[12] Fu Y, Xu Y and Huang T S 2007 Estimating Human Age by Manifold Analysis of Face Pictures and Regression on Aging Features. In: *IEEE International Conference on Multimedia and Expo,* pp 1383-6
[13] Cootes T F, Taylor C J, Cooper D H and Graham J 1995 Active Shape Models-Their Training and Application *Computer Vision & Image Understanding* **61** 38-59
[14] Gunay A and Nabiyev V V 2008 Automatic age classification with LBP. In: *International Symposium on Computer and Information Sciences,* pp 1-4
[15] Eidinger E, Enbar R and Hassner T 2014 Age and Gender Estimation of Unfiltered Faces *IEEE Transactions on Information Forensics & Security* **9** 2170-9
[16] Ylioinas J, Hadid A and Pietikäinen M 2012 Age Classification in Unconstrained Conditions Using LBP Variants. In: *International Conference on Pattern Recognition,* pp 1257-60
[17] Gunay A and Nabiyev V V 2007 Automatic Detection of Anthropometric Features from Facial Images. In: *Signal Processing and Communications Applications, 2007. Siu 2007. IEEE,* pp 1-4
[18] Guo G, Mu G, Fu Y and Huang T S 2009 Human age estimation using bio-inspired features. In: *Computer Vision and Pattern Recognition, 2009. CVPR 2009. IEEE Conference on,* pp 112-9
[19] Cao D, Lei Z, Zhang Z, Feng J and Li S Z 2012 Human Age Estimation Using Ranking SVM. In: *Chinese Conference on Biometric Recognition,* pp 324-31
[20] Guo G, Fu Y, Dyer C R and Huang T S 2008 Image-Based Human Age Estimation by Manifold Learning and Locally Adjusted Robust Regression *IEEE Transactions on Image Processing* **17** 1178-88
[21] Fu Y and Huang T S 2008 Human Age Estimation With Regression on Discriminative Aging Manifold *IEEE Transactions on Multimedia* **10** 578-84
[22] Guo G and Mu G 2013 Joint estimation of age, gender and ethnicity: CCA vs. PLS. In: *IEEE International Conference and Workshops on Automatic Face and Gesture Recognition,* pp 1-6
[23] Yi D, Lei Z and Li S Z 2014 Age Estimation by Multi-scale Convolutional Network. In: *Asian Conference on Computer Vision,* pp 144-58
[24] Ranjan R, Zhou S, Chen J C and Kumar A 2015 Unconstrained Age Estimation with Deep Convolutional Neural Networks. In: *IEEE International Conference on Computer Vision Workshop,* pp 351-9
[25] Zhang A K, Guo B L, Gao C C, Zhao D Z, Sun E M and Yuan F X 2017 Age group classification in the wild with deep RoR architecture. In: *IEEE International Conference on Image Processing,* pp 1587-91
[26] Chen S, Zhang C, Dong M, Le J and Rao M 2017 Using Ranking-CNN for Age Estimation. In: *IEEE Conference on Computer Vision and Pattern Recognition,* pp 742-51
[27] Lin T Y, Goyal P, Girshick R, He K and Dollar P 2017 Focal Loss for Dense Object Detection 2999-3007
[28] Aydogdu M F, Celik V and Demirci M F 2017 Comparison of Three Different CNN Architectures for Age Classification. In: *IEEE International Conference on Semantic Computing,* pp 372-7
[29] Hinton G E, Srivastava N, Krizhevsky A, Sutskever I and Salakhutdinov R R 2012 Improving neural networks by preventing co-adaptation of feature detectors *Computer Science* **3** págs. 212-23
[30] Jia, Yangqing, Shelhamer, Evan, Donahue, Jeff, Karayev, Sergey, Long and Jonathan 2014 Caffe: Convolutional Architecture for Fast Feature Embedding 675-8

IOP Conf. Series: Materials Science and Engineering **428** (2018) 012044 doi:10.1088/1757-899X/428/1/012044

Topology Design and Sliding Mode Controller Design for a Novel High Step-Up DC-DC Converter

Yicheng Liu, JiePing Wang, Hao Chen, Haiyan Tu*

College of Electrical Engineering and Information Technology, Sichuan University, Chengdu 610065, China

*Corresponding author. Email: haiyantu@163.com

Abstract. This paper discusses a high voltage gain boost converter with enhanced conversion efficiency. A passive clamped circuit is applied to the switch of quadratic boost converter. A step-up capacitor and resonant inductor is adopted to connected in series with the clamp capacitor through diodes. Those constitute a voltage multiplier cell that can absorb stray energy to clamp capacity, reduce the stresses of switch and diodes, and promote voltage gain. Furthermore, a sliding mode control (SMC) scheme with constant frequency is addressed for this converter. It is shown that the tracking error can converge to the equilibrium point against input voltage disturbance and load variation. Numerical simulations are given to verified the theoretical considerations.

1. Introduction

Recently, many industrial applications, for example, renewable energy sources, uninterruptible power supply(UPS), electric vehicles [1-3] develop rapidly. Those are increasing the demand of high performance dc-dc converters. In these applications, the static voltage gain and efficiency of the converter need to be high respectively. However, basic boost converter is hard to provide such a high static voltage gain even for an extreme duty cycle which probably leads to high power losses, and a severe reverse-recovery problem [4].

Many researches have devoted to increase static voltage gain and conversion efficiency. Such as, switched inductor, switched capacitor, coupled-inductor, cascaded structure, etc [5-9]. Those can present higher voltage gain than the conventional boost converter. But more switched-inductor or switched capacitor stages are adopted to obtain a large voltage gain, resulting additional cost and complex circuit. The quadratic boost converter using a single active switch is another topology. That can provide a higher voltage gain than the previously mentioned converters and less components are needed. However, the overlarge voltage stress still exists which reduces efficiency and causes additional cost. In some isolated converters and the converters with coupled-inductor [10], the ultra voltage gain is easily obtained with a proper duty cycle, but the leakage inductor of the coupled-inductors and transformers may cause high voltage spikes and induce energy losses. Moreover, the power transformer and coupled inductor's volume and weight are obstacles for the development of a compact converter. Clamp techniques are adopted to reduce voltage stress of power devices in high voltage level applications [11]. A voltage multiplier cell is also alternative choice [12]. This cell can not only improve voltage gain but also alleviate voltage stress.

Due to the variational input voltage of dc-dc converters, a control strategy is necessary to stabilize it against load variations, input voltage disturbances and parameter uncertainties. Many classical linear methods can't guarantee the stabilization or achieve good performance due to the strong nonlinear property of DC-DC converters. Thus, many nonlinear control algorithms have been implemented in DC-DC converters, for example, neural network control, adaptive control, sliding mode control(SMC) [13-15]. SMC is inherently suitable to DC-DC converters because the two switching states of DC-DC converters are similar to switch operation of SMC. Moreover, SMC is characterized by strong robustness, good stability and fast transient response and ease of implementation. Thus, SMC has been widely applied to DC-DC converters. In earlier research, a boundary layer was adopted to alleviate chattering with a proper frequency [16]. However, the switching frequency was still variational. This can cause excessive power loss, EMI problem and filter design complication. To overcome this drawback, a fixed frequency SMC with linear sliding mode was proposed [17-19]. This controller can ensure a constant operation frequency against external disturbance.

This paper aims to improve the output voltage gain and efficiency of boost converter. First, A passive clamped circuit is applied to the switch of quadratic boost converter in [8] to release voltage stress. Second, a step-up capacitor and resonant inductor is connected in series with the clamp capacitor through diodes to improve voltage gain and promote efficiency. Moreover, a fixed frequency SMC scheme with reaching law is designed for this converter.

2. Operational principles of the proposed converter

The simplified equivalent circuit of the proposed converter can be shown by Figure 1. In this figure, Q is switch; L_1, L_2, L_r are input inductors and snubber inductor respectively; C_1, C_o, C_r, C_c are capacitors respectively; D_1, D_2, D_3, D_4, D_o are diodes respectively; R is load. To simplify steady state analysis, many assumptions are as follows:

- All components are ideal.
- L_1, L_2 are large enough so that the converter works under current continues mode (CCM).
- C_1, C_o are also sufficiently large such that the voltage across it is considered to be a constant.

Based on the mentioned assumptions, the relative operating modes are presented by Figure 2-3, and the key theoretical waveforms in one period are shown by Figure 4. This converter mainly includes five operation modes. It restarts the whole cycle and repeats the operation in mode 1 after mode 5 ended.

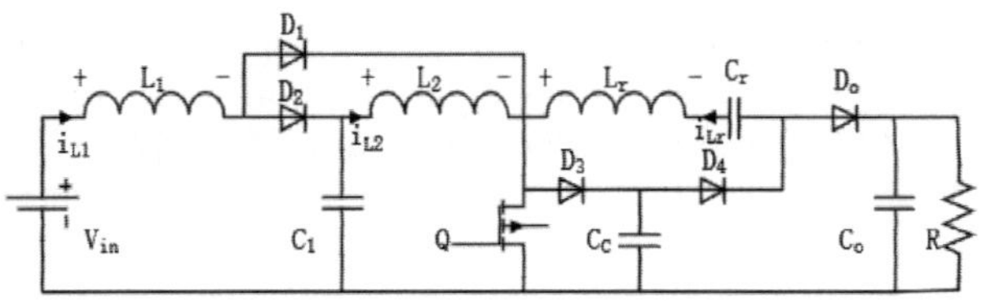

Figure 1. The equivalent circuit of the proposed converter

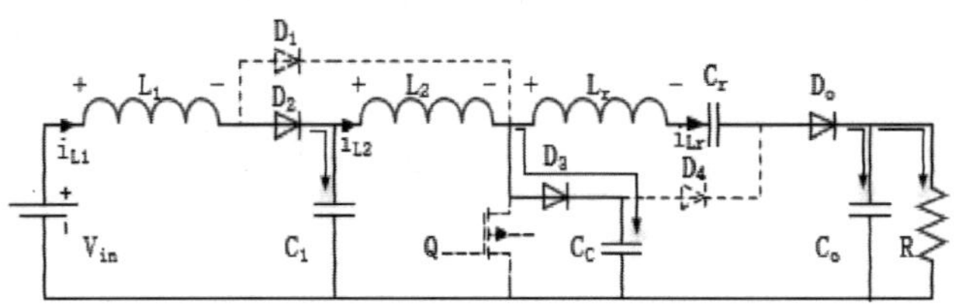

Figure 2. Mode 1

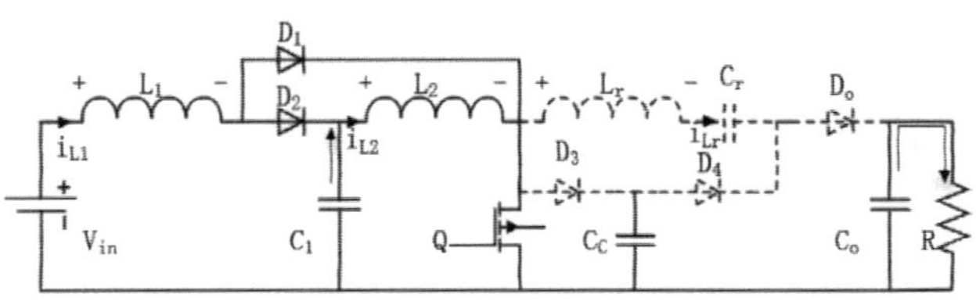

Figure 3. Mode 2

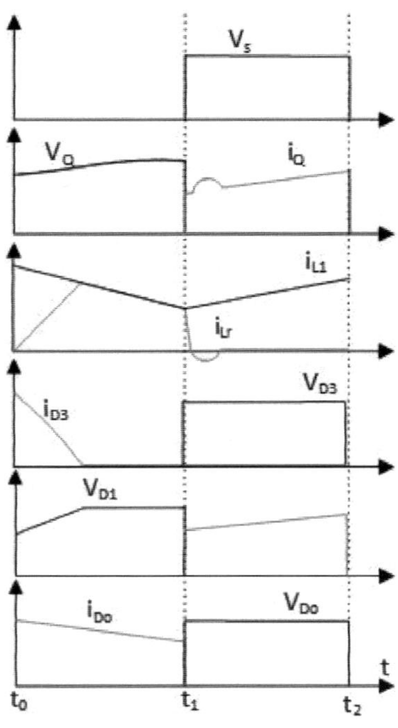

Figure 4. Key theoretical waveforms

1) Mode 1 ([t_0-t_1], Figure 2): at the instant t_0, the switch is turned off, D_1, D_4 are reverse biased, D_2, D_3, D_o are conducted. The energy stored in the input inductor L_1, L_2, is transferred to load R and C_o through D_o, and to C_c through D_3. Meantime, the inductor L_1 releases energy to C_1 through D_2. i_{L_1}, i_{L_2} decrease linearly. The relative functions are given as follows:

$$\frac{di_{L_1}}{dt}=\frac{V_{in}-V_{C_1}}{L_1},\frac{di_{L_2}}{dt}=\frac{V_{C_1}-V_{Cr}}{L_2} \quad (1)$$

$$\frac{dV_{C_1}}{dt}=\frac{i_{L_1}-i_{L_2}}{C_1},\frac{dV_{C_o}}{dt}=\frac{i_{L_2}-V_o/R}{C_o} \quad (2)$$

where i_{L_1},i_{L_2} denote the current flowing through L_1,L_2; $V_{in},V_{C_o},V_{C_1},V_{C_r}$ denote averaged input voltage, output voltage and the voltage across C_1,C_r, respectively.

2) Mode 2([t_1-t_2], Figure 3): at the instant t_1, the switch is turned on. D_3, D_o, D_4 are reverse biased, D_1, D_2 are conducted. The resonant circuit stops working. The DC source and C_1 transfer energy to L_1 and L_2 respectively so that i_{L_1}, i_{L_2} increase linearly until turned off the switch at instant t_2. The relative functions are given as follows:

$$\frac{di_{L_1}}{dt}=\frac{V_{in}}{L_1},\frac{di_{L_2}}{dt}=\frac{V_{C_1}}{L_2} \quad (3)$$

$$\frac{dV_{C_1}}{dt}=\frac{-i_{L_2}}{C_1},\frac{dV_{C_o}}{dt}=\frac{-V_o/R}{C_o} \quad (4)$$

Clearly, C_c can be considered as an output of the quadratic boost converter. C_r is charged by C_c in resonant way in a much shorter period than the switch cycle and the current flowing into C_c is also neglected. Thus, $V_{C_r}=V_{C_c}=V_o/2$ can be obtained. The averaged state-space model of proposed converter operating in CCM can be written as (5):

$$\begin{aligned}\frac{di_{L_1}}{dt}&=\frac{V_{in}-V_{C_1}(1-u)}{L_1},\quad \frac{di_{L_2}}{dt}=\frac{V_{C_1}-V_o(1-u)/2}{L_2}\\ \frac{dV_{C_1}}{dt}&=\frac{i_{L_1}(1-u)-i_{L_2}}{C_1},\quad \frac{dV_O}{dt}=\frac{i_{L_2}(1-u)-V_o/R}{C_2}\end{aligned} \quad (5)$$

where u is the control input which takes '0' for the turned off state and '1' for the turned on state of the switch.

According to steady state of this converter, the DC output voltage gain can be calculated by

$$V_o=\frac{2}{(1-D)^2}V_{in} \quad (6)$$

From (6), it can be seen that the voltage gain of the proposed converter is much higher than that of basic boost converter. Figure 5 shows a voltage gain comparison of the proposed converter with converter in [9] and [10].

The maximum voltage stresses across power devices of the proposed converter and other converters are given in Table 1. This table shows that the voltage stresses of all power devices are relatively lower than others at a certain voltage. Specially, the voltage stresses of the output diode and switch are half of the converter in [9] and [10]. The above analysis shows that the power loss of output diode and switch is reduced and many low-cost components can be applied.

IOP Conf. Series: Materials Science and Engineering **428** (2018) 012044 doi:10.1088/1757-899X/428/1/012044

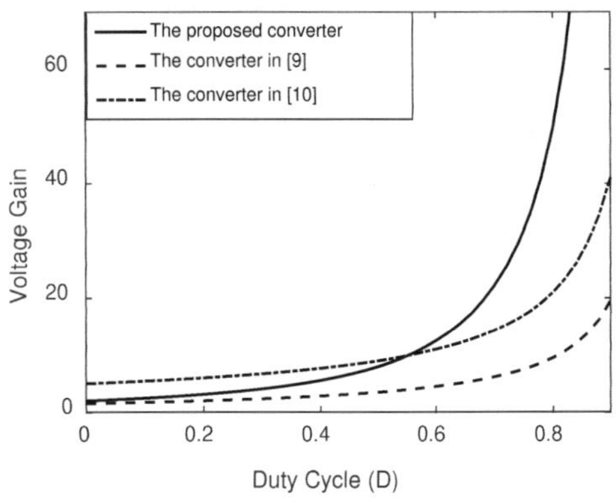

Figure 5. Voltage gain comparison of converters.

Table 1. Power devices voltage stress comparison

Power devices	The proposed converter	The converter in [9]	The converter in [10]
Q	$\frac{V_o}{2}$	$\frac{2V_o}{3+D}$	$\frac{2V_o}{2+nk+nDk}$
D_o	$\frac{V_o}{2}$	$\frac{DV_o}{2(1-D)}$	$\frac{(2+nDK)V_o}{2+nk+nDk}$
D_1	$V_{in}/(1\text{-}D)$		
D_2	$V_{in}/(1\text{-}D)$		
D_3	$V_o/2$		
D_4	$V_o/2$		

3. Sliding mode controller design

In this section, a sliding mode control scheme with fixed frequency is addressed. The output voltage V_O is selected to be the desired control variable. The tracking error can be expressed by:

$$e_1 = V_r - \beta V_O \tag{7}$$

where e_1 is the voltage error; V_r is the reference voltage; β is the voltage feedback network ratio.

The state-space descriptions of the converter operating under CCM is given as follows:

$$\begin{bmatrix} \dot{e}_1 \\ \dot{e}_2 \\ \dot{e}_3 \end{bmatrix} = \begin{bmatrix} 0 & 1 & 0 \\ 0 & -\frac{1}{RC_2} & 0 \\ 1 & 0 & 0 \end{bmatrix} \begin{bmatrix} e_1 \\ e_2 \\ e_3 \end{bmatrix} + \begin{bmatrix} 0 \\ \beta\left(\frac{V_{C_1} - V_o/2}{L_2 C_2}\right) \\ 0 \end{bmatrix} \bar{u} \tag{8}$$

where e_2, e_3 are the voltage error dynamics, and the integral of voltage error, respectively; V_{C_1} is the voltage across C_1; $\bar{u}$ is the inverse logic of u.

A general SMC law adopts a switching function, as follows:

$$u = \begin{cases} 1, & S > 0 \\ 0, & S < 0 \end{cases} \tag{9}$$

where S is the sliding mode and can be chosen as follows:

$$S = \alpha_1 e + \alpha_2 \dot{e} + \alpha_3 \int e \tag{10}$$

α_1, α_2, α_3 represent the sliding coefficients.

To ensure that the state trajectory is maintained within the vicinity of the sliding line, the existence condition derived from Lyapunov's second method must be obeyed:

$$\lim_{S\to 0} S\dot{S} < 0 = \begin{cases} S \to 0^+, \dot{S} < 0 \\ S \to 0^-, \dot{S} > 0 \end{cases} \tag{11}$$

where $\dot{S}$ is the time derivative of S and is calculated as follows:

$$\dot{S} = -\alpha_1 \beta \frac{i_C}{C_O} + \alpha_2 \beta \frac{i_C}{RC_O{}^2} + \alpha_3 e_1 \tag{12}$$

The existence condition is given by:

$$0<\beta L_2\left(\frac{\alpha_1}{\alpha_2}-\frac{1}{RC_o}\right)i_{C_O}-L_2C_o\frac{\alpha_3}{\alpha_2}e_1<\beta\left(\frac{V_o}{2}-V_{C_1}\right) \tag{13}$$

the α_1, α_2, α_3 must be satisfied by (13) with considering the steady value of state variables. Hence employed the control input u, tracking error can converge to zero. However, SM controlled converters generally suffer from significant switching frequency variation when the input voltage or output load is varied [17]. To solve the drawback, an equivalent SMC with constant switching frequency is proposed.

Equating

$$\dot{S}=0 \tag{14}$$

yields the equivalent control input

$$u_{eq}=-\frac{L_2}{(V_O/2-V_{C_1})}\left(\frac{\alpha_1}{\alpha_2}-\frac{1}{RC_o}\right)i_{C_O}-\frac{\alpha_3L_2C_oe_1}{\beta\alpha_2\left(V_O/2-V_{C_1}\right)}+\frac{1}{\left(V_O/2-V_{C_1}\right)} \tag{15}$$

where u_{eq} is a smooth function of the discrete control input u and $0<u_{eq}<1$.

An exponential reaching law is adopted to improve transient response and can be expressed:

$$\dot{S}=-k_1S-k_2sign(S) \tag{16}$$

where k_1, k_2 are positive parameters.

When (16) is solved for u, the total smooth control input:

$$u=-\frac{L_2}{(V_O/2-V_{C_1})}\left(\frac{\alpha_1}{\alpha_2}-\frac{1}{RC_o}\right)i_{C_O}-\frac{L_2C_o\left(\alpha_3e_1+k_1S+k_2sign(S)\right)}{\beta\alpha_2\left(V_O/2-V_{C_1}\right)}+\frac{1}{\left(V_O/2-V_{C_1}\right)} \tag{17}$$

Finally, the total smooth control input u and ramp signal $V_{ramp}=1$ with a certain frequency are fed into a pulsewidth modulator to generate the discrete control input u.

4. Simulation results and discussions

In this section, the simulation of the PWM-based SM controller is presented. Table 2 shows the specification of the proposed converter.

Table 2. main parameter of proposed converter

L_1	L_2	L_r	C_1	C_c	C_r	C_o	R	f_s
$100\mu H$	$150\mu H$	$40\mu F$	$250\mu F$	$15\mu F$	$40\mu F$	$470\mu F$	500Ω	$40kHz$

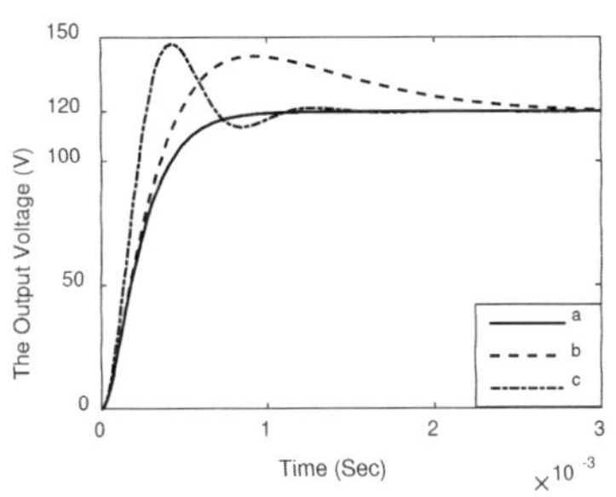

Figure 6. The transient response of the system.

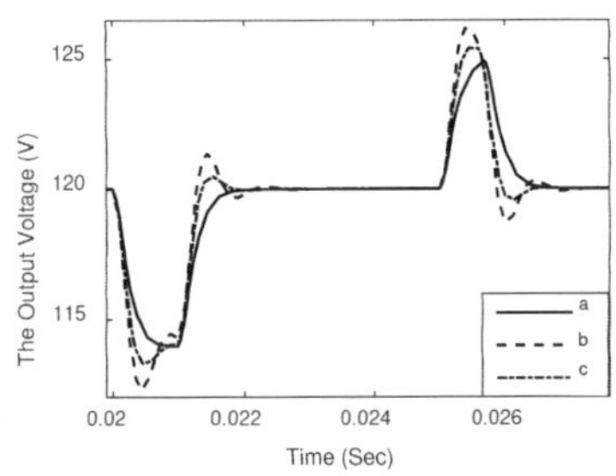

Figure 7. The output voltage waveform versus load variation

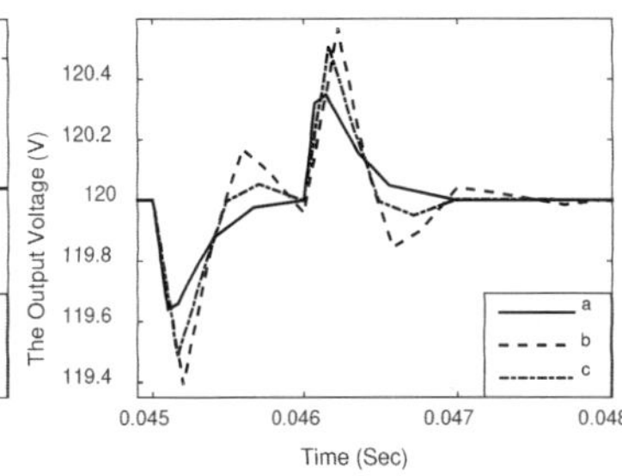

Figure 8. The output voltage waveform versus input variation

Figure 6 illustrates the transient responses comparison of the different SM controllers for the proposed converter. In the figures, 'a' represents output trajectory with the proposed SM controller in (17), 'b' denotes the SM controller in (15), 'c' is the SM controller formulated by the sliding mode in

(10) without integral term respectively. Taking 120 V as the reference, the system transient responses for the proposed SM controller has a settling time of 1 ms and no overshoot, which is relatively better than other responses with 1.5ms and 3ms settling time respectively.

Figure 7 presents the output response versus load disturbance. A sudden load change is introduced at 0.02s, and 0.025s. three responses possess similar settling time around 2ms, but the smallest output voltage variation belongs to the proposed SM controller around 5V. Similarly, figure 8 shows the output response versus input variation, which decreases from 24V to 14V, then return to 24 V. Three output voltages maintain at 120 V around 0.5 V variation and the system for the SM controller in (15) has the longest settling time about 2ms. It can be seen that the proposed controller is relatively superior and robust compared with 'b' controller and 'c' controller for the new converter.

5. Conclusion

This paper has developed a high step-up boost converter with a voltage multiplier. The newly proposed converter has the following features. (1) the output voltage can highly heightened because of the voltage multiplier and serial structure of boost converter. (2) the stray energy also can be recycled into the clamped capacity (C_c) to improve the efficiency. (3) the voltage stresses of the MOSFET and diodes (D_o , D_3 , D_4) are decreased to half of the output voltage so that more economical components can be applied to it. Moreover, a sliding mode control scheme with fixed frequency is designed to stabilized the converter and track reference voltage. The simulation results show that the controlled system can track the reference voltage with short settling time and no overshoot. The highly robustness of the proposed controller against input voltage and load disturbances is also verified.

References

[1] Carvalho, C., & Paulino, N. (2010). A MOSFET only, step-up DC-DC micro power converter, for solar energy harvesting applications. Mixed Design of Integrated Circuits and Systems IEEE. (pp.499-504).

[2] Pahlevaninezhad, M., Das, P., Drobnik, J., Jain, P. K., & Bakhshai, A. (2012). A novel zvzcs full-bridge dc/dc converter used for electric vehicles. IEEE Transactions on Power Electronics, 27(6), 2752-2769.

[3] Lai, C. M., Cheng, Y. H., Hsieh, M. H., & Lin, Y. C. (2018). Development of a bidirectional dc/dc converter with dual-battery energy storage for hybrid electric vehicle system. IEEE Transactions on Vehicular Technology, 67(2), 1036-1052.

[4] Mohan, N., Undeland, T. M., & Robbins, W. P. (2003). Power electronics converter, applications and design. Microelectronics Journal, 28(1).

[5] Forouzesh, M., Shen, Y., Yari, K., Siwakoti, Y. P., & Blaabjerg, F. (2017). High-efficiency high step-up dc-dc converter with dual coupled inductors for grid-connected photovoltaic systems. IEEE Transactions on Power Electronics, PP(99), 1-1.

[6] Ajami, A., Ardi, H., & Farakhor, A. (2015). A novel high step-up dc/dc converter based on integrating coupled inductor and switched-capacitor techniques for renewable energy applications. IEEE Transactions on Power Electronics, 30(8), 4255-4263.

[7] Leipold, D., Dolin, A., & Sheehy, P. (2014). Switched inductor DC-DC converter. US, US8917075.

[8] Leyva-Ramos, J., Ortiz-Lopez, M. G., Diaz-Saldierna, L. H., & Morales-Saldana, J. A. (2009). Switching regulator using a quadratic boost converter for wide dc conversion ratios. Power Electronics Iet, 2(5), 605-613.

[9] Saravanan, S., & Babu, N. R. (2017). Design and development of single switch high step-up dc-dc converter. IEEE Journal of Emerging & Selected Topics in Power Electronics, PP(99), 1-1.

[10] Wong, Y. S., Chen, J. F., Liu, K. B., & Hsieh, Y. P. (2017). A novel high step-up dc-dc converter with coupled inductor and switched clamp capacitor techniques for photovoltaic systems. Energies, 10(3), 378.Zhao, Y., Li, W., Yang, B., & He, X. (2010).

[11] Active clamp boost converter with switched capacitor and coupled inductor. Applied Power Electronics Conference and Exposition (pp.801-806). IEEE.

[12] Beni, S. H., Mirtalaei, S. M. M., Kianpour, A., & Beni, S. A. (2018). Design and improvement of a soft switching high step-up boost converter with voltage multiplier. Iet Power Electronics, 10(15), 2163-2169.

[13] Wang, W., Yi, J. Q., Zheng, Y. L., & Zhao, D. B. (2004). Fuzzy neural network control method for dc-dc converter. Acta Simulata Systematica Sinica, 16(11), 2567-2558.

[14] Babazadeh, A., & Maksimovic, D. (2008). Hybrid digital adaptive control for synchronous buck DC-DC converters. Power Electronics Specialists Conference, 2008. Pesc (pp.1263-1269). IEEE.

[15] Guldemir, H. (2005). Sliding mode control of dc-dc boost converter. Journal of Applied Sciences, 5(3).

[16] Whittington, H. W., Flynn, B. W., & Macpherson, D. E. (2002). Switched mode power supplies: design and construction. Power Engineering Journal, 12(2), 87-87.

[17] Tan, S. C., Lai, Y. M., & Tse, C. K. (2006). A unified approach to the design of pwm-based sliding-mode voltage controllers for basic dc-dc converters in continuous conduction mode. IEEE Transactions on Circuits & Systems I Regular Papers, 53(8), 1816-1827.

IOP Conf. Series: Materials Science and Engineering **428** (2018) 012045 doi:10.1088/1757-899X/428/1/012045

Backstepping Controller Design for Pantograph-Catenary System

Zhi Liu[1], Yicheng Liu[*], Ning Zhou[2], Dong Zou[2] and Haiyan Tu[1]

[1] College of Electrical Engineering and Information Technology, Sichuan University, Chengdu 610065, China;
[2] State Key Laboratory of Traction Power, Southwest Jiaotong University, Chengdu 610031, China

[*]Corresponding author. Email: liuyicheng@scu.edu.cn

Abstract. This paper proposes an active control method to suppress the contact force fluctuation between pantograph and catenary for high-speed trains with external disturbance. First, fitted by using the non-linear least square method, stiffness of catenary is applied to the Pantograph-Catenary system(PCS). Second, a backstepping control strategy for PCS is developed to decrease fluctuation of the contact force. Theoretical analysis shows that the proposed control strategy has strong robustness and disturbance-attenuation ability. The control performances of the closed-loop systems can be shaped as desired by suitably choosing the design parameters. Finally, simulation studies are presented to validate the proposed scheme.

1.Introduction

High speed trains usually acquire current through the PCS. With the speed of the high-speed trains is further improved, the fluctuation of the contact force between the pantograph and the catenary will be more and more violent. There are two methods to solve this problem. One method is to improve the stiffness of the catenary or increase its tension. The other one is to increase the above-mentioned contact force. However, These two methods will lead to huge investment or severe pantograph catenary wear [1]. In order to decrease the fluctuation of the contact force, an uplift force can be adopted through the lower frame of pantograph. If the uplift force is too large, it will cause too much contact force and eventually cause too serious wear of the slide plate. If the uplift force is too small, it will result in too weak contact force and eventually lead to an unnecessary arc [2]. To overcome the above problem, active control method has been put forward and has been receiving significant attentions [3]. Yamashita et al. developed a low-noise active pantograph, and then applied a PID controller or an impedance control method [4]. However, the complete model of a pantograph mechanism for high-speed rail system is complex so that resulting in tuning difficultly the optimal control gains of PID controller [5]. In [6], J Evans et al. discusses several proposals for pantograph active control, taking into account the feedback of the contact force measure. In [7], Corriga et al. employed a frequency-shaping control approach to improve the pantograph system's performance. However, the pantograph system was modeled with a constant stiffness to represent the catenary. In [8] the control strategy is based on Extended Kalman Filter technique, used to get a contact force estimation available for control feedback. Wu and Zhang also suggested setting up an active actuator between upper and lower frame of pantograph, and applying fuzzy logic control technology as control algorithm [9]. A three degree of PCS is established, and a controller based on optimal control strategy is designed in [10]. We can see that the fluctuation of contact force has been significantly improved. However, the contact force has not reached an ideal value, or there is a certain standard deviation with external disturbance.

Combining with backstepping control theory, this paper designs a controller for pantograph catenary. Considering the external disturbance, we can see that the proposed controller has good tracking performance and better robustness. In order to verify the effectiveness of the control method, the latter part also make digital simulation.

2. A model of the PCS

There are many mathematical models of pantograph, such as nonlinear multibody model or simplified mechanical model. But we usually study the control method of PCS, and generally use the lumped-parameters model. There are three reasons for this. First reason, The pantograph adopts the two mass block model, and the simulation results can be compared with the standard. Second reason, the lumped-parameters model is used as the control object. Compared with the multibody model, its dynamic equation is simple and its computation speed is fast. Third reason, nowadays the research of pantograph active control is mainly based on control algorithm, which weakens the influence of pantograph model characteristics on control effect [11]. Therefore, the linear mass model of two degrees of freedom is adopted in this paper [12]. The lumped-parameters model is shown in Figure 1.

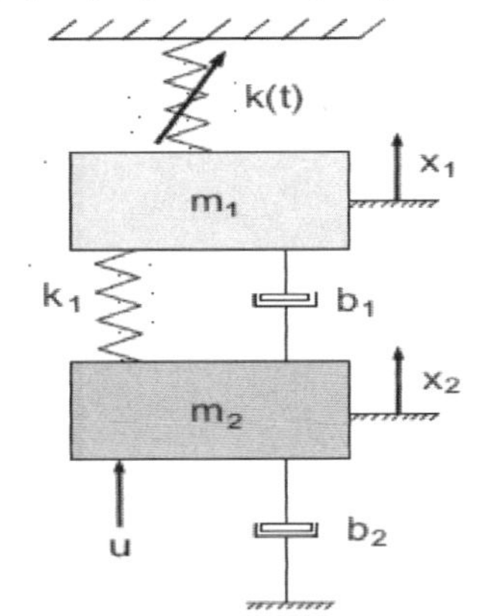

Figure 1. A lumped-parameters model of the Pantograph-Catenary system

The equations of motion of the pantograph-catenary system model shown in Figure 1 can be written as

$$\begin{cases} m_1\ddot{x}_1 - k_1(x_2 - x_1) - b_1(\dot{x}_2 - \dot{x}_1) + k(t)x_1 = 0 \\ m_2\ddot{x}_2 + k_1(x_2 - x_1) + b_1(\dot{x}_2 - \dot{x}_1) + b_2\dot{x}_2 = u + d(t) \end{cases} \tag{1}$$

where, m_1 and m_2 represents the mass of the pantograph head and the quality of the frame, respectively; k_1 stands for the pantograph head stiffness and frame stiffness; b_1 represents the damping of the pantograph head and that of frame ; $x_{i=1,2}$, $\dot{x}_{i=1,2}$ and $\ddot{x}_{i=1,2}$ stands for the displacements, velocities and the acceleration, respectively; u is a controller; $d(t)$ is an external disturbance signal.

$k(t)$ is the equivalent stiffness of the catenary. Stiffness of catenary is fitted by using the non-linear least square method

$$k(t) = k_0(1 + \alpha_1 f_1 + \alpha_2 f_2 + \alpha_3 f_1^2 + \alpha_4 f_3^2 + \alpha_5 f_4^2) \tag{2}$$

$$f_1 = \cos(\frac{2\pi v}{L_1}t), f_2 = \cos(\frac{2\pi v}{L_2}t), f_3 = \cos(\frac{\pi v}{L_1}t), f_4 = \cos(\frac{\pi v}{L_2}t) \tag{3}$$

where k_0 is average stiffness coefficient within span. $\alpha_{i=1,2,3,4,5}$, $L_{i=1,2}$ are the stiffness variation coefficient of catenary and catenary span, and V is the running speed of the locomotive .

In order to design the controller conveniently, we take the following state variables.

$$z(t) = [z_1, z_2, z_3, z_4]^T = [x_1, \dot{x}_1, x_2, \dot{x}_2]^T \tag{4}$$

Taking into account external disturbance, the formula (1) can be written as a state space expression.

$$\begin{cases} \dot{z}(t) = A(t)\, z(t) + Bu(t) + d(t) \\ y(t) = C(t)\, z(t) \end{cases} \tag{5}$$

where $y(t)$ is the contact force and

$$A(t) = \begin{bmatrix} 0 & 1 & 0 & 0 \\ -\frac{k_1 + k(t)}{m_1} & -\frac{b_1}{m_1} & \frac{k_1}{m_1} & \frac{b_1}{m_1} \\ 0 & 0 & 0 & 1 \\ \frac{k_1}{m_2} & \frac{b_1}{m_2} & -\frac{k_1}{m_2} & -\frac{b_1 + b_2}{m_2} \end{bmatrix} \tag{6}$$

$$B = [0 \quad 0 \quad 0 \quad \frac{1}{m_2}]^T \tag{7}$$

$$d(t) = [0 \quad 0 \quad 0 \quad \delta(t)]^T \tag{8}$$

$$C(t) = [k(t) \quad 0 \quad 0 \quad 0] \tag{9}$$

So, we will combine the state space expression and backstepping method to design controller.

3. Controller design

We use backstepping control theory to regulate the contact force. Our aim is to adjust the contact force be kept constant. In order to achieve this goal, we must assume that all the state variables are available for backstepping control. Because our goal is to suppress the fluctuation of the contact force, so we can get a reference signal

$$z_{1r} = \frac{F_c}{k(t)} \tag{10}$$

where F_c is an ideal contact force.

In this purpose, we will build the control law by the three following steps:

Step 1: Let the first tracking error

$$e_1 = z_1 - z_{1r} \tag{11}$$

and considering the Lyapunov function positive definite

$$V_1 = \frac{1}{2} e_1^2 \tag{12}$$

Its time derivative is

$$\dot{V}_1 = e_1 \dot{e}_1 = e_1(\dot{z}_1 - \dot{z}_{1r}) = e_1(z_2 - \dot{z}_{1r}) \tag{13}$$

we found that z_2 is the virtual control, so there will be an error e_2

$$e_2 = z_2 - \beta_1 \tag{14}$$

where β_1 is the virtual control stabilization function to be tracked by

$$\beta_1 = -a_1 e_1 + \dot{z}_{1r} \tag{15}$$

where a_1 is positive real. So, deriving V_1 with respect to time yields

$$\dot{V}_1 = e_1(e_2 - a_1e_1 + \dot{z}_{1r} - \dot{z}_{1r}) \\ = -a_1e_1^2 + e_1e_2 \tag{16}$$

Step 2: Introducing the modified Lyapunov function

$$V_2 = V_1 + \frac{1}{2}e_2^2 \tag{17}$$

Its time derivative can be calculated by

$$\begin{aligned} \dot{V}_2 &= \dot{V}_1 + e_2\dot{e}_2 \\ &= -a_1e_1^2 + e_1e_2 + e_2(\dot{z}_2 - \dot{\beta}_1) \\ &= -a_1e_1^2 + e_1e_2 + e_2(-\frac{k_1+k(\mathrm{t})}{m_1}z_1 - \frac{b_1}{m_1}z_2 + \frac{k_1}{m_1}z_3 + \frac{b_1}{m_1}z_4 - \dot{\beta}_1) \end{aligned} \tag{18}$$

we found that z_4 is the virtual control, so there will be an error e_3

$$e_3 = z_4 - \beta_2 \tag{19}$$

β_2 is the virtual control stabilization function to be tracked by

$$\beta_2 = \frac{m_1}{b_1}(\dot{\beta}_1 - \frac{k_1}{m_1}z_3 + \frac{b_1}{m_1}z_2 + \frac{k_1+k(\mathrm{t})}{m_1}z_1 - e_1 - a_2e_2) \tag{20}$$

where a_2 is positive real. So, deriving V_2 with respect to time yields

$$\begin{aligned} \dot{V}_2 &= -a_1e_1^2 + e_1e_2 + e_2(-\frac{k_1+k(\mathrm{t})}{m_1}z_1 - \frac{b_1}{m_1}z_2 + \frac{k_1}{m_1}z_3 + \\ &\frac{b_1}{m_1}(e_3 + \frac{m_1}{b_1}(\dot{\beta}_1 - \frac{k_1}{m_1}z_3 + \frac{b_1}{m_1}z_2 + \frac{k_1+k(\mathrm{t})}{m_1}z_1 - e_1 - a_2e_2)) - \dot{\beta}_1) \\ &= -a_1e_1^2 - a_2e_2^2 + \frac{b_1}{m_1}e_2e_3 \end{aligned} \tag{21}$$

Step 3: Introducing the Lyapunov function

$$V = V_2 + \frac{1}{2}e_3^2 \tag{22}$$

Its time derivative can be calculated by

$$\begin{aligned} \dot{V} &= \dot{V}_2 + e_3\dot{e}_3 = \dot{V}_2 + e_3(\dot{z}_4 - \dot{\beta}_2) \\ &= -a_1e_1^2 - a_2e_2^2 + \frac{b_1}{m_1}e_2e_3 + e_3(\frac{k_1}{m_1}z_1 + \frac{b_1}{m_2}z_2 - \frac{k_1}{m_2}z_3 - \frac{b_1+b_2}{m_2}z_4 + \frac{1}{m_2}\mathrm{u} + \delta(\mathrm{t}) - \dot{\beta}_2) \end{aligned} \tag{23}$$

At this point, we find that the controller has appeared. The controller is as follows

$$\mathrm{u} = m_2(\dot{\beta}_2 + \frac{b_1+b_2}{m_2}z_4 + \frac{k_1}{m_2}z_3 - \frac{b_1}{m_2}z_2 - \frac{k_1}{m_1}z_1 - \frac{b_1}{m_1}e_2 - (a_3+c)e_3) \tag{24}$$

where a_3 and c is positive real. So, deriving V with respect to time yields

$$\dot{V} = -a_1e_1^2 - a_2e_2^2 - a_3e_3^2 - ce_3^2 + \delta(\mathrm{t})e_3 \tag{25}$$

For convenience,here we make the order Δ

$$\Delta = -ce_3^2 + \delta(\mathrm{t})e_3 \tag{26}$$

$$\Delta \le -ce_3^2 + k\|e_3\| = -ce_3^2 + b\|e_3\|\frac{k}{b} \tag{27}$$

where k and b is positive real, k satisfaction $\|\delta(\mathrm{t})\| \le k$, So

IOP Conf. Series: Materials Science and Engineering **428** (2018) 012045 doi:10.1088/1757-899X/428/1/012045

$$\Delta \le -ce_3^{\ 2} + \frac{1}{2}b^2 e_3^{\ 2} + \frac{k^2}{2b^2}$$

$$\Delta \le \frac{k^2}{2b^2} \tag{28}$$

Then $\dot{V}$ satisfaction

$$\dot{V} \le -a_1 e_1^{\ 2} - a_2 e_2^{\ 2} - a_3 e_3^{\ 2} + \frac{k^2}{2b^2} \tag{29}$$

when k satisfaction $k \ll b$

$$[z_1 - z_{1r}]_{t\to\infty} = 0 \tag{30}$$

4. Simulation results

In order to prove the performance of the proposed controller, numerical simulation has been carried out. The model parameters with the following values:

Catenary: $k_0 = 3600N/m; L_1 = 65m; L_2 = 9m;$

$$\alpha_1 = 0.4665; \alpha_2 = 0.0832; \alpha_3 = 0.2603; \alpha_4 = -0.2801; \alpha_5 = -0.3364$$

Pantograph: $m_1 = 8kg; k_1 = 10000N/m; b_1 = 120Ns/m; m_2 = 12kg; b_2 = 30Ns/m$

Ideal contact force: $F_c = 100N$

The first simulation takes into account the train in 300km/h, and the external signal $\delta(\mathrm{t}) = \sin(2\pi \mathrm{t})$. The first simulation results are found Figure 2. We will eventually find that the contact force will reach a constant value 100N. The second simulation takes into account the train in 100m/s, and the external signal $\delta(\mathrm{t}) = 10\sin(2\pi \mathrm{t})$. The second simulation results are found Figures(3-4). It is also found that the contact force is almost constant 100N. Comparing with [10], the backstepping control method proposed in this paper greatly decreases the fluctuation of contact force under the speed of 400km/h and external disturbance. The simulation results can prove a good tracking effect of contact force at high speed and a strong robustness against interference. Therefore, the effectiveness of the backstepping method is verified.

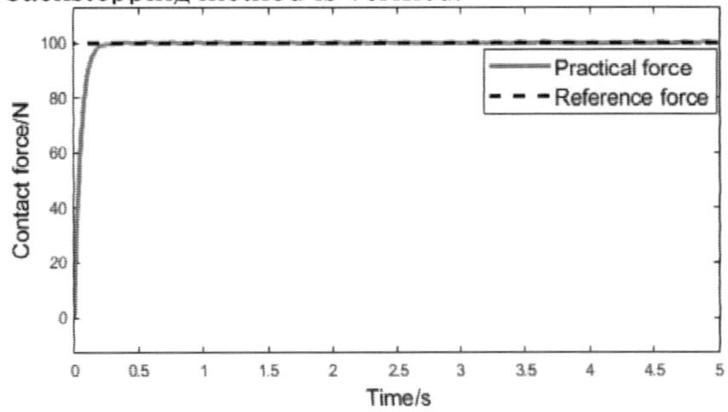

Figure 2. Contact force when the speed is 300km/h

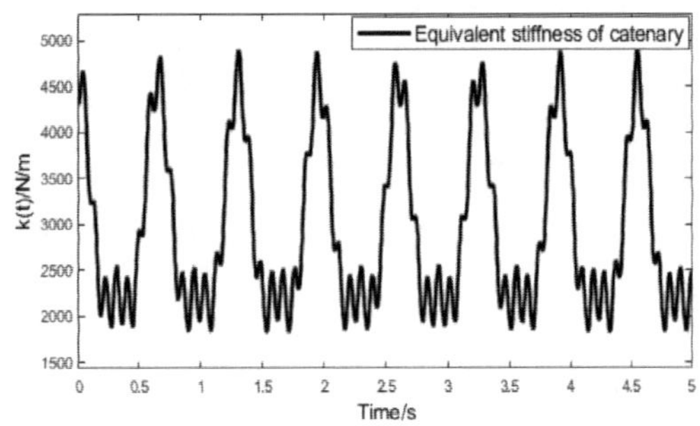

Figure 3. Equivalent stiffness of catenary

5. Conclusion

This paper presents a method for suppressing the contact force fluctuation between pantograph and catenary for high-speed trains using a backstepping active control technology. Theoretical analysis shows that the proposed control strategy has strong robustness and disturbance-attenuation ability. The control performances of the closed-loop systems can be shaped as desired by suitably choosing the design parameters. The simulation results show that: The first simulation, under the speed of 300km/h and external disturbance, the contact force can approximately reach the ideal value. The second simulation, the backstepping control method proposed in this paper greatly decreases the fluctuation of contact force under the speed of 400km/h and external disturbance. Therefore, the controller proposed in this paper has good tracking performance and a better robustness.

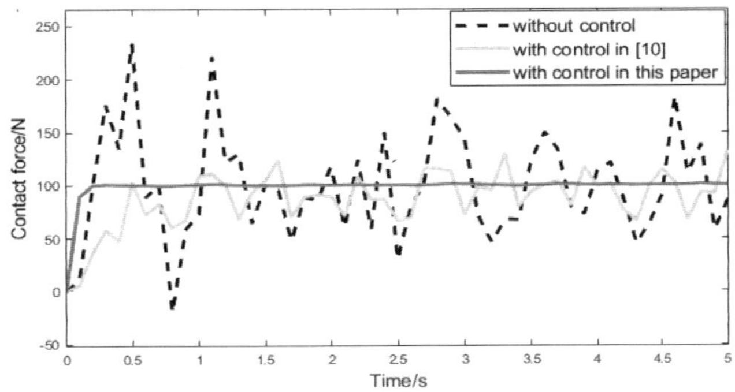

Figure 4. Contact force when the speed is 400km/h

Acknowledgments
This project is supported by the National Key Research and Development Program of China under Grant No. 2016YFB1200506

References

[1] Lu, X., & Liu, Z. (2015). Applicability of active control algorithms for pantographs of high-speed railway. Xinan Jiaotong Daxue Xuebao/journal of Southwest Jiaotong University, 50(2), 233-240.

[2] Pappalardo, C. M., Patel, M., Tinsley, B., & Shabana, A. A. (2015). Contact force control in multibody pantograph/catenary systems. Proceedings of the Institution of Mechanical Engineers Part K Journal of Multi-body Dynamics.

[3] Tieri, R. (2012). Innovative active control strategies for pantograph catenary interaction.Vehicle Engineering.

[4] Yamashita,Y.(2014).Active control of contact force between pantograph and catenary for high-speed trains. Foreign Rolling Stock, 53(1), 28-33.

[5] Visioli, A. (2001).Tuning of pid controllers with fuzzy logic. IEE Proceedings - Control Theory and Applications, 148(1), 1-8.

[6] Evans, J. (1997). Pantograph/catenary dynamics and control. Vehicle System Dynamics, 28(2-3), 159-195.

[7] Corriga, G.,Giua, A., Matta, W., & Usai, G. (1995). Frequency-shaping design of a gain-scheduling controller for pantographs. Decision and Control, 1994. Proceedings of the, IEEE Conference on (Vol.1, pp.393-398 vol.1). IEEE.

[8] Allotta, B., Papi, R., Pugi, L., & Toni, P. (2001). Experimental campaign on a servo-actuated pantograph. Ieee/asme International Conference on Advanced Intelligent Mechatronics, 2001. Proceedings (Vol.1, pp.237-242 vol.1). IEEE.

[9] Wu, X. (2002).Research of active vibration control for the pantograph-catenary system. Journal of Vibration Engineering, 15(1), 36-40.

[10] Guo, J. B., Yang, S. P., & Gao, G. S.(2004).Study on active control of high-speed-train pantographs. Journal of the China Railway Society.

[11] Lu X B, Liu Z G, Song Yang.(2014).Overview of active control of pantograph[J].Journal of transportation and Transportation Engineering.

[12] Ide, C. K., Olaru, S., Rodriguez-Ayerbe,P.,& Rachid, A. (2013). A nonlinear state feedback control approach for a Pantograph-Catenary system. System Theory, Control and Computing (pp.268-273). IEEE.

IOP Conf. Series: Materials Science and Engineering **428** (2018) 012046 doi:10.1088/1757-899X/428/1/012046

Ultrasound Environment Classification Based on Fuzzy Model

Zhili Long, Ronghua He, Xiaobing Zhang and Yuxiang He

Harbin Institute of Technology, Shenzhen, Guangdong Province, China

longworking@163.com

Abstract. The fuzzy model based on the centreline ultrasound signal is pro-posed to solve the poor orientation of ultrasonic sensors in the environment modelling in this paper. The locations of ultrasonic arc feature points were extracted and distinguished. For the problem of poor resolution of split type ultrasonic sensors, fuzzy algorithm is applied to identify the obstacle such as line, corner, or arc structure. Ultrasound environment map based on fuzzy is constructed and the maximum distance and angle deviation is 1.2cm and 3°, showing that the accuracy of obstacle classification based on ultrasound and Fuzzy is improved.

1. Introduction

Because most of the environments that robots face cannot be known in advance, autonomous identification of the environment and mapping is necessary. The environmental modeling based on ultrasonic signal has broad application prospect in indoor service robot since ultrasonic sensor is the earliest development sensor with mature technology and has the advantages of the high precision of small range measurement, no harm to human body, low cost of transforming products, etc.

Research of ultrasonic environment modeling can be divided into three aspects, as grid map[1,2], topological map[3] and geometric feature map[4].To form the geometric feature map, features such as straight lines, corner, and arcs from environments is extracted and combined. The main algorithms of feature extraction include Range of Constant Depth Method(RCD)[5], Arc-Transversal Median(ATM) and Triangulation-Based Fusion (TBF). Many ultrasonic environment modeling in the form of feature maps have emerged in recent years, because the geometric feature map not only has the ability to accurately acquire the necessary information for positioning, but also has a low storage space occupancy rate[6,7]. Heinen[8] used Gaussian mixture model to represent the surrounding environment and proposed a new feature environment map algorithm which has the advantages of small memory usage and high processing speed avoiding the discrete error in fast calculation. Hesham Ismail and Balakumar Balachandran[9] applied revolving measurement data with double ultrasonic sensor and proposed a fusion algorithm of environment feature extraction. The circular arc feature extraction algorithm was combined with the Hough transform-based TBF algorithm and advantages of those algorithms were fully developed. SJ Lee[10] proposed a new ultrasonic feature structure, which uses an ultrasonic data correlation method to extract circular sets of convex corner of objects. The integrated circle center of each set is used as a EKF-SLAM roadmap. The above method can realize the discrimination of angles and arcs, thus the problem of complex indoor environment modeling was solved.

In this paper, by analyzing the geometric features of revolving platform scanning, a fuzzy algorithm is proposed to extract and classify ultrasonic signals. Finally, the environmental modeling based on revolving ultrasonic signals is realized.

IOP Conf. Series: Materials Science and Engineering **428** (2018) 012046 doi:10.1088/1757-899X/428/1/012046

2. Experiment Measurement Based On Ultrasound

In experiment, the ultrasonic signal is collected by a four array ultrasonic sensors, fixed in a rotary motor. Figure 1 shows the experiment measurement platform. The typical distance data of the obstacles is shown in figure 2.

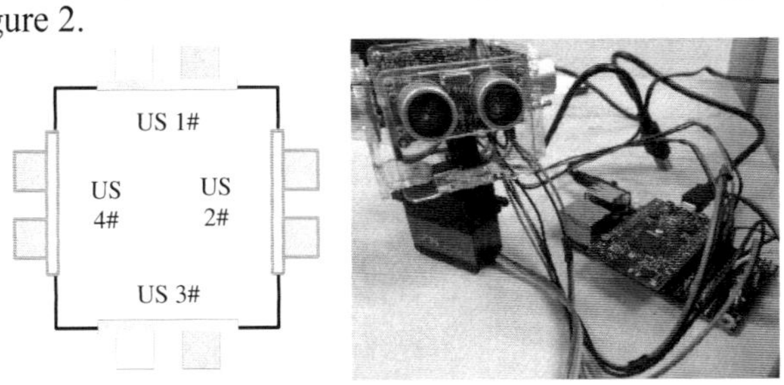

Figure 1. Experiment measurement

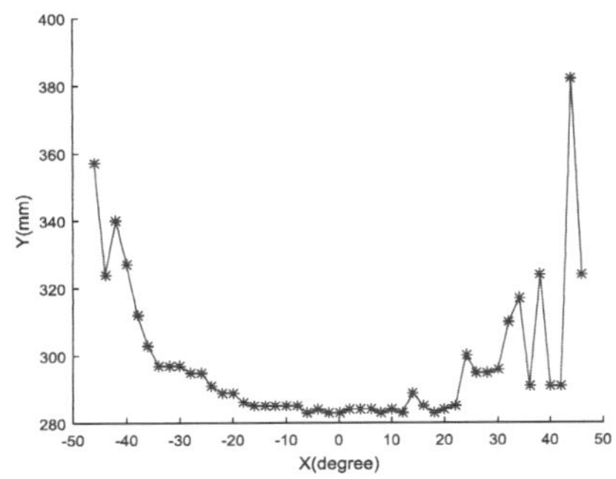

Figure 2. Typical distance data of the obstacles

3. Fuzzy model and application

In actual conditions, the ultrasonic arc fluctuates greatly, and it is difficult to distinguish the source of different feature points. To solve the above problem, a fuzzy algorithm was proposed. The advantage of fuzzy theory is to reduce the data accuracy requirements, and data classification can be achieved through the probability distribution of data and the correlation between data features. As shown in figure 3, the principle of fuzzy algorithm is that by fuzzing the width of ultrasonic arc and fuzzy reasoning with the boundary state of ultrasonic arc and processing it with defuzzification module, the type of geometric feature that can be recognized by computer is obtained finally. The ultrasonic arc width membership can be expressed and the graph is shown in figure 4.

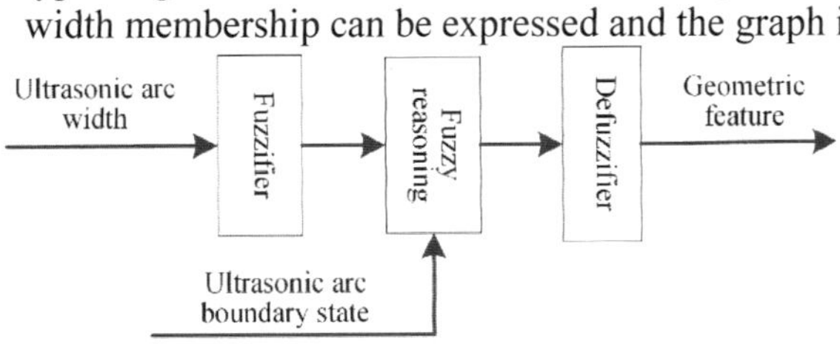

Figure 3. Principle and process of fuzzy model

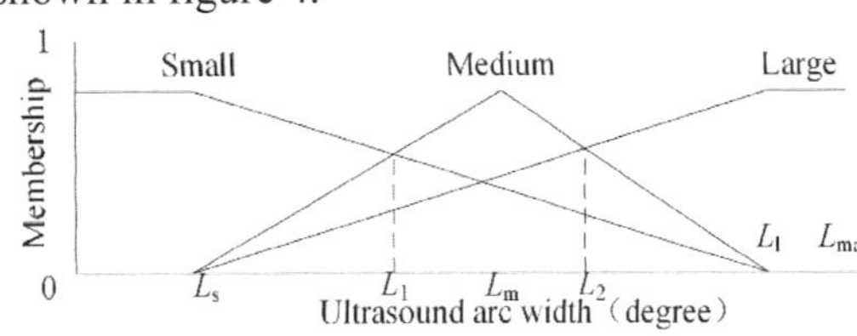

Figure 4. Membership of ultrasound arc width

3.1. Language Conversion

3.1.1. Ultrasonic arc width. According to the ultrasonic arc model, the arc width of the ultrasound is related to the geometric features. In combination with the actual measurement of the ultrasonic sensor, generally, the surface has a maximum arc width, and the arc width of the cylindrical surface is reduced accordingly. The effective arc of convex corner and the concave corner width is the smallest because of the poor recognition ability. The width of the ultrasonic arc can be expressed in table 1.

Table 1. Numerical form of ultrasonic arc width fuzzy state

State	Small	Medium	Large
Value	0	1	2

(2) Boundary of ultrasound arc

The boundary reflects whether the ends of arc are shielded by other ultrasonic arcs. The blocked ultrasonic arc will reduce part of the arc length, thus affecting the judgment of the arc state. The boundary is shown in table 2.

Table 2. Numerical form of boundary states

State	Unobstructed	Left front	Right front	Front
Value	0	1	2	3

(3)Geometric Features

As shown in table 3, geometric features classified include angle, cylinder, plane with value of 0~2.

Table 3. Numerical form of geometric features

State	Corner	Cylinder	Plane
Value	0	1	2

3.2. Data Fuzzification

In order to fuzzify the ultrasonic arc width, it is divided into three categories: large, medium, and small. Three categories of maximum membership threshold corresponding to arc width are set. The critical arc width of different categories is related to the maximum width of the ultrasonic arc.

3.3. Reasoning and defuzzification

3.3.1. Reasoning. The boundary states as the influencing factor of the ultrasonic arc width play a key role in the judgment of geometric features in fuzzy reasoning. A boundary state of 0 means that it is not occluded, a boundary state of 1, 2 means that one side is occluded, and a boundary state of 3 means that both sides are occluded. For each fuzzy arc width, the system has a boundary state corresponding to it. Different boundary states will redefine the types of membership in the current arc width. The reasoning process is as follows. When the arc width is 0, the membership of small width type drops by 0.25 and the membership degree in the rises by 0.25 if the boundary condition is 1, 2. The small membership degree decreases by 0.3 and the large membership degree increases by 0.3 if the boundary condition is 3.When the arc width is 1, the membership degree of medium width type decreases by 0.2 and the large one increases by 0.2 if the boundary condition is 1, 2. The membership of medium decreases by 0.3, and the large membership degree increases by 0.3 if the boundary condition is 3.

*3.3.2. Defuzzification.*When the maximum membership degree is μ_0, the ultrasonic arc comes from the angular feature and the output of the system is 0; when the maximum membership degree is μ_1, the ultrasonic arc comes from the cylindrical feature and the output of the system is 1; when the maximum membership degree is μ_2, the ultrasonic arc comes from the plane feature, and the output of the system is 2.

4. Environment classification with fuzzy algorithm

For the classification of the environmental obstacles, the feature of lines, arcs, and angles need to be identified. After the multi-position measurement data were obtained, the interference is eliminated by preprocessing data points, and the ultrasonic arc clustering with different characteristics is carried out.

4.1. Ultrasound collection and data preprocess

Figure 5 is the schematic diagram of the indoor environment used for modeling in this experiment. There is a cylinder with a radius of 37 mm as an obstacle to detect the ability of the arc modeling.

In experiment, a total of 10 locations were scanned and the data was visualized in the MATLAB. As shown in figure 6, the red circle represents the measurement position of rotating platform, and dotted line represents the platform movement trajectory. The figure shows a large number of discrete points around the distribution environment, due to the poor ability of the sensor to identify the concave corner. The ultrasonic arc representing the concave corner feature is not located in the actual concave position, but in the direction of the concave corner in the form of discrete or small arcs.

In order to reduce the influence of the interference points on the ultrasound arc clustering, the data needs to be filtered. The similar distance points that one's number less than 4 are selected and cleared as the interference point. The removal result is shown in figure 7. It can be seen that the discrete points are cleared and the ultrasonic arcs representing the line features and arc features are displayed clearly.

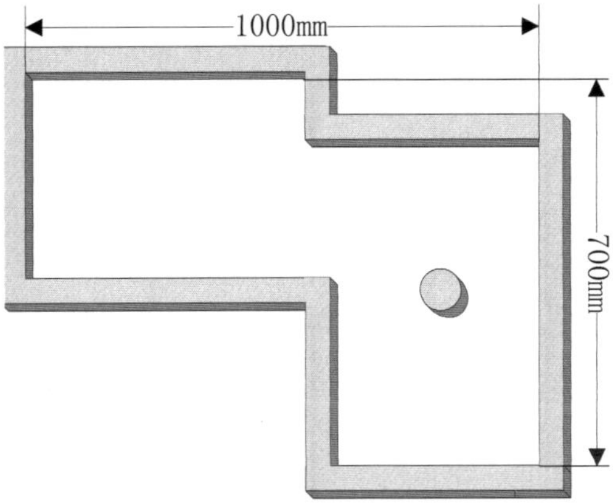

Figure 5. Schematic diagram of the indoor environment

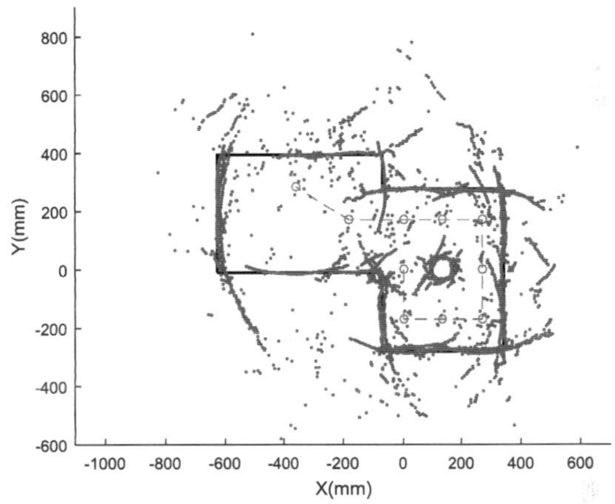

Figure 6. Result of multi-position ultrasound scanning

4.2. Fuzzy feature point acquisition

In experiment, we need to collect many environmental features such as plane, cylinder, convex corner, and use fuzzy algorithm to extract and classify feature points. Firstly, the preprocessed data points are clustered according to the differential distance clustering method. Since there is no interference from outliers, the width of most ultrasonic arcs is regularly distributed according to the source of the features. The width of the plane is the widest, the cylinder is the second, and the angle is the shortest. After clustering, the least squares polynomial fitting is to extract the minimum distance and the result is shown in figure 8.

The clustered ultrasonic arc width and the boundary state are the input of fuzzy classification algorithm, and classified line features, arc features and corner features are the output. The result of fuzzy classification is shown in figure 9.

Data associations are set for the line feature points, arc feature points, and corner feature points related to geometric features. The data association classifies feature points in units of lines, arcs, and points by the geometric association of multi-position features. In experiment, 8 straight line feature point sets, 1 arc feature point set and 1 corner feature point set were formed by classification. Since the corner feature points in the figure that correspond to the secondary measurement intersections are only located at the lower convex corner, the remaining locations do not meet the requirements to be cleared, so there is only one corner feature point set. In addition, the unmatched data in the fuzzy classification is rematched, achieving 100% matching rate in experiment.

4.3. Construction of multi-feature map

The straight-line fitting method is used to process the set of straight line feature points at each position. 8 straight lines are generated, and the intersections between the straight lines are calculated. Least squares curve fitting is performed on the arc feature point set to generate an arc. The nearest point of intersection of the two ends of the line is taken as the position of the optimal corner feature which serves as the two endpoints of the line. Finally, a complete environment map as shown in figure 10 is generated.

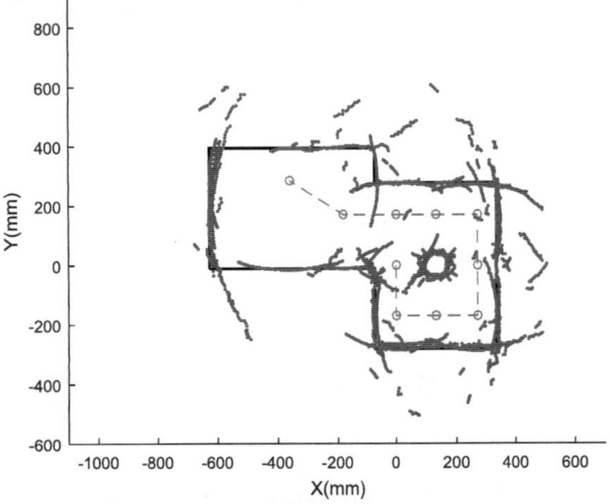

Figure 7. Result after data preprocessing

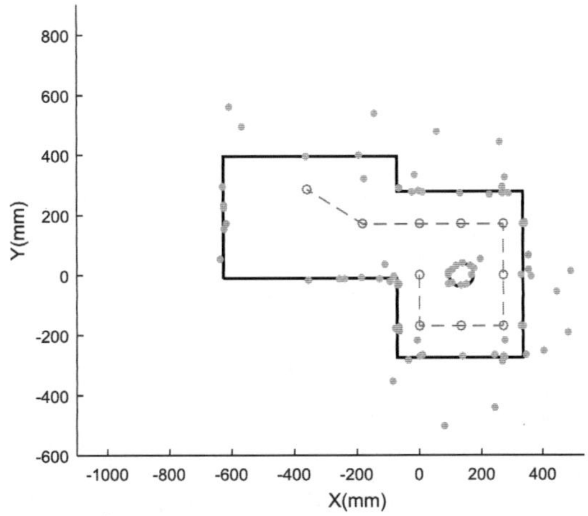

Figure 8. Result of distance minimum point extraction

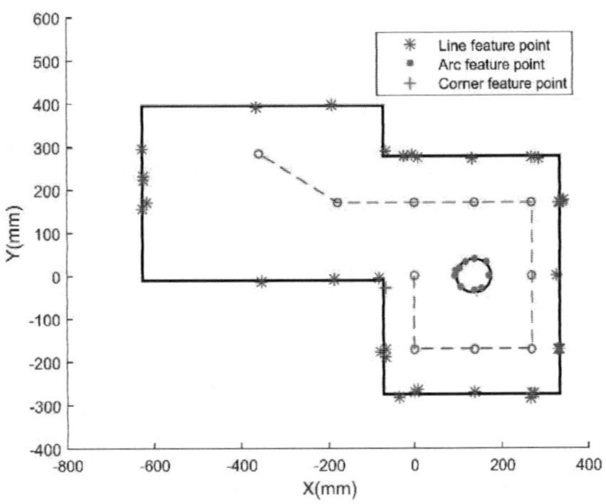

Figure 9. Classification result of fuzzy

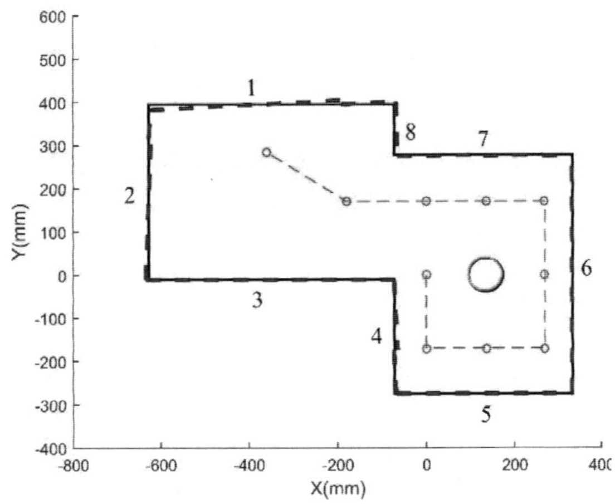

Figure 10. Constructed environment map

By comparing the original boundary with the boundary generated by the fuzzy algorithm, the maximum distance deviation and the angle deviation can be obtained, as shown in table 4. We can see in the farthest measurement distance 1000mm range, the distance deviation of the 1st boundary is the largest, reaching 12.8 mm, and the angle deviation is 3°; the 5th boundary has the smallest distance deviation, only 0.5 mm, and angle deviation is 0°. Because there is only one feature point in the 8th boundary, data association cannot be performed after fuzzy classification. In this case, the distance deviation is 7 mm and the angle deviation is 1°. Therefore, fuzzy algorithm can realize complex map construction with high accuracy and multiple features in the case of split revolving ultrasonic scanning.

Table 4. Deviation result

Boundary index	Maximum distance deviation(mm)	Angle deviation(°)
1	12.8	3
2	6.7	2
3	1.0	0
4	7.8	3
5	0.5	0
6	1.8	0
7	3.8	1
8	7.0	1

5. Conclusion

In this paper, the feature extraction algorithm via ultrasound sensor is proposed. This method solves the problem of feature point location error and feature matching error. A fuzzy extraction algorithm is proposed to solve the problem that the feature location error because of the low resolution and large fluctuation of measurement for ultrasonic sensor. Compared with the physical map, the maximum distance deviation constructed by the ultrasound signal is 1.2 cm and the maximum angle deviation is 3 °, so the accuracy of the fuzzy algorithm is improved to meet with requirements of environmental modeling.

Acknowledgments

This work was supported by the following funds: 1) Nation-al Natural Science Foundation of China (U1713206), 2) Basic Research Plan of Shenzhen (JCYJ20170413112645981, JCYJ20150928162432701), and 3) Shenzhen Technology In-novation Program(JCYJ20170811160003571).

Reference

[1] Blanco JL. A robust, Multi-hypothesis Approach to Matching Occupancy Grid Maps[J]. Robotics, 2013, 31(5): 687-701.
[2] WANG Wei, CHEN Weidong,WANG Yong. Probabilistic Grid Map Based Localizability Estimation for Mobile Robots [J]. Robots, 2012, 34(4): 485-491.
[3] Werner F, Sitte J, Maire F. Topological Map Induction using Neighbourhood Information of Places[J]. Autonomous Ro-bots, 2012, 32(4): 405-418.
[4] Du Zhenjun, Qu Daokui, Xu Fang, Jia Kai. Mobile Robot Map Building Based on Ultrasonic Sensors [C]. Nation-al Conference on access and processing of Information, 2010(S2).
[5] Leanard J, Durrant-Whyte F. Mobile Robot Localization by Tracking Geometry Beacons[J]. IEEE Transactions on Robotics and Automation, 1991(7): 376-382.
[6] Lee SJ, Dong WC, Wan KC, Lim JH, Kang CU. Feature based Map Building using Sparse Sonar Data[C]. IEEE/RSJ International Conference on Intelligent Robots & Systems, 2005, 21(1): 1648 – 1652.
[7] Ismail H, Balachandran B. A Comparison of Feature Ex-traction Algorithms based on Sonar Sensor Data[C]. Asme International Mechanical Engineering Congress & Exposition, 2013: V04AT04A024.
[8] Heinen MR, Engel PM. Incremental Feature-based Map-ping from Sonar Data using Gaussian Mixture Models[J]. Acm Symposium on Applied Computing, 2011: 1370-1375.
[9] Ismail H, Balachandran B. Algorithm Fusion for Feature Extraction and Map Construction From SONAR Data[J]. IEEE Sensors Journal, 2015, 15(11): 6460-6471.
[10] Lee SJ, Song JB. A New Sonar Salient Feature Structure for EKF-based SLAM[C]. IEEE/RSJ International Conference on Intelligent Robots & Systems, 2010, 25(1): 5966-5971.

IOP Conf. Series: Materials Science and Engineering **428** (2018) 012047 doi:10.1088/1757-899X/428/1/012047

Multi-UUV Task Allocation with Limited Communication Range Based on Hoplites Framework

Ning Li[1], Minglong Li[1,*], Da Huang[1] and Wei Yi[2]

[1]The State Key Laboratory of High Performance Computing, College of Computer, National University of Defense Technology
[2]The Artificial Intelligence Research Center, National Innovation Institute of Defense Technology

*huangda1109@163.com

Abstract. Multiple Unmanned Underwater Vehicles(UUVs) can be widely used in marine science expedition, ocean development and ocean engineering. And in the Unmanned Underwater Vehicles cluster, the Multi-Robot Task Allocation with communication range limitations is an important issue. In the past, many works were researched based on perfectly good communication conditions. However, in actual situations, especially underwater environments, communication range between robots is often limited. So in this paper, in order to ensure the communication range between the UUVs as much as possible, based on the Hoplites framework, we first design the communication keep revenue so that the UUVs not only perform the task but also try to ensure the communication range between the UUVs. Second, when there are individual tasks that are particularly far away from all other tasks, a communication relay node needs to be added between the UUV and other UUVs. We regard the communication relay node as a task, which is realized by using the active coordination mechanism of the Hoplites framework. Through the simulation environment of multiple UUVs, we conduct experiments and analysis of the algorithm, and the results show tangible performance gain of our methods over previous methods.

1. Introduction

As humans continue to explore the ocean, the Unmanned Underwater Vehicle（UUV）has become increasingly irreplaceable. The Unmanned Underwater Vehicles can photograph and investigate the geographical environment, ecological environment and hydrological information in the ocean. According to the operator's prior instructions, it can reach places where other detection tools cannot reach and avoid personal injury and waste of resources due to unknown conditions [1] [2]. At the same time, the research on multiple Unmanned Underwater Vehicles has attracted increasing attention in order to be able to perform large-scale tasks and improve the efficiency of task execution [3]. Multiple Unmanned Underwater Vehicles can provide significant advantages for a variety of applications including marine sampling, mapping, surveillance, and communications.

For the multi-UUVs, the most widely studied problem at present is the Multi-Robot Task Allocation(MRTA) [4][5][6]. The MRTA problem refers to assigning multiple tasks to multiple robots according to certain rules. The rules may be non-repetitive allocations, the minimum total time to perform all tasks or the least resources consumed by all robots [7]. Patrolling [8], formations [9] and cooperative transportation [10] all belong to the MRTA. In this paper, we study a task set T that contains multiple tasks $t_1, t_2 \cdots t_n$, where n is the number of tasks, and a UUV set U that contains $uuv_1, uuv_2 \cdots uuv_m$, where m is the number of UUVs. We hope to create a mapping between T and U, and the UUVs can work together to complete all tasks. At the same time, a very important indicator is the limited communication range rc, Throughout the task allocation process, we must ensure that the distance between each UUV and other UUV's is less than or equal to rc, through direct communication or multi-hop communication.

As we know, the underwater environment is quite complicated. It is full of areas that we don't know, and the interference of water with the communication signal is very large. So for multi-UUVs performing missions underwater, maintaining communication with each other is crucial [11]. So in this paper, we study how to use a distributed market-based approach to solve MRTA problem for multi-UUVs with limited communication range.

In the field of MRTA, people have proposed some methods for solving Multi-Robot Task Allocation problem with limited communication range. Wei Hua Sheng et al. [12] studied the exploration problem of multi-robots in unknown environments and proposed a market-based method to try to ensure that the distance of communication among multiple robots is in the limited scope. Although this can give the robots a state of mutual aggregation to a certain extent, it cannot be completely guaranteed. Nguyen et al. [13] studied whether the communication between the leader robot and the base station was interrupted by sensing the RF signal quality. If the RF signal quality is less than a certain threshold, an attached robot is required to relay the radio link. Although this method can form a complete communication chain, this is not a distributed method. If the leader robot is broken, then the entire link is meaningless. Based on the existing methods and their deficiencies, we decided to study the problem of Multi-Robot Task Allocation with communication range limitations based on the Hoplites framework.

The Hoplites framework is a distributed framework based on market-based method [14]. Hoplites framework includes two mechanisms: passive coordination and active coordination. Passive coordination mechanism is used to handle relatively simple task distribution scenarios. It can generate solutions quickly but does not consider the optimality of the global solution. If the situation is more complex, active coordination is needed to obtain better solutions through virtual economic transactions between robots. In this paper, we study the problem of multi-robot task allocation with communication range limitation based on the Hoplites framework. First, by using the market-based approach, we not only consider the revenue of the task and the cost of executing the task, but also the revenue earned by maintaining communications. In other words, when the UUV chooses a task, it will consider the distance between the task and other UUVs and perform comprehensive calculations to obtain the profit gained from the communication maintenance. Finally, the task that has the most comprehensive profit in all tasks is taken as the task that one will perform. However, the limitation of revenue alone is not enough. When we consider that there is a task that is separate from other tasks, we have designed a communication coordination mechanism based on the active coordination mechanism of the Hoplites framework. If an UUV has to perform a task that is out of communication range with all other robots, it can actively cooperate and give high rewards to the UUV that is furthest away from it, and request it to perform tasks of the communication relay node. In this way, a communication subnetwork will be maintained between the UUVs, so that the robots will not lose contact with other robots during executing tasks. At the same time, information out of sync or UUV lost due to out of communication range is also avoided.

2. Proposed Method

In this paper, we propose two levels of algorithms for multiple UUV communication problems with limited range based on Hoplites Framework. Now we have a detailed exposition of the proposed method.

2.1. Framework Overview

IOP Conf. Series: Materials Science and Engineering **428** (2018) 012047 doi:10.1088/1757-899X/428/1/012047

Hoplites framework is a market-based framework that includes passive coodination mechanism and active coordination mechanism. The specific algorithm of the Hoplites framework is shown in Algorithm 1. It is the Hoplites framework of the program running on each UUV. Based on the Hoplites framework, we illustrate the improved algorithm separately.

Algorithm 1: Hoplites Framework

```
currentPlan = Ø
coopAccepted = 0
while NotEmpty(GetTasks())
  if MyTurn(r)
    newPlan = Ø
    Tasks= GetTasks()
    newPlan = PassiveCoordination(Tasks, r)
    R = GetRevenue(newPlan)
    if R <= Threshold
      activePlan,Rc,Plans = activeCoordination(Tasks, r)
      coopAccepted = AskForCoop( Rc, Plans)
      if (coopAccepted) then
        newPlan = activePlan
    currentPlan = newPlan
```

2.2. Communication keep revenue

In order to solve the Multi-Robot Task Allocation problem of multi-UUVs with communication range limitations, and also keep the UUVs in a same communication network as much as possible, we first design a revenue-based communication maintenance mechanism. In other words, when using the Hoplites framework to select the highest-profit tasks, we not only consider the revenue that comes from the completion of the task and the cost of performing the tasks, we also calculate the distance between the UUV and the other UUVs as part of the net profit calculation. That is:

$$P_i = \alpha \bullet R_i + \beta \bullet C_i - \gamma \bullet D_i \tag{1}$$

Where P_i is the net profit obtained from the execution of task i, R_i is the revenue of the task itself, and C_i is the revenue gained from communication maintenance, and D_i is the cost of executing the task i, in this paper we consider it the distance that needed to perform the task. α, β, γ are the adjustment coefficients, which are used to adjust the proportion of R_i, C_i, D_i in the net profit calculation. And

$$C_i = \theta \bullet B \bullet e^{-\frac{d_1}{rc}} + \theta^2 \bullet B \bullet e^{-\frac{d_2}{rc}} + \cdots + \theta^{n-1} \bullet B \bullet e^{-\frac{d_{n-1}}{rc}} \tag{2}$$

In this formula, B is a constant and it is the basic profit that communication maintains. θ is the adjustment coefficient, $d_1, d_2 \cdots d_{n-1}$ is the descending order of the distances between the task i and the tasks currently being performed by other UUVs, and n is the number of UUVs, rc is the maximum communication distance that the UUV can withstand. From this formula, we can see that C_i depends mainly on the UUV closest to this task. This can ensure that when the UUV chooses a task, it prefers to perform tasks that are closer to other UUVs. And finally:

$$P = \max P_i \tag{3}$$

The UUV will choose the task that with the highest total profit in all tasks to execute.

2.3. Communication relay task

Through the method described in *2.2*, we can try to ensure that the UUVs stay in a certain communication range when performing underwater tasks. However, when the degree of dispersion of tasks is particularly large, or when there is a very long distance between a single task and all other tasks, there is no way to guarantee the limited communication distance. At this time, we need to rely on the Hoplites framework, virtualize a communication relay task, and through an active coordination mechanism, request an UUV to perform a communication relay task when an UUV needs to go out of communication with all other UUVs because of the execution of a task. Here we choose to request the UUV that is currently farthest from itself, as this will not destroy the previously complete communication network. It should be noted that although this will allow the requested UUV to walk a little longer, it is worth it for the UUV not to lose communication with the entire robot team. The specific algorithm is as Algorithm 2.

Algorithm 2: Communication relay task

```
Choose the most profitable task T
Distance = get_nearest_distance(T)
if Distance > rc:
      num = get_farthest_uuv()
      active_coordination(T,num)
      waiting for answer()
else:
      execute(T)
```

Here we base on the active coordination mechanism of the Hoplites framework. The essence of the Hoplites framework is a market-based approach. Through a market-based mechanism, purely distributed methods are used to perform the task of buying and selling, to gain the UUV to perform communication relay tasks.

3. Application scenarios

In this paper, our experiments were simulated using UUV Simulator [15]. UUV Simulator extends the open-source robot simulator Gazebo to underwater scenes. It can simulate multiple Unmanned Underwater Vehicles. UUV Simulator is based on the ROS, an open source operating system. It implements simulation and control through a series of plugins that simulate underwater pressure effects and hydrodynamic effects. It also simulates the thrusters, sensors, and external disturbances of Unmanned Underwater Vehicles. The UUV Simulator universal robot simulation platform reuses and extends to the underwater environment, greatly facilitating the simulation of underwater environments. At the same time, we can use RVIZ to visualize data in Gazebo for research and analysis. The Figure 1 is a simulation experiment scenario.

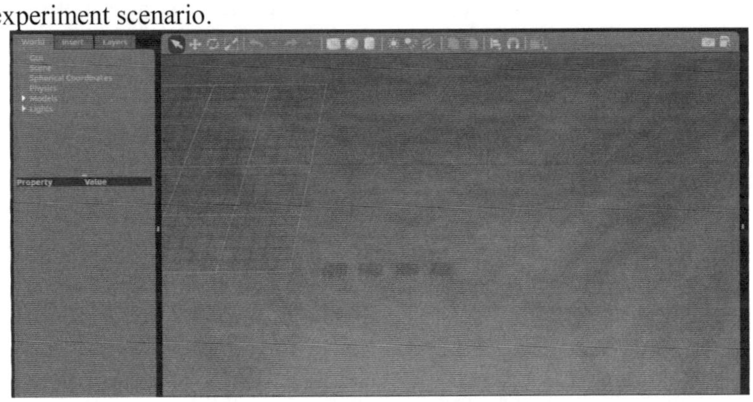

Figure 1. Multiple Unmanned Underwater Vehicles simulation environment UUV Simulator.

In the simulation experiment, we study the patrolling tasks of the Multi-Robot Task Allocation problem. In the underwater environment, we set 10 location points. Ten UUVs need to patrol all task points and cannot repeat. In this process, the communication range between UUVs is limited by a certain distance. Therefore, we have to meet the limitation of the communication range between UUVs through the above-mentioned method and perform all tasks reasonably.

4. Experimental results

For the simulation scenarios applied in the previous section, we discuss and analyse the experimental results of the communications keep revenue the communication relay task.

4.1. Communication keep revenue

In this section, we examine how the addition of communications keep revenue affects the behaviour of team UUVs. We first let the α=1, β=0, γ=1 in formula (1), and the θ in formula (2) is equal to 0.8, the communication range rc is 10 meters, B=R_i=100, *e* is the constant. The movements of the four UUVs are shown in Figures 2, 3, 4. From the figures we can see that since no communication revenue was added, the UUV 1 (the rightmost UUV) moves away from the other three UUVs at the beginning. And due to the follow-up tasks, UUV 1 is further away from the other three UUVs.

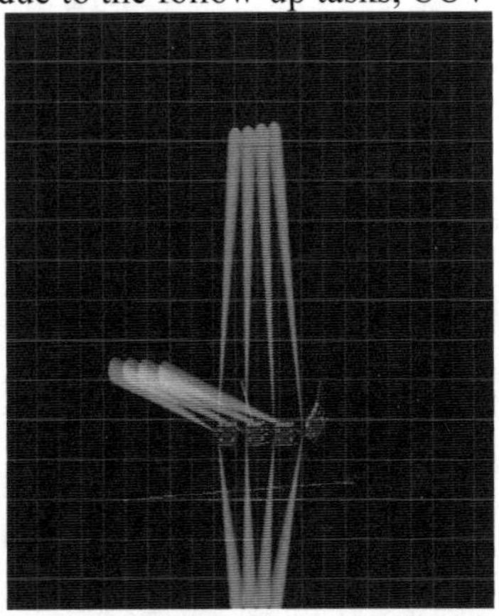

Figure 2. The four UUVs begin to perform tasks.

Figure 3. The rightmost UUV starts to deviate from other UUVs.

Figure 4. The rightmost UUV 1 deviates from the team alone.

Then we let the α=1, β=1, γ=1 in the formula (1), and the rest of the parameters remain unchanged, and the experiment results are shown in Figures 5, 6, and 7. It can be seen from the figures that due to the addition of communication keep revenue, UUV 1 chooses to perform tasks that are closer to other UUVs from the very beginning, and finally, the four robots collectively perform more distant tasks to ensure the UUVs in a relatively close distance, and the communication network is maintained.

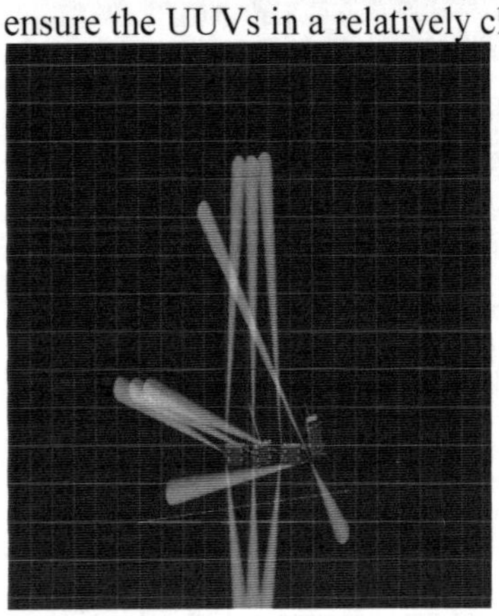

Figure 5. The four UUVs begin to perform tasks and UUV 1 moves in the direction closer to other UUVs.

Figure 6. The rightmost UUV keeps closer with other UUVs than Figure 3.

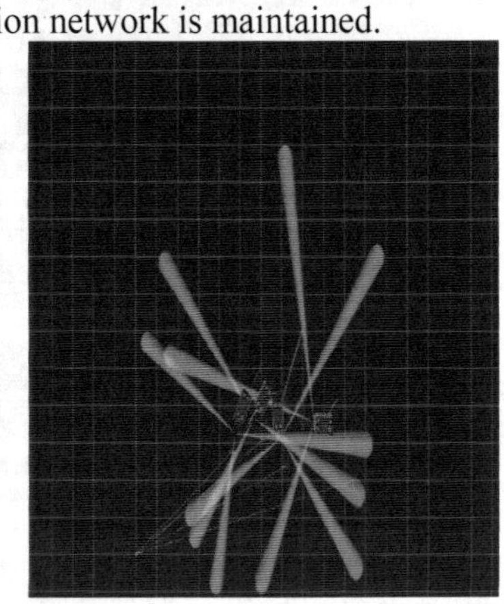

Figure 7. The UUVs have shown a state of closeness, maintaining the communication network.

Finally, we plot the distance between each UUV and other UUVs as a broken line during UUVs movement, as shown in the figure. The abscissa indicates that the UUV performs a task each time.

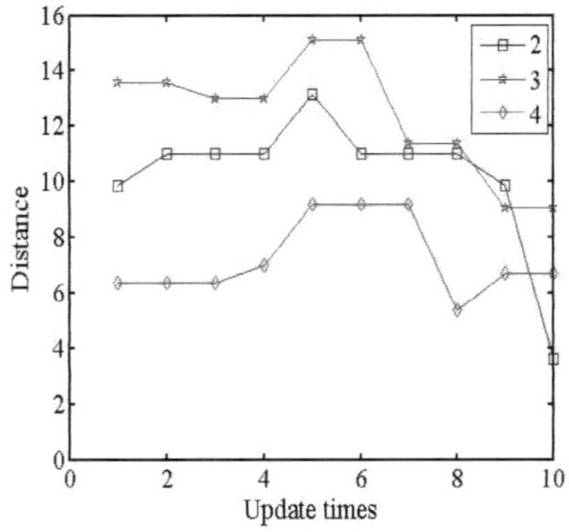

Figure 8. The distance between UUV 1 and other UUVs when not considering communication keep revenue.

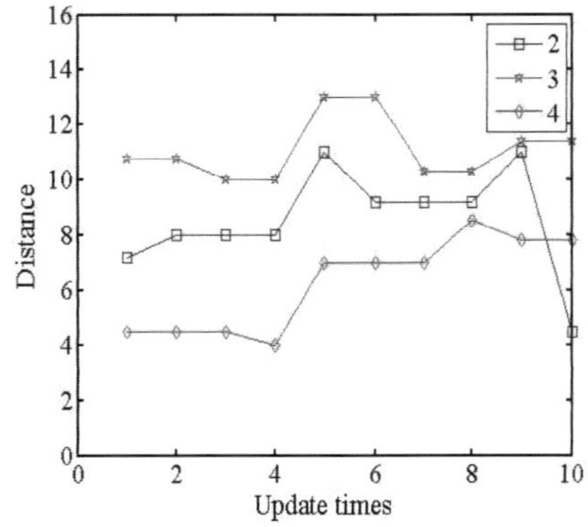

Figure 9. The distance between UUV 1 and other UUVs when considering communication keep revenue.

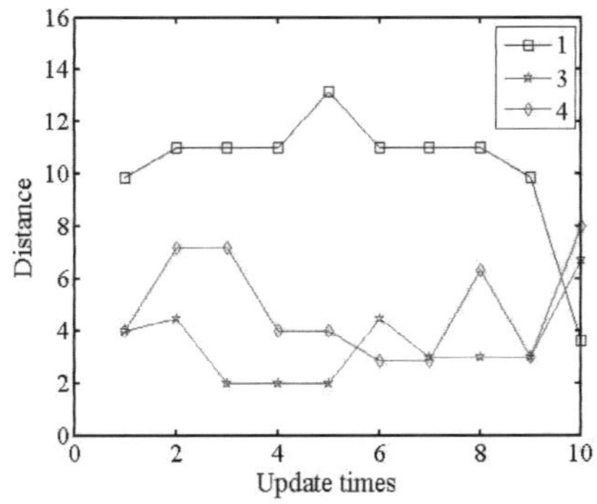

Figure 10. The distance between UUV 2 and other UUVs when not considering communication keep revenue.

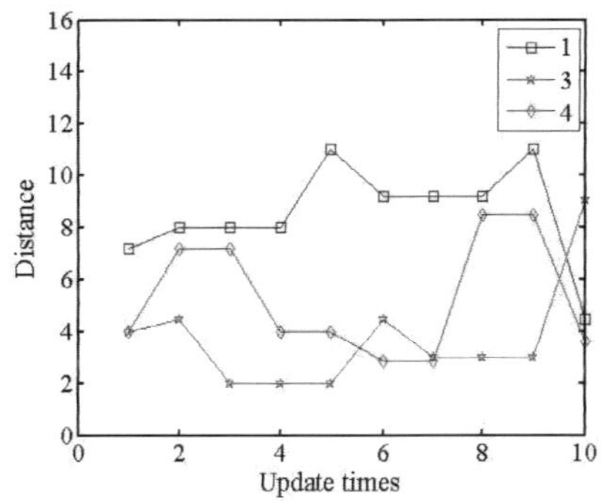

Figure 11. The distance between UUV 2 and other UUVs when considering communication keep revenue.

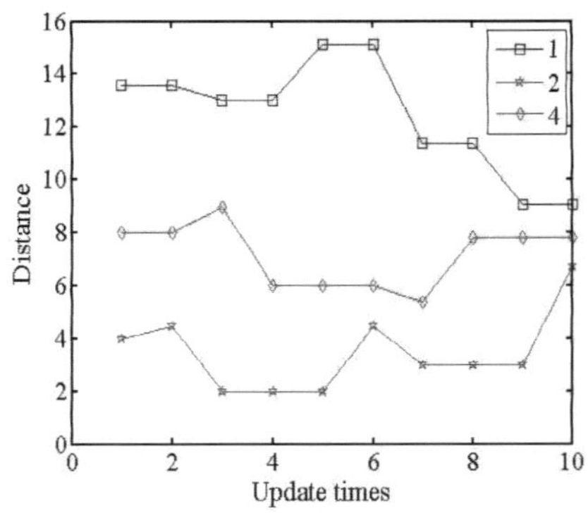

Figure 12. The distance between UUV 3 and other UUVs when not considering communication keep revenue.

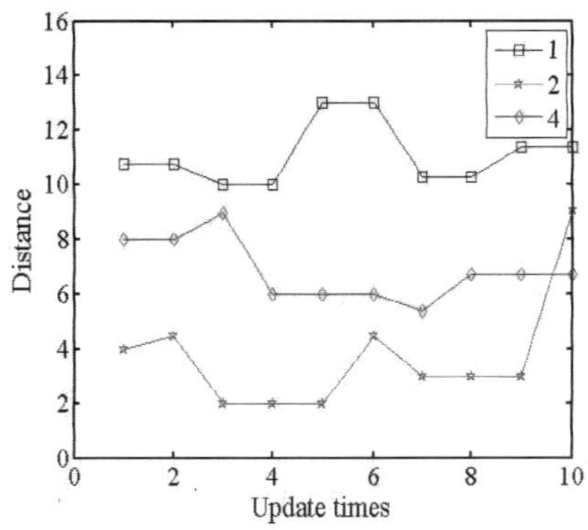

Figure 13. The distance between UUV 3 and other UUVs when considering communication keep revenue.

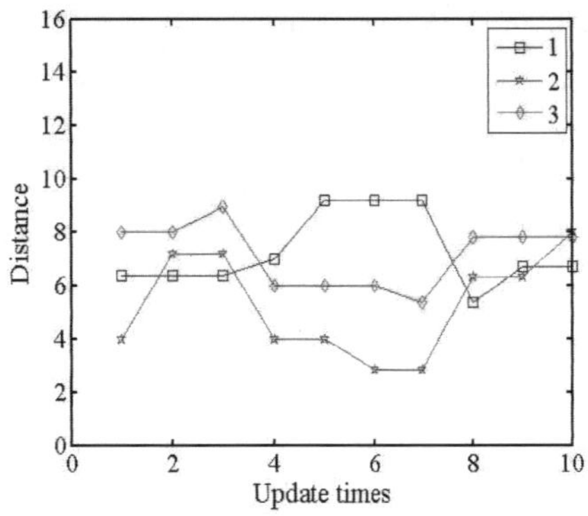

Figure 14. The distance between UUV 4 and other UUVs when not considering communication keep revenue.

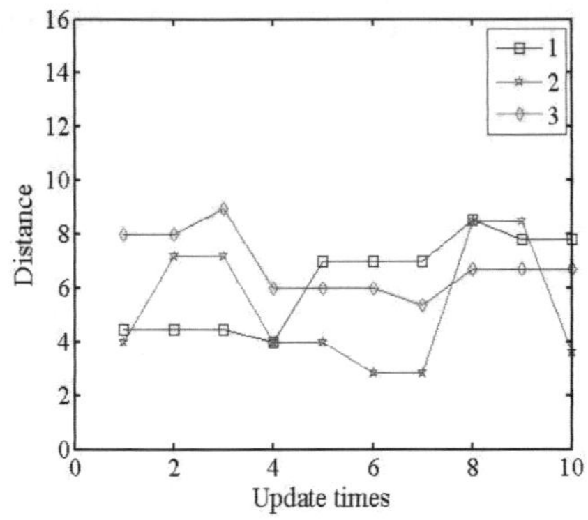

Figure 15. The distance between UUV 4 and other UUVs when considering communication keep revenue.

From the comparison of Figure 8 and 9, 10 and 11, 12 and 13, 14 and 15, it can be seen that the distance between the UUVs that using communication keep revenue is significantly smaller than that of UUVs that have not used, which is very helpful for the communication between UUVs and does not lead to the robots being separated by long distances.

4.2. Communication relay task

In the experiment of the communication relay task, we set a situation where one task and all other tasks are particularly far apart. If an UUV performs this task, it will certainly lose contact with other UUVs where only the communication keep revenue is not enough. Figure 16 shows the initial positions of the four UUVs before executing tasks. And the result of the movements of the UUVs in the experiment are shown in Figure 17 and 18 respectively. From the figures, we can see that the last task performed by UUV 1(the rightmost) is beyond the range of communication. Therefore, UUV 1 triggered the active coordination mechanism and given a high revenue to UUV 3 from the farthest distance to perform communication relay task, UUV 3 received the request, weighed and agreed upon the request to perform communication relay task and restore the entire communication network. It should be noted that we did not consider communication keep revenue in the experiments of communication relay tasks.

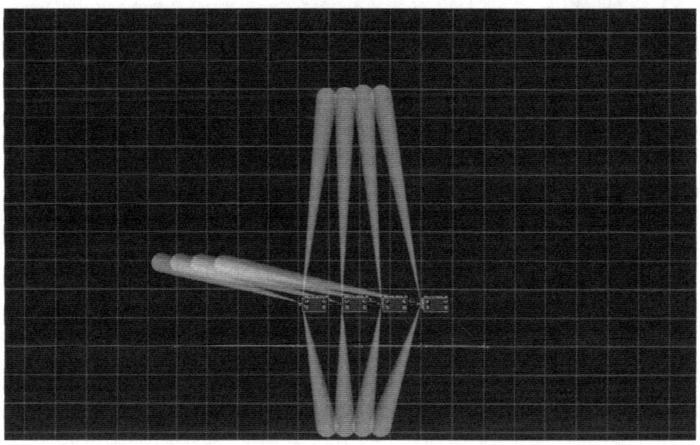

Figure 16. The initial position of four UUVs.

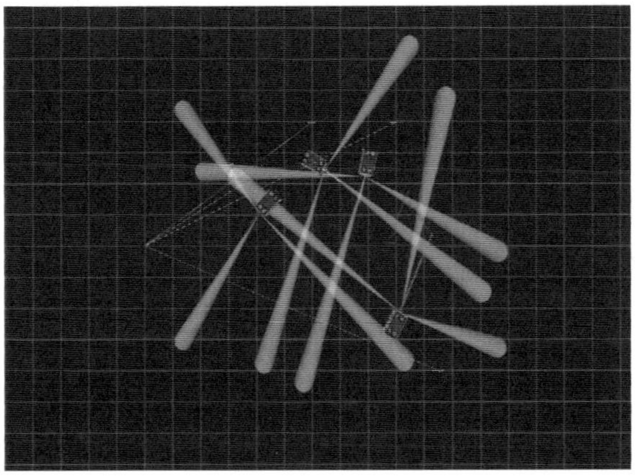

Figure 17. UUV movement not joining a relay task.

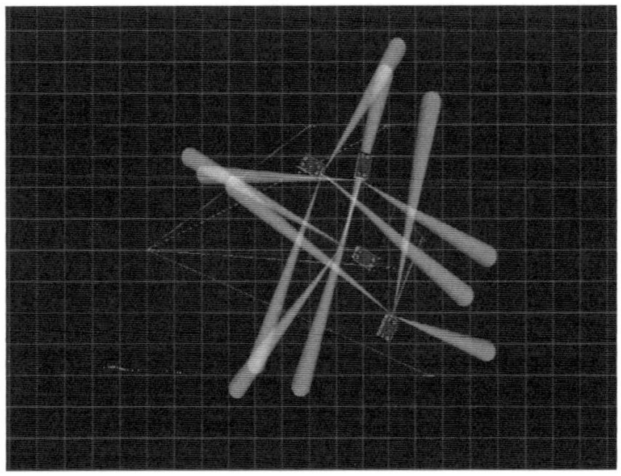

Figure 18. UUV movement after joining a relay task.

5. Conclusions and future work

In this paper, we studied the problem of Multi-Robot Task Allocation with communication range limitation and give two levels of solutions. First of all, we designed a communication keep revenue based on the market-based approach and Hoplites framework. The UUVs appeared to be moving closer due to the inclusion of communication keep revenue, which ensures the scope of communication to a certain extent. Secondly, we designed the communication relay task using the active coordination mechanism of the Hoplites framework. When the distance between the UUV and all other UUVs beyond the communication limit, it restored the entire communication network by actively requesting cooperation from other UUVs. Through simulation experiments, we verified the effectiveness of the algorithm. In the future, we hope to verify our algorithm in physical experiments.

Acknowledgments

This work was partially supported by National Natural Science Foundation of China (Grant No.91648204, No.61532007, No.61601486), The National Key Research and Development Program of China (Grant No.2017YFB1001900), and National Science and Technology Major Project.

References

[1] Rogowski P, Terrill E, Chen J, et al. Observations of the frontal region of a buoyant river plume using an autonomous underwater vehicle[J]. Journal of Geophysical Research, 2014, 119(11): 7549-7567.
[2] Yan Z, Hao B, Liu Y, et al. Movement Control in Recovering UUV Based on Two-Stage Discrete T-S Fuzzy Model[J]. Discrete Dynamics in Nature and Society, 2014: 1-11.
[3] Fiorelli E, Leonard N E, Bhatta P, et al. Multi-AUV Control and Adaptive Sampling in Monterey Bay[J]. IEEE Journal of Oceanic Engineering, 2006, 31(4): 935-948.
[4] M. B. Dias, R. Zlot, N. Kalra, and A. Stentz, "Market-based multirobot coordination: A survey and analysis," Proceedings of the IEEE, vol. 94, no. 7, pp. 1257–1270, 2006.
[5] M. Alighanbari, Y. Kuwata, and J. P. How, "Coordination and control of multiple uavs with timing constraints and loitering," in American Control Conference, 2003. Proceedings of the 2003, vol. 6, pp. 5311–5316, IEEE, 2003.
[6] L. Xu and A. Stentz, "Market-based coordination of coupled robot systems," in Intelligent Robots and Systems (IROS), 2011 IEEE/RSJ International Conference on, pp. 2784–2789, IEEE, 2011.
[7] Alighanbari M, Kuwata Y, How J P, et al. Coordination and control of multiple UAVs with timing constraints and loitering[C]. american control conference, 2003: 5311-5316.

IOP Conf. Series: Materials Science and Engineering **428** (2018) 012047 doi:10.1088/1757-899X/428/1/012047

[8] Soysal O, Sahin E. Probabilistic aggregation strategies in swarm robotic systems[C]. ieee swarm intelligence symposium, 2005: 325-332.
[9] Spears W M, Spears D F, Hamann J C, et al. Distributed, Physics-Based Control of Swarms of Vehicles[J]. Autonomous Robots, 2004, 17(2-3):137-162.
[10] Gross R, Dorigo M. Towards group transport by swarms of robots[J]. International Journal of Bio-inspired Computation, 2009: 1-13.
[11] Xia G, Ban R. Parameter identification to motion model of UUV by extended Kalman Filter [C]// IECON 2017 -, Conference of the IEEE Industrial Electronics Society. IEEE, 2017:6911-6915.
[12] Sheng W, Wang Q, Yang Q, et al. Minimizing data exchange in ad hoc multi-robot networks [C]. international conference on advanced robotics, 2005: 811-816.
[13] Nguyen H G, Pezeshkian N, Gupta A, et al. Maintaining Communication Link for a Robot Operating in a Hazardous Environment[J]. Ans Int.conf.on Robotics & Remote Systems for Hazardous Environments, 2004:28--31.
[14] Kalra N, Ferguson D, Stentz A. Hoplites: A Market-Based Framework for Planned Tight Coordination in Multirobot Teams[C]// IEEE International Conference on Robotics and Automation. IEEE, 2005:1170-1177.
[15] Manhães M M M, Scherer S A, Voss M, et al. UUV Simulator: A Gazebo-based package for underwater intervention and multi-robot simulation[C]// Oceans. IEEE, 2016.

IOP Conf. Series: Materials Science and Engineering **428** (2018) 012048 doi:10.1088/1757-899X/428/1/012048

A Method to Compensate Interaction between Actuator Dynamics and Control Allocator under Incremental Nonlinear Dynamic Inversion Controller

Xiang Li[1], Yi Jiang[1], Jiannan Zhang[2], Shaoyan Shi[1] and Liangyu Zhao[1]

[1]School of Aerospace Engineering, Beijing Institute of Technology, Beijing, 100081, China

[2]Institute of Flight System Dynamics, Technical University of Munich, Munich, 85748, Germany

Corresponding author's e-mail address: jy2818@163.com

Abstract. When designing flight control allocator for an over-actuated flight system, actuator dynamics, generally, are assumed to be ideal since the bandwidths of actuators are much larger than the frequency of aircraft. However, under incremental nonlinear dynamic inversion controller, the interaction between the constrained control allocation frame and actuator dynamics can cause serious consequence. In this work, a method is developed to compensate for actuator dynamics in s-domain analysis. Here, first-order and second-order actuator dynamics are taken into consideration as more complicated actuator dynamics can be combined by these two simplified models. This method solves for gains, which post-processe the output of control allocation frame. The gains are terse in form and are easy to be calculated, which means that it can be implemented into typical flight computer. Simulation results show proposed method can significantly compensate the flaw caused by actuator dynamics.

1. Introduction

Incremental nonlinear dynamic inversion (INDI), which is a nonlinear control law, is established on the assumption that actuator can reflect instantly [1]. The basic ideal of INDI is using Taylor expansion to linearize the control system at every sampling time, therefore it is less dependent on system modelling and shows better robustness compared to nonlinear dynamic inversion [2]. The basic structure of INDI controller is shown in Figure 1.

For an advanced aircraft, it is quite common to be designed as an over-actuated system. Compared to traditional flight system, performance, reliability, and reconfigurability can be improve after extra actuators are utilized [3]. Control allocation algorithms are used to compute aircraft control surface deflections which produce a desired set of generalized moment commands [3]. Typical control allocation equation is,

$$\boldsymbol{u}_{cmd} = \boldsymbol{B}^{-1}\boldsymbol{\tau}_{des} \tag{1}$$

Where, $\boldsymbol{u}_{cmd}$ is actuator command vector. $\boldsymbol{\tau}_{des}$ is desired virtual input vector and $\dot{\boldsymbol{y}}_{des}$ is desired generalized moment vector. Generally, $\boldsymbol{\tau}_{des} = \dot{\boldsymbol{y}}_{des}$. $\boldsymbol{B}$ is effectiveness matrix [3].

For the overall control system, desired goal is that the moments produced by effectors are equal to desired generalized moments, namely,

IOP Conf. Series: Materials Science and Engineering **428** (2018) 012048 doi:10.1088/1757-899X/428/1/012048

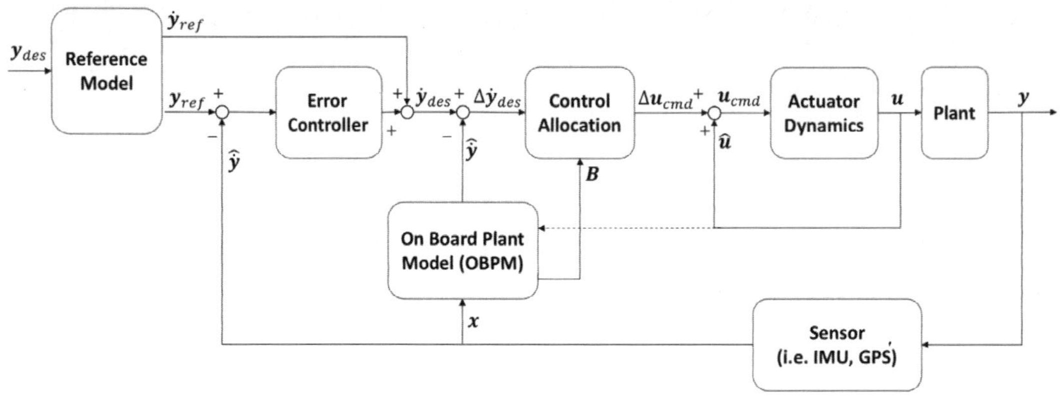

Figure 1. Structure of incremental nonlinear dynamic inversion baseline controller.

$$\boldsymbol{Bu} = \boldsymbol{\tau}_{des} \tag{2}$$

Where, $\boldsymbol{u}$ is actual actuator respond vector. A number of approaches have been developed to ensure that the commands sent to the effectors are physically realizable [3]. Generally, it is reasonable to ignore actuator dynamics on the condition that bandwidths of actuators are normally much larger than the frequencies of the vehicle's rigid body model [4]. But, in some cases, actuator dynamics may have serious influence on the performance of control system if the interactions between constrained control allocators and actuator dynamics are ignored [4]. Bolling has shown that the interaction between first order actuator dynamics and constrained control allocation algorithms can be eliminated by overdriving the actuators [5]. Doman, Oppenheimer and Venkataraman proposed the dynamic control allocation method, which takes effector dynamics into account [6]. Doman and Oppenheimer have proposed a compensation method in discrete analysis [4]. Härkegård posed the control allocation problem as a constrained quadratic program, which provides automatic redistribution of the control effort when one actuator saturates in position [7]. Chen et al. developed another compensation method based on linear matrix inequality approach, which shows good performance on the unconstrained control allocation problem [8].

In this paper, the discrepancy caused by actuator dynamics is presented first under INDI controller. Then compensation gains are deduced for typical two actuator dynamics separately in s-domain analysis. Test system is built to test the performance of proposed compensation method. Simulation results show the proposed method can improve tracking performance.

2. Discrepancy caused by actuator dynamics under INDI controller

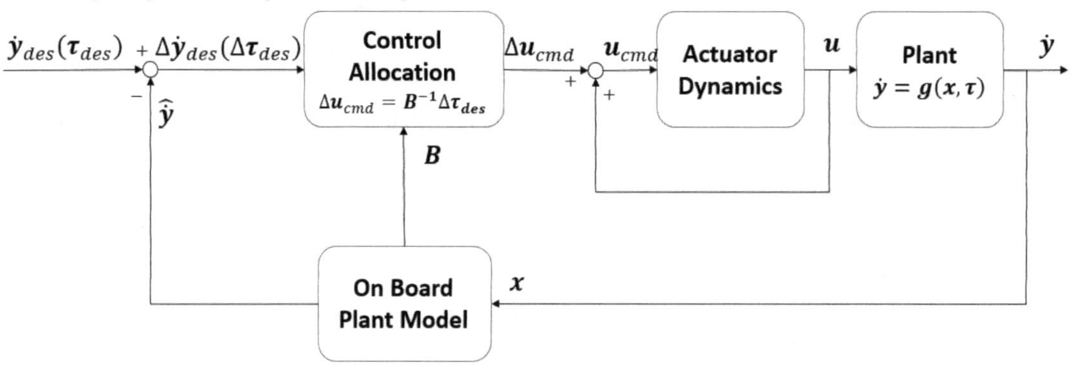

Figure 2. Structure of inner loop of INDI controller

Figure 2 shows the inner loop of flight system under incremental nonlinear dynamic inversion. Control allocation frame allocates the incremental desired generalized moments to each actuator. The basic equation for control allocation to solve is as below,

$$\Delta\boldsymbol{\tau}_{des,m\times1} = \boldsymbol{B}_{m\times n}\Delta\boldsymbol{u}_{cmd,n\times1} \tag{3}$$

Where, $\Delta\boldsymbol{\tau}_{des,m\times1}$ is desired incremental virtual input vector, which is equal to $\Delta\dot{\boldsymbol{y}}_{des}$, namely desired incremental generalized moment vector. $\Delta\boldsymbol{u}_{cmd,n\times1}$ is incremental actuator command vector. Basically, n is bigger than m in an over-actuated system. Equation 3 is different with Equation 2, due to the special structure of INDI controller.

Compared to unconstrained control allocation, constrained control allocation takes into consideration of both effectors absolute position limits and rate limits [9]. In an INDI controller, actuator commands also need constraining to respect the most restrictive of the rate or position limits in a sampling time. Mathematically, for a single actuator, the constraints are defined as set $\boldsymbol{\Omega}$,

$$\boldsymbol{\Omega} = \{\Delta u | \underline{\Delta u} \le \Delta u \le \overline{\Delta u}\} \tag{4}$$

Where,

$$\begin{aligned} \underline{\Delta u} &= \max(u_{min} - u, \dot{u}_{min}\Delta t) \\ \overline{\Delta u} &= \min(u_{max} - u, \dot{u}_{max}\Delta t) \end{aligned} \tag{5}$$

u is current respond of actuator and $\dot{u}_{max}$, $\dot{u}_{min}$ are the maximum rate and minimum rate of actuator, respectively. u_{max}, u_{min} are absolute upper and lower bounds of an actuator, respectively. $\overline{\Delta u}$, $\underline{\Delta u}$ are the incremental upper bound and incremental lower bound respectively in next sampling time and Δt is sampling period of the digital flight control system.

From Figure 2, the control allocation frame of INDI controller does not directly tell effectors to deflect to a desired absolute position, but to deflect a certain degree which is the difference between desired position and current position [1]. Ideal situation is that actuators are infinitely fast, which means $u = u_{cmd}$ or transfer function of actuator is $\boldsymbol{G}_u(s) = 1$. Under this assumption, transfer function from Δu_{cmd} to u can be deduced directly,

$$G_u''(s) = \frac{u}{\Delta u_{cmd}} = \frac{1}{\Delta t \cdot s} \tag{6}$$

In fact, this ideal case can never be achieved due to the existence of actuator dynamics. Here are two figures to present the discrepancy caused by actuator dynamics under INDI controller.

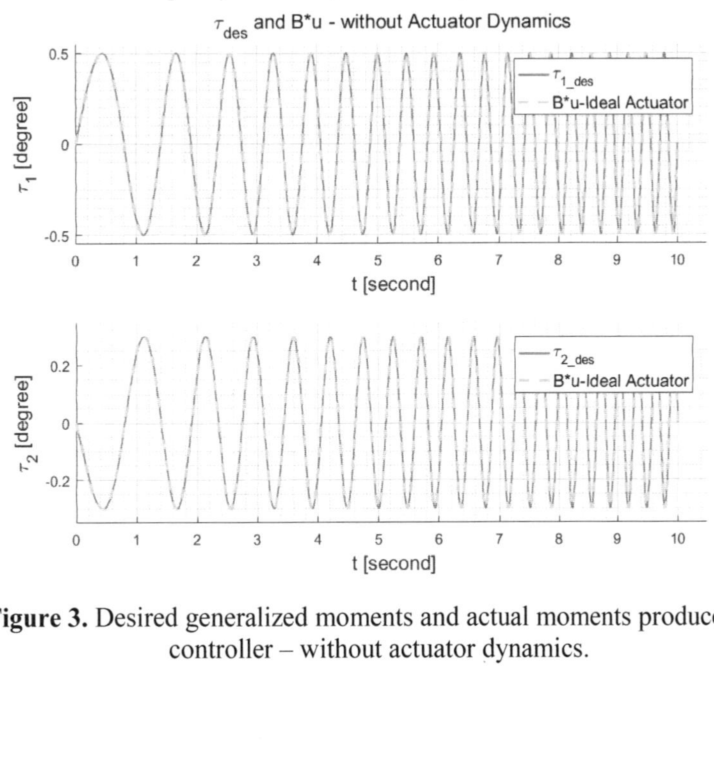

Figure 3. Desired generalized moments and actual moments produced by controller – without actuator dynamics.

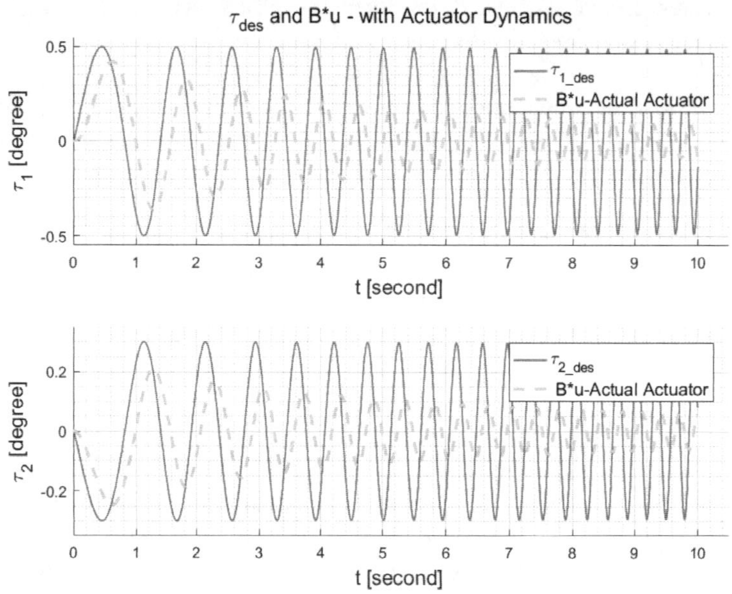

Figure 4. Desired generalized moments and actual moments produced by controller – with actuator dynamics.

In Figure 3, actuators are ignored, the desired result is achieved, which means $\boldsymbol{Bu} = \boldsymbol{\tau}_{des}$. However, in Figure 4, when actuator dynamics are taken into account, it is evident that the tracking error in this case is quite large. Notice that, not only actuator dynamics cause the discrepancy of the system, but the positive feedback, which envelops actuator dynamics in INDI controller, also entails this consequence. Thus, it is obliged to propose a method to compensate this discrepancy.

3. Compensation Method

The objective of compensation method is to modify control allocation framework and actuator dynamics such that $G(s) = u/\Delta u_{cmd} = 1/(\Delta t \cdot s)$. Idea of this method is putting s-domain gains, which post-processe the output of control allocation frame, so that it can directly augment the incremental actuator command. The basic structure is shown as Figure 5,

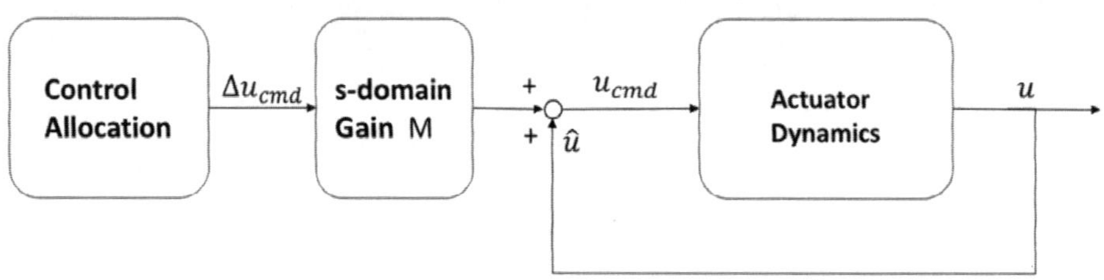

Figure 5. Structure of compensation method

For first-order actuator dynamics, assume transfer function is below,

$$G_{PT1}(s) = \frac{u}{u_{cmd}} = \frac{1}{Ts+1} \tag{7}$$

In Figure 2, transfer function from Δu_{cmd} to u can be deduced before adding a s-domain gain M,

$$G_{PT1}'(s) = \frac{u}{\Delta u_{cmd}} = \frac{1}{Ts} \tag{8}$$

In Figure 5, transfer function from Δu_{cmd} to u can be deduced after adding a s-domain gain M,

$$G''_{PT1}(s) = \frac{u}{\Delta u_{cmd}} = M \cdot G_{PT1}'(s) = \frac{M}{Ts} \tag{9}$$

The objective is to find a gain, M, that modifies the output of control allocation algorithm such that,

$$G''_{PT1}(s) = \frac{1}{\Delta t \cdot s} \tag{10}$$

solving for M yields,

$$M = \frac{T}{\Delta t} \tag{11}$$

Where, compensation gain M is a number and $M > 1$, as the time constant of actuator is much bigger than sampling time. Since M can be computed from the known quantities T and Δt, one can compensate for command increment attenuation using Equation 11 above. For a bank of decoupled first-order actuators with nominal bandwidths of T_i and parameter a_i corresponding values of M_i can be computed using $M_i = T_i/\Delta t$.

For second-order actuator dynamics, assume transfer function is below,

$$G_{PT2}(s) = \frac{u}{u_{cmd}} = \frac{\omega^2}{s^2 + 2\xi\omega s + \omega^2} \tag{12}$$

In Figure 2, transfer function from Δu_{cmd} to u can be deduced before adding a s-domain gain M,

$$G'_{PT2}(s) = \frac{u}{\Delta u_{cmd}} = \frac{G_{PT2}(s)}{1 - G_{PT2}(s)} = \frac{\omega^2}{s^2 + 2\xi\omega s} \tag{13}$$

In Figure 5, transfer function from Δu_{cmd} to u can be deduced after adding a s-domain gain M,

$$G''_{PT2}(s) = \frac{u}{\Delta u_{cmd}} = M \cdot G'_{PT2}(s) = \frac{M \cdot \omega^2}{s^2 + 2\xi\omega s} \tag{14}$$

The objective is to find a gain, M, that modifies the output of the control allocation algorithm such that,

$$G''_{PT2}(s) = \frac{1}{\Delta t \cdot s} \tag{15}$$

solving for M yields,

$$M = \frac{s^2 + 2\xi\omega s}{\Delta t \cdot \omega^2 \cdot s} \tag{16}$$

Where, compensation gain of this case is more complex than that of first order actuator. M is a transfer function whose order of numerator is bigger than the order of the denominator, which is non-causal and non-realistic. Since the expected behavior is an integration, it is desired for the element before the integrator to have the same steady state gain, by using final value theorem,

$$\lim_{s \to 0}\left(s \cdot G''_{PT2}(s)\right) = \lim_{s \to 0}\left(s \cdot \left(\frac{M \cdot \omega^2}{s^2 + 2\xi\omega s}\right)\right) = \frac{1}{\Delta t} \tag{17}$$

Solution of the equation gives,

$$M = \lim_{s \to 0}\left(\frac{s^2 + 2\xi\omega s}{s \cdot \omega^2 \cdot \Delta t}\right) = \frac{2\xi}{\Delta t \cdot \omega} \tag{18}$$

Where, gain M is a number can be calculated easily. Since M can be computed from the known quantities ξ, ω and Δt, one can compensate for command increment attenuation using Equation 20. For a bank of decoupled second-order actuators with nominal natural frequency ω_i and damping ratio ξ_i, corresponding values of M_i can be computed by using $M_i = 2\xi_i/(\Delta t \cdot \omega_i)$.

Referring to Figure 5, for a system with more than one actuator, compensation gain $\boldsymbol{M}$ would be a diagonal matrix with entries along the main diagonal being $M_1, M_2, M_3 \ldots M_n$. Here, M_i would be computed using Equation 11 and Equation 18, corresponding the actuator type and their parameters.

4. Simulation Results

In this section, results from a simulation of a test system displayed in Figure 6 will be shown. Test system is an inner loop of INDI controller for an over-actuated system with 5 actuators. Control allocation algorithm is minimum error redistributed pseudo-inverse (MERPI) proposed by Jin, which is a linear programming algorithm [10]. In the inner loop of INDI controller, the plant block just considers the process from directive $\boldsymbol{\tau}_{des}$ to produce of actual moment $\boldsymbol{\tau}_p$ but neglects transformation from actual moment to the attitude change of flight. For this reason, some dynamics process happened in the outer loop is neglected. Here, sampling time is $\Delta t = 1/200$. The indicator is tracking error between desired generalized moment vector $\boldsymbol{\tau}_{des}$ to actual moment vector $\boldsymbol{\tau}_p$. Here, overall system goal is to achieve $\boldsymbol{u}$ such that the real moment vector $\boldsymbol{\tau}_p$ produced by deflection of effectors can follow $\boldsymbol{\tau}_{des}$ as,

$$\boldsymbol{\tau}_p = \boldsymbol{\tau}_{des} \tag{19}$$

subject to,

$$\boldsymbol{\tau}_p = \boldsymbol{B}\boldsymbol{u} \tag{20}$$

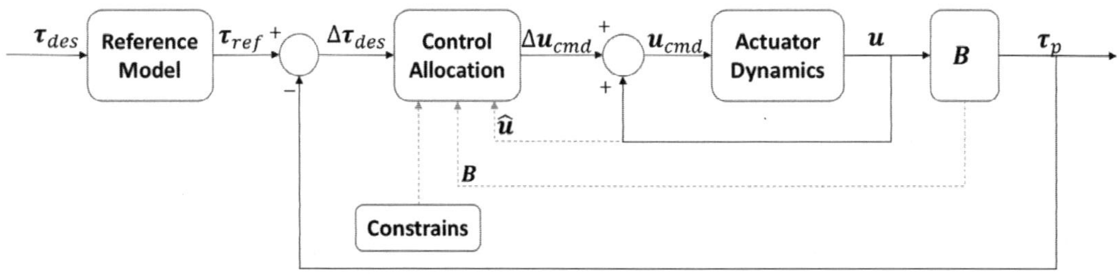

Figure 6. Structure of test system

$\boldsymbol{\tau}$ is chosen as two dimensions' vector as below,

$$\boldsymbol{\tau} = \begin{bmatrix} \tau_1 \\ \tau_2 \end{bmatrix} = \begin{bmatrix} 0.4\sin(0.3\pi t^2 + \pi t) \\ -0.3\sin(0.3\pi t^2 + \pi t) \end{bmatrix} \tag{21}$$

In this simulation, assume value of control effectiveness matrix $\boldsymbol{B}$ is below:

$$\boldsymbol{B} = \begin{bmatrix} 0.3091 & -0.3346 & -0.6043 & 0.2922 & 0.6753 \\ -0.7023 & -0.6807 & 0.4852 & -0.9145 & -0.6983 \end{bmatrix} \tag{22}$$

As for the actuators system block, it contains 5 independent mutually actuators in total. Specifically, the dynamics of each actuator are chosen as below,

$$\begin{aligned} \frac{u_1}{u_{cmd_1}} &= \frac{36}{s^2 + 9.6s + 36} \\ \frac{u_2}{u_{cmd_2}} &= \frac{36}{s^2 + 9.6s + 36} \\ \frac{u_3}{u_{cmd_3}} &= \frac{25}{s^2 + 10s + 25} \\ \frac{u_4}{u_{cmd_4}} &= \frac{5}{5s + 1} \\ \frac{u_5}{u_{cmd_5}} &= \frac{3}{3s + 1} \end{aligned} \tag{23}$$

Each actuator is rate and position limited by the following values,

$$\begin{aligned} \underline{\boldsymbol{u}}^T &= [-3\ \ -2\ \ -4\ \ -5\ \ -3](\text{degree}) \\ \overline{\boldsymbol{u}}^T &= [+1\ \ +5\ \ +4\ \ +5\ \ -3](\text{degree}) \\ \dot{\boldsymbol{u}}^T{}_{min} &= [-3\ \ -5\ \ -2\ \ -10\ \ -4](\text{degree/second}) \\ \dot{\boldsymbol{u}}^T{}_{max} &= [+3\ \ +5\ \ +2\ \ +10\ \ +4](\text{degree/second}) \end{aligned} \tag{24}$$

Proposed method is implemented to test system and comparison are made with the discrete gains in Michael W. Oppenheimer and David B. Doman's paper [4].

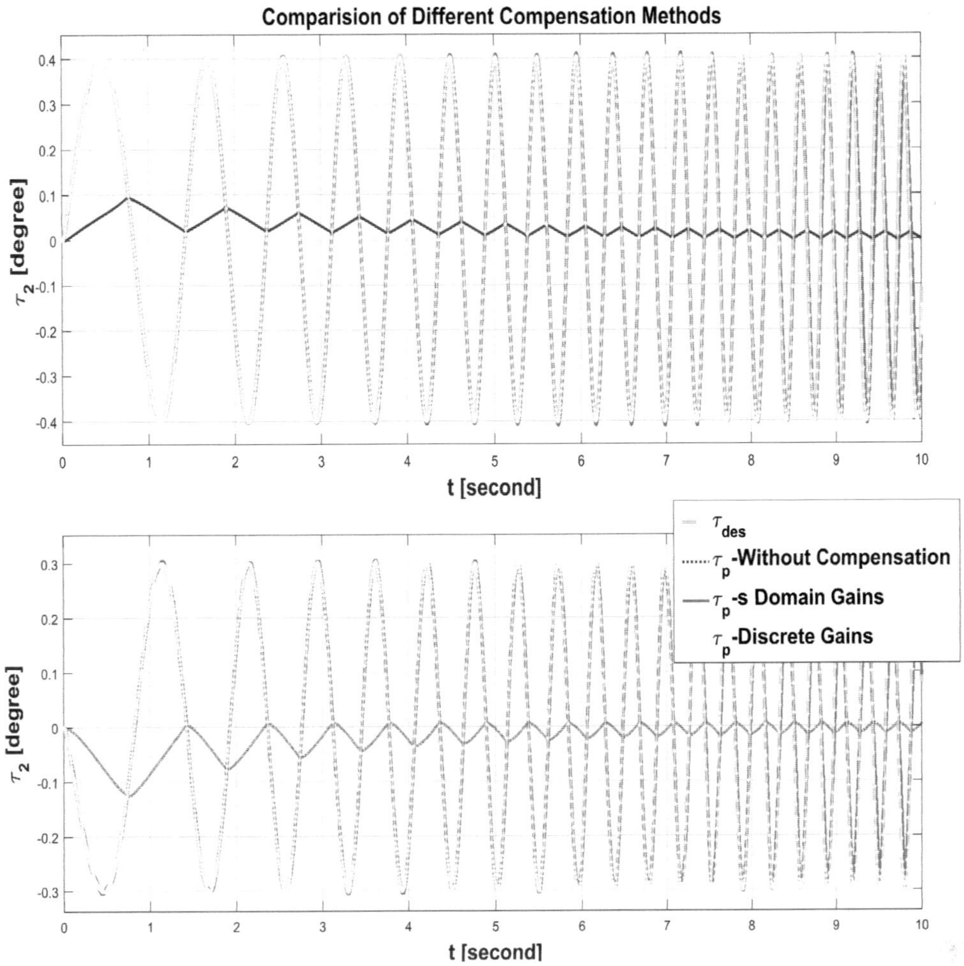

Figure 7. Simulation result

From Figure 7, without augmentation, the original system cannot follow the $\boldsymbol{\tau}_{des}$ definitely and a large error exists between them. Both proposed method and the method in reference [4] can track the reference signal of $\boldsymbol{\tau}_{des}$ closely. Signals of $\boldsymbol{\tau}_p$ under these two compensation methods are almost overlap with signal of $\boldsymbol{\tau}_{des}$. Compared to original system, proposed method significantly improve the tracking performance.

5. Conclusion

Interactions between control allocation algorithm and actuator dynamics degrades the performance of INDI controller. A method is proposed to deal with this problem. Compensation gains are deducted under first-order actuator dynamics and second-order actuator dynamics, separately. Simulation results show that proposed method can reduce the tracking error greatly. There are two advantages of proposed method compared to the compensation method in Reference [4]. First, the form of M is simpler. Especially, for these two typical actuator dynamics, both compensation gains are numbers. Thus, it is quite easy to be calculated. Second, proposed method does not need to know actuators' state of previous step, which is necessary for method in Reference [4]. Therefore, there is no need to put extra estimator to observe actuator.

Reference

[1] Zhang, J. N. 2016 Nonlinear modelling and control of a fighter aircraft. Master thesis, Technical

University of Munich, Germany.

[2] Smeur, E. J., Chu, Q. P., and Croon, G. C. D. 2016 Adaptive incremental nonlinear dynamic inversion for attitude control of micro aerial vehicles. *Journal of Guidance Control & Dynamics,* 39(3), 1-12.

[3] Johansen, T. A., and Fossen, T. I. 2013 Control allocation—a survey." *Automatica,* 49(5), 1087-1103.

[4] Oppenheimer, M. W., and Doman, D. B. 2005. A Method for Compensation of Interactions Between Second-Order Actuators and Control Allocators. *IEEE Aerospace Conference* (pp.1-8). IEEE.

[5] Bolling, J. G. 1997 Implementation of constrained control allocation techniques using an aerodynamic model of an f-15 aircraft.

[6] Venkataraman, R., Oppenheimer, M., and Doman, D. 2004 A new control allocation method that accounts for effector dynamics. 4, 2710-2715 Vol.4.

[7] Härkegård, O. 2012 Dynamic control allocation using constrained quadratic programming. *Journal of Guidance Control & Dynamics,* 27(6), 1028-1034.

[8] Chen, Y., Dong, X., Xue, J., and Wang, F. 2010 Coupling compensation control allocation of dynamic actuator system. *Chinese Journal of Scientific Instrument.*

[9] Ma, J. J. 2009 Research andapplication of control allocation theory for overactuated systems. Ph.D. thesis, National University of Defense Technology, Changsha, China.

[10] Jin, J. 2005 Modified pseudoinverse redistribution methods for redundant controls allocation. *Journal of Guidance Control & Dynamics,* 28(5), 1076-1079.

IOP Conf. Series: Materials Science and Engineering **428** (2018) 012049 doi:10.1088/1757-899X/428/1/012049

Improving Social Responsibility of Artificial Intelligence by Using ISO 26000

Wei-wei Zhao

China National Institute of Standardization,No.4 Zhichun Road, Haidian District, Beijing , China

Zhaoww@cnis.gov.cn

Abstract. The vigorous development of artificial intelligence has had a profound and long-term impact on human production and life. It is a double-edged sword. While letting people enjoy the good life created by new technology, it also allows people to feel its negative effects, such as infringing on human privacy, and bringing new inequalities to human beings. Discussing the social responsibility of artificial intelligence has become a hot topic in academic circles in the past two years. This article starts with adopting the research framework of ISO 26000, comprehensively analyzing the problems of artificial intelligence social responsibility in theory and practice, and putting forward their own thinking. It is concluded that in the age of artificial intelligence, we will proceed from the seven themes of this standard to enhance the social responsibility of artificial intelligence, and ultimately achieve the sustainable development of artificial intelligence by adopting the social responsibility international standard ISO 26000,.

1. Introduction

With the rapid development of artificial intelligence, it has more and more influence on nature and human society. Artificial intelligence (AI) refers to the effect produced by the realization of human minds through computers. It is the description and construction of agents that can obtain perception and execute actions from the environment. The feature of this technology is that they have the ability to automatically learn and acquire knowledge from the data and use this knowledge to help humans achieve their technical goals. With regard to the tremendous social changes that artificial intelligence may bring, people are accompanied by uneasy emotions. In March 2018, Facebook's share price fell sharply because of the fact that Cambridge Analytica, a political consultancy, obtained millions of users' personal information from Facebook. The latest estimate of the affected users is as high as 87 million. The disclosure of this news has caused Facebook to fall into a crisis of trust among users, advertisers, employees, and investors.

We enjoy the conveniences offered by artificial intelligence, at the same time,we need think seriously about the society, the environment, and business ethics challenges that it can bring us, and how we should deal with them. Isaac Asimov set out his famous three laws of robotics in 1942. Richard O. Mason put forward the famous "PAPA" theory in his four Ethical issues in the Information Age[1]. Elon Musk, Stephen Hawking, along with hundreds of other researchers, technology leaders and scientists have joined forces to support the 23 basic principles (Asilomar AI Principles) in 2017 that artificial intelligence should follow in the areas of productivity, ethics, and security. For a number of reasons, awareness about the social responsibility of AI is increasing. However, there is not much literature to discuss artificial intelligence from the perspective of ISO 26000. It provides guidance on the underlying principles of social responsibility, recognizing social responsibility and engaging stakeholders, the core subjects and issues pertaining to social responsibility and on ways to integrate socially responsible behaviour into the organization.

2. Measures

Based on ISO 26000[2],social responsibility means "responsibility of an organization for the impacts of its decisions and activities on society and the environment, through transparent and ethical behaviour that contributes to sustainable development, including health and the welfare of society; takes into account the expectations of stakeholders; is in compliance with applicable law and consistent with international norms of behaviour; and is integrated throughout the organization and practised in its relationships." To define the scope of its social responsibility, identify relevant issues and set its priorities, an organization should address the following core subjects (see Figure 1): organizational governance; human rights; labor practices; the environment; fair operating practices; consumer issues and community involvement and development. This article comprehensively analyze the problems of artificial intelligence on social responsibility in theory and practice by using the research framework of ISO 26000.

Figure 1. Social responsibility core subjects bases on ISO 26000.

3. Social responsibility core subjects of AI

3.1. Organizational governance

Organizational governance is the most crucial factor in enabling artificial intelligence to take responsibility for the impacts of its decisions and activities and to integrate social responsibility throughout the organization and its relationships. In April of 2018, due to a large number of communications challenging the moral bottom line and even breaking the law, some online video platform have been ordered by competent authorities of China to rectify[3]. The platform has exploded to challenge the moral bottom line and pushed it to more users through algorithmic recommendation. In the online video industry, such chaos also abounds.

3.2. Human rights

In October 2017, Saudi Arabia granted robotic Sofia citizenship produced by Hanson Robotics, which is the first robot in history to acquire citizenship. Making the machine aware may be the most essential goal of artificial intelligence research. This is also a source of concern for artificial intelligence. Self-aware artificial intelligence may conflict with humans and rebel against humanity. Artificial intelligence has some humanoid functions. Does it have human rights? What if AI evolves smarter than humans? Will artificial intelligence lose control? If an AI has a goal, it may try to achieve it in a way that humans do not approve of. However, humans cannot stop it. AI can destroy humans without needing to be very evil. Can AI rule humans? How to limit the behavior of artificial intelligence?

3.3. Labor practices
AI simulates and expresses human's wisdom power. It is superior to human working speed, superior to human work accuracy, better than human work attitude, and assists human beings in solving various problems, including dangerous situations and extreme environments. But at the same time, this advantage will also have an adverse effect on humans. First of all, there will be a lot of jobs that will be replaced by robots in the future. All of which will have an impact on the rights of existing workers. Secondly, robots grab people's jobs and AI "worker groups" are forming. How to protect robot rights or how to prohibit the abuse of robots? There are two problems to be solved in the copyright law of AI. Do robots design works with rights? Should rights belong to the machine or to the person who created the machine?

3.4. The environment
The rapid development of artificial intelligence, followed by more and more electronic waste. Electronic waste is an electrical or electronic equipment that is abandoned and unused. Some of them contain lead, arsenic, mercury, cadmium and other harmful substances. These substances are not only very difficult to remove, but also are very toxic and are very harmful to the human body and the environment.

3.5. Fair operating practices
Fair operating practice issues in AI arise in the areas of anti-corruption, fair competition, socially responsible behaviour, relations with other organizations and respect for property rights[4]. How to make artificial intelligence tools better identify false accounts and prevent foreign forces from interfering with elections is a very important issue . This is a new " arms race " .

If you have AI technology, you will have wealth. Tech giants will enjoy rich resources of big data, which may pose a threat to mankind, although it cannot be inferred. Big date's monopoly will become a barrier and a tool for industry giants to pursue their personal interests. It may lead to the accumulation of wealth in the hands of a small number of people, resulting in new inequalities.

3.6. Consumer issues
Some artificial intelligence products are a good kind of auxiliary work for workers, soldiers and disabled people who can not communicate normally because of noisy environment. For the enterprise, after collecting and analyzing the data of the main information, such as the social attribute, the living habit, the consumption behavior and so on, the enterprise perfectly abstracts the complete picture of a user's business as the basic way for the enterprise to apply AI technology in the later stage. For different users to do precision marketing. Enterprises want to obtain higher income, mainly through the way of price discrimination to convert low value users into high value users. Price discrimination means when an enterprise sells goods or services of the same rank and quality to different recipients.

On privacy issue, it is easy to touch and even disclose privacy and secret information by using AI. Therefore, at this stage, AI security concerns may need more attention than to worry about artificial intelligence surpassing human beings.

3.7. Community involvement and development
In addition to offline communities, there are online communities. The leakage of Facebook user data reflects the fact that personal privacy has been embezzled in the era of big data. In the digital age, we often face the sharp opposition between public safety and personal privacy, which has caused a moral and ethical dilemma. In such a dilemma, it is a real dilemma for tech giants with information data to

strike a balance between personal and public responsibility. With the rise of social media, some say privacy is dead. People don't keep their information secret anymore, but they volunteer to share their information online, but it just means people are adopting new definitions and new privacy rules. Keeping information private does not mean keeping it secret, people want to control the information share to whom and how to use the shared information [5].

4. Suggestions

4.1. Responsible attitude

The policy makers, researchers and consumers of products need to take responsibility towards artificial intelligence, because once the artificial intelligence technology is developed, it is unlikely will fall back. We can't make fun of the fate of human beings, let alone use artificial intelligence as a tool to fight against nature. With the development of artificial intelligence, We should amend the legal system more fairly and effectively, and manage the risks associated with artificial intelligence.

Enhancing the social responsibility of scientists. Where the development of artificial intelligence technology go has not got a unified opinion. Technology itself is not good or evil. It is only a tool. The key is to master and use tools. The development of science and technology needs to put the interests of mankind first. The users and developers of artificial intelligence technology should be responsible for the social consequences of their own behavior and strictly fulfill their social responsibility in the early stage of technology development.

4.2. Integrate the social responsibility concept into the artificial intelligence technology

In the development of artificial intelligence technology, it is necessary to introduce a corresponding social responsibility management system to consider the environment, labor rights and interests, and consumers from the beginning, in addition to considering the rationality, science and effectiveness of technology. The core theme of social responsibility based on ISO 26000, such as fair operation practice, will greatly avoid ethical conflicts that conflict with human development. Artificial intelligence is a new technology, and many social responsibility problems have not really emerged in the development. If we wait until these problems are exposed and solved, it may be too late. If we can strength the exchange between scientists in artificial intelligence and experts in social responsibility, integrate social responsibility management into the technological development of artificial intelligence, which will promote the technology of artificial intelligence to be more harmonious, sustainable and human, and ultimately help to alleviate some ethical problems in artificial intelligence technology.

4.3. Legislative norm

The current international has not yet promulgated laws and standards on artificial intelligence aspects of the general social responsibility to restrain and standardize its development, we need to do preventive work, after all, the development of artificial intelligence technology may bring certain risks to human beings. It is the responsibility of all mankind to improve the social responsibility standard of artificial intelligence technology, this is important ,which means to regulate the development of technology, people of all countries need support and cooperation to the ability of mankind, which will make the sustainable development of artificial intelligence technology.

The development of AI can not bring harm to human beings, including human life and the spirit of human. It should respect the right of human privacy . It means that the peace of private life and private information enjoyed by human beings, which are protected according to law from illegal intrusion, knowledge and collection by others. The right to use and make public a personal right. That is, It is the right to ensure that a person does not want to be known by others in private affairs. It involves all aspects of life and is the most advantageous means and excuse for protecting personal interests in modern life.

5. Social responsibility practice of AI

5.1. Organizational governance

Effective governance should be based on incorporating the principles of social responsibility into decision making and implementation on artificial intelligence. Firstly, the AI company may adhere to the full integration of social responsibility principles and concepts into corporate development strategy, production management and corporate culture. Secondly, the AI company may follow some principles, such as accountability, transparency, ethical behaviour, respect for stakeholder interests, respect for the rule of law, respect for international norms of behaviour .Thirdly, the AI company may establish an all-dimensional social responsibility management system that effectively promotes the implementation of corporate social responsibility in achieving economic benefits and social benefits.

Leadership is also critical to effective organizational governance. This is true not only for decision making but also for employee motivation to practice social responsibility and to integrate social responsibility into organizational culture.

5.2. Human rights

When artificial intelligence discovers that even human beings themselves are doing limitless harm to the same kind in a limited human history, how can robots be understood not to harm human beings? And in the face of people who do not speak logic and cannot calculate machines, why does artificial intelligence have to obey unconditionally? So relying entirely on artificial intelligence is unreliable, and humans are ultimately responsible for their destiny. To ensure that humans can control robots. Humans are prohibited from using robots illegally. Prevent human abuse of robots. But at the same time, humans should realize that intelligent robots are essentially machines, but also some of the attributes of intelligent robots Respect is respect for human beings themselves.

5.3. Labor practices

In the era of artificial intelligence, companies must have employees with certain capabilities. Specifically includes creativity and problem solving, information literacy, self-awareness and self-regulation, critical thinking, learning and lifelong learning, civic responsibility and social participation. AI company may cooperate with local universities and colleges, provide internship opportunities and practical training posts, participate in course teaching.

5.4. The environment

First of all, raw materials as far as possible environmental protection materials, choose recyclable materials as product materials. Second, the production process to achieve environmental protection. Finally, the product is green and environmental protection, the formed product is conducive to decomposition or recycling.

5.5. Fair operating practices

We can use artificial intelligence to collect remote identity information and verify the true identity of individuals through face recognition. In fact, no matter how advanced algorithms and technologies are, they need value-oriented guidance. Behind the algorithm is the human, the defect of the algorithm is the defect of the values. The Internet is not an extrajudicial place. Behavior in cyberspace, including business innovation, must be on the rule of law track. For a long time, there has been a view that the Internet should be completely self-disciplined, to survive the fittest by itself. Now it has been found that this is not the case. If this is the case, it will probably affect the values of a generation, especially young people. Minor users are strictly prohibited after identifying minors through AI user portrait technology Open certain authority, optimize algorithm recommendation and report disposal process and other measures to create a good network space.

5.6. Consumer issues

The wide use of artificial intelligence not only brings efficiency and convenience, but also inevitably brings harm to human beings. When artificial intelligence harms humans, is it a technical loophole for the robot creators, or is it the improper use of intelligent machine managers? Perhaps even artificial intelligence systems go beyond the "self-behavior" of the original control scheme? In this case, it is necessary to analyze the cause of the robot's fault behavior in order to find out the main body of tort and share the liability.

5.7. Community involvement and development

Pay attention to the network community. In fact, no matter how advanced algorithms and technologies are, they need value-oriented guidance. Behind the algorithm is the human, the defect of the algorithm is the defect of the values. The Internet is not an extrajudicial place. Behavior in network community, including business innovation, must be on the rule of law track. For a long time, there has been a view that the Internet should be completely self-disciplined, to survive the fittest by itself. Now it has been found that this is not the case. If this is the case, it will probably affect the values of a generation, especially young people. After identifying minors by AI user portrait technology, it is strictly prohibited for minor users to open certain rights, optimize the algorithm recommendation and report and deal with the process and other measures to create. Good cyberspace.

6. Conclusion

There are a lot of problems in theory and practice of artificial intelligence on social responsibility. However, these problems will not be solved in a short time. The goal of human development has shifted from the pursuit of economic to ecological and social equity. If the artificial intelligence companies are fulfill their social responsibilities, which are more capable of enhancing are long-term development, and it is easy to gain a better trust in the public. Today's society must work together to create a good environment for artificial intelligence development.

Acknowledgments

This work was funded by the National Key R&D Program of China under grant No.2017YFF0207603 and No.2016YFF0202502.

References

[1] Mason, R. O. 1986.Four Ethical Issues of the Information Age. MIS QUART, 10, 1, 5-12.
[2] ISO, ISO 26000:2010 - Guidance on social responsibility., 2010.
[3] Wang Sibei HXYJ. 2018. Live consuming light bulb rap to teach drug use by "sucking eye" platform valuation tens of billions - online video industry chaos survey. In xinhuanet
[4] Johnson, Deborah G., and Mario Verdicchio. "AI, agency and responsibility: the VW fraud case and beyond." AI & SOCIETY (2018): 1-9.
[5] Kekulluoglu, Dilara, Nadin Kokciyan, and Pinar Yolum. "Preserving privacy as social responsibility in online social networks." ACM Transactions on Internet Technology (TOIT) 18.4 (2018): 42.

IOP Conf. Series: Materials Science and Engineering **428** (2018) 012050 doi:10.1088/1757-899X/428/1/012050

Fuzzy Lookup Table Controller (FLTC) via Center Average De-fuzzifier (CAD) with a New Efficient Heaviside Search Algorithm (HSA) for Lookup Table Implementation

Chi-Hsu Wang[1], Kar-Chun Hor[2] and Wei-Chuan Chen[2]

[1] Department of Electrical and Computer Engineering, National Chiao Tung University, 1001 University Road, Hsinchu 300, Taiwan, ROC.
[2] College of Electrical and Computer Engineering, National Chiao Tung University, 1001 University Road, Hsinchu 300, Taiwan, ROC.

E-mail: cwang@cn.nctu.edu.tw; s358389@gmail.com

Abstract. This paper proposes an alternate interpretation of the traditional lookup table controller via the Center Average De-fuzzier (CAD) in Fuzzy Set Theory with a set of Disjoint Block Pulse Membership Functions (DBPMFs). Further, a new Heaviside Search Algorithm (HSA) is also proposed to implement the Fuzzy Lookup Table Controller (FLTC) in a more efficient way. The computation complexity of our new HSA is $\boldsymbol{O}(\Sigma R_i)$, where R_i is the number of membership functions for i^{th} input variable in FLTC. Whereas the average computational complexity in traditional Linear Search Algorithm (LSA) is $\boldsymbol{O}((\prod R_i)/2)$. In comparison with LSA, the HSA also reduces the complexity of coding significantly in the same order as the comparison of computation complexities between HSA and LSA. The balancing of Inverted Pendulum System (IPS) is adopted as the benchmark to show the feasibility and efficiency of this new FLTC and HSA.

1. Introduction

It is well-known that Fuzzy Logical Control (FLC), with its broad feasibility, has been widely applied to lots of industrial and system control applications [1-4]. As an important real-time controller in the past few decades, the traditional Lookup Table Techniques (LUTs) [5-7] have been applied effectively in designing real-time controller with the avoidance of massive computation burden. The de-fuzzification of FLC can also be time-consuming to inhibit its real-time applications.

In this paper, by using the Center Average De-fuzzifier (CAD) with the same Disjoint Block-Pulse Membership Functions (DBPMFs) in [7], we can show that the FLC with DBPMFs is also equivalent to the traditional LUT, so that a Fuzzy Lookup Table Controller (FLTC) can be generated to simplify the de-fuzzification process in FLC. Furthermore, a new efficient Heaviside Search Algorithm (HSA) is also proposed to the implementation of FLTC. It is easy to show that the average computational complexity of applying the traditional Linear Search Algorithm (LSA) to FLTC is $\boldsymbol{O}(\prod_{i=1}^{n} R_i/2)$, where R_i is the number of membership functions for i^{th} input variable. However, the computational complexity of HSA for the search of FLTC is only in the order of $\boldsymbol{O}(\sum_{i=1}^{n} R_i)$. In order to show the feasibility of FLC with DBPMFs to the control of nonlinear systems, the balancing of Inverted Pendulum System (IPS) [8-10] is illustrated via the FLTC with HSA.

2. The Fuzzy Lookup Table Controller (FLTC)

A general Fuzzy Logic Controller (FLC) consists of four modules, i.e., fuzzification module, inference engine module, fuzzy rules base module and de-fuzzification module. In this paper, n is the total number of fuzzy input variables for FLC, m is the total number of output variables for FLC, and L is the total fuzzy rules number in the fuzzy rule base. The total fuzzy rules number can be found via the following equation

$$L = \prod_{i=1}^{n} R_i \tag{1}$$

where R_i is the number of membership functions for i^{th} input variable. The fuzzy rules in the fuzzy rule base can be inferred as

$$\begin{aligned} &\text{Rule } j: \text{ IF } x_1 \text{ is } A_1^j \text{ and ... and } x_n \text{ is } A_n^j \\ &\qquad \text{THEN } u_1 \text{ is } W_1^j \text{ and ... and } u_m \text{ is } W_m^j \end{aligned} \tag{2}$$

where j is the fuzzy rule number (j = 1, 2, …, L), A_i^j are the input membership functions for x_i in the jth fuzzy rule, and W_k^j are the output membership functions for u_k (k = 1, 2, …, m) in the jth fuzzy rule. We adopt product inference engine and the center average de-fuzzifier (CAD) as the de-fuzzification method, as shown in the following equation:

$$u_k = \frac{\sum_{l=1}^{L} \bar{y}_k^l \left(\prod_{i=1}^{n} A_i^l(x_i)\right)}{\sum_{j=1}^{L} \left(\prod_{i=1}^{n} A_i^j(x_i)\right)} = \frac{\sum_{l=1}^{L} \bar{y}_k^l S_l}{\sum_{j=1}^{L} S_j}; \quad (k = 1, \ldots, m) \tag{3}$$

$$S_j = \prod_{i=1}^{n} A_i^j(x_i) \tag{4}$$

where $\bar{y}_k^l$ is the center value of the kth output MF in the lth fuzzy rule, S_j is the product of the MFs' mappings ($A_i^j(x_i)$, i =1, 2, …, n)) for all input variables (x_1, …, x_n) in the jth fuzzy rule, and u_k is the kth crisp output of the FLC (k = 1, 2, …, m). It is noted that the conventional FLC is, in general, an online floating point computation process controller, which is time consuming for real time applications. Therefore, the Lookup Table (LUT) method is a useful tool to overcome this obstacle [7]. Hence, a special Fuzzy Lookup Table controller (FLTC) with DBPMFs will be developed and discussed in this section, in order to minimize the online computing resources of a controller. The DBPMF for the lth MF for the x_i input of FLC can be defined as:

$$A_i^l(x) = \begin{cases} 1, & -\dfrac{\sigma_i^l}{2} + c_i^l < x < \dfrac{\sigma_i^l}{2} + c_i^l \\ 0, & \text{otherwise} \end{cases} \tag{5}$$

where σ_i^l and c_i^l are the lth variance (width) and mean (center) of the DBPMF in x_i. The disjoint properties of DBPMFs will be used to construct the FLTC, which is shown as follows:

$$A_i^l(x) A_i^k(x) = \begin{cases} A_i^l(x), & l = k \\ 0, & l \neq k \end{cases} \tag{6}$$

where l and k = 1, 2, …, R_i. The output of the center average de-fuzzifier in (3) can be simplified to a certain center value of an output MFs via the following Theorem 1.

Theorem 1: For the FLC with DBPMFs as the MFs, the FLC will be simplified to a Fuzzy Lookup Table Controller (FLTC), where every entry { u_k | k = 1, …, m} in FLTC is a center value of a certain output membership function (MF). The u_k corresponds to a certain activated fuzzy rule number i is equal to

$$u_k = \bar{y}_k^i \tag{7}$$

where i is a certain activated fuzzy rule number and is decided by the inference process and $\bar{y}_k^i$ is a center value of a certain output MFs corresponded to a certain fuzzy rule.

Proof: Assume that the FLC has n input variables, k outputs, each input variable has R_i DBPMFs and has total L fuzzy rules, therefore, we can rewrite (3) as:

$$u_k = \frac{\overline{y}_k^1 S_1 + \overline{y}_k^2 S_2 + \cdots + \overline{y}_k^L S_L}{\sum_{j=1}^{L} S_j} \tag{8}$$

Since all the DBPMFs are disjoint with each other, there are only single i^{th} rule activated at each time, and due to the fact that the mapping value for every DBPMFs are equal to 1, therefore we have

$$\begin{cases} S_i = 1, & l = i \\ S_i = 0, & l \neq i \end{cases} \tag{9}$$

so we have

$$\sum_{j=1}^{L} S_j = S_i = 1 \tag{10}$$

Thus, we can rewrite (8) as

$$u_k = \frac{\overline{y}_k^1 * 0 + \overline{y}_k^2 * 0 + \cdots + \overline{y}_k^i * 1 + \cdots + \overline{y}_k^L * 0}{1} \tag{11}$$

Hence, we have the following result

$$u_k = \overline{y}_k^i \tag{12}$$

where i is the index of the activated fuzzy rule.

Q.E.D

In order to show the feasibility of FLTC (Theorem 1) for real control purpose, the popular nonlinear IPS will be balanced by the FLTC in the next section.

3. Balancing the IPS via FLTC

This section shows the feasibility of FLTC for balancing the IPS. The mass of the cart is m_c, the pivot point mass at the upper end of the pendulum pole is m, l is half the length of the pendulum pole, g is the gravity acceleration, θ is the angle of the pole from the origin point and u is the control force applied to the cart. We define the first state variable x_1 equals to θ ($x_1 = \theta$) and the second state variable x_2 equals to the angle velocity of the pole $\dot{\theta}$ ($x_2 = \dot{\theta}$). The dynamic equation of IPS can be expressed as:

$$\begin{cases} \dot{x}_1 = x_2 \\ \dot{x}_2 = \dfrac{g \sin x_1 - \dfrac{mlx_2^2 \cos x_1 \sin x_1}{m_c + m}}{l\left(\dfrac{4}{3} - \dfrac{m\cos^2 x_1}{m_c + m}\right)} + \dfrac{\dfrac{\cos x_1}{m_c + m}}{l\left(\dfrac{4}{3} - \dfrac{m\cos^2 x_1}{m_c + m}\right)} u \end{cases} \tag{13}$$

where the parameters are $g = 9.8\ \text{m/s}^2$, $m_c = 1.0\text{kg}$, $m = 0.1\text{kg}$ and $l = 0.5\text{m}$. The control objective is to balance the pole around the origin position by controlling the cart moving forward ($u < 0$) or backward ($u > 0$) via FLTC. The following Example 1 shows the balancing of the IPS via a 5 by 5 FLTC.

Example 1: The 5 by 5 FLTC to balance the Inverted Pendulum System.

In this example, the 5 by 5 FLTC consists of two inputs ($x_1 = e$ and $x_2 = \Delta e$), and each has 5 DBPMFs ($R_1 = R_2 = 5$). The control intervals for x_1 and x_2 are partitioned into five subintervals or five DBPMFs by using $\eta_1^2 = -0.1047$, $\eta_1^3 = -10^{-6}$, $\eta_1^4 = 10^{-6}$ and $\eta_1^5 = 0.1047$ for x_1, $\eta_2^2 = -0.04$, $\eta_2^3 = -10^{-5}$, $\eta_2^4 = 10^{-5}$ and $\eta_2^5 = 0.04$ for x_2, which are shown in Figure 1a and 1b.

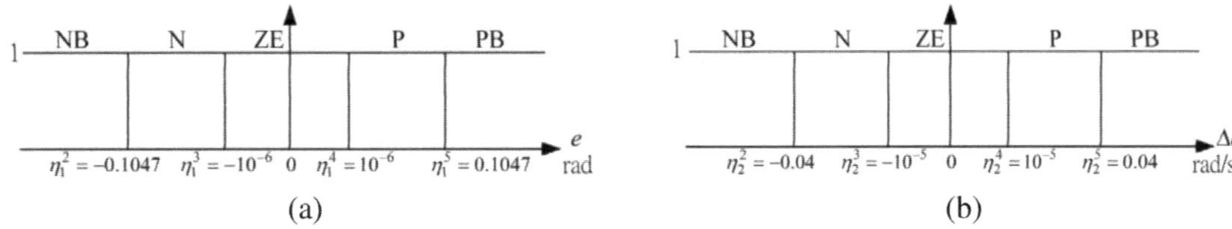

Figure 1. (a) DBPMFs for input variable x_1 (= e). (b) DBPMFs for input variable x_2 (=Δe).

These subintervals also serve as the DBPMFs for the input x_1 and x_2. The boundary values of every subinterval are designed to match the system behaviour of the IPS. Moreover, 7 Gaussian MFs are adopted for output MFs, where the center of each output Gaussian MFs is defined as {NB = -5, NM = -3, NS = -1, ZE = 0, PS = 1, PM = 3, PB = 5}. Therefore, we can further construct a 5 by 5 FLTC (by Theorem 1) with 7 output MFs, which is shown in Figure 2a.

e \ Δe	NB	N	ZE	P	PB
NB	NB (=-5)	NB (=-5)	NM (=-3)	NM (=-3)	NM (=-3)
N	NB (=-5)	NM (=-3)	NS (=-1)	NS (=-1)	NS (=-1)
ZE	NM (=-3)	NS (=-1)	ZE (=0)	PS (=1)	PM (=3)
P	PS (=1)	PS (=1)	PS (=1)	PM (=3)	PB (=5)
PB	PM (=3)	PM (=3)	PM (=3)	PB (=5)	PB (=5)

(a)

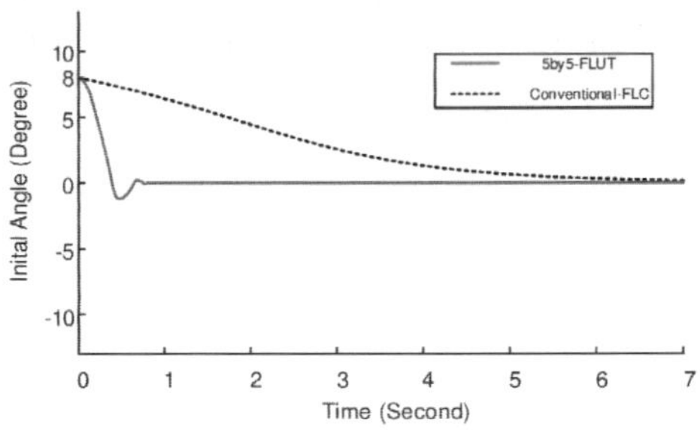

(b)

Figure 2. (a) The FLTC for 5 DBPMFs. (b) The control response of IPS using conventional FLC (black dotted line) and the 5 by 5 FLTC (blue solid line).

We compare the performance of balancing the IPS via 5 by 5 FLTC and the conventional FLC found in [8], where the result is shown in Figure 2b. The initial angle of the IPS is 8 degrees, and the performance of the 5 by 5 FLTC (blue solid line) is better than conventional FLC (black dotted line), where 5 by 5 FLTC takes only 1 second to balance the IPS around the origin and the conventional FLC takes 7 seconds to fully balance the IPS.

4. The Heaviside Search Algorithm (HSA) for FLTC

This section proposes an efficient search algorithm based on the Heaviside function for FLTC to search the corresponding output value. The Heaviside function is expressed as follows:

$$H(\eta)=\begin{cases}0, & \eta<0\\ 1, & \eta\geq 0\end{cases} \tag{14}$$

The Heaviside index for FLTC is expressed as:

$$\Phi(x_i(t))=1+\sum_{j=2}^{R_i} H(x_i(t)-\eta_i^j)\ ;\ \ (i=1,\ldots,n) \tag{15}$$

where $H(\cdot)$ represents the Heaviside function in (14), $x_i(t)$ is the input signal from input x_i at a certain time t, j is the index of DBPMFs and η_i^j is the boundary value for x_i in its jth DBPMF A_i^j. Note that $\eta_i^j < \eta_i^{j+1}$. The Heaviside index $\Phi(x_i(t))$ in (15) returns the accumulative index of the corresponding DBPMF A_i^j. This Heaviside search algorithm can then be performed to find the Heaviside indices for all input values { $x_1(t)$, $x_2(t)$, …, $x_n(t)$ }, and the corresponding control output can be obtained as:

$$u_k = FLTC[\Phi(x_1(t)), x_2(t), \ldots, x_n(t)] \tag{16}$$

where *FLTC* is the n dimension fuzzy lookup table controller and u_k is the control output in the kth (k = 1,…, m) output. The following ***Algorithm 1*** illustrates the Heaviside search algorithm (HSA) for FLTC.

Algorithm 1: Heaviside search algorithm (HSA) for FLTC

Input:

R_i : The number of DBPMFs in each input variable x_i.

$x_i(t)$: Input value from input variable x_i, i=1, …, n.

Output:

$\Phi(x_i(t))$: The Heaviside indices for $x_i(t)$.

Step 1: Define the DBPMFs intervals for all inputs.

For $i=1$ to n
  For $j=2$ to R_i
    Set η_i^j
  End
End

Step 2: Find the all the Heaviside indices $\Phi(x_i(t))$ for every input variables.

$\Phi(x_i(t))=1$
For $i=1$ to n
  For $j=2$ to R_i
    If $(x_i(t)-\eta_i^j)\geq 0$
      $\Phi(x_i(t))=\Phi(x_i(t))+1$
  End
End

Step 3: The output is

$u_k = FLTC[\Phi(x_1(t)),\Phi(x_2(t)),...,\Phi(x_n(t))]$

Step 4. Stop.

5. The comparison of computational and coding complexities between HSA and LSA

This section compares the computational complexities between the HSA and traditional LSA. Assume a FLC has n inputs and m outputs, each input has R_i DBPMFs. Therefore, a FLTC with the dimension equals to $\prod_{i=1}^{n} R_i$ can be constructed. The LSA is roughly shown as follows:

$$\begin{aligned}&\text{if (rule 1) then } u_{1,\dots,m} = \bar{y}_{1,\dots,m}^{1}\\ &\text{else if (rule 2) then } u_{1,\dots,m} = \bar{y}_{1,\dots,m}^{2}\\ &\vdots\\ &\text{else if (rule } L\text{) then } u_{1,\dots,m} = \bar{y}_{1,\dots,m}^{L}\end{aligned} \tag{17}$$

The above (17) shows the fact that we need to prepare a big loop which consists of $L=\prod_{i=1}^{n} R_i$ if-then-else commands in the big loop, which is a cumbersome job for programmers. It is obvious that the computational complexity of LSA is in the order of $O(\prod_{i=1}^{n} R_i/2)$. On the other hand, the HSA for a FLTC with n inputs is shown as follows:

$$\begin{aligned}u_k = FLTC[&\Phi(x_1(t)) = 1 + H_2(x_1(t)-\eta_1^2)+\cdots+H_{R_1}(x_1(t)-\eta_1^{R_1}),\\ &\Phi(x_2(t)) = 1 + H_2(x_2(t)-\eta_2^2)+\cdots+H_{R_2}(x_2(t)-\eta_2^{R_2}),\\ &\vdots \qquad ,\\ &\Phi(x_n(t)) = 1 + H_2(x_n(t)-\eta_n^2)+\cdots+H_{R_n}(x_n(t)-\eta_n^{R_n})]\end{aligned} \tag{18}$$

where k = 1, 2, …, m. The illustration for HSA is shown in Figure 3 and it is obvious that the coding of HSA requires only n small loops to find their respective indices, that only consists of $\sum_{i=1}^{n} R_i$ if-then-else commands, which is much simpler than that of LSA.

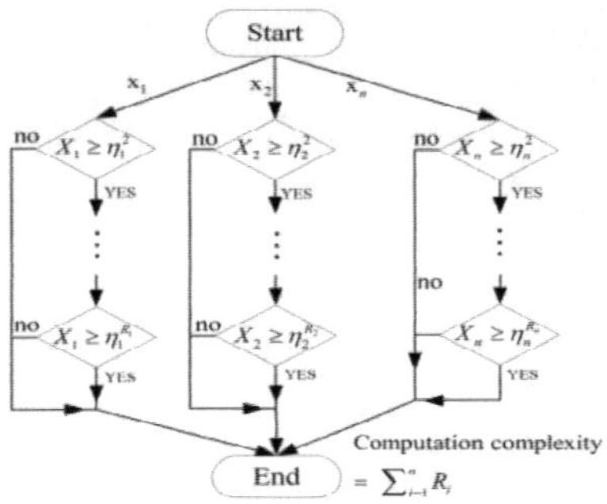

Figure 3. The illustration of FLT using Heaviside searching algorithm.

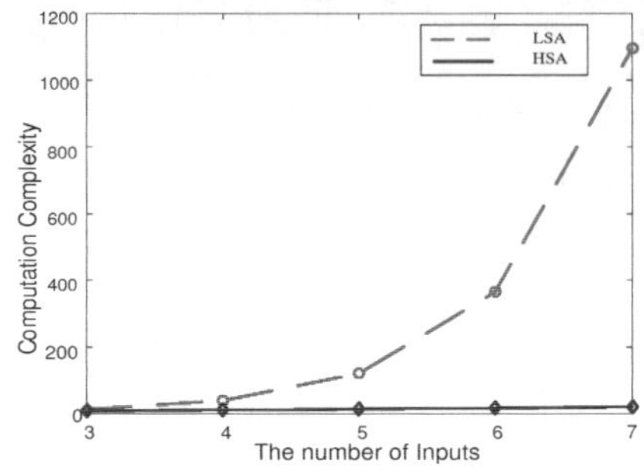

Figure 4. The comparison of computational complexities between HSA and LSA with three DBPMFs for each input.

Note that the number of if-then-else commands in coding is identical to the order of computational complexity. According to (18) and Figure 3, the computational complexity for Heaviside search algorithm is in the order of $\boldsymbol{O}(\sum_{i=1}^{n} R_i)$. The following experiment compares the computational complexities of both HSA and LSA by increasing the number of input from 3 to 7 with fixed 3 DBPMFs for every input x_i (R_i = 3), which is shown in Figure 4. It is obvious to find that the computational complexity of HSA is significant lower than that of LSA while increasing the number of inputs. The table searching time of HSA is generally faster than LSA with an approximate ratio of $(\prod_{i=1}^{n} R_i)/(2*\sum_{i=1}^{n} R_i)$. It is worth mentioning that the coding of HSA is also much simpler compared with traditional LSA, especially for the FLTC with large amount of inputs or DBPMFs.

6. Conclusion

This paper proposes a new interpretation of de-fuzzification via Center Average De-fuzzification when the input MFs are the Disjoint Block Pulse Membership Functions (DBPMFs). This will result in a Fuzzy Lookup Table Controller (FLTC). We also propose a new efficient Heaviside Search Algorithm (HSA) for a much more efficient table search performance. In comparison with the traditional Linear Search Algorithm (LSA) for table search, the search performance of HSA is not only superior, but the coding simplicity is also obtained. The balancing of Inverted Pendulum System (IPS) is adopted as the benchmark to show the feasibility of FLTC. The control performance of adopting a simple 5 by 5 FLTC is much better than that of the traditional Fuzzy Logic Controller (FLC). The comparison of computation complexities for HSA and LSA is also performed. For more inputs, the performance of HSA will be dramatically much better than that of LSA.

Acknowledgments

This work was supported by the Ministry of Science and Technology (MOST) of Taiwan (No. 106-2221-E-009 -062 -).

References

[1] Sugeno M 1985 *Industrial applications of fuzzy control* (Amsterdam, Netherland: Elsevier)

[2] Wang C H and Huang D Y 2013 A new intelligent fuzzy controller for nonlinear hysteretic electronic throttle in modern intelligent automobiles *IEEE Trans. Ind. Electron.* **60**(16) 2332-45

[3] Ali M H, Murata T and Tamura J 2004 The effect of temperature rise of the fuzzy logic-controlled braking resistor on transient stability *IEEE Trans. Power Syst.* **19**(2) 1085-95

[4] Schouten N J, Salman M and Kheir N 2002 Fuzzy logic control for parallel hybrid vehicles *IEEE Trans. Control Syst. Technol.* **10**(3) 460-68

[5] Rad A B, Chan P T, Lo W L and Mok C K 2003 An online learning fuzzy controller *IEEE Trans. Ind. Electron.* **50**(5) 1016- 21

[6] Hwang C L 2008 Microprocessor-based fuzzy decentralized control of 2-D piezo-driven systems *IEEE Trans. Ind. Electron.* **55**(3) 1411-20
[7] Wang C H and Wen J S 2008 On the equivalence of a table lookup (TL) technique and fuzzy neural network (FNN) with block pulse membership functions (BPMFs) and its application to water injection control of an automobile *IEEE Trans. Syst., Man, Cybern. C, Applicat. Rev.* **38**(4) 574-80
[8] Wang L X 1997 *A Course in Fuzzy Systems and Control* (Upper Saddle River, NJ: Prentice Hall)
[9] Jung S, Cho H T and Hsia T C 2007 Neural network control for position tracking of a two-axis inverted pendulum system: Experimental studies *IEEE Trans. Neural Netw.* **18**(4) 1042-48
[10] Huang C H, Wang W J and Chiu C H 2011 Design and implementation of fuzzy control on a two-wheel inverted pendulum *IEEE Trans. Ind. Electron.* **58**(7) 2988-3001

IOP Conf. Series: Materials Science and Engineering **428** (2018) 012051 doi:10.1088/1757-899X/428/1/012051

Motion Control of Manipulators Based on Model-free Adaptive Control

Ziqiao Zhang and Songyi Dian

Department of Automation, College of Electrical Engineering and Information Technology, Sichuan University, Chengdu, Sichuan, China

scudiansy@scu.edu.cn

Abstract. In this paper, the model-free adaptive control (MFAC) method which is based on compact form dynamic linearization (CFDL) of multiple input and multiple output (MIMO) nonlinear systems is applied to the control of Puma560 manipulator. The controller design requires only the input and output data of the system. The simulation results verify the effectiveness of the model-free adaptive control method. Based on the structure of the controller, the function of controller parameters is analysed. By adding external disturbances, the anti-interference of Puma560 based on model-free adaptive control can be verified. The model free adaptive controller is optimized.

1. Introduction

At present, the modern control theory has been continuously developed and perfected. Model-free control theory was proposed and has been gradually developed to solve the problems faced by model-based control. Model-free adaptive control(MFAC) is a data-driven model-free control theory, which only requires the input/output data of system with no need for system information to design controller. It also has easy implementation, small computational burden and strong robustness [1]. MFAC has been widely applied in linear motor control, liquid level control system, robot control system, etc.

The manipulator system is nonlinear, highly coupled and time-varying [2]. Even in well-structured environment, the system is inevitably influenced by structural and non-structural uncertainties. The use of MFAC method can reduce the influence of these uncertainties [3]. It will also achieve data-driven control of manipulators by using online data [4, 5]. In this paper, the MFAC method is applied to Puma560 manipulator system. Simulation results show the effectiveness of the MFAC method.

2. Model-free adaptive control algorithm design

Considering a class of multiple input and multiple output discrete-time nonlinear systems [6]

$$\boldsymbol{y}(k+1)=\boldsymbol{f}\left(\boldsymbol{y}(k),\dots,\boldsymbol{y}(k-n_y),\boldsymbol{u}(k),\dots,\boldsymbol{u}(k-n_u)\right) \tag{1}$$

where $\boldsymbol{u}(k)\in\mathbf{R}^m$ is control input, $\boldsymbol{y}(k)\in\mathbf{R}^m$ is system output at time k; n_y and n_u are unknown integers; $\boldsymbol{f}(\dots)=\left(f_1(\dots),\ \dots,f_m(\dots)\right)^T\in\prod_{n_u+n_y+2}\mathbf{R}^m\mapsto\mathbf{R}^m$ is an unknown nonlinear function.

For nonlinear system (1), assumptions are made as follows:

Assumption 1: the partial derivatives of $f_i(\dots), i=1,\dots,m$, with respect to every component of $\boldsymbol{u}(k)$ are continuous.

Assumption 2: system (1) satisfies the general Lipschitz condition, that is

$$\|\boldsymbol{y}(k_1+1)-\boldsymbol{y}(k_2+1)\| \le b\|\boldsymbol{u}(k_1)-\boldsymbol{u}(k_2)\| \tag{2}$$

for any $k_1 \neq k_2, k_1 \ge 0, k_2 \ge 0$ and $\boldsymbol{u}(k_1) \neq \boldsymbol{u}(k_2)$, b is a positive constant.

When nonlinear system satisfies these assumptions and for any given time k $\Delta\boldsymbol{u}(k) \neq 0$, the compact form dynamic linearization (CFDL) of this system can be written as follows:

$$\boldsymbol{y}(k+1)=\boldsymbol{y}(k)+\boldsymbol{\Phi}_c(k)\Delta\boldsymbol{u}(k) \tag{3}$$

where $\boldsymbol{\Phi}_c(k)=\begin{bmatrix}\phi_{11}(k) & \phi_{12}(k) & \cdots & \phi_{1m}(k)\\ \phi_{21}(k) & \phi_{22}(k) & \cdots & \phi_{2m}(k)\\ \vdots & \vdots & \vdots & \vdots\\ \phi_{m1}(k) & \phi_{m2}(k) & \cdots & \phi_{mm}(k)\end{bmatrix} \in \mathbf{R}^{m\times m}$ is the pseudo Jacobian matrix (PJM) of the system.

Use the following cost function of control input to make estimation of system controlling effect:

$$J\big(\boldsymbol{u}(k)\big)=\|\boldsymbol{y}^*(k+1)-\boldsymbol{y}(k+1)\|^2+\lambda\|\boldsymbol{u}(k)-\boldsymbol{u}(k-1)\|^2 \tag{4}$$

where $\lambda > 0$ is the weighting factor and $\boldsymbol{y}^*(k+1)$ is the desired system output.

Differentiating (4) with respect to $\boldsymbol{u}(k)$, set this derivative to zero to get the following control law,

$$\boldsymbol{u}(k)=\boldsymbol{u}(k-1)+\left(\lambda\boldsymbol{I}+\boldsymbol{\Phi}_c{}^T(k)\boldsymbol{\Phi}_c(k)\right)^{-1}\boldsymbol{\Phi}_c{}^T(k)\big(\boldsymbol{y}^*(k+1)-\boldsymbol{y}(k)\big) \tag{5}$$

Control law (5) requires the calculation of matrix inversion and a simplified control law is introduced to avoid this process,

$$\boldsymbol{u}(k)=\boldsymbol{u}(k-1)+\frac{\rho\boldsymbol{\Phi}_c{}^T(k)\big(\boldsymbol{y}^*(k+1)-\boldsymbol{y}(k)\big)}{\lambda+\|\boldsymbol{\Phi}_c(k)\|^2} \tag{6}$$

where $\rho \in (0,1]$ is the step factor.

Cost function of PJM estimation is given as follows,

$$J\big(\boldsymbol{\Phi}_c(k)\big)=\|\Delta\boldsymbol{y}(k)-\boldsymbol{\Phi}_c(k)\Delta\boldsymbol{u}(k-1)\|^2+\mu\left\|\boldsymbol{\Phi}_c(k)-\widehat{\boldsymbol{\Phi}}_c(k)\right\|^2 \tag{7}$$

where $\mu > 0$ is the weighting factor.

Minimizing cost function (7) and improved projection algorithm can be acquired as follows,

$$\widehat{\boldsymbol{\Phi}}_c(k)=\widehat{\boldsymbol{\Phi}}_c(k-1)+\left(\Delta\boldsymbol{y}(k-1)-\widehat{\boldsymbol{\Phi}}_c(k-1)\Delta\boldsymbol{u}(k-1)\right)\times\Delta\boldsymbol{u}^T(k-1) \times\left(\mu\boldsymbol{I}+\Delta\boldsymbol{u}(k-1)\Delta\boldsymbol{u}^T(k-1)\right)^{-1} \tag{8}$$

An improved algorithm is introduced to avoid the calculation of matrix inversion,

$$\widehat{\boldsymbol{\Phi}}_c(\boldsymbol{k})=\widehat{\boldsymbol{\Phi}}_c(\boldsymbol{k}-\boldsymbol{1})+\frac{\boldsymbol{\eta}\left(\Delta\boldsymbol{y}(\boldsymbol{k})-\widehat{\boldsymbol{\Phi}}_c(\boldsymbol{k}-\boldsymbol{1})\Delta\boldsymbol{u}(\boldsymbol{k}-\boldsymbol{1})\right)\Delta\boldsymbol{u}^T(\boldsymbol{k}-\boldsymbol{1})}{\boldsymbol{\mu}+\|\Delta\boldsymbol{u}(\boldsymbol{k}-\boldsymbol{1})\|^2} \tag{9}$$

where $\eta \in (0,2]$ is a step factor and $\widehat{\boldsymbol{\Phi}}_c(k)=\begin{bmatrix}\hat{\phi}_{11}(k) & \hat{\phi}_{12}(k) & \cdots & \hat{\phi}_{1m}(k)\\ \hat{\phi}_{21}(k) & \hat{\phi}_{22}(k) & \cdots & \hat{\phi}_{2m}(k)\\ \vdots & \vdots & \vdots & \vdots\\ \hat{\phi}_{m1}(k) & \hat{\phi}_{m2}(k) & \cdots & \hat{\phi}_{mm}(k)\end{bmatrix} \in \mathbf{R}^{m\times m}$ is the estimation of PJM $\boldsymbol{\Phi}_c(k)$.

The CFDL-MFAC scheme for MIMO nonlinear system can be built by combining control law (6) with PJM estimation algorithm (9) as follows,

$$\widehat{\boldsymbol{\Phi}}_c(k)=\widehat{\boldsymbol{\Phi}}_c(k-1)+\frac{\eta\left(\Delta\boldsymbol{y}(k)-\widehat{\boldsymbol{\Phi}}_c(k-1)\Delta\boldsymbol{u}(k-1)\right)\Delta\boldsymbol{u}^T(k-1)}{\mu+\|\Delta\boldsymbol{u}(k-1)\|^2} \tag{10}$$

$$\hat{\boldsymbol{\phi}}_{ii}(k)=\hat{\boldsymbol{\phi}}_{ii}(1),\ \text{if}\left|\hat{\boldsymbol{\phi}}_{ii}(k)\right|<b_2 \text{ or } \left|\hat{\boldsymbol{\phi}}_{ii}(k)\right|>\alpha b_2 \text{or sign}\left(\hat{\boldsymbol{\phi}}_{ii}(k)\right)\neq \text{sign}\left(\hat{\boldsymbol{\phi}}_{ii}(1)\right) \tag{11}$$

$$\hat{\boldsymbol{\phi}}_{ij}(k)=\hat{\boldsymbol{\phi}}_{ij}(1),\ \text{if}\left|\hat{\boldsymbol{\phi}}_{ij}(k)\right|>b_1 \text{ or sign}\left(\hat{\boldsymbol{\phi}}_{ij}(k)\right)\neq \text{sign}\left(\hat{\boldsymbol{\phi}}_{ij}(1)\right),\ i\neq j \tag{12}$$

$$\boldsymbol{u}(k)=\boldsymbol{u}(k-1)+\frac{\rho\widehat{\boldsymbol{\Phi}}_c{}^T(k)\big(\boldsymbol{y}^*(k+1)-\boldsymbol{y}(k)\big)}{\lambda+\left\|\widehat{\boldsymbol{\Phi}}_c(k)\right\|^2} \tag{13}$$

where $\lambda > 0$, $\mu > 0$, $\rho \in (0,1]$, $\eta \in (0,2]$; $\hat{\phi}_{ij}(1)$ is the initial value of $\hat{\phi}_{ij}(k)$.

3. Modelling of Puma560 manipulator system

IOP Conf. Series: Materials Science and Engineering **428** (2018) 012051 doi:10.1088/1757-899X/428/1/012051

3.1. Basic structure of Puma560 manipulator system

The system block diagram is shown in figure 1. The desired rotation angle $\boldsymbol{q}^*$, angular velocity $\dot{\boldsymbol{q}}^*$ and angular acceleration $\ddot{\boldsymbol{q}}^*$ are taken as desired signal $\boldsymbol{y}^*$. The actual rotation angle $\boldsymbol{q}$, angular velocity $\dot{\boldsymbol{q}}$ and angular acceleration $\ddot{\boldsymbol{q}}$ are used as actual signal $\boldsymbol{y}$. Desired and actual signals are used as input signals of the controller. These signals are calculated and processed using MFAC method in the controller. Control signal $\boldsymbol{u}$, which is used to control Puma560 manipulator, can be obtained as the output signals of the controller.

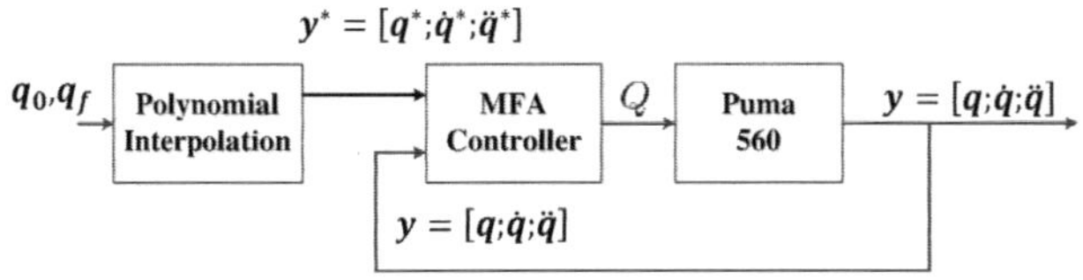

Figure 1. Block diagram of Puma560 manipulator system based on MFAC

The algorithm used by the controller is MFAC algorithm, with no requirement for information of system parameter or structure, establishes a dynamic linearized data model based on input/output data, and carries out real-time data-driven control based on the online data for the manipulator. By establishing the dynamic linearized data model, the driving torque of each joint can be obtained to control the movement of manipulator.

3.2. Controller design

An MFA controller is designed by using the MFAC scheme. Since Puma560 has 6 degrees of freedom, we can determine the dimension of $\boldsymbol{q}^*$, $\dot{\boldsymbol{q}}^*$, $\ddot{\boldsymbol{q}}^*$, $\boldsymbol{y}^*$, $\boldsymbol{q}$, $\dot{\boldsymbol{q}}$, $\ddot{\boldsymbol{q}}$, $\boldsymbol{y}$, $\boldsymbol{u}$:

$$\boldsymbol{q}^*, \dot{\boldsymbol{q}}^*, \ddot{\boldsymbol{q}}^*, \boldsymbol{q}, \dot{\boldsymbol{q}}, \ddot{\boldsymbol{q}}, \boldsymbol{u} \in \mathbf{R}^{6\times1}, \boldsymbol{y}^* = [\boldsymbol{q}^*; \dot{\boldsymbol{q}}^*; \ddot{\boldsymbol{q}}^*] \in \mathbf{R}^{18\times1}, \boldsymbol{y} = [\boldsymbol{q}; \dot{\boldsymbol{q}}; \ddot{\boldsymbol{q}}] \in \mathbf{R}^{18\times1}.$$

The Puma560 manipulator based on MFAC can be written as (1), where $\boldsymbol{u}(k) \in \mathbf{R}^{6\times1}$; $\boldsymbol{y}(k) \in \mathbf{R}^{18\times1}$; $\boldsymbol{f}(\ldots) = \left(f_1(\ldots), \ldots, f_{18}(\ldots)\right)^T \in \prod_{n_u+n_y+2} \mathbf{R}^{6\times1} \mapsto \mathbf{R}^{18\times1}$ is unknown nonlinear function.

The Puma560 system satisfies the Assumption 1 and Assumption 2. The CFDL of this system can be written as,

$$\boldsymbol{y}(k+1) = \boldsymbol{y}(k) + \boldsymbol{\Phi}_c(k)\Delta\boldsymbol{u}(k) \tag{14}$$

where $\boldsymbol{\Phi}_c(k) = \begin{bmatrix} \phi_{1,1}(k) & \phi_{1,2}(k) & \cdots & \phi_{1,6}(k) \\ \phi_{2,1}(k) & \phi_{2,2}(k) & \cdots & \phi_{2,6}(k) \\ \vdots & \vdots & \vdots & \vdots \\ \phi_{18,1}(k) & \phi_{18,2}(k) & \cdots & \phi_{18,6}(k) \end{bmatrix} \in \mathbf{R}^{18\times6}$ is the PJM of the system.

The CFDL-MFAC scheme of this system can be written as,

$$\widehat{\boldsymbol{\Phi}}_c(k) = \widehat{\boldsymbol{\Phi}}_c(k-1) + \frac{\eta\left(\Delta\boldsymbol{y}(k) - \widehat{\boldsymbol{\Phi}}_c(k-1)\Delta\boldsymbol{u}(k-1)\right)\Delta\boldsymbol{u}^T(k-1)}{\mu + \|\Delta\boldsymbol{u}(k-1)\|^2} \tag{15}$$

$$\widehat{\boldsymbol{\phi}}_{ii}(k) = \widehat{\boldsymbol{\phi}}_{ii}(1), \text{ if} \left|\widehat{\boldsymbol{\phi}}_{ii}(k)\right| < b_2 \text{ or } \left|\widehat{\boldsymbol{\phi}}_{ii}(k)\right| > \alpha b_2 \text{or sign}\left(\widehat{\boldsymbol{\phi}}_{ii}(k)\right) \neq \text{sign}\left(\widehat{\boldsymbol{\phi}}_{ii}(1)\right) \tag{16}$$

$$\widehat{\boldsymbol{\phi}}_{ij}(k) = \widehat{\boldsymbol{\phi}}_{ij}(1), \text{ if } \left|\widehat{\boldsymbol{\phi}}_{ij}(k)\right| > b_1 \text{ or sign}\left(\widehat{\boldsymbol{\phi}}_{ij}(k)\right) \neq \text{sign}\left(\widehat{\boldsymbol{\phi}}_{ij}(1)\right), i \neq j \tag{17}$$

$$\boldsymbol{u}(k) = \boldsymbol{u}(k-1) + \frac{\rho\widehat{\boldsymbol{\Phi}}_c^{\,T}(k)(\boldsymbol{y}^*(k+1) - \boldsymbol{y}(k))}{\lambda + \|\widehat{\boldsymbol{\Phi}}_c(k)\|^2} \tag{18}$$

where $\lambda > 0$, $\mu > 0$, $\rho \in (0,1]$, $\eta \in (0,2]$; $\hat{\phi}_{ij}(1)$ is the initial value of $\hat{\phi}_{ij}(k)$.

In addition to determining input and output signals of the controller, the MFAC controller design also needs to initialize input/output signals of the controller, and to determine the weight factors λ, μ, the step factor ρ, η used in the control algorithm.

4. Simulation and analysis

Take initial joint position $q_0 = [0\ 0\ 0\ 0\ 0\ 0]$, final joint position $q_f = \left[\frac{\pi}{4}\ \ -\frac{\pi}{2}\ \ \frac{\pi}{2}\ \ 0\ 0\ 0\ 0\right]$, simulation time $t_{max} = 10s$, time interval $\Delta t = t_{max}/10000 = 0.001s$, weight factors $\lambda = 5$, $\mu = 11$, step factors $\rho = 1$, $\eta = 0.45$. By using kinematics model, the motion of the terminal actuator can be observed and analyzed in the Cartesian space according to the rotation angle of each joint.

4.1. Parameters adjustment

4.1.1 Weighting factors. The influence of weighting factors on control effect is analyzed by changing weighting factors of CFDL-MFAC scheme.

The simulation of λ=0.1,1,5,10,100 and μ=0.1,1,11,50, 200 is carried out respectively. The motion of the terminal actuator is plotting in figure 2 and 3, and the comparative analysis is made.

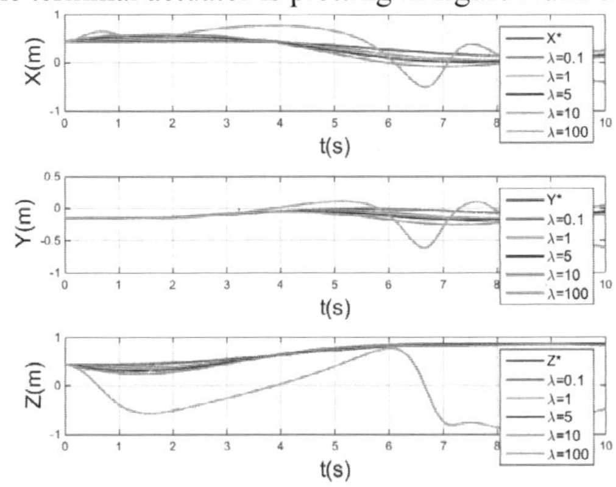

Figure 2. Changing weighting factor λ

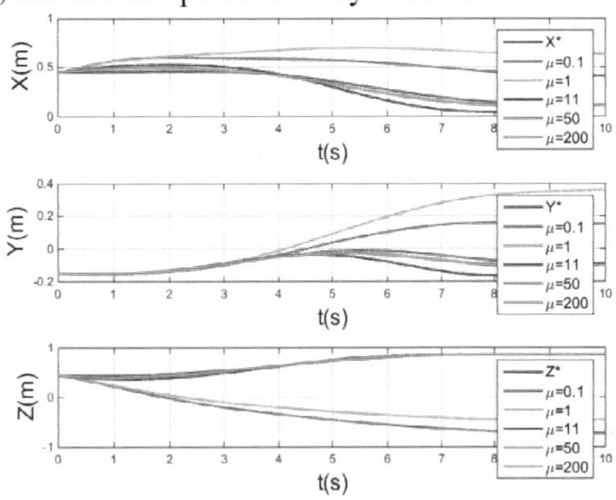

Figure 3. Changing weighting factor μ

From the simulation results, we can see that when λ is too large, the CFDL-MFAC scheme cannot get satisfactory control effect. It can be seen from the control scheme (18) that, when other conditions are the same, the smaller λ is, the smaller $\Delta\boldsymbol{u}(k)$, which means that the change of control input becomes smaller, and this will lead to the decrease of manipulator control's sensibility. When lambda is smaller, $\Delta\boldsymbol{u}(k)$ is mainly dependent on the value of $\widehat{\boldsymbol{\Phi}}_c(k)$, $\boldsymbol{y}^*(k+1)$ and $\boldsymbol{y}(k)$, which means that $\Delta\boldsymbol{u}(k)$ is related to the desired trajectory, the actual trajectory and the structure of the system. For this reason, the change of λ will not have great influence on $\Delta\boldsymbol{u}(\boldsymbol{k})$, and affect control effect.

As for the weighting factor μ, when μ is too small, desired control effect cannot be achieved. It can be seen from the control scheme (15) that, when other conditions are the same, the smaller μ is, the bigger $\Delta\widehat{\boldsymbol{\Phi}}_c(k)$, then $\Delta\boldsymbol{u}$ will decrease which leads to unsatisfactory joint rotation of manipulator. When μ becomes bigger, $\Delta\widehat{\boldsymbol{\Phi}}_c(k)$ becomes smaller and will have little effect on $\Delta\boldsymbol{u}$. Therefore, when μ is big enough, changing μ will have little influence on manipulator control.

Therefore, the selection of weighting factor has effect on the control effect of the system to some extent.

4.1.2 Step factors. Chang the step size factors of CFDL-MFAC scheme ρ, η, respectively. The simulation results are shown in figure 4 and 5.

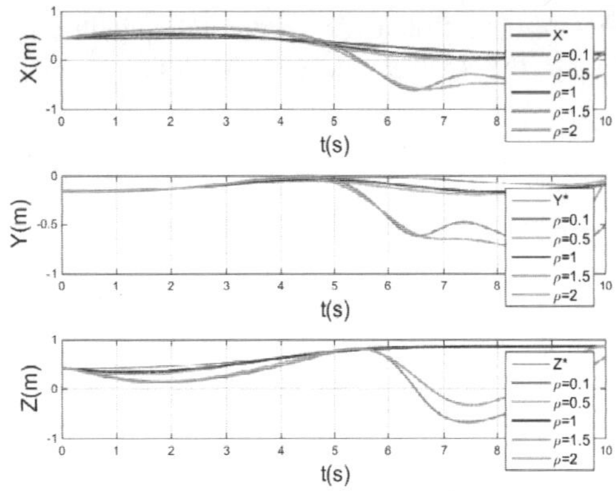

Figure 4. Changing step factor ρ

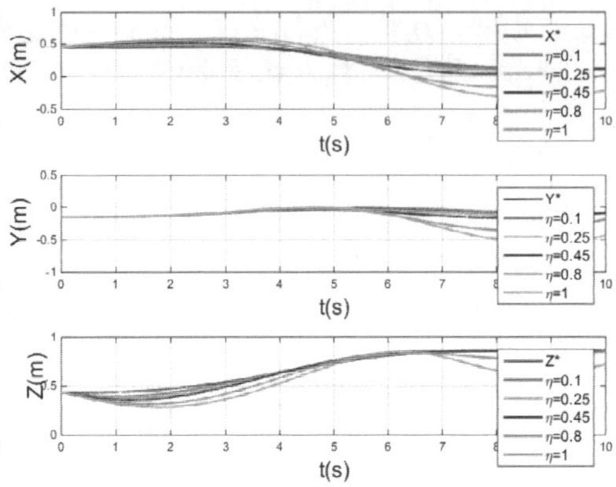

Figure 5. Changing step factor η

As shown in figure 4, the change of ρ will affect the control effect, and when ρ is over 1, the control effect of the controller will be worse. Compared with the desired trajectory, the actual trajectory changes sharply and the stability is poor. It can be seen from the control scheme (18) that, when other conditions are the same, the smaller ρ is, the bigger $\Delta\boldsymbol{u}(k)$, and manipulator control's sensibility will increase, which means that manipulator will react strongly to minor changes in control input or even resulting in a larger fluctuation. When ρ becomes smaller, $\Delta\boldsymbol{u}(k)$ mainly depends on the value of $\widehat{\boldsymbol{\Phi}}_c(k)$, $\boldsymbol{y}^*(k+1)$ and $\boldsymbol{y}(k)$. $\Delta\boldsymbol{u}(k)$ is related to the desired trajectory, the actual trajectory and the structure of the system, so satisfying control effect can be achieved.

As shown in figure 5, the change of η will affect the control effect, and when η is big, the control effect of the controller will be worse. It can be seen from the control scheme (15) that, when other conditions are the same, the bigger η is, the smaller $\Delta\widehat{\boldsymbol{\Phi}}_c(k)$, then $\Delta\boldsymbol{u}(k)$ will increase, which means that manipulator will react strongly to minor changes in control input.

From the control scheme, it can be concluded that the adjustment of the step factor can be used to modify the control effect without affecting the overall trend.

4.2. External disturbance

During simulation process, a short time disturbance, τ_1=[2 2 2 0 0 0]kg·m, is applied to the drive torque in 2-4s. By observing the curve of actual and desired position, and comparing with the curve without disturbance, the anti-interference performance of the Puma560 manipulator under the MFAC scheme is obtained.

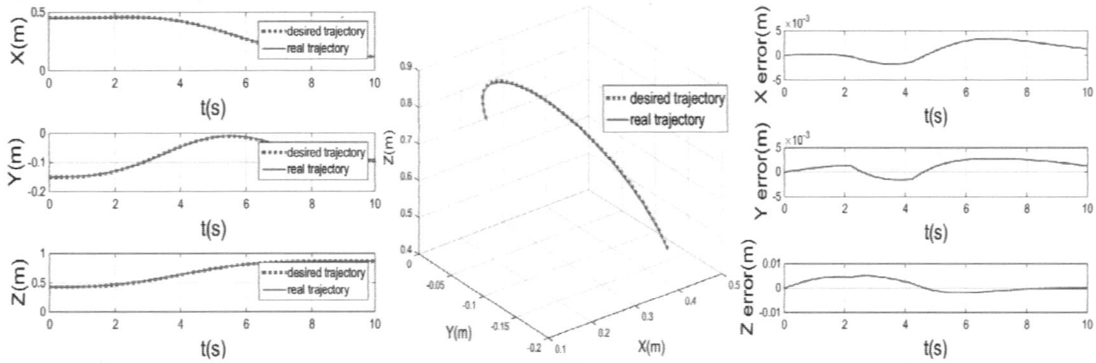

(a) (b) (c)

Figure 6. Simulation results with disturbance τ_1

In the Cartesian space, the influence of disturbance $\boldsymbol{\tau}_1$ on the Puma560 manipulator is observed. The simulation results are shown in figure 6. When $\boldsymbol{\tau}_1$ is applied, the motion of end effector is still stable, and follows the desired trajectory. It can be seen from the coordinate error curve that in the 2-4s, $\boldsymbol{\tau}_1$ produces a small mutation, and the rate of change is not continuous. Therefore, the application of the disturbance, $\boldsymbol{\tau}_1$, has a small interference effect on the motion of the end actuators, but does not affect the smooth motion of the end actuators following the desired trajectory. Though being disturbed by $\boldsymbol{\tau}_1$, the controller can still realize the motion control of the Puma560 manipulator, which shows the robustness of the MFAC algorithm.

4.3. Simplifying the dimension of controller input signal

Considering that the actual joint rotation $\boldsymbol{q}$ is the main factor that determines the trajectory of manipulator, the actual joint speed $\dot{\boldsymbol{q}}$ and acceleration $\ddot{\boldsymbol{q}}$ can be obtained by differential calculation of $\boldsymbol{q}$, so we consider reducing the dimension of input signal of controller to reduce the dimension to satisfy the control effect and improve the efficiency of operation at the same time.

Take $\boldsymbol{q},\dot{\boldsymbol{q}},\boldsymbol{q}^*,\dot{\boldsymbol{q}}^*$ as input signals of the controller. The corresponding calculation and processing of these signals are carried out in the controller by using the MFAC scheme. The driving torque of each joint is obtained as the output signal from the controller. The actual input signal dimension of the controller is 12 dimensions, and the desired input signal dimension is 12 dimensions.

The simulation results are shown in figure 7.

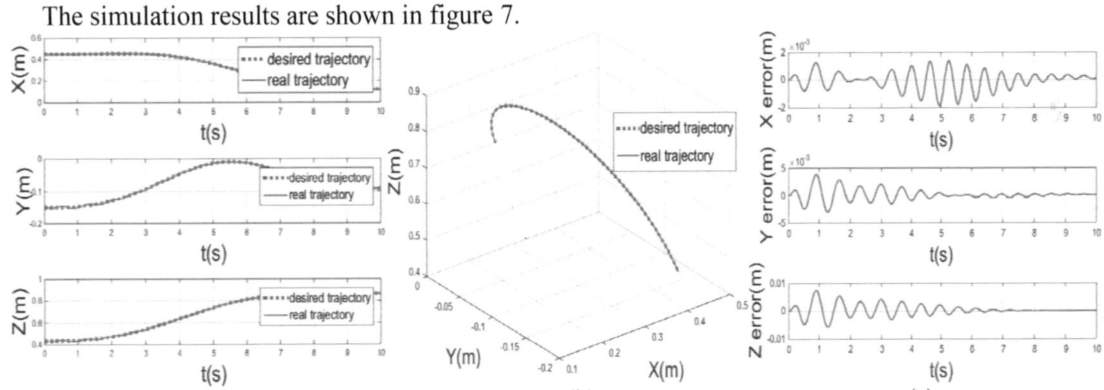

Figure 7. Simulation results with $y = [q; \dot{q}], y^* = [q^*; \dot{q}^*]$

As can be seen from figure 7, the joint rotation signals and speed signals are used as input, and the calculation process is carried out in the controller. The drive torque, as the output of the controller, can drive manipulator to track the desired trajectory well. Compared with the system with joint rotation, speed and acceleration as the input signal of the controller, the tracking error of end actuator is small, and the controller under the simplified method can guarantee the control effect of the original system.

The simulation time of the original system is 8.816931s while the simulation time of the simplified system is 5.189842s. Therefore, simplifying the input signal dimension of the controller can improve the computing speed to some extent.

Using the signal of joint rotation and velocity as the input signal of the controller, the operation speed can be improved and the control effect can also be kept.

5. Conclusion

In this paper, the Puma560 manipulator is used as the research object, and the control method of the manipulator's motion control is studied. The model-free adaptive control method is used to control the motion of the manipulator, which can avoid the unmodeled dynamics caused by the uncertainty of the structure and the unstructured uncertainty. The controller is designed by using the input and output data of the system. During this process, no mathematical model, parameter structure, or other related information of the system are required. Robustness and adaptability of the system can be guaranteed.

Acknowledgments

This work is supported by Fundamental Research. Special Funds for Central Universities (20826041A4133).

References

[1] Z. S. Hou and S. T. Jin, Model Free Adaptive Control: Theory and Applications. Boca Raton, FL, USA: CRC Press, 2013.

[2] D. M. Jia, Manipulator control based on iterative sliding mode algorithm (Master Thesis) Zhengzhou University, Zhengzhou, Henan, China, 2011.

[3] Y.Dai, Research on Intelligent PID control method for Robot Manipulators. (Master Thesis) Shengyang Ligong University, Shengyang, Liaoning, China, 2015..

[4] X. F. Wang, X. Li, J. H. Wang, Active interaction exercise control of exoskeleton upper limb rehabilitation robot using model-free adaptive methods. Acta Automatica Sinica, 2016, 42(12): 1899-1914.

[5] Y. Wang, Y. Q. Yan and L. B. Song, .Study on control method of modular mobile robot based on MFA theory. Journal of Mechanical & Electrical Engineering, Apr. 2012: 400-403

[6] Z. S. Hou and S. T. Jin, "A novel data-driven control approach for a class of discrete-time nonlinear systems," IEEE Trans. Control Syst. Technol., vol.19, no.6, pp. 149–1558, Nov. 2011.

IOP Conf. Series: Materials Science and Engineering **428** (2018) 012052 doi:10.1088/1757-899X/428/1/012052

Jacquard Control System of Warp Knitting Machine Based on Embedded System

LI Ce, MENG Zhuo and SUN Yize

College of Mechanical Engineering, Donghua University, Shanghai 201620, China

Email:ce_li@dhu.edu

Abstract: In order to improve the control precision and speed, jacquard control system of warp knitting machine is designed using Embedded System. Based the overall design of the jacquard control system, the hardware and software are designed respectively. Industrial Ethernet is used to complete the data transmission in the system, aiming to improve the capacity and reliability for high speed data transmission. The synchronous control of the yarn guide pin is based on the distributed clock algorithm. The designed control system can simultaneously control multi-jacquard with faster speed and higher stability.

1. Introduction

With the deepening restructuring of the textile industry, warp knitting machine is developing towards high speed, intelligence and networking[1]. At present, the control system of warp knitting machine mostly adopts SCM or PLC processor. The advantage of SCM is relatively low cost, while the disadvantages are poor anti-interference capability, complex wiring, high failure rate, limited number of jacquard controlling and long development cycle. Compared with SCM, PLC has its advantages in anti-interference ability, working environment, development and maintenance. But its cost is relatively high, which restricts the application in efficient algorithm and intelligent control.

For Jacquard control system of warp knitting machine, the demands of real-time, reliability and data transmission speed are continuously raising. SCM and PLC are difficult to meet the requirements of machining accuracy. Here, system design is based on the embedded microprocessor, and the Linux operation system is used to control the jacquard. The embedded system has flexible clipping property of hardware and software, making it possible to integrate special platform for hardware and software, which results in good real-time, high reliability and low power consumption. For data communication, the industrial Ethernet EtherCAT bus is used. Industrial Ethernet has been widely used in the field of industrial automation, it has high response speed, low CPU occupancy rate and good synchronization performance.[2]Using EtherCAT's distributed clock mechanism, the clock synchronization accuracy between each slave node less than 1 μs can be realized,[3][4] which is very important to the synchronizing action of the guide needle.

The integration of micro processing technology, communication technology and automatic control technology can improve the processing speed of the control system, shorten the development cycle of the system, make the system have better real-time and stability, faster speed, stronger anti-interference ability, lower power consumption and higher scalability. The level of information and automation of the warp knitting machine is improved.

IOP Conf. Series: Materials Science and Engineering **428** (2018) 012052 doi:10.1088/1757-899X/428/1/012052

2. Overall design of the control system

Based on the analysis of the control requirements and working environment, the overall design of the piezoelectric jacquard control system is completed. As shown in Figure 1, the jacquard control system includes IPC, jacquard controller, jacquard repeater, jacquard driver and piezoelectric jacquard guide needle. Considering the large quantity short offset interval and high frequency of the yarn guide needle,[5] the whole control system adopts embedded technology and EtherCAT bus, and distributing clock synchronization algorithm mechanism is used to realize the real-time control, providing high reliability, real-time and anti-jamming capability, large data packet capacity and high transmission speed. which meets the requirements of the high-speed stability of the jacquard control system and can adapt to a poor working environment.

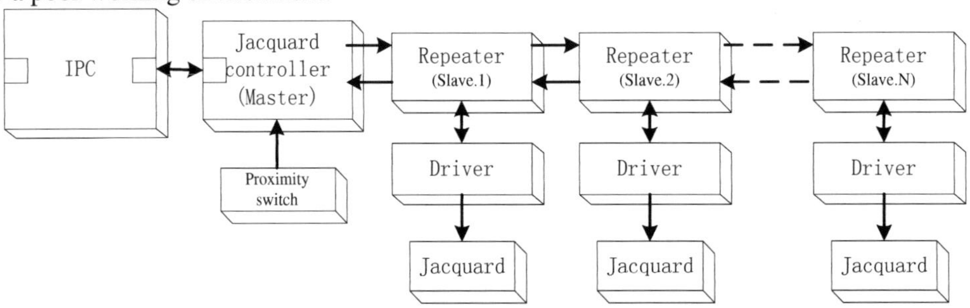

Figure 1. Jacquard control system

In the system, IPC realizes data transmission, parameter configuration, debugging and real-time control of jacquard controller. Master-slave mode is adopted between jacquard controller and repeater. The controller downloads the process data from the industrial control machine and transfers the process data to each jacquard repeater. The repeaters insert the data while extracting, so that the controller can monitor each repeater in real time. The repeater passes the received data to the jacquard drive after analyzing, and the real-time control of the piezoelectric jacquard is realized through the driver. The system uses EtherCAT to complete the bus communication from the industrial field layer to the management layer, which can replace the existing complex bus systems in the industrial control industry. While improving the transmission rate, the number of wires is reduced, and failures due to complicated wiring can also be reduced, contributing to less jacquard errors by malfunctions.

3. Hardware design of the control system

3.1 Design of the controller

The hardware of jacquard controller is mainly composed of processor, memory, communication interface and power supply. The hardware structure is shown in Figure 2. In the control system, the jacquard controller is the main communication station in the entire EtherCAT communication system, so the high performance industrial application processor AM3359 is adopted to meet the high requirement of jacquard controller. AM3359 is a 32-bit ARM core processor based on the ARM Cortex-A8 core developed by TI. The main communication module of the controller is implemented by two Gigabit Ethernet interfaces. The physical layer is connected by AR8031 chip using RGMII interface.

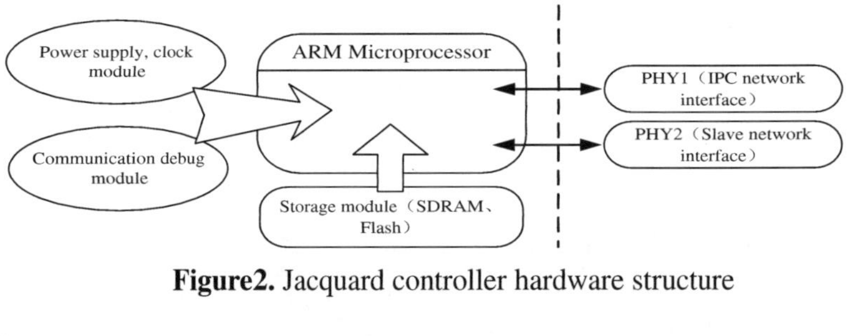

Figure2. Jacquard controller hardware structure

3.2 Design of the repeater

Jacquard repeater is used as the slave station of communication system. Due to the large amount of system data and high communication rate, the jacquard repeater hardware adopts ARM+FPGA architecture. The ARM processor mainly processes EtherCAT data frames, and sends the location instructions and pattern data to FPGA. The output control is realized through FPGA. Figure 3 shows the hardware structure of the jacquard repeater.

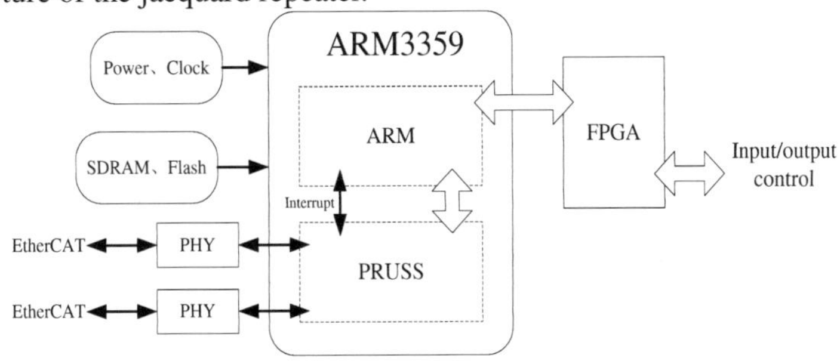

Figure 3. Jacquard repeater hardware structure

Slave processor also chooses AM3359 chip. The microprocessor chip has independent running clock with good application flexibility, and has programmable real-time unit subsystem PRU separated from ARM unit. TI provides PRU-ICSS-based EtherCAT slave routines for free.

FPGA uses the Cyclone III series EP3C16Q240C8N chip produced by Altera Corporation. It combines high performance and low power consumption with relatively low cost. It can also realize high-speed communication and control with multiple jacquard drives at the same time.

ARM and FPGA are connected by 16 bit data bus, three-way control signals and four-channel GPIO. Because ARM chip and FPGA adopt different clock signals, asynchronous FIFO can be used to realize fast real-time data transmission conveniently. Compared to serial bus interface, parallel bus of data transmission can realize high-speed data transmission[5][6]. The three-way control signals include write enable signal (W_EN), read enable signal (R_EN) and chip select signal (NGCS5). The four-channel GPIO can be used as a control command signal for FPGA to realize the control of FPGA state.

4. Software design of the control system

4.1. Design of the master station

As the main station of the system, the jacquard controller completes the analytical communication protocol, and manages the whole communication. The software architecture of jacquard controller is mainly divided into three parts: network device module, application program and master station module, user management module. The software architecture of the master station is shown in Figure 4.

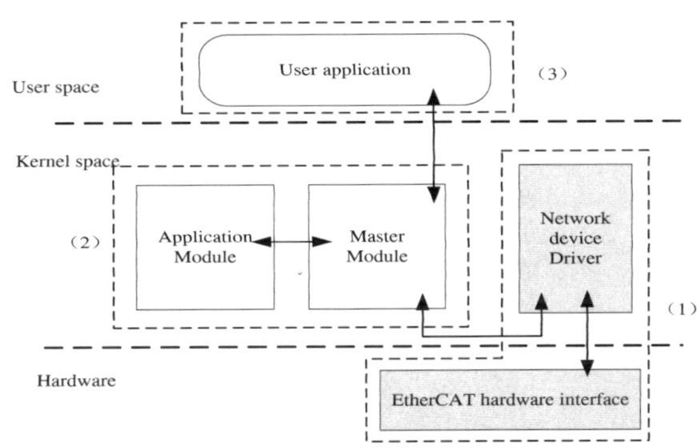

Figure4. Software architecture for master station of control system

The functions implemented by each module are as follows:

(1) Network device module. EtherCAT's network communication function can be realized through common network devices. In kernel layer, it is difficult to realize EtherCAT protocol by common network device driver, so a universal network card driver provided by the open source EtherCAT protocol stack is adopted, which can support EtherCAT master communication without modifying existing drivers.

(2) Application program and master station module. In order to ensure the real-time performance of the controller, both the application and the master module are in the kernel layer. The EtherCAT master module analyzes the protocol, completes the task scheduling, and provides the function interface for the network device and the application program.[7] The application program module is written according to the control request of jacquard repeater, thus realize the automatic control of repeater.

(3) User management module. A variety of commands that can be run at the Linux user tier is provides, enabling direct access to jacquard repeaters and setting.

4.2. Design of the slave station

In the system, jacquard repeater is set as slave station. The entire software architecture of the secondary station will be divided into three layers: the driver layer, the EtherCAT protocol layer and the user application layer. The standard interface function is used to complete the connection between each layer. The driver layer includes all hardware drivers for initialization and operation of PRU, power, memory and corresponding interfaces. Among them, PRU driver includes configuring PRU-ICSS register program, reading and writing the cache function. The EtherCAT protocol layer mainly includes EtherCAT state machine, Mailbox Communication and Process Data Communication for processing EtherCAT communication. The FPGA control and data interaction are realized through the user application layer. The overall architecture of the slave software is shown in Figure 5.

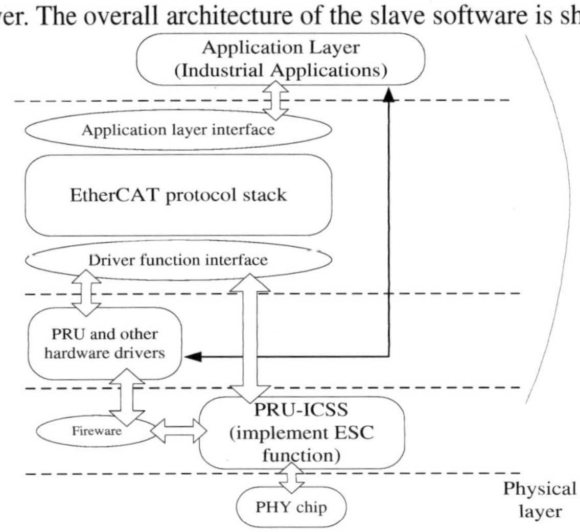

Figure 5. Control System Slave Software Architecture

4.3. Design of the distributed clock and periodic synchronization controlling

The control system is used to synchronize the guide pins, and the precise synchronization control is a critical part of the system. EtherCAT bus has been widely used in industrial control field because of its excellent real-time performance[8]. The distributed clock algorithm can be used to achieve synchronous output control of the jacquard repeaters.

The distributed clock mechanism synchronizes all jacquard repeater to a reference clock. The first repeater with the distributed clock function connected by the jacquard controller is choosed as a reference clock to synchronize the clock of other devices and controllers. The distributed clock synchronization algorithm consists of three parts: transmission delay measurement, offset compensation and clock drift compensation.[9][10] The implementation process of the distributed clock algorithm is shown in Figure 6.

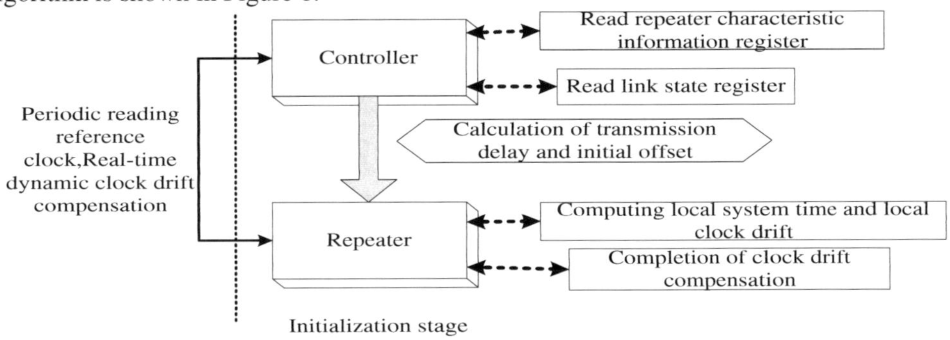

Figure 6. Distributed clock algorithm implementation flow

As shown in Figure 6, the controller initializes the distribution clock synchronization function by reading and writing some configuration registers of the repeater. According to the running stage of the EtherCAT state machine, the controller completes the reading and writing of the repeater corresponding register in the initialization state, making the repeater support the distributed clock and configure the distributed clock related registers. During the configuration process, it is necessary to measure and calculate the transmission delay and the initial clock offset. Under safe running state, the controller begins to synchronize the clock of all repeater, and sends the message to every repeater by sending order addressing command. Through reading current system time from the reference clock repeater and writing the time in other devices, the speed of the local clock is adjusted based on time control loop. In the running state, the controller can periodically send sequential addressing commands along with the process data. By reading the system time of the first repeater and writing to other devices, the dynamic clock drift compensation on the other repeaters can be realized in real time.

5. Conclusion

The system applies embedded control technology. The data transmission and information collection are completed through industrial Ethernet EtherCAT bus. The overall scheme of the control system is given, and the software and hardware design of jacquard controller and jacquard repeater are carried out respectively. The control system has faster transmission speed, larger packet capacity, stronger anti-interference capability and better function expansion. At the same time, the distributed clock mechanism is adopted to meet the speed and accuracy requirements of the control system, and the speed and intelligence level of the warp knitting machine control system are improved.

References

[1] Jiang G M, 2014, The knitting technology: innovation and development J. Jiangsu Textile, (6):1-7.
[2] Chen X, Li D, Wan J, et al. A clock synchronization method for EtherCAT master J. Microprocessors & Microsystems, 2016, 46:211-218.
[3] Cena G, Bertolotti I C, Scanzio S, et al. Evaluation of EtherCAT distributed clock performance[J]. IEEE Transactions on Industrial Informatics, 2012, 8(1):20-29.
[4] Chen H, Song B, Tang X Q. The realization of the EtherCAT precision clock synchronization technology[J]. Combination Machine Tools and Automatic Processing Technology, 2014(5):69-72.
[5] Zhang F H, Xu C H, Hu S F. Design and implement of ARM and FPGA data communication based on Linux [J]. Journal of Hangzhou Dianzi University, 2015(4):26-30.
[6] Li S Q. The Research and simulation of ARM parallel BUS based on FPGA[J]. Electronic Test, 2013, 4(8):22-23

[7] Ma C M, Kang C F, Huang X D, et al. Study of EtherCAT masterbased on Linux[J]. Manufacturing Automation, 2011, 33(8):78-82.

[8] Song Y H, Suk-Lee, Kim H J, et al. Computational delay and real-time performance of industrial network[C]// International Conference on Control, Automation and Systems. 2015.

[9] Xun J, Liu Y Q. Design and application of EtherCAT driver for industrial Ethernet Fieldbus[M]. Beijing University of Aeronautics and Astronautics Press, 2010.

[10] Zhao J, Song S, Liu W, et al. Study on the distributed synchronous acquisition function based on EtherCAT Bus[C]// International Conference on Electrical and Control Engineering. IEEE, 2011:3193-3197.

A Novel 2R Parallel Mechanism for Alt-azimuth Pedestal

C C Xu, C Xue and X C Duan*

Key Laboratory of Electronic Equipment Structure Design, Ministry of Education of China, No.2 South Taibai Road Xi'an, China

*E-mail: xchduan@xidian.edu.cn

Abstract. This paper presents a novel 2R parallel mechanism used as alt-azimuth pedestal, i.e. a 2-DOF (degree-of-freedom) pedestal for supporting and rotating equipment about two perpendicular axes. Its active joints are two sliders equipped with servo control on a circular rail. The synchronous and asynchronous motions of the two sliders lead to the azimuth (rotation about the vertical axis) and pitch (rotation about the horizontal axis) of the rotatable platform, respectively. The DOF computation and kinematics are dealt with firstly, and then the dynamic model of the 2R parallel mechanism is established by using the Lagrange equation. Eventually, a numerical simulation of the typical trajectory motion is carried out, and the actuating torque needed in the actuators for the trajectory tracking motion is worked out by employing the derived inverse dynamic equations, which is consistent with the result obtained from commercial software. This study lays a solid base for the further investigation and engineering implementation of the 2R parallel mechanism.

1. Introduction

An alt-azimuth pedestal is an easy-to-realize two-axis pedestal for supporting and rotating equipment such as radar, radio telescope, solar panel about two perpendicular axes[1], i.e., one vertical axis and one horizontal axis. The rotation about the two axes can adjust the azimuth and altitude of the pointing direction respectively. A conventional alt-azimuth pedestal can be described as follows: There are two active perpendicular axes (implemented with two revolute joints) stacked up. Each axis is driven by a servo motor and gear box. It has the advantages of a simple structure and an easy implementation of control modules. However, the joint error accumulation would be very bulky for supporting a heavy rotational platform because of its serial property of the two axes. As an improved structure, another type of alt-azimuth pedestal has a circular rail and several rollers supporting the weight in the azimuth motion[2]. But the serial configuration of the two axes also can increase the power of the azimuth actuator and total weight. It is thus desirable to propose a novel light structure of an alt-azimuth pedestal particularly for the heavy rotational platform.

From the view point of mechanism development, the advantages such as high rigidity, high precision, large load weight ratio, high speed and acceleration of parallel mechanism satisfy the development trend of modern mechanical and electrical equipment with high precision and high speed [3, 4]. Therefore, the parallel mechanism provides a potential solution to the novel light structure of alt-azimuth pedestals. Due to the fact that different degrees of freedom are needed in practical applications, only a few degrees of freedom are needed in some applications, such as 2, 3, 4 or 5 degrees of freedom. Therefore, the lower-mobility parallel mechanism such like the alt-azimuth pedestal has turned to the focus of attention [5]. A planar 2-DOF drive is constructed by introducing a

IOP Conf. Series: Materials Science and Engineering **428** (2018) 012053 doi:10.1088/1757-899X/428/1/012053

series of branches and a redundant driver in [6]. Gosselin discussed the application of the 2-DOF parallel robot in the space to a vector which can be used for the tasks of two DOFs [7]. A kind of spherical 2-DOF 5R parallel mechanism, which can be used as the positioning device on the spherical surface, is studied by Zhang and Li et al. [8]. Based on the Lagrange equation, the dynamic model of spherical 5R parallel mechanism is established in [9]. However, according to current publications, there exist no 2R parallel mechanism configuration able to provide the 0-360 (or -180 to 180) degree azimuth motion and ordinary pitch motion for the alt-azimuth pedestal.

This paper presents a novel 2R parallel mechanism used as alt-azimuth pedestal, which two active joints are two sliders with independent servo control on a circular rail. The synchronous and asynchronous motions of the two sliders lead to the azimuth and pitch of the rotatable platform, respectively. The DOF computation, kinematics and dynamics are dealt with firstly; moreover a numerical simulation is also carried out to validate the related results.

2. Structural description

As shown in Figure 1, the 2R parallel mechanism based alt-azimuth pedestal includes a vertical column (8) ,which is connected with ground or an external base; a rotatable platform (7) attached to the column by revolute joints (5) and (9); a horizontal rail (1) with the rack on its inner lateral face; two rods (6) with fixed length, one end of each rod connecting the rotatable platform with a 2-DOF revolute joint(4) and the other end of each rod connecting the sliders (2) with a 3-DOF rotational joint (3). There are prismatic joints between the rail and the sliders. A gear pair attached to the slider engages with the rack and is driven by the servo motor and the reducer.

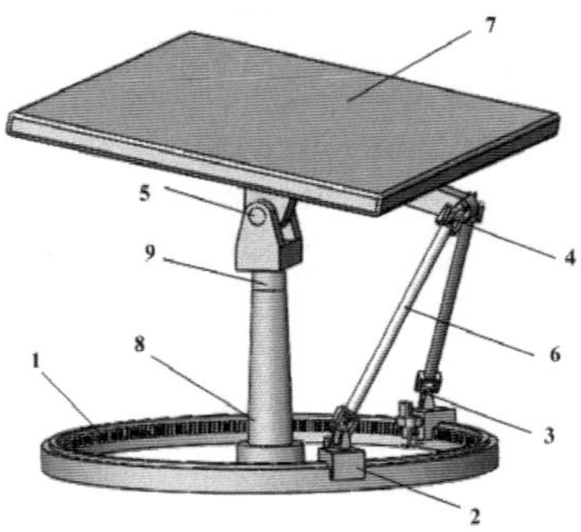

Figure 1. The 2R parallel mechanism based alt-azimuth pedestal.

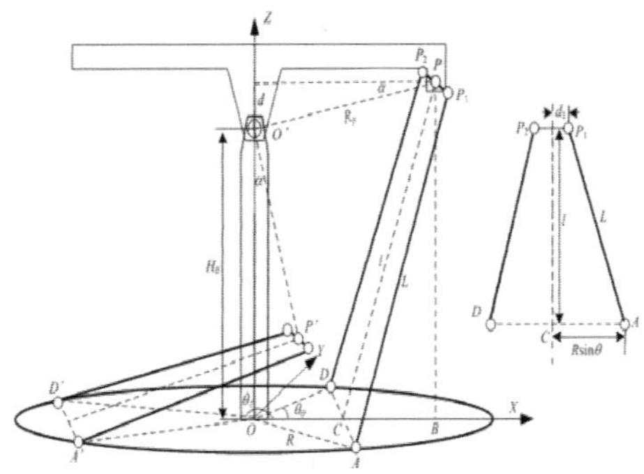

Figure 2. Schematic diagram of the 2R parallel mechanism.

The configuration synthesis of the mechanism is a fundamental process to guarantee the workspace of the rotatable platform. Referring to Figure 2, the directional angles of sliders θ_0 and θ_1 should be determined firstly according to the condition of interference-free when the pitch angle of rotatable platform is 0° and 90°. And then, in order to achieve the pitch angle $\varphi\left(0^\circ \le \varphi \le 90^\circ\right)$ and azimuth angle $\gamma\left(-180^\circ \le \gamma \le 180^\circ\right)$ without interference, the radius of the circular rail R is calculated as follows.

$$R = \frac{H_0\left(R_p \cos\alpha + d\right)}{d_1\left(\sin\theta_0 - \sin\theta_1\right) - d\cos\theta_1 + R_p \cos\alpha\cos\theta_0} \tag{1}$$

where, $\alpha = \arcsin\left(dR_p^{-1}\right)$, d is the height difference of the two revolute joints, d_1 is half of P_1P_2, R_p is the equivalent rotational radius as shown in Figure 2.

The length of the rod L is eventually determined as follows.

$$L = \left[l^2 + (R\sin\theta_0 - d_1)^2\right]^{1/2} \tag{2}$$

where, $l=\left[(H_0+d)^2+(R_p\cos\alpha-R\cos\theta_0)^2\right]^{1/2}$ is the length of equivalent rod *PC*.

The number of DOF can be calculated according to the modified Kutzbach-Grubler formula [10]

$$M=d(n-g-1)+\sum_{i=1}^{g}f_i+\nu-\zeta=6\times(7-8-1)+14=2 \tag{3}$$

3. Kinematic model

3.1. Inverse orientation solution

The directional angles of sliders *D* and *A* shown in Figure 2 are assumed as θ_1 and θ_2, respectively. It is obviously that the positions of *D* and *A* are symmetric when $\gamma=0^\circ$. And at this point, the directional angle of *D*, assumed as $\Delta\theta$, can be obtained as follows by solving the geometrical model.

$$\Delta\theta=\arccos M+\arccos N \tag{4}$$

where, $M=\dfrac{d_1^2+H_0^2+R_p^2+R^2-L^2-2H_0R_p\sin(\varphi-\alpha)}{2R\left[R_p^2\cos^2(\varphi-\alpha)+d_1^2\right]^{1/2}}$, $N=\dfrac{R_p\cos(\varphi-\alpha)}{\left[R_p^2\cos^2(\varphi-\alpha)+d_1^2\right]^{1/2}}$.

Further, the inverse orientation can be derived as follows.

$$\begin{aligned}\theta_1&=\gamma+\Delta\theta\\ \theta_2&=\gamma-\Delta\theta\end{aligned} \tag{5}$$

Figure 3 shows the curves of θ_1 and θ_2 in the case of the pitch angle $\varphi=45^\circ$ and the azimuth angle $0\le\gamma\le45^\circ$. One can observe that the inverse orientation solution is analytical and unique.

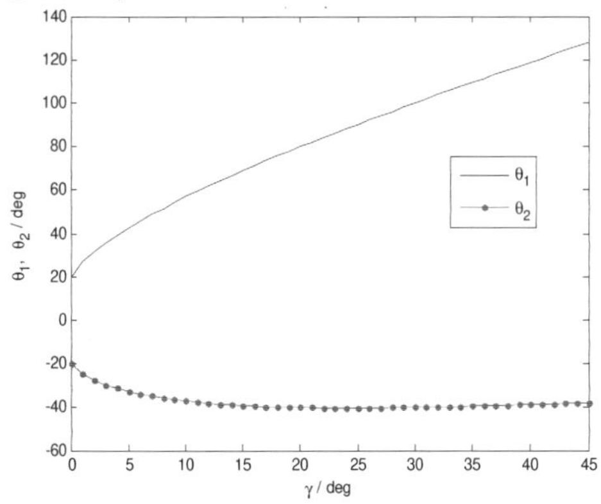

Figure 3. θ_1 and θ_2 vs γ ($\varphi=45^\circ, 0\le\gamma\le45^\circ$).

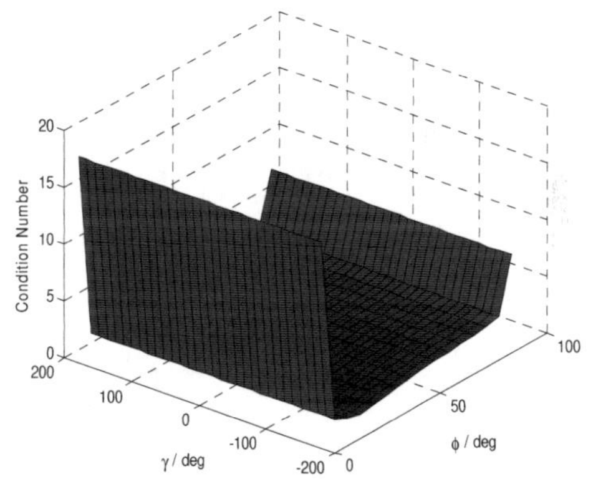

Figure 4. Condition number of the Jacobian matrix.

3.2. Forward orientation solution

Similarly, according to known θ_1 and θ_2, φ and γ can be calculated as follows.

$$\gamma=\frac{\theta_1+\theta_2}{2} \tag{6}$$

$$\varphi=\eta+\alpha-\arccos\lambda \tag{7}$$

where, $\eta=\arccos\dfrac{R\cos\Delta\theta}{\left(R^2\cos^2\Delta\theta+H_0^2\right)^{1/2}}$, $\lambda=\dfrac{d_1^2+H_0^2+R_p^2+R^2-L^2-2d_1R\sin\Delta\theta}{2R_p\left(R^2\cos^2\Delta\theta+H_0^2\right)^{1/2}}$, $\Delta\theta=\dfrac{\theta_1-\theta_2}{2}$.

IOP Conf. Series: Materials Science and Engineering **428** (2018) 012053 doi:10.1088/1757-899X/428/1/012053

3.3. Jacobian matrix

The Jacobian matrix of a mechanism is the time-varying mapping coefficient between input and output generalized velocity. The common dexterity index and singularity evaluation function are also derived from the Jacobian matrix. By differentiating Eq. (5) with respect to time, one can obtain

$$\begin{bmatrix} \dot{\theta}_1 \\ \dot{\theta}_2 \end{bmatrix} = [J] \begin{bmatrix} \dot{\varphi} \\ \dot{\gamma} \end{bmatrix} = \begin{bmatrix} \dfrac{\partial \theta}{\partial \varphi} & 1 \\ -\dfrac{\partial \theta}{\partial \varphi} & 1 \end{bmatrix} \begin{bmatrix} \dot{\varphi} \\ \dot{\gamma} \end{bmatrix} \tag{8}$$

Figure 4 shows the condition number in the domain of $-180^{\circ} \le \gamma \le 180^{\circ}$ and $0^{\circ} \le \varphi \le 90^{\circ}$. One can observe that this 2R parallel mechanism is singularity-free and well conditioned. There is a notable feature that when the pitch angle is fixed, the change of the azimuth angle does not affect its Jacobian matrix condition number, this result agree with our intuition.

4. Dynamic model

The study of mechanism dynamics is to establish the relationship between the driving force and moment of each joint and motion of its output element. Especially, under the condition that the part of the actuator has been given, the force and moment of each driving part and the kinematic pair are solved, which is called dynamic inverse problem [11-13].

4.1. Kinetic energy and potential energy

In order to solve the kinetic energy and potential energy of the mechanism, the mechanism is divided into three parts: the rotational platform, the two rods and the two sliders.

The kinetic energy and potential energy of the rotational platform are

$$K_1 = \frac{1}{2}\left(\boldsymbol{\omega}^T \boldsymbol{I}_s \boldsymbol{\omega}\right), \; P_1 = m_s g z_s \tag{9}$$

where, $\boldsymbol{I}_s$ is the inertia matrix of the coordinate system on the platform with respect to the center of mass, $\boldsymbol{\omega} = \dot{\gamma}\boldsymbol{R}_z(\alpha)\boldsymbol{R}_y(\beta)\vec{x} + \dot{\beta}\boldsymbol{R}_z(\alpha)\vec{y} + \dot{\alpha}\vec{z}$ and m_s are angular velocity and mass of the rotational platform, z_s is the distance from the center of rotational platform to the rail plane.

The kinetic energy and potential energy of the two rods are

$$K_2 = \frac{1}{2}m_1\left(\dot{\boldsymbol{r}}^2_{P_1A} + \dot{\boldsymbol{r}}^2_{P_2D}\right) + \frac{1}{2}J\left(\boldsymbol{\omega}^2_{P_1A} + \boldsymbol{\omega}^2_{P_2D}\right), \; P_2 = m_1 g\left[H_0 - R_p \sin(\varphi - \alpha)\right] \tag{10}$$

where, $\boldsymbol{r}_{P_1A}$ and $\boldsymbol{r}_{P_2D}$ are the position vectors of the rods' centers of gravity, $\boldsymbol{\omega}_{P_1A}$ and $\boldsymbol{\omega}_{P2D}$ are the angular velocity of each rod, m_1 is the mass of the rod, $J = \frac{1}{12}m_1 L^2$.

The kinetic energy and potential energy of the two sliders are

$$K_3 = \frac{1}{2}m_h\left(\dot{\boldsymbol{r}}^2_A + \dot{\boldsymbol{r}}^2_D\right), \; P_3 = 0 \tag{11}$$

where $\boldsymbol{r}_A$ and $\boldsymbol{r}_D$ are the position vectors of D and A as shown in Figure 2, m_h is the mass of the slider.

4.2. Kinetic equation

The total kinetic energy and potential energy of the system are

$$K = K_1 + K_2 + K_3 = \frac{1}{2}\left(\boldsymbol{\omega}^T \boldsymbol{I}_s \boldsymbol{\omega}\right) + \frac{1}{2}m_1\left(\dot{\boldsymbol{r}}^2_{P_1A} + \dot{\boldsymbol{r}}^2_{P_2D}\right) + \frac{1}{2}J\left(\boldsymbol{\omega}^2_{P_1A} + \boldsymbol{\omega}^2_{P_2D}\right) + \frac{1}{2}m_h\left(\dot{\boldsymbol{r}}^2_A + \dot{\boldsymbol{r}}^2_D\right) \tag{12}$$

The total potential energy has the following form

$$P = P_1 + P_2 + P_3 = m_s g\left(H_0 + d\cos\varphi\right) + m_1 g\left[H_0 - R_p \sin(\varphi - \alpha)\right] \tag{13}$$

By substituting Eqs. (12) and (13) into Lagrange dynamic equation $\frac{\mathrm{d}}{\mathrm{d}t}\left(\frac{\partial K}{\partial \dot{\theta}_i}\right) - \frac{\partial K}{\partial \theta_i} + \frac{\partial P}{\partial \theta_i} = \tau_i$

A further simplification yields[14]

$$\begin{cases} \tau_1 = \hat{J}_1\ddot{\theta}_1 + \frac{1}{2}\frac{\partial \hat{J}_1}{\partial \theta_1}\dot{\theta}_1^2 + \frac{\partial \hat{J}_1}{\partial \theta_2}\dot{\theta}_1\dot{\theta}_2 + 4\hat{J}_3\ddot{\theta}_2\dot{\theta}_1 + 2\hat{J}_3\dot{\theta}_2^2\ddot{\theta}_1 + \frac{\partial \hat{J}_3}{\partial \theta_1}\dot{\theta}_1^2\dot{\theta}_2^2 + 2\frac{\partial \hat{J}_3}{\partial \theta_2}\dot{\theta}_1\dot{\theta}_2^3 - \frac{1}{2}\frac{\partial \hat{J}_1}{\partial \theta_1}\dot{\theta}_2^2 + \frac{\partial P}{\partial \theta_1} \\ \tau_2 = \hat{J}_2\ddot{\theta}_2 + \frac{1}{2}\frac{\partial \hat{J}_2}{\partial \theta_2}\dot{\theta}_2^2 + \frac{\partial \hat{J}_2}{\partial \theta_1}\dot{\theta}_1\dot{\theta}_2 + 4\hat{J}_3\ddot{\theta}_1\dot{\theta}_2 + 2\hat{J}_3\dot{\theta}_1^2\ddot{\theta}_2 + \frac{\partial \hat{J}_3}{\partial \theta_2}\dot{\theta}_1^2\dot{\theta}_2^2 + 2\frac{\partial \hat{J}_3}{\partial \theta_2}\dot{\theta}_1\dot{\theta}_2^3 - \frac{1}{2}\frac{\partial \hat{J}_2}{\partial \theta_2}\dot{\theta}_1^2 + \frac{\partial P}{\partial \theta_2} \end{cases} \tag{14}$$

Due to space limitation, $\hat{J}_1$, $\hat{J}_2$ and $\hat{J}_3$ in the Eq. (14) are not listed in detail.

5. Numerical simulation

In order to verify the correctness of the dynamic model of the 2R parallel mechanism, the results of the dynamic analysis software and the inverse solution of the dynamic model are compared. The system parameters used in the simulation are shown in the Table 1. The simulation is carried out by using the typical trajectory motion of the rotational platform, which is specified in Eq. (15).

Table 1. The parameters of the 2R parallel mechanism

Symbol	H_0	R_p	R	L	d	d_1	m_s	m_1	m_h	g
Value	5250	3800	4100	5300	1500	3750	70000	8000	1000	9.8
Unit	mm	mm	mm	mm	mm	mm	kg	kg	kg	N/kg

$$\dot{\varphi} = \begin{cases} 5t, 0 \le t < 2 \\ 10, 2 \le t < 8 \end{cases} \tag{15}$$

$$\dot{\gamma} = 2t, 0 \le t \le 8$$

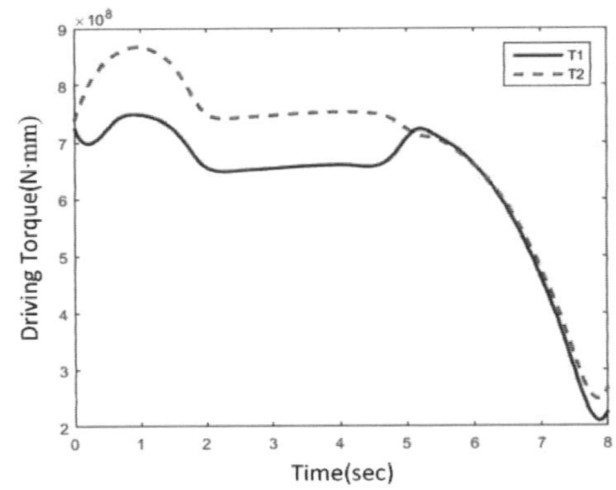

Figure 5. The driving torque of the sliders by calculation.

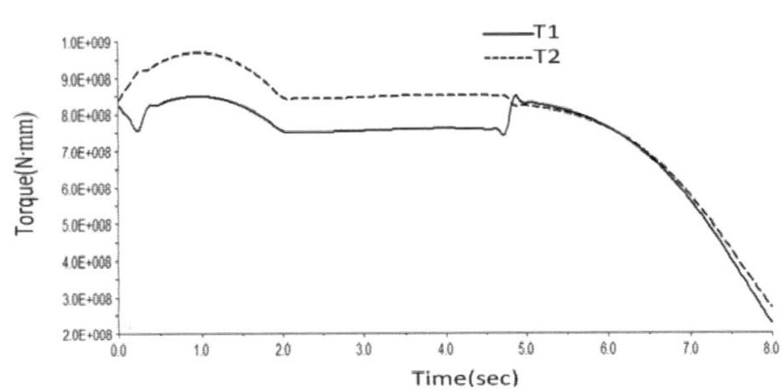

Figure 6. The driving torque of the sliders by simulation.

Figure 5 is the driving torque curves by calculation and Figure 6 is the torque curves by simulation. From the comparison of software simulation and dynamic model calculation, it can be seen the curves basically agrees with each other. The reason for the difference in numerical values arises from the simplification of the kinetic model and the parameter variation such as part mass.

By numerical simulation, the correctness of the dynamic model is verified. In the course of motion planning, we will be able to carry out simulations of more complicated motion, and consider more reasonable motion planning, so as to minimize the required torque.

6. Conclusions

By introducing a novel 2R parallel mechanism, this paper presents an alt-azimuth pedestal for supporting and rotating a rotational platform about two perpendicular axes. Some meaningful conclusions can be drawn as follows.

(1)The active joints of the 2R parallel mechanism are two sliders with servo control on a circular rail. The synchronous and asynchronous motions of the two sliders lead to the azimuth (Rz) and pitch (Rx) of the rotatable platform, respectively.

(2)The solutions to its inverse and forward orientation problem are both analytical and unique. The 2R parallel mechanism is singularity-free and well-conditioned in its large workspace.

(3)Numerical simulation shows the effectiveness of the dynamic model based on Lagrange equation.

This study lays a solid base for the engineering implementation of the 2R parallel mechanism, which can be extended to other related applications.

Acknowledgments

The authors acknowledge the financial support from the 13th Five-year Plan Advanced Research Project under Grant. 30508040102 and the National Natural Science Foundation of China under Grant. 51679398.

References

[1] Shitov V G 1980 Evalution of the precision in setting a telescope with an alt-azimuth mounting *Measurement Techniques* **23**(1) 16–18
[2] Su D G, Wang Y N and Cui X Q 2004 A configuration for future giant telescope *Acta Astronomica Sinica* **28**(3) 356–366
[3] Huang Z and Kong L F 1997 *Theory and Control of Parallel Robot Mechanism* (Beijing: Machinery Industry Press)
[4] Merlet J P 2006 *Parallel Robots* (Netherlands: Springer)
[5] Huang Z *et al* 2010 Lower-mobility parallel robots: theory and applications *Advances in Mechanical Engineering* **2010**(2) 1652–60
[6] Kock S and Schumacher W 1998 A parallel x-y manipulator with actuation redundancy for high-speed and active stiffness applications *Proc. 1998 IEEE Int. Conf. on Robot and Automation* (Leuven) vol 3 pp 2295–2300
[7] Gosselin C *et al* 1994 Analysis of spherical two degree of freedom parallel manipulators *Proc. of the 23rd Biennial Mechanisms Conf.* (Minneapolis) pp 255–262
[8] Zhang L J, Li Y Q and Huang Z 2006 On the kinematics analysis of a 2-degree-of-freedom spherical 5r parallel manipulator *China Mechanical Engineering* **17**(4) 343–346
[9] Bozorgi E R J *et al* 2014 Design, development, dynamic analysis and control of a 2-degree-of-freedom spherical parallel mechanism *2nd RSI/ISM Int. Conf. on Robotics and Mechanics* (New York) pp 445-450
[10] Yang Y Z, Cheng B and Duan X C 2016 Modeling and analysis of two-degrees-of-freedom spherical parallel manipulator *World Sci-Tech R and D* **5** 1054–59
[11] Craig J J 2004 *Introduction to robotics: mechanics and control* (Upper Saddle River: Prentice Hall)
[12] Liu S Z *et al* 2014 Kinematics and dynamics analysis of a three degree-of-freedom parallel manipulator *Journal of Central South University* **21**(7) 2660–66
[13] Zhang D 2010 Parallel Robotic Machine Tools (The United States: Springer)
[14] Liu S Z *et al* 2009 Kinematic and dynamic analysis of a three-degree-of-freedom parallel manipulator *Journal of Mechanical Engineering* **45**(8) 11–17

IOP Conf. Series: Materials Science and Engineering **428** (2018) 012054 doi:10.1088/1757-899X/428/1/012054

Neural Network Based Adaptive Chaotification of Uncertain Robot Manipulators Incorporating Motor Dynamics

Yang Li[*] and Yuxiang Wu

School of Automation Science and Engineering, South China University of Technology, Guangdong, Guangzhou, 510641, China

[*]liyang_scut@163.com

Abstract. Chaotification refers to the problem of generating chaos from an originally non-chaotic system by using a control law. In this paper, an adaptive Radial Basis Function Neural Network (RBF NN) control method is proposed to realize the chaotification of uncertain robot manipulators incorporating motor dynamics. In order to achieve velocity tracking of robot manipulators, the velocity field is introduced as a chaos reference field, and the adaptive RBF NN is used to approximate the unknown nonlinear function of the system. Finally, the uniformly ultimately boundedness of all signals in the closed-loop system is proved via Lyapunov stability theorem, and the effectiveness and feasibility of the proposed control method is verified through the simulation on two-link rigid robot manipulators.

1. Introduction

Chaotification is also called anti-control of chaos which refers to the generation, maintenance and enhancement of chaos[1]. Early researchers believed that chaotic motion can't match the practical application, so they focused on suppressing and eliminating chaos[2]. However, with the deepening of research, it was discovered that it is beneficial to apply chaos in appropriate fields, such as secure communications[3], vibration, mixing and compaction[4], weather prediction[5], biomedical science[6] and etc. In the next, a literature review on chaotification of mechanical systems is given.

In 2003, Y. Gao[7] studied chaotification for the permanent magnet DC motor via delay feedback method, which improved the vibration effect. H. Zargarzadeh[8] proposed a chaotification method for a rigid body by using a variable speed control moment gyroscope as an internal torque generator in 2009. More recently, by using the Lyapunov function framework, an adaptive chaos anti-controller was proposed in [9]. In [10], Javier Moreno achieved the anti-control of chaos for robot manipulators via adaptive neural network algorithm.A great feature of neural networks is that they can estimate nonlinear functions without any prior knowledge on structure of the function to be compensated. Therefore, neural networks have been used to solve many sorts of problems in complex systems and chaos system[11]. In addition, the dynamics model of the robot system incorporating actuators is highly nonlinear, heavily coupled and uncertain, which increase the complexity of controller design. In order to overcome the problems above, important researches have been carried out on control of mechanical systems incorporating actuator dynamics[12]. However, there are few researches on anti-control of chaos for robot manipulators incorporating actuator dynamics.

The main purpose of this paper is to introduce a novel adaptive RBF neural network chaos anti-controller for the uncertain manipulators incorporating motor dynamics. Firstly, a velocity field[13] has been defined as the chaotic reference field for the chaotification of robot manipulators, thus anti-

IOP Conf. Series: Materials Science and Engineering **428** (2018) 012054 doi:10.1088/1757-899X/428/1/012054

control of chaos will be assured if the robot joint velocity uniformly matches the chaotic reference field[14]. The entire design process of adaptive RBF NN algorithm is based on back-stepping method, meanwhile, the RBF NN is used to approximate the unknown nonlinear functions of the system, which has the advantage that the prior knowledge of the robot model is completely ignored.

2. Problem description

2.1. Robot model and its properties

The dynamics in link space of n-link robot manipulators can be expressed as[15]:

$$M(q)\ddot{q}+C(q,\dot{q})\dot{q}+G(q)+F(\dot{q})+\tau_d=\tau \tag{1}$$

Where q denotes $n\times 1$ vector of joint position, $\dot{q}$ denotes $n\times 1$ vector of joint velocity, $\ddot{q}$ denotes $n\times 1$ vector of joint acceleration, $M(q)$ is $n\times n$ symmetric positive definite inertia matrix, $C(q,\dot{q})$ is $n\times n$ centripetal and Coriolis matrix, $G(q)$ is $n\times 1$ vector of gravitational torques, $F(\dot{q})$ is $n\times 1$ vector of friction torque, τ_d denotes $n\times 1$ vector of unknown disturbance, τ denotes $n\times 1$ vector of applied torque inputs. The robot manipulators have properties as follows[15]:

Property 1: The matrix M is uniformly positive definite.

Property 2: The matrix $[\dot{M}(q)-2C(q,\dot{q})]$ is skew symmetric resulting in the following characteristic:

$$x^T\left[\dot{M}(q)-2C(q,\dot{q})\right]x=0,\forall x,q,\dot{q}\in R^n \tag{2}$$

Assumption 1: The disturbance τ_d is bounded. It satisfies $\|\tau_d\|\le d_m$ and d_m is a positive constant.

Assumption 2: Robot joints are driven by DC motors, and the motor dynamics is given by[12]:

$$\begin{cases} L\dot{i}+Ri+K_e\dot{q}_m=u \\ \tau_m=K_T i \end{cases} \tag{3}$$

Where, $\tau_m=\bar{N}\tau$, $q=\bar{N}q_m$, τ_m is $n\times 1$ vector of motor torque, q_m is $n\times 1$ vector of motor angle position, $\dot{q}_m$ is $n\times 1$ vector of motor angle velocity, i is $n\times 1$ vector of motor armature current, L,R,K_e are $n\times n$ diagonal positive definite matrix, respectively representing inductance, resistance, and voltage constant, u denotes $n\times 1$ vector of motor voltage, K_T denotes $n\times n$ motor torque constant matrix, $\bar{N}$ denotes reduction ratio. Let $q=x_1,\dot{q}=x_2,i=x_3$, combining (1) with (3), it can be obtained that:

$$\begin{cases} \dot{x}_1=x_2 \\ \dot{x}_2=M^{-1}[\bar{N}^{-1}K_T x_3-Cx_2-F-G-\tau_d] \\ \dot{x}_3=L^{-1}u+L^{-1}(-Rx_3-K_e\bar{N}^{-1}x_2) \end{cases} \tag{4}$$

2.2. Chaos system

A chaos system can be defined as the one that is locally unstable but with solutions globally bounded. Consider the nonlinear system given by:

$$\dot{x}=f(x) \tag{5}$$

Where $x\in R^l$, $f:R^l\to R^l$. The nonlinear system (5) is chaotic when it satisfies the conditions[10] as follows: 1) the system is extremely sensitive to the initial value; 2) the periodic orbits of system are dense; 3) the system is topological mixing. The nonlinear system (5) can be called chaotic reference field, which will be used in the controller design process of chaotification for the robot manipulators.

2.3. Control objective

Assume that the joint position $q\in R^n$ and the joint velocity $\dot{q}\in R^n$ are measurable. Define the extended coordinate as $\bar{q}=[\bar{q}_1\,\bar{q}_2\cdots\bar{q}_l]^T=[q\;p]^T\in R^l\;(l>n)$, where $p=[p_1\cdots p_{l-n}]^T\in R^{l-n}$. Corresponding to the chaotic reference system (5), the extended coordinate can be expressed as $\dot{\bar{q}} = [\dot{q}\ \dot{p}]^T = f(\bar{q})$. Because $l > n$, we have $f(\bar{q}) = [f_1(\bar{q})\ f_2(\bar{q})]^T \in R^l$, where, $f_1(\bar{q}): R^l \to R^n, f_2(\bar{q}): R^l \to R^{l-n}$, $p \in R^n$ is an internal state obtained by solving:

$$\dot{p} = f_2(\bar{q}), p(0) \in R^{l-n} \tag{6}$$

For the two-link robot manipulators in section 4, it will generate chaotic motion by using jerk-type system[14] as chaotic reference field. Take the extended coordinate as $\bar{q} = [q_1\ q_2\ p_1]^T$, then the chaotic reference field can be given by[14]:

$$f(\bar{q}) = \begin{bmatrix} f_1(\bar{q}) \\ f_2(\bar{q}) \end{bmatrix} = \begin{bmatrix} f_{11}(\bar{q}) \\ f_{12}(\bar{q}) \\ f_{21}(\bar{q}) \end{bmatrix} = a_0 \begin{bmatrix} q_2 \\ p_1 \\ a_1 p_1 + a_2 q_2 + (q_1^2 + a_3) \end{bmatrix} \tag{7}$$

Where $a_0 = 2.0, a_1 = -0.44, a_2 = -2.0, a_3 = -1.0$. In addition, it is assumed that $f(\bar{q})$ is differentiable, and it satisfies $\|f(\bar{q})\| \le v_{m1}$, and $\left\| \frac{\partial f(\bar{q})}{\partial \bar{q}} \right\| \le v_{m2}, \forall \bar{q} \in R^l$, where v_{m1}, v_{m2} are positive constants. Defining the velocity field tracking error as:

$$\varpi(t) = f(\bar{q}(t)) - \dot{\bar{q}}(t) = \begin{bmatrix} f_1(\bar{q}(t)) \\ f_2(\bar{q}(t)) \end{bmatrix} - \begin{bmatrix} \dot{q}(t) \\ \dot{p}(t) \end{bmatrix} \tag{8}$$

In this paper, the control objective is to design a motor voltage u to achieve that: 1) velocity field tracking error $\varpi(t)$ converges to a small neighborhood of zero in a finite time; 2) all the signals in the closed-loop system realize uniformly ultimately boundedness.

3. Controller design and stability analysis

3.1. Controller design

In this paper, an adaptive RBF NN control algorithm is proposed by using back-stepping method for chaotification of uncertain robot manipulators incorporating motor dynamics. The design steps of controller are as follows:

Step 1: Let $e_1 = x_1 - x_d$, $e_2 = x_2 - \alpha_1$, combining velocity reference field proposed in equation (7), and differentiating e_1 we get:

$$\dot{e}_1 = \dot{x}_1 - f_1(\bar{q}) = e_2 + \alpha_1 - f_1(\bar{q}) \tag{9}$$

Where $x_d \in R^n$ is vector of reference position, $f_1(\bar{q}) \in R^n$ is vector of reference velocity. Construct the virtual control variable $\alpha_1 = -K_1 e_1 + f_1(\bar{q})$, and $K_1 \in R^{n\times n}$ is a diagonal positive definite matrix, thus:

$$\dot{e}_1 = -K_1 e_1 + e_2 \tag{10}$$

Step 2: The time derivative of e_2 can be expressed as:

$$\dot{e}_2 = M^{-1}\left[\bar{N}^{-1} K_T x_3 - C x_2 - F - G - \tau_d\right] - \dot{\alpha}_1 \tag{11}$$

Where $\dot{\alpha}_1 = -K_1 \dot{e}_1 + \dot{f}_1(\bar{q})$, thereby the equation (11) can be rewritten as:

$$\dot{e}_2 = M^{-1} \bar{N}^{-1} K_T \left[x_3 + H_1(Z_1)\right] \tag{12}$$

Where $H_1(Z_1) = -\bar{N} K_T^{-1}(C x_2 + F + G + \tau_d + M\dot{\alpha}_1)$. It can be noted that M, C, G, F, τ_d are unknown, and RBF NN is used to approximate the unknown function $H_1(Z_1)$, thus:

$$H_1(Z_1) = W_1^{*T} \varphi(Z_1) + \varepsilon_1 \tag{13}$$

Where $Z_1 = [x_1^T\ x_2^T]$ represents the input vector of NN, W^* represents ideal weight vector, $\varphi(Z_1)$ is basis function, and Gaussian function is taken as basis function of RBF NN in this paper; $\varepsilon_1 \in R^n$ is vector of NN approximation error, and it satisfies $\|\varepsilon_1\| \le \varepsilon_1^*$, where ε_1^* is a small positive constant. Then, let $e_3 = x_3 - \alpha_2$, choosing the virtual control variable $\alpha_2 = -e_1 - K_2 e_2 - \hat{W}_1^T \varphi(Z_1)$, where $K_2 \in R^{n \times n}$ is a diagonal positive definite matrix, $\hat{W}_1$ denotes estimated vector of NN ideal weight vector W_1^*, with estimated error $\tilde{W}_1 = \hat{W}_1 - W_1^*$.Therefore, the $\dot{e}_2$ can be rewritten as bellow:

$$\dot{e}_2 = M^{-1}\bar{N}^{-1}K_T\left[e_3 - e_1 - K_2 e_2 - \tilde{W}_1^T \varphi(Z_1) + \varepsilon_1\right] \tag{14}$$

Step 3: The time derivative of e_3 can be given by:

$$\dot{e}_3 = L^{-1}u + L^{-1}\left(-Rx_3 - K_e \bar{N}^{-1} x_2\right) - \dot{\alpha}_2 \tag{15}$$

Where $\dot{\alpha}_2 = \sum_{i=1}^{2}\left(\frac{d\alpha_2}{dx_i} \times \dot{x}_i\right) + \frac{d\alpha_2}{d\hat{W}_1} \times \dot{\hat{W}}_1 + \frac{d\alpha_2}{dx_d} \times f_1(\bar{q}) + \frac{d\alpha_2}{df_1(\bar{q})} \times \dot{f}_1(\bar{q})$, thus equation (15) can be replaced by:

$$\dot{e}_3 = L^{-1}\left(u + H_2(Z_2)\right) \tag{16}$$

Where $H_2(Z_2) = -Rx_3 - K_e \bar{N}^{-1} x_2 - L\dot{\alpha}_2$. It can be noted that R, K_e, L are unknown, and RBF NN is employed to approximate the unknown function $H_2(Z_2)$, thus:

$$H_2(Z_2) = W_2^{*T} \varphi(Z_2) + \varepsilon_2 \tag{17}$$

Where $Z_2 = \left[x_1^T \; x_2^T \; x_3^T\right]$ represents the input vector of NN, W_2^* represents ideal weight, $\varphi(Z_2)$ is basis function, $\varepsilon_2 \in R^n$ is the vector of NN approximation error, and it satisfies $\|\varepsilon_2\| \le \varepsilon_2^*$, where ε_2^* is a small positive constant. Constructing practical control law, we have:

$$u = -K_3 e_3 - e_2 - \hat{W}_2^T \varphi(Z_2) \tag{18}$$

$$\dot{e}_3 = L^{-1}\left(-K_3 e_3 - e_2 - \tilde{W}_2^T \varphi(Z_2) + \varepsilon_2\right) \tag{19}$$

Where $K_3 \in R^{n \times n}$ is a diagonal positive definite matrix, $\hat{W}_2$ is estimated value of NN weight W_2^*, with estimated error $\tilde{W}_2 = \hat{W}_2 - W_2^*$. The update law of weight is given by:

$$\dot{\hat{W}}_i = \Gamma_i \varphi(Z_i) e_{i+1}{}^T - \sigma_i \Gamma_i \hat{W}_i \; (i = 1, 2) \tag{20}$$

Where $\Gamma_i \in R^{n \times n}$ denotes diagonal positive definite matrix, and σ_i is a small positive constant.

3.2. Stability analysis

The closed-loop system composed of control object (4), the error vector (10) (14) (19), NN weight updating law (20), and practical control law (18) can be expressed as:

$$\begin{cases} \dot{e}_1 = -K_1 e_1 + e_2 \\ \dot{e}_2 = M^{-1}\bar{N}^{-1}K_T\left[e_3 - e_1 - K_2 e_2 - \tilde{W}_1^T \varphi(Z_1) + \varepsilon_1\right] \\ \dot{e}_3 = L^{-1}\left(-K_3 e_3 - e_2 - \tilde{W}_2^T \varphi(Z_2) + \varepsilon_2\right) \\ \dot{\hat{W}}_i = \Gamma_i \varphi(Z_i) e_{i+1}{}^T - \sigma_i \Gamma_i \hat{W}_i \; (i = 1, 2) \end{cases} \tag{21}$$

Theorem: Considering the closed-loop system (21), the given chaotic velocity reference field (7), there are initial conditions $x_1(0) \in \Omega_0$ (Ω_0 is a compact set) and $\hat{W}_i(0) = 0$, we can obtain: 1) all the signals in the closed-loop system (21) are uniformly ultimately bounded; 2) Choosing appropriate design parameters, there exists limited time t_l which can make velocity field error $\varpi(t)$ converges to a small neighborhood of zero, $\forall t \ge t_l$.

Proof 1)**:** Select Lyapunov function as:

$$V = \frac{1}{2}e_1^T e_1 + \frac{1}{2}e_2^T \bar{N} K_T^{-1} M e_2 + \frac{1}{2}e_3^T L e_3 + \frac{1}{2}tr\left(\tilde{W}_1^T \Gamma_1^{-1} \tilde{W}_1\right) + \frac{1}{2}tr\left(\tilde{W}_2^T \Gamma_2^{-1} \tilde{W}_2\right) \tag{22}$$

Combining equation (21) with equation (22), the time derivative of V can be expressed as:

$$\dot{V} = -K_1\|e_1\|^2 - K_2\|e_2\|^2 - K_3\|e_3\|^2 + e_2^T\varepsilon_1 + e_3^T\varepsilon_2 - \sigma_1\tilde{W}_1^T\hat{W}_1 - \sigma_2\tilde{W}_2^T\hat{W}_2 \tag{23}$$

According to the Young inequality, we can obtain:

$$\dot{V} \le -\lambda_{\min}(K_1)\|e_1\|^2 - \lambda_{\min}(K_{e2})\|e_2\|^2 - \lambda_{\min}(K_{e3})\|e_3\|^2 - K_{W1}\|\tilde{W}_1\|^2 - K_{W2}\|\tilde{W}_2\|^2 + \frac{\varepsilon_1^{*2} + \sigma_1\|W_1^*\|^2}{2j_1} + \frac{\varepsilon_2^{*2} + \sigma_2\|W_2^*\|^2}{2j_2} \tag{24}$$

Where $0 < j_1 < 2$, and $0 < j_2 < 2$ are constants, $K_{e2} = K_2 - \frac{j_1}{2}I$, $K_{e3} = K_3 - \frac{j_2}{2}I$, $K_{W1} = \frac{\sigma_1(2-j_1)}{2}$, $K_{W2} = \frac{\sigma_2(2-j_2)}{2}$, and I is 2×2 identity matrix. Selecting appropriate design parameters K_2, K_3, j_1, j_2 to make $K_{e2} > 0, K_{e3} > 0$, inequality (24) can be turned into :

$$\dot{V} \le -\rho V + \sigma \tag{25}$$

$$\begin{cases} \rho = \min\left\{2\lambda_{\min}(K_1), \dfrac{2\lambda_{\min}(K_{e2})}{\lambda_{\max}(\bar{N}K_T^{-1}M)}, \dfrac{2\lambda_{\min}(K_{e3})}{\lambda_{\max}(L)}, \dfrac{2K_{W1}}{\lambda_{\max}(\Gamma_1^{-1})}, \dfrac{2K_{W2}}{\lambda_{\max}(\Gamma_2^{-1})}\right\} \\ \sigma = \dfrac{\varepsilon_1^{*2} + \sigma_1\|W_1^*\|^2}{2j_1} + \dfrac{\varepsilon_2^{*2} + \sigma_2\|W_2^*\|^2}{2j_2} \end{cases} \tag{26}$$

Where $\lambda_{\max}[\bullet]$ and $\lambda_{\min}[\bullet]$ represent the maximum and minimum eigenvalue of the matrix $[\bullet]$. Obviously, because $\|W_1^*\|, \|W_2^*\|$ are constants, and $\varepsilon_1^*, \varepsilon_2^*$ are small constants, choosing sufficiently small σ_1, σ_2, and suitable j_1, j_2 can make σ a small positive constant. According to inequality (25), it can be seen that: when $t \to \infty$, the following inequalities are established:

$$\|e_1\| \le \sqrt{\frac{\sigma}{\lambda_{\min}(K_1)}},\quad \|e_2\| \le \sqrt{\frac{\sigma}{\lambda_{\min}(K_{e2})}},\quad \|e_3\| \le \sqrt{\frac{\sigma}{\lambda_{\min}(K_{e3})}},\quad \|\tilde{W}_1\| \le \sqrt{\frac{\sigma}{K_{W1}}},\quad \|\tilde{W}_2\| \le \sqrt{\frac{\sigma}{K_{W2}}} \tag{27}$$

Therefore, the variable e_1, e_2, e_3, $\tilde{W}_1, \tilde{W}_2$ are uniformly ultimately bounded. Moreover, for all Z, $\varphi(Z)$ is bounded, so combining equation (18) we can know that motor voltage u is bounded too. 2) Solving the inequality (25), we get:

$$0 \le V(t) \le e^{-\rho t} \times (V(0) - \frac{\sigma}{\rho}) + \frac{\sigma}{\rho} \tag{28}$$

Combining equation (22) with inequality (28), we can obtain:

$$\frac{1}{2}\|e_1\|^2 < V(0)e^{-\rho t} + \frac{\sigma}{\rho},\quad \frac{1}{2}\lambda_{\min}(\bar{N}K_T^{-1}M)\|e_2\|^2 < V(0)e^{-\rho t} + \frac{\sigma}{\rho} \tag{29}$$

According to inequality (29), it can be known that: for the given $\gamma_1 > \left(\frac{2\sigma}{\rho}\right)^{\frac{1}{2}}, \gamma_2 > \left(\frac{2\sigma}{\rho\lambda_{\min}(\bar{N}K_T^{-1}M)}\right)^{\frac{1}{2}}$, there exists limit time t_l such that $\|e_1\| < \gamma_1, \|e_2\| < \gamma_2\ \forall t \ge t_l$, where γ_1, γ_2 are small residual sets. By selecting large enough K_2, K_3 and appropriate j_1, j_2, β_1, β_2 can make K_{e2}, K_{e3}, K_{W1}, K_{W2} large enough so that ρ is large enough, and σ is small enough, thereby $\|e_1\|$ and $\|e_2\|$ are small enough $\forall t \ge t_l$, then combining equation(11), it can be seen that $\dot{e}_1$ is small enough too; that is to say, selecting design parameters appropriately allows the velocity tracking error $\dot{e}_1$ converges to a small neighborhood of zero in a finite time. Combined with equation (6), it can be proved that the chaotic velocity field error $\varpi(t)$ converges to a small neighborhood of zero in a finite time. Proof has been completed.

4. Simulation results

In this section, a two-link planar manipulators is taken to carry on the simulation experiment. The dynamics expressions of robot manipulators are[15]:

IOP Conf. Series: Materials Science and Engineering **428** (2018) 012054 doi:10.1088/1757-899X/428/1/012054

$$M(q)=\begin{bmatrix} l_2^2 m_2 + l_1^2(m_1+m_2)+2l_1l_2m_2\cos(q_2) & l_2^2m_2+l_1l_2m_2\cos(q_2) \\ l_2^2m_2+l_1l_2m_2\cos(q_2) & l_2^2m_2 \end{bmatrix} \quad (30)$$

$$C(q,\dot{q})\dot{q}=\begin{bmatrix} -2l_1l_2m_2\sin(q_2)(\dot{q}_1\dot{q}_2+0.5\dot{q}_2^2) \\ l_1l_2m_2\sin(q_2)\dot{q}_1^2 \end{bmatrix} \quad (31)$$

$$G(q)=\begin{bmatrix} l_1(m_1+m_2)g\cos q_1 + l_2m_2g\cos(q_1+q_2) \\ l_2m_2g\cos(q_1+q_2) \end{bmatrix} \quad (32)$$

$$F(\dot{q})=[0.5sign(\dot{q}_1) \quad 0.5sign(\dot{q}_2)] \quad \tau_d=[0.5\sin t \quad 0.5\cos t] \quad (33)$$

Where l_i denotes the length of link $i\,(i=1,2)$, m_i denotes the mass of link $i\,(i=1,2)$, g is gravity acceleration. The parameters are chosen as: $l_1=l_2=1.0m$, $m_1=0.8kg$, $m_2=2.3kg$, $g=9.8kg/m^2$, $L=diag[0.048 \;\; 0.048]\,H$, $R=diag[1.6 \;\; 1.6]\Omega$, $K_e=diag[0.19 \;\; 0.19]V{\bullet}s/rad$. Chaotic reference field is shown in equation (7), and the initial position of robot manipulators is $q_0=[-1 \;\; 0.5]^T$. The first RBF NN has four inputs and two outputs, there are 25 neurons in the hidden layer, $\eta_i=10$, $\Gamma_1=diag\{15\}$, $\sigma_1=0.001$, $\hat{W}_1(0)=0$; the second RBF NN has six inputs and two outputs, there are 25 neurons in the hidden layer, $\eta_i=8$, $\Gamma_2=diag\{20\}$, $\sigma_2=0.001$, $\hat{W}_2(0)=0$; the controller gains are $K_1=diag\{50\}$, $K_2=diag\{15\}$, $K_3=diag\{20\}$. Simulation results are shown as follows:

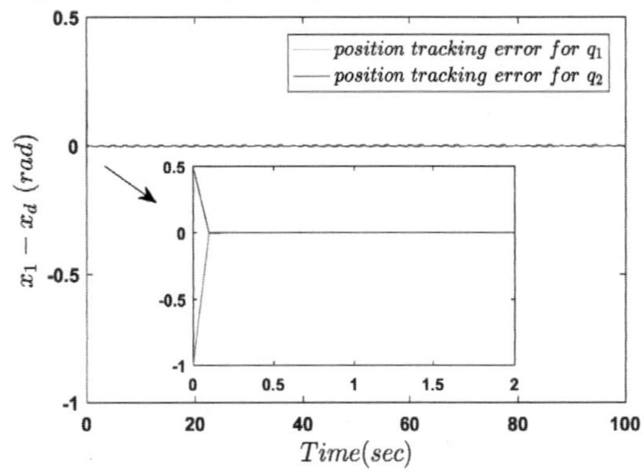

Figure 1: Position tracking error

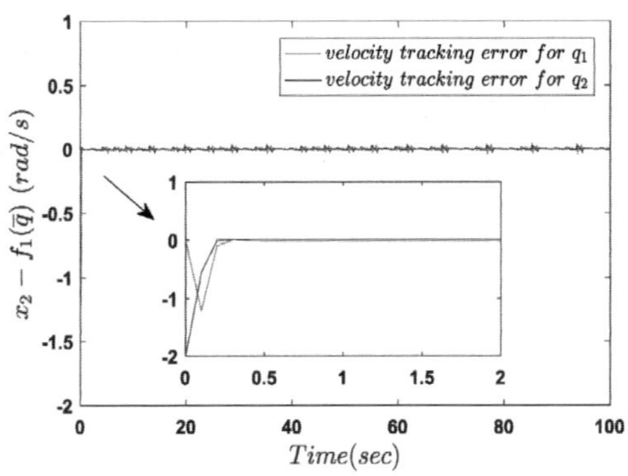

Figure 2: Velocity tracking error

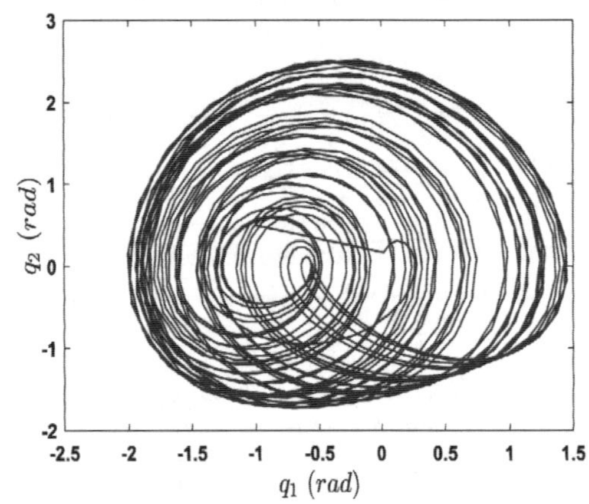

Figure 3: Actual chaos trajectory

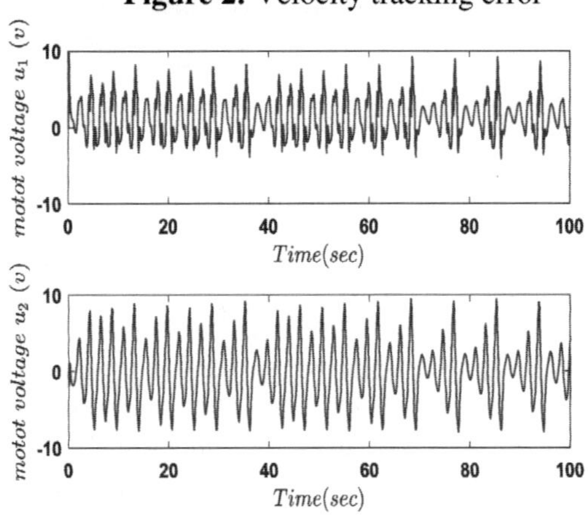

Figure 4: DC motor voltage u_1 and u_2

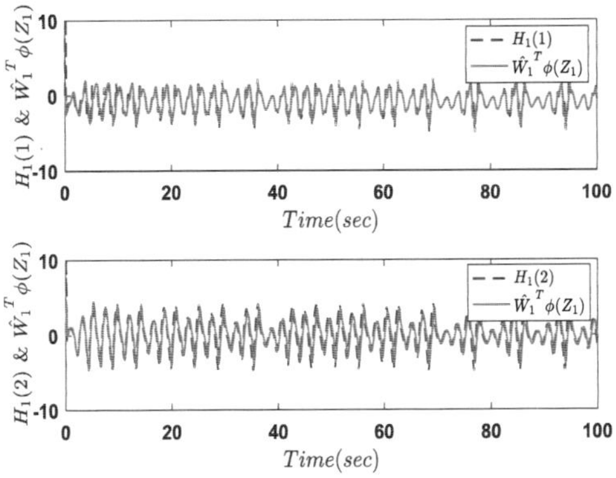

Figure 5 (a): RBF NN approximation to $H_1(Z_1)$

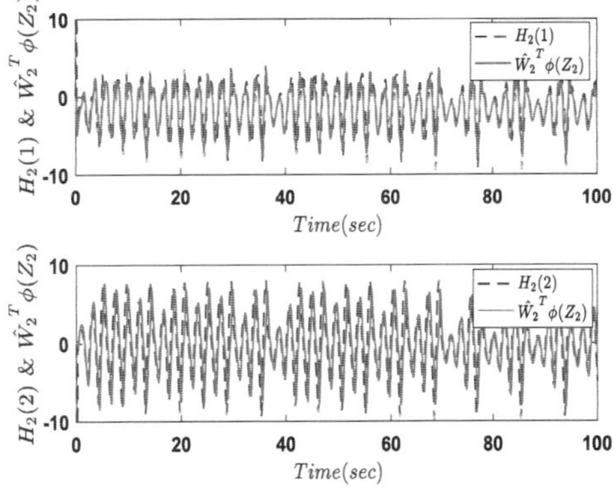

Figure. 5 (b): RBF NN approximation to $H_2(Z_2)$

From figure. 1, it can be seen that: the proposed algorithm can achieve great performance on position tracking for both two links, and the position tracking error converges to a small neighborhood of zero extremely quickly. From figure. 2, it can be seen that: the algorithm we proposed has a good velocity tracking performance for both two links, and the velocity tracking error converge to a small neighborhood of zero in a finite time, in other words, the proposed algorithm achieved control objective of velocity field. Figure. 3 gives the plot of robot manipulators joint position q_1 and q_2, and the chaotic attractor has observed there. Figure. 4 shows the DC motor voltage u_1 and u_2. Figure. 5 gives the approximation curve of RBF NN to the unknown nonlinear function $H_1(Z_1)$ and $H_2(Z_2)$, both two unknown nonlinear functions are approximated accurately.

5. Conclusion

This paper has proposed an adaptive RBF NN control algorithm based on back-stepping method to achieve the anti-control of chaos for the robot manipulators incorporating motor dynamics. Firstly, the velocity field is introduced as a chaos reference field, and RBF NN is employed to realize the unknown dynamic approximation of the system, thus, the proposed design approach does not require any prior knowledge of the robot model. It is found that the proposed algorithm has good performance on velocity tracking and approximation to the unknown nonlinear function, meanwhile, the chaotic attractor has observed successfully. These results make the proposed control scheme be more suitable for practical applications of chaos.

Acknowledgments

The authors would like to acknowledge the financial support from Guangdong Science and Technology Project (2015B010133002, 2017B090910011).

References

[1] Qiang Lai. Chaos Analysis and Research of Its Anti-control [D]. Huazhong University of Science and Technology, 2014.

[2] YunZhong Song, Guangzhou Zhao, Donglian Qi, et al. Reaserch of Chaotification Control [J]. *Journal of Zhejiang University of Technology*, 2007, (03):313-319.

[3] Simin Yu, Lvjin Hu, Chengqing Li. Chaotic Password and Its Application in Multimedia Secure Communication [J]. *Journal of Electronics & Information Technology,* 2016, (03):735-752.

[4] Porta G, Tamellini L, Lever V, et al. Inverse Modeling of Geochemical and Mechanical Compaction in Sedimentary Basins through Polynomial Chaos Expansion [J]. *Water Resources Research*, 2014, 50(12):9414-9431.

[5] Hayashi M, Nagasaka K. Wind Speed Prediction and Determination of Wind Power Output with Multi-area Weather Data by Deterministic Chaos [C]. *International Conference on Advanced Mechatronic Systems. IEEE*, 2014:192-197.

IOP Conf. Series: Materials Science and Engineering **428** (2018) 012054 doi:10.1088/1757-899X/428/1/012054

[6] Yanyu Zhou, Yanchen Du. Realize Anti-Control of Chaos on Heart Beats Using Rossler System [J]. *Journal of Progress in Biomedical Engineering,* 2014, (03):129-132.
[7] Gao Y, Chau K T. Chaotification of Permanent-magnet Synchronous Motor Drives using Time-delay feedback [C]. *IECON. IEEE*, 2003, (01):762-766.
[8] Zargarzadeh H, Motlagh M R J. Anti-Control of Chaos in Rigid Body Motion Using an Internal Torque Source [J]. *IFAC Proceedings Volumes*, 2009, 42(7):349-354.
[9] Javier Moreno–Valenzuela. Lyapunov Function-based Adaptive Chaos Anti Control of Robot Manipulators [C]. *IEEE, International Symposium on Industrial Electronics.* 2014:1221-1226.
[10] Javier Moreno–Valenzuela, Carlos Torres–Torres. Adaptive Chaotification of Robot Manipulators via Neural Networks with Experimental Evaluations [J]. *Neurocomputing*. 2016:56–65.
[11] Meng D. Neural Networks Adaptive Synchronization for Four-dimension Energy Resource System with Unknown Dead Zones [J]. *Neurocomputing,* 2015, 151(03):1495-1499.
[12] Jhan Z Y, Lee C H. Adaptive Impedance Force Controller Design for Robot Manipulator including Actuator Dynamics [J]. *International Journal of Fuzzy Systems*, 2017, 19(03):1-11.
[13] Javier Moreno–Valenzuela, R. Campa, V. Santibez. Velocity Field Control of a Class of Electrically-driven Manipulators. [J] *International Journal of Systems Science.* 2014, 45, 254–270.
[14] Sprott J C. Simple Chaotic Systems and Circuits[J].*American Journal of Physics*,2000, 68:758-763.
[15] Kelly R, Davila V S. *Control of Robot Manipulators in Joint Space* [M]. Springer, London, 2005.

IOP Conf. Series: Materials Science and Engineering **428** (2018) 012055 doi:10.1088/1757-899X/428/1/012055

Improving the Parameters of the Model of the Wheeled Mobile Robot with Constraints by Adaptive Fuzzy Method

Shiqi Wang and Songyi Dian*

College of Electrical Engineering and Information Technology, Sichuan University, Chengdu 610065, China

*scudiansy@scu.edu.cn(Songyi Dian)

Abstract. In the process of motion of the ground, the mobile trolley has branches of motion in the direction of the axis of X , Y and Z. However, considering that the mobile robot only has the ability of adjustment of motion on the axis of X and Y, it is said that the mode of the motion of this kind of mobile robots with constraints is not complete. The method of control used in the past is the constraint method of predictive control. The whole mobile robot can be considered as a system including three inputs and three outputs. The motion of the mobile robot in the direction of the axis of Z can be considered as a kind of disturb from the inputs with a certain effect on the axis of X and Y, but the control of the output is converted to the three parts of control used in the motion on the axis of X and Y and the control of the input of disturbance.The three systems with one independent input and one independent output are considered as the predictive control on the system including three inputs and three outputs. In this paper, the parameters of the predictive model are optimized in the way similar to an adaptive fuzzy control and based on errors and change of errors.

1. Introduction

In the face of a system of motion of the wheeled mobile robot (WMR) with constraints with multiple inputs and multiple outputs, [1] in order to ensure the robustness and accuracy of tracking, the method of traditional predictive control often needs to find a large number of parameters of the state. This method reduces the anti-interference ability of the prediction model, [2] and because the model is built complex, the structure of the model is complicated and the cost is expensive.

In order to ensure the stability of the system, the adaptive fuzzy method is designed to set the parameters of the model, and the stability of the adaptive adjustment with initial experience is guaranteed by the lyapunov rule, which can be widely used to meet the condition of the equation of the motion. [3] The stability of control and optimization of parameters are realized in the system.

The method of robust predictive control is introduced including the theoretical basis and the related simulation in references [4] and [5]. The errors and change of errors are used as the reference for the prediction of the initial value of each step of the prediction model. At the same time, the method of robust predictive control with the traditional constraint is mentioned in the reference [6]. The study of the class of problems involving a lot of tuning parameters, is the usual systems of this method. The desired tracking effect is guaranteed when the required environment for control is ideal. Excessive precision leads to the decrease of the anti-interference ability of the whole model. And the stick model is unhelpful to the tuning of parameters in the actual control. With this consideration, there is the present idea of this paper because the unsuitable parameters can cause the system to producing huge delay and may lead to instability of system. In the tuning of parameters, I refer to the related theory of fuzzy adaptive control [7] , [8] hoping parameters of the predictive model be adjusted in real time, the adjustment of the parameters is based on the size of the interference, the errors and the change of errors of the expected values of three systems compared to the values of the output at the previous time, and the adaptive adjustment of the parameters without the initial experience is often unable to get the desired results.

The fuzzy rule is a set where different initial fuzzy rules are selected for the disturbances of different level, and each initial model is always used. The present rule is determined by the errors and the change of the errors of the predictive model. By adjusting the parameters of the fuzzy rules according to tracking, the most suitable fuzzy rules are determined. The specific adaptive law is established by the lyapunov function related to the errors of the single-output and single-output model and the change of errors related to the errors. The adaptive law that needs to be used is the self-adjustment with the initial experience. This adjustment, comparing to the adaptive setting without initial experience, reduces the number of trial and errors in the actual movement, and has the more rapid tracking effect, and the accuracy of the motion is unchanged. Learning from the references [9] and [10] is the extension of this method.

The structure of the paper, in addition to the first part about the introduction, includes these parts as follows. The second part will be the traditional constraint predictive scheme for the specific model of movement of vehicle and the system of complex control . The third part will give an adaptive fuzzy control of simple sine single tracking on the motion of the specific model. The forth part will compare the adaptive tuning of parameters on the predictive control with the robust prediction scheme without initial experience. The simulation gives the result of the compare. Finally, it is the fifth part of the conclusion of the work.

2. Dynamic model of WMR

2.1. Dynamic error model with constraints on the WMR

In the complex system, the model of WMR is as shown in figure 1.

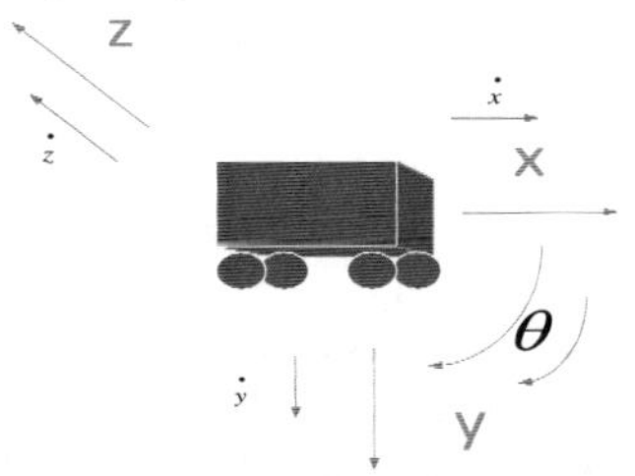

Figure 1. The model of the WMR without the ability of changing the motion on the axis of Z

The robust controller needs to firstly obtain the full dynamic error model of the WMR. The robust tracking controller is designed by using the rolling time domain and linear matrix inequalities. When the robot has uncertain parameters in the dynamic model, the position and angle of the direction and speed are gradually tracked, as long as the next one is known in advance. When the desired position is needed, the predictive model parameters for control can be adjusted according to the deviation between the current position and the predictive position. It can realize the rolling optimization of the predictive model, decomposing the motion of the next motion on the axis of Z to the two parts only including the motion happening on the axis X and Y as shown in figure 2.

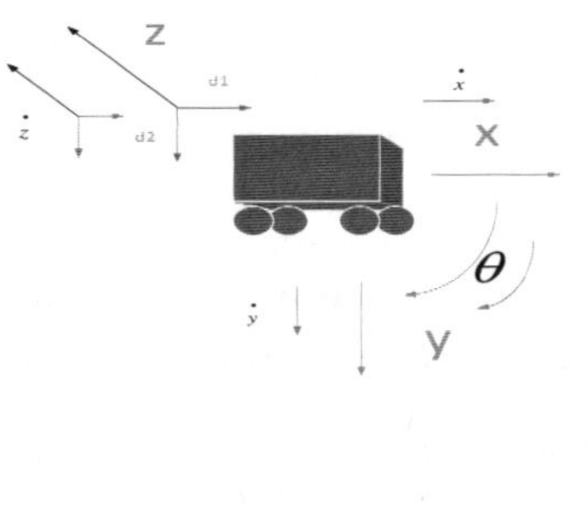

Figure 2. The motion on the axis of Z is decomposed into X and Y

In the formula (1), model of motion is discomposed into three parts on X-Y plane has been shown.

$$x'' = \frac{\lambda}{m}\sin\phi + \beta u_1 \cos\phi$$

$$y'' = -\frac{\lambda}{m}\cos\phi + \beta u_1 \sin\phi$$

$$\phi'' = \alpha u_2$$

$$x'\sin\phi - y'\cos\phi = 0 \tag{1}$$

In the formula (2), lagrange multiplier has been given.

$$\lambda = -m\phi'(x'\cos\phi + y'\sin\phi) \tag{2}$$

2.2. Predictive model of the WMR

Since the coefficient of this predictive model is the contribution of each increment of control compared to the output of predictive control, it will require a large number of solutions, and because of the interaction between three inputs and outputs with initial values at the time of K. The process of the whole dynamic error control system as follow:

- The predictive array and the controlling array are set up respectively, then the trajectory of motion is decomposed to be regarded as discrete-state points, and each state point includes three states of angle, horizontal velocity and vertical velocity, and three parts of predictive control are established respectively.
- According to the trajectory of reference , the value of output on the time K is calculated, and the value of control are stored. The current time K is iterated, and the state transformation matrix P and the output transformation matrix Q are solved on the lyapunov rule.
- Calculating the error between desired trajectory of position and current trajectory of position then correcting the parameters of output matrix, and then calculating the predicted trajectory from output value and the given value in the current time K .
- Adding three discomposed parts to the control matrix, and adding bounded interference.
- To limit the three control states.

In the formula (3) , (4) and (5), the matrix of rolling optimization has been shown.

$$\begin{bmatrix} \hat{y}(k+1) \\ \hat{y}(k+2) \\ \cdots \\ \hat{y}(k+n) \end{bmatrix} = \begin{bmatrix} a_1 & & & 0 \\ a_1 & a_1 & & \\ \vdots & \vdots & \ddots & \\ a_n & a_{n-1} & \cdots & a_{n-m+1} \end{bmatrix} \begin{bmatrix} \Delta u(k) \\ \Delta u(k+1) \\ \vdots \\ \Delta u(k+m-1) \end{bmatrix} + \begin{bmatrix} y_0(k+1) \\ y_0(k+2) \\ \vdots \\ y_0(k+m) \end{bmatrix} \tag{3}$$

$$\Delta u(k) = f(e(k)) \tag{4}$$

$$y_0(k+1) = y_0(k) + e(k) \tag{5}$$

3. Tracking of sine single tracking on the WMR based on adaptive fuzzy control

In the tracking control of the WMR, the fixed parameters of the lyapunov function, the positive diagonal matrix and the error transformation function are determined by the established fuzzy membership function, and the main memberships of five Gauss membership functions are decided on the intervals including ($-\pi/4$, $-\pi/12$), ($-\pi/6$,0), (0, $\pi/6$), ($\pi/12$, $\pi/4$), then initial state and expected tracking sine trajectory are selected .

In the formula (6) , the model of the WMR has been given, in(7), the states and the errors in the time domain have been shown, in (8), the input of adaptive control has been given.

$$x^{(n)} = f(x, x', \ldots, x^{(n-1)}) + g(x, x', \ldots, x^{(n-1)})u$$

$$y = x \tag{6}$$

$$x(k), x'(k), x''(k), \ldots x^{(n-1)}(k)$$

$$e(k), e'(k), e''(k), \ldots e^{(n-1)}(k) \tag{7}$$

IOP Conf. Series: Materials Science and Engineering **428** (2018) 012055 doi:10.1088/1757-899X/428/1/012055

$$u^* = \frac{1}{g(x|\theta_g)}\left[-f(x|\theta_f) + y_m^{(n)} + k^T e\right] \tag{8}$$

Fuzzy adaptive control with initial fuzzy experience, as shown in figure 3.

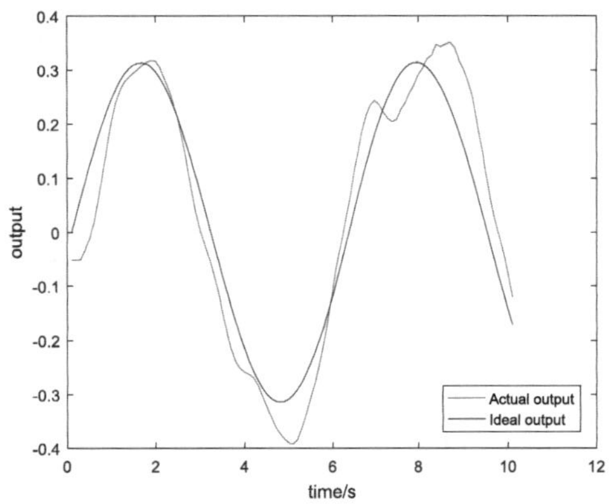

Figure 3. Sine tracking of the WMR

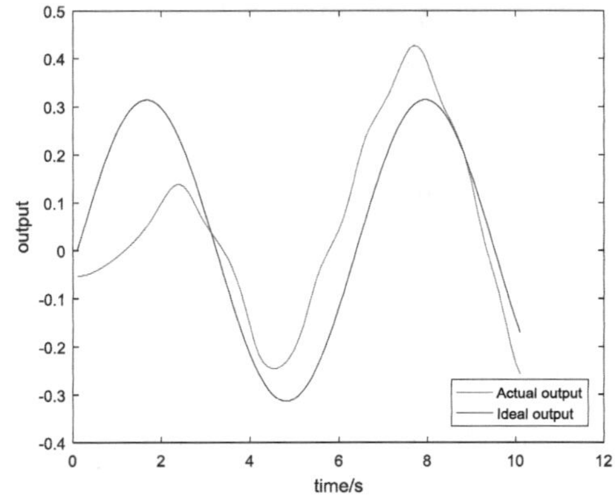

Figure 4. Tracking of the WMR with adjusting parameters

Adjusting the adaptive parameters, the output will be improved, but still basically stable. At this time, even if the initial fuzzy experience is selected a little badly, the control can also achieve stable sine signal tracking of the WMR, as shown in figure 4.

4. Comparison between adaptive scheme and robust predictive scheme

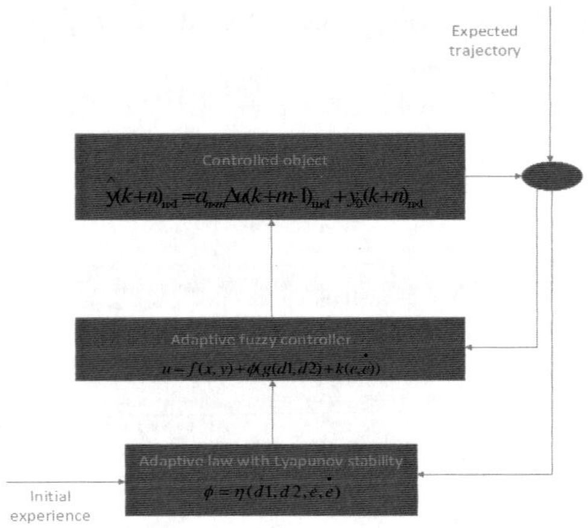

Figure 5. Structure of adaptive control

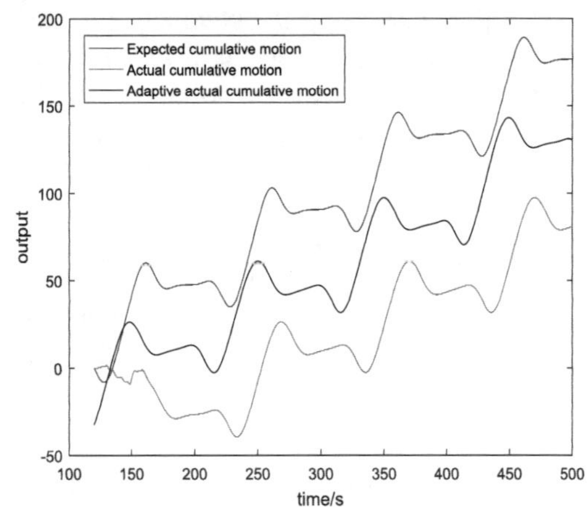

Figure 6. The effect of tracking on the two methods

The figure 5 shows the structure of adaptive control with the initial experience. Based on the errors and the change of errors, in the figure 6, comparing the two methods for the motion control of the prediction model, we can see that adaptive method (black) is quicker to following the expected motion than the robust predictive method (red). With the interfere, the used motion is a helical motion composed of two sinusoidal motions on the axis of X and Y.

In the formula(9), the changing rule of errors and change of errors on X in the helical motion has been given, in(10), lyapunov function guaranteeing the stability has been given, in (11), the adaptive rule of the current chosen fuzzy rules has been given.

$$e^{(n)} = -k^T e + \left[\hat{f}(x|\theta_f) - \mathrm{f(x)}\right] + +\left[\hat{g}(x|\theta_g) - \mathrm{g(x)}\right]u_1$$

$$\Delta = \begin{bmatrix} 0 & 1 \\ -k_2 & -k_1 \end{bmatrix}, b = \begin{bmatrix} 0 \\ 1 \end{bmatrix} \quad (9)$$

$$e' = \Delta e + b\left\{\left[\hat{f}(x|\theta_f) - \mathrm{f(x)}\right] + +\left[\hat{g}(x|\theta_g) - \mathrm{g(x)}\right]u_1\right\}$$

$$V = \frac{1}{2}e^T Pe + \frac{1}{2\gamma_1}(\theta_f - \theta_f^*)^T(\theta_f - \theta_f^*) + \frac{1}{2\gamma_2}(\theta_g - \theta_g^*)^T(\theta_g - \theta_g^*) \quad (10)$$

$$\Delta^{\mathrm{T}} P + P\Delta = -Q$$

$$\theta_f' = -\gamma_1 e^T Pb\xi(x)$$

$$\theta_g' = -\gamma_2 e^T Pb\eta(x)\mathrm{u}_1 \quad (11)$$

Then, figure 7 shows a direct robust predictive mechanism without initial experience on the helical motion with interfere. figure 8 is an adaptive fuzzy control mechanism with initial experience on the helical motion with interfere, and the whole tracking can be clearly seen. Compared to the several errors of the robust predictive mechanism at the start of the tracking, the signals of the helical motion no longer need to be tried many times at the beginning of the control in the adaptive mechanism, and with the initial experience, the method can achieve a better tracking.

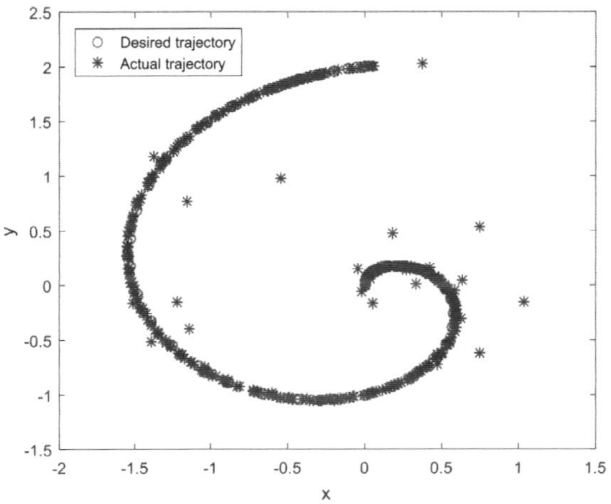

Figure 7. Mechanism without initial experience

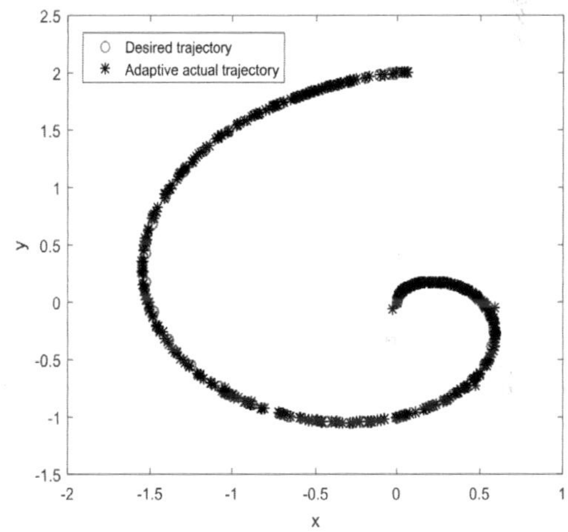

Figure 8. Adaptive mechanism with initial experience

5. Conclusion

The experimental simulation on the constrained WMR shows that the method for control is stable, and the less parameters need to be set, so the cost is lower and the actual tracking effect is ideal and the anti-interference ability is improved. Because of the existence of adaptive law and initial fuzzy experience, the shortage of the control lies in the high requirement in the class of the motion, if the actual motion is beyond the limits of the motion on some direction, there will be a huge deviation on the whole tracking, then because of the limits of adaptive parameter what is set to prevent the adjustment from becoming frequent and violent, it is difficult for the method to return to the needed trajectory. Similar problems are encountered in the simulation, so the initial experience and the possible states of motion in all directions should be taken into account so that the complex trajectory can be perfectly tracked in a certain range. Although the whole adaptive system can be determined to keep stable because of the lyapunov rule, the whole tracking can be influenced deeply by initial parameters, and this is the work that needs to be improved in the future.

Acknowledgments
This work is supported by Fundamental Research Special Funds for Central Universities (20826041A4133).

References

[1] Li, Y., He, L., & Yang, L. M. (2013). Path-following control for multi-axle car-like wheeled mobile robot with nonholonomic constraint. *Ieee/asme International Conference on Advanced Intelligent Mechatronics*(Vol.8212, pp.268-273). IEEE.

[2] Saccon, A. (2005). Minimum Time Maneuver for a Nonholonomic Car with Acceleration Constraints: Preliminary Results. *Intelligent Control, 2005. Proceedings of the 2005 IEEE International Symposium on, Mediterrean Conference on Control and Automation* (pp.1337-1342). IEEE.

[3] Anderson, B. D. O., Agathoklis, P., Jury, E. I., & Mansour, M. (1986). Stability and the matrix lyapunov equation for discrete 2-dimensional systems. *Circuits & Systems IEEE Transactions on, 33*(3), 261-267.

[4] Li, T. S., Tong, S. C., & Feng, G. (2010). A novel robust adaptive-fuzzy-tracking control for a class of nonlinear multi-input/multi-output systems.*IEEE Transactions on Fuzzy Systems, 18*(1), 150-160.

[5] Deng, M., Inoue, A., & Bi, S. (2009). Robust non-linear control and tracking design for multi-input multi-output non-linear perturbed plants. *Iet Control Theory & Applications, 3*(9), 1237-1248.

[6] Wu, T. S., Karkoub, M., Chen, H. S., Yu, W. S., & Her, M. G. (2015). Robust tracking observer-based adaptive fuzzy control design for uncertain nonlinear mimo systems with time delayed states. *Information Sciences, 290*, 86-105.

[7] Chien, C. J., Hsu, C. T., & Yao, C. Y. (2004). Fuzzy system-based adaptive iterative learning control for nonlinear plants with initial state errors. *Fuzzy Systems IEEE Transactions on, 12*(5), 724-732.

[8] Li, Y., Deng, J. M., & Wei, M. Y. (2002). Meaning and precision of adaptive fuzzy systems with gaussian-type membership functions. *Fuzzy Sets & Systems, 127*(1), 85-97.

[9] Boulkroune, A., Tadjine, M., M'Saad, M., & Farza, M. (2010). Fuzzy adaptive controller for mimo nonlinear systems with known and unknown control direction. *Fuzzy Sets & Systems, 161*(6), 797-820.

[10] Erbatur, K., & Kaynak, O. (2010). Use of adaptive fuzzy systems in parameter tuning of sliding-mode controllers. *IEEE/ASME Transactions on Mechatronics, 6*(4), 474-482.

IOP Conf. Series: Materials Science and Engineering **428** (2018) 012056 doi:10.1088/1757-899X/428/1/012056

Optimized Landmark Distribution for Mobile Robot Visual Homing

Xun Ji[1,*] and Qidan Zhu[1]

[1]College of Automation, Harbin Engineering University, Harbin, China

[*]Corresponding author: grady_heu@126.com

Abstract. In this paper, we present an optimized landmark distribution method for mobile robot visual homing technology. To satisfy the equal distance assumption for feature-based homing algorithms, the proposed method eliminates the low-quality features according to the image pixel distance difference and the scale information. Finally, a near-ideal distribution of the landmarks can be obtained. We take the feature-based average landmark vector (ALV) algorithm as the example, and prove the effectiveness of the proposed method by the experiments.

1. Introduction

Visual homing is an attractive technology in the field of autonomous mobile robot navigation [1]. Using visual homing algorithms, the robot compares the panoramic image captured at its current location with the pre-stored goal image captured at the home location, and calculates the home vector to guide the robot to move towards the destination [2]. In contrast to traditional vision-based robot navigation methods (such as visual SLAM [3]), visual homing does not need any localization and mapping, but only the direction controlling the robot's movement. Therefore, visual homing can greatly reduce the calculation amount while maintaining high navigation precision [4].

According to the different input forms, visual homing can be broadly divided into two categories, including image-based homing and feature-based homing. Image-based homing usually uses the image pixels from the panoramic images as landmarks, such as warping [5] and descent in image distance (DID) [6]. Feature-based homing usually uses the scale-invariant features obtained by the feature detection and matching algorithms (SIFT [7] and SURF [8]) as landmarks, such as average displacement vector (ADV) [9] and average landmark vector (ALV) [10].

The above visual homing algorithms have proved their effectiveness to guide the robot to the destination, but almost all the homing methods should obey equal distance assumption, which describes the ideal landmark distribution for visual homing: If all the landmarks are uniformly distributed and located the same distance from the destination, the visual homing methods will exhibit the optimal performance. However, this assumption is unrealistic and always violated in practical applications, irregular landmark distribution will greatly reduce the performance of the visual homing algorithms, thus affecting the robot's actual trajectory.

To cope with the above problem, we propose an optimized landmark distribution method for the feature-based visual homing algorithms in this paper. We take the popular ALV method as the example and consider the SIFT features as the natural landmarks, the landmark distribution can be

IOP Conf. Series: Materials Science and Engineering **428** (2018) 012056 doi:10.1088/1757-899X/428/1/012056

improved to a nearly ideal state according to the image pixel distance difference and the scale information of SIFT, so that the better homing performance can be achieved.

2. Average Landmark Vector (ALV)

ALV is a classic visual homing algorithm inspired by biological navigation. In the ALV algorithm, the robot only memorizes the landmarks instead of the whole panoramic images, the landmarks are reduced to an average form (defined as the average landmark). By processing the average landmark, the two-dimensional home vector can be calculated, pointing from the robot's current location to the destination. Among all the visual homing algorithms, ALV can has the most simple model with the least amount of computation, in the meantime, ALV can also produce a very precise home vector.

The basic model of ALV can be described in figure 1. Taking the current image as an example, C is the robot's current location, L_1, L_2, ..., L_n are the n landmarks (i.e. SIFT features) extracted in the scene. $\overrightarrow{CL_1}$, $\overrightarrow{CL_2}$, ..., $\overrightarrow{CL_n}$ are defined as the unit landmark vectors pointing from the current location to the corresponding landmarks.

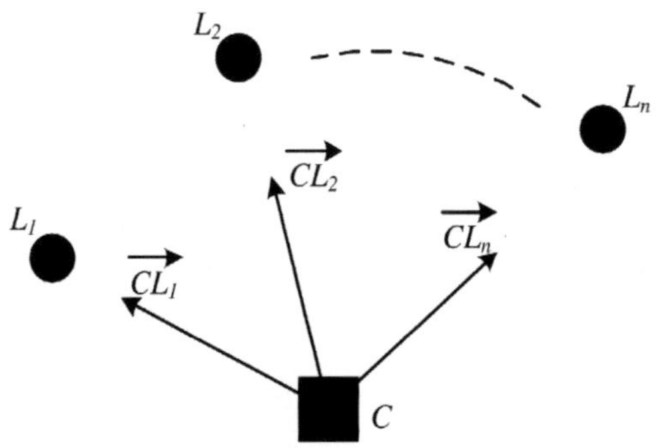

Figure 1. Basic model of ALV

We set the center of the current image as the origin of the image coordinate system, the coordinates of the robot's location can thus be $P_C = (0,0)$, the image coordinates of the ith landmark L_i are defined as $P_i = (x_i, y_i)$. Therefore, the ith landmark vector at C can be computed by:

$$\overrightarrow{CL_i} = \frac{P_i - P_C}{\|P_i - P_C\|} \tag{1}$$

The average landmark at C can then be calculated by averaging all the landmark vectors at C:

$$\overrightarrow{CL} = \frac{1}{n}\sum_{i=1}^{n}\overrightarrow{CL_i} \tag{2}$$

Similarly, the average landmark at H can be calculated by averaging all the landmark vectors at H:

$$\overrightarrow{HL} = \frac{1}{n}\sum_{i=1}^{n}\overrightarrow{HL_i} \tag{3}$$

Finally, the home vector $\boldsymbol{h}$ can be generated by subtracting $\overrightarrow{CL}$ from $\overrightarrow{HL}$:

$$\boldsymbol{h} = \overrightarrow{CL} - \overrightarrow{HL} \tag{4}$$

3. Optimized Landmark Distribution

The proposed method consists of two principles. To make the actual distribution of the landmarks closer to the ideal state, the method optimizes the landmark distribution from two aspects: the image pixel distance and the frequency where landmarks appear in different directions. The models of the two principles are shown in figure 2.

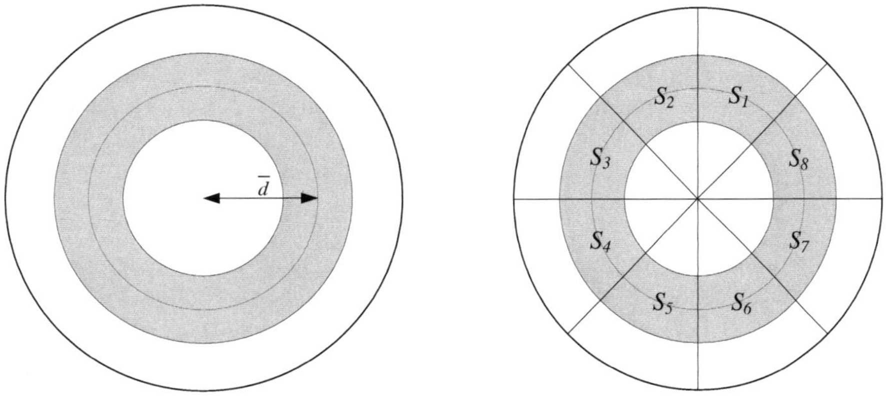

(a) (b)

Figure 2. The models of the principles. (a) The model of Principle 1. The distance from the image center to the red circle is $\bar{d}$. (b) The model of Principle 2.

3.1. Principle 1

As shown in figure 2(a), based on the imaging characteristics of panoramic images, most of the landmarks are extracted and located nearby the horizon circle [11]. However, the image positions of some landmarks are very close or far from the image center, these landmarks will impact on the homing performance. Therefore, Principle 1 is proposed to eliminate such types of the landmarks.

Assume that the image coordinates of the *i*th landmark are defined as (x_i, y_i). The image pixel distance of the *i*th landmark is defined as d_i:

$$d_i = (x_i^2 + y_i^2)^{-1} \tag{5}$$

The average image pixel distance of all the landmarks can be computed by:

$$\bar{d} = \frac{1}{n}\sum_{i=1}^{n} d_i \tag{6}$$

Then, the distance difference of the *i*th landmark can be obtained by:

$$\Delta d_i = \left| d_i - \bar{d} \right| \tag{7}$$

Principle 1 only preserves the landmarks with $\Delta d_i < 0.3\bar{d}$ (the gray area), and removes the landmarks with $\Delta d_i \geq 0.3\bar{d}$ (the white area).

3.2. Principle 2

As shown in figure 2(b), Principle 2 evenly divides the goal image into *N* sector rings, denoted as S_1, S_2, …, S_N, respectively (we set *N* to 8). Assume that *n* landmarks are extracted, and the *i*th sector ring contains n_i landmarks, Principle 2 firstly calculates the average landmarks in each sector ring:

$$\bar{n} = \frac{n}{N} \tag{8}$$

Generally, if each sector ring contains $\bar{n}$ landmarks locating the same distance from the image center, the ideal distribution can be achieved. In Principle 2, the landmark amounts in each sector ring is compared with $\bar{n}$. For the sector rings whose actual landmark amounts are smaller than $\bar{n}$, all the landmarks in these sector rings are preserved without further processing. Conversely, for the sector rings that contain more landmarks, Principle 2 will reduce the landmark amounts in these sector rings based on the scale information.

A phenomenon is stated in [12] that if a certain SIFT feature has a larger value, the distance between the robot and the corresponding landmark is relatively closer. Therefore, when the two images are matched, the two SIFT features in the each matching pair have different scale values, this implies that the represented landmark has different distances to the two capture positions. Although we cannot obtain the exact distance values, we can qualitatively compare the difference of the above two distances.

Since visual homing is a local navigation strategy for a small scale experimental area, for most landmarks, their distances to the robot will not change too much, resulting in small scale changes. Hence, For the sectors containing more landmarks, Principle 2 sorts all the preserved landmarks according to the scale difference from small to large, the landmarks with small scale difference are removed in sequence, and the removed number is the difference between the actual number of the landmarks and $\bar{n}$.

4. Experiments

4.1. Panoramic Image Databases

The experiments were performed based on the panoramic image databases collected at Bielefeld University. The capture environment of the databases is a 4.8m×2.7m indoor spaces, including 170 capture positions, which are spaced 30cm apart. The robot equipped with a panoramic vision system captured the panoramic image at each capture position. We selected two versions of the database, including Original and Chairs. The Original version is the default condition of the environment. The curtains and doors are closed, the lights in the scene are turned off. The Chairs version is based on the Original version, three chairs are randomly set in the environment. Two samples of the panoramic image versions are shown in figure 3.

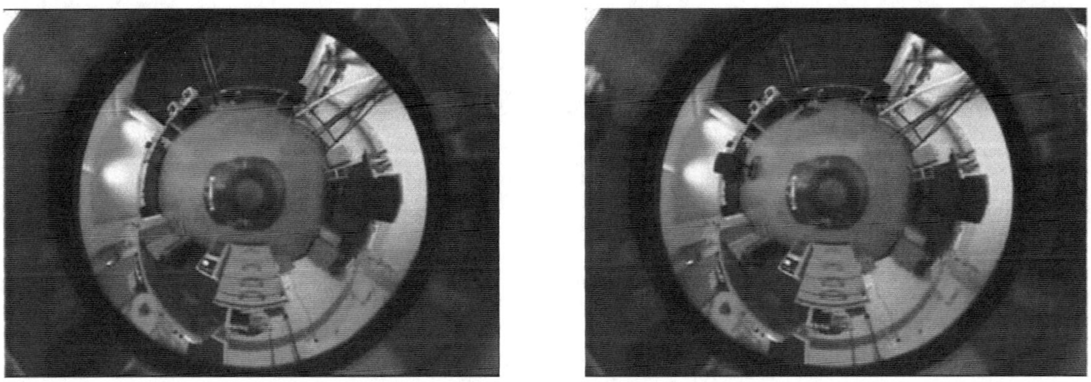

(a) (b)

Figure 3. Two samples of the panoramic image versions.

4.2. Homing Performance

We test the homing performance between the initial and improved ALV algorithms. To intuitively exhibit the effect of the algorithms, we arbitrarily selected a panoramic image from the Original version as the goal image, and the remaining 169 images were considered as all the possible current images. The home vectors were calculated between the goal image and these current images. Figure 4 shows the visualization results of the initial and the improved ALV algorithm. (6,2) was set as the goal location.

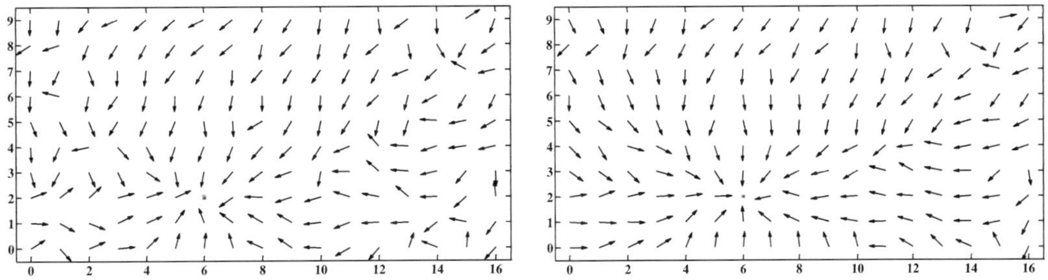

(a) (b)

Figure 4. Visualization results. Red square represents the goal location (6,2). Arrows represent the calculated home vectors. (a) Initial ALV algorithm. (b) Improved ALV algorithm.

4.3. Average Angular Error and Return Ratio

To quantitatively analyze the performance of the above two algorithms, we adopted two widely used performance metrics, called average angular error (*AAE*) and return ratio (*RR*).

AAE indicates the average error between the ideal home vector and the actual home vector. Assume that the direction of the ideal home vector is denoted as α_{ide}, pointing from the current location to the goal location. The direction of the calculated home vector is denoted as α_{act}. The angular error (*AE*) can be computed by:

$$AE(H,C) = |\alpha_{ide} - \alpha_{act}| \tag{9}$$

AAE can thus be calculated by:

$$AAE = \frac{1}{pq}\sum_{k=1}^{pq} AE(H,C_k) \tag{10}$$

Where C_k represents the kth current location. p and q respectively denotes the rows and columns of the capture location. According to the image databases, p is set to 17 while q is set to 10.

RR shows the proportion of all the possible current locations that can successfully return to the destination. A dummy robot is created to simulate the movement according to the calculated home vector, the standard to decide whether a certain current location will become a successful homing location can be presented as follows:

Step 1: Control the robot to move towards its nearest location according to the calculated home vector.

Step 2: If the following two cases happen, jump to Step 4:

Case 1: The robot finally reach the pre-set destination.

Case 2: The robot moves more than 17+10=27 steps, this means the robot has moved more than half the circumference of the environment but still not reach the destination.

Step 3: Continue to perform Step 1.

Step 4: If Case 1 happens, we declare the tested current location is valid. Conversely, if Case 2 happens, we declare the tested current location is failed.

After all the current locations are tested according to the above steps. *RR* can be computed by:

$$RR = \frac{T_s}{T} \tag{11}$$

Where T_s indicates the number of the valid current locations. T indicates the total number of the current locations. According to the image databases, T is set to 169.

Table 1 shows the corresponding *AAE* values, we set (6,2) from the Original version and (8,5) from the Chairs version as the home locations.

Table 1. *AAE* Results (°).

Version	Goal	ALV	Improved
Original	(6,2)	34.504	23.417
Chairs	(8,5)	47.036	34.147

Table 2 and Table 3 shows the *RR* results for the two algorithms based on the two image versions. Three evenly distributed locations are selected as the goal location to obtain a general conclusion.

Table 2. *RR* Results for Original Version

Goal	ALV	Improved
(2,8)	0.663	0.781
(8,3)	0.805	0.828
(15,1)	0.586	0.639

Table 3. *RR* Results for Chairs Version

Goal	ALV	Improved
(2,8)	0.680	0.787
(8,3)	0.799	0.834
(15,1)	0.568	0.598

5. Conclusions

In this paper, we propose two principles to optimize the landmark distribution for visual homing. The landmark ditribution is optimized from the two aspects, including the image pixel distance and the frequency at which landmarks appear in different directions. After the landmarks are filtered, the optimized distribution are more close to the ideal state, satisfying the equal distance assumption. We take the popular ALV algorithm as the example, and perform a series of the experiments based on the panoramic image databases. Results reveal that the landmarks optimized by the proposed principles can make the ALV algorithm exhibit better homing performance than the initial state, the robot can return to the specified destination from more possible locations with more accurate and robust home vectors.

Acknowledgments

This work is partially supported by the National Natural Science Foundation of China (61673129, 51674109).

References

[1] Zhu Q, Liu C and Cai C, A novel robot visual homing method based on sift features, *Sensors*, 2015. vol 15, no.10: pp 26063-26084.

[2] Lee C and Kim D, Local homing navigation based on the moment model for landmark distribution and features, *Sensors*, 2017. vol 17, no.11: pp 2658.

[3] Gamallo C, Mucientes M and Regueiro C V, Omnidirectional visual slam under severe occlusions, *Robotics and Autonomous Systems*, 2015. vol 65, no. C: pp 76-87.

[4] Zhu Q, Liu X and Cai C, Feature optimization for long-range visual homing in changing environments, *Sensors*, 2014. vol 14, no.2: pp 3342.

[5] Franz M O, Schölkopf B, Mallot H A and Bülthoff H H, Where did I take that snapshot? Scene-based homing by image matching, *Biological Cybernetics*, 1998. vol 79, no.3: pp 191-202.

IOP Conf. Series: Materials Science and Engineering **428** (2018) 012056 doi:10.1088/1757-899X/428/1/012056

[6] Zeil J, Hoffmann H I and Chal J S, Catchment areas of panoramic images in outdoor scenes, *Journal of the optical society of America A*, 2003. vol. 20, no.3: pp 450-469.
[7] Lowe D G, Distinctive image features from scale-invariant keypoints, *International Journal of Computer Vision*, 2004. vol. 60, no.2: pp 91-110.
[8] Bay H, Ess A, Tuytelaars T, and Gool L V, Speeded-up robust features, *Computer Vision and Image Understanding*, 2008. vol. 110, no.3: pp 346-359.
[9] Hong J, Tan X, Pinette B, Weiss R and Riseman E M, Image-based homing, *IEEE Control Systems*, 1992. vol. 12, no.1: pp 38-45.
[10] Lambrinos D, Möller R, Labhart T and Pfeifer R, A mobile robot employing insect strategies for navigation, *Robotics and Autonomous Systems*, 2000. vol. 30, no.1-2: pp 39-64.
[11] Zhu Q, Liu C and Cai C, A robot navigation algorithm based on sparse landmarks, *Sixth International Conference on Intelligent Human-Machine Systems and Cybernetics. IEEE*, 2014: pp 188-193.
[12] Churchill D and Vardy A, An orientation invariant visual homing algorithm, *Journal of Intelligent and Robotic Systems*, 2013. vol. 71, no.1: pp 3-29.

IOP Conf. Series: Materials Science and Engineering **428** (2018) 012057 doi:10.1088/1757-899X/428/1/012057

Motion Planning and Simulation of Combined Land-Air Amphibious Robot

Kailin LI[1], BaolingHAN[1], YutingZHAO[1] and Chen ZHU[1]

[1]School of Mechanical Engineering, Beijing Institute of Technology, Beijing, China

Corresponding author e-mail: 190268663@qq.com

Abstract. In order to carry out complex tasks such as disaster rescue and field investigation, this paper designs a land-air amphibious robot that combines the four-legged walking robot and the quadrotor. The robot can not only perform structural separation according to operational requirements in order to realize land and air multi-field investigations. At the same time, the quadrotor can fly freely on the quadruped walking robot body to perform combined operations, providing more possibilities for multi-purpose investigation. The foot trajectory of the four-legged walking robot adopts the Bezier curve, and the center of the robot body is planned to move in a straight line, which can ensure the fluency, stability, and flexibility of the walking motion of the robot. The results of simulation and verification show that the combined land-air amphibious robot has excellent sport characteristics and has popularization and application value.

1. Introduction

In order to successfully carry out a variety of investigation operations under complex terrain and respond flexibly to changing environmental conditions, a land-air amphibious robot combining the advantages of aircraft and foot robots has emerged.The HyTAQ Robot developed by ArashKalantari[1] of Illinois Institute of Technology combines a quadrotor and a cage, four-rotor mode is used for air flight, and ground motion is used to roll the cage by adjusting the attitude of the aircraft.Koji Kawasaki[2] developed a MUWA, a universal land, sea and air robot consisting of four variable-pitch propellers and a peripheral ring. These robots are integrated with the ground robots and cannot be separated and work alone. Moreover, the ground robots do not have speed and acceleration sensorsand cannot accurately control the movement and cannot adapt to a complicated ground environment.

For a combined land-air amphibious robot, the motion planning of the quadrupedal walking robot is particularly important to enhance the adaptability to the ground environment. Currently used methods for gait planning include Wang Lipeng[3] using composite cycloids as the swing-phase plantar trajectory. Using this kind of foot-end trajectory can make the impact of the foot end fall to zero.

2. Land and Air Amphibious Robot Configuration Scheme

The combined configuration scheme combines the functions of a quadrotor and a quadrupedal walking robot to achieve the flexibility of a quadrotor and the superior ground adaptability of a quadrupedal

IOP Conf. Series: Materials Science and Engineering **428** (2018) 012057 doi:10.1088/1757-899X/428/1/012057

walking robot. They can achieve structural separation and perform reconnaissance missions in the land and air space respectively. Simultaneously, four-legged walking robots can be used as ground stations.

Four-rotor aircraft can freely land and assemble on a four-legged walking robot body to perform combined operations. This can accurately control the four-legged walking robot overcomes the complicated and varied ground environment and can simultaneously conduct the detection of land and air. The three-dimensional model isshown in Figure1. The four-rotor aircraft is a standard four-rotor UAV, and the rack is in the form of an X configuration. Taking into account the speed of the walking of the four-legged walking robot, the leg adopts a compound structure of the legs. In this paper, the three degree of freedom leg configuration is adopted, so that the method of adjusting the center of gravity of the robot by adjusting the position of the leg has more choice and is more adaptable to the environment.

3. The foot robot's movement analysis model

The positive kinematics of the robot is a method which is on the basic of knowing the type of each joint of the robot, the size of the adjacent joints and the relative amount of the adjacent joints, and finally determines the posture of the robot end in a fixed coordinate system[4]. According to the configuration scheme of the robot, the D-H method was used to establish the leg 2 coordinate system for the whole robot model,as shown in Figure 2. According to the D-H parameter table, the transformation matrix of the foot-end with respect to the fuselage is obtained, and the coordinates of the foot-end trajectory point are further obtained. This is the solution of forward kinematics.

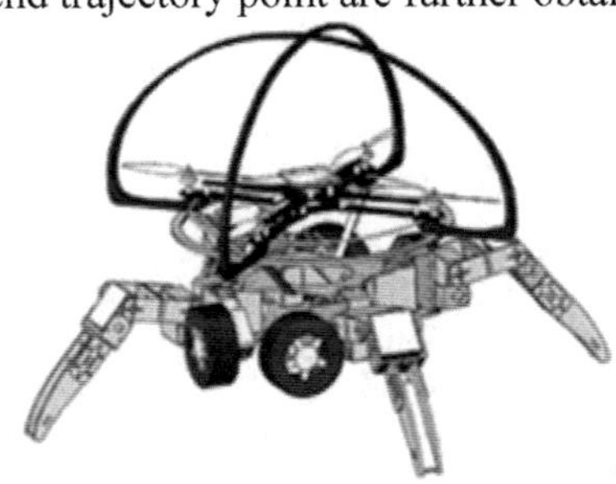

Figure 1.3D model of robot

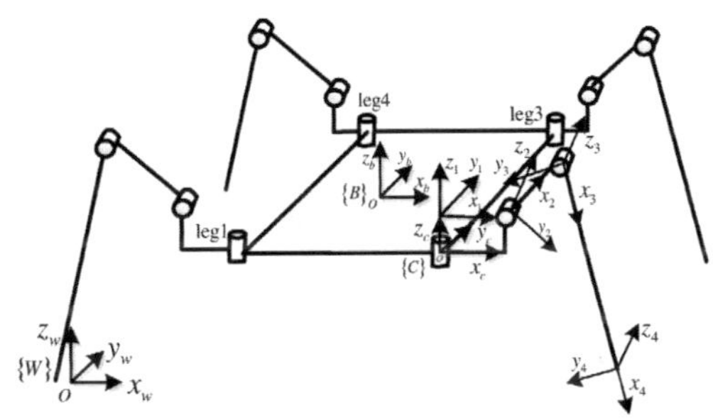

Figure 2. Robot kinematics model

4. Ground robot motion planning

4.1 Gait Planning Analysis

Motion planning is a key technology to ensure robot motion stability. In order to ensure the stability and fluency of their movements, robots can be required to be simple. Slow and stable walking under the terrain. Therefore, the static gait is selected. The preliminary selected gait is the walk gait. The foot trajectory used is the Bezier curve.

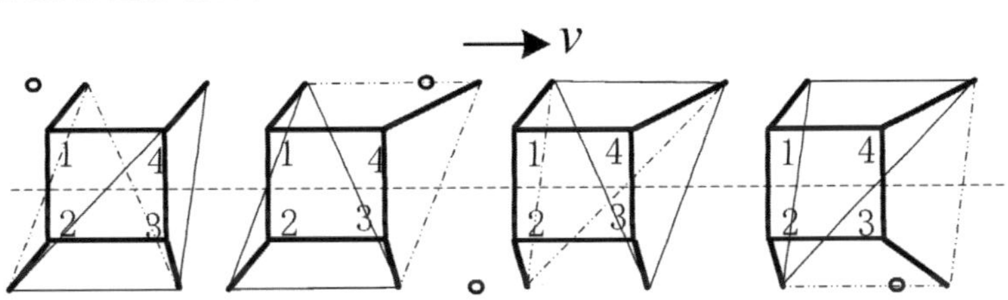

Figure 3.A sequence diagram of a robot

IOP Conf. Series: Materials Science and Engineering **428** (2018) 012057 doi:10.1088/1757-899X/428/1/012057

In the Walk gait plan, the triangle formed by the fulcrum of the three legs of the robot is called the supporting triangle. The central projection method (COP) is used as the stability criterion of the robot.The minimum value of the distance from the center projection point to the triangle edge is called the stability margin. The greater the stability margin, the more stable the robot[5]. When a step is at any moment in the movement and the stability margin of the robot is not zero, it is considered that the step is stable and feasible. For quadruped robots, there are six non-singular static gait steps[6]. Among these six steps, 1-4-2-3 is the best stability margin[7]. Thus, as shown in Figure 3, the order of the steps of 1-4-2-3 is adopted.

4.2 Bessel curve-based planning analysis

The Bessel curve initially determines four points according to the known curve parametric equations, and then connects polygons end to end, and then approximates the polygons by the Bezier formula to obtain the Bezier curve [8]. Its equation is shown in equation (1). In the actual planning, interpolation is used, that is a series of points are selected on the Bezier curve. By controlling the legs of the robot to reach these points, the desired planning is achieved.

$$\begin{aligned} \mathrm{B}(\mathrm{t}) &= \sum_{i=0}^{n} \binom{n}{i} P_i (1-t)^{n-i} t^i \\ &= \binom{n}{0} P_0 (1-t)^n t^0 + \binom{n}{1} P_1 (1-t)^{n-1} t^1 + \ldots + \binom{n}{n} P_n (1-t)^0 t^n, \quad t \in [0,1] \end{aligned} \tag{1}$$

Control the two points to coincide and produce zero speed; the three points of the control point coincide, resulting in zero acceleration[9] is the characteristic of the Bessel curve derived equation. In this paper, the commutation point is added to the trajectory so that when the foot end moves according to the trajectory curve, the impact on the body is reduced as much as possible. The coordinates of 12 interpolation points are selected from the zero point as shown in Table 1. They can satisfy the zero-direction velocity component of the four-legged walking robot when the feet are connected to the ground, and the foot acceleration is as smooth as possible to ensure the height of the legs.

In the table1, S and h represent the step length and step height. According to the robot design scheme, the walking speed in the actual plan is v=0.05m/s,the phase difference between each leg and the first swing leg [*b1,b2,b3,b4*]=[0.25,0.75,0,0.5], the length of each leg segment is *L1, L2*, and *L3*. The foot-end trajectory equations obtained for each leg are shown in equations (2) to (5), and the airframe trajectory is shown in equation (6). The resulting Bezier curve and each interpolation point are shown in Figure 4.

$$\begin{cases} \mathrm{y}_{\mathrm{c}1} = \sum_{i=0}^{11} \binom{11}{i} y_i (1-t_1)^{11-i} t_1^i \\ z_{c1} = \sum_{i=0}^{11} \binom{11}{i} x_i (1-t_1)^{11-i} t_1^i , t_1 = \dfrac{t-b_1T}{(1-\beta)T} \in [0,1] \\ x_{c1} = -a\cos\dfrac{\pi}{4} - n \end{cases} \tag{2}$$

$$\begin{cases} \mathrm{y}_{\mathrm{c}2} = \dfrac{\mathrm{S}_1}{4} + \dfrac{\mathrm{S}_2}{4} + \sum_{i=0}^{11} \binom{11}{i} y_i (1-t_2)^{11-i} t_2^i \\ z_{c2} = \sum_{i=0}^{11} \binom{11}{i} x_i (1-t_2)^{11-i} t_2^i \quad , t_2 = \dfrac{t-b_2T}{(1-\beta)T} \in [0,1] \\ x_{c2} = a\cos\dfrac{\pi}{4} + n \end{cases} \tag{3}$$

$$\begin{cases} y_{c3} = \frac{s_1}{4} - \frac{s_3}{4} + 2m + a\sin\frac{\pi}{4} + \sum_{i=0}^{11}\binom{11}{i} y_i (1-t_3)^{11-i} t_3^i \\ z_{c3} = \sum_{i=0}^{11}\binom{11}{i} x_i (1-t_3)^{11-i} t_3^i \qquad , t_3 = \frac{t - b_3 T}{(1-\beta)T} \in [0,1] \\ x_{c3} = a\cos\frac{\pi}{4} + n \end{cases} \tag{4}$$

$$\begin{cases} y_{c4} = \frac{s_1}{4} + \frac{s_4}{4} + 2m + a\sin\frac{\pi}{4} + \sum_{i=0}^{11}\binom{11}{i} y_i (1-t_4)^{11-i} t_4^i \\ z_{c4} = \sum_{i=0}^{11}\binom{11}{i} x_i (1-t_4)^{11-i} t_4^i \qquad , t_4 = \frac{t - b_4 T}{(1-\beta)T} \in [0,1] \\ x_{c4} = -a\cos\frac{\pi}{4} - n \end{cases} \tag{5}$$

$$\begin{cases} y_b = m + a\sin\frac{\pi}{4} + \frac{1}{4}s + vt \\ z_b = H \\ x_b = 0 \end{cases} \tag{6}$$

Table1. Bayer curve interpolation point

Number	Coordinate(X,Y)mm	Number	Coordinate(X,Y)mm
1	(0,0)	7	(*S*/2,0.926*h*)
2	(-0.4025*S*/2,0)	8	(*S*/2,1.19*h*)
3	(-0.5*S*/2,0.926*h*)	9	(2.516*S*/2,1.19*h*)
4	(-0.5*S*/2,0.926*h*)	10	(2.516*S*/2,1.19*h*)
5	(-0.5*S*/2,0.926*h*)	11	(2.413*S*/2, 0)
6	(*S*/2,0.926*h*)	12	(*S*,0)

5. Virtual Prototyping and Simulation

The virtual prototype technology is a method of using the virtual prototype established in the computer instead of the physical prototype to design and test the product. In order to verify the correctness of the robot gait obtained through Bezier curve programming, The robot virtual prototype model is established by ADAMS software, and simulation analysis is carried out.

First create a model in Solidworks and import it into ADAMS[10]. Define the quality information for each part, then add 12 connectors and motion, as shown in Figure 5. In order to facilitate the analysis and planning effect, add the Maker point to the foot of each leg of the robot. Through the measure, you can see the real-time changes of the values such as the displacement speed of the relevant point.

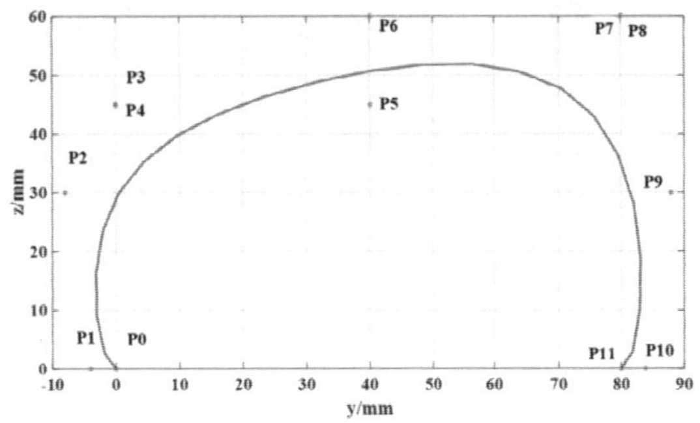

Figure 4. Foot track diagram

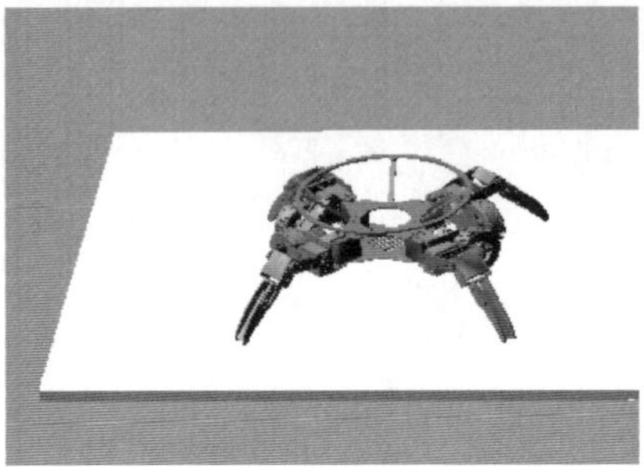

Figure5.ADAMS model

By analyzing the displacement data of the foot-end trajectory, the effect of the planning can be evaluated. The displacements of the center of mass of the robot in the three directions of X, Y, Z in the first 10 cycles are measured, as shown in Figures 6 to 8. Through analysis, it is found that the displacement curve of the robot in the Y direction is relatively smooth, and the speed is gradually increasing, which is consistent with the gradual increase of the step length in the first three cycles of the previous plan. After 4s, it is a diagonal line with constant slope, the corresponding motion is a uniform linear motion, and the fluctuation is very small, indicating that the fuselage follows a straight line and the amount of deflection occurring during the advancement is very small. The maximum X-direction offset of the robot is 2.5mm, indicating that the robot has better stability. The offset of the center-of-mass equilibrium position in the Z direction is about 0.5 mm, mainly because the first three cycles are gradual adjustment phases, and the center of mass does not reach the predetermined position. After stabilization, the change of the center of mass is regular and the equilibrium position is fixed, indicating that the robot maintains linear motion. The ability is better.

Through simulation analysis, it is found that the movement of robot motion planning based on the foot-end trajectory can reach the expected requirements in the virtual environment. The actual movement performance of the robot and the planning are consistent, indicating that this planning method is scientific.

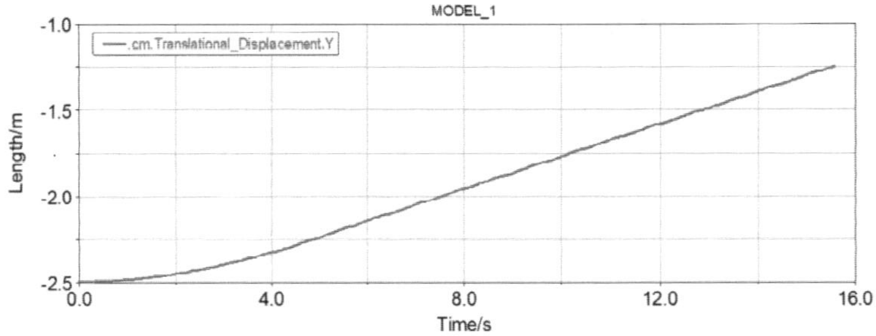

Figure6. Displacement in the y direction of the fuselage

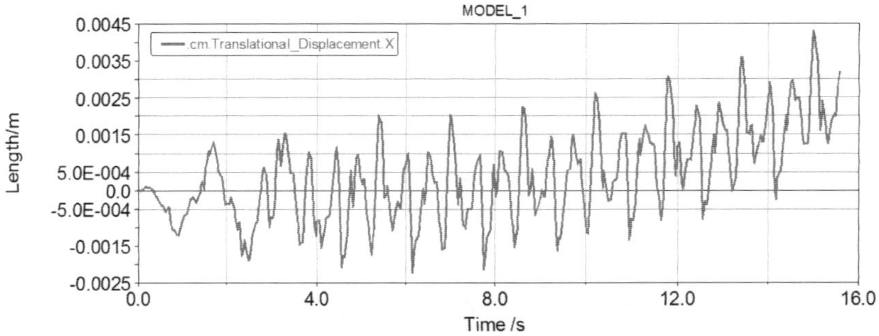

Figure7. Displacement in the x direction of the fuselage

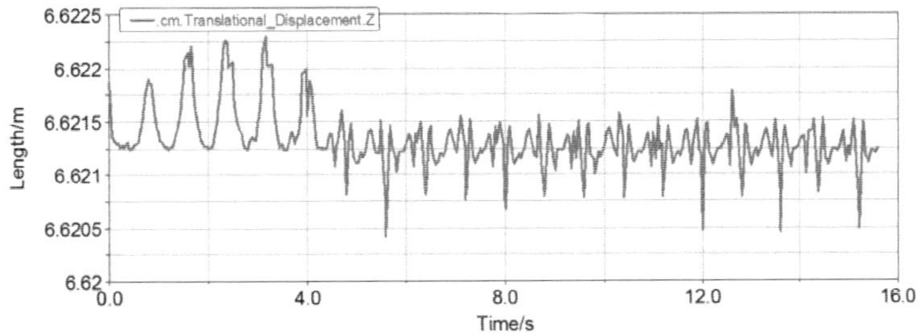

Figure8. Displacement in the z direction of the fuselage

6. Summary

In this paper, a land-air amphibious robot combining a quadrotor and a four-legged walking robot is designed. The robot can not only perform structural separation according to operational requirements in order to realize land and air multi-field investigation, but also the quadrotor is free. Rising and landing on a four-legged walking robot body to perform combined operations provides more possibilities for multi-purpose detection. The separable structural design not only greatly enhances the environmental adaptability of the land-air amphibious robot, but also lays a solid foundation for the future diversified design. The four-legged walking robot's leg has selected a three-degree-of-freedom configuration. The foot trajectory adopts the Bezier curve. The robot body's center of mass is planned to move in a straight line, which can ensure the fluency, stability, and flexibility of the walking robot. The results of simulation and verification show that the combined land-air amphibious robot has excellent sport characteristics and has popularization and application value.

References:

[1] Kalantari A, Spenko M. Design and experimental validation of hytaq, a hybrid terrestrial and aerial quadrotor[C]//Robotics and Automation (ICRA), 2013 IEEE International Conference on. IEEE, 2013: 4445-4450.

[2] Kawasaki K, Zhao M, Okada K, et al. Muwa: Multi-field universal wheel for air-land vehicle with quad variable-pitch propellers[C]//2013 IEEE/RSJ International Conference on Intelligent Robots and Systems. IEEE, 2013: 1880-1885.

[3] WANG Lipeng, WANG Junzheng, WANG Shoukun, et al. Gait control strategy for hydraulic quadruped robot based on foot end trajectory planning algorithm [J]. Journal of Mechanical Engineering,2013,49(1):39-44.

[4] LIU Guanchu, XIONG Jinqi, QIAO Lin, et al. Modeling and algorithm implementation of free walking gait planning for Quadruped Robot[J]. Journal of Xi'an Jiaotong University, 2015(06): 84-89.

[5] FAN Tiantian.Preliminary study on gait planning and motion stability control of small - quadruped robot[D]. Control theory and control engineering of Nanjing University of Aeronautics and Astronautics, 2013:43-46.

[6] LIU Z W, McGhee Robert B,Frank Andrew A. On the stability properties of quadruped creeping gaits [J]. Math-ematical Biosciences,1968,3:331-351.

[7] de Santos P G, Garcia E, Estremera J. Quadru-ped-al Locomotion:an Introduction to the Control of Four-legged Robots[M].Lond on: Springer,2007:59-60.

[8] NA Qi, Research and implementation of motion control technology for Quadruped Robot [D]. School of Mechanical Engineering, Beijing Institute of Technology, 2015:32-35

[9] Hyun D J, Seok S, Lee J, et al. High speed trot-running: implementation of a hierarchical controller using proprioceptive impedance control on the MIT Cheetah [J]. The International Journal of Robotics Research.2014, 33(11): 1417–1445.

[10] MA Zongli, WANG Jianming, LI Hua, et al. Walk gait simulation of a quadruped robot [J], 2013,41(11):146-488+190.

IOP Conf. Series: Materials Science and Engineering **428** (2018) 012058 doi:10.1088/1757-899X/428/1/012058

A New Mitigation Method for Space Robot Charging Effect Based on CNTs

X N Yang[1,2], Y M Liu[1,2], M Li[1], J H Wang[1] and Z H Yuan[3]

[1] Beijing Institute of Spacecraft Environment Engineering, Beijing 100084, P. R. China

[2] Science and Technology on Reliability and Environment Engineering Laboratory, Beijing 100084, P. R. China

[3] School of Materials Science and Engineering, Tianjin University of Technology, Tianjin 300384, P. R. China

Lyming2005@126.com

Abstract. Autonomous space robot will be widely used to complete various space missions in future. Unfortunately electrostatic surface discharge (ESD) induced by space plasma environment may produce one of the most common anomaly in space robots. Liquid Metal Ion Source (LMIS) is one of the effective mitigation methods for space robot charging effect. But as a devices for space-robots charging mitigation equipment, has strict requirements to its weight and volume. The present LMIS can only work at a very high voltage, so the weight and volume of LMIS isn't suitable for space robot. Here we propose that carbon nanotubes (CNTs) can be used as capillary-type emitters in LMIS to meet the challenge. Well-aligned CNT arrays are synthesized on self-ordered hexagonal porous anodic aluminum oxide (AAO) template using chemical vapor deposition (CVD) method. The CNTs obtained are monodispersed and have open tip, which indicates that they have the similar geometry to the present capillary-type emitters in LMIS. So the LMIS made of CNTs can emit ions at a very lower voltage. A new mitigation device for space robot charging effects with very small weight and volume can be developed. The key steps for fabricating CNT emitters are presented.

1. Introduction

With the development of space exploration, autonomous space robot will be widely used to complete various missions. Ongoing human missions to space station have an integrated mix of crew working with robots and supporting autonomous systems on-board spacecraft and in mission control. On orbit servicing tasks of spacecraft will be completed mainly by robots in future. Deep-space exploration missions will require many automation technologies to enable safe missions, become more independent, and enable intelligent autonomous operations. Space robots may play a more and more important role in space missions in future. We can expect in the next few decades that robots designed for space missions will approach or even exceed the performance of the astronauts.

However, there are many unresolved challenges in space robots, one of which is natural environment hazard. Various aspects of the space environment can cause on-orbit spacecraft anomalies. Studies have shown that adverse interactions between the natural space environment and space systems can have deleterious consequences comparable to those caused by human or design errors[1]. Electrostatic surface discharge (ESD) is one of the most common anomaly producing mechanisms in spacecraft. Interactions between hazardous space plasmas and spacecraft surfaces often result in spacecraft charging. Such as in geosynchronous orbit, the plasma environment can cause differential charging of satellite components and lead to ESD on spacecraft. Some mitigation methods have been developed for spacecraft, but as a devices for space-robots charging mitigation equipment, has strict requirements to its weight and volume. Lightweight and miniaturization technology must be developed.

2. Liquid Metal Ion Source (LMIS)

The concept of mitigation of spacecraft charging is defined as a method or design which makes spacecraft charging less severe. Emission of electron or ion beams can control the spacecraft potential effectively. One of the spacecraft charging compensator is Liquid Metal Ion Source (LMIS), which can emit ions beams under high voltage [2-4]. Field emission of ions is a key principle used in LMIS, which refers to the process of using a strong electric field to produce a spray of charged ions and droplets from emitters. If the potential difference between an electrode and a liquid metal surface is sufficiently high, the surface is deformed to an equilibrium shaped so called Taylor cone and then ions and droplets are directly pulled out of the liquid metal surface producing thrust [5, 6].

In LMIS, there is a critical minimum electric potential called onset voltage U_0, below which no ions emission will occur, and the liquid metal gradually distorted into a Taylor cone or a series of Taylor cones with the radius of curvature at the apex of these cones becoming smaller and smaller as the voltage increases [7]. It has been presented experimentally that electric filed is extraordinarily high with a value level of 10^9 Vm^{-1} to extract ions [8]. It is very difficult to pull ions out of the liquid metal.

To lower the applied voltage, several type emitters have been used as field-enhanced structure for LMIS due to their small radius of curvature, such as sharpened needles, capillaries or nozzles, slits, rings. But, the radiuses of curvature of the emitters are usually more than 1μm. The application voltage should be above 5, 000V to emit ions in LMIS [9]. So the weight and volume of LMIS is so high that LMIS can't be used in space robots to mitigate charging effect.

Due to their high aspect ratios and small radii of curvature carbon nanotubes (CNTs) are excellent field emission emitters with high field emission current density at low electric field and stable currents [10-12]. But CNTs used as emitters in LMIS have never been reported. Here a new method to fabricate capillary-emitters used in LMIS is presented. It is shown that the emitters being made of open-end CNTs can emit ions at a much lower voltage about 2, 000V. Such LMIS may be suitable for being used in space robots with much smaller weight and volume.

3. Experimental method

The key processes of fabricating CNT emitters used in LMIS are shown in Figure.1. Firstly, an AAO film was fabricated to be used as template for synthesizing CNTs. When aluminum is anodized as part of an electrolysis cell in the presence of certain weak acid electrolytes, a porous AAO film develops on the surface of the anode. Due to this phenomenon, an AAO template was prepared using a two-step anodization process [13]. A high purity (99.99%) aluminum sheet was ultrasonically washed sequentially with acetone, ethanol, and distilled water. After being annealing in Ar flow at 500℃ for 4h, aluminum sheet was anodized in a 0.3M oxalic acid at 17℃ with a constant supplied voltage of 40V for 3-4 h. Then, the porous alumina film formed was removed by a wet chemical etching in a mixture of phosphoric acid (6 wt%) and chromic acid (1.8 wt%) at 40℃ for 18h. Thereby a relatively ordered indent pattern on the surface of aluminum sheet is prepared. Subsequently, the aluminum sheet was anodized again for 5-6h under the same conditions as in the first anodization process. To separating the AAO film from remaining aluminum film, a wet chemical process is employed to etch the aluminum film by immersing the sample in CuCl2 solution. The barrier layer of AAO film was also removed and the AAO template with high-ordered pores as shown in Figure.1 (A) was obtained.

Secondly, a catalyst-assisted chemical vapor deposition method was used to synthesize well-aligned CNT arrays on self-ordered porous AAO template. The AAO template was immersed in Fe(NO3)3 solution for 10 min, and then washed with distill water. An annealing treatment was carried out in air at 300 ℃ overnight to oxidize the iron nitrate To depositing carbon on the surface of the pore in AAO template, the template was placed in a quartz boat sealed at one end, which was then inserted into a quartz tube mounted in a tube furnace. The furnace was heated to 700 ℃ in an argon flow, then a mixture of Ar/H_2 with a volume ratio of 10:1 was flown for 10 min to activate the catalyst. CNTs were synthesized with C_2H_4 as carbon source for 20min, after which the furnace was cooled down to room temperature.

Finally, the tips of CNTs were exposed by a wet etching process in order to form a similar structure of the emitters in LMIS. After CNTs were synthesized by CVD method, the surface of the AAO template is blanketed with clumps of foam-like amorphous carbon as shown in Figure.1 (B). An oxygen plasma treatment was carried out for 10 min to remove the amorphous carbon (Figure.1 (C)).

Then the template containing CNTs was attached on a ring type substrate. At last, the upper side of the AAO template was partially etching away by dissolving it in 5% wt NaOH solution to expose the tips of CNTs (Figure.1 (D)).

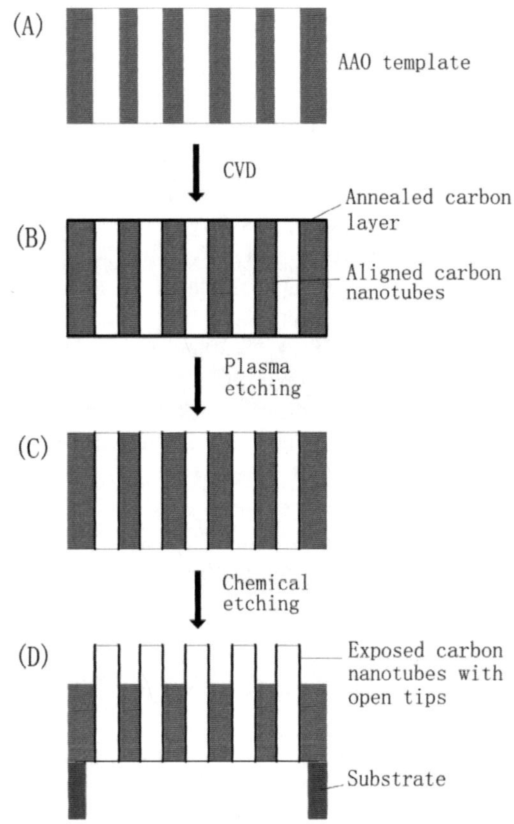

Figure.1 Fabrication steps of CNT emitters

The morphologies of CNTs were characterized by scanning electron microscopy (SEM, JEM-6301F) and transmission electron microscopy (TEM, JEM-200CX).

4. Result and discussion

Figure.2 shows the SEM image of CNTs grown on AAO after chemical etching process. The CNTs are partially exposed with uniform height of about 100nm. The TEM image of CNTs also presents that all CNTs have open tips and large inner diameters (Figure.3). The outer diameters of CNTs are proportional to the pore size in AAO, which indicates that the diameters of CNTs can be controlled by adjusting the pore size in AAO. It had been proved that different pore diameters in AAO could be obtained by varying the anodizing voltage because the pore diameter is proportional to the voltage [14]. The height of CNTs exposed can also been controlled by knowing the etching rate of a particular acid and varying the etching time. So optimizing the geometry of the CNT emitters to enable a better performance in LMIS is possible.

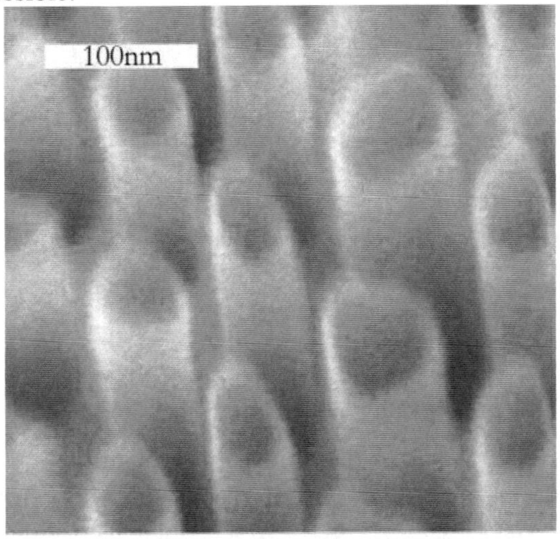

Figure.2 SEM image of CNTs grown on AAO after partially etching away the AAO template

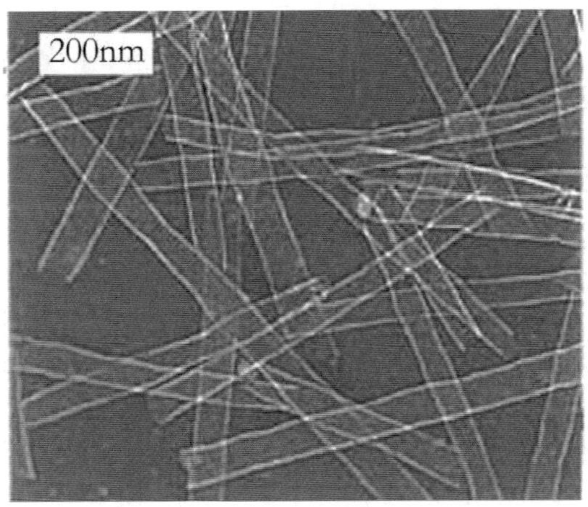

Figure.3 TEM image of CNTs grown on AAO after completely etching away the AAO

The schematics of a LMIS with CNTs as emitters are shown in Figure.4. By capillarity action, liquid metal would fill in CNTs. A voltage is applied between an extractor electrode and CNT emitters, and then a high electric field will extract charged droplets form liquid metal. The droplets are then accelerated by the same voltage and the thrust are produced.

IOP Conf. Series: Materials Science and Engineering **428** (2018) 012058 doi:10.1088/1757-899X/428/1/012058

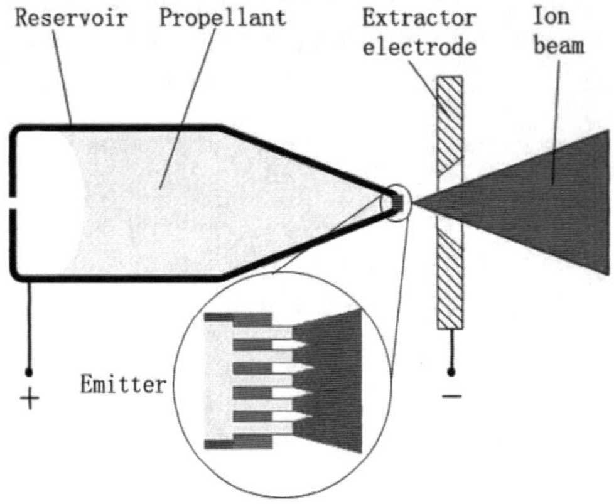

Figure.4 Schematics of a LMIS

To extract charged droplets form liquid metal, a high electric field should be applied on the surface of liquid metal. So a very high voltage usually is used in LMIS. Due to their high aspect ratios and small radii of tip curvature, CNTs are good field emission emitters and have great enhanced field-emission properties. A lower voltage could be enough to extract charged droplets.

Figure.5 shows a simulated result of the electric field intensity between the CNT arrays and the extractor electrode when a voltage of 1000V was applied. The distance between the cathode and the tips of the emitters was about 20 μm. The height of exposed CNTs is 100nm and the distance between two CNTs is 150nm. It is shown that the electric field on the tips of CNTs is about 8.0×10^{4}kV/m, which is much larger than the average field intensity. The electric field has been enhanced by a factor of 1.6. We also calculated the electric field intensity on the tips of the silicon emitters. Similar structure of silicon emitters had been fabricated by MEMS manufacturing techniques [15]. The diameters of the emitters were miniaturized to 10 μm. The height of the silicon emitters was 10 μm, and other parameters didn't change. The result shows that the electric field on the tips of silicon emitters is about 6.4×10^{4}kV/m. So CNT emitters have much better enhanced field emission properties than present emitters in LMIS.

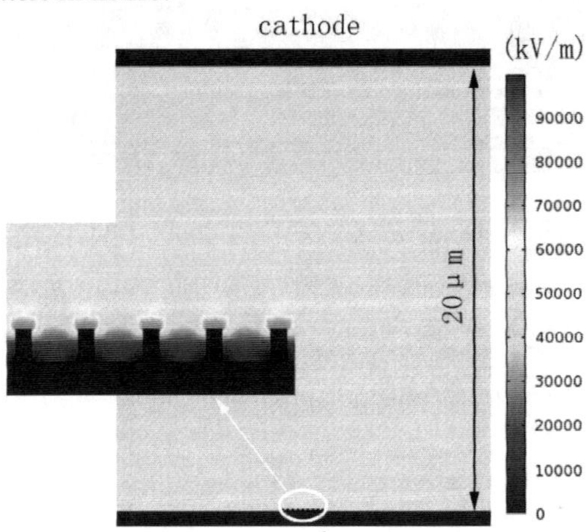

Figure.5 Simulated results of electric field distribution (COMSOL Multiphysics Software)

The ion emission pattern of a prototype sample is shown in Figure.6. When a high voltage of 2000V was applied each time, an emission current occurred simultaneously. The ion beam wasn't very the homogeneous and stable. But it presented that the self-ordered nano-material arrays can be used as emitters in LMIS. With the lower emitting voltage, the weight and volume of LMIS can be reduced.

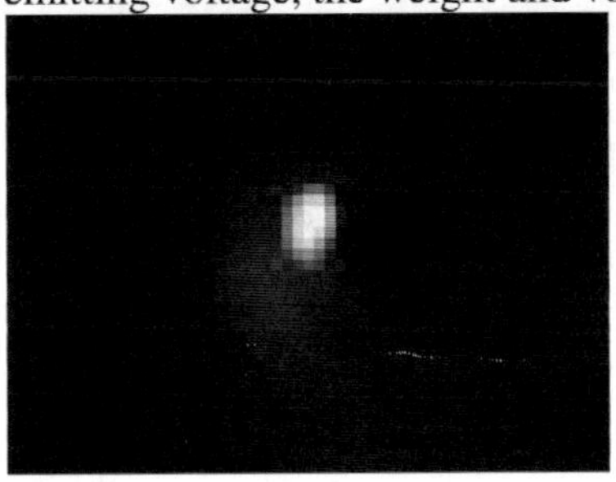

Figure.6 Self-ordered nano-scale emitters during ion emission operation

5. Conclusion

The capillary-type emitters in LMIS should have uniform sizes and distribution. CNTs grown on AAO template are self-ordered and monodispersed with uniform diameter and can be used as capillary-type emitters in LMIS. By this method, small size emitters with diameter of several ten nanometers can be easily realized. With the benefit of the high aspect ratios, the LMIS made of CNTs can emit ions at a very low voltage. So we can develop a new mitigation device for space robot charging effects with very small weight and volume.

References

[1] Garrett H B 1981 *Rev. Geophys. Space Phys.* **19** 577–616
[2] Lai S T 2003 *IEEE Transactions on Plasma Science* **31** 1118-1124
[3] Tajmar M, Scharlemann C, Genovese A, Buldrini N, Steiger W and Vasiljevich I 2009 *Ultramicroscopy* **109** 442–446
[4] Vladimirov V V, Gorshkov V N and Mozyrsky D V 1996 *Appl. Surf. Sci.* 94-94 171-176
[5] Taylor G 1964 *Proc. R. Soc. Lond. A*, **280** 383-397
[6] Marcuccio S and Genovese A 1998 *J. Propul. Power* **14** 774-781
[7] Mair G L R 1980 *Nucl. Instrum. Methods* **172** 567-576
[8] Genoveses A, Tajmar M and Buldrini N 2004 *J. Propul. Powe*r **20** 219-227
[9] Mitterauer J 1987 *IEEE T. Plasma Sci.* **5** 593-598
[10] Rinzler A G, Hafner J H and Nikolaev P 1995 *Science* **269** 1550-1553
[11] De Heer W A, Chatelain A and Ugarte D 1995 *Science* **270** 1179-1180
[12] Bonard J M, Majer F and Stökli T 1998 *Ultramicroscopy* **73** 7-15
[13] Masuda H and Fukuda K 1995 *Science* **268** 1466-1468
[14] Chik H and Xu J M 2004 *Mater. Sci. Eng.* R **43** 103-138
[15] Lenard R, Kravitz S and Tajmar M 2005 *AIP Conf. Proc.* **7** 954-964

IOP Conf. Series: Materials Science and Engineering **428** (2018) 012059 doi:10.1088/1757-899X/428/1/012059

A Lateral Error S-Plane Control for Path Tracking on Autonomous Underwater Vehicle

SUI Haichen[1,2], ZHOU Ziye[3*], JIANG Yanqing[3], CUI Lintao[3] and LI Ye[3]

[1.] Tianjin Research Institute for Water Transport Engineering,M.O.T., Tianjin 300456, China;

[2.] Tianjin Key Laboratory of Surveying and Mapping for Water Transport Engineering, Tianjin Survey and Design Institute for Water Transport Engineering, Tianjin 300456, China

[3.] Science and Technology on Underwater Vehicle Laboratory, College of Shipbuilding Engineering, Harbin Engineering University, 150001, China

Corresponding e-mail: ziye95@foxmail.com

Abstract. This research aims at the motion control on a certain torpedo-type underactuated AUV (Autonomous Underwater Vehicle), using the S (Sigmoid) plane control method to build the heading controller and speed controller, and analyzes the stability of them, then establishes a linear path tracking control system in horizontal plane. In order to improve the performance of the path tracking controller under a large initial disturbance, an asymptotic controller of the heading error is proposed. And to solve the curve path tracking, a curve path tracking controller with a feedforward control idea is proposed. The simulation shows that these two improved controllers have a satisfactory control effects and effectively improved the performance of underwater robot motion control.

1. Introduction

The motion of underwater vehicles has the characteristics of strong non-linearity, mutual coupling between mutil-degrees of freedom, strong environmental disturbances, and it is difficulty to obtain motion models. The underwater vehicle control methods commonly include: neural network control, fuzzy control, sliding mode control, etc. Liu Xuemin[1] proposed an underwater vehicle control method called S-plane, which integrates the idea of fuzzy control and proportional-integral-derivative (PID) control structure, with the advantages of simple structure, less parameters, easy adjustment, and convenient application. It is proved that the S-plane control method is suitable for underwater vehicle motion control[2].

This research uses S-plane control method to construct a lateral error linear path tracking controller, and analyzes the stability of the controller. In combination with the idea of feed-forward control, this linear path tracking controller is applied to the curve path tracking. This research analyzes and compares the control effects under different initial errors, and proves that the superiority of this kind of curve tracking with feed-forward thinking.

2. S-plane control

S-plane control stems from [3]the idea of fuzzy control,it uses a fuzzy control rule table to perform a nonlinear fitting, using a Sigmoid planeto replace the fuzzy rule table. Then continuous expression of the S-plane control is obtained:

$$f = \frac{2}{1+\exp(k_1 e + k_2 \dot{e})} - 1 + \Delta f \tag{1}$$

In the equation(1), k_1, k_2 are the control parameter, e, $\dot{e}$ are the control input, respectively indicate the error and the rate of change of error. f is the control output. To facilitate analysis of e, $\dot{e}$ through normalization, the range of f is [-1, +1], indicates the negative maximum output to the positive output. Δf is the adjustment item to adapt environmental disturbances, factors such as buoyancy changes and currents are considered as fixed interference forces for a period of time. By adjusting the Δf to offset of S-plane, the control output is changed, so as to eliminate the fixed error. The two parameters of the S-plane controller can be adjusted by referring to the PD control[4].

The S-plane control method does not need a model of the control object, and the S-plane control method is actually a kind of non-linear PD control. Therefore, the S-plane control is very suitable for the underwater vehicle, which is nonlinear and difficult to obtain an accurate model[5].

3. Horizontal Linear path Tracking Controller

In the layered control system, the guidance law is an important component of the planning layer and plays an important role in the path tracking of the underactuated AUV[6].For the path tracking control in the horizontal plane, the mechanism of the guidance control of the planning layer is: firstly, according to the geometric principle, the lateral error is obtained from the relative distance between the actual position and the target line; secondly, the control law is based on the Lyapunov stability method, according to the dynamic change of the error, calculate the appropriate target orientation by the S-plane controller, and finally the AUV's movement is converged to the target line[7].

The guidance controller uses the lateral distance error P_e as the control input to calculate the compensation heading angle ψ_c. So that the reference angle of the AUV is:

$$\psi_{ref} = \psi_o + \psi_c(P_e) \tag{2}$$

The error of the heading angle ψ_e can be calculated by the reference heading angle ψ_{ref} and the real heading angle ψ_s

$$\psi_e = \psi_{ref} - \psi_s \tag{3}$$

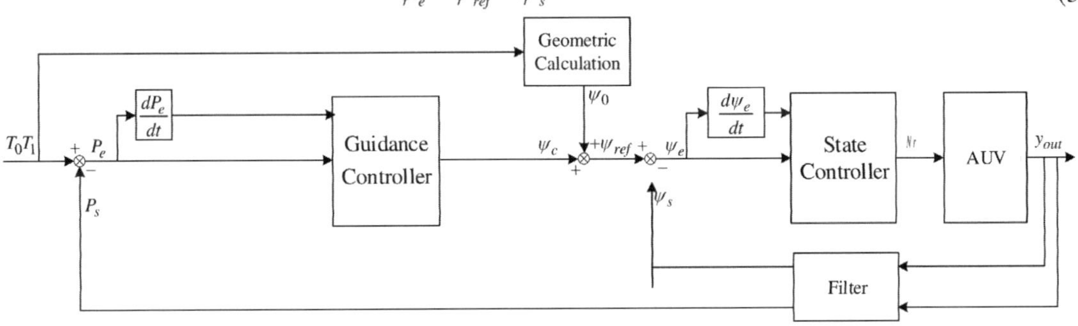

Figure 1. Structure of horizontal straight line tracking control system

The structure of the complete linear path tracking control system in horizontal plane is shown in Figure 1. It is a single input single output (SISO) system. The compensation heading angle ψ_c of the guidance control part is a function of the error P_e.

4. Speed Controller Stability Analysis

First, we designed the S-plane speed controller. The speed control is actually the control of the position change rate, and the angle control is the position control. Let the desired speed be u , and the speed deviation and the change rate of speed deviation are:

$$\begin{cases} e_u = u_{\text{ref}} - u \\ \dot{e}_u = \dot{u}_{\text{ref}} - \dot{u} \end{cases} \tag{4}$$

The discrete form is:

$$f = \sum k[\frac{2}{1+\exp(-k_{\text{p}}e_u - k_{\text{d}}\dot{e}_u)} - 1] \tag{5}$$

Where k is the integration step size.

The S-plane speed controller is derived from the S-plane position controller and can directly use the parameters of the S-plane position controller. In theory, the stability problem can only be qualitatively analyzed[8].

Under the homing and reclamation mission, the AUV usually performs a cruise control during the tracking of the trajectory, $u = u_{\text{ref}}$. The overcome resistance is F, and the controller output is expressed as $f_0 = F / F_{\max}$.

Set $s = -k_p e_u - k_d \dot{e}_u$,the Lyapunov function is: $V(s) = \frac{1}{2}s^2$,so, $\dot{V}(s) = s\dot{s}$,

When $f > f_0$, $\dot{s} > 0$;when $f < f_0$, $\dot{s} < 0$, So, $\text{sgn}(\dot{s}) = \text{sgn}(f - f_0)$.

There are:

$$\text{sgn}(\dot{V}) = \text{sgn}[s \cdot (f - f_0)] = \text{sgn}[s \cdot (\int k\dot{f}dt - f_0)] \tag{6}$$

because of $s \cdot \dot{f} < 0$, $s \cdot (\int k\dot{f}dt - f_0)$,When k is big enough, $\text{sgn}[s \cdot (\int k\dot{f}dt - f_0)] = \text{sgn}[s \cdot \dot{f}] < 0$,So $V(s)$ is gradually stable[9].

5. Linear path tracking with asymptotic control

The path tracking control will face an unstable tracking control condition when the initial steering angle error is greater than $\pi / 3$, and even the path convergence time will rise sharply and even lose control of the AUV. In multi-interference oceans, the heading of AUV sometimes suffers from the ocean currents disturbances, which are accompanied by lateral errors. If the AUV's heading error from the target line is $70° - 90°$, it will lose control stability, and such a control system is not suitable of operating in a varied marine environment. Therefore, an improved method is proposed in which the idea of progressive control is used to control the error of the heading angle of AUV.In the simulation, it isshown that when the initial heading angle error is less than $\pi / 8$, the peak value of the error of the AUV was extremely small, and the convergence time was also small.In this way, $\pi / 8$ is considered as the threshold value of the error of the heading angle. When the absolute value of AUV headingangle error is greater than $\pi / 8$, the target heading is considered as the $\pi / 8$ offseton the target line direction of the AUV heading angle, and the magnitude of the lateral error Pe has no change. When the AUV headingangle error is less than $\pi / 8$, it need not change the value of the heading angle error.In this way, the error of the heading angle at the input of the controller is always limited to $\pi / 8$ or less. A long-term convergence process is decomposed into several short-term convergence processes, achieving stable control and quicker time convergence.

6. Curve path tracking control

For the sake of simplifying the problem, it was decided not to change the linear guidance controller and the state controller, and the linear tracking controller was applied to the curve tracking. A curved path tracking method is proposed to change the tracked target straight line in real time as the curve curvature changes. As shown in the Figure2, the tracking target is selected as the hypothesis line T0L1, and the lateral error between the actual position of the AUV and the hypothesis line T0T1 is obtained as the current control error. The solution of the arbitrary curve lateral error P_e is shown in Figure 2.

As shown in Figure 2, the AUV is at point O and the tracking target curve is C. First, a perpendicular line is made in the x direction of the vertical AUV, intersecting with the target curve C at a point T0, and the slope of the curve C at the point of the tangent line L0 is k0; the idea of introducing the feedforward control takes into account the next change trend of the curve, along with the target curve C, taking a point T2 of 1m-10m after T0, we can also find the slope K2 of C at this point's tangent line L2, and then weighting the two to obtain k1, that is:

$$k_1 = (k_2 + k_3) / 2 \tag{7}$$

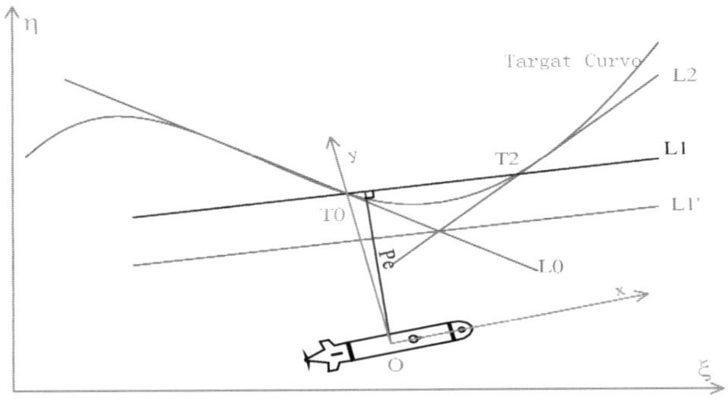

Figure 2. Horizontal position error of arbitrary curves

Then use k_1 as the slope to make a straight line through T0, and find a point T1 on this straight line. Think of this straight line as the straight line traced at that moment. The advantage of this consideration is that it takes into account both the distance and the error of the current position from the target curve C, and the trend of the next trend of the curve. This allows the AUV to approach the target curve C faster, and the precision of the follow-up curve motion of AUV is greatly improved.

In practical tasks, when k1 is determined, different control effects can be achieved by changing the weights of k_2 and k_3, that is:

$$k_1 = (c_2 \cdot k_2 + c_3 \cdot k_3) \cdot (c_2 + c_3)^{-1} \tag{8}$$

The simulation shows that changing the weights c_2 and c_3 will have a certain impact on the quality of the path tracking. Therefore, in the actual situation, the values of c_2 and c_3 should be reasonably determined according to the characteristics of the path.

7. Simulation

7.1 Linear path tracking asymptotic control

The initial position of the AUV is set on the target straight line, and the initial steering angle is changed to compare the different control effects of the S-plane controller and the improved progressive error S-plane controller. When the lateral error Pe<0.5m, the AUV converges on the target path.

It is shown inFigure 3. and Figure 4. that the improved linear tracking controller has greater advantages in rejecting a large initial heading errors, and the convergence time and the convergence distance are approximately linearly positively correlated with the initial heading angle error.

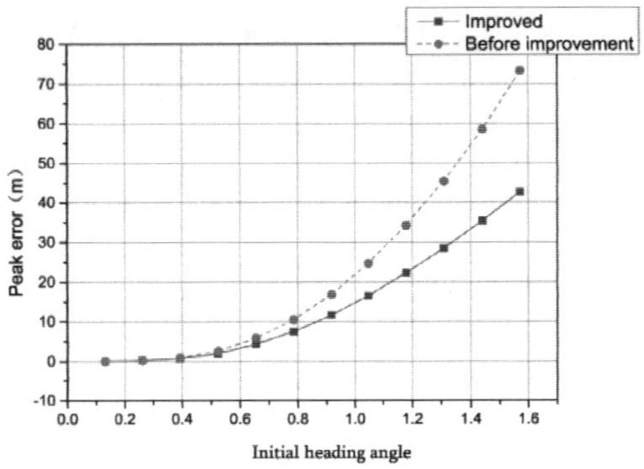

Figure 3. Comparison of error peaks before and after improvement

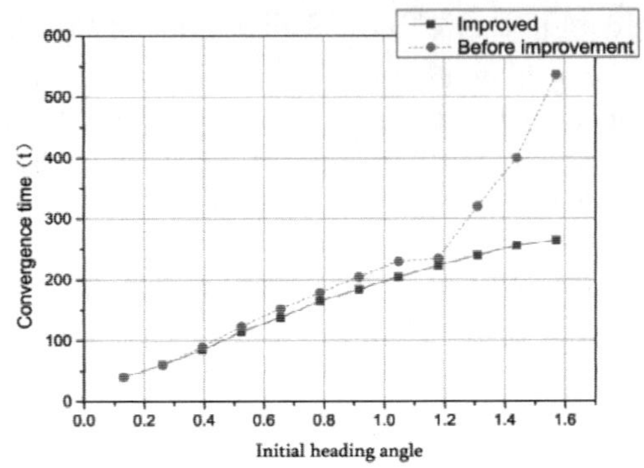

Figure 4. Comparison of convergence time before and after improvement

7.2 Curve path tracking control

The curve path has an amplitude of 100 m and a period of 800 m. The initial position of the AUV is selected to be (0, 120) and the initial angle of the direction is 0,the simulation time is set to 2000s. Take the guidance controller parameters k1 = 2, k2 = 55.2, state controller k1 = 0.1, k2 = 0.02. The following is an image of the AUV along the set cosine curve and the image of the lateral error P_e :

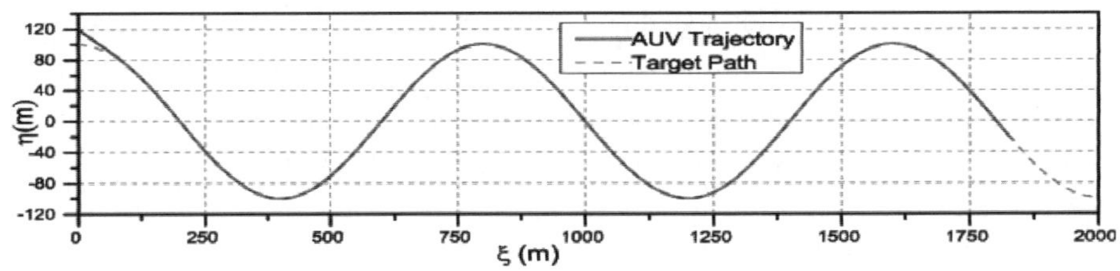

Figure 5. Path tracing of a cosine curve

In Figure 5, Figure 6. and Figure 7, it is shown that the S-plane controller has a very low error in tracking the curve. The lateral distance error varies from -0.1m to 0.1m. At the same time, the turning torque Nr of the AUV is also changing smoothly, which also shows that the S-plane controller under the parameters is adapted to control the motion of the AUV.

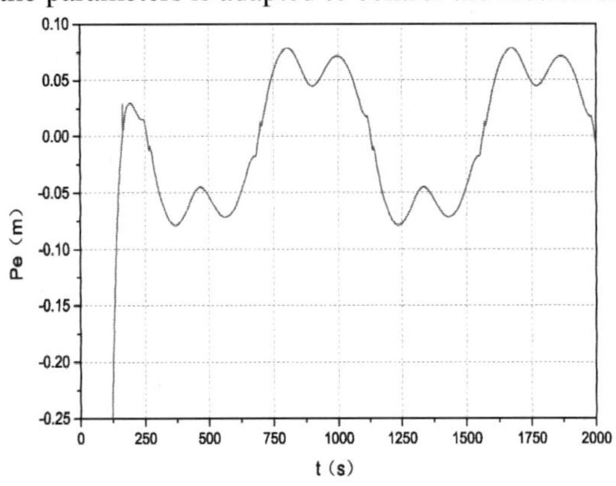

Figure 6. Horizontal Positional Errors of Cosine Curve TrackingPe

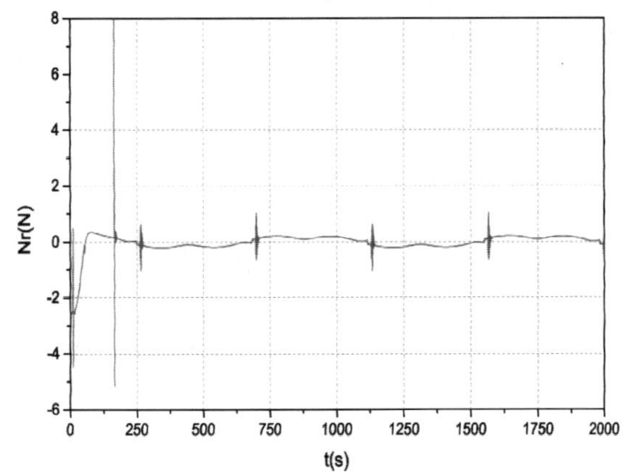

Figure 7. Turning moment tracked by the cosine curveNr

7.3 Curve simulation for different lateral errors

Different initial lateral position errors are selected to research the characteristics and applicability of the curve path tracking system. The following is the result of changing the simulation of a set of initial lateral positions. Figure 8. is the variation of the lateral positional error P_e of the AUV at different initial positions:

It can be seen from Figure8. that when tracing arbitrary curves, the initial lateral position is from (0,50) to (0,170), and after a certain period of time (about 150s-300s), it will converge to almost zero.

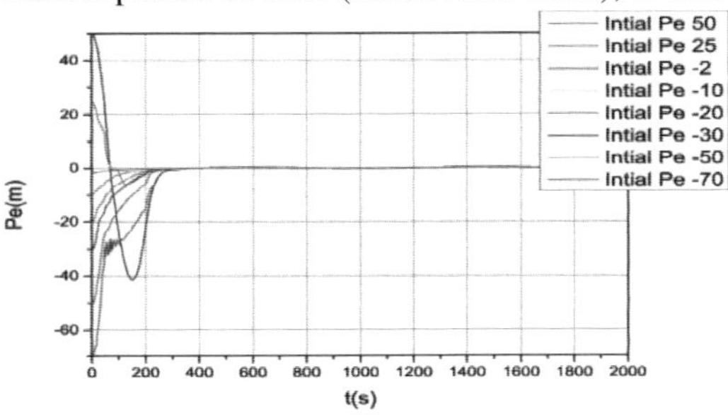

Figure 8. Tracking lateral error of cosine curve path at different initial positions track

8. Conclusion

Practice has proved that the S-plane control method is an effective method for path tracking of underwater vehicle. In this paper, the S-plane control method is used to design a linear path tracking system, and the stability of the controller and speed controller is analyzed. Aiming at the instability at a large initial heading angle error for S-plane controllers, a method of progressive heading angle error control was proposed, so that in the simulation, the convergence time and convergence distance were reduced. Aiming at the curve tracking task, a curve tracking method with feedforward idea is proposed. The simulation results show that this curve tracking method can track the curve and can complete convergence under different initial lateral errors.

Acknowledgments

This work was supported by the special fund project of the basic scientific research service of the Central Public Welfare Research Institute (No. TKS170224), and "Research on underwater emergency search and detection based on AUV " (No. TKS180406), Support Program of China, National Key Research and Development Plan (No. 2017YFC0305700), Financially Supported by Qingdao National Laboratory for Marine Science and Technology (No. QNLM2016ORP0406), the Fundamental Research Funds for the Central Universities (Grant No. HEUCFG201810).

Reference

[1] Liu, X. M., & Yu-Ru, X. U. (2001). S control of automatic underwater vehicles. *Ocean Engineering*.
[2] Yue-Ming, L. I., Pang, Y. J., & Wan, L. (2012). Adaptive s plane control for autonomous underwater vehicle. *Journal of Shanghai Jiaotong University, 1*, 482 - 489.
[3] YANUSHEVSKY R. Modern Missile Guidance[M]. CRC Press, 2008.
[4] SIOURIS G M. Missile Guidance and Control Systems[M]. New York: Springer-Verlag New York, Inc., 2004.
[5] SPEARMAN M L. Historical development of wordwide guided missiles[C]. In: AIAA 16th Aerospace Sciences Meeting, Huntsville, Alabama, USA. 1978.
[6] LIN C L, SU H W. Intelligent control theory in guidance and control system design: An overiview[C]. Proceedings of the National Science Council. 2000,24: 15-30.
[7] MIZUTANI A, CHAHL J S, SRINIVASAN M V. Motion camouflage in dragonflies[J]. Nature, 2003, 423(6): 604.
[8] DRIELS M. Linear Control Systems Engineering[M]. Columbus, Ohio: McGraw-Hill, 2000.
[9] YE L, YONGJIE P, LEI W. Stability analysis on speed control system of anutonomous underwater vehicle[J]. China Ocean Engineering, 2009, 23(2): 345-350.

IOP Conf. Series: Materials Science and Engineering **428** (2018) 012060 doi:10.1088/1757-899X/428/1/012060

Design and Implementation of Four-link Robot Crawler with Variable Structure

ZHAO Yuting[1], Han Baoling[1], Luo Qingsheng[2] and Li Kailing[1]

[1.] School of Mechanical Engineering, Beijing Institute of Technology, Beijing 100081, China

[2.] School of Mechatronical Engineering, Beijing Institute of Technology, Beijing 100081, China

Corresponding author e-mail:zytkite@163.com

Abstract. A kind of crawler-type robot with variable structure based on four-bar linkage was designed to solve the obstacle clearance problem. Firstly, the design scheme of the variable structure tracked robot was proposed, and the virtual prototype was designed. Then the mathematical model between the deformation of the four-bar linkage and the total length of the track was established. the numerical analysis shows that the total length of the track is 0.91% in the deformation process, which is smaller than the elongation of the general rubber track. The feasibility of the scheme was verified theoretically. Finally, a physical prototype experiments were carried out and experiments show that the variable structure crawler could achieve good obstacle performance through deformation.

1. Introduction

In recent years, small variable-structured crawler robots have been playing an increasingly important role in the search for holes and tunnels, inside buildings, and the detection and removal of anti-personnel landmines and other operations. This has aroused the attention of all countries in the world. The company of Inuktun in U.S. has developed a variable structure crawler-type reconnaissance robot (VGTV) [1-3], which can achieve the effect of simultaneous deformation of the crawler during the course of its movement. When the robot body is relatively flat, the robot is in the form of a tank; the relative change in the position of the main driven wheel can make the crawler become a triangular shape, this feature greatly enhances the robot's environmental adaptability and obstacle-obstacle ability. The "Ling-B" explosives robot developed by Shenyang Institute of Automation, Chinese Academy of Sciences, adopts a compound mobile mechanism of wheels, legs, and crawlers [4]. It has made great progress in the research of variable structure robots, but in terms of performance and reliability, compared with similar foreign robots, there is still a certain gap. Based on the deformation principle of the four-bar linkage mechanism, this paper proposes and designs a new type of variable structure double tracked robot. Its related properties and characteristics are systematically studied and in-depth discussions, and the virtual prototype design and the physical prototype movement and implementation are carried out.

2. Structural design

As shown in figure 1, for the track structure with no change in tension, the distance L between the two wheels and the length A of the main arm is a fixed value. The axis of the main arm is located at the

IOP Conf. Series: Materials Science and Engineering **428** (2018) 012060 doi:10.1088/1757-899X/428/1/012060

center point of the connecting line between the two wheels. Assume that the radius R of the wheel and the leading wheel; the angle α between the center line of the main arm and the two wheels and the main arm.

The track perimeter C can be expressed as follows while 0°≤α≤90°:

$$C = L + 2\pi R + \sqrt{A^2 + (\frac{L}{2})^2 - 2ALCOS(\alpha)} + \sqrt{A^2 + (\frac{L}{2})^2 + 2ALCOS(\alpha)} \quad (1)$$

From the analysis of formula (1), it can be seen that C increases with α. However, when the rate of change of C exceeds 5%. When the main arm is lifted, the tensioned track will break and when the main arm falls, the tensioned track begins to fall off. This structure results in a limited range of motion of the main arm and affects the crawler robot function.

To solve the above problem, a four-bar linkage mechanism is used to replace the rigid structure between the connecting rod and the wheel. As shown in figure 2, the four-link variable structure double tracked robot is mainly deformed by a robot body and two symmetrically distributed crawler belts. Device composition. The crawler deformation device is composed of a link mechanism and a swing arm. The track rod end is equipped with a track roller near the ground, and the rear wheel is a passive wheel.

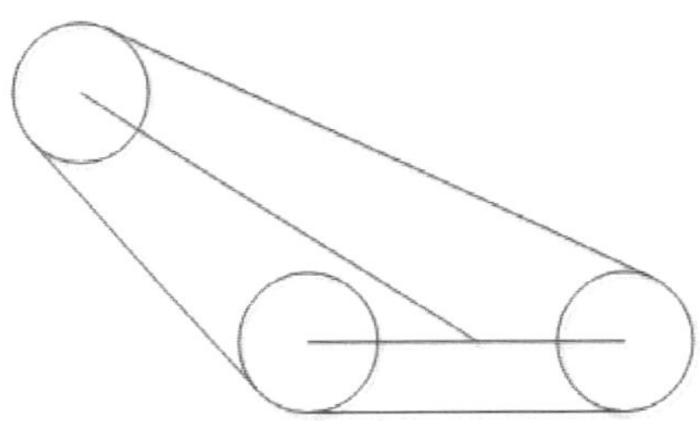

Figure 1. Unchanged track structure without tension.

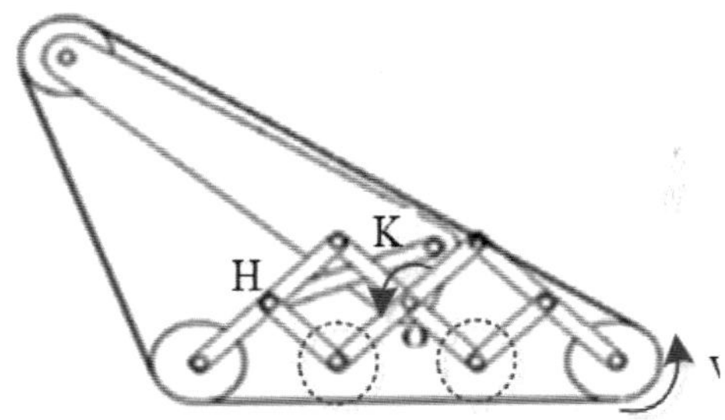

Figure 2. Four-link variable structure crawler structure.

The mechanism adopts a special elastic tension device. When the swing arm is raised and lowered, the swing arm drives the connecting rod HK to pull or push the rotation point O of the four-bar linkage mechanism. Because the track roller is installed at the end of the connecting rod, the main arm swings can be driven four times. The link mechanism zooms to drive the change in distance between the track rollers. While changing the structure, the lengths of crawlers on all sides are changed in real time, and the total track length of the mechanism can be kept constant by the relationship between the lengths of each link.

The advantage of the four-link variable structure double tracked robot is to fully utilize the characteristics of the four-bar linkage mechanism, and one original mover can realize the deformation of the entire mechanism. The structure of this mechanism is simple, and it is convenient to adapt to the relatively complicated working environment such as low clearance through miniaturization. In addition, the robot has good obstacle avoidance and mobility. And connect the rotary joint to zoom in the main arm and linkage mechanism. When the main boom is lifted and rotates around the center of rotation, the four-bar linkage mechanism is also shortened accordingly, which means that L is reduced to accommodate constant-length C.

3. Theoretical analysis of structural design

The main structural principle of the variable structure double tracked robot is shown in figure 3, KD rod, ED rod, and EC rod belong to the swing arm portion, so their relative positional relationship is always the same. The movement of the KD rod of the active part causes a change in the quadrilateral KDMH, which ultimately results in a change in the distance between the track rollers A, M, N, B. From the three states below, the effect of changes in AB pitch on the track perimeter C is analyzed [5].

3.1. Flat state

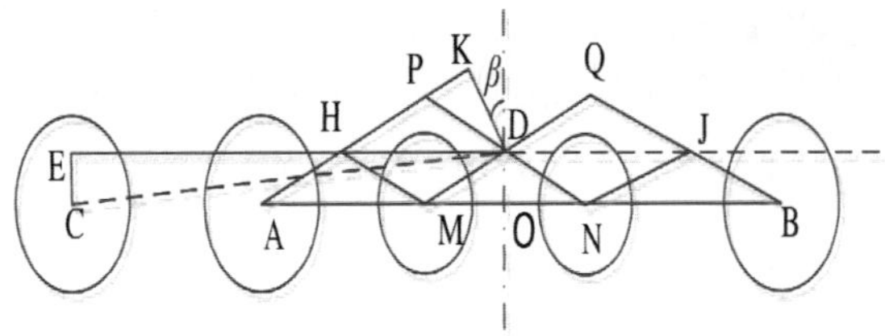

Figure 3. Schematic diagram of the flat state of the robot.

As shown in figure 3, when the ED and the HD coincide with each other, the crawler robot is in a flat state with the lowest height. The theoretical length of the crawler belt is calculated as follows:

In the quadrilateral MDKH, set the angle between the KD and the vertical center line to be β, the length of the HM and the MD equal and known, HK and ED lengths are known, the radius of the track wheel is d, and the coordinate at the K point is D The point is the relative coordinates of the origin. Then, the tangent of the included angle β is equal to the ratio of the abscissa to the ordinate of the K point, which is obtained from the triangular trilateral theorem:

$$COS(\frac{\pi}{2}-\beta)=\frac{HD^2+KD^2-HK^2}{2\times KD\times HD} \tag{2}$$

$$C=L+2\pi R+\sqrt{A^2+(L/2)^2-2ALCOS(\alpha)}+\sqrt{A^2+(L/2)^2+2ALCOS(\alpha)} \tag{3}$$

Thus calculate the length of the HD. Therefore, the theoretical track length is:

$$C(0)=2\times\left(ED+3\times\frac{HD}{2}\right)+\pi\times d \tag{4}$$

3.2 When the robot is deformed to the maximum

As shown in figure 4 when the height of the robot is the highest, the length of the crawler contact with the ground is the shortest.

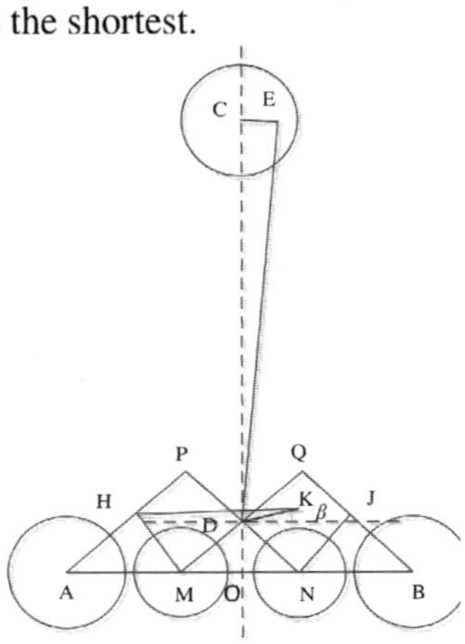

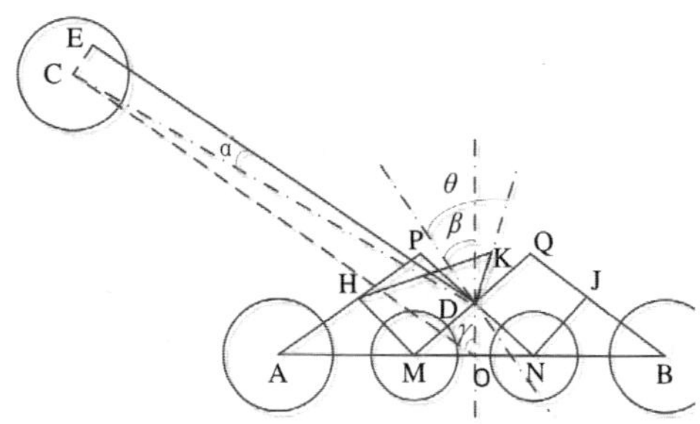

Figure 4. schematic diagram of the robot's highest state.

Figure 5. Schematic diagram of the general state.

Let the angle between the ED and the vertical center line be α. As can be seen from figure 3, CE and ED are perpendicular, so the tangent of α angle is equal to the ratio of CE and ED. Because HD and HK are set to known values, the expression of HD is derived from the triangular trilateral theorem formula:

$$\cos(\pi-\beta+\alpha)=\frac{HD^2+KD^2-HK^2}{2\times KD\times HD} \tag{5}$$

Let the vertical height between point D and AB be h, then h can be represented by HD and DM as:

IOP Conf. Series: Materials Science and Engineering **428** (2018) 012060 doi:10.1088/1757-899X/428/1/012060

$$h = \sqrt{DM^2 - (HD/2)^2} \tag{6}$$

From this we can find the length of the AC:

$$AC = \sqrt{(3 \times HD/2)^2 + (DC + h)^2} \tag{7}$$

The figure shows that AC and BC are equal. So when the height of the variable structure double tracked robot reaches the highest point, the theoretical track length is:

$$C(90) = AB + BC + AC + \pi \times d \tag{8}$$

3.3 General deformation

Figure 5 shows the relationship between the theoretical track length and the DK linkage rotation angle under normal conditions.

Assume that the angle at which the connecting rod DK turns to the normal position when starting from the initial flat state is θ. Then, the relationship between rod KD and angle θ in ΔHDK is:

$$COS(\pi/2 - \beta + \theta) = \frac{HD^2 + KD^2 - HK^2}{2 \times KD \times HD} \tag{9}$$

The angle between CO and the vertical plane is γ, which is known from the cosine law in △COD:

$$COS\gamma = \frac{CO^2 + DO^2 - DC^2}{2CO \cdot DO} \tag{10}$$

Known by △AOC and △BOC:

$$CO^2 + AO^2 + 2AO \cdot CO \cdot COS(\frac{\pi}{2} - \gamma) = AC^2 \tag{11}$$

$$CO^2 + BO^2 + 2BO \cdot CO \cdot COS(\frac{\pi}{2} + \gamma) = BC^2 \tag{12}$$

Therefore, the total theoretical track length obtained is:

$$C(\theta) = AB + BC + AC + \pi \times d \tag{13}$$

Using MATLAB software, in a number of computational experiments, it was found that the position of the K-point was the main factor affecting the deformation of the robot. Referring to figure 3, when the ratio of the abscissa and the ordinate at the K-point was 0.26 with the D-point as the origin, the structure Track length change rate is less than 1%.

In order to adapt to the working environment with a small gap and to increase the obstacle avoidance capacity, the length of the KD rod is 27.9 mm in accordance with the above formula and the simulation of the ratio of the abscissa 7 and the ordinate 27 in the software to the K point. The rod isometric rod is 80mm; the HM and other short rods are 40mm; the HK rod length is 72mm; the DE length is 200mm; the CE length is 15mm; the track wheel diameter d is 50mm.

The initial length is taken in equation (2) when the variable structure robot is in a flat state.

$$HD = 73.7\text{mm}$$

Bring the HD length into equation (4), the theoretical track length is:

$$C(0) = 778.3\text{mm}$$

When the initial length is brought into the maximum height of the variable structure double tracked robot, equation (5) is used to obtain HD=44.4 mm. The length of the theoretical track is calculated by taking the HD length into equation (7):

$$C(90) = 775.4\text{mm}$$

Bring the initial value to the relationship between the theoretical track length and the DK link rotation angle θ in general formulas (9~13). Calculated by MATLAB software and obtained when the θ∈ (0°~ 90°), the length of the theoretical crawler in the process of the lifting rod rotation in the range of 772mm ~ 779mm, so the rate of change is:

$$\eta = \frac{779 - 772}{779} = 0.91\% \tag{14}$$

IOP Conf. Series: Materials Science and Engineering **428** (2018) 012060 doi:10.1088/1757-899X/428/1/012060

Query data found that the general rubber crawler elongation is less than 4%, so the above formula calculates that η is less than 1% meets the requirements, the same track can adapt to changes in the shape of the triangle, thus enabling the robot to adapt to different geographical environment through structural changes.

4.Prototypes and experiments

The robot is virtually assembled in 3D modeling software. The rear cover assembly, the bottom frame assembly and the front cover assembly are assembled to obtain the overall assembly of the variable structure double tracked robot. The rear wheel of the robot is a driving wheel. Through the rotation of the rear wheel, the movement of the crawler is driven to realize the back and forth movement of the robot. By controlling the unilateral motor, the robot turns. In a flat environment, the robot can travel in the flattest manner, as shown in figure 6(a).

When a special variable-structure crawler robot encounters obstacles such as bosses, stairs, and trenches under relatively harsh working conditions [6-7]. By controlling the swing arm motor located in the middle part of the robot, the swing arm motor rotates, and the front cover of the robot is rotated by the gear transmission. Since the connecting rod is connected to the front cover, the rotation of the front cover drives the movement of the connecting rod and realizes four movements. Deformation of the linkage mechanism, thus achieving the deformation of the overall robot. As shown in figure 6(b):

The front cover of the robot swings upwards, and the four-bar linkage shrinks, so that the caterpillar shape is triangular, which creates an angle of attack between the front crawler of the robot and the ground. It can greatly improve the variable structure double track when it meets the environment of the boss and ravine. Obstacle ability of the robot [8]. Figure 6(c) shows the state when the robot has the largest angle of attack:

(a) The flat state (b) The general state (c) The highest state

Figure 6. Virtual Prototype Model.

The assembly of the physical prototype of the four-link variable structure double tracked robot is shown in figure 7, where 7(a) is the state when the robot is flat, and figure 7(b) is the state when the front wheel of the robot is raised to form a large angle of attack.

(a) The flat state (b) The general state

Figure 7. Physical prototype.

The test of the variable structure double tracked robot over the boss is shown in figure 8. where 8(a) shows the state of the robot on the ground. At this point, the contact area of the robot's track with the ground is the largest. When the front side encounters the boss, the robot drives the crawler by swinging the front cover of the robot through the swing arm motor, so that the track forms an angle of attack as shown in figure 8(b). At the same time, the rear drive wheel of the robot rotates, causing the robot to move forward and climb onto the boss. After the upper boss deforms the robot to its initial position, as shown in figure 8(c). After the robot moves, the center of gravity of the robot passes through the edge. Under the pulling of gravity, the robot moves smoothly down the boss and finally passes through the boss. As shown in figure 8(d).

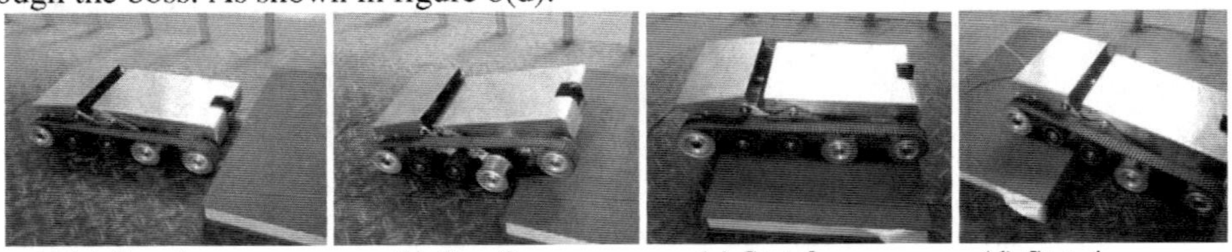

(a) State1 (b) State2 (c) State3 (d) State4

Figure 8. Overbore Experiment.

Experiments show that the tracked robot can achieve the task of overturning the boss by actively adjusting the angle of attack of the track. The four-link zoom mechanism can ensure that the track of the robot stays in a tensioned state without slackening and slipping by adjusting the track wheel spacing.

5. Conclusion

This paper proposes a design method of a four-bar linkage variable structure double tracked robot. Taking the four-link double tracked robot as the research object, a mathematical model is established between the deformation of the four-bar linkage and the change of the meshing length of the track section. The rationality of the scheme was verified from both theoretical calculations and experimental verifications. Through analysis, the following conclusions can be drawn:(1) The robot can actively adjust the pitch of the angle of attack and the track wheel through the linkage zoom mechanism, which effectively ensures the ability of the robot to adapt to the terrain by changing the structure.(2) In the process of adapting the deformation of the zoom mechanism to the terrain, there is no detachment or relaxation of the track and track wheels, and the tension is maintained at all times, which proves the scientific rationality of the proposed mathematical model. The track length deformation is less than 1%, which can greatly improve the tracked robot's obstacle performance.

References:

[1] WANG C C. A novel variable structure theory applied in design for wheeled mobile robots[J]. Artificial Life and Robotics, 2011, 16(3):378-382.

[2] LIN C J,LEE K S. Contour Tracking of a Redundant Robot Using Integral Variable Structure Control with Output Feedback[J]. Journal of Intelligent & Robotic Systems, 2011, 62(2):241-270.

[3] GE L, LI R, YU D, et al. A Nonlinear Adaptive Variable Structure Trajectory Tracking Algorithm for Mobile Robots[M]// Intelligent Robotics and Applications. Springer Berlin Heidelberg, 2008:892-900.

[4] LI Y W, GE S R, ZHU H, et al. Obstacle-surmounting Mechanism and Capability of Four-track Robot with Two Swing Arms[J]. ROBOT, 2010, 32(2):157-165.

[5] SONG M J, ZHANG M L. Kinematics analysis of deformable wheel-leg of multi-sports mode mobile robot[J]. JOURNAL OF MACHINE DESIGN, 2011, 28(5):46-50.

[6] FU Y L, LI H H,ZHANG F H. Design and Kinematic Analysis of Anti-terrorism Robot With Wheel- tracked Variable Structure[J]. Machinery & Electronics,2011, (10):66-69.

[7] LIU Z W, SHI C H, KANG S H, et al. Design and Analysis of the Pose-Varied Wheel-Tracked Robot Travel Mechanism[J]. Machinery Design & Manufacture, 2014(4):175-177.

[8] XU R Q, HAN B L, LUO Q S. Relationship Between Six-track Robot's Structure Parameter and its Performance on Overcoming Obstacles [J].Machinery Design & Manufacture, 2012(7):113-115.

IOP Conf. Series: Materials Science and Engineering **428** (2018) 012061 doi:10.1088/1757-899X/428/1/012061

Multi-objective Identification of UAV Based on Deep Residual Network

Wang JiaQi[1,1], Dai JiYang[1,2], Zhai JinYou[1,3] and Ying Jin[1,4]

[1] School of Information Engineering, Nanchang Hangkong University, Nanchang 330063, China

1196006635@qq.com[1], djiyang@163.com[2], 710644534@qq.com[3] and azx519@foxmail.com[4]

Abstract. Beacuse of the classical RPN (Region Proposal Net) exists the defect of large computation and high time complexity when extracting targets candidate region, a search mode called CRPN (Cascade Region Proposal Network) mode was proposed to ameliorate it in this paper. In order to suppress the degradation phenomenon in deep convolutional neural network training, the residual learning theory was introduced, a novel Mu-ResNet (multi-strapdown deep residual network) was proposed. Combined the Mu-ResNet with CPRN, a network model for multi-target identification of UAV was designed and tested. Compared with the network model that combines ResNet with RPN, the identification accuracy was increased nearly 2 percentage points.

1. Introduction

The convolutional neural network is a branch of the artificial neural network. It has been successfully applied in computer vision, pattern recognition and other fields. The advantages of the convolutional neural network are obvious. Firstly, it processes input datas through convolution operations and down sampling, which and lower the computational complexity. Secondly, its network weights are shared with each other, which can effectively reduce unnecessary weight connections and lower model complexity. Thirdly, the network structure has high degree of scale invariance, rotation invariance and other affine invariance, so the learned features have topological correspondence and robustness.

Imagenet Large Scale Visual Recognition Challenge (ILSVRC) competitions at recent years, the top three participants are both used neural networks algorithms, such as literature [1]. This phenomenon revealed the strong vitality of CNN and its generalization capabilities. Deep convolutional neural networks [2,3] are CNN with deeper level networks, have led a series of breakthroughs for image classification [4,5]. Deep CNN naturally integrate low/mid/advanced functions and categories at end-to-end pattern [6], and the 'levels' of functions can be enriched by stacking more layers (depth). Recent evidences [7,8] reveals that network depth is crucial importance, and the most advanced results of image classification [9,10,11] on the challenging ImageNet dataset [12] all exploited very 'deep' [7] models.

However, with the deepening of the number of network layers, data amount and computation amount are also increasing dramatically which caused the notorious problem of vanishing/exploding gradients[13]. Not only hamper convergence from the beginning but also exposed degradation problem. That means, with the network depth increasing result accuracy gets saturated and then degrades rapidly [14].

IOP Conf. Series: Materials Science and Engineering **428** (2018) 012061 doi:10.1088/1757-899X/428/1/012061

2. Deep Residual Network Construction

The residual learning module is a part of the deep convolutional neural network and it can effectively solve the problem of accuracy degradation. Residual learning does not map the image features successively to the next layer, but directly maps the features to the residual layer[14]. Training deep neural networks is very difficult, however, using the deep residual module to build the network can greatly reduce the amount of training data and achieve deeper network training. The two-layers and three-layers residual learning module learning mechanism was shown in figure 1 and figure 2 respectively.

In this paper, a Strapdown ResNet with multiple shortcut channels was designed, which combines the advantages of the two-layers and three-layers residual modules. The Strapdown ResNet uses a five-layer stack, it's structure and connect model was shown in figure 3. By setting multiple shortcuts to ease the training difficulty and makes it is easier to converge in deep network training. And using dimensionality reduction and convolution to reduced the time complexity of training.

Aiming at the multi-class target images with arbitrary size, UAV target identification system was combined with CNN and residual learning model. The five layers residual learning module was used to build Multi-Strapdown ResNet with different depths.

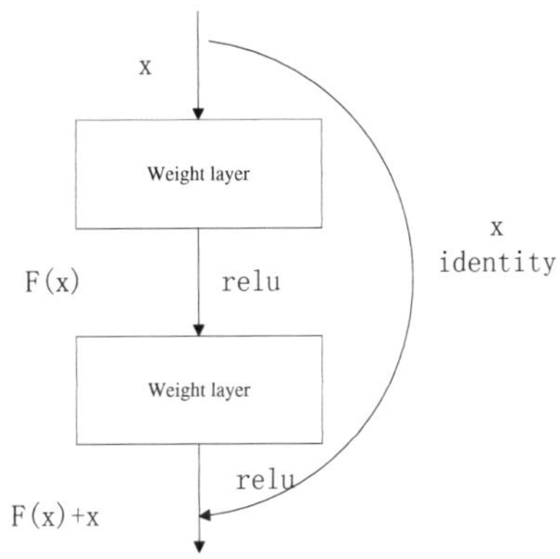

Figure 1. Residual network learning module.

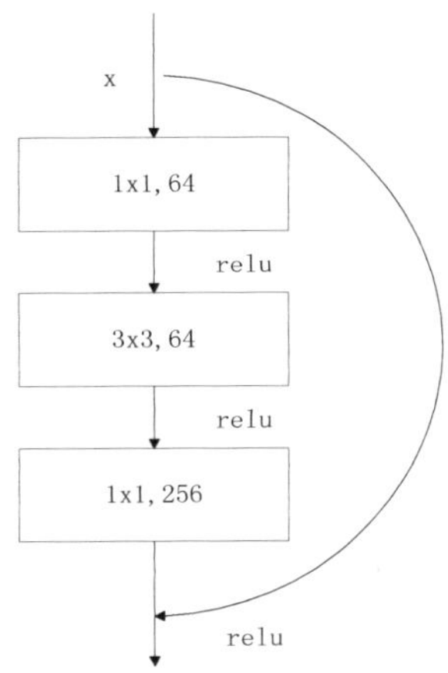

Figure 2. Three layers residual learning module.

3. Target Identification Algorithm

RBG et al. proposed the Regional Proposal Network (RPN) [15] and gives three kinds of area $\{128^2,256^2,512^2\}$ and three scales {1:1,1:2,2:1} which constitute total 9 types of anchors to extract detection areas, as shown in figure 4. Anchor mechanism gives the size of a standard sliding window, then different size windows can obtained according to different multiples and length-to-width ratios.

Beacuse of the large number of search areas in the classical RPN network, the computational complexity and time complexity were high when target candidate areas were extracted. In order to solve this problem, this paper used a three-level CRPN search mode. In the first level CNN, a 32x32 convolution kernel were used, sliding window with 4 step size on the WxH size feature map to get detection window. Due to small search area scale, the first-level CRPN can filter out more than 90% regional windows. Then use non-maximum suppression to eliminate candidate areas with high coincidence rate. This method greatly reduces the computational complexity and time complexity of lower level networks. Next, the target candidate areas obtained through the first level network was adjusted to 64x64 size and input to second-level CRPN to filter out 90% candidate region windows, then the remaining windows passes through the NMS to eliminate the candidate region of high coincidence rate. Finally, the candidate region of the second level CRPN's output was adjusted to 128x128. After the same operation, the background area and the height overlap area are filtered out

IOP Conf. Series: Materials Science and Engineering **428** (2018) 012061 doi:10.1088/1757-899X/428/1/012061

and obtained 300 rectangle candidates in the end. The final candidate region classification layer and window regression layer are implemented with a convolutional layer of 1x1 size. The CRPN model was shown in figure 5.

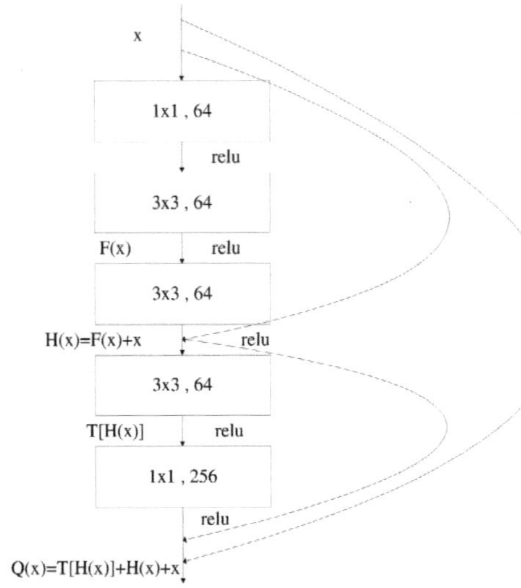

Figure 3. Strapdown Residual Module.

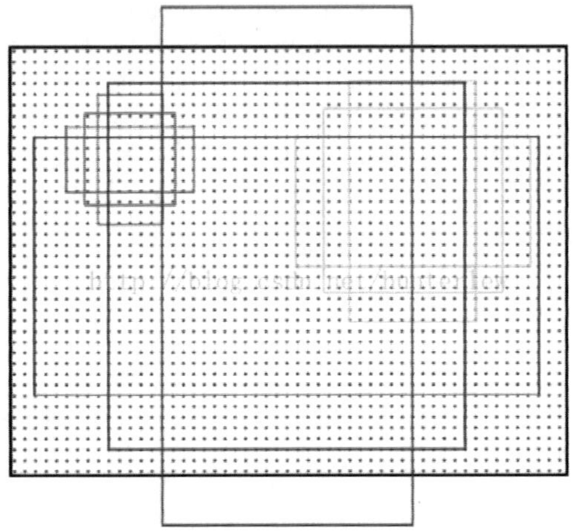

Figure 4. Anchor schematic diagram.

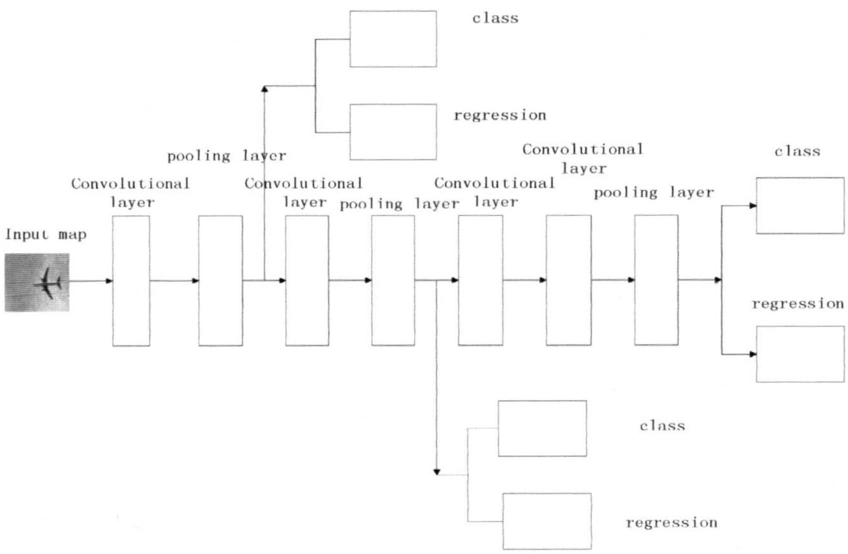

Figure 5. Cascade regional proposal network.

4. Target Identification Experiment

In this paper, the image set for UAV target identification mainly includes 10 kinds of targets, such as aircraft, pedestrians, cars, large buses, ships and flying birds. Thouse are common targets when UAV execution tasks. Data sources mainly come from ImageNet, major open source databases and our own aerial images taken by drones, total 100,000 images. Among them, 70,000 images were used as the training set, and 30,000 images were used as the test set.

In order to study the effect of feature extraction network with different performance parameters on identification results, this paper constructs three kinds of network with different depth for performance test. That are, Mu-ResNet-22, Mu-ResNet-37, Mu-ResNet-57, represent 22/37/57 layers depth residual network separately.

For conduct horizontal comparison, the above three networks are compared with VGG16, ZF and classical residual networks, and the CRPN is combined for performance comparison. Firstly, training on the target training set to obtain each network model, then testing the accuracy of target identification and detecting time consumed for each picture on the test set. The specific results are shown in table 1. From table 1, we can see that the identification accuracy of Mu-ResNet-57 is slightly higher than ResNet-56, and the detection time is less than ResNet-56. The overall performance of Mu-ResNet-22 and Mu-ResNet-37 is also better than VGG16 and ZF networks. Therefore, using the 5-layers residual learning module proposed in this paper to build the network model can improve the performance of the algorithm more effectively.

Table 2 shows the detection curvature of each type target in the UAV data set by combining VGG16, ResNet-56 and Mu-resnet-57 with RPN and CRPN respectively. The proposal parameters of RPN set 2000 can get best effect, according to [15]. The proposal parameters of CRPN in this paper takes 1000. From Table 2, we can see that Mu-ResNet is better than VGG16 and ResNet when classificating airplane, human, automobile and truck. And it's worse than ResNet in classification of ship, bird and horse. The overall performance of CRPN were better than RPN. Through above experiments, considering detection time of each image and detection accuracy, the 57-layer Mu-ResNet combined with CRPN constructed by a 5-layer residual learning module can achieve 90.4 identification rate on the data set and detect each image only needs 0.093s, so this network model is more suitable for UAV target recognition.

Figure 6 shows the test results based on the Mu-ResNet-57 combined with CRPN. The model has good effects on the identification of targets and has good adaptability that it's not influenced by objects size.

Table 1. Comparison of different networks experiment results.

net	times(s)	acuracy
ZF	0.035	79.2%
VGG16	0.116	86.6%
ResNet-56	0.105	89.1%
Mu-ResNet-22	0.076	87.8%
Mu-ResNet-37	0.885	88.2%
Mu-ResNet-57	0.093	90.4%

Table 2. Results of different network models for each category of data set.

net	VGG16 +RPN	VGG16 +CRPN	ResNet-56 +RPN	ResNet-56 +CRPN	Mu-ResNet-57 +RPN	Mu-ResNet-57 +CRPN
mAP(%)	86.8	86.6	89.7	89.1	91.9	90.4
airplane	92.13%	92.36%	94.21%	91.895	93.03%	95.07%
automobile	90.27%	90.55%	93.20%	89.38%	90.06%	94.08%
bird	89.35%	90.10%	92.05%	89.14%	88.31%	89.30%
balloon	94.78%	96.48%	95.11%	93.75%	95.17%	96.14%
cat	86.33%	85.31%	86.31%	87.86%	84.93%	90.11%
dog	93.27%	93.17%	95.07%	93.15%	95.02%	96.22%
human	90.30%	93.09%	94.30%	91.81%	93.34%	94.87%
horse	87.92%	88.16%	91.82%	83.09%	86.79%	90.60%
ship	90.12%	91.07%	91.74%	88.92%	90.88%	93.17%
truck	82.06%	85.30%	86.26%	88.48%	90.12%	88.40%

IOP Conf. Series: Materials Science and Engineering **428** (2018) 012061 doi:10.1088/1757-899X/428/1/012061

Figure 6. Target identification effect diagram.

5. Conclusion

Aiming at the real-time and accuracy requirements of UAV target recognition, based on convolutional neural network, a variety of frameworks of deep residual network models are built for training. Through compared performance of the network, we ultimately find most suitable network for UAV target recognition. Compared with the traditional target recognition algorithm, target recognition algorithm based on depth residual network can avoid artificial error when target feature extraction and greatly improved recognition accuracy. For the situation of targets in complex background the deep residuals network's shortcut feedback mechanism also can effectively reduce the difficulty of network training and enable deeper networks can be effectively trained and used. The target detection using the deep residual network is applied to the UAV target recognition and the comprehensive performance of the accuracy and detection time is superior to the traditional deep convolutional neural network.

However, the target data sets studied in this paper are less. In future research, we need to acquire more and more drone targets data sets, build deeper networks to get better performance and use more efficient feature extraction methods to improve targets detection accuracy while lower detection time.

References

[1] Chang, L., Duarte, M. M., Sucar, L. E., and Morales, E. F. 2012. A bayesian approach for object classification based on clusters of sift local features. Expert Systems with Applications An International Journal, 39(2), 1679-1686.

[2] Lecun, Y., Boser, B., Denker, J. S., Henderson, D., Howard, R. E., and Hubbard, W., et al. 2014. Backpropagation applied to handwritten zip code recognition. Neural Computation, 1(4), 541-551.

[3] Krizhevsky, A., Sutskever, I., and Hinton, G. E. 2012. ImageNet classification with deep convolutional neural networks. International Conference on Neural Information Processing Systems (Vol.60, pp.1097-1105). Curran Associates Inc.

[4] Sermanet, P., Eigen, D., Zhang, X., Mathieu, M., Fergus, R., and Lecun, Y. 2013. Overfeat: integrated recognition, localization and detection using convolutional networks. Eprint Arxiv.

[5] Zeiler, M. D., & Fergus, R. 2014. Visualizing and Understanding Convolutional Networks. European Conference on Computer Vision (Vol.8689, pp.818-833). Springer, Cham.

[6] Srivastava, R. K., Greff, K., and Schmidhuber, J. 2015. Highway networks. Computer Science.

[7] Simonyan, K., and Zisserman, A. 2014. Very deep convolutional networks for large-scale image recognition. Computer Science.

[8] Szegedy, C., Liu, W., Jia, Y., Sermanet, P., Reed, S., and Anguelov, D., et al. 2014. Going deeper with convolutions. 1-9.

[9] Saxe, A. M., Mcclelland, J. L., and Ganguli, S. 2013. Exact solutions to the nonlinear dynamics of learning in deep linear neural networks. *Computer Science*.

[10] He, K., Zhang, X., Ren, S., and Sun, J. 2015. Delving deep into rectifiers: surpassing human-level performance on imagenet classification. 1026-1034.

[11] offe, S., Szegedy, C. 2015. Batch normalization: accelerating deep network training by reducing internal covariate shift. 448-456.

[12] Russakovsky, O., Deng, J., Su, H., Krause, J., Satheesh, S., and Ma, S., et al. 2014. Imagenet large scale visual recognition challenge. *International Journal of Computer Vision*, **115(3)**, 211-252.

[13] X. Glorot, Y. Bengio. Understanding the difficulty of training deep feedforward neural networks. In AISTATS, 2010.

[14] He, K., Zhang, X., Ren, S., and Sun, J. 2015. Deep residual learning for image recognition. 770-778.

[15] Ren, S., He, K., Girshick, R., and Sun, J. 2015. Faster R-CNN: towards real-time object detection with region proposal networks. *International Conference on Neural Information Processing Systems* (Vol.**39**, pp.91-99). MIT Press.

IOP Conf. Series: Materials Science and Engineering **428** (2018) 012062 doi:10.1088/1757-899X/428/1/012062

Research on The Control System of The Friendly Walking Assistant Robot for The Elderly

Xiaojuan Wei

School of Electrical Engineering, Northwest Minzu University , Lanzhou, 730124, China

E-mail: weixiaojuan925@126.com

Abstract. With the intensification of aging population, medical and social insurance system is facing up with the unprecedented pressure. Most of old people need nursing and care. On the one hand, it needs expensive expenditure and manpower. On the other hand, living quality of old people in degree of freedom is greatly reduced. The problem of helping old people to walk has become a great social problem. The self-adaption coordination control of robot is an important indicator of walking assistant robot for the elderly and also the precondition to support and protect old people and the disabled. In order to realize the control of direction for walking assistant robot and get the smaller turning radius and more flexible movement, this paper studied a dual-motor differential control method to realize forward, backward, and cornering of robot. Finally, the experimental results show that the control system of the friendly walking assistant robot for the elderly is rational and feasible, and the designed control system is stable and reliable.

1. Introduction

Similarly to all robots, walking assistant robot for the elderly completes operation under the control system of the robot. As the core constituent part of the robot, the control system performance fundamentally determines good or bad robot performance in today's society and also affects robot development. However, since assistant robot was developed, the control system applied by the assistant robot has had poor cross-country power, low speed of control system or large space, high power and finite processing data, so that it can't satisfy demands of assistant robot with high performance and high speed[1-5]. As a result, this thesis constructed the walking assistant robot for the elderly control system platform with abundant software and hardware resources and fast data processing speed. The core control system applies the fixed-point 32-bit DSP micro-processing chip TMS320F2812 with low power, multiple functions and high cost performance to realize motor control, walking intention decision and identification, so as to realize rapid response of the walking assistant robot for the elderly.

2. The overall system design of the walking assistant robot for the elderly

According to the motion control principle of the walking assistant robot for the elderly, the robot control system designed in this paper applies the combined control mode. It not only realizes the tactile-slip sense drive control of the walking assistant robot for the elderly, but also conducts steering control through joystick[6]. The walking assistant robot for the elderly control system is composed of three parts: one is the power driven and control module, which includes power-driven circuit,

IOP Conf. Series: Materials Science and Engineering **428** (2018) 012062 doi:10.1088/1757-899X/428/1/012062

DSP(TMS320F2812) and control system. The second one is the tactile-slip signal and signal processing circuit of joystick, which is composed of position, voltage, current and fault detection circuits. The third one is the power supply changeover. 24V DC accumulators are changed into the required voltage 15V, 9V, +5V, -5V and 3.3V through DC/DC transformation. The overall structure of the control system is shown in figure 1.

The control system is the core control circuit by using DSP(model TMS320F2812) produced by American Texas Instrument(TI) Company as the control chip, which mainly includes the signal processing circuit to regulate the motor current, tactile-slip sensor and smart action bars, so as to send it to AD conversion module of DSP. The motor rotor position acquisition circuit collects the rotor position signal of two motors and sends to the capture unit of DSP. By judging and confirming the rotor position and rotational speed of the current motor, the breakover sequence of power devices is confirmed. Based on the signals of tactile-slip sensor and action bars, it is transformed into PWM output through the treatment of the corresponding motor control algorithm. Finally, it passes through the drive circuit to drive two brushless DC motors, so as to drive operation of the robot and realize forward, backward, left turn and right turn. At the same time, A/D module of DSP collects the motor's voltage and current signals to realize double closed-loop control of voltage and current. The fault protection circuit constantly conducts fault detection and reports the corresponding fault signals to the PDPINTA/B port of the main control chip. Once there is a fault, PWM signal output should be immediately closed, so as to play a role of protecting motor and controlling system.

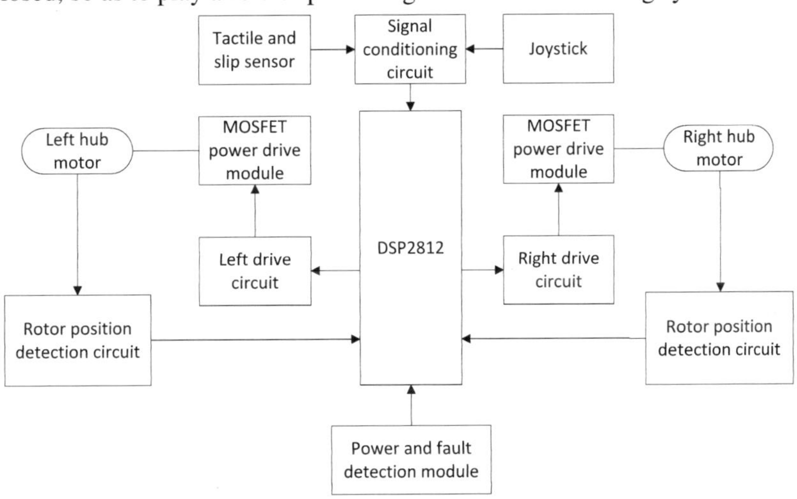

Figure 1. Overall structure diagram of the control system.

3. Control strategy of the walking assistant robot for the elderly

Good or bad control strategy is directly related to the performance of the entire control system. In order to realize stable rotational speed and no static error and have the favorable dynamic property, this control system applies the double closed-loop control of speed loop and current loop. In terms of the control algorithm, the control system requires for the strong generality and timeliness. Also, it has the fault-tolerant capacity and it should be simple and reliable as much as possible under the precondition of satisfying performance indicator[7]. The current PID control is the relatively mature control algorithm and it is advantaged with simple structure, stable performance and high reliability, thus it is widely used in various motion control systems. The target of the system control is the brushless DC motor. PID control can meet control requirements, thus the control system applies PID control algorithm. The double closed loop control of speed ring and current loop designed in this thesis is illustrated in figure 2.

The speed loop is the outer ring for control, which takes charge of making a comparison on the given reference speed and measured speed. The speed regulator is used to do PID regulation on the speed difference, thus the current reference value required by the current loop regulation is obtained.

Next, by comparing with the measured current value, the current regulator conducts PI regulation on the current difference to complete the double closed loop control. The current loop not only can ensure the maximal allowable current as launching the motor, accelerates the response speed of the system, and improves acceleration performance of the motor, but also limits the armature current value and plays a role of rapidly protecting the system as motor overloading or blocking. In this way, it can greatly improve the speed control performance of the system.

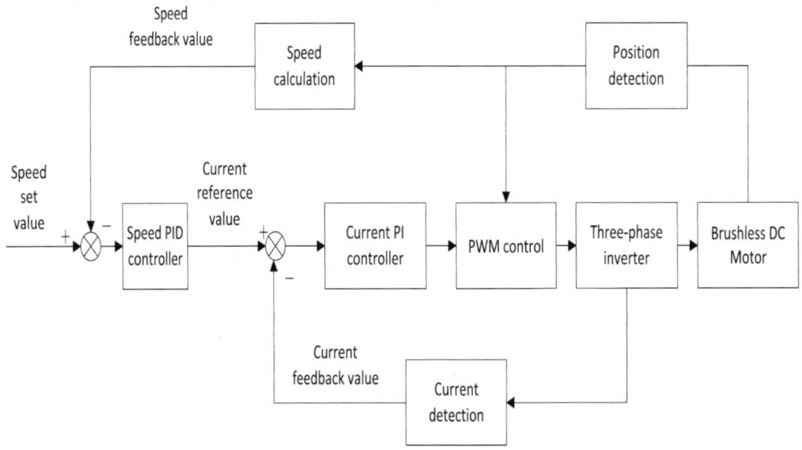

Figure 2. The Double-closed Speed Regulating System of the Brushless DC Motors.

4. Construction and algorithm realization of dual-motor differential control model

4.1. The construction of the dual-motor differential control model

Walking assistant robot for the elderly has the slow travel speed. The maximal speed is 10m/s. Moreover, travel road condition is relatively good, thus Ackerman-Jeantand turning model can be referred. Ackerman-Jeantand model is applied to calculate the speed of each drive wheel in the differential process. Therefore, the control system is based on the theory to realize the speed control for each drive wheel, so as to accurately achieve the purpose of differential control[8]. Considering that the front wheel of the walking assistant robot for the elderly is equipped with the universal wheel, the turning and speed are follow-up, the speed and direction of the walking assistant robot for the elderly can be controlled by controlling the back wheel speed of two back drive wheels. In this way, the dual-motor differential control model that is suitable for the walking assistant robot for the elderly can be constructed, as shown in figure 3. L is the car body length, B is the car body width, d is the half of the car body length; R is the turning radius; θ is the turning angle; Rin is the internal wheel turning radius; and Rout is the external wheel turning radius. Speed and direction of the assistant robot's centroid are used as the control targets to control the movement direction and speed of the assistant robot.

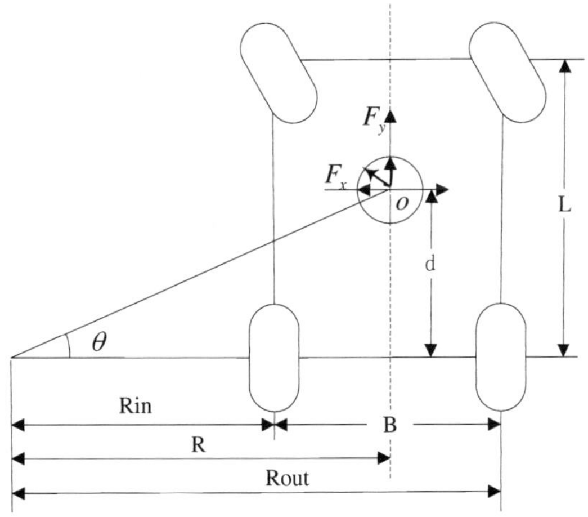

Figure 3. Dual-motor differential control model.

4.2. Realization of the dual-motor differential control algorithm

According to the signal output features, signals are respectively represented as x axis and y axis in the 2D coordinate system. x axis can be used as the set signal of the left and right turning speed for the walking assistant robot for the elderly, while y axis is used as the forward and backward speed set signals of the walking assistant robot for the elderly. As a result, for turning and forward, the operation speed of the walking assistant robot for the elderly is considered as the vector combination of x and y. In other words, the direction is used as the movement direction of the robot. The vector size is considered as the movement speed of the robot.

The output 2D electrical signal and 2D coordinates are reflected in the centroid of the robot, as shown in figure 3. According to the signals of x and y, vector compound is calculated. The amplitude size is considered as the rotational speed Vout of the outer wheel. The computational formula is shown as follows.

$$V_{out} = \begin{cases} \sqrt{F_x^2 + F_y^2} & V_{out} < V_{\max} \\ V_{\max} & V_{out} \geq V_{\max} \end{cases} \quad (1)$$

F_x, F_y —output 2D electrical signals

$V_{\max}$ —the robot's set speed maximum

It can be observed from the geometrical relationship in the figure.

$$R = \frac{d \times F_y}{F_x} \quad (2)$$

$$R_{in} = R - \frac{B}{2} \quad (3)$$

$$R_{out} = R + \frac{B}{2} \quad (4)$$

$$C_{in} = 2\pi R_{in} = 2\pi R - \pi B \quad (5)$$

$$K = \frac{\pi B}{\Delta T} \quad (6)$$

$$V_{in} = \frac{C_{in}}{\Delta T} = \frac{2\pi R}{\Delta T} - \frac{\pi B}{\Delta T} = V_C - K \tag{7}$$

$$V_{out} = \frac{C_{out}}{\Delta T} = \frac{2\pi R}{\Delta T} + \frac{\pi B}{\Delta T} = V_C + K \tag{8}$$

C_{in} and C_{out} is the perimeter(m) of internal and external wheel trajectory; V_{in} and V_{out} is the rotational linear velocity of internal and external wheels(m/s); ΔT is the time required by one week trajectory; V_c is the mean linear velocity of the back wheel(m/s).

The turning radius of the internal and external wheels Rint and Rout can be calculated by the above-mentioned formula.

$$\frac{V_{out}}{V_{in}} = \frac{R_{out}}{R_{in}} \tag{9}$$

5. Experimental verification and result analysis

5.1. The basic functional experiment of the control system

This part correspondingly tested the A-B phase voltage was tested, as shown in figure 4-(a). The A-phase current was detected, as illustrated in figure 4-(b). Through the experimental waveform, it can be observed the designed system can meet the design requirements, showing that the motor is operated well in the drive operation.

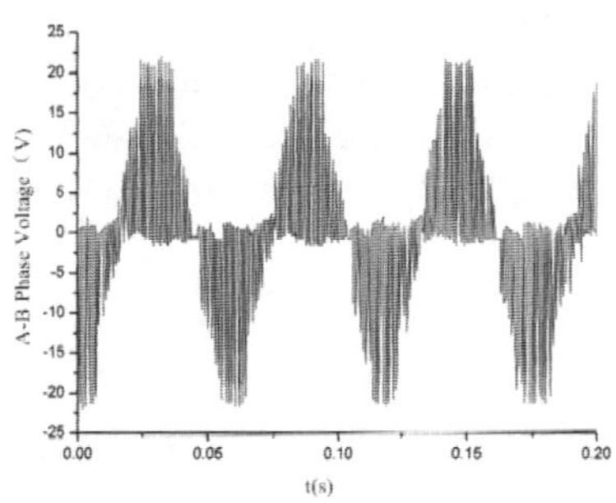

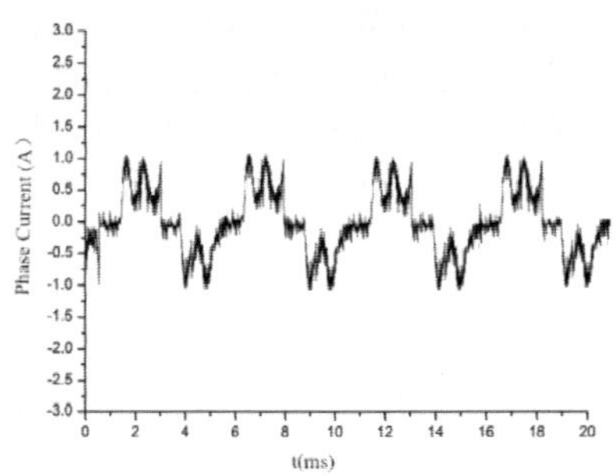

(a) (b)

Figure 4. The normal operation waveform of Brushless DC motors.

5.2. The testing experiment for the operation status of the control system

The testing experiment for the operation status of the control system aims to detect the electrical differential model in the above-mentioned theoretical study and whether the differential algorithm is feasible and correct. This is the key part in this study. To detect the operation status of the control system is to detect the differential performance of the controller. The detection method is shown as follows: the action bars are kept in certain position and detect 2D output voltage to confirm the operation direction of the robot and compare with the PWM wave duty cycle of two motor devices, as shown in figure 5. Next, the practical operation direction of the robot is confirmed. By making a comparison between the direction of the action bars and operational direction of the robot, the differential and control performance can be obtained. For example, when 2D signal of action bars is x=0.500, y=2.499, the ideal operation direction of the robot turns left in situ. The left and right motor's PWM modulating wave is shown in figure 5-(a). The practical operation direction of the robot also turns left around the left vehicle wheel in situ, conforming to the expectation. Other situation analysis is similar, as illustrated in table 1. Based on the experimental data and waveform analysis, it can be observed that the direction of action bars and practical operation direction of the robot are basically identical, reaching the expected design target.

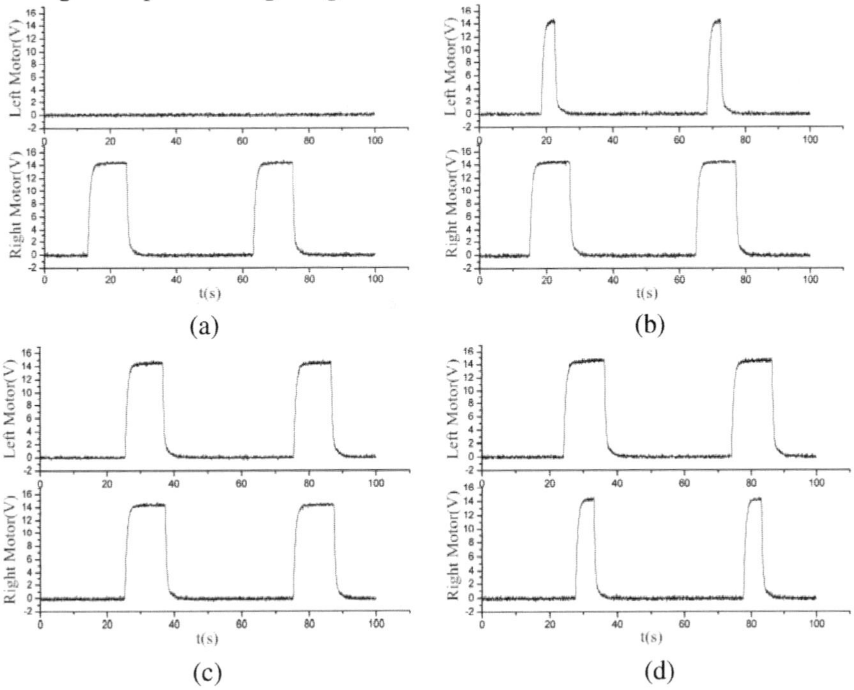

Figure 5. The differential speed waveform of dual motor drive.

Table 1. Experimental analysis data.

x	y	PWM	Direction
0.502	2.499	Figure 5-(a)	Turn left in situ
0.800	3.496	Figure 5-(b)	63 ° to the left
2.498	4.496	Figure 5-(c)	Straight forward
4.343	3.499	Figure 5-(d)	59 ° to the right

6. Conclusions

In this thesis, the walking assistant robot for the elderly control system platform with the abundant hardware and software resources and fast data processing speed was constructed to complete the overall design of the control system and construct the dual-motor differential model that is suitable for the robot, providing the differential realization algorithm. The control system can realize motor control and walking intention detection and identification based on tactile-slip sense, so as to greatly realize the rapid response for the assistant robot. The basic functional experiment of the control system and testing experiment of the operation status could prove that the control system designed is feasible, stable and reliable.

Acknowledgments

This work is supported by Gansu Province Science Foundation for Youths(17JR5RA281), the National Natural Science Foundation of P. R. China (61463046), and Fundamental Research Funds for the Central Universities of Northwest University for Nationalities of China(31920180039, 31920180120).

IOP Conf. Series: Materials Science and Engineering **428** (2018) 012062 doi:10.1088/1757-899X/428/1/012062

References

[1] Steven D, Frank G,et al 2000 *Proc. Int. Conf. on Robotics &Automation* pp 521-524
[2] Feng C, Yong Y and Yunjian G 2007 *Proc. Int. Conf. on Information Acquisition* pp 34-38
[3] Heedon L, Wansoo K, Jungsoo H, and Changsoo H 2012 *International Journal of Precision Engineering and Manufacturing* vol 8 pp 2-5
[4] Kong K, Jeon D 2006 *IEEE ASME Transactions on Mechatronics* vol 2 pp 53-55
[5] Dubowsky S, Genot F, Godding S, etc 2004 Assistive robotics and an ecology of elders living independently intheir homes vol 19 pp 25-59
[6] Xiaodong Z and Xiaojuan W 2011 Detecting System Design of Tactile Sensor for the Elderly-walking assistant & Walking-walking assistant Robot *Key Engineering Materials* vol 455 pp 37-39
[7] Xiaojuan W and Xiaodong Z, and Peng Y 2012 *ICMA* vol 8 pp 2083-2085
[8] Shim H, Hong J, Chung S, et al 2002 *IEEE International Symposium on Industrial Electronics* pp 1553-1556.

Model-Free Recurrent Reinforcement Learning for AUV Horizontal Control

Yujia Huo[1,2,a], Yiping Li[1,b] and Xisheng Feng[1]

[1]State Key Laboratory of Robotics, Shenyang Institute of Automation, Chinese Academy of Sciences, Shenyang, 110016, China
[2]University of Chinese Academy of Sciences, Beijing, 100049, China

Corresponding e-mail: [a]huoyujia@sia.cn; [b]lyp@sia.cn

Abstract. In this paper, aiming at the problems of 2-DOF horizontal motion control with high precision for autonomous underwater vehicle(AUV) trajectory tracking tasks, deep reinforcement learning controllers are applied to these conditions. These control problems are considered as a POMDP (Partially Observable Markov Decision Process). Model-free reinforcement learning(RL) algorithms for continuous control mission based on Deterministic Policy Gradient(DPG) allows robots learn from received delayed rewards when interacting with environments. Recurrent neural networks LSTM (Long Short-Term Memory) are involved into the reinforcement learning algorithm. Through this deep reinforcement learning algorithm, AUVs learn from sequences of dynamic information. The horizontal trajectory tracking tasks are described by LOS method and the motion control are idealized as a SISO model. Tanh-estimators are presented as data normalization. Moreover, AUV horizontal trajectory tracking and motion control simulation results demonstrate this algorithm gets better accuracy compared with the PID method and other non-recurrent methods. Efforts show the efficiency and effectiveness of the improved deep reinforcement learning algorithm.

1. Introduction

Benefiting from the continuous advances in several fields like control, computer, sensor and communication, autonomous underwater vehicles(AUVs) have play irreplaceable rolls in maritime application both in civilian and military areas including oceanographic survey, underwater operation and maritime reconnaissance and surveillance. The autonomy of AUVs has been one of the most critical criteria and been studied for decades by various control theories ranging from traditional techniques to several different artificial neural network-based control architectures.

Conventional approaches focus on controlling an AUV described by one or limited and very few accurate models with hydrodynamic parameters obtained from experiments and achieve variable results. Due to the inertial, buoyancy and hydrodynamic effects, dynamics of AUVs are with strong nonlinearity. AUVs suffer from the surrounding disturbance and the uncertainty when operating underwater. Linear approximations of robot dynamic model are insufficient and nonlinear models are destined for inaccuracy because some parameters are unknown or vary with un-modelled conditions.

With the rapid developments of both computing hardware and artificial intelligence in recent years, artificial agents with learning abilities achieve encouraging performance particularly in the condition of making plans and decisions for complex systems which difficult to establish a set of accurate models. Reinforcement learning(RL) allows agents learn the action when interacting with environment so as to maximize some notion of cumulative reward. In [1], reinforcement learning was applied to an AUV. A Q(λ)-learning algorithm for 3D-navigation control of an AUV URIS[2] and a semi on-line-neural Q-learning(SONQL) in [3] were proposed. For enhancing and accelerating learning in real robotics applications, a tow-step structure actor-critic(AC) reinforcement learning was designed for underwater robotic behaver learning. An open-frame AUV Ictineu was trained on simulation platform approximated to the environment and the experience of policy from simulation remained when agents continuing training in the real world[4]. Adaptive RL was integrated into the adaptive control algorithm and rigorous theoretical analysis is proposed to prove the stability[5]. In other similar motion control application of the UAV including multirotors and helicopters[6,7], RL was also involved.

To improve the performance of agents, deep neural networks with RL has made significant achievements in discrete state/action space, for example playing Atari with the method deep q-networks(DQN)[8]. A algorithm with deterministic policy gradient(DPG) instead of ineffective stochastic policy gradient was proposed in [9] and based on this algorithm, deep deterministic policy gradient(DDPG) was introduced in later work[10]. Application on AUV depth control[11] and heading control[12], DDPG was introduced and compared with PID controller[12], linear quadratic Gaussian integral controller[11] and nonlinear model predictive controller[11].

However, all of these works based on model-free reinforcement learning assume fully observed state[13]. Due to the sample frequency deference and noise of sensors, unobserved variations of systems under control, or state-aliasing caused by function approximation. Systems with these problems are described as Partially-Observable Markov Decision Processes(POMDP). In [14], a RL algorithm with a LSTM recurrent neural network was presented and solve non-Markovian tasks. In [15], recurrent networks replaced the DQN's first fully connected layer [8] and a better performance was got. Multiple LSTM layers approximating the Q-function controlled a real 7-axis articulated robot arm for high precision assembly tasks[16]. [13] presented an algorithm allows the agent to solve POMDP in continuous state/action space primarily considering the DPG and this algorithm gets an effective and satisfying result.

This paper aims to design a controller based on recurrent reinforcement learning for AUV trajectory tracking tasks on horizontal plane when these problems are considered as a POMDP.

The remainder of the paper is organized as follows. In section 2, the underwater dynamic system description is presented. Section 3 explains the composition of controller combining LSTM deterministic policy gradient algorithm. In section 4, simulation study is provided the effectiveness of the algorithm proposed, followed by conclusion in section 5.

2. Problem Formulation

AUVs operate underwater with six degrees of freedoms in three dimensional space but the system with coupled dynamics is difficult to control in practice works. To simplify the controller design, a motion task in 3D space is often decomposed in speed, steering and depth control.

In this paper, we consider the AUV moves and track a desire trajectory in the horizontal plane. Figure 1 illustrates the motion of the AUV in the horizontal plane.

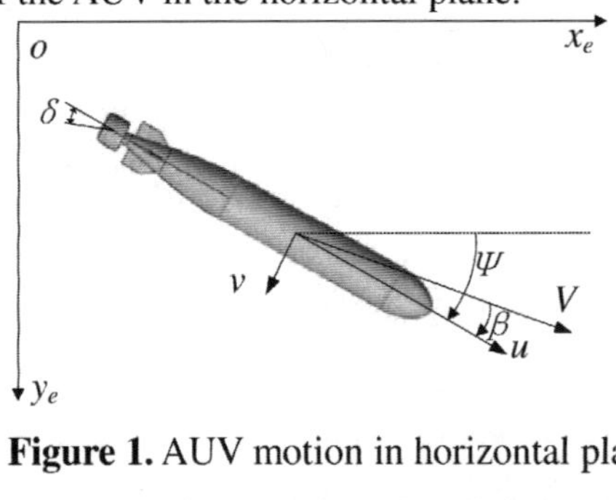

Figure 1. AUV motion in horizontal plane

In Figure 1, a positon coordinate frame $\{OX_eY_e\}$ is defined as the earth-fixed coordinate in the top view and accords to the right-hand system. V is the velocity of the AUV and two decoupled velocities u in surge and v in sway. Heading angel Ψ is defined as the angle between the axis X_e and the nose pointing of AUV. Drift angle is defined as the angle between the velocity V and the velocity u in surge. δ is the angel of the vertical rudder which controls the heading of the robot.

According to [17], an AUV horizontal plane dynamics can be described as follows:

$$\begin{bmatrix} m-Y_{\dot{v}_r} & -Y_{\dot{r}} & 0 \\ -N_{\dot{v}_r} & I_{zz}-N_{\dot{r}} & 0 \\ 0 & 0 & 1 \end{bmatrix}\begin{bmatrix} \dot{v} \\ \dot{r} \\ \dot{\psi} \end{bmatrix} = \begin{bmatrix} Y_{uv}u & (Y_r-m)u & 0 \\ N_{uv}u & N_r u & 0 \\ 0 & 1 & 0 \end{bmatrix}\begin{bmatrix} v \\ r \\ \psi \end{bmatrix} + \begin{bmatrix} Y_\delta u^2 \\ N_\delta u^2 \\ 0 \end{bmatrix}\delta + \begin{bmatrix} Y_{r|r|}r|r| + Y_{v|v|}v|v| \\ N_{r|r|}r|r| + N_{|v|r}|v|r + N_{|v|v}|v|v \\ 0 \end{bmatrix} \tag{1}$$

Where we consider the center of gravity $[x_G, y_G, z_G]^T$ stays at the body-fixed vehicle coordinate system origin and same as the geometric center of the AUV. Then we find it convenient to assume that the sway v is far less than the surge u and a simplified discrete-time dynamic model for heading control in the horizontal plane can be expressed as follows:

$$\begin{bmatrix} r(k+1) \\ \psi(k+1) \end{bmatrix} = T_s M^{-1}\left[\left(\begin{bmatrix} N_r u & 0 \\ 1 & 0 \end{bmatrix} + \frac{M}{T_s}\right)\begin{bmatrix} r(k) \\ \psi(k) \end{bmatrix} + \begin{bmatrix} N_\delta u^2 \\ 0 \end{bmatrix}\delta + \begin{bmatrix} N_{r|r|}r|r| + N_{|v|r}|v|r + N_{|v|v}|v|v \\ 0 \end{bmatrix}\right] \tag{2}$$

Where M is the is

$$\begin{bmatrix} I_{zz}-N_{\dot{r}} & 0 \\ 0 & 1 \end{bmatrix} \tag{3}$$

And T_s is the sampling time. Thus, a dynamic model function which describe a map from action rudder angle δ to states yaw r and heading angle ψ.

In trajectory tracking tasks, line-of-sight method is employed to detect a target position along the desire trajectory. When operating underwater, the heading angle ψ_{AUV}, the target position (x_{target}, y_{target}) and the actual position (x_{AUV}, y_{AUV}) are known. And

$$x_t = x_{AUV} - x_{target};\quad y_t = y_{AUV} - y_{target} \tag{4}$$

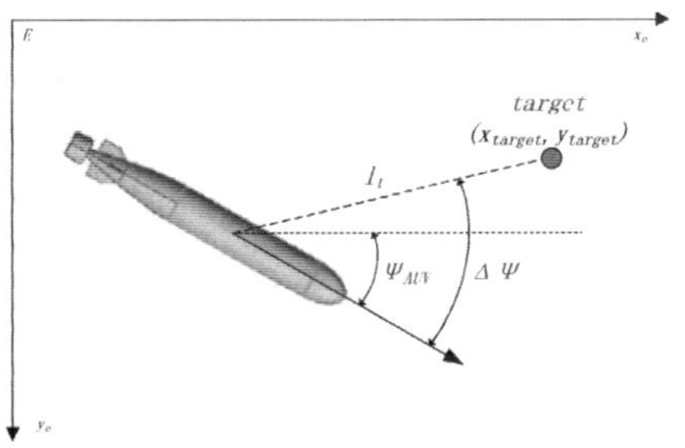

Figure 2. Line of sight(LOS) guidance approach

Figure 2 illustrates the line-of-sight guidance approach. Where $\Delta\Psi$ means the angle between AUV heading angle and the angle towards the target position. Now we have

$$l_t = \sqrt{x_t^2 + y_t^2} = \sqrt{(x_{AUV} - x_{target})^2 + (y_{AUV} - y_{target})^2};\quad \Delta\psi = \psi_{AUV} - \psi_{target} = \psi_{AUV} - \arctan\left(\frac{y_{AUV} - y_{target}}{x_{AUV} - x_{target}}\right) \tag{5}$$

Thus, the target position can be describe as $(l_t, \Delta\psi)$ in a polar coordinate.

As the problem of control in a 3D space has been separated into a steering control, diving control and speed control. For the control problem in the horizontal plane, we only consider the relationship between the heading angle rate r, its accumulation form - heading angle Ψ and the action rudder angle δ. The velocity in surge u is often set as a constant. The desire trajectory is composed of a sequence of target spots $(x^1_{target}, y^1_{target}, x^2_{target}, y^2_{target}, \cdots)$ and these target spots are equidistant along with the trajectory.

3. LSTM Deterministic Policy Gradient

As mentioned in introduction, reinforcement learning has produced several achievements but all these works assume fully observe state. In this section, a deep reinforcement learning algorithm with current neural networks is introduced. Since the algorithm has an actor-critic structure based on DPG[18] with an actor neural network and a critic neural network synonyms for the policy and value function. An off-policy deterministic actor-critic method is also presented in this section.

3.1 Partially Observable Markov Decision Process

A Markov decision process(MDP) is a stochastic process which satisfies the Markov property. When interacting with the environment at each of a sequence of discrete time steps, $t = 0,1,2,......$, the agent receives representation of the environmental state, $s_t \in \mathcal{S}$, where $\mathcal{S}$ is the set of possible states, and on that basis chooses an action, $a_t \in \mathcal{A}(S_t)$,where $\mathcal{A}(s_t)$ is the set of actions available in the state s_t [19]. At each step after action, consequently, the agent receives a numerical reward, $r_{t+1} \in \mathcal{R} \subset \mathbb{R}$, and transfer itself in new state s_{t+1}. Thus a MDP is described $(s_t, a_t, r_t, p(s_{t+1}=s'|s_t=s, a_t=a))$, where $p(s_{t+1}=s'|s_t=s, a_t=a)$ is transition probability that action a in state s at time t will lead to state s' at time $t+1$.

But in the real world, the full states of the system being needed are rarely provided to the agent or even perceived by sensors. In other words, it is rare that the Markov property holds well in the real world. A Partially Observable Markov Decision Process(POMDP) captures the environmental dynamics of the real world more effectively by explicitly acknowledging that the sensations that the agent receives are only partial glimpses of the underlying system state[15]. Then, a POMDP is described as (s, a, p, r, Ω, O) with 6 components, and the s, a, p, r are described as states, actions, transitions and rewards as before. But the agent no longer receives the states directly from the system, instead an observation $o \in \Omega$ is received. And this observation is based on the probability distribution $o \sim O(s)$.

3.2 Off-Policy Deterministic Actor-Critic

For a MDP, the performance can be described and evaluated as an expectation with the conditional probability density at $\pi_\theta(s_t, a_t)$ as following:

$$J(\pi_\theta) = \int_S \rho^\pi(s) \int_A \pi_\theta(s,a) r(s,a) \mathrm{d}a\mathrm{d}s = \mathbb{E}_{s\sim\rho^\pi, a\sim\pi_\theta}[r(s,a)] \tag{6}$$

where we denote the discounted state distribution by

$$\rho^\pi(s') = \int_S \sum_{t=1}^{\infty} \gamma^{t-1} p_1(s) p(s \to s', t, \pi) \mathrm{d}s \tag{7}$$

For this performance function(6), the basic goal is to maximize the performance function with the optimal policy. And the policy gradient is widely utilized for a continuous state/action tasks. By updating the parameters θ with the performance gradient(8), the algorithm finds the optima parameterized with θ, and at the same time, maximizing the time cumulative reward function.

$$\nabla_\theta J(\pi_\theta) = \int_S \rho^\pi(s) \int_A \nabla_\theta \pi_\theta(a|s) Q^\pi(s,a) \mathrm{d}a\mathrm{d}s = \mathbb{E}_{s\sim\rho^\pi, a\sim\pi_\theta}[\nabla_\theta \log \pi_\theta(a|s) Q^\pi(s,a)] \tag{8}$$

$$\theta \leftarrow \theta + \alpha \widehat{\nabla_\theta J(\theta)} \tag{9}$$

where $\alpha \widehat{\nabla_\theta J(\theta)}$ is a stochastic approximation of the true gradient of the performance function and $Q^\pi(s,a)$ is the action-value function but not easy to be estimated.

As an extended method based on the policy gradient, the actor-critic has a better performance. The actor-critic contains two eponymous components. The *actor* adjusts the stochastic policy $\pi_\theta(s)$ with parameters θ by stochastic gradient of the performance function(8). Meanwhile, as the action-value function $Q^\pi(s,a)$ hard to estimate, an approximate policy evaluation algorithm such as a temporal-difference(TD) method is introduced to *critic* and estimate the action-value function $Q^w(s,a) \approx Q^\pi(s,a)$ with parameters ω.

To improve the actor-critic algorithm, a deterministic policy replaces the stochastic policy, because the gradients of deterministic policy can be estimated more efficiently without a problematic integral over the action space [18].And it has been proved that the deterministic policy gradient is a limiting case of the stochastic policy gradient theorem.

Unlike the stochastic policy as mentioned before, the actor updates the parameter θ of the deterministic policy $\mu_\theta(s)$. Since the we choose a deterministic policy, an off-policy algorithm is needed to allow the agent to explore and consequently to learn the evaluation. Then, the updating for an off-policy deterministic actor-critic with an off-policy algorithm Q-learning in *critic* and a deterministic policy gradient in *actor* can be described as following:

$$\delta_t = r_t + \gamma Q^w(s_{t+1}, \mu_\theta(s_{t+1})) - Q^w(s_t, a_t) \tag{10}$$

$$w_{t+1} = w_t + \alpha_{critic}\delta_t \nabla_w Q^w(s_t, a_t) \tag{11}$$

$$\theta_{t+1} = \theta_t + \alpha_{actor}\nabla_\theta \mu_\theta(s_t)\nabla_a Q^w(s_t, a_t)\Big|_{a=\mu_\theta(s)} \tag{12}$$

Where, δ_t is the TD error of Q-learning used for critic.

3.3 LSTM Deterministic Policy Gradient

As mentioned above, the states cannot be received by the agent. Instead, the agent only indirectly observes the Markov decision process through the observations. Then there involves a history $h_t = (o_1, a_1, o_2, a_2, \cdots, o_{t-1}, a_{t-1}, o_t)$. Same as the goal of off-policy deterministic actor-critic algorithm, we tend to maximize the deterministic policy performance function(13) re-described for POMDP.

$$J = \mathbb{E}_\tau[\sum_{t=1}^{\infty} \gamma^{t-1} r(s_t, \mu(h_t))] \tag{13}$$

In the case of partial observability, the optimal policy and the estimated action-value function are both considering the observation-action h_t with states as a sequence of history, instead of just states of one step for fully observable MDP. To solve the POMDP, feedforward networks policy and evaluation in DPG are replaced with neural networks(LSTM) in practice, which allow the agent to learn from the information preserved from past. Thus, a new policy gradient is obtain as following by replacing $\mu(s)$ and $Q(s, a)$ with (h) and $Q(h, a)$.

$$\nabla_\theta J = \mathbb{E}_\tau[\sum_t \gamma^{t-1} \frac{\partial Q^\mu(h_t, a)}{\partial a}\Big|_{a=\mu^\theta(h_t)} \frac{\partial \mu^\theta(h_t)}{\partial \theta}] \tag{14}$$

where trajectory $\tau = (s_1, o_1, a_1, s_2, o_2, a_2 \cdots)$.

Based on ideas from the DQN[8] method, a relay buffer is also introduced which can improve the data efficiency and stability. Meanwhile, target networks are appended to the algorithm. Two copies of the evaluation function Q and the policy are involved, with parameters *w'* and θ'. *w'* and θ' are updated as mirror of *w* and θ with some delay. This Asynchronous update also enhances the stability of the neural networks.

LSTMDPG algorithm:

Initialize critic network $Q^w(a_t, h_t)$ and actor $\mu^\theta(h_t)$ with parameters w and θ

Initialize target network $Q^{w'}(a_t, h_t)$ and $\mu^{\theta'}(h_t)$ with weight $w' \leftarrow w$ and $\theta' \leftarrow \theta$

Initialize replay buffer R

for episode i=1 to max episode M **do**:

  Initialize empty history memory h_0

  for j=1 to T **do**:

    Observe o_t

    Append observation and action to previous history memory $h_t \leftarrow h_{t-1}, a_{t-1}, o_t$

    Choose action $a_t = \mu(h_t|\theta)$

  end for

  Store the sequence $(o_1, a_1, r_1, \cdots, o_T, a_T, r_T)$ to R

  Sample a minibatch $(o_1^n, a_1^n, r_1^n, \cdots, o_T^n, a_T^n, r_T^n)_{n=1-N}$ with shape N episodes from R

  Update $h_t^n = (o_1^n, a_1^n, \cdots, o_{t-1}^n, a_{t-1}^n, o_t^n)$

Compute target values $(\delta_t^n, \cdots, \delta_T^n)$ for each sample episode by LSTM

$$\delta_t^n = r_t^n + \gamma Q^{w'}(h_{t+1}^n, \mu^{\theta'}(h_{t+1}^n)) - Q^w(h_t^n, a_t^n)$$

Calculate critic network and actor network update with BPTT[20]

$$\Delta w = \frac{1}{NT} \sum_n \sum_t \delta_t^n \frac{\partial Q^w(h_t^n, a_t^n)}{\partial w}$$

$$\Delta \theta = \frac{1}{NT} \sum_n \sum_t \delta_t^n \frac{\partial Q^w(h_t^n, \mu^\theta(h_t^n))}{\partial a} \frac{\partial \mu^\theta(h_t^n)}{\partial \theta}$$

Update actor and critic with Adam[21]

Update the target networks with the learning rate α_{critic} and α_{actor}

$$w' \leftarrow \alpha_{critic} w + (1 - \alpha_{critic}) w'$$

$$\theta' \leftarrow \alpha_{actor} \theta + (1 - \alpha_{actor}) \theta'$$

end for

4. Simulation

In this section, results of simulation for the AUV trajectory tracking tasks are presented. Primary parameters of AUV in simulation are listed. The simulation platform is coded in Python3.5 on Ubuntu16.04 system and TensorFlow is in use for deep neural networks implementation.

In comparison with the LSTM deterministic policy gradient(LSTMDPG), a PID controller and a simple deep deterministic policy gradient(DDPG)[10] controller are designed for simulation of AUV trajectory tasks.

In LSTMDPG algorithm, both the actor network and critic network are LSTM recurrent neural networks with 4 layers and the number of hidden layer units is 200. In the practice, we set the discount factor γ as 0.99, learning rate α as 0.001 and the batch size as 64.

In DDPG algorithm, as the problem we aim to solve is in a continuous state/action space, it is enough that both the actor network and the critic network are fully connected. The numbers of the units in the hidden layers of both actor and critic networks are 64. And discount factor γ, learning rate α, the batch size and the activation function of the neural networks are each set as 0.99, 0.001, 64 and Relu.

It is appropriate to define the input as an error form like (7). Both in the LSTMDPG and DDPG, the action δ and state $\Delta\psi$, l and r are normalized by the tanh-estimators which are robust and highly efficient. The normalization is given by:

$$x_k' = \tanh\left(\frac{x_k - \mu}{\sigma}\right) \tag{15}$$

where, μ and σ are the mean and standard deviation estimates. Standard deviations of states $\Delta\psi \in$ (-180°,180°], r and the action $\delta \in$ [-45°,45°] that transferred into neural networks for training are each set as $\sigma_{\Delta\psi} = 80; \sigma_r = 4; \sigma_\delta = 20, \sigma_l = 10$, and the normalizations are shown in figure 3.

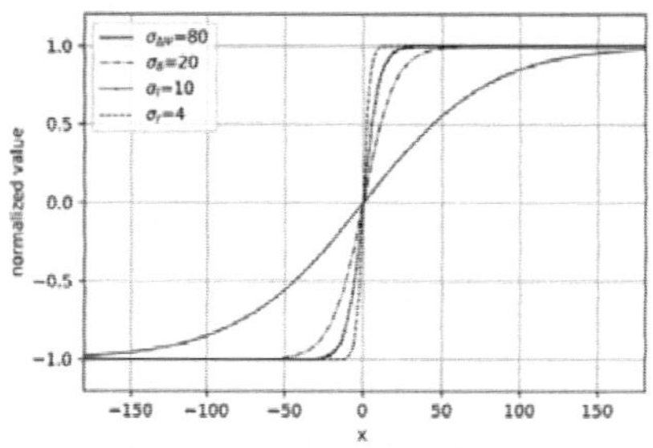

Figure 3. The normalization of states and action

Then for reinforcement learning of the trajectory tracking tasks on the horizontal plane, the reward that agent receives at each step is set as the quadratic sum of the states and actions with different weights:

$$reward(k) = -[\Delta\psi(k)^2 + r(k)^2 + l(k)^2 + 0.1 * \delta(k)^2] \tag{16}$$

And during training the AUV, the states of the agent are reset as (*Δψ*=45°+*dis*, *l* =*10*, *r=0*) at the beginning of each episode, where *dis* is a random disturbance.

For another comparison, a PID controller is also designed and the controller can be described as following and the factors of PID are set as (0.6,1.5,0.08):

$$\delta(t) = K_p \Delta\psi(t) + K_d \frac{\mathrm{d}\Delta\psi(t)}{\mathrm{d}t} + K_I \int_0^t \Delta\psi(\tau)\mathrm{d}\tau \tag{17}$$

Random disturbances appended to the observation are described by a Gaussian distribution *N(0,0.05)*.

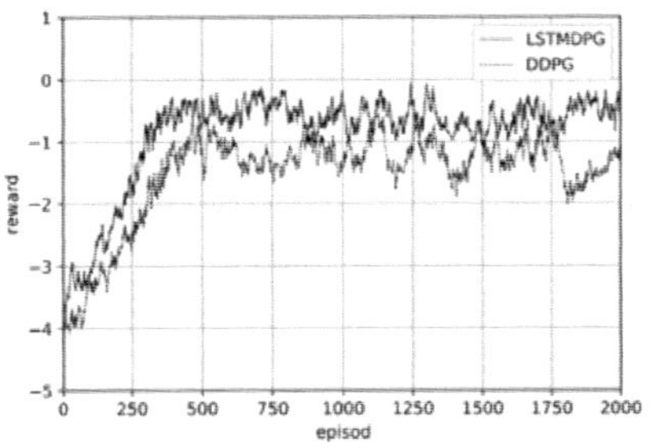

Figure 4. Rewards of LSTMMPG and DDPG.

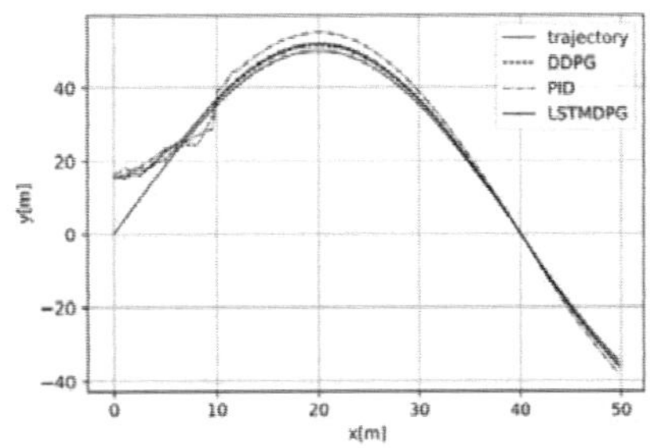

Figure 5. Trajectory tracking with different algorithm.

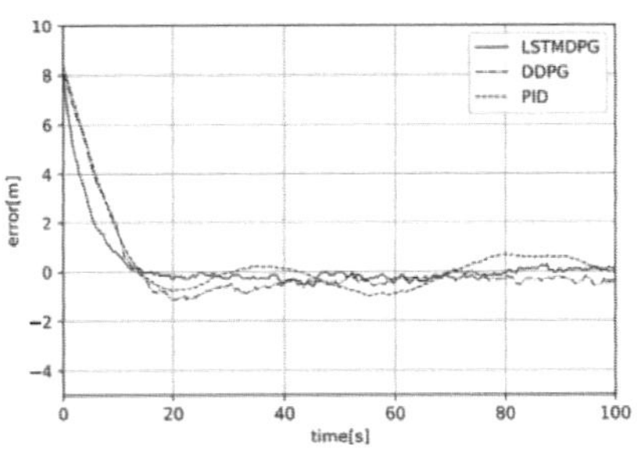

Figure 6. Error of trajectory tracking with different algorithm.

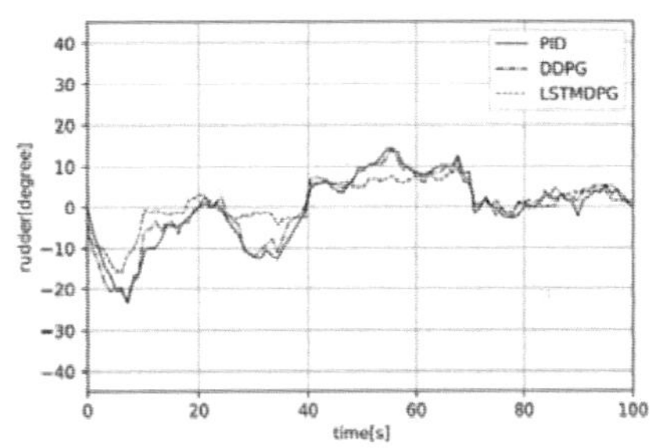

Figure 7. Rudder action of different algorithm.

We set a maximum of 2000 episodes to train both DDPG and LSTMDPG, and the rewards in figure 4 illustrate LSTMDPG has accelerated convergence speed. Furthermore, the LSTMDPG has a better performance with bigger rewards. Figure 5 and figure 6 show the trajectory and the error on the horizontal plane when agents are operating in the tracking task. It can be seen that LSTMDPG has a fast convergence with little steady-state error when compared with a DDPG controller and a PID controller. And figure 7 is shows the rudder action of the AUV.

Simulations illustrate that a LSTMDPG controller can solve the trajectory tracking tasks with a satisfying performance which is assumed as a POMDP

5. Conclusion

In this paper, a trajectory tracking task assumed as a partially observable Markov decision process is solved by a recurrent reinforcement learning. Extended from deterministic policy gradient, LSTM recurrent neural networks are involved and replace the feedforward networks both in actor and critic. This method with recurrent neural networks allow agent to learn effectively, and is validated in simulation compared with a PID controller and DDPG controller.

IOP Conf. Series: Materials Science and Engineering **428** (2018) 012063 doi:10.1088/1757-899X/428/1/012063

References

[1] Gaskett C 1999 Reinforcement Learning applied to the control of an Autonomous Underwater Vehicle *Proc. Aust. Conf. Robot. Autom.*

[2] Carreras M, Batlle J and Ridao P 2001 Hybrid coordination of reinforcement learning-based behaviors for AUV control *Proc. 2001 IEEE/RSJ Int. Conf. Intell. Robot. Syst. Expand. Soc. Role Robot. Next Millenn. (Cat. No.01CH37180)* **3** 1410–5

[3] Carreras M, Yuh J, Batlle J and Ridao P 2005 A Behavior-Based Scheme Using Reinforcement Learning for Autonomous Underwater Vehicles *IEEE J. Ocean. Eng.* **30**

[4] El-Fakdi A and Carreras M 2013 Two-step gradient-based reinforcement learning for underwater robotics behavior learning *Rob. Auton. Syst.* **61** 271–82

[5] Cui R, Yang C, Member S S S, Li Y, Member S S S, Sharma S, Member S S S, Li Y and Member S S S 2017 Adaptive Neural Network Control of AUVs With Control Input Nonlinearities Using Reinforcement Learning *Ieee Trans. Syst. Man, Cybern. Syst.* 1–11

[6] Bou-Ammar H, Voos H and Ertel W 2010 Controller design for quadrotor UAVs using reinforcement learning *Proceedings of the IEEE International Conference on Control Applications* pp 2130–5

[7] Ng A Y, Coates A, Diel M, Ganapathi V, Schulte J, Tse B, Berger E and Liang E 2006 Autonomous inverted helicopter flight via reinforcement earning *Springer Tracts Adv. Robot.* **21** 363–72

[8] Mnih V, Silver D and Riedmiller M Playing Atari with Deep Reinforcement Learning *NIPS* 1–9

[9] Conditions A R 2014 Deterministic Policy Gradient Algorithms : Supplementary Material *Icml* **1**

[10] Lillicrap T P, Hunt J J, Pritzel A, Heess N, Erez T, Tassa Y, Silver D and Wierstra D Continuous control with deep reinforcement learning *arXiv Prepr. arXiv1509.02971, 2015.*

[11] Wu H, Song S, You K and Wu C 2018 Depth Control of Model-Free AUVs via Reinforcement Learning *IEEE Trans. Syst. Man, Cybern. Syst.* 1–12

[12] Yu R, Shi Z, Huang C, Li T and Ma Q 2017 Deep Reinforcement Learning Based Optimal Trajectory Tracking Control of Autonomous Underwater Vehicle *36th Chinese Control Conf.* 4958–65

[13] Heess N, Hunt J J, Lillicrap T P and Silver D 2015 Memory-based control with recurrent neural networks *NIPS Deep Reinf. Learn. Work.* 1–11

[14] Bakker B 2003 Reinforcement learning with long short-term memory *In Advances in Neural Information Processing Systems 15 (NIPS-2002* pp 1475–82

[15] Hausknecht M and Stone P 2015 Deep Recurrent Q-Learning for Partially Observable MDPs 29–37

[16] Inoue T, De Magistris G, Munawar A, Yokoya T and Tachibana R 2017 Deep Reinforcement Learning for High Precision Assembly Tasks

[17] Feldman J 1979 DTNSRDC Revised Standarrd Submarine Equations of Motion 12

[18] Silver D, Lever G, Heess N, Degris T, Wierstra D and Riedmiller M 2014 Deterministic Policy Gradient Algorithms *Proc. 31st Int. Conf. Mach. Learn.* 387–95

[19] Sutton R S and Barto A G 1998 *Reinforcement learning: an introduction.* vol 9

[20] Werbos P J 1990 Backpropagation Through Time: What It Does and How to Do It *Proc. IEEE* **78** 1550–60

[21] Kingma D P and Ba J Adam: A Method for Stochastic Optimization *arXiv Prepr. arXiv1412.6980, 2014.*

IOP Conf. Series: Materials Science and Engineering **428** (2018) 012064 doi:10.1088/1757-899X/428/1/012064

A Design of Magic Cube Robot Based on STM32

Guoyou Zhao[1], Haiyan Tu[1] and Yicheng Liu[*, 1]

[1] College of Electrical Engineering and Information Technology, Sichuan University, Chengdu 610065, China;

[*]Corresponding author: E-mail: liuyicheng@scu.edu.cn

Abstract: This paper proposed a new project plan to build a magic cube robot. Under the precondition of ensuring to convenient to move and carry, we use external HD camera to collect magic cube's surface color information and then sent it to a PC equipped with MATLAB to handle. So that we can get the optimal path to restore magic cube. As for image processing, we chose K-means to gain the magic cube's original statue, in terms of path planning, we selected dimensionality reduction method, in general, the number of steps were about 20. According to the final result, the time of restoring process will be about 1 minute and which proved the project plan to be viable.

1. Introduction

Robot is an intelligent product with high technology content and energy. It arouses people's great interest and enthusiasm for research. Magic cube robot is such a mechanical design which combines system control and machine vision. The basic calculation of the solution of the magic cube is given in literature [1], but there were too many steps of the algorithm which will make the process of turning the manipulator to be long. According to literature [2], the design of the manipulator is low cost and small, but the magic cube turns slowly, and the stability is not strong. This paper proposes a design of a low grade, small size, stable performance and strong popularity which can Restore the magic cube fast. According to [3], based on layer-by layer algorithm and OpenGL 3D technology, a demo system is realized, which application platform the users to resolve the magic cube problem cube problem manully. And [4] embedded image processing system based on ARM integrates all the functions of image capturing, processing and dis-playing takes on the character of small size and low power consumption. The system designed in this paper is based on the processor of S3C2410 with the ARM9 core, which has the operation system modified from the ARM-Linux and the software based of the V4L library. In [5] this paper presents a novel reconfigurable modular robots(changeful magic cube robots) that also shares characteristics with self-reconfigurable and self-assembly and swarm robots. each module of changeful magic cube robots can be free to move and can be self-assembled to form a magic cube structure with the other modules. According to [6] aiming at the problem of Rubik's cube recovered by dual-arm robot, this paper designed a system of novel fast solving Rubik's cube. The system consists of image acquisition, image processing and mechanical arm control. First of all, Droid Cam which is an android mobile phone software is used to acquire Rubik's cube image, then the image information is transmitted to PC. Secondly, the library function provided by LabVIEW software is utilized to analyze and recognize the image, then the result is sent to the STM32 processor. In paper [7], Aim at the process detecting system in nowadays of real-time need, designing a low cost and moderate function image processing system. The system use ARM7(S3C44B0X) as core with circuit realize image processing function, multi-communication interface constitute image transmission channels, SD card interface acquire image data. For [8], this paper presents a novel reconfigurable modular robots(Changeful Magic Cube Robots) that also shares characteristics with self-reconfigurable and self-assembly and swarm robots based on the Infineon XC886 microcontroller. Each module of Changeful Magic Cube Robots can be free to move and can be self-assembled to form a magic cube's structure with the other modules. According to [9], The architecture and simulation technology of RACS are also discussed. OpenGL and Visual C++ based modelling method has been used to establish the platform after considering the function of RACS. And for [10], a control system based on a 2-DOF manipulator and the image obtained through a camera, the coordinates of the manipulator's end executor are resolved by analyzing the image of motion area. This coordinates are subtracted with the anticipant coordinates to get an error as a compensation of the manipulator control system, then the full close-loop control of the system is realized and the problem of low control precision brought by half close-loop control is resolved.

2. General scheme design of the magic cube robot

The one based STM32 with two self is such a robot which can let any a three order magic cube to be reductive. It involves the visual tracking, the data collection, magic cube decoupling algorithm, manipulator control system. The feature of the robot is that it is fully controlled by STM32, small in size, simple in structure and cost low. Figure 1 is the robot control system, including control system, vision system, manipulator system.

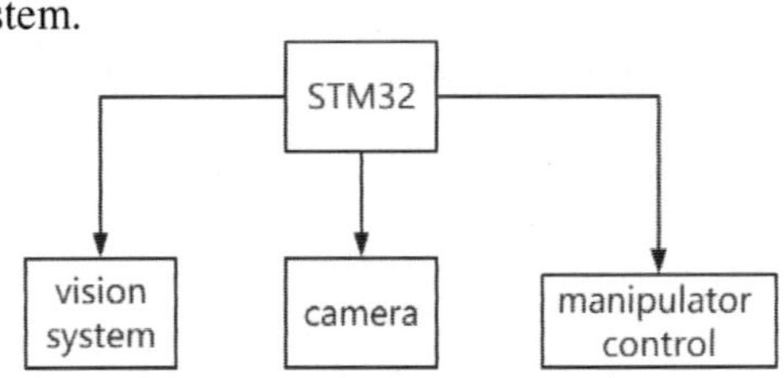

Figure 1. robot control system

3. Hardware Design

It is very important for the magic cube robot to design hardware, which would decide whether the robot run fluently.

3.1. Mechanical structure design

Mechanical structure is the most basic part of the magic cube robot. The main task of mechanical structure is to achieve the magic cube clip, holding, single side rotation and the whole rotation of it. In this paper, the design of magic cube robot with two degrees of freedom mechanism, which can not only cost less but also let the structure to be most brief. As the following Figure 2, the two arms are parallel to each other and have arms in the same plane. Such structure of the machine claw the magic cube up and down to reduce the influence of gravity on the process of solving the magic cube.

Figure 2. mechanical structure of magic cube robot

3.2. Electrical system design

The electric control system of the magic cube robot mainly includes control system system, communication system. Figure3 gives a detailed description of the controller connection which include a camera, steering gear, and serial communication, mechanical arm and so on.

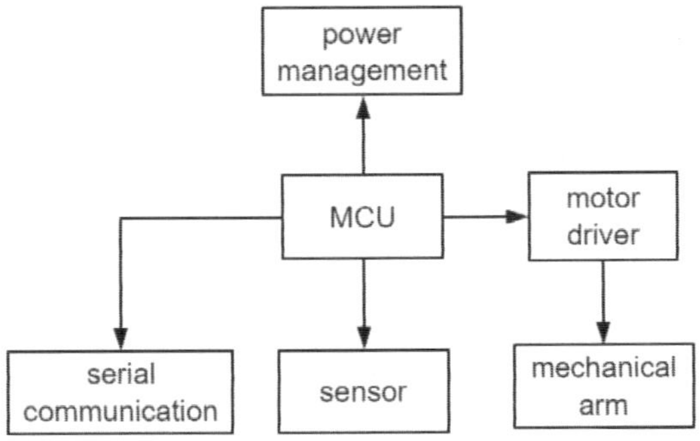

Figure 3. control system function

(1)control system: Choosing STM32F407ZG with main frequency 168Mhz and enough pins to meet the needs , which was made by ST semiconductor company.

(2)sensor: We mainly use camera to collect the color information, the process requires higher anti-interference capability for data transmission so that the capacitor filter circuit is designed in this circuit. By the design data transmission and the power supply lines do not interfere with each other, which makes the collected RGB data to be reliable.

(3)communication system: Communication system adopts a serial communication, which is used to do communications between the STM32 main control center and motor controller so that we can sure that the operator group we made advanced can be operated fluently.

4. Algorithm design

The software design part of magic cube robot mainly includes vision subsystem design, manipulator action group controscdx l and path optimization algorithm.

4.1. Vision processing algorithm

In this design we use a computer outside the camera to get the color information of the floor of the magic cube. Once get it, those information will be sent to PC to analysis, then the results will be be sent to control center so that STM32 will know how to operate the mechanical arm. The diagram is shown in Figure 4.

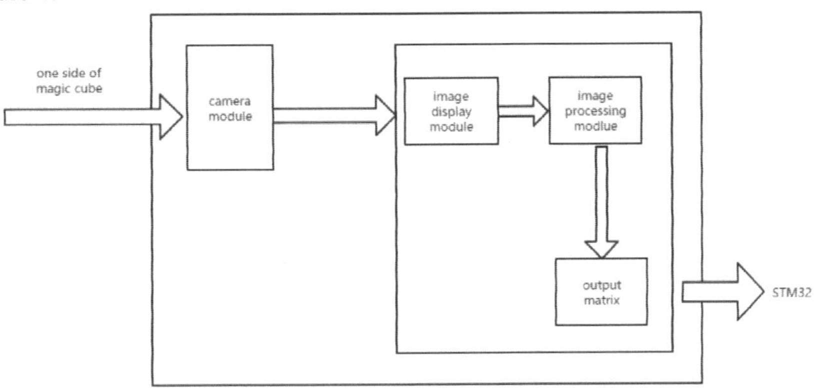

Figure 4. the diagram of processing image

As the result of the fact that the common type cameras we can buy on the market(such as OV7670, OV2460) are poor with pixel of the equivalent and too sensitive with light, in order to finish

IOP Conf. Series: Materials Science and Engineering **428** (2018) 012064 doi:10.1088/1757-899X/428/1/012064

restoration in shortest time and reduce the time of image analysis as soon as possible so we choose the high performance and high definition camera to connect to computer.

This design not any has optimized the use of hardware, but also the algorithm of image analysis. Once PC get the information from camera, their RGB value will be checked. Base it, we define such law:1 for white, 2 for yellow, 3 for red, 4 for orange, 5 for blue, 6 for green. Then for the color collected by camera, checking it from the center of the color block to outside by K-means. So that we can classify the color block to one of a number from 1 to 6. The reason why we sue K-means instead of classification is that the latter need to define category advance, which are too sensitive with outer factor such as light so it will be difficult to analysis its data. K-mean can select several data object as the center of clustering from various of data object. As for the rest of the data object could be adjusted to the most similar group.

4.2. Manipulator control algorithm

The every step used to describe mechanical arm take is called action group. There are 18 groups of action group are set in total, and they are named R, U, F, D, B, L, E, S, M, Ri, Ui, Fi, Di, Bi, Li, Ei, Si, Mi, as Figure 5 and Figure 6 showed.

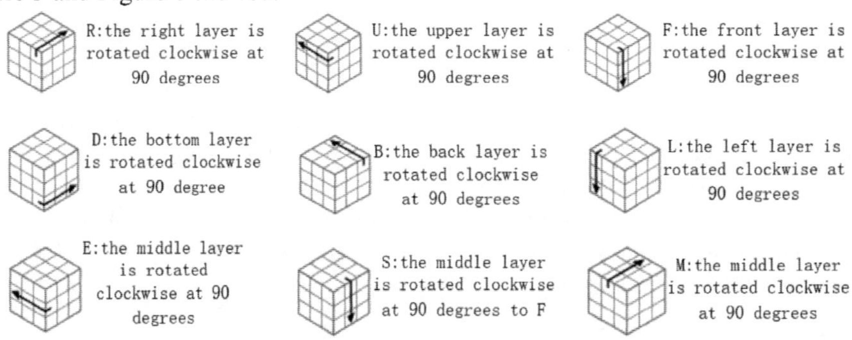

Figure 5. clockwise rotation action group

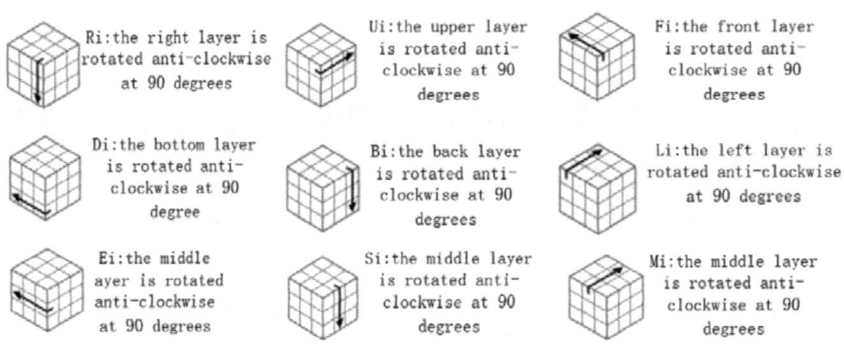

Figure 6. anti-clockwise rotation action group

4.3. Path optimization algorithm

In fact, in the magic cube rotation algorithm, there are whole magic cube rotation and a part of magic cube rotation. In consideration of its complexity, we need to simplify it to avoid too many unnecessary steps.

Step1:First, make the magic cube algorithm to be restored by the conventional magic cube algorithm, and record the instruction queue.

Step2:X, Y, Xi, Yi are used as transformation operators, transform the instruction sequence through string operation and record the instruction length n.

Step3:Check whether two adjacent characters are dual instructions, or can three adjacent instructions be turned into a reverse instruction, at the same time, simplify the instruction queue and update the instruction length n by operating character string.

Step4:repeat Step3, if the instruction length n is a static value, output the current instruction set.

5. Software design

As for robot system, the hardware circuit to robot is equivalent to a body to a human, it is the same that software to robot is equipment to a head to a human, because all the function of information processing and controlling depends on software. On this text, software is mainly designed to communicate with PC by Bluetooth and control mechanical arm to execute steps we wanted. The software process is designed as following Figure 7 showed.

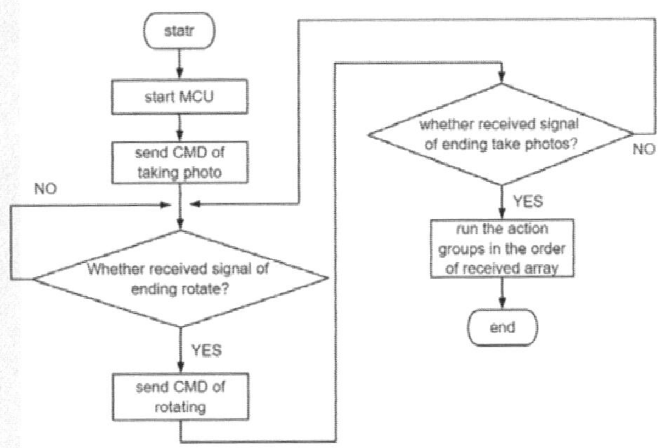

Figure 7. software design process

6. Concluding remarks

Based on the embedded processor, the whole magic cube robot is designed and completed. According test, the magic cube robot could restore magic cube by average 2 minutes. The robot has the advantages of low cost, small volume, stable performance, strong popularity and strong adaptability to the environment.

7. Direction of improvement

According to existing schemes, the text designed a two degree of freedom intelligent robot which can restore a disordered magic cube automatically in 30 steps and in 3 minutes. Because of limited time and energy and so on, the robot remains many things could be do better, such as the following aspects:

(1) Image information processing program runs on a PC, which made the whole system to be clumsy and inconvenient. It will be better to remove it to MCU for better flexibility and adaptability.

(2) The algorithm of restoring a magic cube could be more efficient and so robot could restore a magic cube in shorter time.

(3) Replace the motor the text used with such the motor which has feedback angle information, and the motor will run more accurately.

Acknowledgement

This research was supported by research grants from Science and technology support program of Sichuan Province (Grant No.: 2018KF013)

Reference

[1] Zhang Y, Changjun G U, Wang F, et al. Control System Design of Magic Cube Robot[J]. Journal of Shenyang Ligong University, 2016.

[2] Huang M, Si-Liang L U, Kong F R. Design and Implementation of Vision Subsystem of Rubik's

IOP Conf. Series: Materials Science and Engineering **428** (2018) 012064 doi:10.1088/1757-899X/428/1/012064

Cube Robot[J]. Machinery & Electronics, 2013.
[3] He Z Y, Ben K R. Automatic resolving algorithm and realization of magic cube based on OpenGL[J]. Journal of Harbin Institute of Technology, 2004, 36(7):893-895.
[4] Jia J J. A Embedded Image Processing System Based On ARM[J]. Microcomputer Information,2009.
[5] Jie Z, Guo L, Cai Z. Design of changeful magic cube robots′ control system[J]. Microcomputer & Its Applications, 2011.
[6] Sheng Q, Yongjun D U, Luo F, et al. Research of Solution of Rubik's Cube and Implemented by a Robot Based on STM-32[J]. Research & Exploration in Laboratory, 2017.
[7] Lin W S, Zhong-Shen L I, Hong J. Design and implementation of embeded image data processing system base on ARM[J]. Journal of Fuzhou University, 2008.
[8] Guo L F, Jie Z C, Cai Z H. Design of Changeful Magic Cube Robots Based on XC886[J]. Computer Knowledge & Technology, 2011.
[9] an Y J, Zhu Q D, Jiang C. OpenGL Based Experimental Platform for Simulation of Robot Arm Control System[J]. Computer Simulation, 2006, 23(8):252-257.
[10] Wang J N. 2-DOF Manipulator Control System Based on Visual Location[J]. Machinery & Electronics, 2008.

IOP Conf. Series: Materials Science and Engineering **428** (2018) 012065 doi:10.1088/1757-899X/428/1/012065

On Locomotion Control Using Position Feedback Only in Traversing Rough Terrains with Hexapod Crawling Robot

Petr Čížek and Jan Faigl

Computational Robotics Laboratory, Faculty of Electrical Engineering (FEE), Czech Technical University in Prague, Technicka 2, 166 27, Czechia

E-mail: cizekpe6|faiglj@fel.cvut.cz

Abstract. In this paper, we report on our results on improved locomotion control of small and affordable hexapod crawling robot using only position feedback from the utilized servo motors. Multi-legged robots represent complex mechanical systems with many degrees of freedom from which they can benefit in traversing rough terrains. However, the crucial ability of multi-legged robots is maintaining stable locomotion over irregularities of the terrain which makes the locomotion control complex and requires reliable and timely detection of the leg contact point with the ground. Such detection may require additional sensory equipment which can increase the cost of the multi-legged platform. Therefore, we focus on exploiting capabilities of nowadays intelligent servo motors with position feedback to develop a minimalistic set up in which the robot uses solely the position feedback of the servo motors to sense the ground reaction force. The first achievements enable a small hexapod crawling robot to navigate rough terrains using stable pentapod gait, where only one leg moves at a time, and five legs support the robot. Later on, we improved the locomotion control to enable faster locomotion using three simultaneously moved legs in the so-called tripod motion gait. This paper reports on further advancements with a faster control loop enabled by hardware based acceleration of the communication latency with the utilized Dynamixel AX12 servo motors that improve the locomotion capabilities of the robot. The reported results indicate the robot locomotion with the used adaptive motion gait is speeded up by a factor of 1.4 with the same stability in traversing the rough terrain of the experimental laboratory mock-up.

1. Introduction

Multi-legged robots can overcome terrains that are impassable by wheeled or tracked platforms at the cost of complex mechanical construction and locomotion control. The crucial property that enables to traverse rough terrains with the minimal possible stress on the robot construction is timely and reliable detection of the leg(s) contact with the ground and obstacles. If foot-contact events are not appropriately detected, the robot construction is put under a stress because of increased load and torque at the joints. Moreover, the stability of the robot posture can be lost, and high torque values in the joints may overheat or even damage the actuators. Therefore sensory feedback is utilized to supply tactile information, and thus negotiate the terrain and cope with its irregularities. However, additional sensory equipment may increase the cost and complexity of the robotic platform.

In our previous work [1], we focus on minimalistic approach of the locomotion control for hexapod crawling robot to enable traversing rough terrains using only the position feedback provided by the utilized servo motors. Such a minimalistic setup is suitable for affordable robots with minimal sensory equipment as the robot consists only from the servo motors that form the robot legs, the body frame, and a control board, see Figure 1. The idea of the developed adaptive motion gait [1] is based on the interpolation of the leg motion into small steps during the swing-down phase of the moving leg and evaluating the difference between the expected position and real position of the actuator read from the servo motor. However, the number of simultaneously controlled actuators (that also provide the required feedback for the control) limits the speed of the locomotion because of the limited bandwidth of the communication channel between the control board and the servo motors. The communication issues have been addressed by developed model of the actuator controller that enables to predict the position of the actuator, and thus compensates the experienced communication latency. Usage of the predicted position improves the reliability of the ground detection and also enables usage of the so-called tripod gait, where three legs provide support and the other three legs are simultaneously moving in the swing phase.

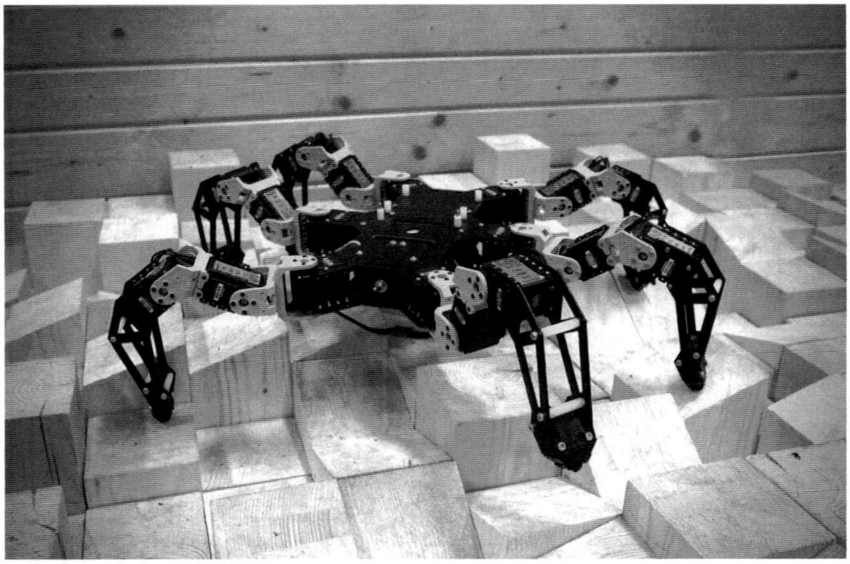

Figure 1. Utilized hexapod crawling robot with 18 Dynamixel AX-12A servo motors and Odroid XU-4 control board.

Based on our first results, we consider the influence of the bandwidth and latency of the readings important to the reliability of the ground detection, the achievable speed of the locomotion, but also to the smoothness of the locomotion itself. Therefore we investigate the influence of the control loop period to the locomotion of hexapod crawling robot in the experimental setup with our hexapod robot (shown in Figure 1) traversing rough terrains of the experimental laboratory mock-up. In this paper, we report on the achieved results and found insights because we believe that such an evaluation is of the significant importance to provide condensed insights on influencing factors of the locomotion stability and design guidelines for developing affordable robotic platforms capable of traversing rough terrains.

The rest of the paper is organized as follows. Section 2 overviews different architectures and approaches to locomotion control of multi-legged robots. The problem statement within the context of the hardware used and the used groundwork is introduced in Section 3. A discussion how latency can affect the locomotion stability is provided in Section 4. The achieved results of the experimental evaluation are reported in Section 5. Concluding remarks are in Section 6.

2. Related work

Locomotion control of multi-legged robots is a problem studied for many years, and therefore, there can be found numerous contributions in the literature. Deployment of the multi-legged platforms in rough terrains is mainly enabled by improvement (and also cost reduction) of the necessary sensory equipment that allows for sensing and negotiating the terrain. Two complementary types of sensors can be used in the locomotion control: the exteroceptive and proprioceptive sensors. The exteroceptive sensors allow taking measurements remotely without a direct visitation of the particular location, and the already reported deployments with hexapod robots include LIDARs [2, 3], stereo [4], and RGB-D cameras [5, 6]. These sensors are used to build the spatial awareness of the robot by creating a local map of the environment that is then utilized to steer the robot to avoid obstacles [4] or select individual foothold positions [7]. However, online built maps tend to suffer from artifacts and inaccuracies causing foothold selection to be unreliable. Besides, further processing techniques might be necessary for, e.g., soft terrains, to distinguish vegetation and the actual supportive soil [8]. Hence, additional information is desirable to improve the accuracy of the build maps [9].

The proprioceptive sensing such as the tactile feedback aims to detect the moment of the leg contact with the ground. A straightforward approach is to place a contact sensor at the leg foot tip [10]. Another method is to estimate or measure ground-reaction forces or joint torques. The joint torques can be measured directly at each joint [11], but they can be estimated using a linear model of the servo motor current [12]. Besides, other existing approaches employ strain-gauges [13] and force-sensitive resistors [14] to build a more affordable solution than using expensive force-torque sensors [15] to measure the ground reaction force at the leg foot-tip. But all these approaches rely on the particular servo motors accompanied with the selected sensor, which increase the cost and complexity of the robot construction and needed electronics. A minimalistic approach based on additional compliant actuator added on each leg is proposed in [16] to measure the ground reaction force in the kinematic chain. Using the ground reaction force, Palankar et al. [17] propose a force threshold-based position controller for crawling over irregular terrains.

In [1], we consider the same affordable servo motors as Palankar et al., but we propose an even more minimalistic approach that directly utilizes position feedback of the active actuators without the need of additional compliant actuators as in [17]. Thus, the proposed approach does not rely on any additional sensory equipment, and it solely uses the actuators that form the hexapod robot legs. Therefore, the developed locomotion control enables an affordable solution for the multi-legged robot capable of locomotion over rough terrains. However, the stability of the motion is influenced by the communication bandwidth and latency between the control board and the utilized servo motors. Therefore, we investigate the effect of the control loop period on the overall performance of the locomotion, and the achieved results are reported in the following sections of this paper.

3. Position-based Adaptive Locomotion

The problem addressed in this paper is to show an influence of the control loop period, e.g., defined by the communication bandwidth and latency, of the adaptive motion gait to the performance of the locomotion in crawling rough terrains. In particular, the reported experimental evaluation is based on our groundwork [1] that proposes a minimalistic control system for a hexapod crawling robot using only the position feedback of the utilized servo motors. In this section, we briefly introduce the underlying adaptive locomotion control that is essential for the elaboration of the control loop period influence on the locomotion performance.

3.1. Hexapod Robot Platform

The utilized affordable hexapod crawling robot is assembled from off-the-shelf components, and it consists of six electrically actuated legs, each with three joints, attached to the trunk which hosts the control board. Each leg follows a yaw-pitch-pitch kinematic chain, where each joint is motorized with the Dynamixel AX-12A intelligent servo motor with position feedback. All servo motors are connected in a daisy chain over a half-duplex serial bus with a bandwidth up to 1 MBaud and communicate using a master-slave protocol with the control board. Each servo motor has a P-type position controller and provides the control unit with its current position on demand. In the default configuration, the robot dimensions are approx. 45×40 cm, see Figure 1.

3.2. Adaptive Locomotion Control

The utilized adaptive locomotion control [1] split the motion of the legs into swing phase and stance phase. The individual legs alternate the phases according to the prescribed motion gait [18], and the legs in the swing phase are reaching new footholds while the legs in the stance phase support the robot body. In the employed locomotion control, the swing phase is further divided into three individual parts with the ground contact detection only in the swing-down phase of the motion as it is visualized in Figure 2a. Moreover, the robot motion is divided into reaching new footholds and body leveling parts because of the proposed ground detection using the position feedback.

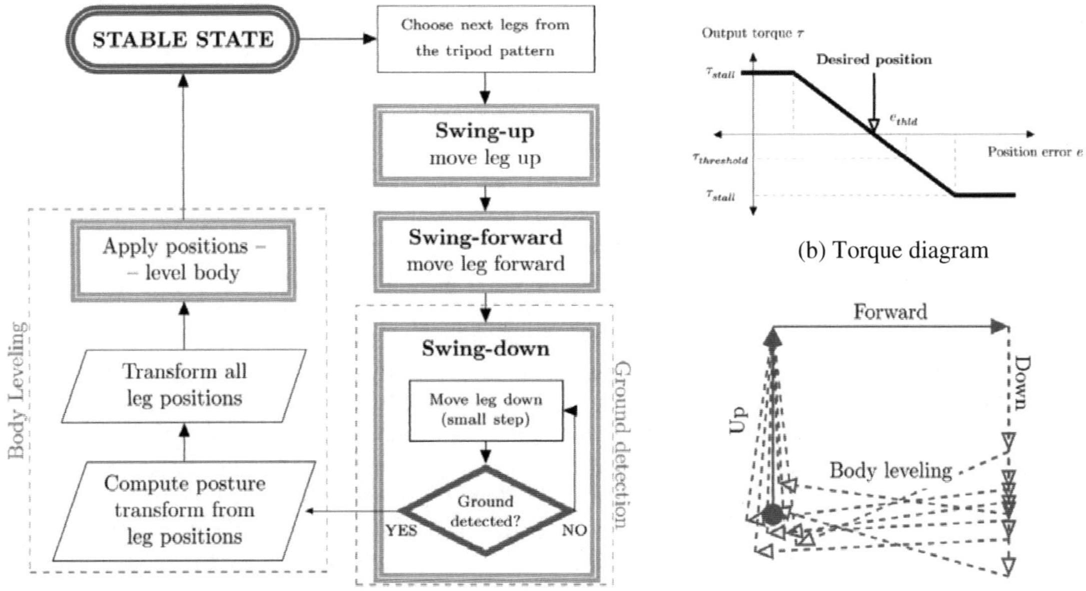

(b) Torque diagram

(a) Locomotion control scheme

(c) Regular gait

Figure 2. (a) Schema of the adaptive locomotion control [1] with the divided swing phase with the ground detection and body leveling. The ground is detected only in the swing-down phase of the leg motion using position error ϵ and its relation to the joint torque τ of the utilized servo motor shown in (b) torque diagram. (c) Leg trajectory in the hexapod body reference frame.

The ground detection is based on the relation of the position error and the torque that acts on the joint as it is depicted in Figure 2b. When the swinging leg touches the ground, the ground reaction force causes an increase of the load applied to the individual leg joints. Therefore the position error is increased, and if it reaches a specific threshold, the leg contact with the ground can be detected [1]. Once all legs in the swing-down phase reach the ground, and thus the foothold positions are acquired, the robot body is moved into an equilibrium position to distribute the weight of the robot evenly among its legs, which increases the robot stability. This motion is called body leveling by which the robot reaches a stable position from which it can easily move in any direction. As the equilibrium position is always given by the new footholds, it makes the whole robot thrust forward. Therefore the resulting trajectory of the leg in the robot body reference frame during several gait cycles may look like in Figure 2c.

The essential part of the timely detection of the foot contact with the ground is an interpolation of the swing-down trajectory of the leg into small incremental steps $\Delta\theta$ that are given according to

$$\Delta\theta = \frac{\theta_{fin} - \theta_{init}}{t_{des}/t_{con}} \qquad (1)$$

where θ_{init} and θ_{fin} are the initial position and the desired final position of the joint, respectively, t_{des} is the requested time to finish the motion, and t_{con} is the control cycle period, which is limited by the minimum period in which the servo motors can be polled for the position of the joint and which

IOP Conf. Series: Materials Science and Engineering **428** (2018) 012065 doi:10.1088/1757-899X/428/1/012065

depends on the used communication chain. The ground is detected by the adaptive controller as follows.

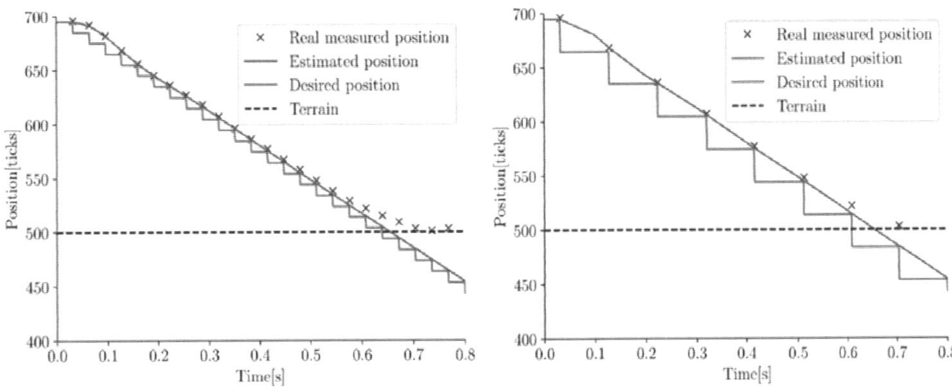

(a) Small interpolation step (b) Large interpolation step

Figure 3.Visualization of the servo trajectory interpolation for settings $\theta_{init} = 700$, $\theta_{fin} = 300$, $t_{des} = 1$ s and (a) $t_{con} = 32$ ms, and (b) $t_{con} = 96$ ms.

In each incremental step of the swing-down motion, the actuators of the currently swinging legs (given according to the motion gait) are set with a new position and their current position is fetched by the controller. The trajectory interpolation for a single actuator swinging between the current position $\theta_{init} = 700$ and the final desired position $\theta_{fin} = 300$ that is expected to be reached in $t_{des} = 1$ s is shown in Figure 3. Note, the reported values of the position are in ticks because the measured angular position of the actuator is discretized. Given the control cycle period t_{con}, each interpolation step corresponds to a single write and a single read of the servo position. When the leg touches the ground, the ground-reaction force starts to act on the robot leg causing the torque in the joints is increased that further propagate into the position error, which is in the case of the utilized Dynamixel AX12 servo motors, approximately proportional to the torque according to the curve depicted in Figure 2b. Hence, if the read position differs in more than ϵ threshold ticks from the expected estimated position of the joint, i.e., the position in which the servo should be at the given time without any external influence, the foot-strike is detected, and the leg motion is stopped.

The length of the interpolation step significantly influences the reaction time of the controller. For longer steps, there are fewer readings throughout the leg swing-down phase, and thus a higher ground reaction force can arise which can further cause, e.g., overheat of the servo motors, but it primarily affects the stability of the locomotion. An example of a large interpolation step for a longer period t_{con} is visualized in Figure 3b. The large interpolation step causes an increased latency of the ground detection process as it makes a longer time to satisfy the threshold condition.

4. Ground Detection, Latency, and Motion Stability

Sensory feedback is necessary to maintain the stability of the multi-legged robot in crawling rough terrains. In the utilized adaptive locomotion control [1], the position feedback of the servo motors is sufficient for the timely stopping of the leg movement in the case of the foot-strike event. However, the position readout latency and other related communication and signal processing factors may further significantly influence the stability of the robot during locomotion over rough and irregular terrains. The main identified influencing factors are as follows.

Robot topology influences the placement and connection of the sensors, and thus it influences the information gain that can be obtained by reading data from the particular sensor. In the utilized adaptive locomotion control [1], the actuators are the sensors that are connected in a tree-like topology with a single communication channel to the control board. The servo motors are connected via a half-duplex serial interface which has to be utilized for both setting the desired pose and reading the current pose. The adaptive locomotion control uses three sub-phases of the swing phase that is split into up, forward, and down parts. Together with the design of the robot leg, it is sufficient to read the position information only from the middle servo of the leg (called femur) as the main portion of the ground reaction force acts on to this servo during the swing-down phase.

Latency of the communication channel is mainly influenced by the utilized communication interface, but it is also related to the topology and connection of the sensors. Besides, the particular communication protocol and the delays caused by both hardware and software level may cause significant latency issues. For example, it is reported that the used Dynamixel AX12 servo motor suffers from 16 ms period latency for serial communication under Windows and Linux-based operating systems [19]. Hence, real-time operating systems, tailored drivers or dedicated hardware solutions might be necessary to speed up the communication further and overcome the boundary conditions, which may further increase the cost of the complete solution for a multi-legged robot capable of traversing rough terrains. The latency of the communication channel is directly expressed by the control cycle period t_{con} as the maximum reading speed influences the overall trajectory interpolation process.

Motion gait prescribes the order in which the individual legs are moved. When multiple legs are swinging simultaneously, e.g., for the tripod gait with three legs in the swing phase, synchronized transmission of the desired poses and readings of the current poses for all the involved servo motors is necessary. Managing multiple servo motors and sensors has a multiplicative effect on the overall latency, and thus the control cycle period t_{con}. Therefore it is desirable to write all the requested commands to the servo motors synchronously, but the readouts from the servos can be performed sequentially because of, e.g., half-duplex serial interface; hence, the control cycle period has to be prolonged appropriately.

Locomotion speed can be considered as the user-defined parameter t_{des}; however, it cannot be set arbitrarily as it should cope with the communication constraints imposed by the above influencing factors. Besides, too low t_{des} may influence the reliability of the ground detection, and therefore, the appropriate value has to be set according to the expected performance of the locomotion.

4.1. Ground Detection and Control Cycle Period

An example of the influence of the communication latency given as the control cycle period t_{con} on the servomotor motion is visualized in Figure 4., where a development of the position error between the desired and the actual position of the joint is shown for $t_{con} \in \{4\text{ ms}, 16\text{ ms}, 32\text{ ms}, 48\text{ ms}\}$. The values of t_{con} have been selected as the representatives of the real latencies in the used hexapod crawling robot as follows.

- 16 ms is the period of the serial communication reported for Linux-based operating systems [19].
- 32 ms as the double of the communication period in the case of reading two servos in every interpolation step.
- 48 ms is the triple of the communication period needed in the tripod gait where three positions are read in each interpolation step because of simultaneous moving of three legs.
- 4 ms represents the minimal control cycle period for tripod gait achieved by the hardware accelerated communication.

For each $\mathrm{t_{con}}$, the same setup with a single leg that swings down and reaches a ground has been utilized, see Figure 4a. The leg, i.e., the femur joint, is initially set to the position $\theta_{\mathrm{init}} = 700$ and it is requested to reach the position $\theta_{\mathrm{fin}} = 300$ in $\mathrm{t_{des}} = 1$ s with the obstacle corresponding to the joint angle $\theta = 500$ that corresponds approximately to the time 0.6 s, when the leg hits the ground.

The results in Figure 4b indicate that shorter interpolation steps induced by the shorter $\mathrm{t_{con}}$ reduce the overall torque in the leg as the servo P-type controller does not have to handle large jumps in the desired position. Moreover, a short interpolation step allows for faster detection of the ground contact event. The overall effects of the lower latency (shorter control cycle period) on the actual locomotion control are discussed in the following section.

(a) Single leg setup

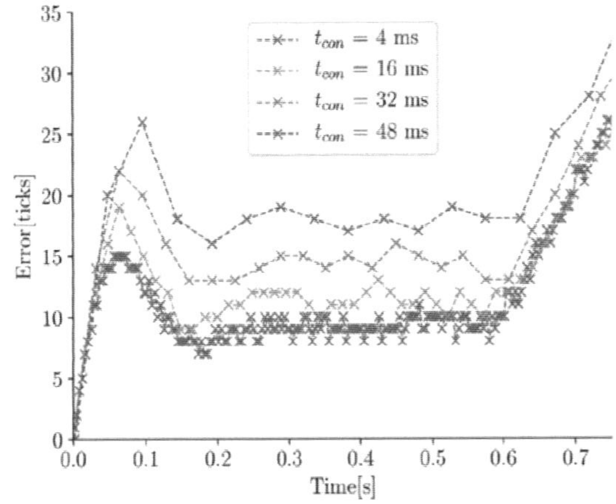

(b) Position errors for different control cycle period t_{con}

Figure 4. (a) Single leg of the hexapod crawling robot in experiment measuring the effect of sensor reading latency on the terrain contact detection; (b) Visualization of measured position errors for experiment with different values of t_{con} and fixed obstacle that is hit at around 0.6 s.

5. Experimental Results

In this section, we report on the experimental evaluation of the adaptive locomotion controller [1] with the settings of the control cycle period $t_{con} \in \{4\text{ ms}, 16\text{ ms}, 32\text{ ms}, 48\text{ ms}\}$ to demonstrate an effect of the communication latency. The selected periods correspond to the adaptive tripod gait with and without the hardware acceleration of the communication with the servos. The evaluation is organized in two experiments. The first experiment is focused on the stability of the platform and reliability of the ground detection for the given t_{con} and requested locomotion speed defined by the period t_{des} according to (1). The second experiment benchmarks the performance of the controller in crawling a rough terrain mockup consisting of wooden blocks with the identical dimensions 10×10 cm, but with variable height and slope, see Figure 1.

5.1. Locomotion Stability

The stability of the platform and reliability of the ground detection is measured by observing the body motion during the locomotion on the flat surface. The stability is measured as smoothness of the locomotion using an inertial measurement unit (IMU) attached to the robot trunk. In a case of the miss detection of the ground contact, the hexapod robot trembles which cause vibrations measured by the IMU. Therefore, the body orientation (pitch and roll angles) and the linear acceleration in the vertical direction (Acc_z) are measured as the stability and reliability indicators during a course of 20 gait cycles of the tripod gait for the control cycle period $t_{con} \in \{4\text{ ms}, 48\text{ ms}\}$ and $t_{des} \in \{0.5\text{ s}, 1\text{ s}\}$. For each setup, a suitable error threshold ϵ for the ground detection has been determined experimentally as the median value of the position error (described in Section 4.1 with the evolution of the error depicted in Figure 4b) increased by three additional ticks. The indicators are measured at 400 Hz using the XSens MTi-30 Attitude Heading Reference System (AHRS). Then, the mean value of each variable is subtracted from the readings, which cancels systematic amplitude displacements, and the five-number summary for all the indicators is calculated.

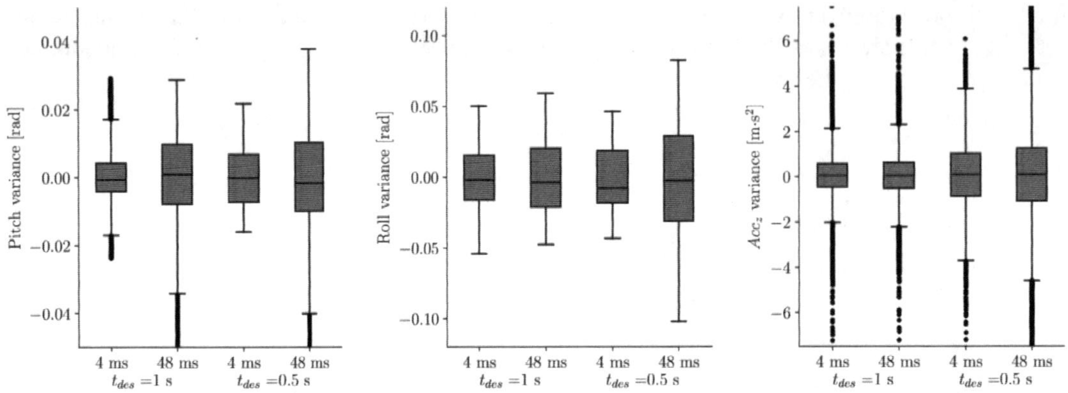

Figure 5. Stability and reliability indicators of the ground-detection according to t_{con} and t_{des}.

The results for the individual measured indicators and each particular setup are visualized in Figure 5, where it can be noticed that in all the cases, the ground detection with the shorter control period $t_{con} = 4$ ms provides smoother locomotion in comparison to the longer period $t_{con} = 48$ ms, and thus it seems to be more reliable. Besides, it can be seen that for the roll and pitch variances, the period $t_{con} = 4$ ms yields better stability even for faster locomotion in comparison to the adaptive tripod gait without the hardware acceleration of the communication.

5.2. Rough Terrain Traversing

The performance of the adaptive locomotion controller in traversing the rough terrain mockup has been experimentally verified to test dependency of the robot ability to crawl over rough terrains with different control cycle period t_{con} and locomotion speeds defined by t_{des}. Also in this case, the tripod gait is considered with $t_{con} \in \{4\text{ ms}, 48\text{ ms}\}$ and with the locomotion speeds $t_{des} \in \{0.5\text{ s}, 1\text{ s}\}$. For each setup, the robot has been requested to traverse the experimental 2.1 m long mockup ten times. The average values of the time to traverse the mockup and the achieved velocities are reported in Table 1 and snapshots of the robot traversing the mockup are presented in Figure 6. The velocity has been estimated using an external visual localization system based on [20] running with 25 Hz according to $v = ds/dt$ where dt is a fixed time window of 5 s and ds is calculated from the robot's trajectory.

Table 1. Rough Terrain Traverse Times and Velocities

	$t_{des} = 1$s		$t_{des} = 0.5$s	
	$t_{con} = 4$ ms	$t_{con} = 48$ ms	$t_{con} = 4$ ms	$t_{con} = 48$ ms
Traverse time [s]	58.6	62.3	47.3	83.1
Velocity $\times 10^{-2}[\text{m} \cdot \text{s}^{-1}]$	3.6	3.4	4.7	2.9

The results indicate that using a short period of the control cycle enables faster but still stable locomotion. Overall, the speed of the locomotion in rough terrains is improved about 1.4 times in comparison to the original controller proposed in [1], which is represented by $t_{con} = 48$ ms and $t_{des} = 1$ s.

Figure 6. Snapshots of the robot traversing the experimental mockup with the rough terrain.

6. Conclusion

In this paper, we report on the experimental evaluation of the control cycle period influence to the locomotion of the hexapod crawling robot traversing rough terrains. The evaluation is based on measuring locomotion stability with different periods and locomotion speeds. Besides, we also report that by lowering the feedback control period (e.g., the latency), the overall torque in the servos is lower, and the stability of the locomotion in both flat and rough terrains is increased. Last but not least, we show that no additional sensory equipment is necessary for an affordable robot to overcome structured terrain and only the position feedback from the used servo motors is sufficient.

Acknowledgments

This work has been supported by the Czech Science Foundation (GAČR) under research project No. 18-18858S. The support of grant No. SGS16/235/OHK3/3T/13 to Petr Čížek is also gratefully acknowledged. The authors acknowledge the support of the OP VVV funded project CZ.02.1.01/0.0/0.0/16_019/0000765 "Research Center for Informatics".

References

[1] Mrva J and Faigl J 2015 *10th International Workshop on Robot Motion and Control (RoMoCo)* pp 240–245

[2] Belter D and Skrzypczyński P 2011 *Journal of Field Robotics* **28** 497–528

[3] Kalakrishnan M, Buchli J, Pastor P and Schaal S 2009 *IEEE/RSJ International Conference on Intelligent Robots and Systems (IROS)* pp 167–172

[4] Stelzer A, Hirschmüller H and Görner M 2012 *The International Journal of Robotics Research* **31** 381–402

[5] Belter D, Łabecki P and Skrzypczyński P 2013 *IEEE International Conference on Safety, Security, and Rescue Robotics (SSRR)*

[6] Čížek P, Masri D and Faigl J 2017 *IEEE International Conference on Intelligent Robots and Systems (IROS)* pp 4096–4101

[7] Belter D, Labecki P and Skrzypczyński P 2016 *Journal of Field Robotics* **33** 337–370

[8] Bradley D M, Chang J K, Silver D, Powers M, Herman H, Rander P and Stentz A 2015 *IEEE International Conference on Intelligent Robots and Systems (IROS)* pp 1144–1151

[9] Walas K 2012 Robot Motion and Control 2011 *(Lecture Notes in Control and Information Sciences vol 422)* ed Kozlowski K (Springer London) pp 137–146

[10] Lewinger W A and Quinn R D 2010 *IEEE/RSJ International Conference on Intelligent Robots and Systems (IROS)* pp 3386–3391

[11] Winkler A, Havoutis I, Bazeille S, Ortiz J, Focchi M, Dillmann R, Caldwell D and Semini C 2014 *IEEE International Conference on Robotics and Automation (ICRA)* pp 6476–6482

[12] Bjelonic M, Kottege N and Beckerle P 2016 *IEEE/RSJ International Conference on Intelligent Robots and Systems (IROS)* pp 2042–2049

IOP Conf. Series: Materials Science and Engineering **428** (2018) 012065 doi:10.1088/1757-899X/428/1/012065

[13] He Zhang, Rui Wu, Changle Li, Xizhe Zang, Xuehe Zhang, Hongzhe Jin and Jie Zhao 2017 *Sensors* **17** 1514
[14] Walas K 2013 *Pomiary, Automatyka, Robotyka* **17** 283–287
[15] Walas K 2013 *Journal of Automation, Mobile Robotics & Intelligent Systems* 718–23
[16] Palankar M 2013 *A distributed local-leg feedback algorithm for robust walking on uneven terrain* Ph.D. thesis University of South Florida
[17] Palankar M and Palmer L 2015 *Autonomous Robots* **38** 301–316
[18] Chen W, Ren G, Zhang J and Wang J 2012 *Journal of Intelligent & Robotic Systems* **67** 255–270
[19] Jung R W ROBOTIS-GIT/DynamixelSDK cited 2018-05-30 URL https://github.com/ROBOTIS-GIT/DynamixelSDK/blob/master/c%2B%2B/src/dynamixel_sdk/port_handler_linux.cpp
[20] Olson E 2011 *IEEE International Conference on Robotics and Automation (ICRA)* pp 3400–3407

IOP Conf. Series: Materials Science and Engineering **428** (2018) 012066 doi:10.1088/1757-899X/428/1/012066

Kinematics Modelling and Co-simulation Research of a Novel Electrical Optical Cable Pruning Robot Using V-REP and MATLAB

Chao Li[1, b], Y F Yang[2] and S Y Dian[1, a]

[1.] Department of Automation, College of Electrical and Information, Sichuan University, Chengdu 610065, China

[2.] Quzhou Electric Power Bureau of ZheJiang Province, Quzhou 324002, China

[3.] Corresponding Author: [a]scudiansy@scu.edu.cn; [b]lcqwe147@163.com

Abstract. In this paper, a new type of serial pruning robot with five degree of freedom is developed to solve the issues of power cable inspection and maintenance. First, the kinematic model of the manipulator is established by Denavit-Hartenbreg (D-H) method, and its kinematic forward and inverse solution is obtained. Then, the kinematic co-simulation is executed by MATLAB and V-REP, the Monte Carlo method is used to analyze the working space of the pruning robot, and the workspace of the manipulator is obtained; To ensure the pruning process runs smoothly, trajectory planning simulation in the workspace is taken; Finally, the co-simulation results show that the kinematic model is valid, each joint and end effector control effect is good, and the pruning robot can achieve flexibly pruning branches and other obstacles in the range of 1 to 1.5 meters, which provides theoretical and experimental basis for follow-up research of pruning robot control problem.

1. Introduction

The inspection work of power optical fiber lines [1] is an important guarantee for the safety of electrical transmission. Traditional line inspection approaches include aerial surveys and ground-based visual inspections, etc. These approaches require the maintenance personnel of the grid to have a high accumulation of operating experience. Due to high intensity labor, low work efficiency, low detection accuracy of inspection personnel, traditional methods have obvious shortcomings. With the continuous development of robot technology, robot inspection and troubleshooting methods have gradually become a research hotspot [2-3]. The literature [4-5] introduced the transmission line inspection robot developed by Hibot in 2008 called Expliner, this robot can achieve the functions that automatic balance adjustment, lift arm to avoid obstacle, etc; The literature [6] introduces power transmission line inspection robot designed by Mexico Institute of Technology and Industrial Development, in which, the kinematics and dynamics models of the robot are established, then the robot model and the joint control algorithm of PD+G are verified by simulation. The literature [7] has designed a double boom patrol inspection robot that can step over obstacles such as shock hammers, linear clamps, and pressure cube through climbing mechanisms; Above research results are of positive significance, but they are all aimed at the inspection of high-voltage transmission lines, and there are few studies on inspection of electric optical cable.

IOP Conf. Series: Materials Science and Engineering **428** (2018) 012066 doi:10.1088/1757-899X/428/1/012066

The co-simulation of robot kinematics control system is one of the hot areas in the study of kinematics and dynamics of robot complex control model [8-9]. The current co-simulation system is mainly based on ADAMS/MATLAB virtual prototype technology [10-11]. V-REP is a kinematics and dynamics simulation platform widely used in robot virtual simulation [12]. It can be combined with MATLAB to build an independent simulation environment, and the results of simulation are more realistic and closer to the actual system.

In this paper, we are inspired by the work of traditional transmission lines mentioned above, for the inspection and pruning problem of electrical optical cable, the kinematics simulation experiment of the robot is carried out using the method of V-REP and MATLAB co-simulation. In our design, a 5-DOF serial pruning robot is mounted on the bottom of a 3-boom wheel-type patrol robot to maximally prune the branches near the cable. The kinematics model of the pruning robot is used to obtain the forward and inverse solutions of the robot. The joint simulation results verify the correctness of the kinematic model, and the Monte Carlo method combined with the MATLAB/V-REP co-simulation system is used to verify the reachable workspace of each joint of the robot.

2. Pruning robot

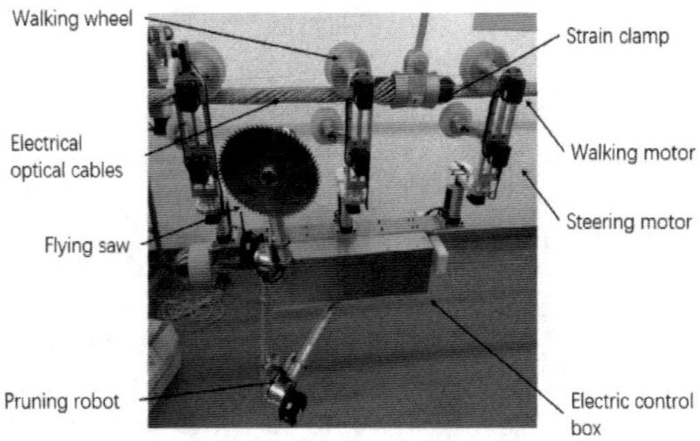

Figure 1. Robot body structure design.

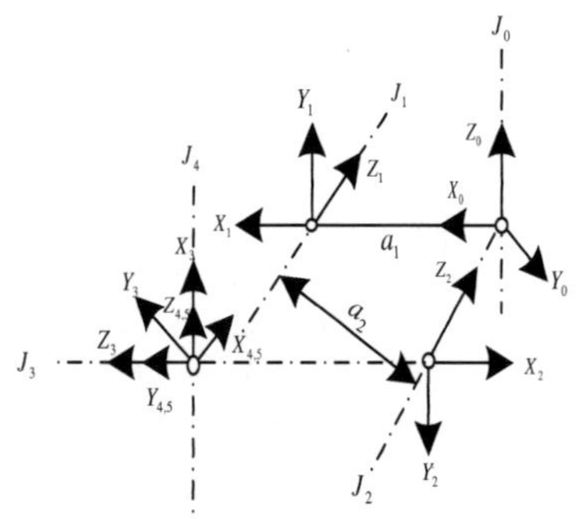

Figure 2. D-H coordinate system of robot.

As shown in Figure 1, the top of the robot is a three-boom wheeled patrol robot. The design of the upper-lower wheels allows the robot to walk on the single-fiber cable not to fall off. And through opening and closing movement of the upper-lower wheels, robot can step across the cable line. the middle of the robot is an electric control box for controlling the rotation of the walking wheel and each joint motor, while the left half is equipped with a high-definition wide-angle camera for monitoring the entire walking and pruning process.

The developed 5-DOF serial pruning manipulator was mounted under the robot, and manipulator's end effector is a fly saw, and the branches near the optical cable are cut along the line by the pruning manipulator, at the same time with the autonomous obstacle-free operation[8].

3. Positive and negative kinematics analysis of pruning robot

3.1. Positive kinematic analysis

First we established a link-pole coordinate system for pruning robots as [9], The following D-H parameters are derived from the coordinate system diagram Figure 2.

Table 1. Pruning robot D-H parameters

Link number	Joint angle θ_i	Twist angle α_i	Length of link a_i	Link offset d_i

1	θ_1	$90°$	$750cm$	$0cm$
2	$\theta_2-180°$	$0°$	$750cm$	$0cm$
3	$\theta_3+90°$	$90°$	$0cm$	$0cm$
4	$\theta_4+90°$	$-90°$	$0cm$	$0cm$
5	θ_5	$0°$	$0cm$	$0cm$

From the D-H parameter table 1, we can obtain the link transformation matrix ${}^{i-1}_{i}T$ of the coordinate system {i} relative to the coordinate system {i-1}, among them:

$$\begin{aligned}{}^{i-1}_{i}T &= Rot(z,\theta_i)Trans(0,0,d_i)Trans(a_i,0,0)Rot(x_i,\alpha_i)\\ &= \begin{bmatrix}\cos\theta_i & -\sin\theta_i\cos\alpha_i & \sin\theta_i\sin\alpha_i & a_i\cos\theta_i\\ \sin\theta_i & \cos\theta_i\cos\alpha_i & -\cos\theta_i\sin\alpha_i & a_i\sin\theta_i\\ 0 & \sin\alpha_i & \cos\alpha_i & d_i\\ 0 & 0 & 0 & 1\end{bmatrix}\end{aligned} \tag{1}$$

substituting data from table 1, and in following formula $s_i=\sin\theta_i$, $c_i=\cos\theta_i$:

$${}^{0}_{1}T=\begin{bmatrix}c_1 & 0 & s_1 & 750c_1\\ s_1 & 0 & -c_1 & 750s_1\\ 0 & 1 & 0 & 0\\ 0 & 0 & 0 & 1\end{bmatrix}\ {}^{1}_{2}T=\begin{bmatrix}-c_2 & s_2 & 0 & -750c_2\\ -s_2 & -c_2 & 0 & -750s_2\\ 0 & 0 & 1 & 0\\ 0 & 0 & 0 & 1\end{bmatrix}\ {}^{2}_{3}T=\begin{bmatrix}-s_3 & 0 & c_3 & 0\\ c_3 & 0 & s_3 & 0\\ 0 & 1 & 0 & 0\\ 0 & 0 & 0 & 1\end{bmatrix}\ {}^{3}_{4}T=\begin{bmatrix}-s_4 & 0 & -c_4 & 0\\ c_4 & 0 & -s_4 & 0\\ 0 & -1 & 0 & 0\\ 0 & 0 & 0 & 1\end{bmatrix}\ {}^{4}_{5}T=\begin{bmatrix}c_5 & -s_5 & 0 & 0\\ s_5 & c_5 & 0 & 0\\ 0 & 0 & 1 & 0\\ 0 & 0 & 0 & 1\end{bmatrix}$$

From the above link coordinate system transformation matrix, a homogeneous transformation matrix of the reference coordinate system of the robot end-effector relative to the base coordinate system can be obtained:

$${}^{0}_{5}T={}^{0}_{1}T\,{}^{1}_{2}T\,{}^{2}_{3}T\,{}^{3}_{4}T\,{}^{4}_{5}T=\begin{bmatrix}n_x & o_x & a_x & p_x\\ n_y & o_y & a_y & p_y\\ n_z & o_z & a_z & p_z\\ 0 & 0 & 0 & 1\end{bmatrix} \tag{2}$$

where n, o, a denote the vector coordinates of the x, y, and z axes of the coordinate system where the actuator is located relative to the reference coordinate system; p denotes the coordinates of the origin of the coordinate system where the actuator is located relative to the coordinates of the reference coordinate system. And the posture equation of the end effector is as follows:

$$\begin{cases}n_x = c_5c_4s_1 + s_4c_1c_2s_3 - c_1c_3s_2 - s_5c_1s_2s_3 + c_1c_2c_3\\ n_y = -c_5c_1c_4 - s_4c_2s_1s_3 - c_3s_1s_2 - s_5s_1s_2s_3 + c_2c_3s_1\\ n_z = -s_5c_2s_3 - c_3s_2 - c_5s_4c_2c_3 + s_2s_3\\ o_x = -s_5c_4s_1 + s_4c_1c_2s_3 - c_1c_3s_2 - c_5c_1s_2s_3 + c_1c_2c_3\\ o_y = s_5c_1c_4 - s_4c_2s_1s_3 - c_3s_1s_2 - c_5s_1s_2s_3 + c_2c_3s_1\\ o_z = s_4s_5c_2c_3 + s_2s_3 - c_5c_2s_3 - c_3s_2\\ a_x = c_4c_1c_2s_3 - c_1c_3s_2 - s_1s_4\\ a_y = c_1s_4 + c_4c_2s_1s_3 - c_3s_1s_2\\ a_z = -c_4c_2c_3 + s_2s_3\end{cases} \tag{3}$$

the positional equation of the end effector is:

$$\begin{cases}p_x = 750c_1 + 750c_1c_2\\ p_y = 750s_1 + 750c_2s_1\\ p_z = -750s_2\end{cases} \tag{4}$$

3.2. Inverse kinematic analysis

The inverse kinematics solution is based on the end position and pose given by (2). The joint angles of each joint of the robot are obtained. There are many methods for obtaining inverse solutions, including analytical method, projection method, and Pieper method. We use analytical methods to solve the inverse kinematics of the robot here. The left side of the equation (2) is multiplied by ${}^{0}_{1}\boldsymbol{T}^{-1}$,

$$\begin{aligned} {}^{0}_{1}\boldsymbol{T}^{-1}\,{}^{0}_{5}\boldsymbol{T} &= \begin{bmatrix} c_1 & s_1 & 0 & -750 \\ 0 & 0 & 1 & 0 \\ s_1 & -c_1 & 0 & 0 \\ 0 & 0 & 0 & 1 \end{bmatrix} \begin{bmatrix} n_x & o_x & a_x & p_x \\ n_y & o_y & a_y & p_y \\ n_z & o_z & a_z & p_z \\ 0 & 0 & 0 & 1 \end{bmatrix} \\ &= \begin{bmatrix} n_x c_1 + n_y s_1 & o_x c_1 + o_y s_1 & a_x c_1 + a_y s_1 & p_x c_1 + p_y s_1 - 750 \\ n_z & o_z & a_z & p_z \\ n_x s_1 - n_y c_1 & o_x s_1 - o_y c_1 & a_x s_1 - a_y c_1 & p_x s_1 - p_y c_1 \\ 0 & 0 & 0 & 1 \end{bmatrix} \\ &= {}^{1}_{2}\boldsymbol{T}\,{}^{2}_{3}\boldsymbol{T}\,{}^{3}_{4}\boldsymbol{T}\,{}^{4}_{5}\boldsymbol{T} = \begin{bmatrix} E_{11} & E_{12} & E_{13} & 750c_2 \\ E_{21} & E_{22} & E_{23} & -750s_2 \\ E_{31} & E_{32} & E_{33} & 0 \\ 0 & 0 & 0 & 1 \end{bmatrix} = \boldsymbol{A}_1 \end{aligned} \tag{5}$$

from the above $A_1(3,4)$:

$$p_x s_1 - p_y c_1 = 0 \tag{6}$$

solution: $\theta_1 = \arctan(\frac{p_y}{p_x})$

In the same way, we get (7) from $A_1(2,4)$:

$$-750\sin\theta_2 = p_z \tag{7}$$

solution: $\theta_2 = -\arcsin\frac{p_z}{750}$

similarly we get solutions $\theta_3 \sim \theta_5$ from $A_2(3,3)$, $A_2(3,1)$, $A_2(1,1)$ and $A_2(2,1)$:

$$\theta_3 = \arccos\frac{(-n_x c_1 c_2 - n_z s_2 - n_y c_2 s_1)^2 - c_5^2 s_4^2}{-n_z s_4 c_5 c_2 + n_x s_4 c_5 c_1 s_2 + n_y s_4 c_5 s_1 s_2 + n_x s_5 c_1 c_2 + n_z s_5 s_2 + n_y s_5 c_2 s_1},\ \theta_4 = \arcsin(a_y c_1 - a_x s_1),\ \theta_5 = \arccos\frac{n_x s_1 - n_y c_1}{c_4}.$$

4. Co-simulation using V-REP and MATLAB

4.1. Kinematics simulation and workspace analysis under MATLAB

The positive and inverse kinematics of the kinematics are calculated and verified in MATLAB. The values of the positive solutions for each joint are shown in Table 2:

Table 2. Each joint angle and inverse solution angle of Pruning Robot

θ_1	θ_2	θ_3	θ_4	θ_5	θ_1'	θ_2'	θ_3'	θ_4'	θ_5'
$\pi/6$	$\pi/6$	0°	$\pi/6$	0°	30.00°	30.00°	0°	30.00°	0.218°

solves the position matrix of the robot's end effector ${}^{0}_{5}\boldsymbol{T}$:

$${}^{0}_{5}\boldsymbol{T} = \begin{bmatrix} 0.2165 & -0.75 & -0.625 & 87.02 \\ -0.875 & 0.433 & 0.2165 & 50.24 \\ 0.433 & 0.5 & 0.75 & -375 \\ 0 & 0 & 0 & 1 \end{bmatrix} \tag{8}$$

for the inverse solution, select the end positions as above to obtain the joint angles as shown in Table 2 $\theta_1' \sim \theta_5'$. It can be seen from the range of changes in the joint angles in Table 2, Table 3,

Figure 3 and Figure 4 that the angle changes in the positive and negative solutions are basically the same; the MATLAB simulation results show that the forward kinematics and inverse kinematics sought in this paper are correct.

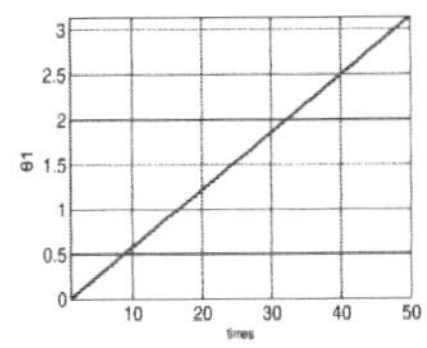

Figure 3. Range of joint angle θ_1 in inverse solution

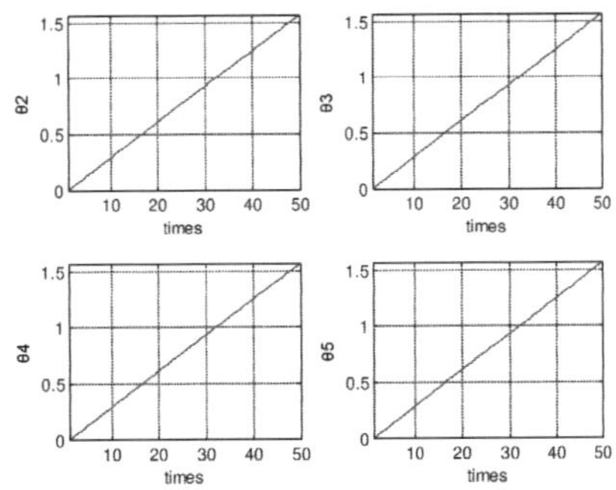

Figure 4. Range of joint angle θ_2~θ_5 in inverse solution

The Monte-Carlo Method [13], also called statistical simulation method, is a numerical method that solves mathematical problems by means of random sampling (pseudo-random number). It is written in MATLAB as a computer program, and the output 5000 points simulate 3D space, and the workspace of the end effector of the pruning robot is as follows, where x: lateral distance, y: vertical distance, z: working height:

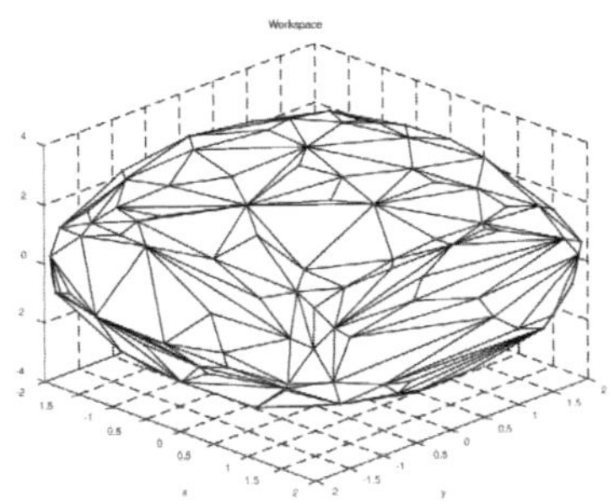

Figure 5. 3D figure of robot's reachable workspace

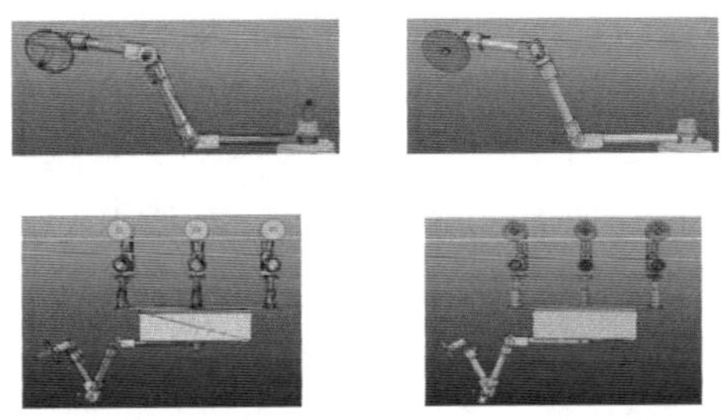

Figure 6. Dynamic modelling and joint definition of Pruning Robot in V-REP

As shown in Figure 5, the simulated workspace conforms to the actual space of each joint. The envelope of the workspace expresses the actual position that can be reached by the manipulator intuitively. The simulation results are in accordance with the actual expectations; the manipulator can reach ±1.5 meters work point in the axes of *x*, *y*, completely achieved the function of flexibly handling obstacles such as branches in the range of 1 to 1.5 meters near the optical cable.

4.2. Establishment of V-REP Model

In order to validate the kinematics model and reachable workspace of the pruning robot, the three-dimensional motion simulation was performed with the example of patrolling-pruning motion performed by the robot in the V-REP platform.

Firstly, the 3D model of the pruning robot was created in SolidWorks and imported into V-REP. Based on the shape construction and joint definition method [14], a three-dimensional V-REP model with dynamic characteristics was established to build a robot simulation experiment scenario. See Figure 6, where: (a) Dynamic model of the pruning robotic; (b) Visible model of the pruning robotic; (c) Dynamic model of the three-boom-armed wheeled robot and the pruning robot; (d) Visible model of the three-boom-armed robot and pruning robot.

4.3. Co-simulation of kinematics with MATLAB/V-REP

V-REP uses the Remote API to communicate with MATLAB [15] so that simulations can be run directly on MATLAB and data can be transferred simultaneously between the two programs. Using the D-H parameter table of the pruning robot proposed above, the PD controller [16] was used to control the V-REP servo motor (i.e. the above defined rotation joint) in MATLAB to complete the robot motion planning in the 3D simulating environment. Table 3 shows the initial pose and the target pose of the robot:

Table 3. 6-DOF pose of Pruning robot

	x	*y*	*z*	(α, β, γ)
initial	1.3	1.69	0.8	(10.38°-19.54°54.86°)
target	1.32	1.73	0.81	(0°0°0°)

Where x, y, z, α, β, γ represent the abscissa, ordinate, and vertical coordinates of the centroid of the end effector in the world coordinate system, and the corresponding attitude angles respectively.

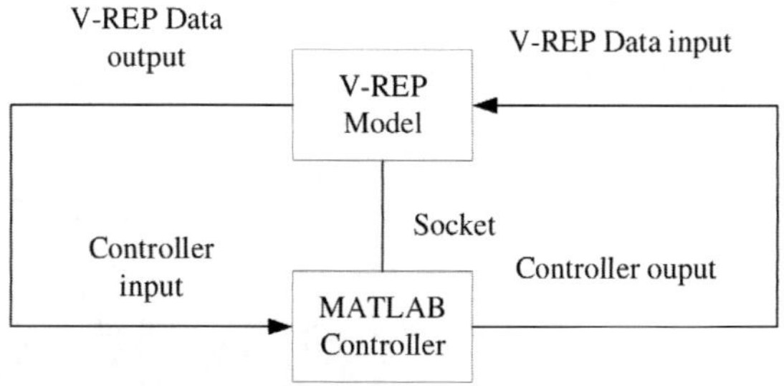

Figure 7. Robot body structure design.

The joint angles acquired on the V-REP platform during the movement of the robot were compared with the inverse solutions obtained by the analytical method in MATLAB. See Figure 8.

Figure 7 shows the two-way communication connection between V-REP and MATLAB via SOCKET. Among them, the PD controller output control quantity designed in MATLAB, the pruning robot joint motor in V-REP rotates under control action, and then current pose of end fly saw is sent back to MATLAB so that it loops until the end fly saw reaches the target point.

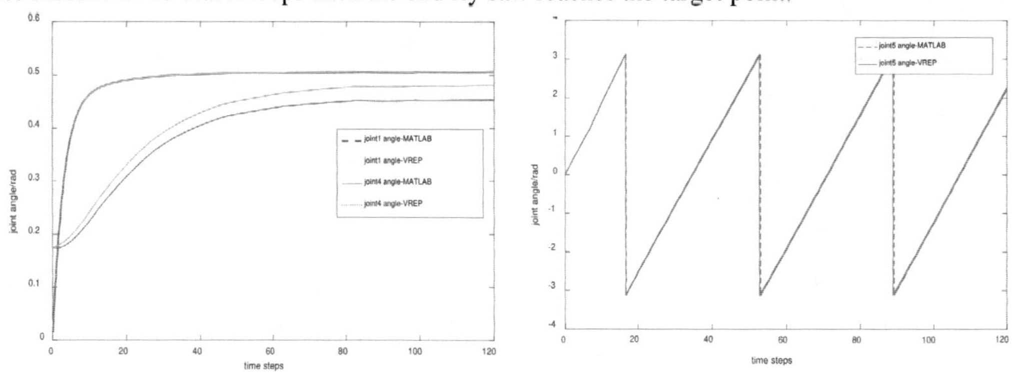

(a) (b)

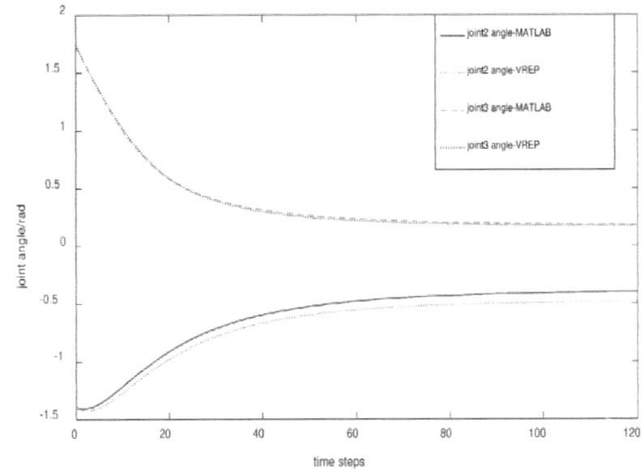

(c)

Figure 8. (a), (b), (c) show comparison of every joint angle output between MATLAB and VREP

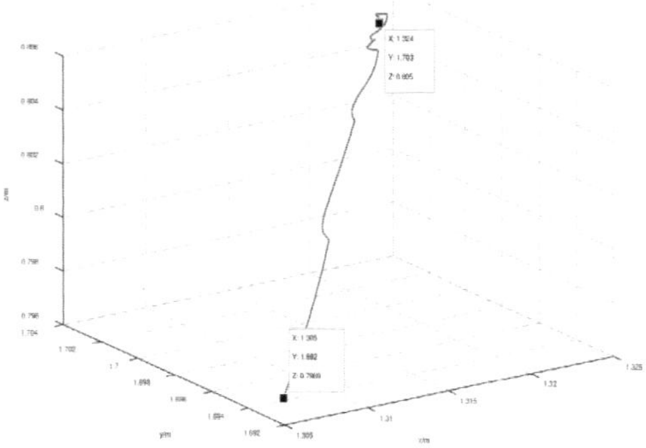

Figure 9. Trajectory planning of end effector

5. Result analysis

Figure 8 shows the rotation angles of the joints of the pruning robot in the motion planning process. It can be seen from the figure that the inverse joint angle calculated based on the kinematics model established above meets the actual motion simulation results of V-REP, indicating that the kinematics model is effective and reliable. Practically speaking, from Figure 9 we can see that the actual target position of the terminal trajectory is (1.324, 1.703, 0.805, 0°, 0°, 0°). The robot trajectory planning in 3D space based on PD controller and D-H parameter method can achieve smoothly movement from a predetermined starting point to a target point. Define the trajectory planning error as e:

$$e = \left\| \frac{P_o - P_d}{P_d} \right\| \tag{9}$$

Where $\boldsymbol{P}_0$ represents the actual target pose vector of the robot and $\boldsymbol{P}_{\mathbf{d}}$ represents the set target pose vector. By calculating the actual target pose and set value error e less than or equal to 1.6%, the trajectory planning results show that the kinematics co-simulation system established by MATLAB and V-REP is stable and reliable, the motion control effect is good, and it has high control accuracy.

6. Conclusion

In this paper, a new type of 5-DOF serial pruning robot kinematic model and co-simulation system is established. Its main innovations are:

1) The kinematics model of a 5-DOF serial pruning robot was established. In order to verify the accuracy of model, dynamics and kinematics simulation software V-REP was introduced and MATLAB combined with kinematics simulation and trajectory planning simulation experiments. The results show that the established kinematics model is accurate, the constraints are reasonable, and the scope of work accords with the actual situation;

2) Different from the MATLAB-based kinematics simulation method used in [13], etc. this paper uses the co-simulation of V-REP and MATLAB to establish the kinematics and dynamics model in V-REP, and the control algorithm is solved by MATLAB, by their respective advantages, established a new method of power optical cable pruning robot co-simulation based on multi-software, dynamics, kinematics and control algorithms decoupled, the final simulation results show that the method can be used for complex robot system, with good reliability and accuracy;

3) The new robot kinematics co-simulation method proposed in this paper can predict various performance indicators of the system before the physical prototype is manufactured, thus well overcoming the drawbacks of the traditional mechanical design process, improving the research and development efficiency. Therefore, it has certain guiding significance for the prototype design and

manufacture of power cable pruning robot, and can also provide theoretical basis and experimental platform for the research of follow-up control problems.

4) Compared with the ADAMS/MATLAB co-simulation system used in [9], [10], and [11], V-REP relies on good interface characteristics, accurate and high-quality physics simulation engine, which combined with MATLAB for establishing robot co-simulation system, taking into account the actual mechanical design and control algorithms need to be continuously and collaboratively iterative with product updates, can increase the speed of product updates, has better stability and high efficiency.

Acknowledgments

The National Natural Science Foundation of China is gratefully acknowledge for their financial support.

References

[1] Liang Z X, Zhang X D, Wei M H. Power communication cable maintenance and external damage preven-tion meatures. Telecommunications for Electric Power System. 2011; 32(06): 67—72.

[2] Qin X Y, Wu G P, Ye X H, et al. A novel method to reconstruct overhead High-Voltage power lines using cable in-spection robot LiDAR data. Remote Sensing. 2017; 9(7): 753—774.

[3] Dian S Y, Chen L, Tan J Y. Gain scheduled dynamic surface control for a class of un-deractuated mechanical systems using neural network disturbance observer. Neu-rocomputing. 2018; 275: 1998—2008.

[4] Paulo D, Michele G. Expliner from proto-type towards a practical robot for inspection of high-voltage lines. Proceedings of 2010 1st International Conference on Applied Robot for the Power Industry. Montreal, 2010, pp. 1—6.

[5] Paulo D, Michele G. Expliner-robot for inspection of transmission lines. Proceedings of IEEE International Conference on Robotics and Automation. Pasadena, 2008, pp.3978—3984.

[6] Zhou F Y, Wang J D, Li Y B, et al. Control of an inspection robot for 110KV power transmission lines based on expert system design methods. Proceedings of the 2005 IEEE Conference on Control Applications. Toronto, 2005, pp.1563—1568.

[7] Wen Xuefeng, Dian Songyi, Dong Han, et.al. Study on climbing mechanism of du-al-arm robot for power Line. Machinery Design & Manufacture, 2013; (7): 168—170.

[8] Su C, Wang Y L, Liu J T. Simulation analysis of virtual platform of close coupling multiple robots combined lifting system. Journal of System Simula-tion, 2015; 27(12): 2981—2987.

[9] Zhao Yucheng, Jia Shuheng, Dun Wentao. Development and application of vir-tual prototyping technology based on ADAMS. Agriculture Network Information. 2013; (08): 40—42+72.

[10] Lu Zhongzheng, Ge Zhengyang, Ding Wei. Co-simulation quadruped weeding robot using ADAMS and MATLAB. Mechanical Science and Technology. 2016; 35(03): 375—380.

[11] Chen Gang, Zhou Qicai, Wu Jing, et.al. Research for co-simulation of 7-DOF ma-nipulator based on ADAMS and MATLAB. Journal of System Simulation. 2017; 29(01): 99—106.

[12] Rohmer E, Singh S P N. and Freese M. V-REP: a versatile and scalable robot sim-ulation framework. 2013 IEEE/RSJ International Conference on Intelligent Robots and Systems, Tokyo, 2013, pp.1321—1326.

[13] Cai D, Xie C X, Zhang T, et.al. Study on workspace analysis and simulation of 6-DOF painting robot based on monte-carlo method. Machinery Design & Manufacture. 2009; (3): 161—162.

[14] Shi Junjun. The control method and ex-perimental research on dynamic stability of the multi-legged mobile robot system. Shengyang: Shenyang Ligong University, 2016.

[15] Ciszewski M, Mitka L, Buratowski T. Modeling and simulation of a tracked mo-bile inspection robot in MATLAB and V-REP software. Journal of Automation, Mobile Robotics & Intelligent Systems. 2017; 11(2): 5-11.

[16] Yang Xiao. Robotic manipulator PD control driven by virtual unmodeled dynamic. Shengyang: Northeastern University, 2014

IOP Conf. Series: Materials Science and Engineering **428** (2018) 012067 doi:10.1088/1757-899X/428/1/012067

Trajectory Planning for Periodic Operation of Autonomous Excavating Robot Based on Time-Jerk Synthetic Optimum

Yaguang Zhu[1,2*], Liang Zhang[1], Wanjin Guo[1] and Zhengcang Chen[1]

[1]Key Laboratory of Road Construction Technology and Equipment of MOE, Chang'an University, Xi'an, 710064, China.
[2]State Key Laboratory of Fluid Power and Mechatronic Systems, Zhejiang University, Hangzhou, 310028, China.

Email: zhuyaguang@chd.edu.cn

Abstract. Considering the problem of trajectory planning of redundant degree of freedom in autonomous operation for excavating robot, a method of trajectory planning for periodic operation of autonomous excavating robot based on time-jerk synthetic optimum was proposed in this paper. Through a kinematic model establishment and forward/inverse kinematics analysis for working mechanism, the trajectory planning based on Hermite interpolation method was carried out. Taking the time-jerk as objective function, a mathematical model of nonlinear programming under the inequality constraints was established, and Genetic Algorithm was used to find out the globally optimal solution. The rapidity and stability of working process of excavating robot is effectively enhanced and it has an important significance to realize the intelligentization of construction machinery.

1. Introduction

In order to realize periodically autonomous operation of the excavating robot, research on the trajectory planning is essential. At recent, there are many optimization methods for the robot trajectory planning[1-4]. Excavating robot is a robot with three degrees of freedom in a planar form, can realize the autonomous motion, obstacle avoidance and other special tasks, but its trajectory planning is more complicated than non-redundant structure. For the common foundation pit digging operation, the robot need to complete excavation point fixing from the start position, then full bucket lifting, rotations, unloading, empty bucket returning and posture adjustment, and at last return to the start position for the next cycle[5]. To enhance at the same time of the work efficiency enhancement of periodic excavation, sudden changes of velocity and acceleration will be inevitable, and so as the joint impact, which eventually decrease service life of the component. The approach to improve the work efficiency effectively and reduce the impact of each joint and further prolong the service life is critical to enhance the comprehensive performance of excavator. As mentioned above, a method of trajectory planning for periodic operation of autonomous excavating robot with redundant degree of freedom based on time-jerk synthetic optimum was proposed in this paper. Then the dual factors of the time and the impact after optimization are analyzed, the result shows that rapidity and stability of the working process of excavating robot is effectively enhanced.

2. Kinematic analysis of excavating robot

2.1. The establishment of D-H model of excavating robot's working device

Figure 1(a) is the photograph of backhoe excavator. The working device consists of boom, arm and bucket joints[6]. To realize autonomous operation, the bucket should keep a certain angle with the ground to realize the horizon digging or autonomous excavation ditch. The tip trajectory of the working device is composed of the movement of three joints when the rotation is not considered. It can be regarded as a planar three bar manipulator. The manipulator, which is made up of four connecting rods with three rotating joints in series, is fixed to the base of excavating robot. In this paper, the trajectory planning of the excavator is carried out by using the combined movement mode. The boom hydraulic cylinder, arm hydraulic cylinder and bucket hydraulic cylinder are cooperated with each other. The digging action is driven by the arm rotating around the hinge point of the boom and the boom hydraulic cylinder is also conducted to control the motion trajectory.

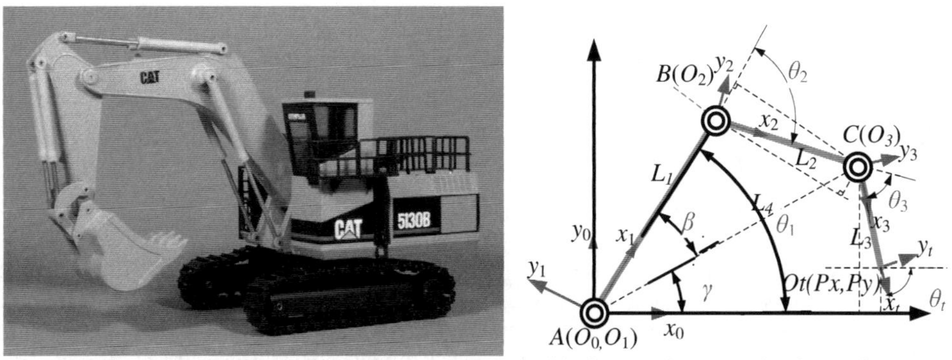

(a) Photograph of Backhoe Excavator (b)Three Joints Mechanism of Excavating

Figure 1. Joints Mechanism of Excavating Robot.

Figure 1(b) is the schematic diagram of the three joints mechanism of excavating robot. The kinematic model of the working device is established according to the Denavit-Hartenberg method[7]. Reference coordinate system of the base is fixed in the connection between the boom and the base of the excavator, and this coordinate system and the coordinate of the connecting rod 1 is set to be coincident. The coordinates, and of connecting rod 2, connecting rod 3 and a bucket tip are established respectively.

2.2. Analysis of kinematics

In this paper, a planar robotic excavator with three degrees of freedom is studied, which is one of the important basic models of robot with redundant degree of freedom. Firstly, the kinematic model of the working device is established, and the forward and inverse kinematics analysis is carried out. In forward kinematics solution, working device of each joint angle and the length of the connecting rod are known, and the position and posture of bucket tip relative to the base coordinate system can be solved[8]. According to the transformation matrix between the two adjacent coordinate. So the solution of the inverse kinematics is obtained

$$\theta_1 = \arctan\frac{Py - l_3\sin\theta_t}{Px - l_3\cos\theta_t} - \arcsin\frac{l_2\sin\theta_2}{l_4}, \theta_2 = \arccos(\frac{l_4^2 - l_1^2 - l_2^2}{2l_1 l_2}), \theta_3 = \theta_t - \theta_1 - \theta_2 . \quad (1)$$

3. Analysis of Working space

In order to ensure the expect target workspace of the bucket tip is completely enveloped the actual movement space, analysis of working space for excavating robot is carried out. The actual movement space of the bucket tip is drawn by the Monte Carlo method[9]. The three joint angles (θ_1, θ_2, θ_3) of the boom, arm and bucket contain a large number of random samples. Position and posture of the bucket tip are deduced by the forward kinematics. Then the approximate solution of this problem is obtained by using the matlab program. Then the corresponding random points of the bucket tip are obtained. Figure 2 is the working space of excavating robot.

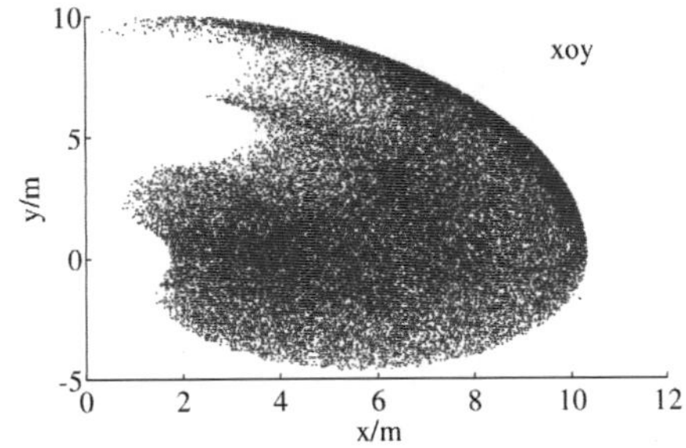

Figure 2. Working Space of Excavating Robot.

According to the map of excavator workspace, it can be convenience to select the target position within the reach of bucket and improve work efficiency of excavating robot. In general, when the Monte Carlo method is used to analyze the working space of excavator, the more random points are given and the results are more close to the truth. However, due to the large amount of computation in the general engineering practice, it will be acceptable if the working space of excavator is solved within a certain error range.

4. Trajectory planning based on time-jerk synthetic optimum

4.1. Trajectory planning based on the spline curve function

The trajectory planning for excavating robot is carried out in the joint space[10]. Of course, cubic polynomials, quintic polynomials or other curves, all can be applied on the proposed the optimization method. But we just show the process of the time-jerk synthetic optimization. So the five groups of cubic spline functions have been used[11], which can make excavator complete the expected action through the set target point in a cycle. Six points of the bucket tip in the working space are given for the spline curve fitting. The method is to represent the joint angle as the cubic functions of time. Joint velocity expression is obtained by solving first order derivative of cubic polynomial, the expression of joint acceleration is achieved by solving the two order derivative, and joint impact function is obtained by solving the three order derivative. So the expression of the joint trajectory for excavator is

$$Q(t)=\theta_1+\dot{\theta}_1 t+\left[\frac{3(\theta_2-\theta_1)}{t_2^2}-\frac{2}{t_2}\dot{\theta}_1-\frac{1}{t_2}\dot{\theta}_2\right]t^2+\left[-\frac{2}{t_2^3}(\theta_2-\theta_1)+\frac{1}{t_2^2}(\dot{\theta}_2+\dot{\theta}_1)\right]t^3 . \qquad (2)$$

4.2. Mathematical model based on time-jerk synthetic optimum

Normally, the excavating robot must meet the requirements of rapidity and stability in the working process. The travelling times between work points affect its working efficiency and the joint impacts affect the vibration amplitude and frequency of the excavator. Considering the influence of the two factors on the performance of the excavator, a mathematical model based on time-jerk characteristic is established. Five groups of cubic polynomials are used in this paper. Excavator completes a work cycle and the bucket tip travels from the beginning to the end through 6 via-points in total. Specified time interval sequences are $[t_1,t_2]$, $[t_2,t_3]$, $[t_3,t_4]$, $[t_4,t_5]$, $[t_5,t_6]$, the length of time h_i between the two adjacent time points and the total travelling time are

$$h_i=t_{i+1}-t_i \quad (i=1,2,3,4,5),\; T=\sum_{i=1}^{5}h_i=h_1+h_2+h_3+h_4+h_5 . \qquad (3)$$

Since the cubic polynomials cannot guarantee continuous of acceleration at the middle point, the joints of excavator will produce flutters and impacts in the transition points of the adjacent trajectories in the digging process. The maximum-minimum impact method is used to satisfy the dynamic constraint conditions, and make the sum of maximum impact values of each joint minimum. The expression of impact function is

$$J = \min[\sum_{j=1}^{3} \max_{t\in[0,t_j]} |\dddot{Q}_{j,i}(t)|] \quad (i = 1,2,3,4,5), \tag{4}$$

where t is the time length of adjacent two segments, $\dddot{Q}_{j,i}(t)$ is the impact function of robot joint. Its expression is

$$\dddot{Q}_{j,i}(t) = 6[-\frac{2}{h_i^3}(\theta_{i+1} - \theta_i) + \frac{1}{h_i^3}(\dot{\theta}_{i+1} + \dot{\theta}_i)]. \tag{5}$$

To consider operation efficiency and jerk restriction simultaneously for excavating robot, the weighted sum function is introduced in this paper. The time weight K_T and the impact weight K_J are set in the different working conditions. Taking these into account, the mathematical model of trajectory planning based on time-jerk synthetic optimum for excavating robot is given

$$\min\{K_T \sum_{i=1}^{5} h_i + K_J [\sum_{j=1}^{3} \max_{t\in[0,t_j]} |\dddot{Q}_{j,i}(t)|]\}, \tag{6}$$

among them, $i = 1,2,...5; j = 1,2,3$; K_T is the weight of the movement time; K_J is the weight of the impact value. Different optimization models can be obtained by selecting different K_T and K_J. The parameters are optimized by Genetic Algorithm

4.3. Constraint condition

In this paper, constraints of the excavating robot are established by limiting the maximum rotational speed and angular acceleration of each joint. Such as velocity expressions

$$\max |\dot{Q}_{j,i}(t)| \le VC_j \quad (i = 1,2,...5; j = 1,2,3), \tag{7}$$

where $t \in [0, h_i]$, VC_j represents the maximum speed of the joint. $\dot{Q}_{j,i}(t)$ is a quadratic curve which appears the maximum speed at $t = 0, t = h_i, t = t^*$ $(t^* \in [0, h_i])$. t^* is the point, where the two derivative is zero, that is $\ddot{Q}_{j,i}(t^*) = 0$.

Similarly, it can get acceleration constraints, such as the equation

$$\max |\ddot{Q}_{j,i}(t)| \le AC_j \quad (i = 1,2,...5; j = 1,2,3), \tag{8}$$

where $t \in [0, h_i]$, AC_j represents the maximum acceleration of the joint. $\ddot{Q}_{j,i}(t)$ is a linear function and thus the maximum value of acceleration occurs at $t = 0$ or $t = h_i$.

Here, Genetic Algorithm is used to solves the optimization problem of excavating robot trajectory. The population size n is set to 100 and the initial population is generated. Objective function $f = K_T * f_1 + K_J * f_2$ is used as the fitness function ($f_1 = \sum_{i=1}^{5} h_i$, $f_2 = \min[\sum_{j=1}^{3} \max_{t\in[0,t_j]} |\dddot{Q}_{j,i}(t)|]$). The probability of initial crossover ξ is 0.5, the number of elite preservation m is 4, the penalty factor ρ is 2, the fitness of each individual in each generation is calculated, and the genetic operation is carried out on the chromosome. According to the genetic strategy, the selection, crossover and mutation are used to the group to form the next generation. The lower bound of the boundary condition is set to $[0.01, 0.01, 0.01, 0.01, 0.01]$, and the upper bound is $[5,5,5,5,5]$. The iteration number is 100, and the stagnation algebra is 80. The performance of groups will be judged to show whether to the results meet the predetermined value. If not, return to modify the genetic parameters. By genetic algorithm, the fitness function value is less than the optimal 10^{-6} and stops about 42 generations.

Different optimizing results can be obtained by using different time weights and impact weights. The time optimal trajectory planning mathematical model is obtained when time weight is zero. After optimization, h_i=[4.724,5,3.931,3.531,5], f=0.6606 Although the maximum impact value of each joint is reduced and smaller fitness results are achieved, but the travelling time of excavator is too long, which cannot reach the expected requirements. The impact optimal trajectory planning mathematical model is obtained when impact weight is zero and here h_i=[1.933,0.697,1.047,0.737,0.892], f=6.3067. It can be seen that the time interval is very short, but does not have the impact optimization. To sum up, in order to guarantee that the excavating robot is working smoothly and efficiently, it should be adequately taken the dual influence of time and impact into consideration by selecting the appropriate weights. The weight factor and are different under different working conditions, but this paper mainly focuses on the optimization method of the excavator in a certain condition. Thus relatively reasonable trajectory planning mathematical model of excavating robot can be obtained.

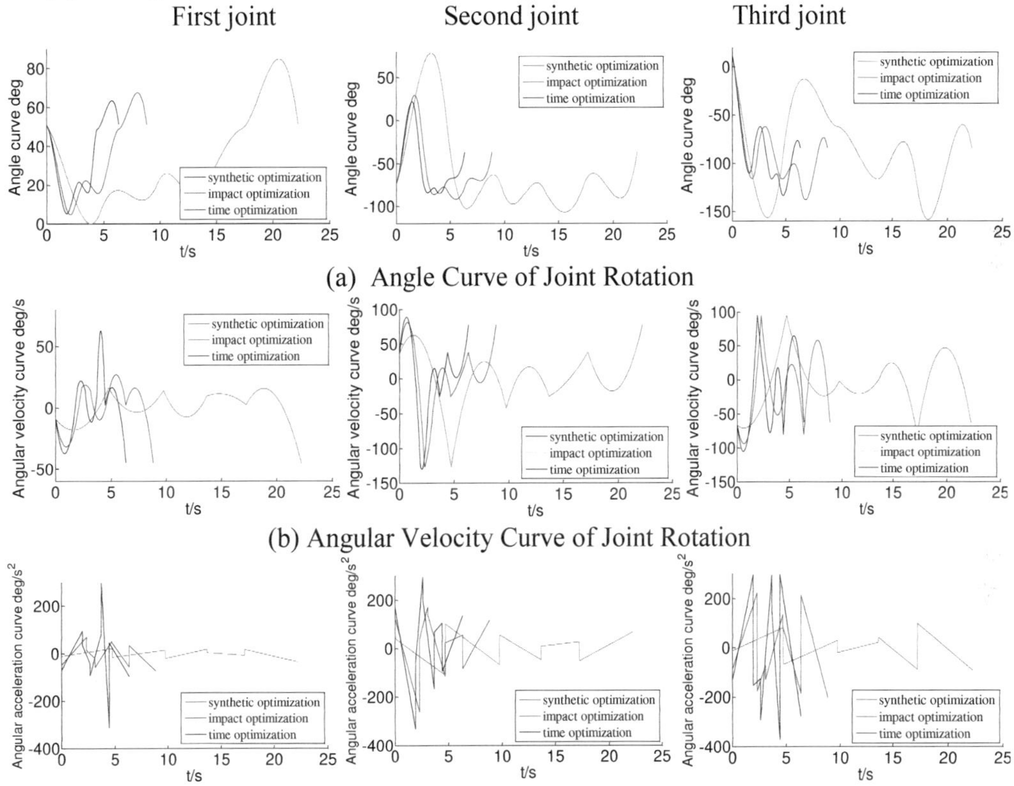

(a) Angle Curve of Joint Rotation

(b) Angular Velocity Curve of Joint Rotation

(c) Angular Acceleration Curve of Joint Rotation

Figure 3. Excavating Robot Trajectory Optimization

5. Analysis of results

Figure 3 is the optimal trajectory of excavating robot. The joint angle, angular velocity and angular acceleration of each joint at each moment are shown. The rotation angle curves of impact optimization are gentle and smooth, and the impact value of the joint in the working process is about 0.6606, and the total time of a work cycle needs 22.186s. But the rotation angle of each joint has exceeded the determination of boundary value, and the cycle time, which is close to the maximum value of 25s, is too long for the efficiency of the excavator in a cycle and cannot meet the requirements in the periodically independent digging. The operate time of the angle curves of time optimization in a working cycle is about 6s, and the joint impact value is about 6.3067 after optimization. It can be seen that the total operation time of this model is the shortest, but it generates a larger mutation for the angular velocity and angular acceleration of each joint. Therefore, the impact value of the joint for the excavator working mechanism is increased, and the service life is reduced, and the working requirement is difficult to meet. The total time of the synthetic optimization curve is about 9s, the rotation angle of each joint is in the normal range and joint impact value is about 11.6977. Through the comprehensive comparison of the parameters of the two curves, the optimization process of the latter approach not only improves the working efficiency of the excavator, but also the angle curve, angular velocity and angular acceleration curve are all within the set boundary. The maximum impact value is also within the allowable range and it can realize periodically efficient operation in a certain working condition.

6. Conclusions

This method contributes to control the motion trajectory of the working mechanism in the periodically autonomous operation both in the time and impact. In a certain sense, it not only improves the speed of movement, but also reduces the joint impacts. Therefore, the work efficiency and the motion stability of the excavating robot are significantly improved. Of course, the proposed method still have the problems about the regulation of weight coefficients about time and impact, the incomplete convergence and slow convergence rate in the simple Genetic Algorithm, but it also has important significance to promote the overall performance for the excavating robot.

Acknowledgments

This study is supported by the National Natural Science Foundation of China (No. 51605039), Shaanxi Postdoctoral Scientific Research Project (No. 2016BSHYDZZ26) and Open Foundation of the State Key Laboratory of Fluid Power Transmission and Control (No. GZKF-201610), Xi'an Science and Technology Project for Talented Personnel Service Enterprise in Colleges and Universities (No. 2017088CG/RC051(CADX001)).

References

[1] Baza SA,Tondu B.Minimum Time On-line Joint Trajectory Generator Based on Low Order Spline Method for Industrial Manipulators. Robotics and Autonomous, 1999, 29 (4): 257-268.
[2] Shin KG, McKay ND. Minimum-time Control of Robotic Manipulators with Geometric Path Constraints. IEEE Transactions on Automatic Control, 1985,AC30(6): 531-541.
[3] Martin B J, Bobrow J E. Minimum Effort Motions for Open Chain Manipulators with Task Dependent End-effected Constraints. IEEE International Conference on Robotics & Automation, 1997, 18 (2): 213-224.
[4] Kyriakopoulos K J, Saridis G N. Minimum Jerk Path Generation. Proceedings of the 1988. IEEE International Conference on Robotics and Automation, Philadelphia, 1988: 364-369.
[5] Masakazu H, Watanabe H, Kazuo F. Digging Control System for Hydaulic Excavator[J]. Mechatronics. 2001, 11(6): 665-676.
[6] Hui Shao, Hiroshi Yamamoto, et al. Automatic Excavation Planning of Hydraulic Excavator[C]. Springer- Verlag Berlin Heidelberg (ICIRA), 2008 (Part II): 1201-1211.
[7] Moosavian S A A, Pourreza A, Alipour K. Kinematics and Dynamics of a Hybrid Serial-parallel Mobile Robot[J]. IEEE International Conference on Robotics & Automation, 2009, 16(1): 1358-1363.
[8] Sun Z, He D, Zhang W J. A Systematic Approach to Inverse Kinematics of Hybrid Actuation Robots[J]. Oncogene, 2012,33(1): 300-305.
[9] Metropolis N, Rosenbluth A W, Rosenbluth M N, et al. Equation of State Calculations by Fast Computing Machines[J]. Journal of Chemical Physics, 2004, 21(6): 1087-1092.
[10] Egerstedt M, Martin C F. Optimal Trajectory Planning and Smoothing Splines[J]. Automatica, 2001, 37(7): 1057-1064.
[11] Zhao X, Zhang C, Yang B, et al. Adaptive Knot Placement Using a GMM-based Continuous Optimization Algorithm in B-spline Curve Approximation[J]. Computer-Aided Design, 2011, 43(6): 598-604.

IOP Conf. Series: Materials Science and Engineering **428** (2018) 012068 doi:10.1088/1757-899X/428/1/012068

Design of Automatic TV Inspection Equipment for CPR1000 Steam Generator 9th Tube Support Plate

Wei-rang Wu*, Yu-chun Tao, Tong Yu, Qing-sen Zhao and Xun-zong Yang

Technical Support Department, Suzhou Nuclear Power Research Institute, Suzhou, China

*wuweirang@cgnpc.com.cn

Abstract. With steam generator running more and more years, sludge will accumulate on the surface of 9th tube support plate, causing broached holes clogging and heat transfer efficiency decreasing. In order to know the situation of sludge deposit and provide evidence of judging whether soft chemical cleaning is necessary, the 9th tube support plate automatic TV inspection equipment is required and needs to be developed based on the upper structure of CPR1000 steam generator. In this paper, the application surroundings and requirements were analyzed. The inspection method, general design, magnetic crawler, probe characteristics and mockup tests were described, especially how to insert the probe into bundles to achieve inspection scope was introduced.

1. Introduction

Steam generator is the vital component of the primary loop in NPP (abbreviation of "nuclear power plant"), which has the role of preventing radiation leaking between the primary loop and the secondary loop[1-3].As steam generator running for longer time, sludge accumulates more on the tubesheet, TSPs (abbreviation of "tube support plate")and tube walls. If sludge is not removed on time, it can affect heat transfer efficiency and main gas pressure. Considering CPR1000(abbreviation of "China Pressurized Reactor 1000 unit") steam generator inspection access, only tubesheet, TSP1 and TSP9 can be carried out TV inspection. By far, tubesheet inspection has already been done during every outage[4] and TSP1 inspection needs to be done if necessary; however, TSP9 TV inspection has never been performed because it is much more difficult to design and apply such an equipment. According to operation feedback from EDF,TSP9 is much more likely to deposit sludge than others except tubesheet[5-6]. Whether TSP9 needs to be cleaned by high pressure water or soft chemical lancing is determined based on TSP9 broached holes clogging. If clogging is over 30 percent, high pressure lancing or chemical lancing is necessary[7]. To obtain clogging information,TSP9 TV inspection is becoming more and more important for steam generator running more than 15 years, which means automatic TV inspection equipment needs to be developed for the purpose above.

In Europe, the TSP9 inspection job is mainly done by a French specialized company using one magnetic crawler inside upper wrapper to climb up and down. The crawler carries a probe that is used for capturing clogging video and travels along wrapper circumference to find the inspected bundle for probe going in and out of bundle. The probe movement is done manually by operator outside of manhole, which is shown in figure 1[8].

Figure 1. TSP9 inspection crawler used in Europe

For CPR1000 steam generator, there has been no such an automatic TV inspection equipment because of developing difficulties, especially with all automatic function which can have more operation convenience and less radiation doses. It is proven to be feasible by enough investigations and performance tests. Therefore, it is worthy to develop it. In this paper, working surroundings were introduced and design difficulties were analyzed. The inspection method, equipment components and properties were mainly focused.

2. Application Analysis

Before developing the automatic equipment, it is necessary to summarize performance parameters and analyze the difficulties based on inspection object and structure surroundings.

2.1. Inspection Object

The main aim of TSP9 inspection is to check clogging of broached holes which number is 8948 in total with square arrangement in the pitch of 27.43mm. The sample inspection view is shown in figure 2, and the result obtained by the newly designed automatic equipment should be as clear as that.

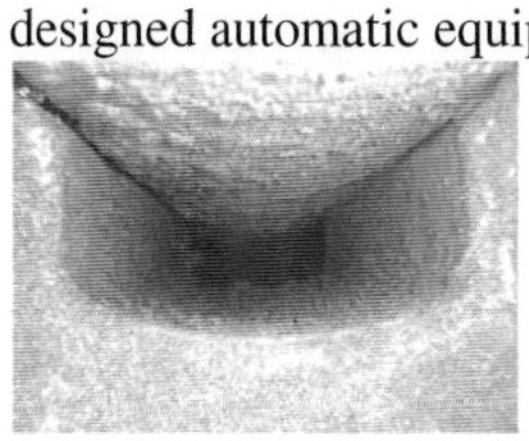

Figure 2. Sample inspection view

2.2. Inspection Surroundings

In order to make sure the designed equipment can be applied on site successfully, it is very important to know every structural detail inside steam generator above TSP9,which is shown in figure 3.

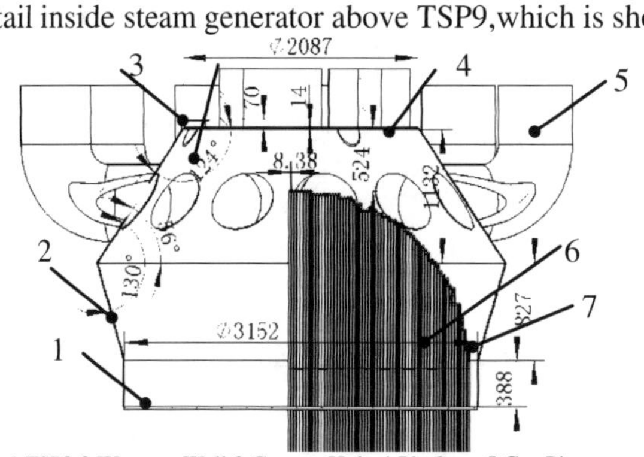

1.TSP9;2.Wrapper Wall;3.Camera Hole;4.Platform;5.Gas Pipe;
6.Tubes;7. Ends of Anti-vibration Bar

Figure 3. Structure drawing of steam generator inside above TSP9 with half tubes

In order to make probe reach the upper surface, the crawler needs to enter inside wrapper through camera hole which is the only access. There are 3 camera holes in total on the platform where the operator places crawler inside. Through the drawing, it can be seen that another 3 characteristics which will bring new challenges to design work. Firstly, there are 12 gas pipes around the wrapper, and the biggest width between next inner holes for crawler passing through is only 207mm. Secondly, there is an edgefold with 130°on the wrapper side wall, which increases risk of falling down when crawler crossing the corner. Thirdly, the distance between wrapper inner wall and its nearest tube wall is only 40mm in the height of within 388mm over TSP9,which is impossible for a crawler carrying probe to travel around. Therefore, crawler must stay where the height is over 388mm. Moreover, in order to make operator receive less radiation, the water level inside steam generator must be over the top of the bend tubes, which means inspection work should be done under water.

3. Equipment Design

3.1. General Design

To achieve the inspection work and satisfy the above requirements, general equipment design is shown in figure 4.

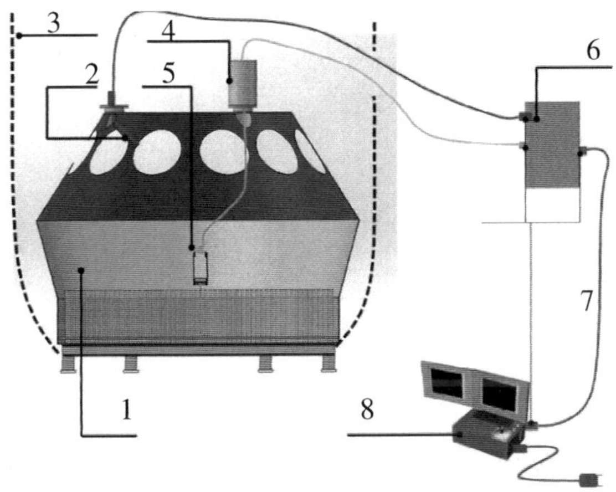

1.TSP9 Mockup;2. Surveillance Camera; 3. Steam Generator Wall; 4. Push-Puller Device; 5.Crawler;6.Site Control Box;7. Composite Cable;8.Monitoring System

Figure 4. Schematic drawing of TSP9 automatic TV inspection equipment

The equipment is composed of the following parts:

- Crawler. It travels around wrapper wall inside with 4 magnetic wheels, with adjusting probe head position to enter bundles easily and driving probe in and out of the inspected bundle. It also has 2 side cameras, 1 front camera and 1 position sensor to assist crawler movement and inspection work.
- Push-Puller Device. It has function of pushing or pulling crawler cable based on crawler position and it is working automatically with the order from monitoring system. It plays an important role of making no cable twisting with anti-vibration bar ends and crawler passing through the space between gas pipe holes.
- Surveillance Camera. It is used to check water level and assists in driving crawler. It is installed on another camera hole.
- Site Control Box. It is placed outside steam generator. It is used to receive orders to control all motors movement, acquire and process signals of encoders or sensors.
- Monitoring System. It has communication with site control box. Operator uses it to move crawler and probe, supervise crawler position, check inspection video and so on.
- Composite Cable. It is used to connect among crawler, push-puller device, site control box and monitoring system. It is used to transfer power, signal, audio ,video between linked components.

IOP Conf. Series: Materials Science and Engineering **428** (2018) 012068 doi:10.1088/1757-899X/428/1/012068

3.2. Crawler

Crawler is the key component of the whole equipment. It is composed of body, flexible nose, front camera , side camera and 4 wheels. Its length is 350mm,its width is 160mm,its thickness is 120mm and its weight is 6.5Kg.The body is divided into two layers vertically and its material is aluminum. The upper layer places probe roller, probe driving mechanism and front camera. The lower layer places 8 motors, control board, and 2 side cameras. There are 4 motors for driving wheels independently and the other 4 motors for driving flexible nose in 4 directions through stretching molybdenum wires inside flexible nose. The crawler structure is shown in figure 5 and lower layer arrangement is shown in figure 6.

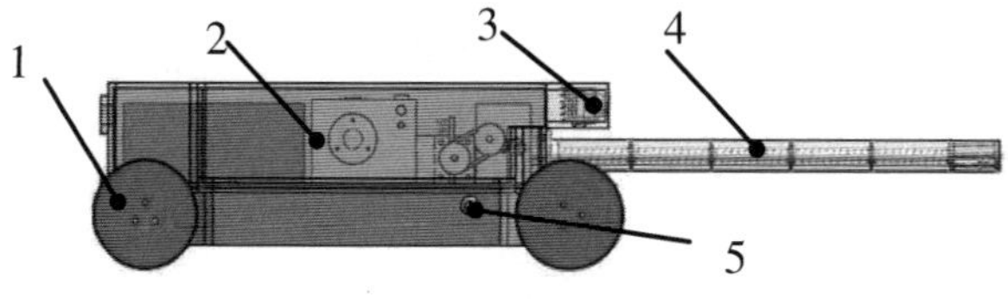

1. Wheel; 2.Body;3.Front Camera;4.Flexible Nose;5.Side Camera

Figure 5. Schematic drawing of crawler

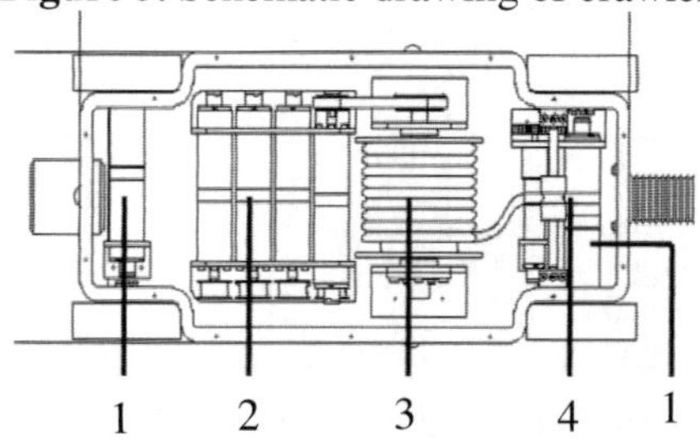

1. Wheel Motor; 2. Molybdenum Wire Motor;3.Probe Roller;4.Probe

Figure 6. Schematic drawing of layer arrangement

To make sure crawler can pass the 130°edgefold smoothly, crawler forces when crossing corner are analyzed and wheel magnetic force is calculated by the following formula.

$$\mathbf{F} = (1.6\times \mathbf{F}n-54)\times\cos50° - 2\times F_2 + 1.6\times Fn \qquad (1)$$

In Equation(1), **F** is required driving force, **F**n is normal pressure of magnetic wheel, $\mathbf{F}_2$ is the magnetic force in the negative X axis. Wheel diameter is 70mm and friction factor between wheels and wrapper wall is 0.8.

On the other side, software ANSYS is adopted to calculate F_2 . It can be calculated that every wheel needs driving torque with 0.907N·m and every chosen motor type is Maxon DCX22S GBKL12V+GPX26A 172:1 which can provide 1.646N·m,which is 1.81 times of required torque. It means that crawler has enough power to climb wrapper wall and cross 130°corner with no risk of falling down. The force diagram is shown in figure 7.

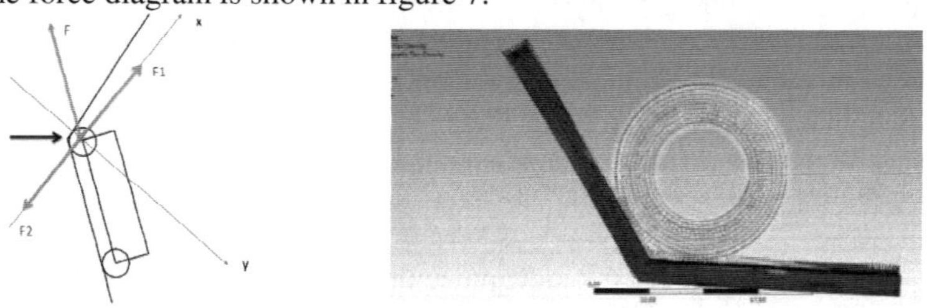

(a) Force Diagram When Crossing the Corner (b)Magnetic Force Calculation by ANSYS

Figure 7. Force diagram and calculation

Concerning the probe, it should have required balance between enough rigidness to move in the bundle and softness to be rolled into the roller. After enough experiments, the chosen body is 6mm in diameter and has a specially designed head ,which is shown in figure 8.

The probe is moving inside flexible nose which can adjust bending angle through stretching molybdenum wires by motors. There are 4 wires inside flexible nose at the location 0°,90 °,180°,270° to control 4 directions respectively. The flexible nose is made of several articulations with one rotating free degree. Articulations are connected one by one through special links. The wires are placed in plastic tubes for protection and easy movement and the outside surface of flexible nose is made by knitted mesh grid. Flexible nose plays an important role in guarantee of inserting probe into the bundles successfully ,which structure is shown in figure 9.

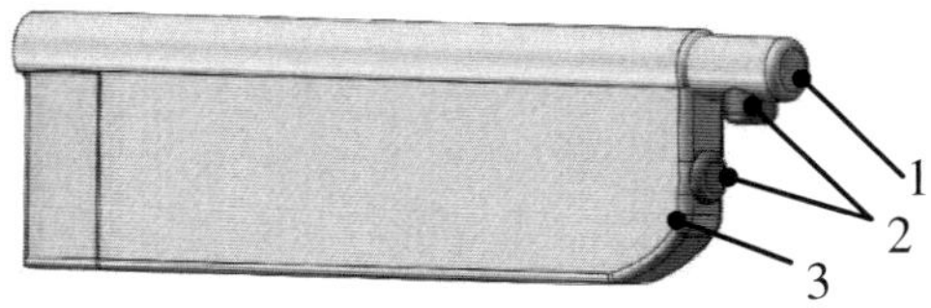

1.Body;2. Camera with LED Lights;3.Support Leg

Figure 8. Picture of probe head

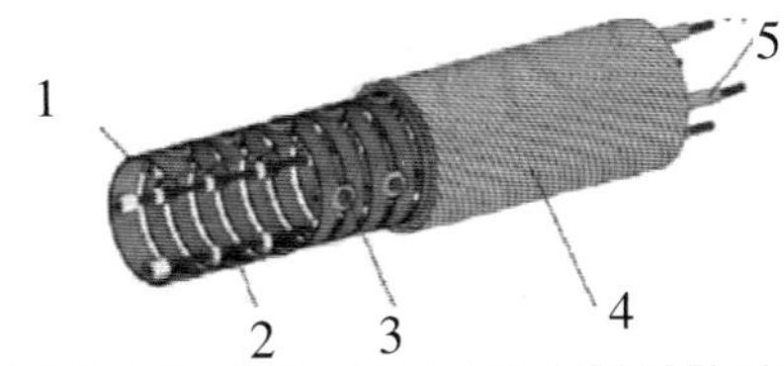

1.Articulation;2.Wire;3.Link;4.Mesh Grid;5.Plastic Tube

Figure 9. Schematic drawing of flexible nose

3.3. Monitoring System

The monitoring system is developed based on ROS framework under Linux system. It is displayed by industrial computer with the latest Intel Core processor .As is shown in figure 10, the computer has double screens. The left screen is showing information of probe video, clogging record, map for marking the inspected area; the right screen is showing information of crawler cameras, surveillance camera,crawler positon and crawler driving interface.

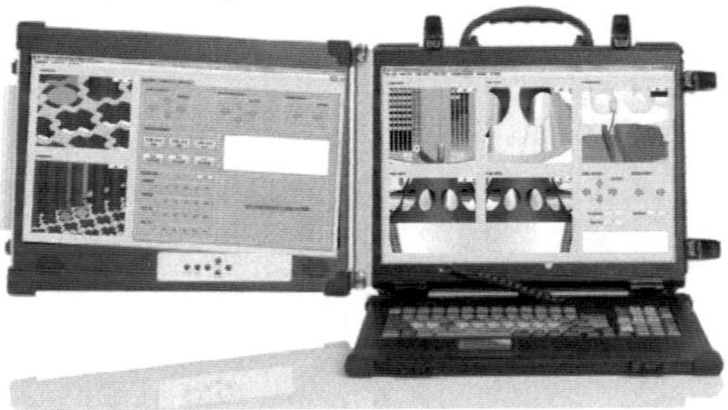

Figure 10. Picture of monitoring system

4. Mockup Test

Before application on site, the equipment needs to be done a lot of tests on mockup to make sure it is coincident with steam generator and it can achieve the required inspection work. The tests include crawler climbing, probe head adjusting ,probe going in and out of bundles, probe articulation, water proof and so on. The equipment is proven to be qualified by passing such tests. It can move to the destination with no falling down, drive the probe into bundles, record video when probe taking back by operating monitoring system. It has a good performance of sealing and video quality, which is shown in figure 11.

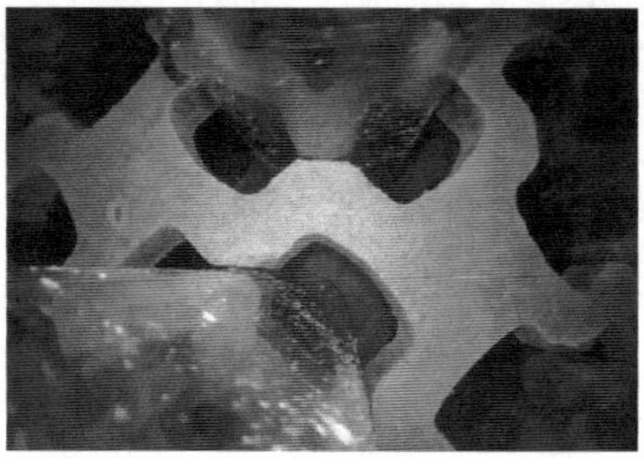

Figure 11. Picture of inspection results on mockup

5. Conclusion

In this paper, the importance and application of TSP9 automatic TV inspection equipment was introduced and application surroundings and requirements were summed up to guide design work. Finally, inspection method and design scheme were given priority to describe, especially crawler structure. Generally speaking, there are mainly 4 following points about equipment development:

1. The scheme of TSP9 inspection equipment with all automatic function is proven to be achievable by enough tests and design.
2. The general design is suitable for TSP9 automatic inspection with no blind spots .The crawler with 4 magnetic wheels and 1 flexible nose can cross the edgefold with 130° easily and safely, adjust probe head to aim at the inspected bundle. The crawler movement for switching inspected bundles is done automatically based on arithmetic.
3. The probe is supportive for integrating into the crawler and performing outstanding inspection effect.
4. Such an equipment has already passed mockup test and will be applied on site, which will provide design and operation feedback for similar designs and applications.

References

[1] Berge P H and Figuet J M.The need for clean steam generators[J]. *Nucl Energy*,2013,32(2):115-116.
[2] Deng Jing-shan, Liu Cheng-chao, Cheng Bao-liang.Nuclear power plant steam generator secondary side sludge lancing system[J].*Cleaning World*,2015,31(21):42-43.
[3] Liao Chang-bin and Dai Bing. Research on sludge lancing technology of steam generator secondary side[J].*SiliconValley*,2013(14):58-60.
[4] Zhou Guo-zheng,Yan An-he and Xie Jie. Development and application of the secondary side inspection technology of CPR1000 unit steam generator[J].*Nondestructive Testing*,2014,36(9):49-50.
[5] Jovica Riznic. Steam Generators for Nuclear Power Plants.Woodhead,Duxford,United Kingdom, ISBN: 978-0-08-100894-2.
[6] Wei-rang Wu,Yu-chun Tao,Xi-ming Tang,Tong Yu and Yue Zhong. “Research on Automatic TV Inspection Technology of Steam Generator Secondary Side Center Lane.” *Proceedings of 2016 IEEE Far East NDT New Technology & Application Forum (FENDT),*Held in Nanchang,China,June 22-24.
[7] Ding Xun-shen. Mechanical cleaning and chemical cleaning for steam generators in nuclear power station [J].*Chemical Cleaning*,1999, 15(5):20-26.
[8] Wang Li-quan, Wu Jian-rong and Liu Yu-long. “Structure Design of Steam Generator Inspection Robot.” *Proceedings of the 2nd International Symposium on Symbiotic Nuclear Power Systems for 21st Century(ISSNP2008).*

Virtual Experiments Design for Robotics Based on V-REP

Mingzuo Xie [1,2,*], Dongxu Zhou[1], Yong Shi[2] and Ruiqing Jia[1]

[1]School of Mechanical Electronic & Information Engineering, China University of Mining and Technology (Beijing), Beijing100083, CHINA
[2]Department of Mechanical Engineering, Stevens Institute of Technology, NJ 07030, USA

*mingzuoxie@126.com

Abstract. Compared with traditional robot laboratory, virtual robot laboratory integrates programming, modeling and controlling, which can assist related learning of robot courses more efficiently and directly under the condition of saving time and equipment cost. A concept of virtual robotics laboratory based on virtual robot platform V-REP is proposed in this paper. The laboratory is divided into coordinate transformation, positive and inverse kinematics, dynamics and trajectory planning. At the same time, virtual teaching tasks of comprehensive course design for robotics is also proposed to improve students' capacity for designing virtual robots comprehensively and improve teaching quality.

1. Introduction

With the advent of the era of artificial intelligence and rapid growth of robotics, increasingly jobs are being replaced by robots. Robots will accelerate to improve social productivity in several future years to create efficiency and convenience for human life in the future.

In recent years, more and more undergraduate colleges start to set up related majors of robots to satisfy great demands of all walks of life on talents in the aspect of robots. However, as robot equipment is relatively expensive and teachers are undertrained, it is difficult for many colleges to set up a robot experiment course. Aimed at this problem, plenty of robot simulation software is developed, such as MATLAB Robotics Toolbox [1], RoboAnalyzer[2], RoKiSim[3], ARTE[4], and others. There are also many virtual robotics laboratories developed for robot learning, such as a virtual experiment framework of robotics based on Qt and Ogre [5], and ROBOLAB based on Java[6]. The numerous software is few designed for robotics basic knowledge teaching. Some software for teaching and learning are lack of systemic and completeness. The biggest difficulty of robotics teaching at present is how to show mathematical physics meanings of these concepts clearly and combine theoretical knowledge and experimental courses. V-REP is a virtual simulation platform of robots developed rapidly in recent years. Although it does not have direct teaching functions, its abundant expansion capability provides a better tool for us.

A concept of virtual robotics laboratory based on V-REP is proposed in this paper, which explains all basic parts of robotics systematically. It can help students learn and understand basic knowledge concepts of robots better and quickly, thus improving teaching quality of robotics.

2. V-REP and Virtual Robotics Laboratory Architecture

2.1 V-REP

V-REP is a strong robot 3D integrated development environment, which has several universal calculation modules (inverse kinematics, physics/dynamics, collision detection, minimum distance calculation, path planning), distributed control architecture (control scripts of unlimited number, thread or non-thread), and several extension mechanisms (plug-in, client application programme and so on.) [7]. It provides multiple functions and can make integration and combination easily by detailed API and script function.

2.2 Robot model

Some typical robot arms are adapted for robot model of virtual robotics laboratory. Only arm robots are involved and mobile robots are not involved in initial stage. Then it is considered to add related parts of mobile robots in subsequent plans. Model of virtual laboratory mainly include palletizing robot arm, SCARA, DELTA, Puma560. Teachers and students also can use model in V-REP or customize models belonging to them. Plan of virtual experiment is mainly introduced and discussed by 3-DOF robotic manipulator in this paper.

The prototype of the manipulator is from ABB palletizing robot. Reconfiguration design is implemented aimed at robotics teaching. The experimental manipulator mainly includes a workbench and mechanical arm. The experimental scene of the manipulator is mainly used for picking and placing and writing or drawing. The manipulator is a typical 3-DOF series mechanical arm shown in Figure 1, which has strong representation in primary learning of robotics.

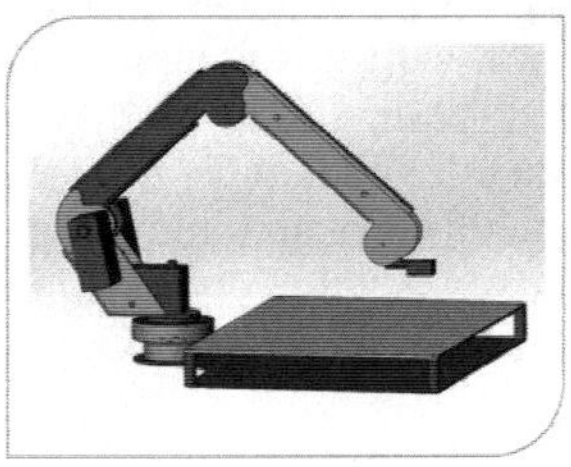

Figure 1. 3-DOF robotic manipulator

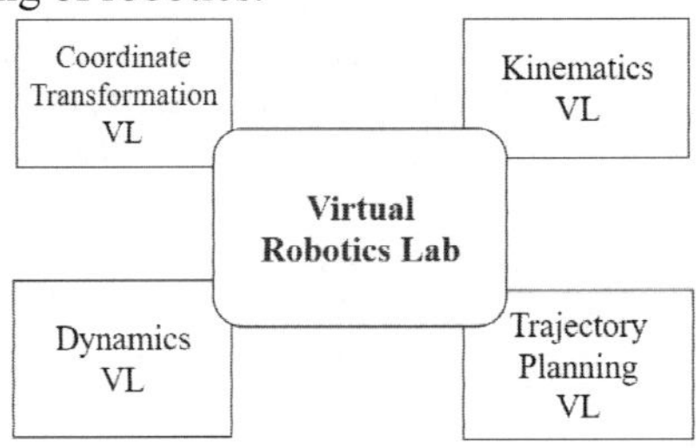

Figure 2. The architecture of virtual robotics laboratory

2.3 The architecture of Virtual Robotics Laboratory

The whole virtual laboratory of robotics is constituted by multiple V-REP document assembly. The contents include important basis chapters in robotics, include coordinate transformation, kinematics, dynamics, trajectory planning. V-REP is a platform supporting multiple systems. The whole software has small capacity and is easy for installation. Document of V-REP is generally several hundred to thousands of KB, which can be used flexibly. These documents can be linked to all parts of textbook respectively and uploaded to cloud for downloading and use by teachers and students in future supporting textbooks. The structure of the whole laboratory is shown in Figure 2.Each laboratory basically involves important knowledge points of the part of robotics. Table 1 shows details.

Table 1. Virtual labs and basic knowledge of robotics

Virtual Lab	Coordinate transformation laboratory	Kinematics laboratory	Dynamics laboratory	Trajectory planning laboratory
Main knowledge concepts	Coordinate transformation Homogeneous transformation Homogeneous transformation matrix Euler Angles Representation RPY Angles Representation Axis / Angle Representation	D-H Parameters Forward Kinematics Inverse Kinematics Velocity Analysis	Inverse Dynamics Forward Dynamics Equations of Motion Euler-Lagrange Formulation Newton-Euler Formulation	Joint Space Planning Cartesian Space Planning Point-To-Point Control Continuous Path Control

Each subsidiary virtual laboratory is an independent file (Figure 3). The interface is designed by OpenGI. Script programming language LUA is adopted to realize programming. LUA programming does not have much syntax requirements and is ideal for robot beginners to implement algorithms. Path planning plug-in OMPL of the robot arm is also involved in path planning laboratory.

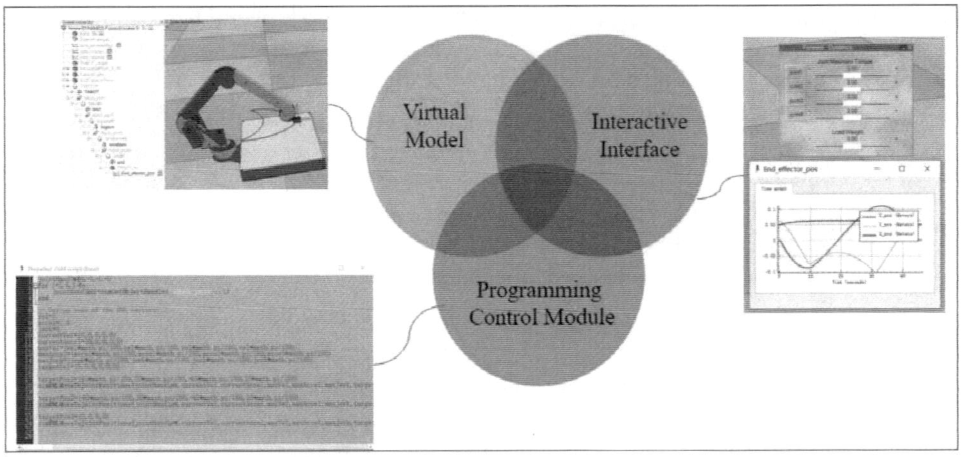

Figure 3. Structure of each virtual robotics laboratory

2.4 Application of virtual robotics laboratory

Application of virtual robotics laboratory is mainly divided into four phases. Each phase is aimed at different teaching purposes from the shallow to the deep to improve students' understanding of robotics and flexible application of basic knowledge step by step, as shown in Figure 4.

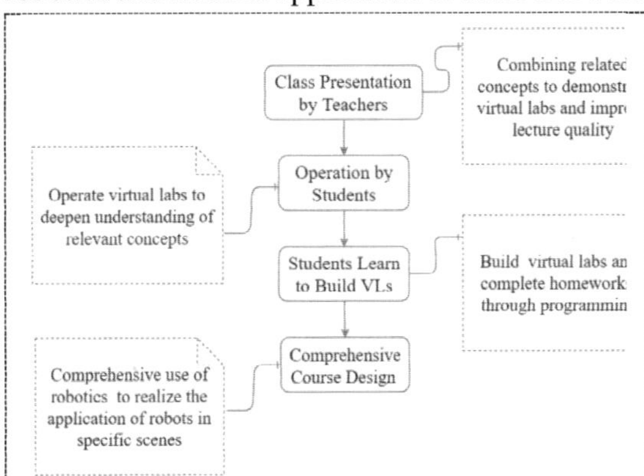

Figure 4. Application of virtual robotics laboratory

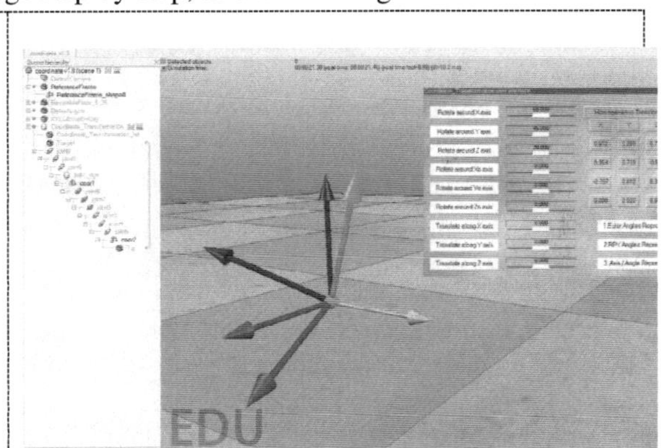

Figure 5. The interface of the coordinate transformation virtual lab

3. Coordinate transformation Virtual lab of robotics

3.1 Introduction to Virtual laboratory of coordinate transformation

Virtual laboratory of coordinate transformation of robotics is composed of a series of coordinate frames and joints. Rotation and translation of coordinate frames are realized by adding rotating joints and prismatic joints between two coordinate frames. OpenGI embedded in V-REP is adopted for UI interface to control coordinate frame model and calculate homogeneous coordinate transformation matrix. Virtual laboratory of coordinate transformation assists students in learning mathematical foundations of coordinate transformation of robotics by vivid and immersing three-dimensional scene, simple and direct operation and quick matrix calculation. The interface of the laboratory is shown in Figure 5. As coordinate transformation has different parameterization method, 3 kinds of different

IOP Conf. Series: Materials Science and Engineering **428** (2018) 012069 doi:10.1088/1757-899X/428/1/012069

calculation methods of homogeneous coordinate transformation matrix are set up here, respectively Euler angle representation method, RPY angle representation method and spindle/angle representation method. Programming module is realized by script document. The main function is to realize interaction between coordinate model and UI: including data transmission, matrix calculation and so on. The coordinate transformation virtual laboratory can perform various translation, rotation, and compound transformations. The rotation transformation can rotate around the current coordinate frame as well as around the fixed coordinate frame, as shown in figure 6.

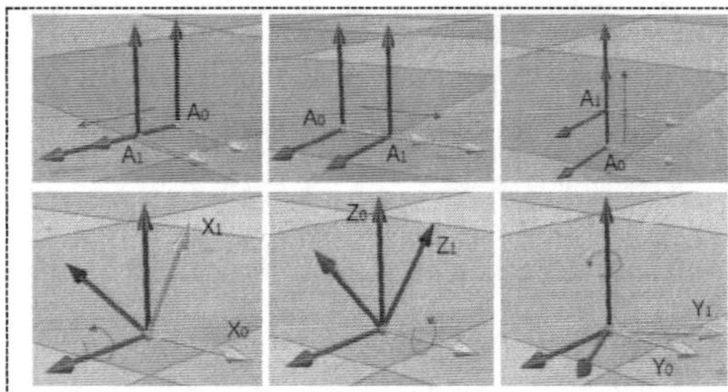

Figure 6. Elementary translations and rotations demonstration

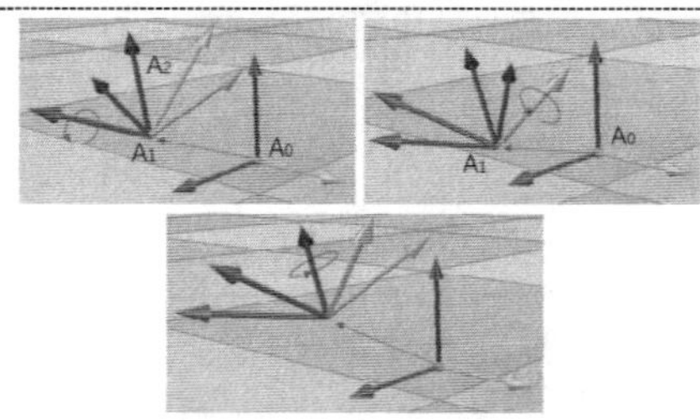

Figure 7. Compound transformation demonstration

Three upper pictures in Figure 6 represent translation along the X axis, Y axis and Z axis respectively, three lower pictures represent rotation around X axis, Y axis and Z axis of current coordinate frame respectively. Visualization and dynamization of coordinate transformation can be realized only by dragging corresponding slider (or entering some numerical values). The compound transformation is shown in Figure 7.

3.2 Teaching demonstration of coordinate transformation

In order to test the experimental results, a coordinate transformation based on the PRY angle representation is shown here:

1.Rotate 60° along the X -axis of the fixed coordinate system (the result is the coordinate frame X1Y1Z1)

2. Rotate 45° along the Y-axis of the fixed coordinate system (the result is the coordinate frame X2Y2Z2)

3. Rotate 30° along the Z-axis of the fixed coordinate system (the result is the coordinate frame X3Y3Z3)

Start the simulation by typing 60°, 45°, and 30° in Rotate around X axis, Rotate around Y axis, and Rotate around Z axis. You can see the changes in Figure 8.

After that, click 2.RPY Angles Representation and the homogeneous transformation matrix will be immediately calculated in the UI as shown in Figure 9.

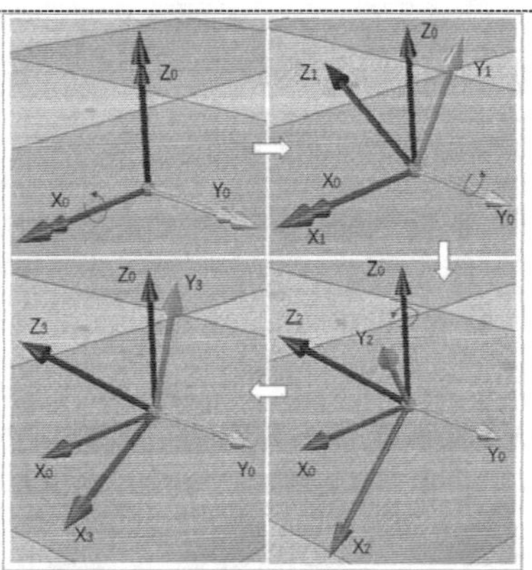

Figure 8. Teaching demonstration

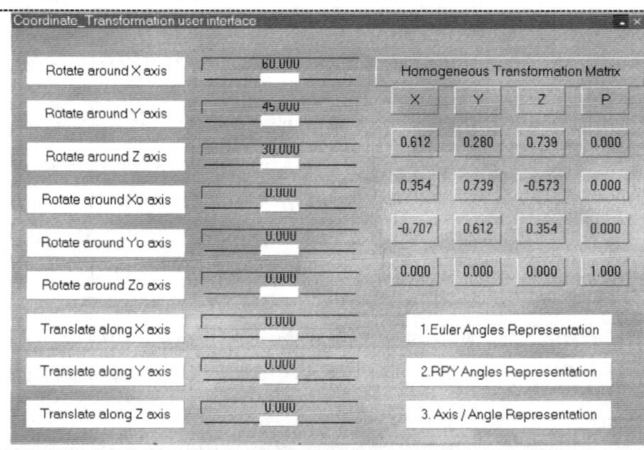

Figure9. Homogeneous transformation matric calculation

4. Kinematics Virtual Lab

4.1 Introduction to kinematics virtual lab

Kinematics is one of the very important basic contents of robotics. Planning of kinematics will directly influence the later dynamics modeling and trajectory planning. In traditional teaching, robotics theory knowledge is usually only explained through textbooks, which is difficult to master these concepts quickly for students. Virtual laboratory of kinematics displays typical robots by visual models so that teachers can show forward and inverse kinematics concepts through simple operations. Students can understand quickly and establish kinematics concept thinking quickly at the same time through the display of D-H coordinate, real-time calculation of matrix transformation and explanation of simple robot algorithm. Robotic kinematics mainly includes forward kinematics and inverse kinematics. In the virtual laboratory, these two situations can be demonstrated separately by the operation of the slider. In the whole process, the D-H coordinate system where all joints are located can also be displayed, and the understanding of kinematics can be deepened through coordinate transformation. After demonstrating the virtual lab, the teacher can explain the corresponding forward and inverse kinematics algorithm through the script.

4.2 Teaching demonstration of kinematics virtual lab

4.2.1 Forward kinematics: Joint positions can be changed by dragging slider in three joints respectively (or inputting angles value through textbox). Positions of end-effector are calculated under drive of script program in real time and manipulator will move to corresponding positions at the same time. As shown in Figure 10, the three joints of the robot arm rotate 30°, 15°, and -15°, respectively.

4.2.2 Inverse kinematics: Set up the robot end-effector position by dragging the X, Y, and Z sliders (or entering the Cartesian space coordinates via the textbox). Then, the joint position will be calculated in real time by the inverse kinematic algorithm of the script program. At the same time, the robot will move to the corresponding posture. The Figure 11. shows the end-effector moving to Cartesian coordinates (0.11, -0.03, 0.04).

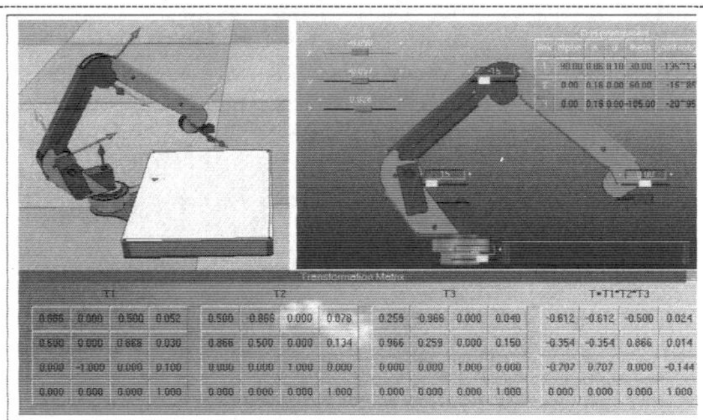

Figure 10. Forward kinematics demonstration

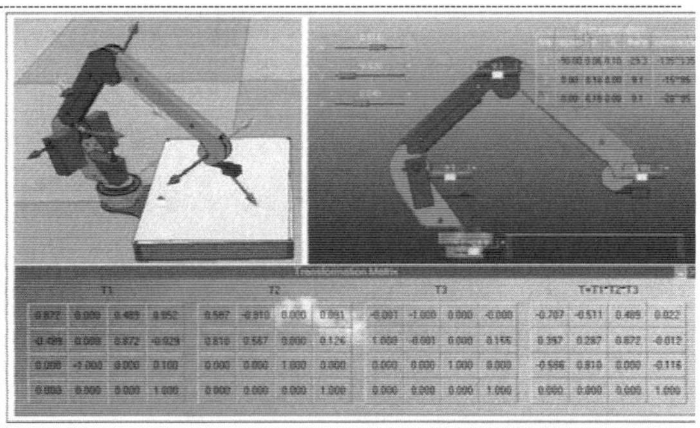

Figure 11. Inverse kinematics demonstration

5. Dynamics Virtual Lab

5.1 Introduction to dynamics virtual lab

The robot's motion characteristics under the influence of various joint forces and moments are studied in dynamics. Firstly, the trajectory control accuracy of robotic manipulator or end-effector depends on the accuracy of dynamics model; secondly, when the robot is constructed, its motion control characteristics can be simulated using dynamics; conversely, results of robot dynamics research can provide basis for robot structure optimization. Similar to kinematics, there are forward and inverse dynamics in robot dynamics. Forward dynamics studies the law of motion of robotic end-effector under the action of various joint forces and moments, and inverse dynamics studies the required joint

IOP Conf. Series: Materials Science and Engineering **428** (2018) 012069 doi:10.1088/1757-899X/428/1/012069

control forces and torques for a given motion trajectory. It can be seen that inverse dynamics plays a key role in real-time control of robots. Virtual laboratory scene of dynamics mainly includes robot model, UI interface, program script and some graphs. Where UI interface mainly realizes maximum torque of controlling all joints and weight of controlling object. Position of each end-effector, all joint angular velocities, joint angular accelerations, torque and other parameters are displayed in the graphs. Dynamics algorithms are implemented in program scripts. Here are two dynamic algorithms for students to learn: Euler-Lagrange formulation and Newton-Euler formulation.

5.2 Teaching demonstration of dynamics virtual lab

It is usually difficult for students to understand the meaning of dynamics analysis in the teaching of robotics. They usually think that kinematics can control manipulator, so why still should learn dynamics. The demonstration of the virtual dynamics laboratory can enable students to have a clearer understanding of dynamics.

5.2.1 Inverse dynamics: In inverse dynamics, firstly, trajectory of end-effector is needed to be set up in inverse dynamics, then joint positions, velocities, accelerations will be calculated through inverse dynamics. Torques and forces of each joint required in different position will be calculated by inverse dynamics and shown in different graphs with curves. Maximum control force or torque of joint required by us can be obtained by graphs, which can provide reference for robot structure design. The Figure 12 is two examples with a track of straight line and a circular arc, and recorded joint positions, velocities, accelerations and joint torques are shown in the graphs.

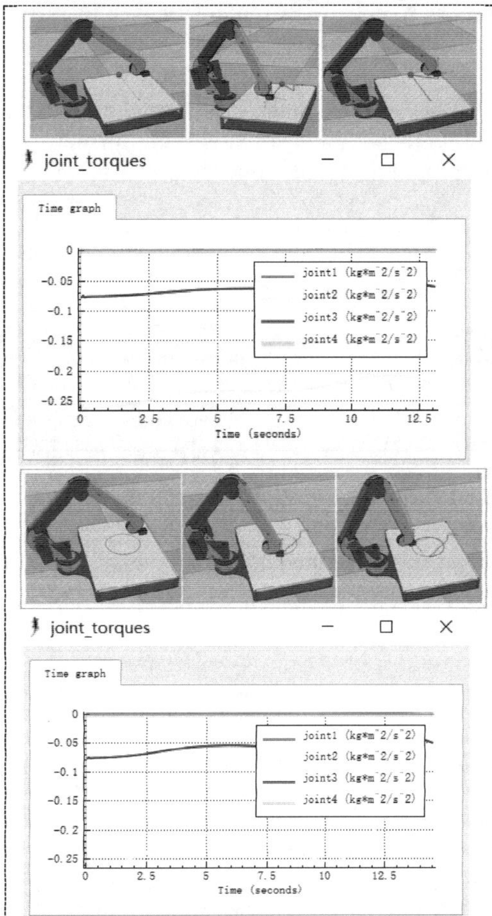

Figure 12. Inverse dynamics demonstration

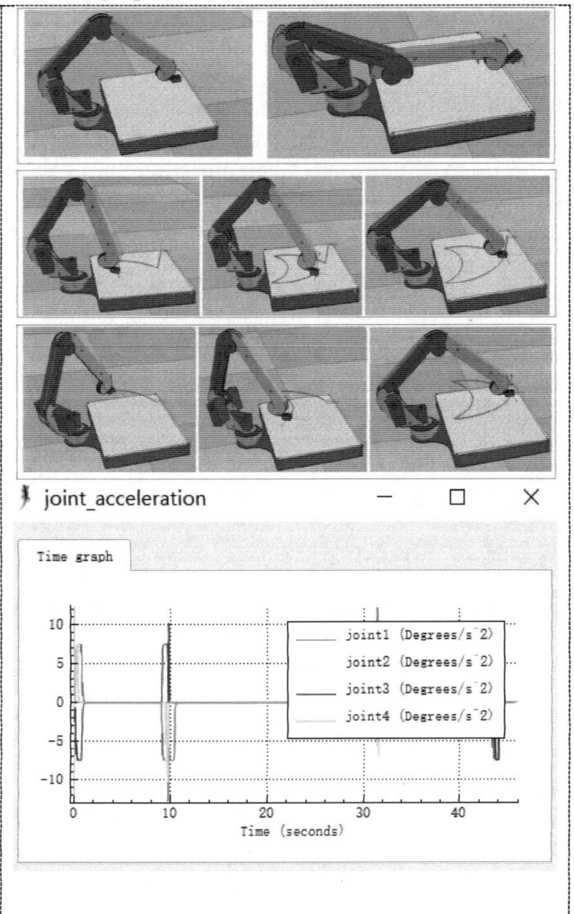

Figure 13. Forward dynamics demonstration

End-effector movement condition can be observed under the condition that different joint torques are given for manipulator operation after above virtual experiment.

5.2.2 Forward dynamics: Maximum torque of each joint is set up and movement rules of the end-effector are observed and studied in positive dynamics. Firstly, the torque of all joints is set to be 0 N·m, as shown in the Figure13. It can be seen that robot arm will fall down due to gravity. When dragging and increasing maximum torque of each joint, for example 0.1 N·m, the end-effector is dragging on the bench. After analyzing the above parameters, it was found that the cause of dragging was mainly due to the insufficient torque of the joint 2. When the joint 2 was set up approximately 0.3 N·m, we can observe that the manipulator is successfully operated. Each joint position, velocity, acceleration are calculated and recorded in graphs in the entire procedure(only the joint acceleration is listed here).

6. Trajectory Planning Virtual Lab

6.1 Introduction to trajectory planning virtual lab

Trajectory planning is the description of robots performing tasks in robots. All the previous foundations are prepared for trajectory planning. The trajectory planning virtual laboratory mainly presents related basic concepts by displaying several typical planning methods. Here, the virtual laboratory shows joint space-based trajectory planning and Cartesian space-based trajectory planning. Point to point (PTP) control and continuous path (CP) control are introduced for trajectory control in trajectory planning.

The virtual laboratory for trajectory planning can realize the motion planning of the virtual robot arm through simple programming and operation. In the virtual environment, the end-effector motion trajectory can be displayed, and the position, angular velocity, and angular acceleration of each joint can be graphically displayed. The Author has presented a complicated function of writing and drawing in the form of welded connection, which is a presentation of trajectory planning.

In the virtual laboratory for trajectory planning, we have used kinematic calculation module of V-REP, used and integrated Reflexxes Motion Library type IV C++. Programming is implemented by some simple APIs.

6.2 Teaching demonstration of trajectory planning

6.2.1 Trajectory planning based on joint space: Trajectory planning is performed directly in the joint space without knowing the entire motion path equation. This is accomplished by specifying the joint displacement function of joints in the motion space. In terms of the virtual model for robot Puma560, the terminal trajectory of robot, and positions, velocities and accelerations of joints are recorded respectively in the graphs as follows: The first thing needs to be defined is the initial and final joint position. Three positions are given in the Table 2. the robot will move to a certain position and then back to the initial position. The entire movement is shown in Figure 14.

Table 2. Three Positions

joint	Joint1	Joint2	Joint3	Joint4	Joint5	Joint6
Initial Pos	0°	0°	0°	0°	0°	0°
Pos 1	45°	45°	45°	30°	15°	0°
Final pos	0°	0°	0°	0°	0°	0°

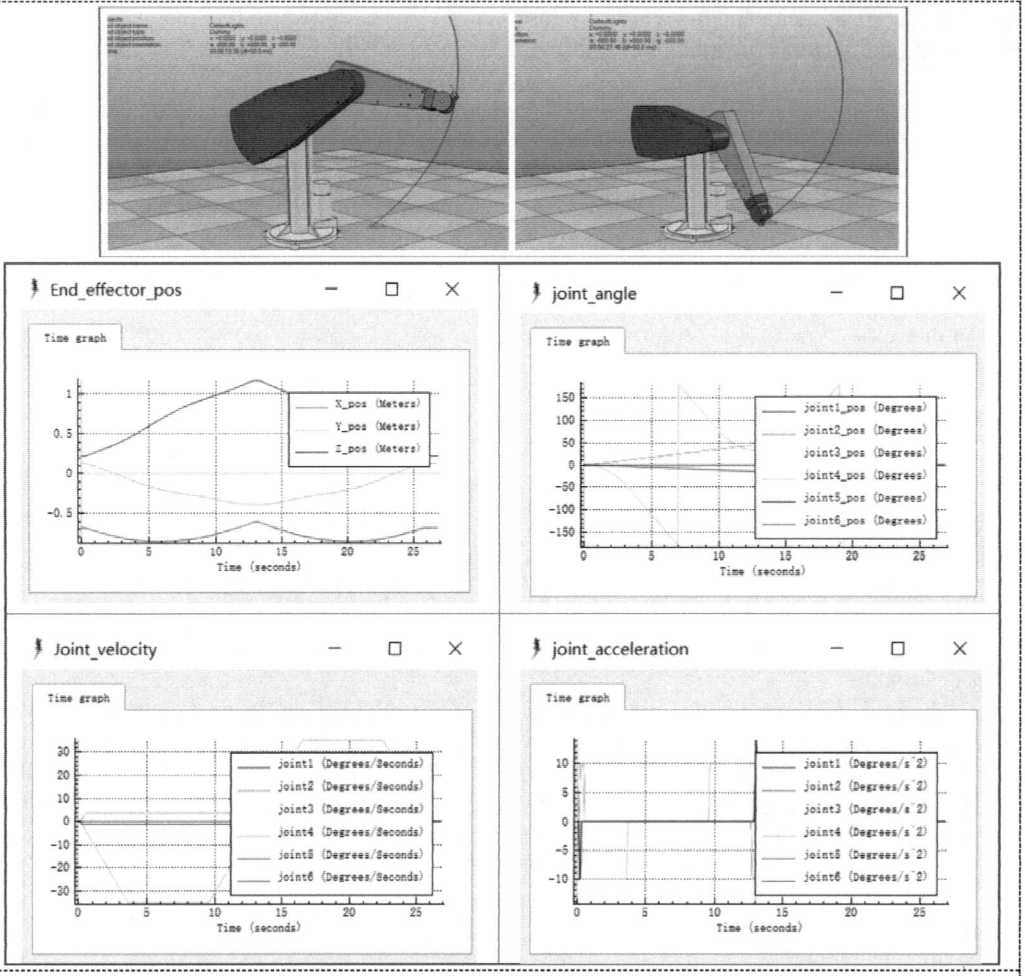

Figure 14. The scene of trajectory planning laboratory for Puma560

6.2.2 Trajectory Planning Based on Cartesian Space: Also known as Orthographic Space Trajectory Planning. The motion path is known, and the key points are converted into joint space by inverse kinematics to control the joint rotation. Trajectory planning of Delta manipulator based on Cartesian space is shown in Figure 15. Pictures show the scenarios of motion along the arched trajectory and arc trajectory.

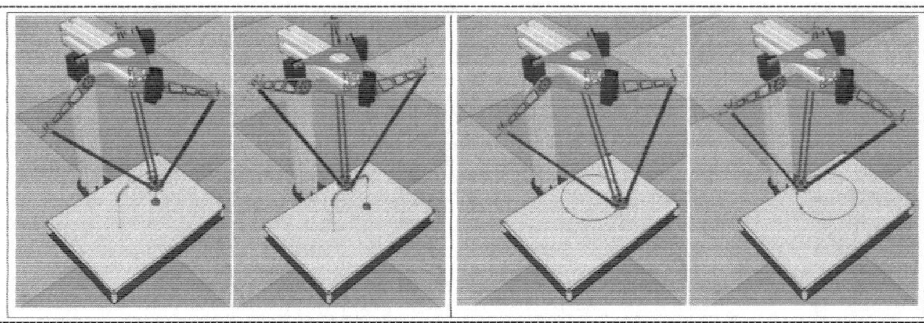

Figure 15. The scene of trajectory planning laboratory for Delta

6.2.3 Point to point control (PTP): The manipulator has to move from an initial to a final joint configuration in a given time; In this case, the end-effector path is of no concern [8]. In the middle of the trajectory, there are only joint geometric constraints, maximum speed and acceleration constraints.

In the trajectory planning virtual laboratory, Dummy can be added as a virtual trajectory point, and then the inverse kinematics algorithm is used to calculate the joints vectors corresponding to the point. As shown in the Figure 16, it is desirable for the robot arm to reach these four positions, but do not care about the intermediate process. Adds the geometric constraints, maximum velocities and acceleration constraints of the robotic joint programmatically. Here, we added a spot weld head to show the trajectory of the tip on the platform. The result was shown in Figure 16.

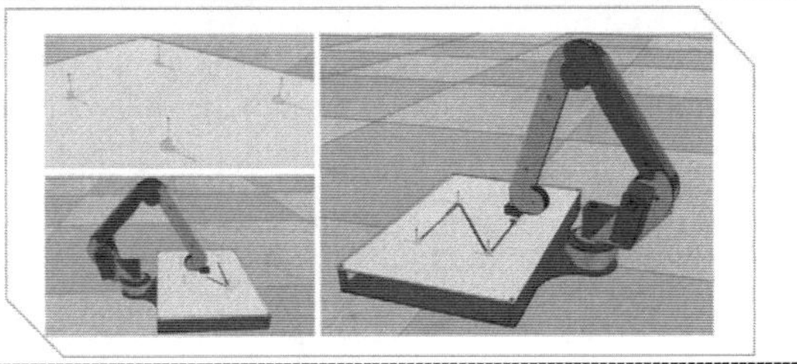

Figure 16. The scene of PTP control

6.2.4 Continuous trajectory control (CP): The end-effector has to follow a prescribed and controlled path within a certain accuracy range in strict accordance with the predetermined trajectory and speed, and it is required that the speed be controllable, the trajectory be smooth, and the movement be performed smoothly and completely. In the trajectory planning virtual laboratory, students can perform continuous path planning by adding a path or generate a continuous path by importing a path file in CVS format. Then, the end-effector is programmed to follow the path movement. The Figure 17. shows generated paths and the motion of continuous trajectory control in the virtual laboratory of trajectory planning.

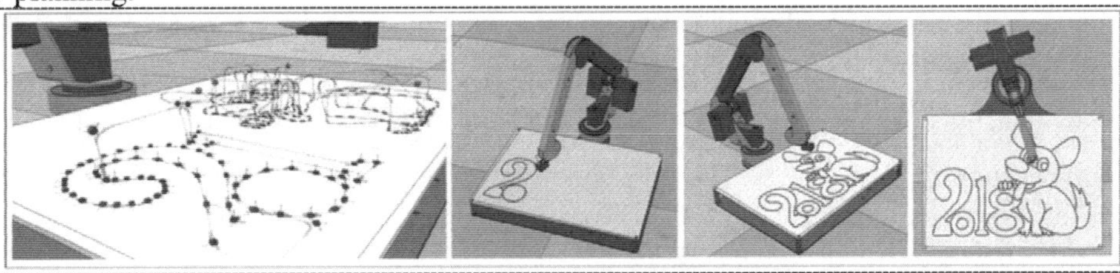

Figure 17. The scene of CP control

7. Robotics Course Comprehensive Design Experiment Plan

In order to test and reinforce students' basic knowledge and primarily develop their comprehensive design capability for the practical robotic application scenario after they have learnt basic courses on robotics, we have proposed comprehensive design of robotics courses, such as implementing cooperation between robot palletizer and conveyor belt to realize pick and place task in assembly line, and robots palletizing. In comprehensive design, overall knowledge framework of robotics will be considered and basic knowledge of each part will be integrated into a design in a way that reinforces basic knowledge, trains the capacity to utilize basic knowledge, and implements high-quality teaching objectives of robotics by the idea learning by doing or learning by projects. Results of comprehensive design of robotics courses of China University of Mining and Technology (Beijing) are shown in Figure 18 and Figure 19.

Figure 18. Robotics Course Comprehensive Design

Figure 19. Robotics Course Comprehensive Design

8. Conclusion

In the paper, a planning scheme for V-REP-based robotics teaching virtual experiments was proposed. Main basic knowledge about robotics involved is demonstrated and explained for teaching in the virtual laboratory established by the model of typical manipulators. The robotic virtual laboratory established covers robot arm of palletizer, Puma560, Delta, SCARA and other typical robot arms. It is suitable for primary learning on robotics. The robotic virtual laboratory shows the obscure concepts of robotics in a manner of simple interaction and in the form of virtual simulation. The virtual laboratory will save more funds for experimental equipment for more colleges and universities, save teaching space and improve teaching efficiency. Besides, V-REP is selected as the experimental platform. On the one hand, V-REP has a good modularity function, which can provide convenience for subsequent integrated design and subject research. It is also a good tool for subject research on robots and can provide a good platform for prototype design and algorithm verification to prepare for future scientific research. The extensive expansion capabilities of V-REP will provide more powerful advantages in robot teaching and subject research. The robotic virtual laboratory will also help teachers improve the quality of teaching with more functions and help students learn robotics-related courses.

References

[1] Corke P I. 1996 A robotics toolbox for MATLAB *IEEE Robotics & Automation Magazine* pp 24-32.
[2] Sadanand R, Joshi R P and Chittawadigi R G et al. 2016 Virtual Experiments for Integrated Teaching and Learning of Robot Mechanics Using RoboAnalyzer. *Robotics and Factories of the Future.*
[3] RoKiSim – available at: http://www.parallemic.org/RoKiSim.
[4] A. Gil, O. Reinoso, J. M. Marin, L. Paya and J. Ruiz 2015 Development and deployment of a new robotics toolbox for education *Comput. Appl.Eng. Educ.* pp 443–454.
[5] Fei Ma and Ruiqing Jia 2017 Study and Design of Virtual Laboratory in Robotics-Learning. 2017 *International Conference on Applied Mechanics and Mechanical Automation.*
[6] S. Kucuk, Z. Bingul 2010 An off-line robot simulation toolbox *Comput.Appl. Eng. Educ.*18 pp 41–52.
[7] Rohmer E, Singh S P N and Freese M 2013 V-REP: A versatile and scalable robot simulation framework *International Conference on Intelligent Robots and Systems.* IEEE, 2013: pp 1321-1326.
[8] Saha S K 2001 *Introduction to Robotics* (Beijing: China Machine Press) p300.

IOP Conf. Series: Materials Science and Engineering **428** (2018) 012070 doi:10.1088/1757-899X/428/1/012070

AUV Tunnel Tracking Method Based on Adaptive Model Predictive Control

Chenkai Jia[1], Zhou Chen[2], Guocheng Zhang[1,*], Lei Shi[2], Yushan Sun[1] and ShuoPang[1]

[1] Science and Technology on Underwater Vehicle Laboratory. Harbin Engineering University. Harbin, China
[2] Zhejiang Design Institute of Water Conservancy & Hydro-electric Power. Hangzhou, China

* Zhangguocheng168@126.com

Abstract. In this paper, a model predictive path tracking controller is designed for the path tracking of the autonomous underwater vehicle (AUV). Firstly, Serret-Frenet coordinate system is introduced, and the path tracking error system model is established in this coordinate system. Based on Lyapunov theory and Backstepping method, the forward heading of AUV is conversed to the target value. Then, a nonlinear model predictive control design constraint path following control law of the AUV based on rolling to solve constrained optimization problems, satisfy the constraint of extension control input, to hit the target heading control. In the end, AUV-T, an underwater robot developed by Harbin engineering university, was used to test the proposed control law. The results showed that the controller showed good tracking effect.

1. Introduction

With the application of autonomous underwater vehicle (AUV) in ocean research and development, path tracking control has become one of the important technologies of AUV motion control. Path tracking is an expectation trajectory that controls AUV to track a course requirement and performance constraint, and is independent of time. Is the research background of this paper is tunnel detection, AUV in the tunnel and tunnel side wall fixed distance within the same horizontal plane along the tunnel track, and through the AUV stem of the front camera and midship section to install 5 groups of lights and cameras to see inside the tunnel environment, thus achieve the purpose of the inspection on tunnel inside detection.

The serret-frenet coordinate system is one of the more classical path tracking control methods, which is widely used in the research of the path tracking control of robots and unmanned ships. There are some literatures[1-7] have studied the path tracking control problem of the motion control system, but it is not comprehensive enough to consider the constraint of AUV physical actuator during the design of the controller. While the model predictive control (MPC) is an advanced intelligent feedback control method, it through to the definition of some time domain in the future on the prediction and forecast input is presented to solve constrained optimization problems to get the control input, the mechanism itself make it has a good ability to deal with constraints explicitly [8-12].

Based on the above research results, this article will first introduce Serret - Frenet frame, and path tracking error system model in this system, and based on the Lyapunov theory and Backstepping method will advance of the AUV heading converge to the target. Then, a nonlinear model predictive control design constraint path following control law of the AUV based on rolling to solve constrained optimization problems, satisfy the constraint of extension control input, thus to control target heading, achieve the result of path tracking.

2. Problem description

In this paper, the AUV is fully driven due to the motion of only the horizontal direction, and the force of x direction and y direction and the control moment of the heading direction are respectively present. The horizontal kinematics equation.is:

$$\dot{x} = u\cos\psi - v\sin\psi \tag{1}$$

$$\dot{y} = u\sin\psi + v\cos\psi \tag{2}$$

$$\dot{\psi} = \mathrm{r} \tag{3}$$

The rotation Angle of {B} around the Z axis is obtained by the new coordinate system {W}. Under {W}, the AUV motion equation can be expressed as:

$$\dot{x} = v_t\, cos\, \psi_W \tag{4}$$

$$\dot{y} = v_t\, sin\, \psi_W \tag{5}$$

$$\dot{\psi_W} = r + \dot{\beta} \tag{6}$$

For the dynamic equation of AUV, based on the literature [13], the AUV horizontal motion and vertical plane motion decouple, so it can ignore the vertical swing and pitch motion. In addition, the AUV-T cross section of the tunnel underwater robot developed by Harbin engineering university is an ellipse, which can be neglected. The above assumptions are combined. The dynamic equation of AUV is as follows:

$$m(\dot{u} - vr) = X_{rr}r^2 + X_{\dot{u}}\dot{u} + X_{vr}vr + X_{uu}u^2 + X_{vv}v^2 + T_x \tag{7}$$

$$m(\dot{v} - ur) = Y_{\dot{r}}\dot{r} + Y_{\dot{v}}\dot{v} + Y_{ur}ur + Y_{v|r|}v|r| + Y_{uu}u^2 + Y_{uv}uv + Y_{v|v|}v|v| + T_y \tag{8}$$

$$I_z\dot{r} = N_{\dot{r}}\dot{r} + N_{r|r|}r|r| + N_{\dot{v}}\dot{v} + N_{ur}ur + N_{v|r|}v|r| + N_{uu}u^2 + N_{uv}uv + N_{v|v|}v|v| + M_{T_z} \tag{9}$$

Among them, m is the quality of AUV, $X_{(\cdot)}$, $Y_{(\cdot)}$ and $Z_{(\cdot)}$ is the hydrodynamic coefficients, I_z is the moment of inertia, T_x and T_y is control, M_{T_z} is to control the torque.

3. AUV path tracking control

We need to study the kinematics model of AUV in frame {F}.

$$\dot{s}_1 = -\dot{s}(1 - c_c y_1) + v_t\, cos\, \psi_e \tag{10}$$

$$\dot{y}_1 = -c_c\dot{s}s_1 + v_t \sin\psi_e \tag{11}$$

$$\dot{\psi_e} = r + \dot{\beta} - c_C\dot{s} \tag{12}$$

Where $c_c(s)$ is the curvature of the expected path, and s is the parameter variable of the expected path.

The AUV kinematics model is formula (1) - (3), and the path tracking error model in serret-frenet coordinate system is formula (10) - (12). Based on literature [14], the Angle of sight navigation method, for a given expected path S and the expected surge speed u_d of AUV, design control as follows, for any initial Q position, make the AUV tracking heading error converges to converge to zero.

First of all, choose the Lyapunov function as $V_1 = \frac{1}{2}(\varphi_e - \delta)^2$, and design the control as follows:

$$r = \dot{\delta} - \dot{\beta} - k_1(\psi_e - \delta) + c_C\dot{s} \tag{13}$$

Where k_1 is the controller gain, then there is：

$$\dot{V}_1 = -k_1(\psi_e - \delta)^2 \leq 0 \tag{14}$$

Under the guidance of the navigation Angle δ, AUV, the trajectory convergence:

$$\{\Omega_1 | (x, y) \in R^2, \psi_e = \delta\} \tag{15}$$

After obtaining the AUV target heading, the design path tracking prediction controller is designed to control the heading of the target. Basic principle of path tracking controller is shown in figure 1.

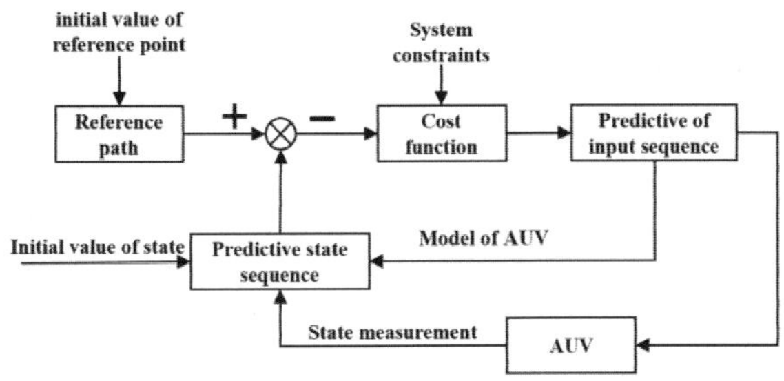

Figure 1. A schematic diagram of AUV path tracking control based on MPC.

Based on literature [15], the cost function of defining the constraint optimization problem at each moment t_k is:

$$J\left(x_{tk},\bar{e}(\cdot),\bar{\theta}(\cdot),\bar{u}(\cdot),\bar{v}(\cdot)\right)=\int_{t_k}^{t_k+T_P}F\left(\bar{e}(\tau),\bar{\theta}(\tau),\bar{u}(\tau),\bar{v}(\tau)\right)d\tau \quad (16)$$

Where $F(e,\theta,u,v)=[e^T \quad \theta]Q[e \quad \theta]^T+[u \quad v]R[u \quad v]^T$ and $E(x,z)=[x^T \quad z^T]P[x^T \quad z^T]^T$ are called the state cost function and the terminal cost function. $Q \in R^3$ and $R \in R^2$ are respectively the state cost weight matrix and the input cost weight matrix. $T_p$is prediction horizon,$\bar{e}(\cdot)$and$\bar{u}(\cdot)$ is the error and input of prediction horizon. $e(\cdot),u(\cdot)$ is the error and input of the actual system, $\bar{v}(\cdot)$ is the virtual input of the control parameters. The constraint optimization problem solved by t_k at each moment is:

$$min \ J\left(x(t_k),\bar{e}(\cdot),\bar{\theta}(\cdot),\bar{u}(\cdot),\bar{v}(\cdot)\right) \quad (17)$$

$$\forall\tau \in \left[t_k,t_k+T_p\right] \quad (18)$$

$$\dot{\bar{x}}(\tau)=f\left(\bar{x}(\tau),\bar{u}(\tau)\right),\bar{x}(t_k)=x(t_k) \quad (19)$$

$$\dot{\bar{z}}(\tau)=\begin{pmatrix}0 & 1\\ 0 & 0\end{pmatrix}\bar{z}(\tau)+\begin{pmatrix}0\\ 1\end{pmatrix}v(\tau) \quad (20)$$

$$\dot{\bar{e}}(\tau)=\eta(\tau)-p\left(\bar{z}(\tau)\right) \quad (21)$$

$$\bar{\theta}(\tau)=\bar{z}_1,\bar{u}(\tau)\in U \quad (22)$$

$$\bar{z}(\tau)\in z,\bar{v}(\tau)\in V \quad (23)$$

For constrained optimization problems (17-23), the starting position of AUV can be used as the initial value of AUV prediction model. In order to ensure that AUV can converge to the reference path as soon as possible, the path parameters corresponding to the nearest path reference point can be selected as the initial point of the path prediction model, namely:

$$z_0=[\theta_{min},p(\theta_{min})]^T \quad (24)$$

$$\theta_{min}=arg\,min\,\|\eta_9-[\theta,p(\theta)]^T\| \quad (25)$$

At the same time, the path parameter state constraint domain can be set to ensure that $\dot{\bar{\theta}}\geq 0$, that is, the forward orientation of the guaranteed path. The optimal input of constraint optimization problem (16) at time t_k is:

$$\left(u_k^*(\cdot),v_k^*(\cdot)\right)=\underset{\bar{u}(\cdot),\bar{v}(\cdot)}{argmin}J\left(x(t_k),\bar{e}(\cdot),\bar{\theta}(\cdot),\bar{u}(\cdot),\bar{v}(\cdot)\right) \quad (26)$$

If the sampling period of AUV path tracking system is δ, namely $t_k=k\delta$, through the system at time t model to calculate the system input and the state and to solve the constrained optimization problem (17-23) optimized input solution is $u_k^*(\cdot)$, when applied to the system input is:

$$(u_k,v_k)=\left(u_k^*(t-t_k;x(t_k)),v_k^*(t-t_k;x(t_k))\right),\forall t\in[t_k,t_{k+1}) \quad (27)$$

The optimal input process (26) for each time t is applied to the control system only (27). Meanwhile, the time domain is predicted from $[t_k,t_k+T_P]$ to $[t_{k+1},t_{k+1}+T_P]$, and the constraint path planning problem of t_{k+1} time is obtained.

To sum up, the constraint path tracking controller algorithm is:

IOP Conf. Series: Materials Science and Engineering **428** (2018) 012070 doi:10.1088/1757-899X/428/1/012070

1) obtain the initial value of path parameters z_0 by solving formula (24-25);
2) at the sampling time, t_k measure system state x_{t_k}, solve the optimization problem equation (17-23), and obtain the optimized solution formula (26);
3) apply equation (27) to the system and roll the predicted time domain to the next moment, when the sampling time is t_{k+1};
4) let k=k+1, back to 2).

The MPC algorithm can guarantee the stability of the algorithm by means of zero terminal constraint or terminal state constraint and terminal cost function. Therefore, when designing the MPC controller in the engineering practice, the terminal state constraint is often dropped, and the control system is restrained by the longer prediction time domain.

4. Experimental analysis.

In order to prove the effectiveness of the tracking controller proposed, the auv-t robot developed by Harbin engineering university was used in this section to conduct experiments in a canal in Hangzhou.

About test, has provided about a quarter of voltage, ensure AUV in section 1 of the lateral speed all the way, the robot began to trace the initial heading Angle $\psi = -60°$, tracking distance S = 3 m. The experimental site and the schematic diagram are shown in figure 2 and figure 3.

Figure 2. Actual scene of the test site.

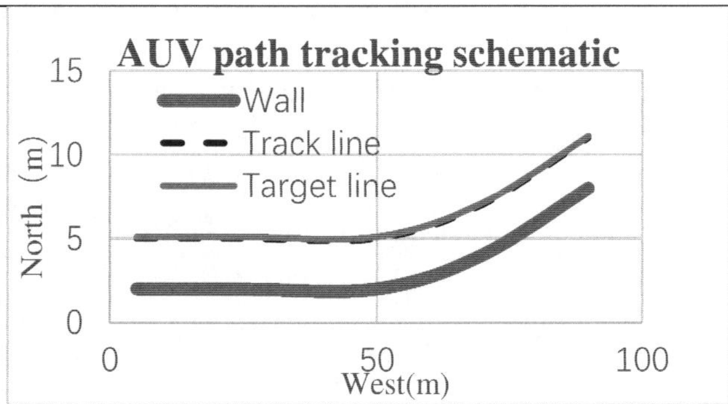

Figure 3. AUV path tracking schematic.

Figure 4, Figure 5 and Figure 6 are respectively the heading change, distance change and voltage change of the AUV path tracking.

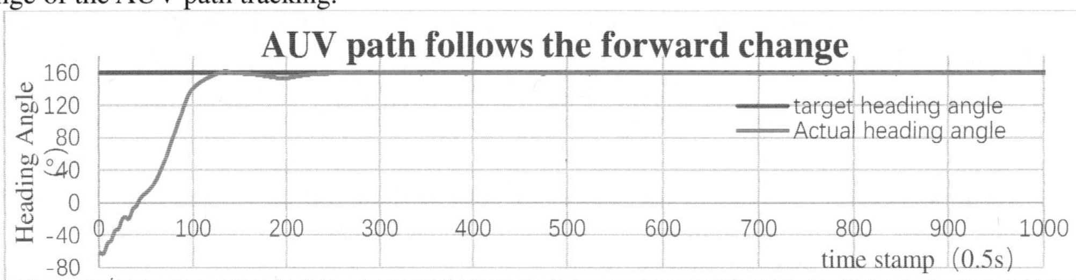

Figure 4. AUV path follows the forward change.

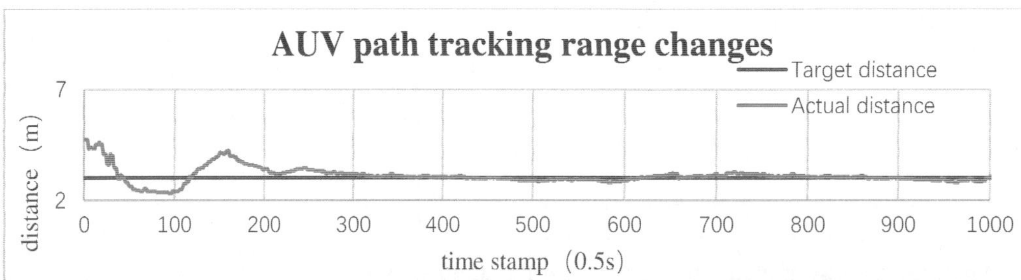

Figure 5. AUV path tracking range changes.

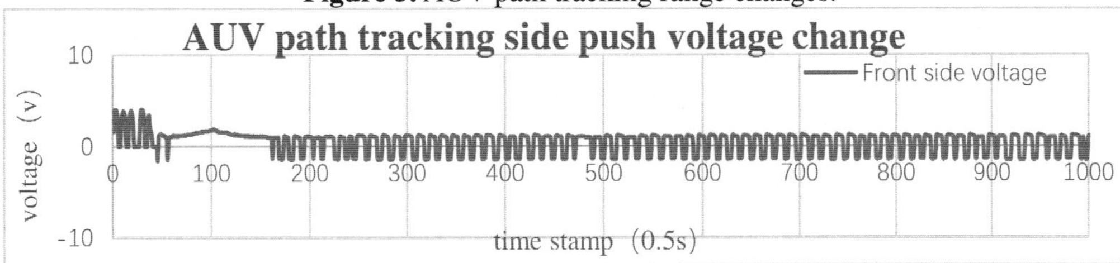

Figure 6. AUV path tracking side push voltage change.

5. Conclusion

In this paper, the path tracking of the plane motion AUV is studied. Firstly, the adaptive control method based on the Lyapunov theory and Backstepping technique is proposed in the serret-frenet coordinate system, and the target heading of AUV is determined. Then, a nonlinear MPC is used to track the reference path of AUV under constraint conditions. During the design of the controller, the freedom of control process is added, and the reference model is introduced to predict the controller. The last test in the outfield lake, a simulated tunnel environment on the wall is distance tracking, finally the analysis of experimental data proved that the design of the underactuated AUV path tracking predictive controller has good tracking performance.

Acknowledgments

This research was part of the project (Grant No. 51779057, 51709061) supported by Natural Science Foundation of China. Great thanks are addressed to them by the research team.

Reference

[1] Lapierre L, Soetanto D. Nonlinear path-following control of an AUV [J]. Ocean Engineering, 2007, 34: 1734-1744.

[2] Lapierre L, Paseoal A. Nonlinear path following with application to the control of autonomous underwater vehicle [J]. Proc. IEEE Conf. Dec. Control, Maui, Hawaii, USA, 2003, 1256-1261.

[3] Lapierre L, Jouvencel B. Robust nonlinear path-following control of an AUV [J]. IEEE Journal of Ocean Engineering, 2008,33(2):89-102

[4] Do K D, Pan J. State-and output-feedback robust path-following controllers for underactuated ships using Serret-Fernet frame [J]. Ocean Engineering, 2004,31(5-6):587-613

[5] Jia H M, Cheng X Q, et al. Three-dimensional path tracking control for an underactuated AUV based on discrete-time sliding mode prediction[J] Control and Decision, 2011,26(10):1452-1458.

[6] Song J, Xu X M, Autopilot controller designing for subsea robot in the dive plane [J]. Journal of Shanghai Jiaotong University,1997,31(3):24-27.

[7] Wang H J, Chen Z Y, Jia H M, et al. Three-dimensional path-following control of underactuated unmanned underwater vehicle using feedback gain backstepping [J]. Control Theory and Applications, 2014, 31(1):66-77.

[8] Xi Y G, Li D W, Lin S, Model predictive control-status an challenges [J]. Acta Automatica Sinica, 2013 ,39(3):222-236(in Chinese).

[9] Mayne D Q, Model predictive control: recent developments and future promise [J]. Automatica, 2014,50(12):2967-2986.

[10] Ghaemi R, Oh S, Sun J. Path following of a model ship using model predictive control with experimental verification [C]. //American Control Conference, June 30-July 2, 2010, Baltimore, Maryland, Baltimore, Maryland: IEEE, 2010: 5236-5241.

IOP Conf. Series: Materials Science and Engineering **428** (2018) 012070 doi:10.1088/1757-899X/428/1/012070

[11] Li Z, Sun J, Oh S. Path following for marine surface vessels with rudder and roll constraints: an MPC approach [C]. //American Control Conference, June 10- 12, 2009, St. Louis, Missouri . St. Louis, Missouri: IEEE, 2009: 3611-3616.
[12] Zheng H, Negenborn R R, Lodewijks G. Trajectory tracking of autonomous vessels using model predictive control [C]. //Proceedings of the 19th IFAC World Congress. Cape Town, South Africa: Elsevier, 2014: 8812-8818
[13] Jiang X S, Feng X S, Wang L T. Unmanned Underwater Vehicles [M]. Liaoning Science and Technology Publishing House, 2002.
[14] Pettersen K Y, Lefeber E. Way-point tracking control of ships [J]. Proc. of 40th IEEE Conference on Decision and Control. 2001, 940-945.
[15] Ferreau H J, Ortner P, Langthaler P, et al. Predictive control of a real-world diesel engine using an extended online active set strategy [J]. Annual Reviews in Control, 2007, 31(2):293-301.

IOP Conf. Series: Materials Science and Engineering **428** (2018) 012071 doi:10.1088/1757-899X/428/1/012071

Design and Implementation of a New Climbing Robot for High Voltage Transmission Tower

Wentao Jiang[1,3], Ronghai Liu[2], Yingchun Yang[2], Shuting Wan[1], Xinliang Guo[2], Zonghan Jiao[2] and Hongjiang Yu[2]

[1] Department of Mechanical Engineering, North China Electric Power University, Baoding 071003, China
[2] Electric Power Research Institute, Yunnan Power Grid Co., Ltd. , Kunming 650217, China
[3] Graduate Workstation of North China Electric Power University & Yunnan Electric Power Grid Corporation, Kunming 650217, China

Email: 192186182@qq.com, 23697835@qq.com

Abstract. A new type of high voltage transmission tower climbing tower robot is designed. Firstly, the structure of the transmission tower is analyzed to determine that the robot manipulator is a five-degree-of-freedom structure. Secondly, a manipulator capable of automatically positioning and tightening the angle steel surface was designed according to the L-shaped angle steel of the tower. The manipulator adopts electromagnetic absorption and grasping and tightening the angle steel. Thirdly, the theoretical calculation of climbing steps is carried out to determine the maximum step of the torsional climbing gait. Then STM32 SCM is used as a carrier to carry out various functional modules to achieve effective control of the robot's rotation, movement and positioning. Finally, the finite element simulation of the robot model was performed to analyze the deformation and stress of each component during the climbing process, and the rationality of the structure was verified. In addition, the experiment of tower climbing was carried out with the experiment iron tower in the outdoor. The results show that the robot manipulator is tightened and reliable, the weight bearing is large, the degree of freedom is not redundant, the climbing speed is fast, and the movement of obstacle crossing can be realized smoothly.

1. Introduction

In recent years, the power loss caused by accidents of high-voltage transmission towers has become increasingly serious. In order to reduce tower accidents, the power grid company regularly arranges artificial tower inspections. However, this detection method is inefficient and has a large potential safety hazard and is affected by the natural environment [1].

At present, a large number of researches on tower climbing robots have been carried out in China, and many representative robots have been designed [2-5]. Although the designed robot meets climbing requirements through theoretical simulation, there are many problems in the actual climbing process. For example, the lack of grasping force of the manipulator is easy to cause the slipping accident, the heavy weight of the complicated structure, the positioning accuracy is poor, the low efficiency of the climbing tower and poor obstacle avoidance.

In this paper, a five-degree-of-freedom self-reconfigured tower-climbing robot with simple structure and fast and reliable design is designed. The advantage is that the manipulator uses an electromagnet to adsorb the L-shaped angle steel and tighten it with a hand, and the clamping force is great to prevent falling. Secondly, the self-reconfigurable structure method makes the structure simple and has little load and no redundant degree of freedom. Finally, a torsional climbing gait makes the stepping speed faster and the obstacles more advantageous.

2. Robot mechanism design

2.1. Field environment and requirements

The transmission line tower is made up of single equilateral L angle steel or combined L angle steel by electric welding or bolt connection. The main body of the tower is composed of four L-shaped angles of the main material. The connection between the oblique material and the main material is directly connected with bolts or a gusset plate method. The tower main material and the main material are directly connected via the outer L-shaped angle steel or the L-shaped angle steel plus the substrate. In addition, the cross arm and tower body are mainly connected with additional gusset plates.

The function of climbing obstacle can be achieved by crossing the L angle steel and bolts at the junction of the main material and the main material, the joint plate of the junction of the inclined material and the main material, and the vertical edge of the oblique material at the direct connection between the inclined material and the main material. As shown in Figure 1.

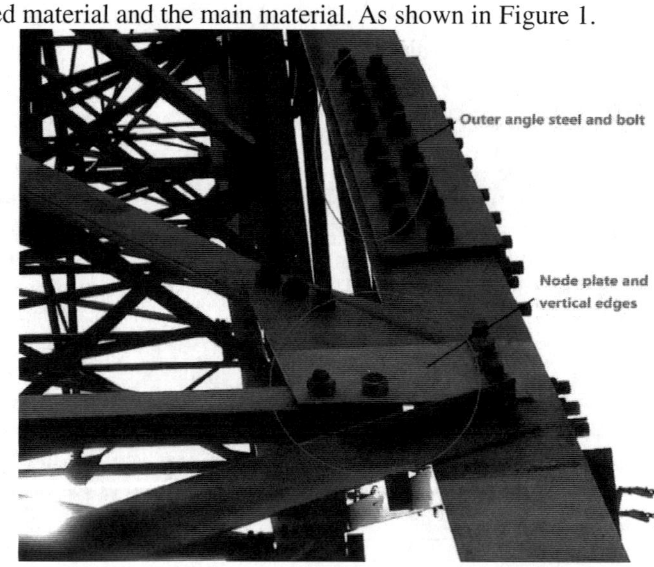

Figure 1. High-voltage transmission tower obstacles.

2.2. Robot Structure Design

The robot structure is designed to ensure that the robot can reach any position in the three-dimensional truss, and to meet climbing and obstacle avoidance functions. In addition, we must also ensure that the structure is simple, compact, flexible, and of low weight.

There are two feasible schemes to select the degree of freedom to meet the above features, including five degrees of freedom and six degrees of freedom [6]. According to the working environment of the robot, the symmetry of the structure and the difficulty of solving the inverse kinematics, the structure diagram of the robot is determined as shown in Figure 2 below.

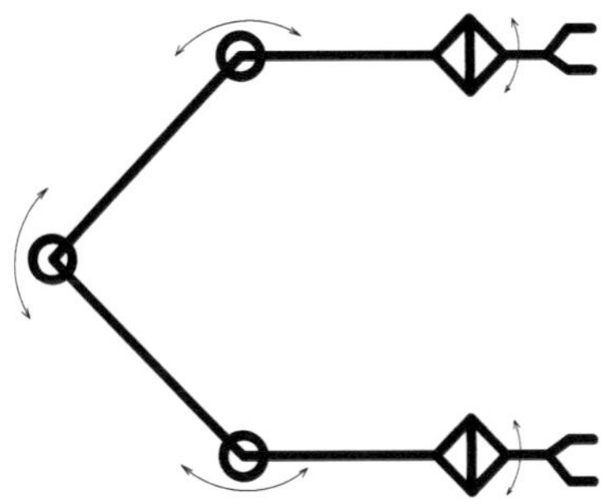

Figure 2. Robot structure diagram.

From the above figure, the robot structure consists of four arms and two manipulators, and there are five degrees of freedom. Figure 3 below shows a three-dimensional model of the robot.

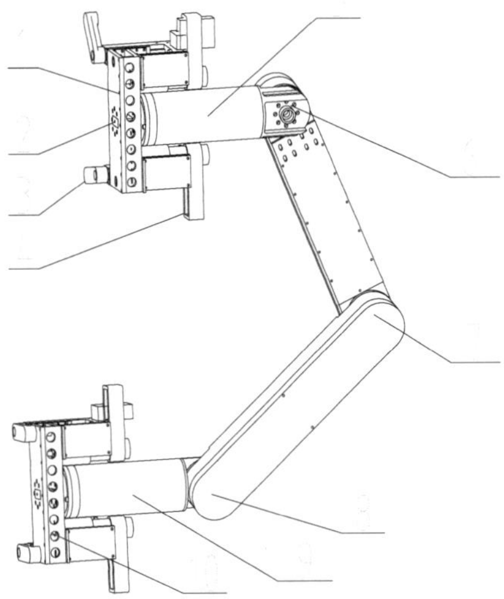

Figure 3. 3D model of robot (1. Manipulator 1 2. Photoelectric sensor 3. Hand grabbing 4. Ranging sensor 5. Shaft 1 6. Shaft2 7. Shaft 3 8. Shaft 4 9. Shaft 5 10. Manipulator 2).

As can be seen from the above figure, the end robot arm and the manipulator are connected by an I-joint module, and the robot arm is connected by an L-joint module. Each joint module is composed of a driving motor and a transmission mechanism, in which the motor provides power, and the transmission mechanism is used to change the torque of the power machine to meet the requirements of the working machine. By analyzing and calculating the output torque of each module joint of the robot, the transmission plan is determined as shown in Table 1.

Table 1. Transmission scheme of joint module.

Module name	Motor model	Motor power	Primary transmission	Secondary transmission	Three-stage transmission
Type I	Maxon re25	200W	Planet gear (1:14)	Harmonic reducer (1: 80)	
Type L	Maxon re25	200W	Turbine worm (1: 120)	Timing belt (1:2)	Harmonic reducer (1: 80)

The above servo motors adopt the motor driver with the advantages of input pulse and direction signal for stepping mode control and high precision PWM signal speed control to ensure that each joint is an independent electromechanical system for convenient control. In addition, in order to achieve precise positioning and ensure motor rotation accuracy, the motor uses a line number of 100 encoders. Calculated from the above table, I joint module can achieve a 1:1120 reduction ratio, and the rotation accuracy is about 0.32 degrees. L module transmission ratio is 1:1280, its motor rotation accuracy is high, about 0.002 degrees. The structural models of I and L modules are shown in Figures 4 and 5 respectively.

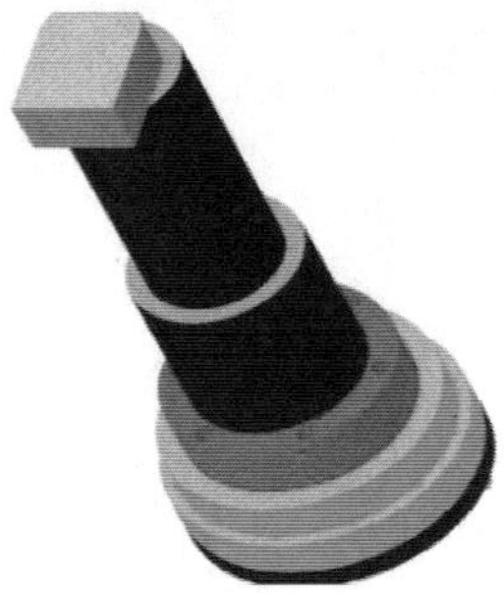

Figure 4. I joint module.

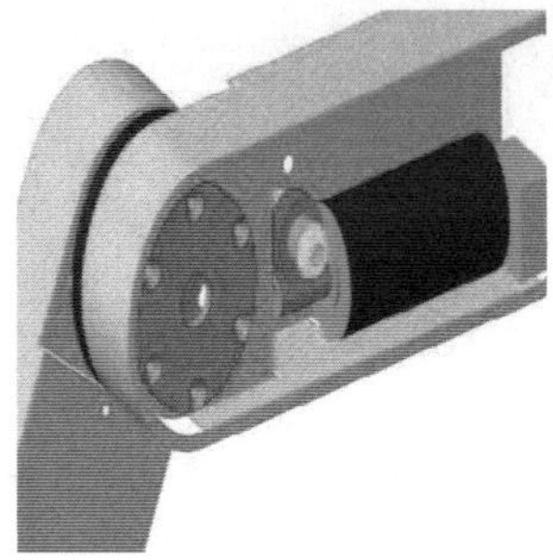

Figure 5. L joint module.

2.3. Manipulator Design
The effective design of the manipulator is the key to safe and reliable climbing of the crawler robot. The manipulator not only needs to be able to reliably grasp the L-shaped angle steel support robot, but also has an operating function, so its structure is more complicated [7-9]. In the process of robot climbing along the tower main material, only one robot grips the L-shaped angle steel in most cases. Therefore, on the one hand, the structural self-optimization is required to reduce the self-weight of the robot. On the other hand, the clamping mechanism must provide sufficient clamping force to ensure that the robot does not fall or slip during climbing. In addition, the robot must be able to accurately position the target angle to avoid erroneous operation.

To meet the above requirements, the manipulator model of climbing tower robot is designed as shown in Figure 6 below.

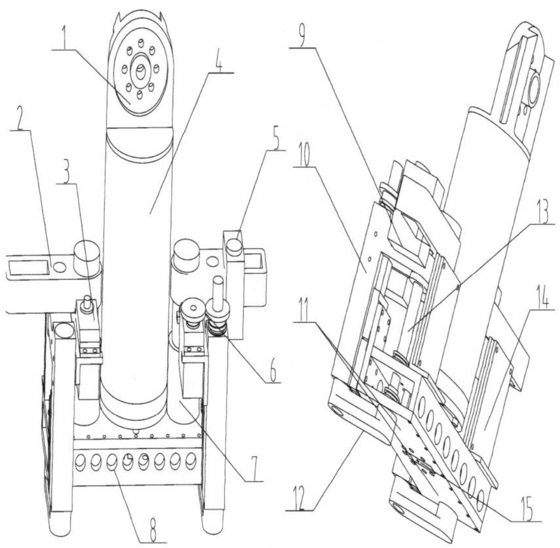

Figure 6. 3D model of manipulator (1. Flange plate 2. Equipped platform 3. Steering gear 4. I module shell 5. Fixed frame 6. Transmission gear 7. DC motor 8. Hollow structure 9. Ranging sensor 10. Moving rbar11. Robot underside 12. Hand grabbing 13. Screw 14. Power Supply 15. Photoelectric sensor).

2.3.1. Manipulator structure In this design, L-shaped angle steel with a length of 15cm and a thickness of 1.5cm is used as a tightening object for the manipulator. Through analyzing the characteristics of the L-angle steel, a gripping method of tightening the vertical face of the angle steel is adopted.

The manipulator adopts a rectangular hollow structure with long 17cm and wide 6.5cm, which not only reduces weight, but also has a large contact area between the lower surface and the angle steel vertical surface, thus increasing the friction force. In addition, in order to achieve the tightening action, each robot has a grip that has two degrees of freedom of rotation and vertical movement. The rotational freedom of the grip is controlled by the steering engine to ensure the rotation direction of the grip. The vertical degree of freedom of hand grasping is realized by a moving bar of hand grasping, a guide rail is arranged on the moving bar, and a slide block is arranged on the mechanical hand body, and both constitute a guide rail slider mechanism. In addition, the nut fixed on the moving rod constitutes a ball screw drive with a screw having only one rotational degree of freedom installed on the manipulator. When the direct current motor drives the screw to rotate through the gear drive, the nut on the moving rod moves vertically relative to the screw. At this time, the hand grasping mounted on the moving bar moves up and down.

2.3.2. Manipulator positioning. The robot body is connected with the end of the I-joint module to achieve a 360-degree rotation in the horizontal direction, and is connected to the L joint module of the robot arm through the flange plate to ensure that it can realize the vertical direction swing. The robot body is connected to the I-joint module through the flange of the lower end of the I-joint module, and the servo motor provides the driving force to realize the 360-degree rotation of the robot. Secondly, the manipulator's swing is achieved by the rotation of the L-joint module. In addition, the manipulator reaches any position of the three-dimensional space through the rotational cooperation of each L-joint module and I-joint module.

As shown in Figure 3, the photoelectric sensor is installed at the center position of the bottom of the manipulator, and the two ranging sensors and cameras are arranged on the platform shown in 2 in Figure 4. The projection of the photoelectric sensor and the two ranging sensors on the bottom of the manipulator is collinear. When the manipulator 1 is moving, the manipulator 2 is always in a tightening state. When the manipulator moves to the target position, the photoelectric sensor on the manipulator 1 opens and detects, and the shaft 5 is rotated. When the photoelectric sensor fails to detect the angle steel, the shaft 5 stops rotating, which is the edge of the angle steel. Secondly, when rotating the shaft 1, when the distance sensor does not detect the angle steel, the point is the other edge of the angle steel. These two points ensure that the edge of the manipulator body is parallel to the edge of the angle steel. Then slowly rotate the shaft 5 so that the robot moves 3cm inward relative to the angle of the steel angle, and then rotate the shaft 2 so that the two ranging sensors measure the same distance to adjust the bottom surface of the manipulator to be parallel to the angle steel surface. Finally, rotate the shaft 3 slowly to contact the bottom surface of the manipulator and the angle steel surface. The servo is energized to rotate the grip to the lower surface of the angle steel, and the torque is provided by the DC motor, and the grip on the moving bar is driven to move upwards through the gear transmission and the ball screw transmission until the angle steel is tightened.

3. Robot climbing tower gait

From the above analysis of the working environment, we can see that in order to obtain a larger climbing step and effective obstacle avoidance effect. The robot uses the gait as shown in Figure 7 below.

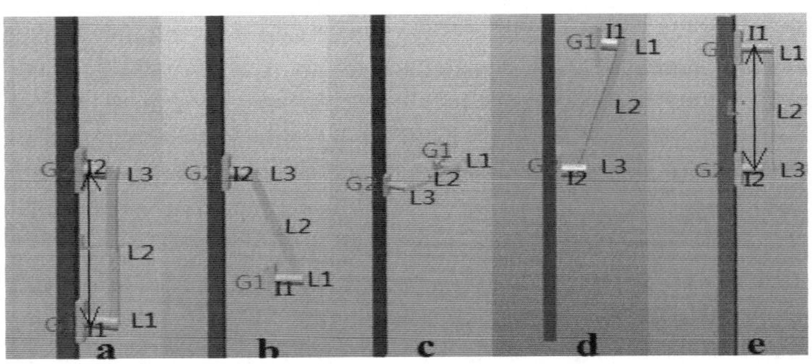

Figure 7. Climbing gait(G1 and G2 are manipulators, L1, L2, and L3 are transmission joints, and L and L' are gait lengths. a, b, c, d, e are climbing gaits).

The climbing process is as follows:

(a) In the initial position, G1 and G2 are clamped. G1 loosens slowly and G2 continues to clamp and supports the robot alone.

(b) L3 opens slowly, G1 moves outwards, leaving the bar completely to avoid interference with the bar when the robot is twisting.

(c) When I2 is rotated 180°, the robot is twisted for half a turn and re-aligned with the rod.

(d) L3 rotates in reverse until it returns to the initial angle.

(e) G1 clamps and the robot completes a climbing cycle. L' can be adjusted by controlling the angles of L1, L2, and L3.

When the robot adopts a twisting gait, the sequence is alternately switched. The distance of one movement cycle is: $0.78L<\Delta L<L$. The climbing step of this gait is larger than other gaits [10-11].

4. Control flow of robot

The design uses STM32 microcontroller to control the driver, motor, sensor, encoder and other modules [12]. The MCU sends the feedback signal to the upper computer through the wireless transmission module. The upper computer sends new instructions to the MCU to control the lower computer and completes the movement of the robotic system. The robot control flow is shown in Figure 8.

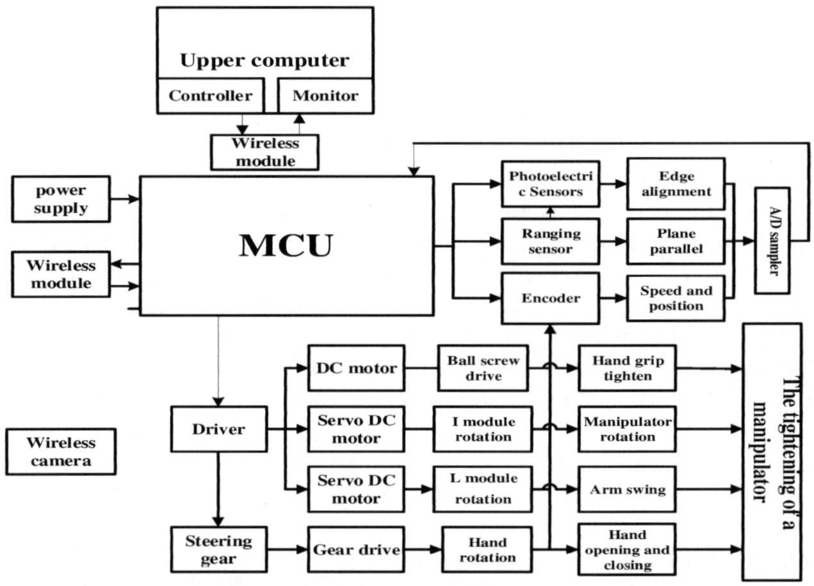

Figure 8. Robot control flow chart.

As illustrated in Figure 8, the power supply is powered by a single chip microcomputer, and both upper computer and MCU are equipped with wireless communication module. The upper computer monitor observes the climbing tower status of the robot in real time through the wireless camera, and the controller sends the instruction to the MCU through the wireless module. The MCU receives the instruction through the signal line and sends it to the motor driver, sensor and encoder. The encoder will detect the position and speed information of the motor and feed it back to the MCU in time. The MCU sends a new instruction command to the driver according to the feedback signal, and the driver can control the rotation speed and direction of the servo motor according to the received new instruction. Similarly, the photoelectric sensor and the distance measuring sensor begin to work after receiving the instruction from the single-chip microcomputer. During the work, the measurement data is continuously fed back to the single-chip microcomputer. The single-chip microcomputer judges according to the feedback signal and re-sends a new instruction to the sensor. The steering gear and DC motor on the manipulator are controlled by the driver which receives the single-chip instruction, without feedback link. In addition, the wireless camera communicates with the SCM through the wireless module. After the wireless camera sends the detected image signal to the SCM, the SCM sends the image signal to the upper computer monitor through the wireless module.

5. Structure simulation and prototype manufacture

5.1. Finite element analysis

In order to calculate the force and deformation of each gait in the climbing process of the robot, the ANSYS Workbench finite element analysis was performed on the deformation, strain and stress of each gait of the climbing tower [13]. The following figures 9, 10, 11 and 12 are simulated cloud charts for each step in the climbing cycle of the climbing tower robot.

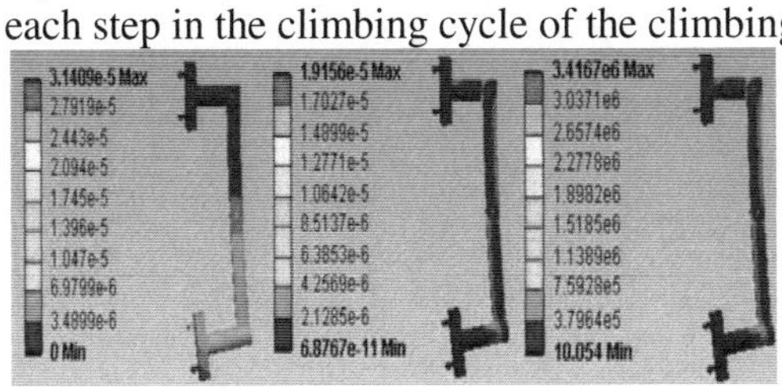

Figure 9. Gait a deformation, strain, stress cloud.

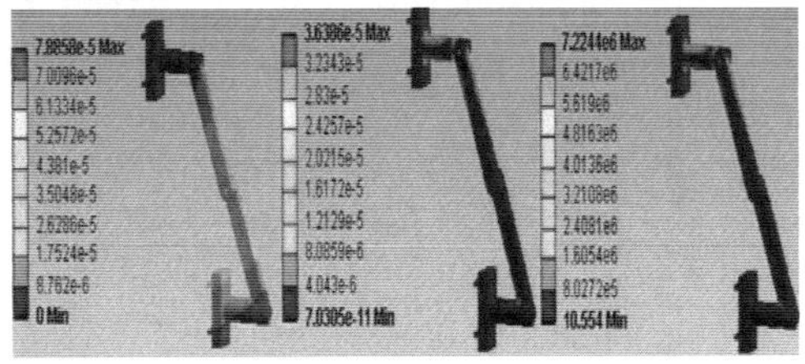

Figure 10. Gait b deformation, strain, stress cloud.

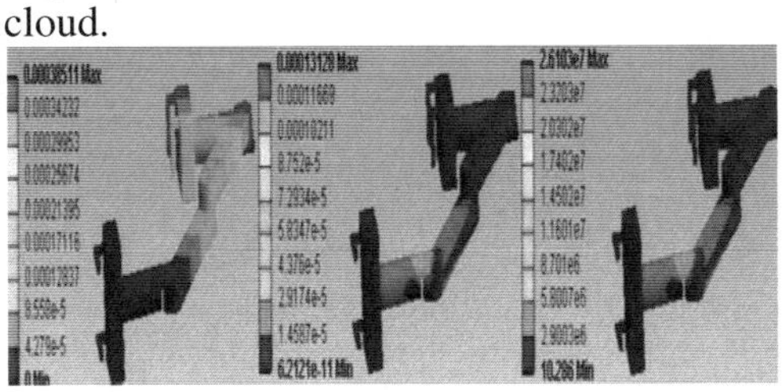

Figure 11. Gait c deformation, strain, stress cloud.

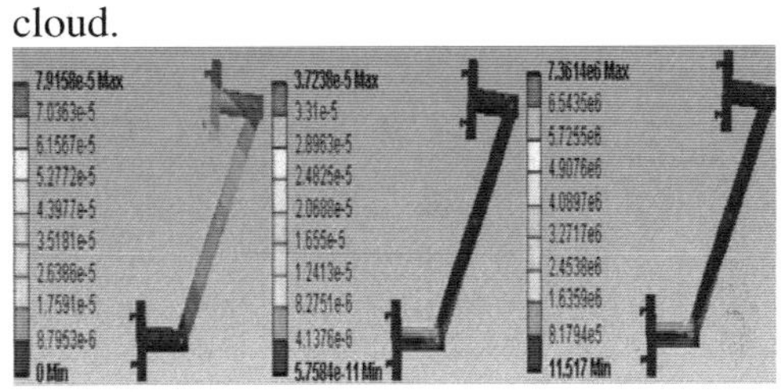

Figure 12. Gait d deformation, strain, stress cloud.

According to Figure 9, under the action of self-weight, the manipulator G1 is slightly deformed, strain and stress occur in both the manipulators and the joints, and the change is not obvious. Therefore, the manipulator G1 needs to be opened slowly to avoid large vibration deformation. It is known from figure 10 that when the mechanical arm starts, the manipulator G2 is slightly deformed under the self-weight, the maximum deformation is small, the strain and stress are mainly concentrated in the stress concentration parts. Therefore, the stress concentration phenomenon should be avoided as far as possible. As can be seen from Figure 11, when the arm is twisted by 90°, G2 tightens the angle steel to support the entire mechanical structure and the deformation of the end is large. Stress and strain are mainly concentrated in the L joint module. Therefore, the strength and rigidity of the material at the site should be strengthened, and the structural self-weight should be reduced. In addition, it can be found from the figure 11 that the maximum stress and strain values are small and are all within the design range. From Figure 12, when the mechanical arm is twisted by 180°, the manipulator G1 deforms within a certain range under its own weight, and the strain and stress are also concentrated at the L joint module, and the maximum stress and strain values are within the design range.

From the above analysis, it can be known that the deformation of the robot is mainly distributed on the suspended end manipulator, and the strain and stress are mainly concentrated on the rotating pair. Since the deformation amount, the stress and the strain are within the allowable range of the design, the structure of the robot is stable and reliable, and the climbing action can be completed. In addition, to ensure that the gravity of the robot itself is small, it is particularly important to ensure the reliability of each rotating pair and the stiffness and strength of the materials at each joint.

5.2. Prototype Experiment

In order to verify the operation of the robot under the actual working environment, the climbing ability of the tower-climbing robot was tested on the oblique material and the main material of the high-voltage transmission tower.

Firstly, the ability of the robot to climb the tower on the high voltage transmission tower is verified.

IOP Conf. Series: Materials Science and Engineering 428 (2018) 012071 doi:10.1088/1757-899X/428/1/012071

The two manipulator are tightened on the high voltage transmission tower, and the climbing experiment is carried out according to the gait described above. The climbing process is shown as shown in Figure 13 below.

Figure 13. Prototype experiment 1.

Secondly, in order to verify the climbing ability of the robot on the main material of the high voltage transmission tower, the experiment of the robot crawling along the main material of the tower is carried out. The process of climbing the tower is shown as shown in Figure 14 below.

Figure 14. Prototype experiment 2.

Through the above experiments, it is found that when the upper computer sends a movement instruction, the climbing robot can climb up along the L-shaped angle ridge line, indicating that this design scheme is feasible. When one end of the manipulator is tightened, it is enough to support the entire fuselage. The torsional climbing gait can effectively cross the tower obstacle. Through the above experiments, the upper computer can send the movement instructions, and the climbing robot can climb up along the L angle steel ridge line, which shows that the design plan is feasible. Experiments show that when the manipulator is tightened at one end, it is enough to support the entire fuselage and meet the load requirement. When the robot climbs the tower, each joint module rotates flexibly, and the manipulator can accurately tighten the surface of the angle steel, and the effect of twisting gait is better.

6. Conclusion

(1)The five-degree-of-freedom mechanical arm structure not only reduces its own weight, but also ensures that there is no redundancy in the degree of freedom.

(2)The manipulator achieves accurate positioning by means of photoelectric and distance measuring sensors. By gripping the angle steel to provide a larger clamping force, the manipulator can support the entire mechanical mechanism to maintain the balance of the robot after grasping the clamp.

(3)The torsional climbing gait step can effectively avoid the obstacles on the tower. In addition, the step is larger and the climbing speed is faster.

(4)Through the simulation analysis of each gait, the deformation of the robot is mainly concentrated in the end manipulator, and the stress and strain are mainly concentrated in the rotating pair and each joint. The values of the parameters meet the actual requirements. The simulation results show that under the premise of ensuring the robot's self-gravity, it is very important to ensure the reliability of each rotating pair and to strengthen the strength and rigidity of each joint.

(5)The prototype experimental results show that the robot can effectively complete climbing obstacles, verifying the rationality of robot design and the feasibility of climbing gaits.

References

[1] Ma W J. *Research on safe operation and maintenance system of overhead transmission line* [D]. Shandong University,2015.(in Chinese)
[2] Jiang L. *Research on double claw modular bionic climbing robot* [D] . South China University of Technology,2012.(in Chinese)
[3] Zhou X L, Lu X L, Zhao S P, Cao Z H. *Pose error analysis of climbing robot for electric power tower* [J] . Mechanical design and manufacture,2013, (11):179-182. (in Chinese)
[4] Huang L, Luo Z J, Xu W Q, Wang X, Wang C Y. *Design, simulation and control of a climbing tower robot* [J]. Journal of engineering design, 2015, 22(05): 476-481. (in Chinese)
[5] Li B Y. *Research on power tower climbing robot for power transmission line* [D]. Harbin Institute of Technology, 2016.(in Chinese)
[6] M. Tavakoil, L. Marques, A. T. De Almeida. *3DClimber: Climbing and Manipulation Over 3D Structures* [J].The Journal of Mechatronics, 2010, 21(2001): 48-62.
[7] Zhao Y Y, Fang Y, Nie J W. *Kinematics research of five degree of freedom robot arm* [J]. science and technology and engineering, 2009,9 (06): 1571-1573+1583.
[8] Zhang N, He Y C, Wang N. Design, *research and application of manipulator for robotic manipulators* [J]. mechanical design, 2013,30 (03): 17-20+61.
[9] Wang G Z, Li X, Deng P, Wu W H. *Obstacle surmounting ability and clamping mechanism analysis of a new type of cable climbing robot* [J] . Mechanical design and manufacture, 2016,(06):30-33.(in Chinese)
[10] Zhu X G. *Gait and path planning of biped robot* [D]. North China Electric Power University, 2012.
[11] Liu J. *Gait analysis and dynamics study of biped walking robot* [D] . East China University of Science and Technology, 2012.(in Chinese)
[12] Zhou J.*STM32 single chip microcomputer principle and hardware circuit design* [J]. digital technology and application, 2015 (11): 1.
[13] Han J, Fang L, Sun J P, Lv W F, Yang Y C. *Collaborative simulation and optimization of complex assembly based on Pro/e and ANSYS Workbench* [J] . Mechanical design and manufacture, 2010,(01):190-192. (in Chinese)

IOP Conf. Series: Materials Science and Engineering **428** (2018) 012072 doi:10.1088/1757-899X/428/1/012072

Research on Kinematics Simulation of Continuum Robot Based on ADAMS

Yueling Zhang, Lin Chen and Songyi Dian

Department of Automation, College of Electrical Engineering and Information Technology, Sichuan University, Chengdu, Sichuan, China

scudiansy@scu.edu.cn

Abstract. As a prevalent biological inspired robot, continuum robots has good bending performance and excellent flexibility. In this paper, a continuum robot is designed, then the kinematic model is established to analyze the motion characteristics of the robot. Secondly, the simulation model of the continuum robot is conducted in ADAMS to verify the feasibility of the established kinematic model and to simulate its bending ability. At the same time, it also showed that ADAMS software had value in building a flexible body model.

1. Introduction

Continuum robot has received the widespread attention. It uses the "bionic" design concept to mimic the "invertebrate" flexible design of animals' structure, and there aren't any rigid joints or rigid links. Compared with the rigid robot, the continuum robot has excellent flexibility and good bending performance. It can regulate its shape flexibly according to the obstacles and working condition. Then make it owes strong adaptability to unstructured and limited environment[1]. So continuum robots have a wide application prospect. Walker designed a continuum robot based on the elephant nose and achieved certain load capacity[2]. Simaan developed a continuum robot for throat surgery, which can be applied to medical treatment[3].

However, continuum robot still has unsolved problems, ranging from material selection, structural design, mechanical analysis, control and perception. The development process of continuum robot based on the virtual prototype technology can reduce the cost greatly[4]· Automatic Dynamic Analysis of Mechanical Systems (ADAMS) is one of the famous virtual prototype analysis softwares. In this paper, it is used to test the designed continuum robot's motion characteristics.

2. Structural design of continuum robot

The structure of designed continuum robot is shown in figure 1. It consists of a main skeleton, four drive lines (secondary skeleton), and some disks. The material of the skeleton is the NiTi alloy, and that of the disk is lightweight aluminum to make the robot lighter. There are five small holes in each disk. The central main hole makes the disk fix to the main skeleton. The sub-holes around the main hole that are distributed at 90 degrees to each other. The diameter of the main skeleton is 1 mm, the diameter of the driving line is 0.6 mm, the diameter of the disk is 12 mm, and the diameter of the hole corresponds to the diameter of the flexible body that passes through. When the drive lines are driven, the corresponding drive lines produce displacement and the robot can be driven to achieve continuum bending motion.

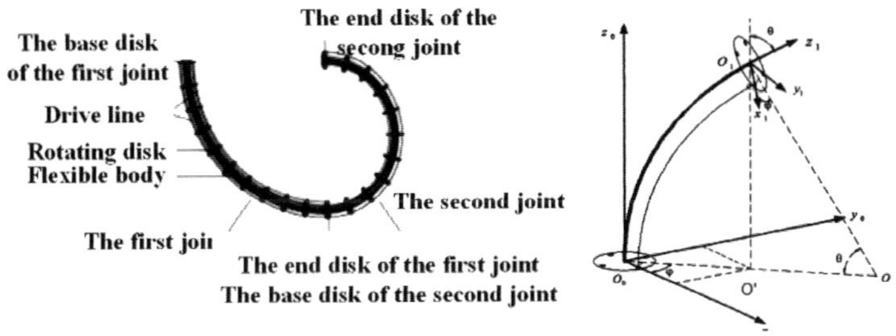

Figure 1. Illustration and the geometric model of the continuum robot[5]

3. Kinematic analysis of continuum robot

Geometric analysis[5] is used for kinematic analysis of the continuum robot. The flexible body is considered as a smooth continuum arc with the same curvature, and its axial deformation is ignored[5]. The single-segment's geometric model is given as in figure 2.

Let the center point of the base disk be the origin of the base coordinate frame$\{O_0\}$, let the center point of the end disk be the origin of the terminal coordinate frame $\{O_1\}$. The coordinate frames are shown as in figure 1. The bending angle is represented by θ, the rotation angle is represented by φ, and the radius of the minor skeleton dividing circle are represented by r.$\{O_1\}$ can be seen as the result of translation and rotation transformation by $\{O_0\}$.A homogeneous transformation matrix expressed as equation (1) is used to represent the coordinate transformation matrix.

$$R = \begin{bmatrix} n_x & o_x & a_x & p_x \\ n_y & o_y & a_y & p_y \\ n_z & o_z & a_z & p_z \\ 0 & 0 & 0 & 1 \end{bmatrix} \tag{1}$$

where $n=[n_x,n_y,n_z]^{\mathrm{T}}$, $o=[o_x,o_y,o_z]^{\mathrm{T}}$, $a=[a_x,a_y,a_z]^{\mathrm{T}}$ is the attitude matrix. $p=[p_x,p_y,p_z]^{\mathrm{T}}$ is the positional matrix .

Use the geometric method , the homogeneous transformation matrix can be obtained as:

$$\begin{aligned} {}^0_1R &= Trans\left[\frac{l}{\theta}c\varphi(1-c\theta),\ \ \frac{l}{\theta}s\varphi(1-c\theta)\ \ \frac{l}{\theta}s\theta\right]\cdot Rot(z,\varphi)\cdot Rot(y,\theta)\cdot Rot(z,-\varphi) \\ &= \begin{bmatrix} c^2\varphi c\theta + s^2\varphi & c\varphi s\varphi c\theta - c\varphi s\varphi & c\varphi s\theta & \frac{l}{\theta}c\varphi(1-c\theta) \\ c\varphi s\varphi c\theta - c\varphi s\varphi & s^2\varphi c\theta + c^2\varphi & s\varphi s\theta & \frac{l}{\theta}s\varphi(1-c\theta) \\ -c\varphi s\theta & -s\varphi s\theta & c\theta & \frac{l}{\theta}s\theta \\ 0 & 0 & 0 & 1 \end{bmatrix} \end{aligned} \tag{2}$$

From the established forward kinematic model equation (2), the pose matrix of the end coordinate frame relative to the base coordinate frame can be obtained as follows:

$$p_x = \frac{l}{\theta}\cos\phi(1-\cos\theta),\quad p_y = \frac{l}{\theta}\sin\phi(1-\cos\theta)\quad a_z = \cos\theta \tag{3}$$

According to equation (3), the bending angle and rotation angle can be obtained as:

$$\begin{cases} \theta = \arccos a_z \\ \varphi = \arctan\left(\dfrac{p_y}{p_x}\right) \end{cases} \tag{4}$$

The change of drive length can be expressed as $\Delta l = l - l_i$, where l is the initial length of the drive line, l_i is the length of instantaneous moment in the movement, Δl_i is the elongation of the drive line.the transformation relationship between the drive variation and the angle can be obtained as:

$$\Delta l_1 = -\Delta l_3 = r\theta\cos\phi, \quad \Delta l_2 = -\Delta l_4 = r\theta\sin\phi \tag{5}$$

Then the inverse solution is:

$$\begin{cases} \varphi = \arctan(\Delta l_2 / \Delta l_1) \\ \theta = \sqrt{\Delta l_1^{\ 2} + \Delta l_2^{\ 2}} / r \end{cases} \tag{6}$$

From equation (1), the vector expression of the end position corresponding to bending angle and rotation angle can be obtained as:

$$p = [p_x, p_y, p_z]^{\mathrm{T}} = F(x) \tag{7}$$

where $x = [\theta, \varphi]^{\mathrm{T}}$. Simultaneously, differentiating equation (7) about θ and φ respectively yields the speed matrix equation, which can be given as:

$$\dot{p} = \begin{bmatrix} \partial p_x / \partial\theta & \partial p_x / \partial\varphi \\ \partial p_y / \partial\theta & \partial p_y / \partial\varphi \\ \partial p_z / \partial\theta & \partial p_z / \partial\varphi \end{bmatrix} = J(x)\dot{x} \tag{8}$$

4. Kinematic analysis of continuum robot

This paper use the method of establishing a rigid body first and then establishing a flexible body. Taking the reference point on the established rigid body as the end point for establishing the flexible body, the operation of moving the flexible body can be avoided. In order to modify the model in the design easily, parametric modeling is used when building the model, which can shorten the design time and improve the efficiency of the design work.

4.1. Disk Parametric Modeling Process

4.1.1. Create variables. Drop down the "Create" option in the menu bar and click on the "Design Variable". The key to parameterize disk model is to parameterize appropriate point. The coordinates of the points are expressed as design variables, the distance between the center of each hole and the center of the disk is denoted by the design variable RA_1. The value of the radius of the disk is represented by (1+RA_1), and the thickness of the disk is represented by the design variable RA_2. Set the standard value for RA_1 to 5, the standard value of RA_2 is 0.5.

4.1.2. Parametric modeling disks. Select the cylinder model in the "Object" option in the toolbar. When certain length and cross-sectional radius are required, enter the design variable in the space corresponding to the length and radius. Check the "√" symbol before the space to indicate that the input value will be applied. Next, set up four parameterized geometric points on the disk, and their coordinates are represented by (RA_1, 0, 0), (0, RA_1, 0), (-RA_1, 0, 0), (0, -RA_1, 0).

4.1.3. Create circular disk holes. It is necessary to create holes on the four parameterized geometric points and the center of the disk. Select the "Drill" icon in the "Features" area in the "Object" option. Set the radius of holes to 0.05cm and the depth to (RA_2), When using the mouse to select the center point it can causes deviations, right-click on the hole, in the menu that appears, select “Modify”,In the empty column that sets the center of the hole, specify the center of the hole as point POINT_1 in the form of (LOC_RELATIVE_TO({0,0,0}, POINT_1)). The hole center is designated point POINT_1.

4.2. Establishing A Flexible Body

4.2.1. Add material settings. Drop down the "Create" option in the menu bar and click on the "New Material", in the dialog that appears, set the relevant information for the material like Young's modulus, Poisson's ratio and density.

4.2.2. Build a flexible body. The stretch method is used to create a liner flexible body with two known points as the endpoints. First, the disk is set up, then the point on the disk can be directly used as the endpoint to create a flexible body. By setting the center line, the stretching path in the global coordinate system is defined, and the radius of the cross section is set according to the diameter of the hole. The unit size will affect the accuracy of the flexible body and the calculation speed during simulation. When the unit setting is larger, the solver will calculate faster, but the accuracy of the simulated flexible body will decrease. When setting the attachment points,take the two Marker points created in defining center line as outer joins. Clicking the OK button then the flexible body will be successfully created.

4.3. Adding System Constraints

There are some constraint among each components. The constraint determines the connection relationship, the relationship between physics and mechanics, the degree of freedom among objects, and the way two objects move relative to each other. In ADAMS, different connects are used to express different constraint relationships.First set fixed constraints, click "Connect" in the menu bar, and select the "fixed" icon in the connectors bar of the toolbar that appears. Use the fixed to fix the main skeleton flexible body and the disks at two ends,and select the parameterized geometry point as the outer joints of the fixed. Use the fixed to connect the top of the drive line with the disk, and the drive lines at both sides of the rotary disk are connected together by fixed into a whole.

4.4. Adding Drivers

The drive is to constrain the two components to allow them to move according to a certain law. Observe the movement of the robot in a plane with a rotation angle of 45 degrees. It is necessary to first establish a translation between the bottom end of the flexible body and the ground, and then add the motion to the translation, then the bending motion of the robot is achieved by setting the parameters of the motion. To make the rotation angle achieve 45 degrees, the four motion functions should be set to the same value, two of the signs are positive, while others are negative.

4.5. Adding Contact

The drive line comes into contact with the inner wall of the disk hole during bending deformation, accompanied by the deformation of the part and the change of the contact force. Therefore, establishing the model in ADAMS requires a contact between the hole and the flexible body of the drive line. And the flexible body needs to contact with the rigid body by adding a dummy part. Otherwise, an error will occur. In this case, the dummy part is constructed by adding a cylinder and setting its mass and inertia matrix to zero. Use fixed to fix the dummy part on the flexible body, in the simulation, the flexible body will contact the rigid disk through the dummy part, therefore, it's necessary to add contact between dummy parts and the disk. Click the "Contact" icon in the "force" area of the toolbar, in the dialog that appears, select the dummy part and the disk as the I Solid and J

Solid to complete the setting of contact. The larger the bending angle to be achieved, the more dummy parts need to be added, and add more dummy parts, the simulation will be more accurate.

5. Kinematic analysis of continuum robot

5.1. Model Simulation

Because there is gravity in the actual application, it't necessary to add a gravity environment in the simulation.After the model is set up, use the model checking tool to check for errors such as redundant constraints. If there is no error, click the "Run Interactive Simulation" icon under the simulation area to start simulate. Select the representative moment to display the simulation process as follows:

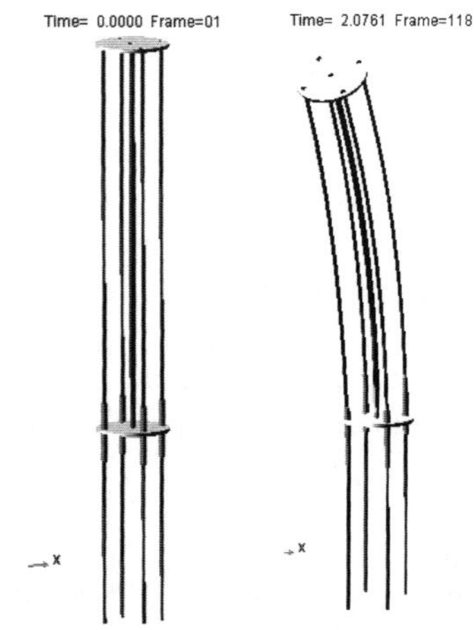

Figure 2. Pose diagram of the single robot

After the simulation is completed, enter the post-processing module to process the data. Select the data source by setting the source and object, then choose the characteristic in the characteristic bar which includes the point's running trajectory, velocity, acceleration, angular velocity, angular acceleration, and other information.

5.2. Analysis of Simulation Results

Because the end coordinate system of the kinematics equation is based on the end disk of the continuum robot, the center of mass of the end disk is used as the analysis object to analyze the motion position, velocity, acceleration and angular velocity. The simulation results are as follows:

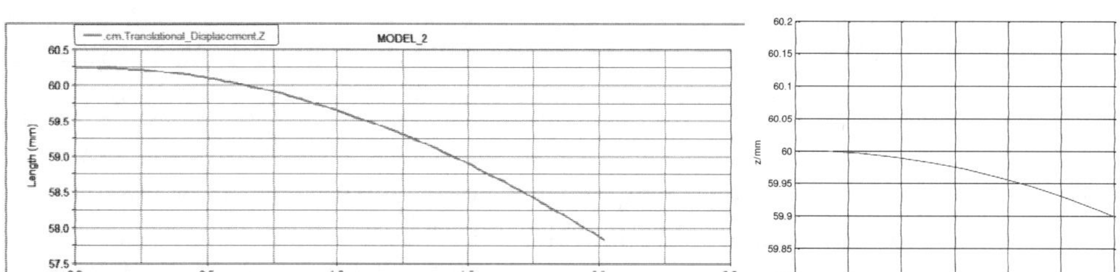

Figure 3. The position curve of the end disk's centroid in the Z direction (in ADAMS and MATLAB)

Figure 3 shows the position of the centroid of the end disk in the Z direction. In order to verify the feasibility of the kinematics equation proposed in section II, the position of the kinematics equation in the Z direction was simulated in MATLAB, compared with the figure of Z position in ADAMS, it can be seen that the movements of the two figures tend to be the same, although there are errors, the trends are generally the same, so the kinematic law established in section III is basically correct. However, it should also be noted that as the bending angle increases, the error between the two also gradually increases. The reason for this phenomenon is that the derivation of kinematics is based on ideal conditions. In order to reduce the error, more realistic conditions must be taken into account in the formula to further improve it.

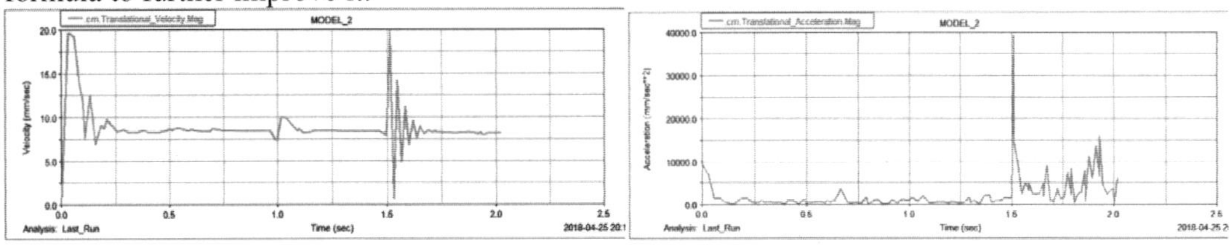

Figure 4. The velocity and acceleration curve of the end disk's centroid

Figure 4 are the velocity and acceleration changes of the centroid of the end disk. It can be seen from the figure that the continuum robot has a certain transition time from the initial state to a more stable running state. After that, it can basically achieve uniform speed operation. When the speed fluctuates greatly appears at 1.5 s, analyze the length of dummy part in the model, it should be the fluctuation caused by the dummy part when it runs at the junction with the hole. The speed in the steady running state is basically consistent with the speed calculated by the formula, which shows that the kinematic relationship established in section II is basically correct, and also illustrates the feasibility of using ADAMS software to build a flexible body model, this software is of great help in the study of flexible study.

6. Conclusion

The model of continuum robots was established in the virtual prototype software ADAMS, rigid and flexible components were established during the process, the connection relationship was implemented with constraints such as connectors and contacts. The model was simulated and the motion of the continuum robot was observed. Finally, image analysis is performed on the motion characteristics data obtained from the simulation, and compared with the images obtained based on the derived kinematics equations, the correctness and rationality of the deduced kinematics model could be verified. At the same time, it also showed that ADAMS software had value in building a flexible body model.

Acknowledgments

The This work is supported by Fundamental Research Special Funds for Central Universities (20826041A4133).

References

[1] Guo Wei, Xiao Tao, Hu Haiyan, et al. Research and Simulation on the continuity of a continuous intestinal robot [J]. mechanical and electronic,2010(7):63-67.

[2] Walker I D,Hannan M W. A novel 'elephant's trunk' robot [C]// IEEFJASME International Conference on Advanced Intelligent Mechatronics. Piscataway,NJ,USA: IEEE,1999:410415.

[3] Simaan N, Taylorr,Flint P. A dexterous system for laryngeal surgery[C]// Proceedings of the 2004 IEEE International Conference on Robotics and Automation Piscataway:IEEE Press,2004:351-357.

[4] Li Yun. Optimization design of clamping mechanism based on ADAMS [J]. mechanical manufacturing and automation.2010, 35(6):35-37.

[5] Liu Yanghui. Motion simulation and optimization of a continuous abdominal endoscopic robot [D]. Guangdong University of Technology,2013.

IOP Conf. Series: Materials Science and Engineering **428** (2018) 012073 doi:10.1088/1757-899X/428/1/012073

Analysis on Attitude Fluctuation in Vertical Descent of Multi-rotor UAVs

Li Chenglong[1], Yang Zuqiang[2] and Gou Jiangchuan[3]

[1,3] College of Air Traffic Management, Civil Aviation Flight University of China, No.46, Section 4, Nanchang Road, Guanghan City, Sichuan Province, China, 618307
[2] Information Science Academy of China Electronics Technology Corporation, No. 4 Building, No. 36, North Road, Haidian District, Beijing, China, 100086

Corresponding author's e-mail: lcl@zju.edu.cn

Abstract. In continuous vertical descent, multi-rotor UAVs' attitude will meet the problem of attitude fluctuation and tends to divergence. The problem comes from the flight controller instead of vortex-ring state. Based on an eight-rotor UAV's nonlinear model in hovering mode, the control law is analysed by using transfer function. The first reason is that the inner ring and outer ring bandwidths of the PID cascade controller do not match well. And the other one is that the lower linear operating point and lower average rotor speed during descent lead to less thrust and torque per propeller, which make it more difficult to resist wind disturbances. By adjusting the position loop's parameters in descent, the eight-rotor UAV reduced the requirement of horizontal position accuracy, and the stability of horizontal attitude control increased, which finally obtained a better performance.

1. Introduction

Multi-rotor UAVs have the ability to hover and descend vertically. But in certain application, for example, multi-rotor UAVs descending vertically from a height more than 1000 meters, it was found that the eight-rotor UAV was easy to encounter attitude fluctuations in continuous descent. For this phenomenon, one probable reason is that multi-rotor UAVs may encounter the vortex-ring state, another reason could be the unsuitable flight controller.

For the first reason, if a multi-rotor UAV descends vertically and rapidly, the rotors may fall into turbulence, in which it will suffer from severe thrust fluctuation. And this unsteady aerodynamic state usually leads to severe attitude fluctuation on multi-rotor UAVs. In engineering practice, reference[1] gives a reasonable criterion of vortex-ring state boundary for helicopter. Reference[2] proposes a method of descending with forward flight for multi-rotor UAVs, which can help reduce the probability of encountering the vortex-ring state. But this method needs a wide horizontal airspace for continuous forward flight. Actually, according to the practical boundary in reference [1], we limit the UAV's maximum descent rate at 2 m/s to avoid being caught in vortex-ring state. But there are still attitude fluctuations in vertical descent experiment. So, the vortex-ring state can be excluded as the main factor. And we turn to consider if the problem comes from the flight controller.

Most multi-rotor UAVs' flight controller use cascade PID control law in four control channels (pitch, roll, yaw and height) [3, 4]. PID controller is robust and adaptable [5] which permits the object deviating from original linear working point. So it is possible to use the same parameters for different flight modes (hovering, forward flight and vertical descent) of multi-rotor UAVs. However, a multi-rotor UAV's linear working point will change in different rates of descent. For this case, using the same set of control law is convenient in engineering practice but may not achieve a perfect performance for different flight modes [6]. Especially complex atmospheric environment, like wind shear, can cause oscillation when the flight controller has insufficient control margin. And it will ultimately lead to a loss of control of multi-rotor UAVs.

2. Model Linearization and Control Law Design for Multi-rotor UAVs

2.1. Multi-rotor UAVs' Model and Control Law Design

In a project, the nonlinear model of a centrosymmetric eight-rotor UAV was established. Based on the dSpace system, a Hardware-in-loop simulation platform was built to roughly test PID parameter and verify the program logic for the flight controller. In this process, the most difficult thing is to get each rotor's dynamic nonlinear model. It can be solved by referring to the method of modelling a helicopter's main rotor [7].

To simplify the control law design, we linearize the nonlinear model of the eight-rotor UAV at hovering working point. The high-order minor terms of the linearized expression represent for interactions between the four control channels, which can be ignored. So, multi-rotor UAVs' four channels can be controlled in a decoupled manner.

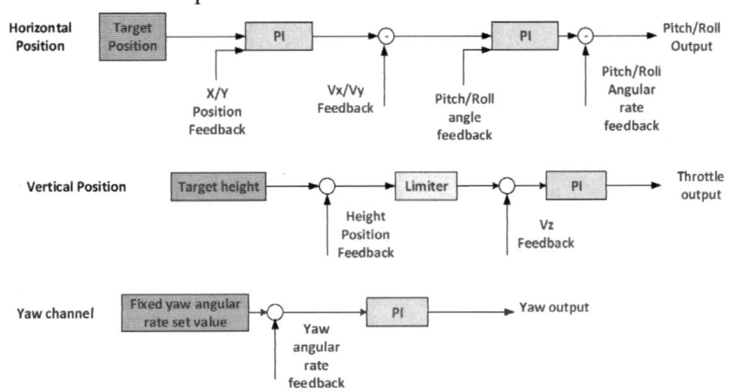

Figure 1. Control law of multi-rotor UAVs in hovering mode

As shown in figure 1, in pitch and roll channels, the eight-rotor UAV uses the same control law during the vertical descending and hovering. Inner ring is attitude loop and it uses PI controller with angular rate damping. The outer ring is position loop and it uses PI controller with velocity damping. So, there is a two-stage PID cascade controller in pitch or roll channels. For the vertical channel, the UAV adopts a PI controller in the command of descent rate whose setting value is guided by altitude position. Finally, four channels' output value will be normalized and converted into rotor speed setting value for each propeller.

2.2. Analyzing Control Structure in Transfer Function

Using the MATLAB linearization kit, the linearized expressions of four channels are obtained and the open-loop transfer function for the angular rate in roll/pitch channel can be written as equation (1).

$$Gqlon(s) = \frac{308.7s^2 + 3.365s}{s^3 + 6.351s^2 + 0.06911s + 0.6396} \tag{1}$$

Based on the transfer function of $Gqlon(s)$, the structure of the inner-loop can be constructed and is shown in figure 2. Then the attitude loop's transfer function can be represented by equation (2). By zero-pole analysis, equation (2) can be approximately equivalent to a second order system, written in equation (3).

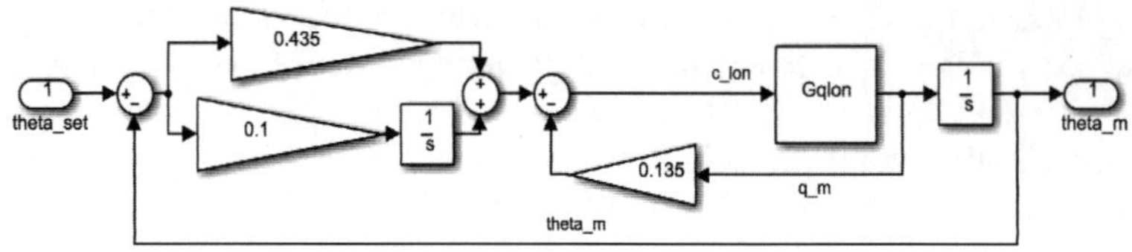

Figure 2. attitude-loop structure

$$Gtheta(s)=\frac{134.3s^2+1.772s+0.003365}{s^4+48.02s^3+134.8s^2+2.412s+0.003365} \quad (2)$$

$$G_\theta(s)=\frac{133.9}{s^2+48.01s+133.9} \quad (3)$$

Comparing the two equation (2) and (3) in bode diagram (Figure 3), it demonstrates that the function $G_\theta(s)$ can be a substitute for $Gtheta(s)$. That is to say, the inner ring (attitude closed loop) of multi-rotor UAVs can be equivalent to a second-order system.

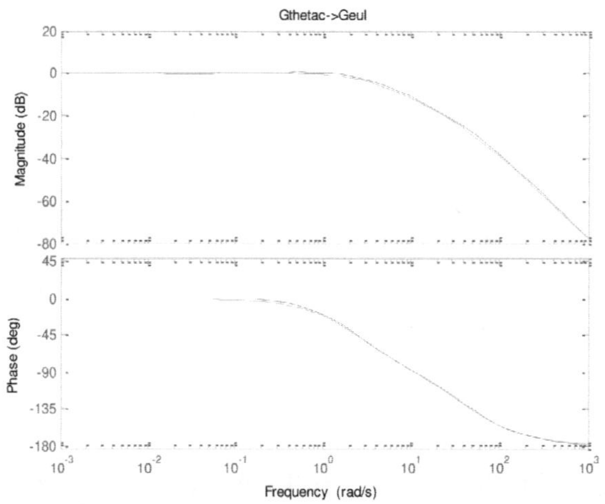

Figure 3. Bode plotting comparison between Gtheata(s) & Gθ(s)

Based on the transfer function of $G_\theta(s)$, the outer-loop is constructed in figure 4 and its transfer function is expressed in equation (4), which can be figured out that the closed-loop system has a damping oscillation period in 3.228 seconds.

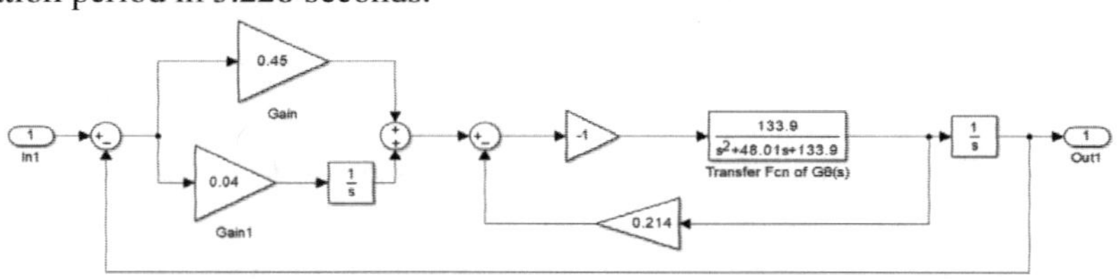

Figure 4. Position closed loop controller structure

$$Gpc(s)=\frac{53.56s+5.356}{s^4+48.01s^3+105.3s^2+53.56s+5.356} \quad (4)$$

2.3. Attitude Tracking Curve Comparison between Hovering and Descent Mode

The control law depicted in section 2.2 is used to control the eight-rotor UAV in different flight modes. The figure 5 shows that the real pitch value keeps tracking the setting value closely in hovering mode in real test. The figure 6 is the pitch curve at a descent rate of 2 m/s, which records the severe attitude oscillation in pitch channel.

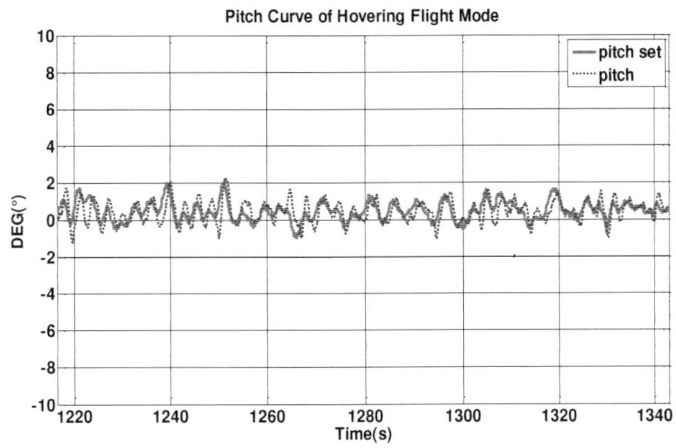

Figure 5. Pitch curve of hovering mode

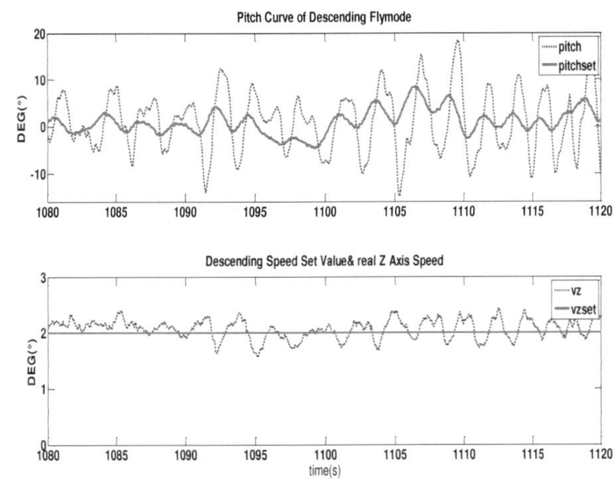

Figure 6. Pitch curve and vertical velocity curve in descent

3. Improvement of Control Law in Descent

3.1. Analysis of Attitude Oscillation in Descent

The attitude tracking performance in hovering mode is good, and this shows that there is little possibility of inner ring causing fluctuation problems. So the attitude fluctuations in descent are mainly attributed to two aspects.

The first reason is that the inner ring and outer ring's bandwidths do not match well. Position loop output is the external incentive of attitude loop. When the external incentive's frequency is close to the inner ring's inherent frequency, it will cause resonate in attitude curve. And if the external incentive frequency is higher than inner loop's bandwidth, it will cause the inner controller's phase delay which can also lead to a drastic attitude oscillation. In engineering practice of PID controller, the inner ring bandwidth should normally be five times that of the outer ring. Based on the transform function, we figure out the attitude loop and the position loop's bandwidths, which are 2.69 rad/s and 0.95 rad/s respectively. To some extent, this indicates that the improper position loop's PID controller parameters cause attitude fluctuation.

Secondly, average rotor speed of the eight-rotor UAV is lower than that in hovering mode. The change of the throttle operating point in descent has a different linear working point. However, each rotor's thrust and torque are lower than that in hovering mode. When using the PID parameters adapted to hovering mode to control the descent process, the eight-rotor UAV is more susceptible to the effects of atmospheric disturbances in descent, resulting in attitude fluctuations.

3.2. Experiment Result

After analysis, the proportion item of outer ring's PI controller was decreased and the integration item was nearly removed. The eight-rotor UAV did a vertical descending experiment again. Figure 7 gives the pitch curve in an experiment after adjusting parameters. And one can see that the attitude curve tracking characteristic is improved in figure 7. Also, Z-axis velocity is less affected by attitude fluctuation.

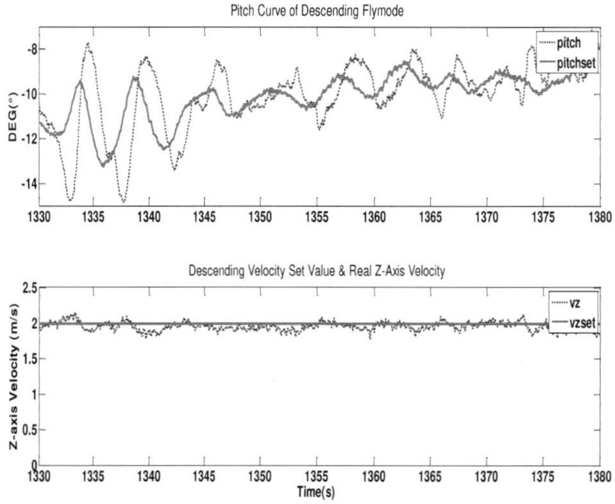

Figure 7. Pitch curve and vertical velocity curve in descent after adjusting controller parameters

4. Conclusion

This work studies the problem of attitude fluctuation in descent of multi-rotor UAVs. From the experiment result, one can see that the mismatch bandwidths and inappropriate controller parameters caused the problem. Using PID cascade controller, the horizontal position precision can be reduced in exchange for the stability of horizontal attitude control. After adjusting the position loop's parameters, the UAV achieve a better stability in continuous descent.

Acknowledgements

This work was supported by Sichuan Science and Technology Program (grant 2018JY0394), Scientific Research Fund of Civil Aviation Flight University of China(grant J2018-30), Joint Funds of the National Science Foundation of China and the Civil Aviation Administration(grant U1733105) and Science and Technology Program of Sichuan, China (grant 2017GZ0047).

References

[1] Lu Yang G Z. Practical Boundary in Helicoptor Vortex-ring State[J]. Transactions of Nanjing University of Aeronautics & Astronau. 2003(01): 6-11.

[2] Chenglong L, Zhou F, Jiafang W, et al. A vortex-ring-state-avoiding descending control strategy for multi-rotor UAVs[C]. 2015.

[3] Hoffmann G, Huang H, Waslander S, et al. Quadrotor Helicopter Flight Dynamics and Control: Theory and Experiment[C]. 2007.

[4] Zulu A, John S. A Review of Control Algorithms for Autonomous Quadrotors[J]. Open Journal of Applied Sciences. 2014, 04(14): 547-556.

IOP Conf. Series: Materials Science and Engineering **428** (2018) 012073 doi:10.1088/1757-899X/428/1/012073

[5] BAI Yongqiang, LIU Hao, Zongying S. Robust Flight Control of Quadrotor Unmanned Air Vehicles[J]. ROBOT. 2012, 34(5): 519-524.
[6] Zihao S. LPV Flight Control of Multi-rotor Control[D]. Zhejiang University, 2015.
[7] Chenglong L. Flight Stability and Automatic Control of Multirotor UAVs in a Wide Range Task[D]. Zhejiang University, 2016.

IOP Conf. Series: Materials Science and Engineering **428** (2018) 012074 doi:10.1088/1757-899X/428/1/012074

Dynamic Analysis of a New Type of Asymmetrical Parallel Mechanism Based on Lagrange Method

Quan Li, Bin Li and XinhuaZhao

Tianjin Key Laboratory for Advanced Mechatronic System Design and Intelligent Control, Tianjin University of Technology, Tianjin 300191, China

Corresponding to: Bin Li (cnrobot@tjut.edu.cn)

Abstract. In this article, a novel over constrained 2-RRU&RSR parallel manipulator is proposed and analyzed. This mechanism is composed of a moving platform and a base, which connected by two same configuration RRU legs and one RSR leg. The article uses the 2RRU&RSR parallel mechanism that can achieve 2R1T (two rotations and one translational motion) as the object. Based on the known kinematics forward solutions, the dynamics study was carried out. The dynamic model of the parallel mechanism was established using the Lagrange method and then it can be obtained that the dynamic inverse solution. Then ADAMS software was used to build a virtual prototype of this mechanism and the dynamics simulation was performed by using Lagrange method. Under the trend of a certain trajectory, the law of driving torque changes with time, paving the way for subsequent trajectory planning and control.

1. Introduction

The rapid development of the machinery industry and the wide application of less-degree-of-freedom agencies have prompted the optimization and upgrading of parallel institutions. One type of 2R1T three-degree-of-freedom parallel mechanism has been studied by more and more people.

The parallel robot has high rigidity and symmetry, low cost, small occupied space, and easy to realize the characteristics of its spatial motion [1,2,3]. Up to now, a great deal of related research works in this area has been described in literature. Xie F. G introduced a systematic type synthesis method for 2R1T LMPM based on the Grossmann Line Geometry, and discussed the typical application of this type parallel manipulators in five-axis machine tools in [4]. 3-RPS parallel manipulator is a typical 2R1T LMPM, which was firstly introduced by [5], the kinematic characteristics analysis, instantaneous motions analysis, kinematic synthesis, direct singular positions analysis, kinematics, dimensional synthesis and singularity analysis, etc., of the parallel manipulator were presented in [6,7,8]. Fu Tie studied the configuration of the 2R1T parallel mechanism using the Lie group theory systematically [9]. Among them, the non-fully symmetrical parallel mechanism has obvious advantages over other symmetrical mechanisms in some special applications and can play a decisive role [10]. The advantages of the parallel mechanism are prominent. However, due to the different closed-loop structures, there is a nonlinear coupling of spatial mechanisms with multiple degrees of freedom, resulting in complex forward and inverse solution of the mechanism, rich singularity, small working space, and difficult control. There are several methods for the dynamic modelling of the parallel mechanism [11,12]: Newton Euler's method, Lagrange method, and D'Alembert's principle. The method used in this paper is the Lagrange method to study the dynamics of 2RRU&RSR. This method is easy to model and easier to read than the equation using the vector force relationship.

In this paper, a novel over constrained 2R1T 2RRU&RSR parallel manipulator is proposed. This mechanism is composed of a moving platform and a base, connected by two same structure RRU legs and one RSR leg. Three revolute joints fixed on the base are set to be the driven joints. The arrangement of this text is planned as follows. In the section 2, the form of the institution is described in detail. In the section 3, kinematics problems, including the analysis of the reverse position of the manipulator are solved by calculations. The fourth section is about the dynamic modelling analysis of the organization. In Section 5, the virtual dynamic modelling and simulation of 2RRU&RSR is described. In section 6, the conclusions of the 2RRU&RSR research are described.

2. Structure characteristics of the mechanism

In figure 1, the 2-RRU&RSR parallel mechanism consists of a movable platform and a fixed platform, connected by three limbs: two RRU limbs and one RSR limb. The mechanism is symmetric with respect to RSR limb but not fully symmetrical. It is important to note that the two RRU limbs share a universal joint at the moving platform, which is named as a Hook joint "U". The parallel mechanism is driven by three rotating joints "R" connected to the fixed platform as driving joints.

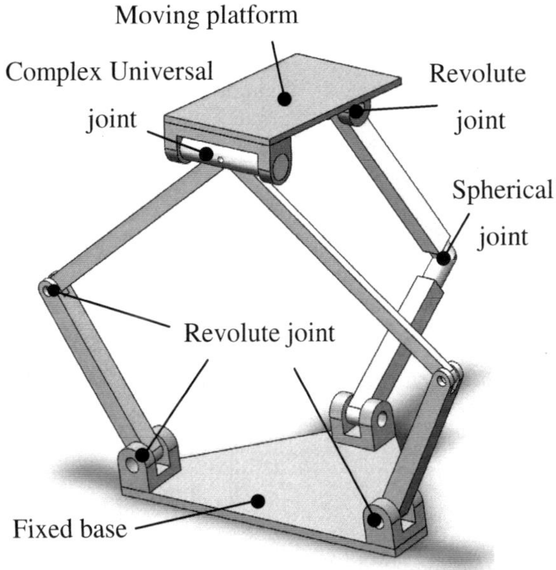

Figure 1 3D model of 2-RRU&RSR

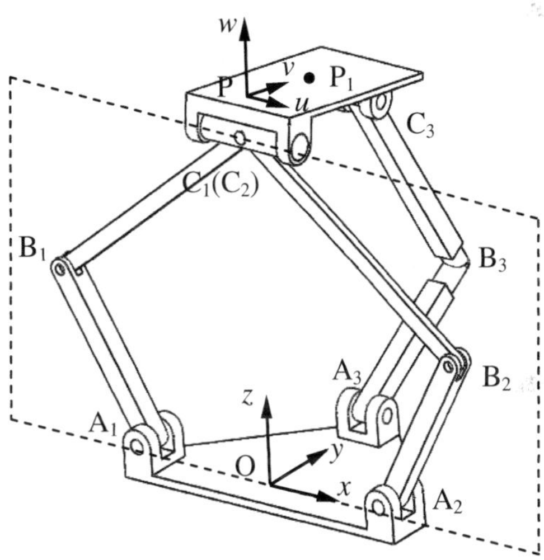

Figure 2 The structural diagrams of 2-RRU&RSR

In Figure 2, A_i, B_i and C_i $(i=1,2,3)$ represent the center of each joint of motion pairs on each limb respectively. The basic coordinate frame $O\text{-}xyz$ is established on the fixed platform. The X axis is along the OA_1 direction and the Y axis along the OA_3 direction and Z axis is vertically upward. The dynamic coordinate system $P\text{-}uvw$ is established on the moving platform. The V axis is located in the moving platform along the C_1C_3 direction. U axis located on moving platform and perpendicular to V axis. The W axis is perpendicular to the moving platform.

The RPY Euler angle $(\psi \quad \theta \quad \phi)$ representing the attitude of the moving platform represents the angle of rotation of the moving platform around the fixed coordinate system x, y, z.

$$\boldsymbol{R}=\begin{bmatrix} c\theta c\phi & s\psi s\theta - c\psi c\theta s\phi & c\psi s\theta + s\psi c\theta s\phi \\ s\phi & c\psi c\phi & -s\psi c\phi \\ -s\theta c\phi & s\psi c\theta + c\psi s\theta s\phi & c\psi c\theta - s\psi s\theta s\phi \end{bmatrix} = \begin{bmatrix} u_x & v_x & w_x \\ u_y & v_y & w_y \\ u_z & v_z & w_z \end{bmatrix} = [\boldsymbol{u} \quad \boldsymbol{v} \quad \boldsymbol{w}] \tag{1}$$

Here "s" and "c" refers to sin and cos.

3. Kinematic analysis of the mechanism

As shown in the figure 2, under the base coordinate system $O-xyz$, the point C_1 can be expressed by vector $r=\begin{pmatrix}x & y & z\end{pmatrix}^{\mathrm{T}}$,

$$\boldsymbol{r}=\overrightarrow{OA_i}+\overrightarrow{A_iB_i}+\overrightarrow{B_iC_1} \quad i=1,2; \qquad \boldsymbol{r}=\overrightarrow{OA_i}+\overrightarrow{A_iB_i}+\overrightarrow{B_iC_i}-\overrightarrow{C_iC_1} \quad i=3 \tag{2}$$

Since $\overrightarrow{A_iB_i}$ and $\overrightarrow{B_iC_1}$ are perpendicular to one of the axis of the U joint at the same time, the two limbs and the intersection point C_1 of the branch chain are confined to the same plane and apart of this mechanism becomes a planar five-bar mechanism. Therefore, according to structural features, the following equations are obtained:

$$\boldsymbol{r}^T c_i=0, \quad i=1,2 \tag{3}$$

Here, $c_1=c_2=(0 \quad 1 \quad 0)^{\mathrm{T}}$

Similarly, for the RSR limb, the constraint imposed by the revolute joint restricts both $\overrightarrow{A_iB_i}$ and $\overrightarrow{B_iC_i}$ to be normal to the unit vector $\overrightarrow{c_3}$ of the revolute joint axis. Thus, taking the dot product with $\overrightarrow{c_i}$ on both sides of equation (3), leads to:

$$\left(\boldsymbol{r}+\overrightarrow{C_iC_1}\right)^T c_i=0, \quad i=3 \tag{4}$$

Here, c_3=$\boldsymbol{R}c_{30}$, $c_{30}=\begin{pmatrix}1 & 0 & 0\end{pmatrix}^T$.

It should be noted that since two RRU limbs can only move in a Oxz plane, thus $u_y=0$.

Equations (3) and (4) lead to

$$x=-a\sin(\psi)\sin(\theta), y=0, z=z \tag{5}$$

$$\phi=0 \tag{6}$$

The forward position analysis of the mechanism is relative to the determination of the mobile platform pose given the position of C_1.

In accordance with the geometric constraint of the mechanism, since two RRU legs can only move in the plane Oxz, the u axis of the moving frame *P-uvw* can only moves in the same plane. So the constraint equation can be deduced as

$$A_0\cos\varphi+B_0\sin\varphi-C_0=0 \tag{7}$$

Where

$A_0=2q_{12}(x_{B2}-x_{B1}), B_0=2q_{12}(z_{B2}-z_{B1}), C_0=q_{12}^2+L_{B1B2}^2-q_{22}^2$, $L_{B_1B_2}=\sqrt{(x_{B_1}-x_{B_2})^2+(z_{B_1}-z_{B_2})^2}$,

$x_{B1}=-b-q_{11}\cos(\pi-\alpha_1), z_{B1}=q_{11}\sin(\pi-\alpha_1)$,

$x_{B2}=b+q_{21}\cos(\pi-\alpha_2), z_{B2}=q_{21}\sin(\pi-\alpha_1)$

Where q_{ij} are the distance of $\|A_iB_i\|$, $\|B_iC_i\|$ and $\|A_iC_i\|$.

Expanding equation (7), yields

$$\varphi=2\arctan\left(\frac{B_0\pm\sqrt{A_0^2+B_0^2-C_0^2}}{A_0+C_0}\right) \tag{8}$$

Through equation (8), yields

$$x_{C1}=x_{B1}+q_{12}\cos(\varphi), z_{C1}=z_{B1}+q_{12}\sin(\varphi) \tag{9}$$

In the same time, using the cosine theorem of the triangle $C_1C_3B_3$ and the geometric relationship of the mechanism, yields

$$\frac{\left|\overrightarrow{B_3C_1}\bullet\overrightarrow{C_1C_3}\right|}{\left|\overrightarrow{B_3C_1}\right|\left|\overrightarrow{C_1C_3}\right|}=\frac{\left|\overrightarrow{B_3C_1}\right|^2+a^2-q_{32}^2}{2a\left|\overrightarrow{B_3C_1}\right|} \tag{10}$$

Implementing necessary addition and subtraction, yields

$$A\cos^2(\psi)+B\cos(\psi)+C=0 \tag{11}$$

Where $A=E^2+D^2, B=-2KE, C=K^2-D^2+\frac{D^2x_{C1}^2}{a^2}, E=-a\left[b+q_{31}\cos(\pi-\alpha_3)\right],$

$E=-a\left[b+q_{31}\cos(\pi-\alpha_3)\right], D=a\left[q_{31}\sin(\pi-\alpha_3)-z_{C1}\right], K=-\frac{L^2+a^2-q_{32}^2}{2}-x_{C1}^2,\ L=\overrightarrow{B_3C_1}$

Expanding equation (11), yields

$$\psi=arc\cos\left(\frac{-B\pm\sqrt{B^2-4AC}}{2A}\right) \tag{12}$$

Next, substituting equation (12) into equation (5) yields

$$\theta=arc\sin\left(-\frac{x_{C1}}{a\sin(\psi)}\right) \tag{13}$$

Lastly, given a set of $(\alpha_1\quad\alpha_2\quad\alpha_3)$, x, y, z, ψ and θ can be solved using equations (9), (12) and (13).

Above all, the coordinate of the center of the moving platform can be written as the following formula:

$$^{O}P_1=R\ ^{P}P_1+{}^{O}P \tag{14}$$

By importing parameters including Euler angles and limb lengths into the MATLAB software, the coordinates of the moving platform can be calculated immediately. Because the result of MATLAB is too long, it is very difficult to express integrity in the text.

4. Dynamic Modelling and Analysis of 2RRU&RSR

This article only considers the dynamic and potential energy of the organization's dynamic platform and its various branches, ignoring friction. The mechanism is divided into two systems to calculate: the dynamic platform and three limbs, and finally all the kinetic energy and potential energy are added together to obtain the total kinetic energy and potential energy of the system. Since this mechanism has three degrees of freedom: translation along the Z axis and rotation around the X axis and the Y axis, the pose parameters of the moving platform are:

$$p=(Z_{P_1}\quad\psi\quad\theta) \tag{15}$$

Here Z_{P_1} is the coordinate of the moving platform in the fixed platform coordinate system, ψ and θ are the angles that the movable platform rotates around the X axis of the fixed platform and the angles that the rotating platform rotates about the V axis.

First, consider the kinetic energy and potential energy generated by the moving platform, ω is the angular velocity of the moving platform. Then the kinetic energy of the moving platform is:

$$E_h=\frac{1}{2}m{v_{P_1}}^T v_{P_1}+\frac{1}{2}\omega^T I_o\omega=\frac{1}{2}(m{\dot{Z}_{P_1}}^2+\omega^T I_o\omega) \tag{16}$$

Where m represents the quality of the moving platform; v_{P_1} is the linear velocity vector of the center of the platform and I_O is the inertial matrix of the motion platform relative to the inertia coordinate system, and I_o=RIR^T, R is the transformation matrix of the primary axis of the branch relative to the fixed reference frame, I is the moment of inertia of the coordinate system of the orbiting platform.

Then, the kinetic energy and potential energy of the three limbs of the parallel mechanism are calculated respectively. The mass of the connecting rod can be assumed uniform, and that the center of mass is unchanged during the movement.

The second step is the calculation of the kinetic energy and potential energy of two RRU limbs and one RSR limb. Each limb has two rods. The mass of the links is m_{ij}, where i represents the number of branches and j represents the lower rod and the upper rod. In order to simplify the process, this article makes the following assumptions: The rod connected to the fixed platform is called the lower rod, and the length is l_1; the rod connected to the moving platform is called the upper rod, and the length is l_2.

According to the spatial geometric relationship, the coordinates of the second R joints B_1, B_2 and B_3 can be obtained as follows:

$$\begin{cases} x_{B_1} = -b - l_1\cos(\pi-\alpha_1) \\ y_{B_1} = 0 \\ z_{B_1} = l_1\sin(\pi-\alpha_1) \end{cases}, \begin{cases} x_{B_2} = b + l_1\cos(\pi-\alpha_2) \\ y_{B_2} = 0 \\ z_{B_2} = l_1\sin(\pi-\alpha_2) \end{cases}, \begin{cases} x_{B_3} = 0 \\ y_{B_3} = b + l_1\cos(\pi-\alpha_3) \\ z_{B_3} = l_1\sin(\pi-\alpha_3) \end{cases} \tag{17}$$

So the potential energy of lower robs and the upper robs can be written as follows:

$$P_{i1} = \frac{1}{2}m_a g Z_{B_i}, P_{i2} = \frac{1}{2}m_b g(Z_{B_i} + Z_{C_i}) \quad (i = 1,2,3) \tag{18}$$

Then the kinetic energy of RRUs and RSR branches are solved by using the following formula:

$$E_{ij} = \frac{1}{2}(v_{ij}^T m v_{ij} + \omega_{ij}^T I_{ij}\omega_{ij}) \quad (i = 1,2,3; j = 1,2) \tag{19}$$

The linear velocity of the connecting rod can be obtained by finding the vector diameter of the connecting rod center of mass, $v_{ij} = \dot{r}_{ij}$. R_{ij} is the center of the connecting rod. Therefore, the linear velocity component of the connecting rod along the x axis, y axis, and z axis can be calculated by deriving the coordinates of the center of mass of the connecting rod respectively.

The vectors of the upper and lower links of three branches' centers are:

$$r_{11} = \frac{1}{2}\begin{bmatrix} x_{B_1} - b \\ 0 \\ z_{B_1} \end{bmatrix}, r_{12} = \frac{1}{2}\begin{bmatrix} x_{B_1} + x_{C_1} \\ 0 \\ z_{B_1} + z_{C_1} \end{bmatrix}; \quad r_{21} = \frac{1}{2}\begin{bmatrix} x_{B_2} + b \\ 0 \\ z_{B_2} \end{bmatrix}, r_{22} = \frac{1}{2}\begin{bmatrix} x_{B_2} + x_{C_2} \\ 0 \\ z_{B_2} + z_{C_2} \end{bmatrix};$$
$$r_{31} = \frac{1}{2}\begin{bmatrix} 0 \\ y_{B_3} + b \\ z_{B_1} \end{bmatrix}, r_{32} = \frac{1}{2}\begin{bmatrix} x_{B_3} + x_{C_3} \\ y_{B_3} + y_{C_3} \\ z_{B_3} + z_{C_3} \end{bmatrix} \tag{20}$$

R_{ij} is a coordinate transformation matrix between a link system and a fixed coordinate system, and I_c is a moment of inertia around a link coordinate system.

To establish the coordinate system of the rod, the z axis is perpendicular to the platform, the y axis coincides with the axis of rotation, and the x axis is determined by the right-handed helix rule. According to the spatial geometric relations, the angular velocity components of the connecting rod around the x, y and z axis can be obtained by deriving the time around the x, y and z axis:

Angular speed of lower robs and upper robs are:

$$\begin{cases}\omega_{ij_x} = \dot{\alpha}_i \\ \omega_{ij_y} = 0 \quad (i=1,2,3; j=1), \\ \omega_{ij_z} = 0\end{cases} \quad \begin{cases}\omega_{ij_x} = \dot{\alpha}_i + 2l_2\dot{l}_2 / \sqrt{4l_1^2l_2^2 - l_1^2} \\ \omega_{ij_y} = 0 \quad\quad (i=1,2; j=2), \\ \omega_{ij_z} = 0\end{cases}$$
$$\begin{cases}\omega_{ij_x} = \dot{\alpha}_i + 2l_2\dot{l}_2 / \sqrt{4l_1^2l_2^2 - l_1^2} \\ \omega_{ij_y} = 0 \quad\quad (i=3; j=2) \\ \omega_{ij_z} = -\dot{x}_{C_i} / x_{C_i} + \dot{y}_{C_i} / (y_{C_i} + b)\end{cases} \tag{21}$$

Substituting these parameters into equation (19) yields the kinetic energy of the three branches.

The total potential energy and the total kinetic energy of the system are:

$$P = P_h + P_{11} + P_{12} + P_{21} + P_{22} + P_{31} + P_{32}; \quad E = E_h + E_{11} + E_{12} + E_{21} + E_{22} + E_{31} + E_{32} \tag{22}$$

5. Dynamic Simulation of the Mechanism

Establish the three-dimensional model of the mechanism, introduce the established mechanism model into ADAMS, and simulate and analyze the dynamics of the organization. The modules in the ADAMS software use the Lagrange equation in the theory of multi-body system dynamics to establish a system dynamics model that can perform static, kinematic, and dynamic analysis on virtual mechanical systems, output displacement, velocity, acceleration, and force curves and so on.

Simulate the forces that drive the joints of each branch of the mechanism. The structural parameters of the parallel robot are in table 1:

Table 1. Structural parameters of the mechanism.

Institutional parameters	l_1	l_2	a	b
Length (mm)	85	120	70	64

Assuming that the initial position of the moving platform is where the angle θ is always 0. Its trajectory is a circular motion around its own center with a radius of $20mm$. The motion trajectory of the moving platform center is: $x = 20\cos\frac{\pi}{2}t, y = 20\sin\frac{\pi}{2}t$.

Where t is the time parameter. According to the given structural parameters, ADAMS is used for simulation analysis of driving torques of this mechanism.

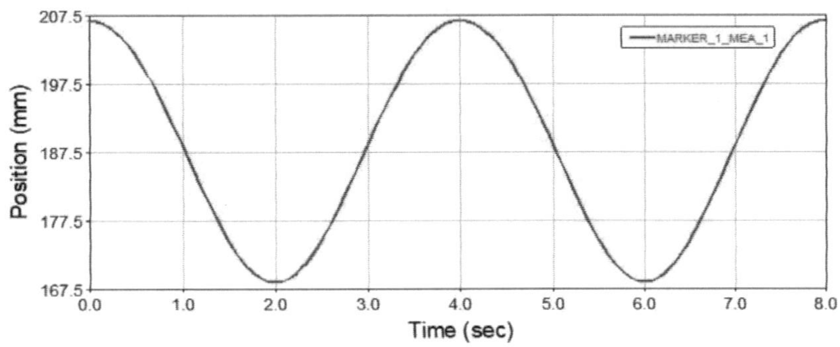

Figure 3. The curve of position change over time.

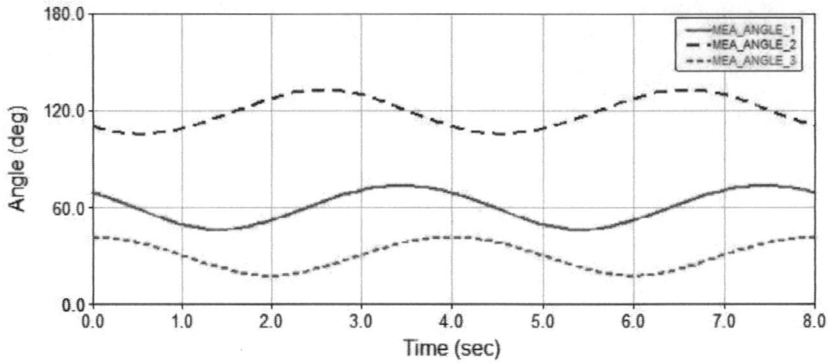

Figure 4. Angle changes with time in different trajectories.

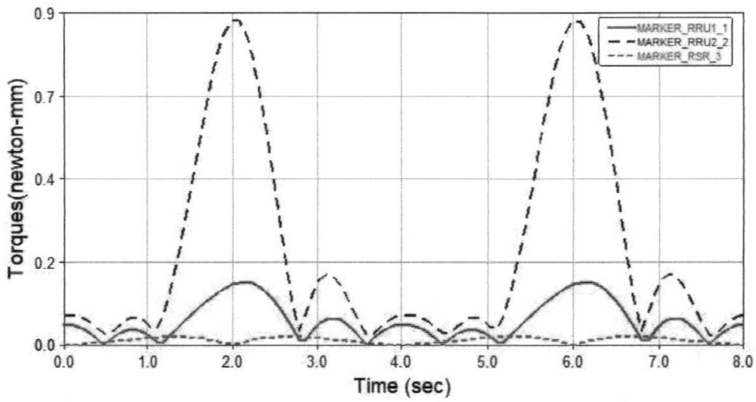

Figure 5. Torque changes with time in different trajectories.

Firstly, it can be seen from figure 3 that the curves of the position change of the moving platform. Secondly, the three driving angles changing in 8s and the driven robs move according to the sine curve in the figure 4. Thirdly, in figure 5, the driving torque of one of one RRU branch is always greater than that of the other RRU branch, which is equivalent to the larger torque of the two branches acting as a driven branch. It can be inferred that when the direction of the trajectory in the plane is opposite, the torque of the smaller branch will become the driven branch. From this simulation, when the sign of the angle changes, the force torque in the RSR branch also suddenly changes.

6. Conclusion

In this paper, (1) According to the characteristics of the 2RRU and RSR parallel robots, kinematics analysis is performed to obtain the positive solution of the motion platform, which lays a foundation for dynamic analysis. (2) The rigid body dynamic model is established by the Lagrange function balance method, and the corresponding displacement and force curves of the system are obtained through ADAMS virtual simulation. (3) Reasonably predict the behaviour of the organization and prepare for the next study.

The above analysis has very important significance for the optimization design and engineering application of this mechanism. It can also be used for dynamic analysis of similar parallel mechanisms.

Acknowledgments

This work was supported by National Key R&D Program of China (No.2017YFB1303502) and Intelligent Manufacturing Technology Major Project of Tianjin (16ZXZNGX00070).

References

[1] Merlet, J. P. 2010 Parallel robots. Solid Mechanics & Its Applications, 128, 2091-2127.
[2] Huang, Zhen 1997 Parallel Robotics Theory and Control. Machinery Industry Press.
[3] Hunt, K. H. 1983 Structural kinematics of in-parallel-actuated robot-arms. *Trans.asme J.of Mechanisms Transmissions & Automation in Design*, 105(4), 705-712.
[4] Yuan T 2012 Theoretical Analysis and Research of Asymmetrical 2-RRUR/2-RPUR Parallel Mechanism. Yanshan University.
[5] Huang, Z., & Fang, Y. F. 1996 Kinematic characteristics analysis of 3 DOF in-parallel actuated pyramid mechanisms. *Mechanism & Machine Theory*, 31(8), 1009-1018.
[6] Han, S. K., & Tsai, L. W. 2003 Kinematic synthesis of a spatial 3rps parallel manipulator. *Journal of Mechanical Design*, 125(1), 92-97.
[7] Rao, N. M., & Rao, K. M. 2009 Dimensional synthesis of a spatial 3-rps parallel manipulator for a prescribed range of motion of spherical joints. *Mechanism & Machine Theory*, 44(2), 477-486.
[8] Xie, F., Liu, X. J., & Li, T. 2013 Type synthesis and typical application of 1t2r-type parallel robotic mechanisms. *Mathematical Problems in Engineering*,2013,(2013-9-22), 2013(9), 497-50
[9] Fu, T., Han, X., Zhu, Y. Y., & Ding, H. S. 2014 Type synthesis of 2r1t parallel mechanism based on lie group theory. *Transactions of Beijing Institute of Technology*, 34(11), 1106-1109 and 1119.
[10] Yiu, Y. K., Cheng, H., Xiong, Z. H., & Liu, G. F. 2001 On the dynamics of parallel manipulators. *IEEE International Conference on Robotics and Automation*, 2001.Proceedings (Vol.4, pp.3766-3771 vol.4). IEEE.
[11] Park, F. C., Bobrow, J. E., & Ploen, S. R. 1995 A Lie group formulation of robot dynamics. *Sage Publications*, Inc.
[12] Li, J., & Wang, J. 2003 Inverse kinematic and dynamic analysis of a 3-dof parallel mechanism. *Chinese Journal of Mechanical Engineering*, 16(1), 54-58.